AF324085

Ordinary Differential Equations

Linear and Nonlinear Systems, Dynamical Systems and Applications

Ordinary Differential Equations

Linear and Nonlinear Systems, Dynamical Systems and Applications

Julián López-Gómez

Instituto de Matemática Interdisciplinar,
Universidad Complutense de Madrid, Spain

Andrea Tellini

Universidad Politécnica de Madrid, Spain

NEW JERSEY • LONDON • SINGAPORE • BEIJING • SHANGHAI • TAIPEI • CHENNAI

Published by

World Scientific Publishing Co. Pte. Ltd.

5 Toh Tuck Link, Singapore 596224

USA office: 27 Warren Street, Suite 401-402, Hackensack, NJ 07601

UK office: 57 Shelton Street, Covent Garden, London WC2H 9HE

Library of Congress Control Number: 2025041113

British Library Cataloguing-in-Publication Data
A catalogue record for this book is available from the British Library.

ORDINARY DIFFERENTIAL EQUATIONS
Linear and Nonlinear Systems, Dynamical Systems and Applications

ISBN 978-981-98-1154-0 (hardcover)
ISBN 978-981-98-1240-0 (paperback)

For any available supplementary material, please visit
https://www.worldscientific.com/worldscibooks/10.1142/14267#t=suppl

Desk Editors: Aanand Jayaraman/Rok Ting Tan

Typeset by Stallion Press
Email: enquiries@stallionpress.com

*In studiis puto me hercules melius esse res ipsas intueri et harum
causa loqui, ceterum verba rebus permittere, ut qua duxerint, hac
inelaborata sequatur oratio. "Quid opus est saeculis duratura
componere? Vis tu non id agere, ne te posteri taceant? Morti natus
es, minus molestiarum habet funus tacitum! Itaque occupandi
temporis causa, in usum tuum, non in praeconium aliquid simplici
stilo scribe; minore labore opus est studentibus in diem." Rursus ubi
se animus cogitationum magnitudine levavit, ambitiosus in verba est
altiusque ut spirare ita eloqui gestit et ad dignitatem rerum exit
oratio; oblitus tum legis pressiorisque iudicii sublimius feror et ore
iam non meo.*

Lucio Anneo Seneca (*De Tranquillitate Animi*, 1.13–14)

En los estudios creo más acertado atender a la substancia de las
cosas, subordinando a ella las palabras, de manera que la oración
siga al pensamiento de forma no elaborada. "¿Qué necesidad hay de
componer obras que perduren siglos? ¿Pretendes que la posteridad
no te silencie? Has nacido para morir. ¡Un funeral silencioso causa
menos molestias! Así pues, escribe en estilo sencillo para ocupar tu
tiempo en tu provecho, no en tu popularidad; estudiar pensando en
el presente requiere menor esfuerzo." Pero en cuanto mi espíritu se
eleva empujado por la grandeza de sus pensamientos, nuevamente
se torna altivo en sus palabras y ansía hablar al ritmo que respira;
olvidándome entonces de mis reglas prescritas, me subo a las
alturas, y dejo de ser yo quien habla por mi boca.

In my literary studies I think that it is better view things as they are, and to speak of them on their own account, and as for words, to trust to things for them, and to let one's speech simply follow whither they lead. "What need is there to compose something that will last for centuries? Will you not give up striving to keep posterity from being silent about you? You were born for death; a silent funeral is less troublesome! And so, to pass the time, write something in simple style, for your own use, not for praise; they that study for the day have less need to labour." Then again, when my mind has been uplifted by the greatness of its thoughts, it becomes ambitious of words, and with higher aspirations it desires higher expression, and language issues forth to match the dignity of the theme; forgetful then of my rule and of my more restrained judgement, I am swept to loftier heights, using a language that is not my own.

(Translation adapted from https://www.loebclassics.com/view/ seneca_younger-de_tranquillitate_animi/1932/pb_LCL254.209.xml)

Preface

The enormous success of Differential Equations in Science and Technology is based on the fact that many mathematical models of real-world systems are formulated in terms of Differential Equations. Astonishingly, the same mathematical models can mimic the evolution of several physical, chemical, economical, or ecological systems. Thus, their solutions allow us to make predictions in a huge variety of real-world systems, some of which might be almost intractable empirically. For this reason, the theory of Differential Equations and its applications are imperative in many branches of Applied Sciences and Engineering.

This textbook introduces some fundamentals of the theory of Ordinary Differential Equations and Dynamical Systems in a rigorous and self-contained way, adopting some slightly new perspectives coming from Nonlinear Analysis. Moreover, it applies the theory to analyze, in a practical way, a series of significant models from different areas, which may be of interest for students in several disciplines. This book germinated from the lecture notes by López-Gómez on *Elements of Ordinary Differential Equations* and *Differential Equations*, prepared for the second and third years, respectively, of the bachelor's degree in Mathematics, which have been taught since 1996 at the Complutense University of Madrid. Some of its contents have also been taught by Tellini in the course on *Ordinary Differential Equations I* of the bachelor's degree in Mathematics at the Polytechnic University of Madrid since 2022.

Most classic textbooks on Ordinary Differential Equations focus on second-order equations from Classical Mechanics, in the spirit of

19th century Mathematics. Nevertheless, during the 20th century, diffusive equations proved to have many applications in Ecology, Economy, Physics, Chemistry, Biology, and Engineering. Although their multidimensional counterparts consist of Partial Differential Equations, when restricted to one dimension, they reduce to Ordinary Differential Equations. Studying these one-dimensional prototypes has been a pillar facilitating later developments in the design of strategies to tackle their multidimensional counterparts, leading to significant advancements in nonlinear differential equations during the 20th century. For this reason, we have decided to include their analysis here.

Other main novelties are some contents that are not covered in most classical books in this field, such as the analysis of linear systems with holomorphic coefficients, Kamke's theorem about the method of sub- and supersolutions for systems of nonlinear ordinary differential equations, and the application of phase portrait techniques for constructing global bifurcation diagrams of solutions for important classes of one-dimensional boundary value problems closely related to the diffusive logistic equation.

This book consists of three main parts: Part 1, "Linear Systems", treats the theory of linear equations and systems, including the cases of periodic and holomorphic coefficients; Part 2, "Nonlinear Systems", covers the abstract theory of first-order nonlinear systems and systems of equations of arbitrary order; Part 3, "Dynamical Systems", introduces the fundamentals of the theory of Continuous Dynamical Systems, with preliminary insights into some more advanced topics in this field.

Part 1 consists of the following three chapters:

- Chapter 1. First-order linear systems.
- Chapter 2. First-order linear systems with constant coefficients.
- Chapter 3. First-order linear systems with holomorphic coefficients.

Chapter 1 presents the general theory for linear systems

$$u' = A(t)u + B(t), \tag{0.1}$$

where $A(t)$ is a matrix of order N with coefficients

$$a_{ij} \in \mathcal{C}([\alpha, \beta]; \mathbb{R}) \quad \text{for all } i, j \in \{1, \ldots, N\},$$

for some $\alpha, \beta \in \mathbb{R}$, $\alpha < \beta$, and $B \in \mathcal{C}([\alpha, \beta]; \mathbb{R}^N)$ is a continuous vector-valued function. Associated with the linear system (0.1), for every $t_0 \in [\alpha, \beta]$ and $u_0 \in \mathbb{R}^N$, we also consider the initial value problem

$$\begin{cases} u' = A(t)u + B(t), \\ u(t_0) = u_0, \end{cases} \qquad (0.2)$$

and its associated integral operator

$$\mathcal{K}h(t) := u_0 + \int_{t_0}^{t} (A(s)h(s) + B(s))\ ds, \quad t \in [\alpha, \beta],$$

defined for all $h \in \mathcal{C}([\alpha, \beta]; \mathbb{R}^N)$, whose fixed points are in one-to-one correspondence with the solutions of (0.2). As a consequence of the *superposition principle*, or, in other words, the linearity of this system, the set of solutions of (0.1), denoted by Σ_B in this book, is an affine subspace of $\mathcal{C}^1([\alpha, \beta]; \mathbb{R}^N)$. Indeed, for every $h_1, h_2 \in \Sigma_0$ and $\lambda, \mu \in \mathbb{R}$, the superposition principle establishes that

$$\lambda h_1 + \mu h_2 \in \Sigma_0.$$

Moreover, for every $u \in \Sigma_B$, we have that $u + h \in \Sigma_B$ if and only if $h \in \Sigma_0$. Thus, Σ_B is an affine manifold. On the other hand, as an application of the contractive mapping theorem to the integral operator $\mathcal{K}$, the existence and the uniqueness of a global solution for problem (0.2) is obtained. This allows us to define the mapping that sends the initial datum u_0 to the unique solution of (0.2), denoted by $u(t; t_0, u_0)$, which establishes an affine isomorphism between $\mathbb{R}^N$ and the set of solutions of (0.1), Σ_B. As a consequence, Σ_B is an N-dimensional affine manifold. Actually, it is linear if $B = 0$. Then, Chapter 1 characterizes whether or not a given set of N solutions of Σ_0 generates Σ_0, and it adapts the abstract theory to deal with systems of linear equations of arbitrary order. Finally, in Chapter 1, we construct the general solutions of linear equations of arbitrary order with constant coefficients, focusing in particular on the special role played by second-order differential equations in describing harmonic motions in Classical Mechanics, which has a huge interest in Applied Sciences and Engineering.

Chapter 2 constructs the general solution of a linear system of N first-order equations with constant coefficients through the exponential matrix, which is calculated from the Jordan canonical form of the matrix of the system. Indeed, when $A(t) \equiv A$ is a constant matrix, the unique solution of (0.2) is given by the variation of constants formula

$$u(t; t_0, u_0) = e^{(t-t_0)A} u_0 + \int_{t_0}^{t} e^{(t-s)A} B(s)\, ds,$$

where, by definition,

$$e^{zA} = \sum_{n=0}^{\infty} \frac{z^n}{n!} A^n$$

is the exponential matrix of zA, for all $z \in \mathbb{C}$. In this chapter, we recall all the results from Linear Algebra which are needed to compute the Jordan canonical form and, in turn, the exponential matrix of zA.

Chapter 3 analyzes linear systems of the form

$$u' = A(z)u + B(z),$$

where the coefficients of $A(z)$ and $B(z)$ are holomorphic functions on some simply connected subdomain, Ω, of $\mathbb{C}$, focusing in particular on the special case of convex Ω. Indeed, in such a case, the variation of constants formula can still be derived and provides us with a rather explicit formula for the solutions of the system. In this context, one can develop the solutions in terms of the Taylor series and try to establish some relations on the corresponding coefficients in order to represent the solutions. These contents were initially designed by López-Gómez for his course on *Functions of one Complex Variable and Differential Equations* in the third year of the old degree in Mathematics at the Complutense University of Madrid.

Then, Chapter 3 solves some fundamental differential equations from the point of view of applications in Science and Engineering, including those of Airy, Hermite, and Bessel, as well as the Schrödinger equation of the one-dimensional linear harmonic oscillator. Other important equations are introduced in the list of exercises in this chapter, such as the Laguerre and Legendre equations, and the Schrödinger equation of the hydrogen atom. A classical reference in this field, which probably has not yet received the deserved attention, is Friedrichs' lecture notes at the Courant Institute collected in 1965.

Part 2 consists of the following five chapters:

- Chapter 4. An introduction to nonlinear differential equations.
- Chapter 5. Cauchy–Lipschitz theory.
- Chapter 6. High-order nonlinear equations.
- Chapter 7. Peano theory.
- Chapter 8. Method of sub- and supersolutions.

Chapter 4 analyzes some important classes of nonlinear scalar problems,

$$\begin{cases} u' = f(t, u), \\ u(t_0) = u_0, \end{cases} \tag{0.3}$$

where $f \in \mathcal{C}((a, b) \times \mathbb{R}; \mathbb{R})$, with $a, b \in [-\infty, +\infty]$, $a < b$, $t_0 \in (a, b)$, and $u_0 \in \mathbb{R}$. In particular, it studies the logistic equation, the Bernoulli equation, and the Riccati equation, as well as equations with separated variables, i.e. equations for which

$$f(t, u) = F(t)G(u),$$

together with other special, but important, classes of differential equations, such as homogeneous and exact differential equations. The integration of many of them can be reduced to the integration of an equation with separated variables, where the implicit function theorem guarantees that, as long as $G(u_0) \neq 0$, problem (0.3) has a unique local solution. A series of pivotal non-uniqueness examples in Chapter 4 show that, though the constant $u := u_0$ solves (0.3) if $G(u_0) = 0$, in such a case, problem (0.3) might possess multiple solutions.

Essentially, Chapter 4 presents a collection of simple examples which allow us to test all abstract results throughout the rest of this book.

Chapter 5 is devoted to the Cauchy–Lipschitz theory on the existence and uniqueness of solutions for initial value problems related to a general class of nonlinear systems. In the first part, given $\alpha, \beta \in \mathbb{R}$, with $\alpha < \beta$, and a natural number $N \geq 1$, we consider a continuous vectorial function $f \in \mathcal{C}([\alpha, \beta] \times \mathbb{R}^N; \mathbb{R}^N)$ satisfying a *global Lipschitz condition*, uniformly with respect to the time variable, t. This means that there exists a constant $L \geq 0$ such that

$$\|f(t, x) - f(t, y)\| \leq L\|x - y\| \tag{0.4}$$

for all $x, y \in \mathbb{R}^N$ and $t \in [\alpha, \beta]$. Under these conditions, we show that, for every $t_0 \in [\alpha, \beta]$ and $u_0 \in \mathbb{R}^N$, the initial value problem

$$\begin{cases} u' = f(t, u), \\ u(t_0) = u_0, \end{cases} \tag{0.5}$$

possesses a unique solution which is globally defined in $[\alpha, \beta]$. The simplest condition ensuring (0.4) is the existence and continuity of all partial derivatives

$$\frac{\partial f_i}{\partial u_j} \in \mathcal{C}([\alpha, \beta] \times \mathbb{R}^N; \mathbb{R}), \quad 1 \leq i, j \leq N,$$

together with the existence of a constant $C > 0$ such that

$$\left| \frac{\partial f_i}{\partial u_j}(t, u) \right| \leq C \quad \text{for all } (t, u) \in [\alpha, \beta] \times \mathbb{R}^N \text{ and } 1 \leq i, j \leq N.$$

In particular, in the case of linear systems

$$f(t, u) = A(t)u + B(t), \quad (t, u) \in [\alpha, \beta] \times \mathbb{R}^N,$$

we have that

$$\frac{\partial f_i}{\partial u_j}(t, u) = a_{ij}(t).$$

Thus, it is apparent that, if $A(t)$ is continuous in $[\alpha, \beta]$, such a function $f(t, u)$ is globally Lipschitz in u, uniformly in $t \in [\alpha, \beta]$. Therefore, the main theorem in Chapter 1, related to the existence and uniqueness of a global solution for the linear initial value problem (0.2), can be seen as a particular case of the main theorem in Chapter 5. Actually, as in Chapter 1, the main theorem in Chapter 5 follows from applying the contractive mapping theorem to the integral operator associated with problem (0.5), which is

$$\mathcal{K}h(t) := u_0 + \int_{t_0}^{t} f(s, h(s)) \, ds$$

for all $t \in [\alpha, \beta]$. This entails that, for every $h_0 \in \mathcal{C}([\alpha, \beta]; \mathbb{R}^N)$, the $\mathcal{K}$-iterates (also known as *Picard–Lindelöf iterates*)

$$h_n = \mathcal{K}h_{n-1} = \mathcal{K}^2 h_{n-2} = \cdots = \mathcal{K}^n h_0 \tag{0.6}$$

converge to the unique solution of (0.5), uniformly in $[\alpha, \beta]$.

Later in this chapter, the global Lipschitz condition (0.4) is replaced with a weaker one of local nature, which will allow us to obtain the local existence and uniqueness for (0.5). The rest of Chapter 5 analyzes how the local solution can be extended to a unique maximal (non-extensible) solution, and ascertains the possible limiting behaviors of the maximal solutions. Finally, the sensitivity of the maximal solution and its interval of definition with respect to variations in u_0 and f is analyzed.

In Chapter 6, given a natural number $N \geq 1$, an open subset $\Omega \subset \mathbb{R}^N$, and a scalar function $g \in \mathcal{C}(\mathbb{R} \times \Omega; \mathbb{R})$, we study the nonlinear scalar differential equation of Nth order,

$$u^{N)} = g\big(t, u, u', u'', \dots, u^{N-1)}\big), \tag{0.7}$$

as well as its associated initial value problem,

$$\begin{cases} u^{N)} = g\big(t, u, u', u'', \dots, u^{N-1)}\big), \\ \big(u(t_0), u'(t_0), u''(t_0), \dots, u^{N-1)}(t_0)\big) = U_0, \end{cases}$$

with $t_0 \in \mathbb{R}$ and $U_0 \in \Omega$, in the special case when g satisfies a Lipschitz condition with respect to its variables in Ω, uniformly on compact subsets of $t \in \mathbb{R}$. We begin by deriving the first-order system associated with the Nth-order equation (0.7), and then we establish a bijection between the set of solutions of (0.7) and the set of solutions of the associated first-order system. This allows us to transfer the theory developed in Chapter 5 for general first-order systems to Nth-order scalar equations and, at a later stage, to general systems of scalar equations of arbitrary order with Lipschitz continuous nonlinearities. Finally, Chapter 6 applies the abstract theory to the simple gravity pendulum and the diffusive logistic equation. The first model is pivotal in Classical Mechanics. The second one is the basis on which the mathematical theories of Population Dynamics and Environmental Sciences, among other fields, have been developed. As an application of the theory in Chapter 6, we derive the so-called Keller–Osserman condition for characterizing the existence of positive solutions for the singular problem

$$\begin{cases} u'' = g(u) & \text{in } (-L, L), \\ u(-L) = u(L) = +\infty, \end{cases}$$

for all $L > 0$, where the boundary condition is meant in a limiting sense.

Chapter 7 begins with the Peano theory, in which the general existence theory for (0.5) with $f \in \mathcal{C}([\alpha, \beta] \times \mathbb{R}^N; \mathbb{R}^N)$, not necessarily Lipschitz continuous in u, is developed. The examples given in Chapter 4 show how, in the context of the Peano theory, one cannot expect to get uniqueness for (0.5). The existence of a local solution in this more general framework follows from the Ascoli–Arzelà theorem—possibly, the first theorem in Functional Analysis—which characterizes the compact subsets of the space of continuous vectorial functions $\mathcal{C}([\alpha, \beta]; \mathbb{R}^N)$.

As a consequence of this theorem, a given subset of the set of solutions of the system $u' = f(t, u)$ is a compact subset of $\mathcal{C}([\alpha, \beta]; \mathbb{R}^N)$ if and only if it is closed and bounded in $\mathcal{C}([\alpha, \beta]; \mathbb{R}^N)$. Indeed, any compact set in a metric space is closed and bounded. Conversely, any bounded subset of the set of solutions of $u' = f(t, u)$ must be equicontinuous, as the derivatives of its elements are uniformly bounded. Thus, thanks to the Ascoli–Arzelà theorem, it is a relatively compact subset of $\mathcal{C}([\alpha, \beta]; \mathbb{R}^N)$. Therefore, this theorem plays a similar role in $\mathcal{C}([\alpha, \beta]; \mathbb{R}^N)$ as the theorem of Bolzano and Weierstrass in $\mathbb{R}^N$. For this reason, it is a pivotal result in the theory of Differential Equations.

Due to the lack of uniqueness for (0.5), in the context of the Peano theory, the set of solutions of (0.5) is far from being well-ordered with the order relation *being an extension of a solution to a larger interval*. As a consequence, in this framework, one must resort to the Kuratowski–Zorn lemma, which is equivalent to the axiom of choice in the Zermelo–Fraenkel set theory, to establish the existence of a maximal (non-extensible) solution for (0.5).

After discussing the global behavior of the maximal solutions, Chapter 7 presents a celebrated example given by Müller to illustrate that, in the context of the Peano theory, one cannot expect, in general, the $\mathcal{K}$-iterates (0.6) to stabilize to some solution of (0.5). Finally, Chapter 7 delivers a theorem by Kneser establishing that, in the setting of the Peano theory, the reachable set of points, at a fixed time, by the solutions of (0.5) is a closed and connected subset of the phase space, $\mathbb{R}^N$. In other words, when (0.5) has not a unique solution, it has a non-trivial topological continuum of solutions.

Chapter 8 begins by introducing the fundamental concepts of sub- and supersolution for (0.5). A subsolution is a function $\underline{u} \in \mathcal{C}([\alpha, \beta]; \mathbb{R}^N)$ such that $\underline{u} \leq \mathcal{K}\underline{u}$ in $[\alpha, \beta]$. Similarly, a supersolution is a function $\overline{u} \in \mathcal{C}([\alpha, \beta]; \mathbb{R}^N)$ such that $\mathcal{K}\overline{u} \leq \overline{u}$ in $[\alpha, \beta]$, where $\mathcal{K}$ is the integral operator associated with (0.5).

The main theorem in Chapter 8 establishes that, if $f(t, u)$ is cooperative, in the sense that $u \mapsto f(t, u)$ is non-decreasing for all $t \in [\alpha, \beta]$, and (0.5) admits a pair of sub- and supersolutions in $[\alpha, \beta]$, say $\underline{u}$ and $\overline{u}$, with $\underline{u} \leq \overline{u}$ in $[\alpha, \beta]$, then the $\mathcal{K}$-iterates from $\underline{u}$ and $\overline{u}$ approximate the minimal and maximal solutions, respectively, of problem (0.5) in the interval $[\underline{u}, \overline{u}]$. This is an important theorem by Kamke, which has a variety of counterparts in several different contexts in Mathematics. Then, Kamke's theorem is used in some illustrative applications to estimate the existence time and/or the blow up time of the solutions of (0.5) when f is increasing in u. Another application of Kamke's theorem in Chapter 8 allows us to prove Wintner's extensibility theorem and Osgood's uniqueness criterium.

Part 3 consists of the following three chapters:

- Chapter 9. Some paradigmatic Dynamical Systems.
- Chapter 10. Newtonian planar conservative systems.
- Chapter 11. Non-conservative systems.

Chapter 9 introduces the theory of Dynamical Systems by constructing the admissible phase portraits of the linear planar system with constant coefficients

$$\begin{cases} u' = au + bv, \\ v' = cu + dv, \end{cases} \qquad (0.8)$$

where $a, b, c, d \in \mathbb{R}$ satisfy $ad \neq bc$, which entails that $(u, v) = (0, 0)$ is the unique equilibrium of (0.8). The phase portrait consists of the set of trajectories $(u(t), v(t))$, $t \in \mathbb{R}$, where (u, v) is a maximal solution of the system. Based on the theory in Chapter 2, this system can be easily integrated, and its trajectories are easily found for all value of the parameters a, b, c, d. Next, Chapter 9 studies some fundamental properties of the trajectories of general autonomous systems of the form $u' = f(u)$, with $f \in \mathcal{C}^1(\mathbb{R}^N; \mathbb{R}^N)$. Then, based on these properties, we ascertain the dynamics of the non-negative solutions of the Lotka–Volterra predator-prey model

$$\begin{cases} u' = \lambda u - uv, \\ v' = -\mu v + uv, \end{cases} \tag{0.9}$$

where $\lambda > 0$ and $\mu > 0$. Finally, Chapter 9 constructs the phase portrait of the competing species model

$$\begin{cases} u' = \lambda u - uv, \\ v' = \mu v - uv. \end{cases} \tag{0.10}$$

The analysis of the dynamics of the symbiotic model

$$\begin{cases} u' = -\lambda u + uv, \\ v' = -\mu v + uv, \end{cases} \tag{0.11}$$

is proposed in Exercise 6 of Chapter 9. Note that (0.9), (0.10) and (0.11) provide us with the simplest examples in which the differential equations $u' = mu$ and $v' = nv$, with $m, n \in \mathbb{R}$, can be coupled in a nonlinear way. In Chapter 9, it will be shown that these systems are conservative, in the sense that there is a function $E : \mathbb{R}^2 \to \mathbb{R}$ such that

$$E(u(t), v(t)) = E(u_0, v_0) \quad \text{for all } t \in I,$$

where I is the maximal interval of definition of the solution of the corresponding system with $u(0) = u_0 \geq 0$ and $v(0) = v_0 \geq 0$. Thus, it is possible to construct the phase portrait of each of these models by analyzing the set

$$\mathcal{C}_{(u_0, v_0)} := \left\{ (u, v) \in \mathbb{R}^2 : E(u, v) = E(u_0, v_0) \right\}$$

through any point (u_0, v_0) of the first quadrant. The set $\mathcal{C}_{(u_0, v_0)}$ is usually called an *integral curve* through (u_0, v_0). By construction, the integral curves consist of a union of trajectories of solutions that are separated from each others by equilibria.

Chapter 10 focuses on Newtonian equations of the form

$$-u'' = f(u),$$

where $f \in \mathcal{C}^1(\mathbb{R};\mathbb{R})$. By multiplying this differential equation by $u'(t)$ and setting $v := u'$, it is easily seen that the solution of the initial value problem

$$\begin{cases} -u'' = f(u), \\ u(0) = u_0, \quad u'(0) = v_0, \end{cases} \tag{0.12}$$

satisfies

$$\frac{v^2(t)}{2} + \int_0^{u(t)} f(s)\, ds = \frac{v_0^2}{2} + \int_0^{u_0} f(s)\, ds \quad \text{for all } t \in I, \tag{0.13}$$

where I denotes the maximal interval of the definition of the solution of (0.12). In Classical Mechanics $\frac{v^2}{2}$ represents the kinetic energy of a body with mass $M = 1$,

$$\varphi(\xi) := \int_0^{\xi} f(s)\, ds$$

can be regarded as the potential energy, and (0.13) establishes the conservation of the total energy induced by the Newtonian force $-f(u)$:

$$E(u, v) := \frac{v^2}{2} + \varphi(u) = \frac{v^2}{2} + \int_0^u f(s)\, ds.$$

Hence,

$$v(t) = \pm\sqrt{2\left(E_0 - \varphi(u(t))\right)} \quad \text{for all } t \in I,$$

where

$$E_0 := \frac{v_0^2}{2} + \int_0^{u_0} f(s)\, ds$$

is the total energy. Therefore, the trajectory of the solution of (0.12) lies on the integral curve through (u_0, v_0)

$$\mathcal{C}_{(u_0, v_0)} := \left\{ (u, v) \in \mathbb{R}^2 : v = \pm\sqrt{2\left(E_0 - \varphi(u)\right)} \right\},$$

which consists of a union of trajectories of solutions of $-u'' = f(u)$ separated by constant solutions of the form $(u, u') = (\omega, 0)$, with $f(\omega) = 0$. Naturally, the phase portrait of $-u'' = f(u)$ can be easily constructed from these integral curves, which can be represented by analyzing the potential energy, as will be discussed in the first two sections of Chapter 10. Then, the dynamics of the free pendulum is constructed and discussed, as well as the dynamics of the diffusive equation

$$-u'' = \lambda u - u^3. \tag{0.14}$$

Actually, in Section 10.4, once the dynamics of (0.14) has been constructed, we will use it to find all the types of solutions of (0.14), as well as to characterize the existence of positive solutions for the boundary value problem

$$\begin{cases} -u'' = \lambda u - u^3 & \text{in } (0, L), \\ u(0) = u(L) = 0. \end{cases} \tag{0.15}$$

The interest of these solutions lies in the fact that they provide us with the asymptotic profiles of positive solutions of the following nonlinear energy transmission problem associated with the one-dimensional heat equation:

$$\begin{cases} \frac{\partial u}{\partial t} - \frac{\partial^2 u}{\partial x^2} = \lambda u - u^3 & \text{for all } t > 0 \text{ and } x \in [0, L], \\ u(0, t) = u(L, t) = 0 & \text{for all } t > 0, \\ u(\cdot, 0) = u_0 & \text{in } [0, L], \end{cases} \tag{0.16}$$

where $u_0 \gneq 0$ represents the initial distribution of temperatures. It turns out that, regardless the value of the function u_0, the unique solution of (0.16), $u_\lambda(t; u_0)$, satisfies

$$\lim_{t \uparrow +\infty} u_\lambda(t; u_0) = \begin{cases} 0 & \text{if } \lambda \le \left(\frac{\pi}{L}\right)^2, \\ \theta_\lambda & \text{if } \lambda > \left(\frac{\pi}{L}\right)^2, \end{cases}$$

where θ_λ denotes the unique positive solution of (0.15). Problem (0.16) can be also viewed as a one-dimensional reaction-diffusion model.

Subsequently, the results on positive solutions for (0.15) will allow us to construct all non-zero solutions of (0.15). Chapter 10 concludes by repeating the analysis done for (0.14) for the parametric differential equation

$$-u'' = \lambda u + u^2.$$

This equation has a multidimensional counterpart,

$$-\Delta u = \lambda u + u^2,$$

where Δ is the Laplace operator in $\mathbb{R}^N$,

$$\Delta = \sum_{j=1}^{N} \frac{\partial^2}{\partial x_j^2}.$$

It turns out that, for any given bounded open subset $\Omega \subset \mathbb{R}^N$ with a sufficiently regular boundary, denoted by $\partial\Omega$, the number and structure of the positive solutions of the boundary value problem

$$\begin{cases} -\Delta u = \lambda u + u^2 & \text{in } \Omega, \\ u = 0 & \text{on } \partial\Omega, \end{cases}$$

might depend on the geometrical and topological properties of Ω in a rather hidden way, as well as on the spatial dimension N. Thus, when dealing with these multidimensional models, it is imperative, as a first step, to analyze their one-dimensional prototypes by using the techniques and tools introduced in Chapters 9 and 10. Actually, this is essentially the unique solvable case.

Finally, Chapter 11 discusses some more advanced topics in the theory of Dynamical Systems that were developed for treating *non-conservative systems*, i.e. for systems where the techniques in Chapters 9 and 10 cannot be applied. Precisely, Chapter 11 begins by discussing the concepts of stability of motion, as introduced by Lyapunov, and proving the stability and instability theorems by Lyapunov for equilibria of autonomous systems, not necessarily planar. These techniques, based on the variational (or linearized) system at the given equilibrium, are imperative when the integral curves of the system are unknown. Lyapunov's theorems are then complemented with a rather detailed discussion of the theorem by Grobman and Hartman about the topological equivalence of systems at hyperbolic equilibria, i.e. equilibria where the linearized system has no eigenvalue with a vanishing real part.

Then, Chapter 11 introduces two pivotal concepts in the theory of Dynamical Systems: the α-limit and ω-limit sets of an orbit. The orbit is the set of points of $\mathbb{R}^N$ covered over time by the solution of a first-order system of ordinary differential equations. The ω-limit set of an orbit is the set of limiting points of the orbit as $t \uparrow +\infty$, while its α-limit set is the set of limiting points of the orbit as $t \downarrow -\infty$. After discussing some of their invariance and topological properties, we give a self-contained proof of the Poincaré–Bendixson theorem for autonomous planar systems. According to it, any ω-limit set of a bounded non-periodic orbit without equilibria must correspond to a periodic solution, i.e. a closed orbit in the phase plane. In such a case, the bounded solution must approach the periodic solution spiraling toward it, from its interior or its exterior.

Later, Chapter 11 applies this abstract theory to analyze the dynamics of the non-negative solutions of the Lotka–Volterra competition model

$$\begin{cases} u' = \lambda u - u^2 - buv, \\ v' = \mu v - cuv - v^2, \end{cases}$$

where λ, μ, b and c are positive constants.

Finally, a periodic solution is constructed in a predator-prey system with saturation as an application of the Poincaré–Bendixson theorem, and the dynamics of the pendulum with friction is constructed. This is a paradigmatic example of a non-conservative physical model where energy is dissipated through friction.

We point out that the three parts of the book can essentially be covered independently to design the content of subjects according to the needs of each bachelor's or master's degree. Of course, it is also possible to combine the contents of the different parts. In addition, one can focus more on the theory, or, alternatively, on its applications and examples, depending on the character of the course and the needs of each student/professor.

Each chapter ends with a list of exercises of variable difficulty, which complement the theory and examples developed in the book, as well as with some final comments, including historical remarks or references to complementary materials that may be useful both for students and their professors.

López-Gómez wants to express his deepest gratitude to the students who have attended the courses he taught at the Complutense

University of Madrid during the last two decades, particularly Carlos Mora Corral, David Gómez Castro, Juan José Madrigal Martínez, Luis Maire, Sergio Fernández Rincón, Eduardo Muñoz Hernández, Alejandro Sahuquillo Cruz, Andrés Méndez Hernández, and Lucía Mondaray Travieso. The interaction with such brilliant students has been for him an important source of motivation and inspiration to develop and polish many contents of this book, which, consequently, is indebted to them. One does not teach the same material one year after the other, as the contents of successive courses experience small mutations every year and, occasionally, this happens because of the influence and mathematical abilities of the students who propose brilliant and appealing questions. Teaching requires knowledge of the precise level that your students can reach with the teachers' daily work; it is a psychological act of synthesis and transmission of knowledge. This author also wants to express his deepest gratitude to Alfredo Somolinos, former Ph.D. student of Burton, who taught him the curse of *Ordinary Differential Equations* at the Complutense University of Madrid during the academic year 1978–1979, and later introduced him to the applications of the abstract theory in Population Dynamics and Biology.

Andrea Tellini is deeply grateful to Professor Fabio Zanolin for transmitting him the passion for Differential Equations over several courses in Mathematical Analysis, Differential Equations, and Dynamical Systems, which the professor taught him at the University of Udine and the Scuola Superiore dell'Università degli Studi di Udine from 2005 to 2010, as well as during the supervision of different works and academic projects. Most of the knowledge of this author about the topics of this book is essentially due to Fabio Zanolin, one of the world's leading experts in the theory of Differential Equations, Dynamical Systems, and Topology. He would also like to thank Professors Sebastiano Sonego and Dikran Dikranjan for what he has learned about Mathematical Physics and Topology, for their deep professional knowledge in these subjects, and for their open-mindedness to consider his interests in Differential Equations. Last, but not least, he wants to acknowledge all the members of the DEG1 research group (https://deg1.uniud.it/), who, over the years, have opened to him new insights in the world of Differential Equations while always having good times together.

Our research in the field of Ordinary Differential Equations and Dynamical Systems has been extraordinarily facilitated by our

Research Grants funded by the Ministry of Science and Technology of Spain (in particular, PID2021-123343NB-I00) and by the Institute of Interdisciplinary Mathematics of Complutense University of Madrid. Andrea Tellini has also been supported by the Ramón y Cajal program RYC2022-038091-I, funded by MCIN/AEI/10.13039/501100011033 and by the FSE+.

Finally, we want to thank Eduardo Muñoz Hernández for finding some misprints in a preliminary version of Chapters 1–3. We have done our best to reduce the number of typos/errors/misprints in the whole book. However, as we are humans, a few will most likely remain. If the reader would like to point them out to us, please email us at jlopezgo@ucm.es and/or andrea.tellini@upm.es. We will really appreciate it!

Julián López-Gómez and Andrea Tellini

Madrid, October 11th 2024

About the Authors

Julián López-Gómez has served as a guest professor and delivered master and doctoral level courses at several institutions worldwide, including: the Mathematical Institute of the University of Zurich (Switzerland); the National Center for Theoretical Sciences of Taiwan, formerly based in Hsinchu (Taiwan); Xi'an University of Technology (P. R. China); the Departments of Mathematics at the University of Trieste (Italy) and the University of Brasilia (Brazil); the National Institute for the Development of the Chemical Industry at Santa Fe, the University of Buenos Aires, and the University of Bahía Blanca (R. Argentina); and the University of Los Andes at Mérida (Venezuela). He has supervised 15 Ph.D. theses, authored around 220 research papers—most published in leading journals—and 10 books, and has edited 5 volumes, including the *Proceedings of the 10th AIMS Conference* (Madrid 2014, the largest conference on Differential Equations ever held, with over 2,300 participants). He has delivered invited talks at 88 conferences, co-organized 33 international scientific events, and served as a member of the editorial boards of several journals, including *Advanced Nonlinear Studies, Nonlinear Analysis, Rendiconti dell'Istituto di Matematica dell'Università di Trieste* (RIMUT), *Abstract and Applied Analysis, Journal of Applied Mathematics, Mathematics,* and *Discrete and Continuous Dynamical Systems.* He has also prepared reports for over 80 international institutions and scientific interdisciplinary journals.

His research interests span a multitude of fields, including partial differential equations, ordinary differential equations, numerical analysis, nonlinear analysis, operator theory, complex analysis, topology, and geometry, as well as a number of applications in Chemistry, Physics, Engineering, and Population Dynamics.

Prof. López-Gómez has undertaken scientific tours in Argentina (1988; Santa Fe, Córdoba, San Juan, Rosario, Buenos Aires, and La Plata), South Korea (2011; Seoul, Pohang, and Busan), Taiwan (2013; Taipei, Hsinchu, and Kaohsiung), Japan (2023; Tokyo, Meiji, Waseda, Ibaraki, Hiroshima, and Fukuoka), and Brazil (2024; Rio de Janeiro, Salvador de Bahía, Belém do Pará, and Brasilia), and has delivered seminars at mathematical institutes worldwide, including at Kyoto, Hiroshima, Kitakyushu, and Fukuoka (Japan); Beijing, Xi'an, Tianjin, and Taiyuan (China); Banff, Toronto, and Winnipeg (Canada); Madison (WI), San Antonio (TX), Miami (FL), Orlando (FL), and Dallas (TX), all in the U.S.; and Sydney (Australia), as well as at the Courant Institute of New York University, where he had the opportunity to discuss mathematics with L. Nirenberg and S. R. S. Varadhan, Abel prize recipients.

He has collaborated with many influential mathematicians, including H. Amann, E. N. Dancer, P. Omari, P. Pucci, F. Zanolin, L. Véron, and P. H. Rabinowitz, former president of the Section of Mathematics of the U.S. National Academy of Sciences. Since December 2021, P. Omari, F. Zanolin, and P. H. Rabinowitz have been members of the research team of his latest research grant from the Ministry of Science and Innovation of Spain.

He has been included in Stanford University's list of the top 2% most influential scientists, and according to the organizers of his most recent course in Brasilia: "He is an influential mathematician, internationally recognized for his profound contributions to nonlinear analysis, bifurcation theory and their applications in partial differential equations."

Andrea Tellini is a Ramón y Cajal Research Fellow at Polytechnic University of Madrid, where he also teaches a course on Ordinary Differential Equations I as part of the bachelor's degree in Mathematics. After completing his bachelor's and master's studies at the University of Udine (Italy) as a student of the *Scuola Superiore* of the same university, he obtained his Ph.D. at Complutense University of Madrid in 2013, under the supervision of Julián López-Gómez and Marcela Molina-Meyer. He subsequently held post-doc positions in Paris at the Centre d'Analyse et de Mathématique Sociales (Paris), under the supervision of Henri Berestycki, and at the Autónoma University of Madrid, under the supervision of Carlos Mora-Corral, before obtaining a position at the Polytechnic University of Madrid, where he has been working since 2018.

He has authored 18 research papers on differential equations, both ordinary and partial, where he has investigated the existence and multiplicity of solutions for reaction-diffusion problems related to variants of the logistic equation, complementing the analytical results with numerical simulations. He has supervised several bachelor's and master's theses related to the analysis and numerical simulations of models involving differential equations. Currently, he is co-supervising two Ph.D. students, together with Julián López-Gómez.

Contents

Part 2 Nonlinear Systems   207

4. An Introduction to Nonlinear Differential Equations   209

5. Cauchy–Lipschitz Theory   273

Part 1
Linear Systems

Chapter 1

First-Order Linear Systems

This chapter studies the *linear*, or *affine*, first-order system of ordinary differential equations

$$u' = A(t)u + B(t), \tag{1.1}$$

as well as the associated *initial value problem*

$$\begin{cases} u' = A(t)u + B(t), \\ u(t_0) = u_0, \end{cases} \tag{1.2}$$

where, for some fixed $\mathbb{K} \in \{\mathbb{R}, \mathbb{C}\}$, $\alpha, \beta \in \mathbb{R}$ with $\alpha < \beta$, and $t_0 \in [\alpha, \beta]$,

$$A(t) := (a_{ij}(t))_{1 \le i,j \le N} = \begin{pmatrix} a_{11}(t) & \cdots & a_{1N}(t) \\ \vdots & \ddots & \vdots \\ a_{N1}(t) & \cdots & a_{NN}(t) \end{pmatrix}, \quad t \in [\alpha, \beta],$$

is a matrix of order $N \ge 1$ whose coefficients are continuous functions, $a_{ij} \in \mathcal{C}([\alpha, \beta]; \mathbb{K})$, for all $i, j \in \{1, \ldots, N\}$, and

$$B(t) := (b_i(t))_{1 \le i \le N} = \begin{pmatrix} b_1(t) \\ \vdots \\ b_N(t) \end{pmatrix}, \quad t \in [\alpha, \beta],$$

3

with $b_i \in \mathcal{C}([\alpha, \beta]; \mathbb{K})$ for all $i \in \{1, \ldots, N\}$. In (1.1) and (1.2), u is

$$u := \begin{pmatrix} u_1 \\ \vdots \\ u_N \end{pmatrix} = (u_1, \ldots, u_N)^T,$$

so T, in the context of vectors, means "transposition", and u' denotes the vector $(u_1', \ldots, u_N')^T$, where $'$ indicates the derivative with respect to the independent variable t. In (1.2), u_0 is a given point of $\mathbb{K}^N$, *the initial datum*. Initial value problems such as (1.2) are also often referred to as *Cauchy problems.*

In this book, a solution of system (1.1) in the interval $[\alpha, \beta]$ is a vectorial function $u \in \mathcal{C}^1([\alpha, \beta]; \mathbb{K}^N)$ such that

$$u'(t) = A(t)u(t) + B(t) \quad \text{for all } t \in [\alpha, \beta].$$

Obviously, $u'(\alpha)$ denotes the right derivative of u at $t = \alpha$, while $u'(\beta)$ denotes the left derivative of u at $t = \beta$.

If a solution u of (1.1) in addition satisfies $u(t_0) = u_0$, then it is said to solve the Cauchy problem (1.2). Note that $u \in \mathcal{C}^1([\alpha, \beta]; \mathbb{K}^N)$ if and only if $u_j \in \mathcal{C}^1([\alpha, \beta]; \mathbb{K})$ for all $j \in \{1, \ldots, N\}$.

Once an integer $N \geq 1$ is fixed, for any given norm $\|\cdot\|$ in the Euclidean space $\mathbb{K}^N$, it is well known that any other norm, $\|\|\cdot\|\|$, is equivalent to $\|\cdot\|$, i.e. there exist two positive constants, $0 < C_1 < C_2$, such that

$$C_1\|x\| \leq \|\|x\|\| \leq C_2\|x\| \qquad \text{for all } x \in \mathbb{K}^N.$$

The most common norms in $\mathbb{K}^N$ are the p–norms, $\|\cdot\|_p$, defined as

$$\|u\|_p := \left(\sum_{j=1}^{N} |u_j|^p \right)^{\frac{1}{p}}, \qquad u \in \mathbb{K}^N,$$

for all $1 \leq p < +\infty$, and

$$\|u\|_\infty := \max_{1 \leq j \leq N} |u_j|, \qquad u \in \mathbb{K}^N.$$

The norm $\|\cdot\|_2$ is the Euclidean norm. Indeed, when $\mathbb{K} = \mathbb{C}$, we have that

$$\|u\|_2 = \sqrt{\sum_{j=1}^{N} |u_j|^2} = \sqrt{\sum_{j=1}^{N} u_j \bar{u}_j},$$

where, for every $x, y \in \mathbb{R}$, $\bar{u}_j = x - iy$ denotes the complex conjugate of $u_j = x + iy$. Thus,

$$\|u\|_2 = \sqrt{\langle u, u \rangle},$$

where $\langle \cdot, \cdot \rangle$ denotes the inner product (a positive definite Hermitian form) of $\mathbb{C}^N$:

$$\langle u, v \rangle := \sum_{j=1}^{N} u_j \bar{v}_j.$$

Throughout most of this book, we work with the 1-norm $\| \cdot \|_1$. So, unless otherwise specified, we simply denote

$$\|u\| := \|u\|_1 = \sum_{j=1}^{N} |u_j| \quad \text{for all } u \in \mathbb{K}^N.$$

As it will be apparent in the sequel, due to the equivalence of norms, the choice of the norm does not alter the qualitative nature of the results of this book. It only changes the quantitative ones, but this is irrelevant.

According to the notation introduced above, for any given subinterval $J \subset \mathbb{R}$ and every integer $N \geq 1$, we denote by $\mathcal{C}(J; \mathbb{K}^N)$ the set of continuous functions $u : J \to \mathbb{K}^N$ and by $\mathcal{C}^1(J; \mathbb{K}^N)$ the set of functions $u : J \to \mathbb{K}^N$ such that $u'(t) \in \mathbb{K}^N$ exists for all $t \in J$ and $u' \in \mathcal{C}(J; \mathbb{K}^N)$. $\mathcal{C}(J; \mathbb{K}^N)$ and $\mathcal{C}^1(J; \mathbb{K}^N)$ can be regarded as infinite-dimensional linear vector spaces with the standard sum of functions and their multiplication by scalars in the field $\mathbb{K}$.

1.1 Structure Properties of the Solutions of $u' = Au + B$

Usually,

$$u' = A(t)u + B(t)$$

is referred to as an *inhomogeneous* first-order linear system, while

$$u' = A(t)u$$

is known as the associated *homogeneous* linear system. Throughout this chapter, we denote by Σ_B the set of solutions of $u' = A(t)u + B(t)$

in $[\alpha, \beta]$. So, Σ_0 denotes the set of solutions of the associated homogeneous system, $u' = A(t)u$, in $[\alpha, \beta]$. The following result establishes that Σ_0 is a linear subspace of the vector space $\mathcal{C}^1([\alpha, \beta]; \mathbb{K}^N)$, while Σ_B is an affine subspace in the direction of Σ_0. To differentiate the solutions of Σ_0 and Σ_B, in this book, we often denote by u, or v, the elements of Σ_B and by h those of Σ_0.

Lemma 1.1 (Superposition principle). *The following hold:*

(a) *for every $h_1, h_2 \in \Sigma_0$ and $\lambda, \mu \in \mathbb{K}$, $\lambda h_1 + \mu h_2 \in \Sigma_0$;*
(b) *for every $u, v \in \Sigma_B$ and $h \in \Sigma_0$, $u - v \in \Sigma_0$ and $h + u \in \Sigma_B$.*

Proof.	(a) Let $h_1, h_2 \in \Sigma_0$. Then, thanks to the linearity of the derivative, for every $\lambda, \mu \in \mathbb{K}$ and $t \in [\alpha, \beta]$,

$$(\lambda h_1 + \mu h_2)'(t) = \lambda h_1'(t) + \mu h_2'(t)$$
$$= \lambda A(t)h_1(t) + \mu A(t)h_2(t)$$
$$= A(t)\left(\lambda h_1(t) + \mu h_2(t)\right).$$

Thus, $\lambda h_1 + \mu h_2 \in \Sigma_0$.

(b) Now, suppose $u, v \in \Sigma_B$. Then,

$$(u - v)'(t) = u'(t) - v'(t)$$
$$= A(t)u(t) + B(t) - (A(t)v(t) + B(t))$$
$$= A(t)\left(u(t) - v(t)\right)$$

for all $t \in [\alpha, \beta]$. Hence, $u - v \in \Sigma_0$. Moreover,

$$(h + u)'(t) = h'(t) + u'(t)$$
$$= A(t)h(t) + A(t)u(t) + B(t)$$
$$= A(t)\left(h(t) + u(t)\right) + B(t)$$

for all $h \in \Sigma_0$ and $t \in [\alpha, \beta]$. So, $h + u \in \Sigma_B$. This ends the proof.	$\square$

By induction, it readily follows that any linear combination of functions in Σ_0 remains in Σ_0. The fact that any linear combination of solutions of $u' = A(t)u$ also solves $u' = A(t)u$ is known as the *superposition principle* since, in the context of differential equations, *linear combinations of solutions* were already known as *superpositions*, while the concept of *linear combination* was coined in the nineteenth century. So, the concept of superposition in the context of differential equations is much older than the more abstract one of linear combination. In this book, we use the concepts of linear combinations and superpositions synonymously.

According to Lemma 1.1(a), Σ_0 is a linear subspace of $\mathcal{C}^1([\alpha, \beta]; \mathbb{K}^N)$, while, due to Lemma 1.1(b), Σ_B is an affine subspace of $\mathcal{C}^1([\alpha, \beta]; \mathbb{K}^N)$ parallel to Σ_0. Indeed, the difference of the two points $u, v \in \Sigma_B$ is a free vector, i.e. $u - v \in \Sigma_0$, and the sum of a free vector $h \in \Sigma_0$ and a point $u \in \Sigma_B$ provides us with another point, $u + h$, which belongs to Σ_B.

These structural properties play a pivotal role in the theory of linear systems.

The main goal of this chapter is to establish that Σ_0 and Σ_B are isomorphic to $\mathbb{K}^N$. Once that done, we will construct affine reference systems in Σ_B starting from any given basis of $\mathbb{K}^N$.

1.2 Linear Scalar Equation

The linear scalar equation is the simplest ordinary differential equation. It reads

$$u' = a(t)u + b(t),$$

where $a, b \in \mathcal{C}([\alpha, \beta]; \mathbb{K})$. For this equation, the following fundamental existence and uniqueness result holds.

Theorem 1.2. *Suppose $a, b \in \mathcal{C}([\alpha, \beta]; \mathbb{K})$. Then, for every $t_0 \in [\alpha, \beta]$ and $u_0 \in \mathbb{K}$, the initial value problem*

$$\begin{cases} u' = a(t)u + b(t), \\ u(t_0) = u_0, \end{cases} \tag{1.3}$$

possesses a unique solution in $[\alpha, \beta]$, $u(t; t_0, u_0)$, which is given by

$$u(t; t_0, u_0) = e^{\int_{t_0}^{t} a(\sigma)\, d\sigma} u_0 + \int_{t_0}^{t} e^{\int_{s}^{t} a(\sigma)\, d\sigma} b(s)\, ds \tag{1.4}$$

for all $t \in [\alpha, \beta]$.

Proof. We start by finding a solution to the homogeneous equation $h' = a(t)h$ such that $h(t_0) = 1$. Since any solution must be continuous, there exists $\tilde{t} > 0$ such that

$$h(t) \neq 0 \quad \text{if } t \in J := \left[t_0 - \tilde{t}, t_0 + \tilde{t}\right] \cap [\alpha, \beta].$$

Thus,

$$\frac{h'(t)}{h(t)} = a(t) \quad \text{for all } t \in J,$$

or, equivalently,

$$\frac{d}{dt} \operatorname{Log} h(t) = a(t) \quad \text{for all } t \in J$$

(here, and in the rest of the book, we denote by Log the *principal value* of the complex logarithm, which reduces to the *natural logarithm*, denoted by ln, for positive real numbers). Hence, for every $t \in J$, integrating the previous relation in (t_0, t) yields

$$\operatorname{Log} \frac{h(t)}{h(t_0)} = \int_{t_0}^{t} a(\sigma)\, d\sigma,$$

and, since $h(t_0) = 1$,

$$h(t) := e^{\int_{t_0}^{t} a(\sigma)\, d\sigma} \quad \text{for all } t \in J.$$

Actually, this argument works as long as $h(t)$ is non-zero, leading to the explicit expression

$$h(t) := e^{\int_{t_0}^{t} a(\sigma)\, d\sigma}, \qquad t \in [\alpha, \beta],$$

which is non-zero for all $t \in [\alpha, \beta]$. Thus, this function satisfies $h' = a(t)h$ in the whole interval $[\alpha, \beta]$, and $h(t_0) = 1$. Next, we look for a solution u to the inhomogeneous differential equation in (1.3) of the form

$$u := hv. \tag{1.5}$$

This change of variable transforms the linear scalar equation $u' = a(t)u + b(t)$ into

$$h'v + hv' = ahv + b$$

or, equivalently,

$$ahv + hv' = ahv + b.$$

Consequently, (1.5) transforms (1.3) into

$$\begin{cases} v' = b/h, \\ v(t_0) = u_0, \end{cases}$$

whose unique solution is obtained by direct integration and is given by

$$v(t) = u_0 + \int_{t_0}^{t} e^{-\int_{t_0}^{s} a(\sigma)\, d\sigma} b(s)\, ds, \qquad t \in [\alpha, \beta].$$

Therefore, by (1.5),

$$\begin{aligned} u(t) &= e^{\int_{t_0}^{t} a(\sigma)\, d\sigma} \left(u_0 + \int_{t_0}^{t} e^{-\int_{t_0}^{s} a(\sigma)\, d\sigma} b(s)\, ds \right) \\ &= e^{\int_{t_0}^{t} a(\sigma)\, d\sigma} u_0 + \int_{t_0}^{t} e^{\int_{t_0}^{t} a(\sigma)\, d\sigma} e^{-\int_{t_0}^{s} a(\sigma)\, d\sigma} b(s)\, ds \\ &= e^{\int_{t_0}^{t} a(\sigma)\, d\sigma} u_0 + \int_{t_0}^{t} e^{\int_{t_0}^{t} a(\sigma)\, d\sigma - \int_{t_0}^{s} a(\sigma)\, d\sigma} b(s)\, ds \\ &= e^{\int_{t_0}^{t} a(\sigma)\, d\sigma} u_0 + \int_{t_0}^{t} e^{\int_{s}^{t} a(\sigma)\, d\sigma} b(s)\, ds \end{aligned}$$

is the unique solution of (1.3). $\qquad\square$

Note that, when $a(t)$ is constant, say $a \in \mathbb{K}$, for all $t \in [\alpha, \beta]$, (1.4) becomes

$$u(t; t_0, u_0) = e^{a(t-t_0)} u_0 + \int_{t_0}^{t} e^{a(t-s)} b(s) \, ds.$$

Also, observe that functions of the form $h(t)C$, where C is a constant, cannot solve $u' = a(t)u + b(t)$, unless $b = 0$, because they solve the homogeneous equation. The main idea of the proof of Theorem 1.2 lies in replacing the constant C by an unknown function, $v(t)$, in order to determine all possible solutions of $u' = a(t)u + b(t)$. For such a reason, this method, which goes back to Lagrange, is called the Lagrange *method of variation of constants*. Similarly, (1.4) is usually referred to as the *variation of constants formula*.

As a particular case of Theorem 1.2, we obtain that the function

$$h(t; t_0, u_0) := e^{\int_{t_0}^{t} a(\sigma) \, d\sigma} u_0, \qquad t \in [\alpha, \beta], \tag{1.6}$$

provides us with the unique solution to the Cauchy problem

$$\begin{cases} h' = a(t)h, \\ h(t_0) = u_0. \end{cases}$$

Indeed, (1.4) reduces to (1.6) when $b = 0$. Note that, when $\mathbb{K} = \mathbb{R}$, $h(t; t_0, u_0) > 0$ for all $t \in [\alpha, \beta]$ if $u_0 > 0$, $h(t; t_0, u_0) < 0$ for all $t \in [\alpha, \beta]$ if $u_0 < 0$, and $h(\cdot; t_0, u_0) = 0$ in $[\alpha, \beta]$ if $u_0 = 0$. Here, as in the rest of this book, we denote with a $\cdot$ the argument of the considered function.

Similarly, owing to Theorem 1.2, we infer from (1.4) that

$$u(t; t_0, 0) = \int_{t_0}^{t} e^{\int_{s}^{t} a(\sigma) \, d\sigma} b(s) \, ds, \qquad t \in [\alpha, \beta], \tag{1.7}$$

is the unique solution of

$$\begin{cases} u' = a(t)u + b(t), \\ u(t_0) = 0. \end{cases}$$

By (1.6) and (1.7), the variation of constants formula, (1.4), can be equivalently expressed in the form

$$u(\cdot; t_0, u_0) = h(\cdot; t_0, u_0) + u(\cdot; t_0, 0). \tag{1.8}$$

Observe that expression (1.8) can be regarded as a superposition, in the spirit of Lemma 1.1, with the difference that here one has to take into account the initial data as well. Consequently, as a by-product of Theorem 1.2, the following result holds.

Theorem 1.3. *The solution operator $\mathcal{S}_0 : \mathbb{K} \to \Sigma_0$, defined as*

$$\mathcal{S}_0(x) := h(\cdot\,; t_0, x) = e^{\int_{t_0}^{\cdot} a} x,$$

is a linear isomorphism. Thus, the solution operator $\mathcal{S}_b : \mathbb{K} \to \Sigma_b$, defined as

$$\mathcal{S}_b(x) := u(\cdot\,; t_0, x) = \mathcal{S}_0(x) + u(\cdot\,; t_0, 0),$$

is an affine isomorphism. In particular,

$$\dim \Sigma_b = \dim \Sigma_0 = 1.$$

Therefore, Σ_0 is a vectorial line and Σ_b is an affine line parallel to Σ_0 passing through the point $u(\cdot\,; t_0, 0)$.

Proof. By definition, for every $\lambda, \mu, x, y \in \mathbb{K}$,

$$\mathcal{S}_0(\lambda x + \mu y) = e^{\int_{t_0}^{\cdot} a}(\lambda x + \mu y) = \lambda e^{\int_{t_0}^{\cdot} a} x + \mu e^{\int_{t_0}^{\cdot} a} y = \lambda \mathcal{S}_0(x) + \mu \mathcal{S}_0(y)$$

and, hence, $\mathcal{S}_0$ is linear. Moreover, since the exponential is non-zero, if $\mathcal{S}_0 x = 0$, necessarily $x = 0$, and so $\mathcal{S}_0$ is an injection. Finally, it is surjective because, by the uniqueness established in Theorem 1.2, for every $w \in \Sigma_0$,

$$\mathcal{S}_0(w(t_0)) = h(\cdot\,; t_0, w(t_0)) = w.$$

The fact that $\mathcal{S}_b$ establishes an affine isomorphism is a direct consequence of (1.8) because (1.8) entails

$$\mathcal{S}_b = \mathcal{S}_0 + u(\cdot\,; t_0, 0).$$

This ends the proof. $\square$

According to Theorem 1.3, Σ_0 is spanned by any non-zero solution of $h' = a(t)h$. In particular, for every $t_0 \in [\alpha, \beta]$,

$$\Sigma_0 = \operatorname{span}\left[e^{\int_{t_0}^t a} \right].$$

Hence, for every *particular solution*, $p(t)$, of $u' = a(t)u + b(t)$,

$$\Sigma_b = \operatorname{span}\left[e^{\int_{t_0}^t a} \right] + p(t).$$

In other words, the parametric equation of the straight line Σ_0 is

$$e^{\int_{t_0}^t a} x, \qquad x \in \mathbb{K},$$

while, for every $p \in \Sigma_b$, the parametric equation of Σ_b is

$$e^{\int_{t_0}^t a} x + p(t), \qquad x \in \mathbb{K}.$$

In the context of ordinary differential equations, these *parametric equations* are usually referred to as the *general solutions* of the differential equations $h' = a(t)h$ and $u' = a(t)u + b(t)$, respectively.

As already remarked above, when $a(t)$ is a constant, $a \in \mathbb{K}$, it is easy to see that $h(t; t_0, u_0) = e^{a(t-t_0)} u_0$, $t \in \mathbb{R}$, is the unique solution of the problem

$$\begin{cases} h' = ah, \\ h(t_0) = u_0. \end{cases}$$

As a consequence of this observation and the structural properties that we have established for the linear equation, we now determine the unique solution of the problem

$$\begin{cases} u' = au + b(t), \\ u(t_0) = u_0, \end{cases}$$

for some specific choices of $b(t)$, without needing to use, if possible, the method of variation of constants.

Case 1. $a \neq 0$ and $b(t) = b$, a constant. In this case, $u = -b/a$ provides us with a constant solution of $u' = au + b$. Thus, there exists $x \in \mathbb{K}$ such that

$$u(t; t_0, u_0) = e^{a(t-t_0)}x - \frac{b}{a}, \qquad t \in \mathbb{R}.$$

Evaluating at $t = t_0$ yields

$$x - \frac{b}{a} = u_0$$

and, therefore,

$$u(t; t_0, u_0) = e^{a(t-t_0)}\left(u_0 + \frac{b}{a}\right) - \frac{b}{a}, \qquad t \in \mathbb{R}.$$

Case 2. $a \in \mathbb{K}$, $b(t) = e^{\omega t}$, for some $\omega \in \mathbb{K} \setminus \{a\}$. In this case, the linear scalar equation

$$u' = au + e^{\omega t}$$

admits a particular solution of the form $p(t) = me^{\omega t}$ for some $m \in \mathbb{K}$. Indeed, the constant m shall be chosen so that

$$m\omega e^{\omega t} = ame^{\omega t} + e^{\omega t}$$

and, hence, $m = \frac{1}{\omega - a}$. Therefore, since Σ_0 is the vectorial line in the direction of e^{at}, the general solution of $u' = au + e^{\omega t}$ is given by

$$u(t) = e^{at}x + \frac{e^{\omega t}}{\omega - a}, \qquad x \in \mathbb{K}.$$

In order to determine the unique solution of the differential equation such that $u(t_0) = u_0$, one must impose

$$e^{at_0}x + \frac{e^{\omega t_0}}{\omega - a} = u_0$$

and, hence,

$$x = e^{-at_0}\left(u_0 - \frac{e^{\omega t_0}}{\omega - a}\right).$$

Therefore,

$$u(t; t_0, u_0) = e^{a(t-t_0)} \left(u_0 - \frac{e^{\omega t_0}}{\omega - a} \right) + \frac{e^{\omega t}}{\omega - a}, \qquad t \in \mathbb{R}.$$

Case 3. $a \in \mathbb{K}$ and $b(t) = e^{at}$. In contrast to Case 2, the differential equation

$$u' = au + e^{at}$$

cannot admit a solution of the form me^{at} with $m \in \mathbb{K}$ because such a function solves the associated homogeneous equation. Hence, in order to obtain the general solution in this case, one shall use the method of variation of constants: the change of variable

$$u(t) = e^{at} v(t)$$

transforms the differential equation into

$$ae^{at} v + e^{at} v' = ae^{at} v + e^{at},$$

or, equivalently, $v' = 1$, whose general solution is $v(t) = t + x$ with $x \in \mathbb{K}$. Therefore,

$$u(t) = e^{at}(t + x) = e^{at} x + te^{at}, \qquad t \in \mathbb{R},$$

where $x \in \mathbb{K}$, provides us with the general solution of $u' = au + e^{at}$. Since $e^{at} x$ solves the homogeneous equation, by Lemma 1.1, the function te^{at} solves the inhomogeneous equation, which is not obvious at a first glance.

In order to determine the unique solution, $u(t; t_0, u_0)$, of the associated Cauchy problem

$$\begin{cases} u' = au + e^{at}, \\ u(t_0) = u_0, \end{cases}$$

we should find the unique value of x such that

$$e^{at_0}(x + t_0) = u_0,$$

which is given by

$$x = e^{-at_0} u_0 - t_0.$$

Consequently,

$$u(t; t_0, u_0) = e^{a(t-t_0)} u_0 + (t - t_0) e^{at}, \qquad t \in \mathbb{R}.$$

The lesson we learn from this example is that, when it is difficult to find a particular solution, $p(t)$, varying the coefficients as we have done here is always a good try.

Observe that the method of variation of constants not only provides us with the solutions to problem (1.3) when $a, b \in \mathcal{C}([\alpha, \beta]; \mathbb{K})$, but it also gives the formula for a solution, for instance, when $a(t)$, or $b(t)$, has a finite number of discontinuities in $[\alpha, \beta]$. Of course, in these cases, the concept of a solution to (1.3) has to be generalized: we require the solution to belong to $\mathcal{C}^1(J; \mathbb{K})$ for every interval J where both a and b are continuous, and, in addition, to $\mathcal{C}([\alpha, \beta]; \mathbb{K})$. Actually, in such a framework, it is possible to show that the solution of a given initial value problem is unique. Indeed, it can be constructed by properly gluing, in a progressive way, the unique solution in the continuity intervals J, whose existence and uniqueness are given by Theorem 1.2.

As an example, consider

$$b(t) := \begin{cases} 1 & \text{if } t \geq 0, \\ 0 & \text{if } t < 0. \end{cases}$$

The variation of constants formula gives that the unique solution of

$$\begin{cases} u' = u + b(t), \\ u(0) = u_0, \end{cases}$$

should be given by

$$u(t; 0, u_0) = e^t u_0 + \int_0^t e^{t-s} b(s) \, ds = \begin{cases} e^t u_0 + e^t - 1 & \text{if } t \geq 0, \\ e^t u_0 & \text{if } t < 0. \end{cases}$$

Indeed, the function $u_+(t) := e^t(u_0 + 1) - 1$ solves the problem

$$\begin{cases} u' = u + 1, \\ u(0) = u_0, \end{cases}$$

in $[0, \infty)$, while $u_-(t) := e^t u_0$ solves

$$\begin{cases} u' = u, \\ u(0) = u_0, \end{cases}$$

in $(-\infty, 0]$. In addition, we have $u_+(0) = u_-(0) = u_0$, but

$$u'_-(0) = u_0 < u_0 + 1 = u'_+(0).$$

Therefore, the solution $u(t; 0, u_0)$ is not differentiable at $t = 0$. Naturally, such a discontinuity of u' at $t = 0$ is provoked by the discontinuity of $b(t)$ at that point.

We conclude this section by solving the problem

$$\begin{cases} u' = 2tu + te^{t^2}, \\ u(0) = x, \end{cases} \tag{1.9}$$

for a generic $x \in \mathbb{R}$. First, we solve the associated homogeneous problem

$$\begin{cases} h' = 2th, \\ h(0) = 1. \end{cases}$$

According to Theorem 1.2, $h(t) = e^{t^2}$ provides us with the unique solution to this problem for all $t \in \mathbb{R}$. Thus, the change of variable

$$u(t) := e^{t^2} v(t)$$

transforms problem (1.9) into

$$\begin{cases} e^{t^2} v' = te^{t^2}, \\ v(0) = x, \end{cases}$$

which can be equivalently expressed as

$$\begin{cases} v' = t, \\ v(0) = x. \end{cases}$$

As the unique solution of this problem is $v(t) = \frac{t^2}{2} + x$, the unique solution of (1.9) is

$$u(t) = e^{t^2} \left(\frac{t^2}{2} + x \right), \qquad t \in \mathbb{R}.$$

1.3   General Existence Theorem for $N \geq 1$

This section generalizes Theorem 1.2 to cover the general multidimensional case $N \geq 1$. When $N > 1$, it is not possible to proceed as in Section 1.2 to solve the homogeneous system explicitly. Actually, in most cases, finding an explicit formula for the solution is impossible, and another approach must be followed. The main result can be stated as follows.

Theorem 1.4. *Suppose*

$$a_{ij}, \; b_i \in \mathcal{C}([\alpha, \beta]; \mathbb{K}) \quad \text{for all } 1 \leq i, j \leq N.$$

Then, for every $t_0 \in [\alpha, \beta]$ and $u_0 \in \mathbb{K}^N$, problem (1.2) possesses a unique solution, $u(t; t_0, u_0)$, in the interval $[\alpha, \beta]$.

The proof of Theorem 1.4 relies on the following result.

Lemma 1.5. *Let $u \in \mathcal{C}([\alpha, \beta]; \mathbb{K}^N)$. Then, the following assertions are equivalent:*

(a) $u \in \mathcal{C}^1([\alpha, \beta]; \mathbb{K}^N)$ *solves the initial value problem (1.2);*
(b) $u \in \mathcal{C}([\alpha, \beta]; \mathbb{K}^N)$ *satisfies the integral equation*

$$u(t) = u_0 + \int_{t_0}^{t} [A(s)u(s) + B(s)] \, ds \quad \text{for all } t \in [\alpha, \beta], \quad (1.10)$$

where the integral is understood component-wise.

Proof. Suppose (a). Then, integrating $u' = A(t)u + B(t)$ yields

$$u(t) - u(t_0) = \int_{t_0}^{t} u'(s) \, ds = \int_{t_0}^{t} [A(s)u(s) + B(s)] \, ds$$

for all $t \in [\alpha, \beta]$, which implies (1.10) because $u(t_0) = u_0$.
   Conversely, suppose (b). Then, since

$$s \mapsto A(s)u(s) + B(s) \in \mathbb{K}^N$$

is continuous in $[\alpha, \beta]$, by the fundamental theorem of calculus, the function $\int_{t_0}^{t} [A(s)u(s) + B(s)] \, ds$ is a differentiable function such that

$$\frac{d}{dt} \int_{t_0}^{t} [A(s)u(s) + B(s)] \, ds = A(t)u(t) + B(t) \quad \text{for all } t \in [\alpha, \beta].$$

Thus, it follows from (1.10) that u is differentiable and

$$u'(t) = A(t)u(t) + B(t) \quad \text{for all } t \in [\alpha, \beta],$$

which in particular implies that $u \in \mathcal{C}^1([\alpha, \beta]; \mathbb{K}^N)$. Moreover, evaluating (1.10) at t_0 yields $u(t_0) = u_0$. Therefore, u solves (1.2). $\square$

According to Lemma 1.5, the solutions of the Cauchy problem (1.2) are the fixed points of the *integral operator*

$$\mathcal{K} : \mathcal{C}([\alpha, \beta]; \mathbb{K}^N) \to \mathcal{C}([\alpha, \beta]; \mathbb{K}^N),$$

defined, for every $u \in \mathcal{C}([\alpha, \beta]; \mathbb{K}^N)$, by

$$\mathcal{K}u(t) := u_0 + \int_{t_0}^{t} [A(s)u(s) + B(s)]\, ds, \qquad t \in [\alpha, \beta]. \qquad (1.11)$$

In terms of the integral operator $\mathcal{K}$, we will prove Theorem 1.4 by showing that $\mathcal{K}$ possesses a unique fixed point. This will be a direct consequence of the contractive mapping theorem applied to $\mathcal{K}$ in the Banach space $\mathcal{C}([\alpha, \beta]; \mathbb{K}^N)$ equipped with an appropriate norm. In the rest of this section, we introduce the concepts needed to state and prove the contractive mapping theorem, up to arriving at the proof of Theorem 1.4 in Section 1.3.4.

1.3.1 *Space of continuous functions*

Throughout the rest of this book, for any given $\alpha < \beta$, $\mathbb{K} \in \{\mathbb{R}, \mathbb{C}\}$, and every integer $N \geq 1$, we consider the $\mathbb{K}$-vector space $\mathcal{C}([\alpha, \beta]; \mathbb{K}^N)$ of the continuous functions $u : [\alpha, \beta] \to \mathbb{K}^N$ equipped with the maximum norm

$$\|u\|_\infty := \max_{t \in [\alpha, \beta]} \|u(t)\|, \qquad u \in \mathcal{C}([\alpha, \beta]; \mathbb{K}^N), \qquad (1.12)$$

where $\|\cdot\|$ denotes any (given) norm of $\mathbb{K}^N$. Since $t \mapsto \|u(t)\|$ is continuous and $[\alpha, \beta]$ is compact, $\|u\|_\infty$ is well defined. Moreover, $\|\cdot\|_\infty$ is a norm because:

- $\|u\|_\infty \geq 0$ for all $u \in \mathcal{C}([\alpha, \beta]; \mathbb{K}^N)$, and $u = 0$ if $\|u\|_\infty = 0$;
- $\|\lambda u\|_\infty = |\lambda| \|u\|_\infty$ for all $\lambda \in \mathbb{K}$ and $u \in \mathcal{C}([\alpha, \beta]; \mathbb{K}^N)$;
- for every $u_1, u_2 \in \mathcal{C}([\alpha, \beta]; \mathbb{K}^N)$, the following inequality holds

$$\|u_1 + u_2\|_\infty \leq \|u_1\|_\infty + \|u_2\|_\infty,$$

referred to as the *triangle inequality*.

Thus,

$$\mathcal{C}([\alpha,\beta];\mathbb{K}^N) := \left(\mathcal{C}([\alpha,\beta];\mathbb{K}^N), \|\cdot\|_\infty\right)$$

is a normed $\mathbb{K}$-vector space. It is infinite-dimensional because, for every $e \in \mathbb{K}^N$, $e \neq 0$, the set of monomials $\{t^n e\}_{n\in\mathbb{N}}$ is a system of independent continuous functions in any compact interval $[\alpha,\beta]$, with $\alpha < \beta$. Indeed, if for some $n \geq 1$, there exists $(c_0, c_1, \ldots, c_n) \in \mathbb{K}^{n+1}$ such that

$$c_0 e + c_1 t e + c_2 t^2 e + \cdots + c_n t^n e = 0 \quad \text{for all } t \in [\alpha,\beta],$$

then, since $e \neq 0$,

$$c_0 + c_1 t + c_2 t^2 + \cdots + c_n t^n = 0 \quad \text{for all } t \in [\alpha,\beta].$$

Hence, differentiating n times with respect to t yields $n!c_n = 0$, which implies $c_n = 0$. Thus,

$$c_0 + c_1 t + c_2 t^2 + \cdots + c_{n-1} t^{n-1} = 0 \quad \text{for all } t \in [\alpha,\beta],$$

and reiterating the previous argument leads to

$$c_{n-1} = c_{n-2} = \cdots = c_0 = 0,$$

which establishes the linear independence of

$$\{e, te, t^2 e, \ldots, t^n e\}$$

for all $n \geq 1$ and $e \in \mathbb{K}^N \setminus \{0\}$.

The following result shows that $\mathcal{C}([\alpha,\beta];\mathbb{K}^N)$, equipped with the maximum norm, is a Banach space.

Theorem 1.6. *$\mathcal{C}([\alpha,\beta];\mathbb{K}^N)$, equipped with the maximum norm, is a Banach space, i.e. for any given Cauchy sequence, $\{u_n\}_{n\geq 1}$, in $\mathcal{C}([\alpha,\beta];\mathbb{K}^N)$, there exists $u \in \mathcal{C}([\alpha,\beta];\mathbb{K}^N)$ such that*

$$\lim_{n\to\infty} \|u_n - u\|_\infty = 0.$$

Proof. Let $\{u_n\}_{n \geq 1}$ be a Cauchy sequence in $\mathcal{C}([\alpha, \beta]; \mathbb{K}^N)$ with respect to the maximum norm. By the definition of Cauchy sequence, for every $\varepsilon > 0$, there exists $n_0 = n_0(\varepsilon) \in \mathbb{N}$ such that

$$\|u_{n+m} - u_n\|_\infty \leq \varepsilon \quad \text{for all } n \geq n_0 \text{ and } m \geq 1.$$

Thus, owing to (1.12),

$$\|u_{n+m}(t) - u_n(t)\| \leq \varepsilon \text{ for } n \geq n_0,\, m \geq 1 \text{ and } t \in [\alpha, \beta]. \quad (1.13)$$

Consequently, for every $t \in [\alpha, \beta]$, $\{u_n(t)\}_{n \geq 1}$ is a Cauchy sequence in $\mathbb{K}^N$. Hence, since $\mathbb{K}^N$ is complete, there exists $u(t) \in \mathbb{K}^N$ such that

$$\lim_{n \to \infty} u_n(t) = u(t).$$

In particular, letting $m \to \infty$ in (1.13) gives

$$\|u(t) - u_n(t)\| \leq \varepsilon \quad \text{for all } n \geq n_0 \text{ and } t \in [\alpha, \beta]. \quad (1.14)$$

It remains to prove that $t \mapsto u(t)$ is continuous in $[\alpha, \beta]$. Indeed, since u_{n_0} is continuous in $[\alpha, \beta]$ and $[\alpha, \beta]$ is compact, u_{n_0} is uniformly continuous in $[\alpha, \beta]$. Thus, for any given $\varepsilon > 0$, there exists $\delta = \delta(\varepsilon) > 0$ such that

$$\|u_{n_0}(t + h) - u_{n_0}(t)\| \leq \varepsilon \quad \text{if } t, t + h \in [\alpha, \beta], \text{ with } |h| \leq \delta. \quad (1.15)$$

Consequently, by (1.14) and (1.15), we find that

$$\|u(t + h) - u(t)\| \leq \|u(t + h) - u_{n_0}(t + h)\|$$
$$+ \|u_{n_0}(t + h) - u_{n_0}(t)\| + \|u_{n_0}(t) - u(t)\| \leq 3\varepsilon$$

provided $t, t + h \in [\alpha, \beta]$, with $|h| \leq \delta$. Therefore, $u \in \mathcal{C}([\alpha, \beta]; \mathbb{K}^N)$. Finally, taking the maximum in $t \in [\alpha, \beta]$ in (1.14) shows that

$$\|u_n - u\|_\infty \leq \varepsilon \quad \text{for all } n \geq n_0.$$

Equivalently,

$$\lim_{n \to \infty} \|u_n - u\|_\infty = 0,$$

which completes the proof of the theorem. $\qquad\square$

In practice, when a sequence of continuous functions, $\{u_n\}_{n\geq 1}$, converges in the Banach space $\mathcal{C}([\alpha,\beta];\mathbb{K}^N)$, equipped with the maximum norm, to a continuous function, u, we say that

$$\lim_{n\to\infty} u_n = u \quad \text{uniformly in } [\alpha,\beta].$$

This is why the topology induced by the norm $\|\cdot\|_\infty$ in the vector space $\mathcal{C}([\alpha,\beta];\mathbb{K}^N)$ is often referred to as the *topology of the uniform convergence* of continuous functions.

Figure 1.1 illustrates the geometrical meaning of the uniform convergence of continuous functions: for every $\varepsilon > 0$, there is $n_0 = n_0(\varepsilon) \in \mathbb{N}$ such that the graphs of the functions u_n, for all $n \geq n_0$, lie in an ε-neighborhood of the graph of u in $[\alpha,\beta]$. Naturally, in general, the smaller $\varepsilon > 0$, the larger $n_0(\varepsilon)$ may need to be chosen to achieve this situation.

In this book, besides the norm $\|\cdot\|_\infty$, for every $t_0 \in [\alpha,\beta]$ and $\eta > 0$ we will also consider the norms

$$\|u\|_\eta := \max_{t\in[\alpha,\beta]} \left\{ e^{-\eta|t-t_0|}\|u(t)\| \right\}, \quad u \in \mathcal{C}([\alpha,\beta];\mathbb{K}^N). \quad (1.16)$$

They will be extremely useful in making global estimates in a compact way, as it will become apparent in the proof of Theorem 1.4. It is straightforward to see that $\|\cdot\|_\eta$ is a norm in $\mathcal{C}([\alpha,\beta];\mathbb{K}^N)$. So, the

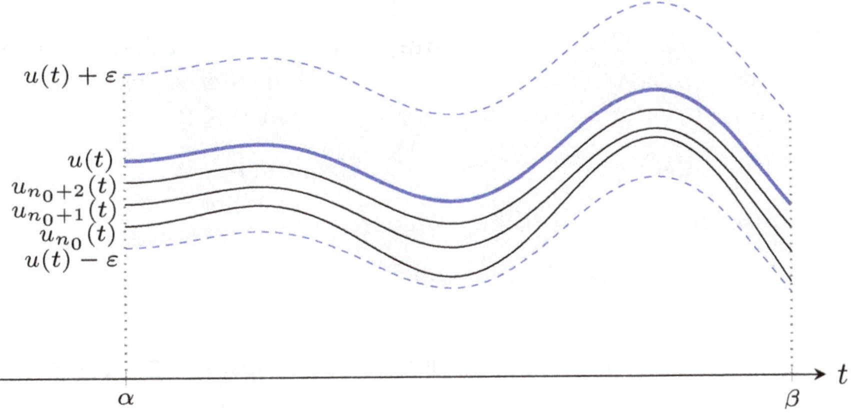

Fig. 1.1. Uniform convergence of continuous functions in $[\alpha,\beta]$.

technical details are left as an exercise (see Exercise 10 of Chapter 1). Moreover, for every $u \in \mathcal{C}([\alpha, \beta]; \mathbb{K}^N)$ and $t \in [\alpha, \beta]$,

$$e^{-\eta|t-t_0|}\|u(t)\| \leq \|u(t)\| \leq \|u\|_\infty,$$

and hence, taking the maximum in $t \in [\alpha, \beta]$ yields

$$\|u\|_\eta \leq \|u\|_\infty.$$

Similarly, for every $t \in [\alpha, \beta]$,

$$\|u(t)\| = e^{\eta|t-t_0|}e^{-\eta|t-t_0|}\|u(t)\| \leq e^{\eta(\beta-\alpha)}\|u\|_\eta$$

and, therefore,

$$\|u\|_\infty \leq e^{\eta(\beta-\alpha)}\|u\|_\eta.$$

Consequently, the norms $\|\cdot\|_\infty$ and $\|\cdot\|_\eta$ are equivalent. In particular, the topologies they induce in the vector space $\mathcal{C}([\alpha, \beta]; \mathbb{K}^N)$ coincide, as each open ball with respect to $\|\cdot\|_\infty$ contains an open ball with respect to $\|\cdot\|_\eta$, and vice versa. Thus, $\mathcal{C}([\alpha, \beta]; \mathbb{K}^N)$ is also complete when equipped with the norm $\|\cdot\|_\eta$.

1.3.2 *Contractive mapping theorem*

In this section we introduce the concept of *contractive mapping*, or simply *contraction*, and state and prove the pivotal *contractive mapping theorem*. Actually, most of the results in Parts 1 and 2 of this book are elegant consequences of this theorem.

Definition 1.7 (Contractive mapping). In a normed space X, an operator $\mathcal{K} : X \to X$ is said to be a *contractive mapping*, or, simply, a *contraction*, if there exists a constant $0 \leq \theta < 1$ such that

$$\|\mathcal{K}u - \mathcal{K}v\| \leq \theta\|u - v\| \quad \text{for all } u, v \in X. \qquad (1.17)$$

In such a case, $\mathcal{K}$ is said to be a θ-contraction.

It should be noted that (1.17) entails the continuity of $\mathcal{K}$ and provides us with the estimate

$$\text{dist}(\mathcal{K}u, \mathcal{K}v) \leq \theta \, \text{dist}(u, v) < \text{dist}(u, v) \quad \text{for all } u, v \in X, \, u \neq v.$$

So, $\mathcal{K}$ shortens, or contracts, the distances between different points of X, which explains the reason of the name "contraction".

Theorem 1.8 (Contractive mapping theorem). *Suppose X is a Banach space and $\mathcal{K} : X \to X$ is a θ-contraction, with $0 \leq \theta < 1$. Then, $\mathcal{K}$ admits a unique fixed point in X, x^*, i.e. $\mathcal{K}x^* = x^*$. Moreover, for every $x_0 \in X$, the sequence of iterates initialized at x_0,*

$$x_n := \mathcal{K}x_{n-1}, \qquad n \geq 1, \tag{1.18}$$

approximates, as $n \to \infty$, the unique fixed point x^*, i.e.

$$\lim_{n \to \infty} x_n = x^*.$$

Proof. Given an arbitrary $x_0 \in X$, we claim that the sequence of iterates $\{x_n\}_{n \geq 0}$, defined through (1.18) and initialized in x_0, is a Cauchy sequence in X. Indeed, for every $n, m \geq 1$, it follows from (1.17) and (1.18) that

$$\|x_{n+m} - x_n\| = \|\mathcal{K}x_{n+m-1} - \mathcal{K}x_{n-1}\| \leq \theta \|x_{n+m-1} - x_{n-1}\|.$$

Hence, by iterating this estimate, it becomes apparent that

$$\|x_{n+m} - x_n\| \leq \theta^n \|x_m - x_0\| \quad \text{for all } m, n \geq 1.$$

Consequently, combining this estimate with the triangle inequality yields

$$\begin{aligned}
\|x_{n+m} - x_n\| &\leq \|x_{n+m} - x_{n+m-1}\| + \|x_{n+m-1} - x_{n+m-2}\| \\
&\quad + \cdots + \|x_{n+2} - x_{n+1}\| + \|x_{n+1} - x_n\| \\
&\leq \left(\theta^{n+m-1} + \theta^{n+m-2} + \cdots + \theta^{n+1} + \theta^n\right) \|x_1 - x_0\| \\
&\leq \frac{\theta^n}{1 - \theta} \|x_1 - x_0\|
\end{aligned}$$

for all $n, m \geq 1$. Therefore, for every $\varepsilon > 0$, there exists an integer $n_0 = n_0(\varepsilon) \geq 1$ such that

$$\|x_{n+m} - x_n\| \leq \frac{\theta^n}{1 - \theta} \|x_1 - x_0\| \leq \varepsilon$$

for all $n \geq n_0$ and $m \geq 1$. Consequently, $\{x_n\}_{n \geq 1}$ is a Cauchy sequence in X. As we are assuming that X is complete, because it is a Banach space, there exists $x^* \in X$ such that

$$\lim_{n \to \infty} x_n = x^*.$$

By the continuity of $\mathcal{K}$, letting $n \to \infty$ in (1.18) yields

$$\mathcal{K}x^* = x^*.$$

So, x^* is a fixed point of $\mathcal{K}$. It remains to establish the uniqueness of the fixed point. Indeed, if $x^* \neq y^*$ are two fixed points, then

$$\|x^* - y^*\| = \|\mathcal{K}x^* - \mathcal{K}y^*\| \leq \theta\|x^* - y^*\| < \|x^* - y^*\|,$$

which is impossible. The proof is complete. $\qquad\qquad\qquad\square$

1.3.3   *Technical preliminaries for the proof of Theorem 1.4*

Throughout the rest of this book, we denote by $\mathcal{L}(\mathbb{K}^N)$ the vector space of all linear endomorphisms of $\mathbb{K}^N$; it is isomorphic to $\mathbb{K}^{N^2}$, as well as to the vector space $\mathcal{M}_N(\mathbb{K})$ of all the square matrices of order N with entries in the field $\mathbb{K}$. Moreover, we denote by $\mathrm{Iso}(\mathbb{K}^N)$ the set of $A \in \mathcal{L}(\mathbb{K}^N)$ that are isomorphisms. Once a norm $\|\cdot\|$ has been fixed in $\mathbb{K}^N$, for every $A \in \mathcal{L}(\mathbb{K}^N)$, we denote by

$$\|A\|_{\mathcal{L}(\mathbb{K}^N)} := \max\{\|Ax\|: \|x\| = 1\}$$

the *operator norm* induced by $\|\cdot\|$ in the vector space $\mathcal{L}(\mathbb{K}^N)$. We leave as an exercise to prove that such a quantity indeed defines a norm in $\mathcal{L}(\mathbb{K}^N)$ (see Exercise 13 of Chapter 1). Here, we only observe that it is well defined because any linear operator is continuous and, hence, it is bounded on the unit sphere of $\mathbb{K}^N$, which is a compact subset of $\mathbb{K}^N$. By definition, we have

$$\|Ax\| \leq \|A\|_{\mathcal{L}(\mathbb{K}^N)}\|x\| \quad \text{for all } x \in \mathbb{K}^N.$$

Actually, the operator norm is the lowest constant satisfying such an inequality. Indeed, for every $x \in \mathbb{K}^N \setminus \{0\}$,

$$\|Ax\| = \left\|A\frac{x}{\|x\|}\right\|\|x\| \leq \|A\|_{\mathcal{L}(\mathbb{K}^N)}\|x\|,$$

and, since $\|A\|_{\mathcal{L}(\mathbb{K}^N)} = \|A\tilde{x}\|$ for some $\tilde{x} \in \mathbb{K}^N$ with $\|\tilde{x}\| = 1$,

$$\|A\tilde{x}\| = \|A\|_{\mathcal{L}(\mathbb{K}^N)} = \|A\|_{\mathcal{L}(\mathbb{K}^N)}\|\tilde{x}\|,$$

which establishes the minimality of $\|A\|_{\mathcal{L}(\mathbb{K}^N)}$ in the above inequality.

Naturally, besides $\| \cdot \|_{\mathcal{L}(\mathbb{K}^N)}$, one can also define the following p-norms in $\mathcal{M}_N(\mathbb{K})$:

$$\|A\|_p := \left(\sum_{i,j=1}^{N} |a_{ij}|^p \right)^{\frac{1}{p}}, \qquad A \in \mathcal{M}_N(\mathbb{K}),$$

for all $1 \le p < +\infty$, or

$$\|A\|_\infty := \max_{1 \le i,j \le N} |a_{ij}|, \qquad A \in \mathcal{M}_N(\mathbb{K}).$$

Nevertheless, all of them are equivalent since

$$\dim \mathcal{L}(\mathbb{K}^N) = \dim \mathcal{M}_N(\mathbb{K}) = \dim \mathbb{K}^{N^2} = N^2 < \infty.$$

Observe that, in general, if $\|A\|_{\mathcal{L}(\mathbb{K}^N)}$ denotes the operator norm induced by the $\| \cdot \|_p$ norm of $\mathbb{K}^N$, $1 \le p < +\infty$, $\|A\|_{\mathcal{L}(\mathbb{K}^N)}$ differs from $\|A\|_p$ defined above (see Exercise 14 of Chapter 1). In the rest of this book, when no ambiguity is possible, we briefly denote $\|A\|$ instead of $\|A\|_{\mathcal{L}(\mathbb{K}^N)}$.

In the last part of this section, we assume that $\alpha \in [-\infty, +\infty)$ and $\beta \in (-\infty, +\infty]$ satisfy $\alpha < \beta$ and that $w \in \mathcal{C}((\alpha, \beta); \mathbb{K}^N)$, where $\mathbb{K} \in \{\mathbb{R}, \mathbb{C}\}$. Our last goal is to establish the inequality

$$\left\| \int_{t_0}^{t} w(s)\, ds \right\| \le \int_{\min\{t_0,t\}}^{\max\{t_0,t\}} \|w(s)\|\, ds \qquad (1.19)$$

for every $t_0, t \in (\alpha, \beta)$, where $\| \cdot \|$ denotes the $\| \cdot \|_1$-norm of $\mathbb{K}^N$. Throughout this book, such a relation will be referred to as the *Minkowski integral inequality*. To show it, consider

$$w := (w_1, \ldots, w_N)^T.$$

Then, by definition, we have that

$$\int_{t_0}^{t} w(s)\, ds = \left(\int_{t_0}^{t} w_1(s)\, ds, \ldots, \int_{t_0}^{t} w_N(s)\, ds \right)^T,$$

and

$$\left\| \int_{t_0}^{t} w(s)\, ds \right\| = \sum_{j=1}^{N} \left| \int_{t_0}^{t} w_j(s)\, ds \right| \le \sum_{j=1}^{N} \int_{\min\{t_0,t\}}^{\max\{t_0,t\}} |w_j(s)|\, ds$$

$$= \int_{\min\{t_0,t\}}^{\max\{t_0,t\}} \sum_{j=1}^{N} |w_j(s)|\, ds = \int_{\min\{t_0,t\}}^{\max\{t_0,t\}} \|w(s)\|\, ds,$$

as claimed.

1.3.4   *Proof of Theorem 1.4*

In this section, we derive Theorem 1.4 from Theorem 1.8 by proving that the integral operator $\mathcal{K}$ associated with the Cauchy problem (1.2) establishes a contraction in the Banach space $X := \mathcal{C}([\alpha, \beta]; \mathbb{K}^N)$, equipped with the norm

$$\|u\|_\eta := \max_{t \in [\alpha,\beta]} \left\{ e^{-\eta|t-t_0|} \|u(t)\| \right\}, \qquad u \in \mathcal{C}([\alpha, \beta]; \mathbb{K}^N),$$

for a certain value of η, whose choice is postponed until the end of the proof.

As mentioned earlier, we consider, in $\mathbb{K}^N$, $\| \cdot \| = \| \cdot \|_1$. By the definition of $\mathcal{K}$ (see (1.11)), for every $u, v \in \mathcal{C}([\alpha, \beta]; \mathbb{K}^N)$ and $t \in [\alpha, \beta]$,

$$e^{-\eta|t-t_0|} \|\mathcal{K}u(t) - \mathcal{K}v(t)\| = e^{-\eta|t-t_0|} \left\| \int_{t_0}^{t} A(s)[u(s) - v(s)]\, ds \right\|.$$

Moreover, by the Minkowski integral inequality (1.19),

$$\left\| \int_{t_0}^{t} A(s)[u(s) - v(s)]\, ds \right\| \le \int_{\min\{t_0,t\}}^{\max\{t_0,t\}} \|A(s)[u(s) - v(s)]\|\, ds.$$

Thus,

$$e^{-\eta|t-t_0|} \|\mathcal{K}u(t) - \mathcal{K}v(t)\| \le e^{-\eta|t-t_0|} \int_{\min\{t_0,t\}}^{\max\{t_0,t\}} \|A(s)[u(s) - v(s)]\|\, ds$$

$$\le e^{-\eta|t-t_0|} \int_{\min\{t_0,t\}}^{\max\{t_0,t\}} \|A(s)\|\|u(s) - v(s)\|\, ds.$$

By the equivalence of norms in $\mathbb{K}^{N^2}$, there exists a constant $C > 0$ such that

$$\|A(s)\| := \|A(s)\|_{\mathcal{L}(\mathbb{K}^N)} \leq C \|A(s)\|_1 = C \sum_{i,j=1}^{N} |a_{ij}(s)| \leq C \sum_{i,j=1}^{N} \|a_{ij}\|_\infty.$$

Thus, setting

$$L := C \sum_{i,j=1}^{N} \|a_{ij}\|_\infty,$$

we find that, for every $t \in [\alpha, \beta]$,

$$e^{-\eta|t-t_0|} \|\mathcal{K}u(t) - \mathcal{K}v(t)\|$$

$$\leq L e^{-\eta|t-t_0|} \int_{\min\{t_0,t\}}^{\max\{t_0,t\}} e^{\eta|s-t_0|} e^{-\eta|s-t_0|} \|u(s) - v(s)\| \, ds$$

$$\leq L e^{-\eta|t-t_0|} \|u - v\|_\eta \int_{\min\{t_0,t\}}^{\max\{t_0,t\}} e^{\eta|s-t_0|} \, ds.$$

On the other hand, by distinguishing the cases $t > t_0$ and $t < t_0$, a direct calculation gives

$$\int_{\min\{t_0,t\}}^{\max\{t_0,t\}} e^{\eta|s-t_0|} \, ds = \frac{e^{\eta|t-t_0|} - 1}{\eta} < \frac{e^{\eta|t-t_0|}}{\eta}.$$

Consequently, for every $t \in [\alpha, \beta]$,

$$e^{-\eta|t-t_0|} \|\mathcal{K}u(t) - \mathcal{K}v(t)\| \leq \frac{L}{\eta} \|u - v\|_\eta,$$

and hence, taking the maximum in $t \in [\alpha, \beta]$ yields

$$\|\mathcal{K}u - \mathcal{K}v\|_\eta \leq \frac{L}{\eta} \|u - v\|_\eta.$$

Therefore, by choosing $\eta > L$, it becomes apparent that $\mathcal{K}$ is a $\frac{L}{\eta}$-contraction and Theorem 1.8 completes the proof of Theorem 1.4. $\qquad\square$

To conclude this section, we work in an unbounded interval for t: for example, we suppose that $a_{ij}, b_i \in \mathcal{C}(\mathbb{R}; \mathbb{K})$ for all $1 \leq i, j \leq N$.

Then, for every integer $n \geq 1$, by applying Theorem 1.4 with $\alpha = \alpha_n := t_0 - n$ and $\beta = \beta_n := t_0 + n$, problem (1.2) possesses a unique solution, $u_n(t; t_0, u_0)$, in the compact interval $[\alpha_n, \beta_n] = [t_0 - n, t_0 + n]$. Moreover, by uniqueness, whenever $n < m$, we have that

$$u_m(t; t_0, u_0) = u_n(t; t_0, u_0) \quad \text{for all } t \in [t_0 - n, t_0 + n].$$

Thus, u_m is a proper *extension* of u_n to the interval $[t_0 - m, t_0 + m]$. Therefore, the following limit is well defined

$$u(t; t_0, u_0) := \lim_{n \to \infty} u_n(t; t_0, u_0)$$

for all $t \in \mathbb{R}$, and it provides us with the unique solution of (1.2) in $\mathbb{R}$. Consequently, we have proved the following result.

Corollary 1.9. *Suppose a_{ij}, $b_i \in \mathcal{C}(\mathbb{R}; \mathbb{K})$ for all $1 \leq i, j \leq N$. Then, for every $t_0 \in \mathbb{R}$ and $u_0 \in \mathbb{K}^N$, problem (1.2) possesses a unique global solution, $u(t; t_0, u_0)$, in $\mathbb{R}$.*

To conclude this section, we point out that, in the proof of Theorem 1.4, with all the elements of the Cauchy problem (1.2) fixed, we had chosen η such that $\mathcal{K}$ was a contraction. Changing perspective, if we fix η and all the elements of the Cauchy problem (1.2) except $[\alpha, \beta]$, it is always possible to make $\mathcal{K}$ a contraction, provided the interval $[\alpha, \beta]$ is sufficiently small. This will become clearer with the theory developed in Chapter 5.

1.3.5   *Picard–Lindelöf iterates*

In the proof of Theorem 1.4 given in the previous section, we have only used the existence and uniqueness of the fixed point given by Theorem 1.8. In this section, instead, we will take advantage of the convergence of the iterates (1.18) to derive a possible constructive way to approach the solution of the Cauchy problem (1.2). In the context of the operator $\mathcal{K}$ defined in (1.11), the iterates (1.18) read, for every $t \in [\alpha, \beta]$,

$$h_n(t) := u_0 + \int_{t_0}^{t} [A(s)h_{n-1}(s) + B(s)] \, ds, \qquad n \geq 1, \quad (1.20)$$

and are referred to as *Picard–Lindelöf iterates*. Theorem 1.8, together with the equivalence of the norms $\|\cdot\|_\eta$ and $\|\cdot\|_\infty$, allows us to obtain the following result.

Corollary 1.10. *Under the assumptions of Theorem 1.4, for every function $h_0 \in \mathcal{C}([\alpha, \beta]; \mathbb{K}^N)$, the sequence of Picard–Lindelöf iterates defined in (1.20) converges in $\mathcal{C}([\alpha, \beta]; \mathbb{K}^N)$, equipped with the $\|\cdot\|_\infty$ norm, i.e. uniformly in $[\alpha, \beta]$, to the unique solution $u(t; t_0, u_0)$ of the Cauchy problem (1.2).*

Rather astonishingly, the Picard–Lindelöf iterates of h_0 converge uniformly in $[\alpha, \beta]$ toward $u(t; t_0, u_0)$, independently of the initial function $h_0 \in \mathcal{C}([\alpha, \beta]; \mathbb{K}^N)$.

In general, it is not possible to compute $\lim_{n\to+\infty} h_n(t)$ explicitly to obtain an expression for the solution of the Cauchy problem (1.2). As an example where such a calculation is instead possible, consider $N = 1$, $A(t) = 1$, and $B(t) = 0$, for all $t \in [\alpha, \beta]$. In this case, the associated integral operator, $\mathcal{K}$, is given by

$$\mathcal{K}h(t) := u_0 + \int_{t_0}^{t} h(s)\, ds, \qquad t \in [\alpha, \beta].$$

A rather natural choice for h_0 is $h_0(t) = u_0$, $t \in [\alpha, \beta]$, because in such a case, h_0 coincides with $u(t; t_0, u_0)$ at least for $t = t_0$. Then, by definition,

$$h_1(t) := \mathcal{K}h_0(t) = u_0 + \int_{t_0}^{t} u_0\, ds = (1 + t - t_0)\, u_0,$$

$$h_2(t) := \mathcal{K}h_1(t) = u_0 + \int_{t_0}^{t} u_0\, (1 + s - t_0)\, ds$$

$$= \left(1 + t - t_0 + \frac{(t - t_0)^2}{2}\right) u_0,$$

and a simple induction argument shows that, for every $n \geq 1$ and $t \in [\alpha, \beta]$,

$$h_n(t) = \sum_{j=0}^{n} \frac{(t - t_0)^j}{j!}\, u_0$$

(here and throughout this book, unless otherwise specified, we use the convention $0^0 = 1$). Actually, as there is no limitation on the size of the interval $[\alpha, \beta]$,

$$\lim_{n \to \infty} h_n(t) = e^{(t-t_0)} u_0$$

uniformly on any compact subinterval of $\mathbb{R}$, which indeed provides us with the unique solution of

$$\begin{cases} u' = u, \\ u(t_0) = u_0. \end{cases}$$

Actually, this global convergence of the Picard–Lindelöf iterates always holds when all the coefficients $a_{ij}(t)$ and $b_i(t)$ are defined and continuous for all $t \in \mathbb{R}$. This can be obtained by reasoning as in the proof of Corollary 1.9.

1.4   Dimensions of Σ_0 and Σ_B

As already mentioned above, we denote by $u(\cdot; t_0, u_0)$ the unique solution of (1.2), while $h(\cdot; t_0, u_0)$ is the unique solution of the homogeneous problem

$$\begin{cases} h' = A(t)h, \\ h(t_0) = u_0. \end{cases}$$

The following result is a counterpart of Theorem 1.3 for linear systems of ordinary differential equations.

Theorem 1.11. *Under the assumptions of Theorem 1.4, the solution operator* $\mathcal{S}_0 : \mathbb{K}^N \to \Sigma_0$, *defined by*

$$\mathcal{S}_0(x) := h(\cdot; t_0, x), \qquad x \in \mathbb{K}^N,$$

is a linear isomorphism. Thus, the solution operator $\mathcal{S}_B : \mathbb{K}^N \to \Sigma_B$

$$\mathcal{S}_B(x) := u(\cdot; t_0, x), \qquad x \in \mathbb{K}^N,$$

is an affine isomorphism. In particular,

$$\dim \Sigma_B = \dim \Sigma_0 = N.$$

Proof. Suppose λ, $\mu \in \mathbb{K}$, and x, $y \in \mathbb{K}^N$. Since Σ_0 is a linear subspace, the linear combination, or superposition,

$$h(t) := \lambda h(t; t_0, x) + \mu h(t; t_0, y), \qquad t \in [\alpha, \beta],$$

lies in Σ_0. Moreover,

$$h(t_0) = \lambda x + \mu y.$$

Thus, by the uniqueness obtained as an application of Theorem 1.4,

$$h(t; t_0, \lambda x + \mu y) = \lambda h(t; t_0, x) + \mu h(t; t_0, y) \quad \text{for all } t \in [\alpha, \beta].$$

In other words,

$$\mathcal{S}_0(\lambda x + \mu y) = \lambda \mathcal{S}_0(x) + \mu \mathcal{S}_0(y).$$

Consequently, $\mathcal{S}_0$ is linear. Suppose $\mathcal{S}_0(x) = 0$. Then, $h(\cdot; t_0, x) = 0$ and, in particular, $x = h(t_0; t_0, x) = 0$. Thus, $\mathcal{S}_0$ is injective. Moreover, for every $h \in \Sigma_0$, we have that

$$h = h(\cdot; t_0, h(t_0)) = \mathcal{S}_0(h(t_0));$$

therefore, $\mathcal{S}_0$ is a linear isomorphism. Finally, thanks to Lemma 1.1 and the uniqueness given by Theorem 1.4, we have that

$$u(t; t_0, x) = h(t; t_0, x) + u(t; t_0, 0)$$

for all $t, t_0 \in [\alpha, \beta]$ and $x \in \mathbb{K}^N$. In other words,

$$\mathcal{S}_B = \mathcal{S}_0 + u(\cdot; t_0, 0).$$

Hence, $\mathcal{S}_B$ is an affine isomorphism. $\qquad\qquad\square$

As any linear isomorphism transforms bases into bases, the previous result ensures that, if $\{e_1, \ldots, e_N\}$ is a basis of $\mathbb{K}^N$, i.e.

$$\mathbb{K}^N = \text{span}\,[e_1, \ldots, e_N],$$

then, for every $t_0 \in [\alpha, \beta]$,

$$\Sigma_0 = \mathrm{span}\,[h(\cdot; t_0, e_1), \dots, h(\cdot; t_0, e_N)],$$

i.e. $\{h(\cdot; t_0, e_1), \dots, h(\cdot; t_0, e_N)\}$ forms a basis for the set of solutions of the homogeneous system. In addition, if

$$u_0 = \sum_{i=1}^{N} c_i e_i,$$

then

$$h(\cdot; t_0, u_0) = \sum_{i=1}^{N} c_i h(\cdot; t_0, e_i) = (h(\cdot; t_0, e_1) \cdots h(\cdot; t_0, e_N)) \begin{pmatrix} c_1 \\ \vdots \\ c_N \end{pmatrix},$$

where $(h(\cdot; t_0, e_1) \cdots h(\cdot; t_0, e_N))$ denotes the matrix whose columns are the vectors $h(\cdot; t_0, e_i)$. This kind of matrix is known as a *fundamental matrix* and is analyzed in the following section.

As a final comment, we point out that, once we know how to determine a basis for Σ_0, in order to determine an affine reference system in Σ_B, it is sufficient to find a particular solution to the inhomogeneous system (1.1). This will be done in Section 1.6 through the variation of constants formula for linear systems, which generalizes the one for linear scalar equations presented in Section 1.2.

1.5   Fundamental Matrices. Jacobi–Liouville Formula

Throughout the rest of this book, for any given Banach space X, we denote by $\mathcal{M}_N(X)$ the set of square matrices of order N with entries in X. The concept introduced in the following definition is pivotal in the theory of linear systems of differential equations. Actually, it is older than the concept of basis in linear algebra.

Definition 1.12 (Fundamental matrix of solutions). A matrix $\Phi \in \mathcal{M}_N(\mathcal{C}^1([\alpha, \beta]; \mathbb{K}))$ is said to be a *matrix of solutions* of $h' = A(t)h$, or, simply, to solve $h' = A(t)h$ in $[\alpha, \beta]$ if

$$\Phi'(t) = A(t)\Phi(t) \quad \text{for all } t \in [\alpha, \beta].$$

A matrix of solutions Φ of $h' = A(t)h$ in $[\alpha, \beta]$ is said to be a *fundamental matrix of solutions* if it solves such a system and if, in addition, the columns of Φ, which are vectors of $\left(\mathcal{C}^1([\alpha,\beta];\mathbb{K})\right)^N$, form a basis of Σ_0.

Observe that a matrix $\Phi(t)$ with columns $h_1(t),\ldots,h_N(t)$, which we denote by

$$\Phi(t) := (h_1(t)\cdots h_N(t)),$$

satisfies $\Phi' = A\Phi$ in $[\alpha,\beta]$ if and only if $h_j \in \Sigma_0$ for all $j \in \{1,\ldots,N\}$. Indeed,

$$\left(h_1'(t)\cdots h_N'(t)\right) = \Phi'(t) = A(t)\Phi(t) = (A(t)h_1(t)\cdots A(t)h_N(t))$$

holds for all $t \in [\alpha,\beta]$ if and only if $h_j \in \Sigma_0$ for all $j \in \{1,\ldots,N\}$.

According to Theorem 1.11, we already know that, given a basis $\{e_1,\ldots,e_N\}$ in $\mathbb{K}^N$,

$$\Phi(t) := (h(t;t_0,e_1)\cdots h(t;t_0,e_N)) \tag{1.21}$$

is a fundamental matrix of solutions of $h' = A(t)h$ in $[\alpha,\beta]$. Conversely, any fundamental matrix of solutions,

$$\Phi(t) = (h_1(t)\cdots h_N(t)), \qquad t \in [\alpha,\beta],$$

can be expressed in the form given by (1.21) for some basis of $\mathbb{K}^N$ by setting

$$e_j := h_j(t_0) \quad \text{for all } j \in \{1,\ldots,N\}.$$

Indeed, for every $t \in [\alpha,\beta]$, we have that

$$\begin{aligned}
\Phi(t) &= (h_1(t)\cdots h_N(t)) \\
&= (h_1(t;t_0,h_1(t_0))\cdots h_N(t;t_0,h_N(t_0))) \\
&= (h_1(t;t_0,e_1)\cdots h_N(t;t_0,e_N)),
\end{aligned}$$

where $\{h_1(t_0),\ldots,h_N(t_0)\}$ is a basis of $\mathbb{K}^N$. This is guaranteed by the following characterization, which establishes a simple criterion to determine whether or not a matrix of solutions is a fundamental matrix of solutions.

Theorem 1.13. *Let $h_j \in \Sigma_0$, $j \in \{1,\ldots,N\}$, and consider the matrix of solutions of $h' = A(t)h$ in $[\alpha,\beta]$*

$$\Phi(t) = (h_1(t)\cdots h_N(t)), \qquad t \in [\alpha,\beta].$$

Then, the following assertions are equivalent:

(a) Φ *is a fundamental matrix of solutions of* $h' = A(t)h$ *in* $[\alpha, \beta]$*;*
(b) *there exists* $t_0 \in [\alpha, \beta]$ *such that* $\det \Phi(t_0) \neq 0$*;*
(c) $\det \Phi(t) \neq 0$ *for all* $t \in [\alpha, \beta]$*.*

In particular, when any of the previous conditions holds,

$$\mathbb{K}^N = \mathrm{span}[h_1(t), \ldots, h_N(t)] \quad \text{for all } t \in [\alpha, \beta].$$

Proof. (a) $\Rightarrow$ (c). Suppose that there exists $t_0 \in [\alpha, \beta]$ such that $\det \Phi(t_0) = 0$. Then, there is $(c_1, \ldots, c_N) \in \mathbb{K}^N \setminus \{0\}$ such that

$$\sum_{j=1}^{N} c_j h_j(t_0) = 0.$$

Consequently, the function $h(t)$ defined by

$$h(t) := \sum_{j=1}^{N} c_j h_j(t), \qquad t \in [\alpha, \beta],$$

provides us with a solution of $h' = A(t)h$ in $[\alpha, \beta]$ such that $h(t_0) = 0$. By Theorem 1.4, $h = 0$. Thus,

$$\sum_{j=1}^{N} c_j h_j(t) = 0 \quad \text{for all } t \in [\alpha, \beta].$$

In particular, $\det \Phi(t) = 0$ for all $t \in [\alpha, \beta]$ and $\{h_1, \ldots, h_n\}$ are linearly dependent in Σ_0. So, Φ cannot be a fundamental matrix.

(c) $\Rightarrow$ (b). This implication is immediate.

(b) $\Rightarrow$ (a). If Φ is not a fundamental matrix, then there exists $(c_1, \ldots, c_N) \in \mathbb{K}^N \setminus \{0\}$ such that

$$\sum_{j=1}^{N} c_j h_j = 0.$$

In other words,

$$\sum_{j=1}^{N} c_j h_j(t) = 0 \quad \text{for all } t \in [\alpha, \beta].$$

Thus, $\{h_1(t), \ldots, h_n(t)\}$ are linearly dependent in $\mathbb{K}^N$ for all $t \in [\alpha, \beta]$. So,

$$\det \Phi(t) = 0 \quad \text{for all } t \in [\alpha, \beta],$$

which completes the proof. $\qquad\qquad\qquad\qquad\qquad\qquad\qquad\qquad\quad\square$

Actually, given any solution matrix, Φ, the following result holds, which provides us with another proof for the equivalence of statements (b) and (c) in Theorem 1.13.

Theorem 1.14 (Jacobi–Liouville formula). *Suppose Φ is a solution matrix of $h' = A(t)h$ in $[\alpha, \beta]$. Then, for every $t, t_0 \in [\alpha, \beta]$,*

$$\det \Phi(t) = e^{\int_{t_0}^{t} \operatorname{tr} A(s)\,ds} \det \Phi(t_0), \tag{1.22}$$

where $\operatorname{tr} A(t)$ is the trace of the matrix $A(t)$,

$$\operatorname{tr} A(t) = \sum_{j=1}^{N} a_{jj}(t), \qquad t \in [\alpha, \beta].$$

Identity (1.22) is referred to as the Jacobi–Liouville *formula.*

Proof. Throughout this proof, we set

$$\Phi(t) = \begin{pmatrix} h_{11}(t) & \cdots & h_{1j}(t) & \cdots & h_{1N}(t) \\ \vdots & \ddots & \vdots & \ddots & \vdots \\ h_{i1}(t) & \cdots & h_{ij}(t) & \cdots & h_{iN}(t) \\ \vdots & \ddots & \vdots & \ddots & \vdots \\ h_{N1}(t) & \cdots & h_{Nj}(t) & \cdots & h_{NN}(t) \end{pmatrix}, \qquad t \in [\alpha, \beta].$$

Naturally, the function

$$\varphi(t) := \det \Phi(t), \qquad t \in [\alpha, \beta],$$

is of class $\mathcal{C}^1([\alpha, \beta]; \mathbb{K})$. Moreover, by induction on N, it is easily seen that

$$\varphi'(t) = \sum_{i=1}^{N} \begin{vmatrix} h_{11}(t) & \cdots & h_{1j}(t) & \cdots & h_{1N}(t) \\ \vdots & \ddots & \vdots & \ddots & \vdots \\ h'_{i1}(t) & \cdots & h'_{ij}(t) & \cdots & h'_{iN}(t) \\ \vdots & \ddots & \vdots & \ddots & \vdots \\ h_{N1}(t) & \cdots & h_{Nj}(t) & \cdots & h_{NN}(t) \end{vmatrix} \tag{1.23}$$

for all $t \in [\alpha, \beta]$ (see Exercise 17 of Chapter 1). On the other hand, since $\Phi' = A\Phi$ on $[\alpha, \beta]$, we find that

$$h'_{ij} = \sum_{k=1}^{N} a_{ik} h_{kj}, \qquad i, j \in \{1, \ldots, N\},$$

and hence, substituting in (1.23) yields

$$\varphi' = \sum_{i=1}^{N} \begin{vmatrix} h_{11} & \cdots & h_{1j} & \cdots & h_{1N} \\ \vdots & \ddots & \vdots & \ddots & \vdots \\ \sum_{k=1}^{N} a_{ik}h_{k1} & \cdots & \sum_{k=1}^{N} a_{ik}h_{kj} & \cdots & \sum_{k=1}^{N} a_{ik}h_{kN} \\ \vdots & \ddots & \vdots & \ddots & \vdots \\ h_{N1} & \cdots & h_{Nj} & \cdots & h_{NN} \end{vmatrix} .$$

Since the determinant is a multilinear map, we have

$$\varphi' = \sum_{i=1}^{N} \sum_{k=1}^{N} a_{ik} \begin{vmatrix} h_{11} & \cdots & h_{1j} & \cdots & h_{1N} \\ \vdots & \ddots & \vdots & \ddots & \vdots \\ h_{k1} & \cdots & h_{kj} & \cdots & h_{kN} \\ \vdots & \ddots & \vdots & \ddots & \vdots \\ h_{N1} & \cdots & h_{Nj} & \cdots & h_{NN} \end{vmatrix} .$$

Therefore, since all these determinants vanish if $k \neq i$, we find that

$$\varphi' = \sum_{i=1}^{N} a_{ii} \begin{vmatrix} h_{11} & \cdots & h_{1j} & \cdots & h_{1N} \\ \vdots & \ddots & \vdots & \ddots & \vdots \\ h_{i1} & \cdots & h_{ij} & \cdots & h_{iN} \\ \vdots & \ddots & \vdots & \ddots & \vdots \\ h_{N1} & \cdots & h_{Nj} & \cdots & h_{NN} \end{vmatrix} = \operatorname{tr} A \cdot \varphi.$$

Consequently, by Theorem 1.2, we obtain that, for every $t, t_0 \in [\alpha, \beta]$,

$$\varphi(t) = e^{\int_{t_0}^{t} \operatorname{tr} A(s)\, ds} \varphi(t_0),$$

which completes the proof. $\qquad\square$

The following result collects some useful properties of fundamental matrices.

Proposition 1.15. *The following properties hold true.*

(a) *Let Φ be a fundamental matrix of solutions of $h' = A(t)h$ in $[\alpha, \beta]$. Then, for every invertible matrix $C \in \mathcal{M}_N(\mathbb{K})$, the matrix*

$$\Psi(t) := \Phi(t)C, \qquad t \in [\alpha, \beta],$$

is also a fundamental matrix of solutions of $h' = A(t)h$ in $[\alpha, \beta]$.
(b) *Let Φ_1 and Φ_2 be two fundamental matrices of solutions of $h' = A(t)h$ in $[\alpha, \beta]$. Then, there exists an invertible matrix $C \in \mathcal{M}_N(\mathbb{K})$ such that*

$$\Phi_2(t) = \Phi_1(t)C \quad \text{for all } t \in [\alpha, \beta].$$

Proof. (a) Since

$$\Psi'(t) = \Phi'(t)C = A(t)\Phi(t)C = A(t)\Psi(t)$$

for all $t \in [\alpha, \beta]$, $\Psi(t)$ is a matrix of solutions of $h' = A(t)h$. Moreover,

$$\det \Psi(t) = \det \Phi(t) \det C \neq 0$$

for all $t \in [\alpha, \beta]$. Thus, by Theorem 1.13, Ψ is a fundamental matrix of solutions of $h' = A(t)h$ in $[\alpha, \beta]$.

(b) If Φ_1 and Φ_2 are two fundamental matrices of solutions of $h' = A(t)h$ in $[\alpha, \beta]$, their columns provide us with two bases for Σ_0. Therefore, the constant matrix C is nothing more than the matrix of the change of basis between their columns. Next, we provide a detailed self-contained proof of this fact. Suppose that

$$\Phi_1(t) = (h_{1,1}(t) \cdots h_{1,N}(t)), \quad \Phi_2(t) = (h_{2,1}(t) \cdots h_{2,N}(t)), \quad t \in [\alpha, \beta],$$

for some $h_{i,j} \in \Sigma_0$, $j \in \{1, \dots, N\}$, $i \in \{1, 2\}$. Since $\{h_{1,1}, \dots, h_{1,N}\}$ is a basis of Σ_0 and $h_{2,j} \in \Sigma_0$ for all $j \in \{1, \dots, N\}$, there exist N^2 coefficients $c_{ij} \in \mathbb{K}$, $i, j \in \{1, \dots, N\}$, such that, for every $t \in [\alpha, \beta]$,

$$h_{2,j}(t) = c_{j1}h_{1,1}(t) + c_{j2}h_{1,2}(t) + \cdots + c_{jN}h_{1,N}(t), \quad j \in \{1, \dots, N\}.$$

In other words,

$$h_{2,j}(t) = (h_{1,1}(t) \cdots h_{1,N}(t)) \begin{pmatrix} c_{j1} \\ \vdots \\ c_{jN} \end{pmatrix} = \Phi_1(t) \begin{pmatrix} c_{j1} \\ \vdots \\ c_{jN} \end{pmatrix}, \quad j \in \{1, \dots, N\}.$$

Thus, for every $t \in [\alpha, \beta]$, we find that

$$\Phi_2(t) = (h_{2,1}(t) \cdots h_{2,N}(t)) = \left(\Phi_1(t) \begin{pmatrix} c_{11} \\ \vdots \\ c_{1N} \end{pmatrix} \cdots \Phi_1(t) \begin{pmatrix} c_{N1} \\ \vdots \\ c_{NN} \end{pmatrix} \right)$$

$$= \Phi_1(t) \begin{pmatrix} c_{11} & \cdots & c_{N1} \\ \vdots & \ddots & \vdots \\ c_{1N} & \cdots & c_{NN} \end{pmatrix} = \Phi_1(t)C,$$

where $C \in \mathcal{M}_N(\mathbb{K})$ is the matrix

$$C := \begin{pmatrix} c_{11} & \cdots & c_{N1} \\ \vdots & \ddots & \vdots \\ c_{1N} & \cdots & c_{NN} \end{pmatrix}.$$

Since $\Phi_2(t) = \Phi_1(t)C$ for all $t \in [\alpha, \beta]$, it becomes apparent from Theorem 1.13 that

$$0 \neq \det \Phi_2(t) = \det \Phi_1(t) \det C$$

for all $t \in [\alpha, \beta]$. Therefore, $\det C \neq 0$. The proof is complete. $\qquad \square$

We conclude this section with the following observation that will be important for obtaining the variation of constants formula. It essentially establishes that we can always take, if needed, a fundamental matrix of solutions Φ satisfying that $\Phi(t_0)$ is the identity matrix of $\mathbb{K}^N$.

Remark 1.16. As a consequence of Proposition 1.15(a), if Ψ is a fundamental matrix of solutions of $h' = A(t)h$ in $[\alpha, \beta]$, then, for every $t_0 \in [\alpha, \beta]$, the matrix

$$\Phi(t) := \Psi(t)\Psi^{-1}(t_0), \qquad t \in [\alpha, \beta],$$

is also a fundamental matrix of solutions of $h' = A(t)h$. Moreover,

$$\Phi(t_0) = I,$$

where $I := I_{\mathbb{K}^N}$ denotes the identity matrix of $\mathbb{K}^N$.

1.6 Variation of Constants Formula

As any linear isomorphism transforms bases into bases, according to
Theorem 1.11, if $\{e_1, \ldots, e_N\}$ is a basis of $\mathbb{K}^N$, it becomes apparent
that

$$\Sigma_0 = \mathrm{span}\,[h(\cdot; t_0, e_1), \ldots, h(\cdot; t_0, e_N)]$$

for all $t_0 \in [\alpha, \beta]$. Thus, setting

$$h_j := h(\cdot; t_0, e_j), \qquad j \in \{1, \ldots, N\},$$

we have that

$$\Psi(t) = (h_1(t) \cdots h_N(t)), \qquad t \in [\alpha, \beta],$$

is a fundamental matrix of solutions of $h' = A(t)h$ in $[\alpha, \beta]$. In particular, the parametric equations of Σ_0—which represent the general
solution of $h' = A(t)h$—can be expressed as

$$x_1 h_1(t) + x_2 h_2(t) + \cdots + x_N h_N(t) = \Psi(t)x, \quad x = (x_1, \ldots, x_N) \in \mathbb{K}^N.$$

In other words, $\Psi(t)x$ provides us with the set of solutions of $h' = A(t)h$ as x varies in $\mathbb{K}^N$. Moreover, by Remark 1.16,

$$\Phi(t) := \Psi(t)\Psi^{-1}(t_0), \qquad t \in [\alpha, \beta],$$

provides us with a fundamental matrix of solutions such that

$$\Phi(t_0) = I.$$

Adapting the strategy of the proof of Theorem 1.2, in order to find the solution of

$$\begin{cases} u' = A(t)u + B(t), \\ u(t_0) = u_0, \end{cases} \tag{1.24}$$

we perform the change of variable

$$u(t) = \Phi(t)v(t), \qquad t \in [\alpha, \beta]. \tag{1.25}$$

Substituting (1.25) in (1.24) yields

$$u_0 = u(t_0) = \Phi(t_0)v(t_0) = v(t_0)$$

and

$$\Phi'(t)v(t) + \Phi(t)v'(t) = A(t)\Phi(t)v(t) + B(t).$$

Thus, since $\Phi' = A\Phi$, it becomes apparent that $u(t)$ solves (1.24) if and only if $v(t)$ satisfies

$$\begin{cases} \Phi(t)v'(t) = B(t), \\ v(t_0) = u_0. \end{cases} \tag{1.26}$$

As $\Phi(t)$ is a fundamental matrix of solutions of $h' = A(t)h$, it is invertible for all $t \in [\alpha, \beta]$ by Theorem 1.13, and hence,

$$v'(t) = \Phi^{-1}(t)B(t) \quad \text{for all } t \in [\alpha, \beta].$$

Thus, by integrating in (t_0, t), we find that

$$v(t) = v(t_0) + \int_{t_0}^{t} \Phi^{-1}(s)B(s)\,ds = u_0 + \int_{t_0}^{t} \Phi^{-1}(s)B(s)\,ds$$

is the unique solution of (1.26). Therefore, by substituting $v(t)$ in (1.25), it becomes apparent that

$$u(t; t_0, u_0) = \Phi(t)u_0 + \int_{t_0}^{t} \Phi(t)\Phi^{-1}(s)B(s)\,ds \tag{1.27}$$

provides us with the unique solution of (1.24). This formula is often referred to as the *variation of constants formula* for (1.24). Note that

$$h(t; t_0, u_0) = \Phi(t)u_0, \qquad u(t; t_0, 0) = \int_{t_0}^{t} \Phi(t)\Phi^{-1}(s)B(s)\,ds,$$

and hence,

$$\Sigma_B = \Sigma_0 + \int_{t_0}^{t} \Phi(t)\Phi^{-1}(s)B(s)\,ds.$$

Specifically,

$$u(t;t_0,0) = \int_{t_0}^{t} \Phi(t)\Phi^{-1}(s)B(s)\,ds, \qquad t \in [\alpha,\beta],$$

provides us with a point of the affine subspace Σ_B, i.e. it is a particular solution of (1.1). Moreover, since

$$\Phi(t)\Phi^{-1}(s) = \Psi(t)\Psi^{-1}(t_0)\left[\Psi(s)\Psi^{-1}(t_0)\right]^{-1}$$
$$= \Psi(t)\Psi^{-1}(t_0)\Psi(t_0)\Psi^{-1}(s) = \Psi(t)\Psi^{-1}(s),$$

we also have that

$$u(t;t_0,0) = \int_{t_0}^{t} \Psi(t)\Psi^{-1}(s)B(s)\,ds.$$

Therefore,

$$\mathcal{R} := \left\{ h_1(t), \ldots, h_N(t); \int_{t_0}^{t} \Psi(t)\Psi^{-1}(s)B(s)\,ds \right\}$$

is an *affine reference system* of the affine subspace Σ_B of $\mathcal{C}^1([\alpha,\beta];\mathbb{K}^N)$. In other words, the *parametric equations* of Σ_B— which provide us with the *general solution* of $u' = A(t)u + B(t)$—are given by

$$x_1 h_1(t) + \cdots + x_N h_N(t) + \int_{t_0}^{t} \Psi(t)\Psi^{-1}(s)B(s)\,ds,$$

with $x_1, \ldots, x_N \in \mathbb{K}$, which can be equivalently expressed as

$$\Psi(t)x + \int_{t_0}^{t} \Psi(t)\Psi^{-1}(s)B(s)\,ds, \qquad x \in \mathbb{K}^N. \tag{1.28}$$

In Chapter 2, where we assume that the entries of $A(t)$ are constant, i.e. $A \in \mathcal{M}_N(\mathbb{K})$, we will define the exponential matrix

$$e^{tA} := \sum_{n \geq 0} \frac{t^n A^n}{n!}$$

(here, and in the rest of the book, we use the convention $M^0 = I$ for all $M \in \mathcal{M}_N(\mathbb{K})$, where I denotes the identity matrix of order N), and we will show that it provides us with a fundamental matrix of solutions of $h' = Ah$. In this particular case, the variation of constants formula (1.28) reduces to

$$e^{tA}x + \int_{t_0}^{t} e^{(t-s)A} B(s)\, ds,$$

which, not surprisingly, has the same formal expression as the general solution of the single scalar equation

$$u' = Au + B(t)$$

in the very special case when $A \in \mathbb{K}$ and $B \in \mathcal{C}([\alpha, \beta]; \mathbb{K})$.

1.7   Linear Equations of Arbitrary Order

As usual in this chapter, we consider $\alpha, \beta \in \mathbb{R}$, with $\alpha < \beta$, and $\mathbb{K} \in \{\mathbb{R}, \mathbb{C}\}$. Given three arbitrary continuous functions, $a_0, a_1, b \in \mathcal{C}([\alpha, \beta]; \mathbb{K})$, the second-order differential equation

$$u'' = a_0(t)u + a_1(t)u' + b(t) \tag{1.29}$$

is referred to as the *second-order linear differential equation*.

The reader should be aware that, despite its simple appearance, there does not exist a general integration method for this equation, unless the coefficients $a_0(t)$ and $a_1(t)$ take some special form (e.g. when both are constants or real analytic functions).

Similarly, given another continuous function $a_2 \in \mathcal{C}([\alpha, \beta]; \mathbb{K})$, the third-order differential equation

$$u''' = a_0(t)u + a_1(t)u' + a_2(t)u'' + b(t) \tag{1.30}$$

is referred to as the *third-order linear differential equation*.

More generally, given a natural number $N \geq 1$ and $N+1$ continuous functions,

$$a_j \in \mathcal{C}([\alpha, \beta]; \mathbb{K}), \quad j \in \{0, \ldots, N-1\}, \qquad b \in \mathcal{C}([\alpha, \beta]; \mathbb{K}),$$

the Nth-order differential equation

$$u^{N)} = \sum_{j=0}^{N-1} a_j(t)u^{j)} + b(t) \tag{1.31}$$

is referred to as the *Nth-order linear differential equation* with coefficients $a_j(t)$, $j \in \{0, \ldots, N-1\}$, and inhomogeneity $b(t)$. A solution of (1.31) in $[\alpha, \beta]$ is defined as a function $u \in \mathcal{C}^N([\alpha, \beta]; \mathbb{K})$ such that

$$u^{N)}(t) = a_0(t)u(t) + a_1(t)u'(t) + a_2(t)u''(t) + \cdots + a_{N-1}(t)u^{N-1)}(t) + b(t)$$

for all $t \in [\alpha, \beta]$.

Throughout the rest of this book, we denote by $\mathcal{C}^N([\alpha, \beta]; \mathbb{K})$ the set of functions $u : [\alpha, \beta] \to \mathbb{K}$ with N derivatives in $[\alpha, \beta]$ such that $u^{N)}$ is a continuous function in $[\alpha, \beta]$. It is a vector space because it is closed for the sum of functions and multiplication by the scalars of $\mathbb{K}$. It is an infinite-dimensional vector space because, as shown in Section 1.3.1, $\{t^n\}_{n \in \mathbb{N}}$ provides us with a set of independent functions in $\mathcal{C}^N([\alpha, \beta]; \mathbb{K})$.

Throughout this section, as usual, we denote by Σ_b^e the set of solutions of (1.31) in the interval $[\alpha, \beta]$, with the super-index "e", emphasizing the fact that we are dealing with solutions of an "equation". The linear equation with $b = 0$,

$$u^{N)} = a_0(t)u + a_1(t)u' + a_2(t)u'' + \cdots + a_{N-1}(t)u^{N-1)}, \qquad (1.32)$$

is referred to as the *homogeneous equation* associated with (1.31). Naturally, Σ_0^e will denote the set of solutions of (1.32).

As in the proof of Lemma 1.1, it is straightforward to show that, for every $\lambda, \mu \in \mathbb{K}$ and $h_1, h_2 \in \Sigma_0^e$,

$$\lambda h_1 + \mu h_2 \in \Sigma_0^e.$$

Indeed,

$$(\lambda h_1 + \mu h_2)^{N)} = \lambda h_1^{N)} + \mu h_2^{N)}$$

$$= \lambda \sum_{j=0}^{N-1} a_j(t) h_1^{j)} + \mu \sum_{j=0}^{N-1} a_j(t) h_2^{j)}$$

$$= \sum_{j=0}^{N-1} a_j(t)(\lambda h_1^{j)} + \mu h_2^{j)}) = \sum_{j=0}^{N-1} a_j(t)(\lambda h_1 + \mu h_2)^{j)}.$$

Thus, Σ_0^e is a linear subspace of the linear vector space $\mathcal{C}^N([\alpha, \beta]; \mathbb{K})$. Similarly, it is easily seen that, for every $u, v \in \Sigma_b^e$ and $h \in \Sigma_0^e$,

$$u - v \in \Sigma_0^e \quad \text{and} \quad h + u \in \Sigma_b^e.$$

Therefore, Σ_b^e is an affine subspace of $\mathcal{C}^N([\alpha, \beta]; \mathbb{K})$ parallel to Σ_0^e.

1.7.1 *The associated first-order linear system*

The fact that all linear equations can be equivalently written, in a canonical way, as linear first-order systems extraordinarily simplifies the study of linear scalar equations of arbitrary order. From the point of view of Newtonian mechanics, fixing the initial position, $u(t_0)$, and the initial velocity, $u'(t_0)$, of a particle moving according to the differential equation (1.29) should determine the value of $u(t)$ for all values of $t \in [\alpha, \beta]$. This fundamental result in deterministic Physics can be easily derived by applying Theorem 1.4 to the first-order system associated with (1.29), which can be easily constructed by setting

$$\begin{pmatrix} u_1 \\ u_2 \end{pmatrix} := \begin{pmatrix} u \\ u' \end{pmatrix}, \tag{1.33}$$

where u is assumed to be a solution of the linear second-order equation (1.29). Differentiating identity (1.33) with respect to t yields

$$\begin{pmatrix} u_1 \\ u_2 \end{pmatrix}' = \begin{pmatrix} u_1' \\ u_2' \end{pmatrix} = \begin{pmatrix} u' \\ u'' \end{pmatrix} = \begin{pmatrix} u' \\ a_0(t)u + a_1(t)u' + b(t) \end{pmatrix}$$

and hence, using again (1.33) shows that

$$\begin{pmatrix} u_1 \\ u_2 \end{pmatrix}' = \begin{pmatrix} u_2 \\ a_0(t)u_1 + a_1(t)u_2 + b(t) \end{pmatrix}.$$

Equivalently,

$$\begin{pmatrix} u_1 \\ u_2 \end{pmatrix}' = \begin{pmatrix} 0 & 1 \\ a_0(t) & a_1(t) \end{pmatrix} \begin{pmatrix} u_1 \\ u_2 \end{pmatrix} + \begin{pmatrix} 0 \\ b(t) \end{pmatrix},$$

which is the first-order system associated with the second-order differential equation (1.29). Similarly, in order to determine the first-order system associated with the third-order scalar equation (1.30), we introduce the variables

$$\begin{pmatrix} u_1 \\ u_2 \\ u_3 \end{pmatrix} := \begin{pmatrix} u \\ u' \\ u'' \end{pmatrix}. \tag{1.34}$$

Then, differentiating with respect to $t \in [\alpha, \beta]$ and using (1.34) shows that

$$\begin{pmatrix} u_1 \\ u_2 \\ u_3 \end{pmatrix}' = \begin{pmatrix} u_1' \\ u_2' \\ u_3' \end{pmatrix} = \begin{pmatrix} u' \\ u'' \\ u''' \end{pmatrix} = \begin{pmatrix} u' \\ u'' \\ a_0(t)u + a_1(t)u' + a_2(t)u'' + b(t) \end{pmatrix}$$

for every solution u of (1.30). Thus,

$$\begin{pmatrix} u_1 \\ u_2 \\ u_3 \end{pmatrix}' = \begin{pmatrix} u_2 \\ u_3 \\ a_0(t)u_1 + a_1(t)u_2 + a_2(t)u_3 + b(t) \end{pmatrix}$$

or, equivalently,

$$\begin{pmatrix} u_1 \\ u_2 \\ u_3 \end{pmatrix}' = \begin{pmatrix} 0 & 1 & 0 \\ 0 & 0 & 1 \\ a_0(t) & a_1(t) & a_2(t) \end{pmatrix} \begin{pmatrix} u_1 \\ u_2 \\ u_3 \end{pmatrix} + \begin{pmatrix} 0 \\ 0 \\ b(t) \end{pmatrix},$$

which is the first-order system associated with the third-order scalar equation (1.30). More generally, for every $u \in \Sigma_b^e$, we consider the vectorial function whose entries are the derivatives of u up to order $N - 1$, i.e.

$$\begin{pmatrix} u_1 \\ u_2 \\ \vdots \\ u_{N-1} \\ u_N \end{pmatrix} := \begin{pmatrix} u \\ u' \\ \vdots \\ u^{N-2)} \\ u^{N-1)} \end{pmatrix}. \tag{1.35}$$

Then, differentiating with respect to t and using (1.31) and (1.35) yield

$$\begin{pmatrix} u_1 \\ u_2 \\ \vdots \\ u_{N-1} \\ u_N \end{pmatrix}' = \begin{pmatrix} u_2 \\ u_3 \\ \vdots \\ u_N \\ \sum_{j=0}^{N-1} a_j u^{j)} + b \end{pmatrix} = \begin{pmatrix} u_2 \\ u_3 \\ \vdots \\ u_N \\ \sum_{j=0}^{N-1} a_j u_{j+1} + b \end{pmatrix}.$$

In other words, for every $u \in \Sigma_b^e$, the vectorial function formed by the derivatives of u up to order $N - 1$,

$$\mathcal{D}(u) := \begin{pmatrix} u \\ u' \\ \vdots \\ u^{N-2)} \\ u^{N-1)} \end{pmatrix}, \tag{1.36}$$

provides us with a solution in $[\alpha, \beta]$ to the linear first-order system

$$U' = A(t)U + B(t), \tag{1.37}$$

where

$$U = (u_1, u_2, \ldots, u_{N-1}, u_N)^T, \qquad B(t) = (0, 0, \ldots, 0, b(t))^T, \tag{1.38}$$

and, for every $t \in [\alpha, \beta]$,

$$A(t) = \begin{pmatrix} 0 & 1 & 0 & \cdots & 0 & 0 \\ 0 & 0 & 1 & 0 & \cdots & 0 \\ \vdots & \ddots & \ddots & \ddots & \ddots & \vdots \\ \vdots & \ddots & \ddots & \ddots & 1 & 0 \\ 0 & \cdots & \cdots & 0 & 0 & 1 \\ a_0(t) & a_1(t) & a_2(t) & \cdots & a_{N-2}(t) & a_{N-1}(t) \end{pmatrix}. \tag{1.39}$$

Throughout this book, system (1.37) will be referred to as the *first-order linear system associated with the Nth-order linear equation* (1.31). Note that $b = 0$ if and only if $B = 0$. In analogy with the notation introduced above, we denote by Σ_B^s the set of solutions of the associated system (1.37). In particular, Σ_0^s denotes the set of solutions of the associated homogeneous system $U' = A(t)U$. The super-index "s" should remind us that we are dealing with solutions of a "system".

1.7.2 *Fundamental properties*

The following result is extremely useful.

Theorem 1.17. *The map $\mathcal{D} : \Sigma_b^e \to \Sigma_B^s$ defined by (1.36) establishes a bijection between Σ_b^e and Σ_B^s. Actually, $\mathcal{D} : \Sigma_0^e \to \Sigma_0^s$ is a linear isomorphism. Therefore,*

$$N = \dim \Sigma_0^e = \dim \Sigma_0^s,$$

and $\mathcal{D}$ transforms the bases of Σ_0^e into the bases of Σ_0^s.

Proof.   We already know that $\mathcal{D}(u) \in \Sigma_B^s$ for each $u \in \Sigma_b^e$. Suppose $\mathcal{D}(u) = \mathcal{D}(v)$. Then, (1.36) implies $u = v$. Thus, $\mathcal{D}$ is injective. Let

$$U = (u_1, \ldots, u_N)^T \in \Sigma_B^s.$$

Then, from the first $N - 1$ differential equations of the linear system (1.37), we find that

$$u_1' = u_2, \quad u_2' = u_3, \quad u_3' = u_4, \quad \ldots, \quad u_{N-1}' = u_N,$$

and hence,

$$u_2 = u_1', \quad u_3 = u_1'', \quad u_4 = u_1''', \quad \ldots, \quad u_N = u_1^{N-1)}.$$

Thus, according to (1.36), $U = \mathcal{D}(u_1)$. Moreover, it follows from the last differential equation in (1.37) that

$$u_1^{N)} = u_N' = \sum_{j=0}^{N-1} a_j u_{j+1} + b = \sum_{j=0}^{N-1} a_j u_1^{j)} + b.$$

Therefore, $u_1 \in \Sigma_b^e$ and $\mathcal{D}$ is surjective. Consequently, $\mathcal{D}$ is a bijection.

As the differentiation is a linear operator, from (1.36) we have that, for every $\lambda, \mu \in \mathbb{K}$ and $h_1, h_2 \in \Sigma_0^e$,

$$\mathcal{D}(\lambda h_1 + \mu h_2) = \lambda \mathcal{D}(h_1) + \mu \mathcal{D}(h_2).$$

Thus, the map $\mathcal{D} : \Sigma_0^e \to \Sigma_0^s$ is linear, and hence, it is a linear isomorphism. This completes the proof. $\square$

By combining Theorems 1.4 and 1.17, the following fundamental result holds.

Corollary 1.18. *For every $t_0 \in [\alpha, \beta]$ and $U_0 \in \mathbb{K}^N$, the Cauchy problem*

$$\begin{cases} u^{N)} = \sum_{j=0}^{N-1} a_j(t) u^{j)} + b(t), \\ \mathcal{D}(u)(t_0) = \big(u(t_0), u'(t_0), u''(t_0), \ldots, u^{N-1)}(t_0) \big)^T = U_0, \end{cases} \tag{1.40}$$

possesses a unique solution, $u(t; t_0, U_0)$, in the interval $[\alpha, \beta]$. Moreover, $u(t; t_0, U_0)$ must be the first component of the unique solution, $U(t; t_0, U_0)$, of the associated Cauchy problem

$$\begin{cases} U' = A(t)U + B(t), \\ U(t_0) = U_0, \end{cases} \tag{1.41}$$

where $B(t)$ and $A(t)$ are given in (1.38) and (1.39), respectively.

When $b = 0$, the unique solution of (1.40) will be denoted by

$$h(t; t_0, U_0), \qquad t \in [\alpha, \beta].$$

Similarly, we denote by $H(t; t_0, U_0)$, $t \in [\alpha, \beta]$, the unique solution of

$$\begin{cases} U' = A(t)U, \\ U(t_0) = U_0. \end{cases}$$

Naturally, the following counterpart of Theorem 1.11 holds.

Theorem 1.19. *The solution operator $\mathcal{S}_0^e : \mathbb{K}^N \to \Sigma_0^e$ defined by*

$$\mathcal{S}_0^e(x) := h(\cdot; t_0, x), \qquad x \in \mathbb{K}^N,$$

is a linear isomorphism. Thus, the solution operator $\mathcal{S}_b^e : \mathbb{K}^N \to \Sigma_b^e$ defined by

$$\mathcal{S}_b^e(x) := u(\cdot; t_0, x), \qquad x \in \mathbb{K}^N,$$

is an affine isomorphism.

Proof. The fact that $\mathcal{S}_0^e$ is a linear isomorphism is a direct consequence of the fact that

$$\mathcal{S}_0^e = \mathcal{D}^{-1} \circ \mathcal{S}_0^s,$$

where we denote by $\mathcal{S}_0^s : \mathbb{K}^N \to \Sigma_0^s$ the solution operator

$$\mathcal{S}_0^s(x) := H(\cdot; t_0, x), \qquad x \in \mathbb{K}^N,$$

associated with the first-order system $U' = A(t)U$, and $\mathcal{D}^{-1} : \Sigma_0^s \to \Sigma_0^e$ denotes the projection operator on the first component, i.e. the inverse of the map $\mathcal{D}$ defined in (1.36), which is a linear isomorphism thanks to Theorem 1.17. As, owing to Theorem 1.11, $\mathcal{S}_0^s$ is a linear isomorphism, it becomes apparent that $\mathcal{S}_0^e$, as a composition of linear isomorphisms, is a linear isomorphism too.

Finally, as for every $U_0 \in \mathbb{K}^N$,

$$u(\cdot; t_0, U_0) = h(\cdot; t_0, U_0) + u(\cdot; t_0, 0),$$

we have that

$$\mathcal{S}_b^e = \mathcal{S}_0^e + \mathcal{S}_b^e(0)$$

and, therefore, $\mathcal{S}_b^e$ is an affine isomorphism. $\qquad\square$

According to Theorem 1.19, whenever

$$\mathbb{K}^N = \operatorname{span}[e_1, \ldots, e_N],$$

it becomes apparent that, for every $t_0 \in [\alpha, \beta]$,

$$\Sigma_0^e = \operatorname{span}[h(\cdot; t_0, e_1), \ldots, h(\cdot; t_0, e_N)].$$

Therefore, thanks to Theorem 1.17, we can also infer that

$$\Sigma_0^s = \operatorname{span}[\mathcal{D}(h(\cdot; t_0, e_1)), \ldots, \mathcal{D}(h(\cdot; t_0, e_N))].$$

In other words,

$$W := (\mathcal{D}(h(\cdot; t_0, e_1)) \cdots \mathcal{D}(h(\cdot; t_0, e_N)))$$

is a fundamental matrix of solutions of $U' = A(t)U$. More generally, given any basis, $\{h_1, \ldots, h_N\}$, of Σ_0^e, by Theorem 1.17, the matrix

$$W(\cdot; h_1, \ldots, h_N) := (\mathcal{D}(h_1) \cdots \mathcal{D}(h_N)) = \begin{pmatrix} h_1 & h_2 & \cdots & h_N \\ h_1' & h_2' & \cdots & h_N' \\ \vdots & \vdots & \ddots & \vdots \\ h_1^{N-1)} & h_2^{N-1)} & \cdots & h_N^{N-1)} \end{pmatrix}$$

is a fundamental matrix of solutions of $U' = A(t)U$, which is called the *Wronskian of the solutions* h_j, $j \in \{1, \ldots, N\}$.

Note that, since the trace of the matrix $A(t)$ defined by (1.39) equals $a_{N-1}(t)$, according to Theorem 1.14, we have that, for every $t, t_0 \in [\alpha, \beta]$,

$$\det W(t) = e^{\int_{t_0}^{t} a_{N-1}(s)\,ds} \det W(t_0).$$

The best general strategy to solve the Cauchy problem (1.40) from a given basis of Σ_0^e, $\{h_1, \ldots, h_N\}$, consists of adopting the methodology detailed in Section 1.6, which is based on the method of varying coefficients, with $u_0 := U_0$, B given by (1.38) and

$$\Psi(t) = W(t; h_1, \ldots, h_N), \qquad t \in [\alpha, \beta],$$

in order to solve the associated Cauchy problem (1.41) through the variation of constants formula (1.27). Once $U(t; t_0, U_0)$ is determined, its first component, $u(t; t_0, U_0)$, provides us with the unique solution of (1.40). The following section describes this whole process in a simple prototypical model that is pivotal from the point of view of the application of differential equations in Science and Engineering.

1.8 Second-Order Equations with Constant Coefficients

The main goal of this section is to obtain the general solution of the linear second-order differential equation

$$u'' = a_0 u + a_1 u' + b(t), \tag{1.42}$$

where $a_0, a_1 \in \mathbb{C}$ and $b \in \mathcal{C}([\alpha, \beta]; \mathbb{C})$ are arbitrary. First, we shall get a basis for the associated homogeneous equation

$$u'' = a_0 u + a_1 u'. \tag{1.43}$$

As the coefficients a_0 and a_1 are constant, it is rather natural to look for solutions to (1.43) of the form

$$u(t) = e^{\lambda t}$$

for some exponent $\lambda \in \mathbb{C}$ to be determined. Indeed, by substituting $u(t) = e^{\lambda t}$ in (1.43) and simplifying, it becomes apparent that this function solves (1.43) if and only if

$$\lambda^2 - a_1\lambda - a_0 = 0, \tag{1.44}$$

which explains why (1.44) is called the *characteristic equation* of (1.43). Actually, by introducing the associated differential operator

$$P(D) := D^2 - a_1 D - a_0, \qquad \text{where } D = \frac{d}{dt},$$

it is apparent that

$$P(D)e^{\lambda t} = P(\lambda)e^{\lambda t}.$$

Thus, $e^{\lambda t}$ solves (1.43) if and only if $P(\lambda) = 0$, which provides us with (1.44). The polynomial

$$P(z) = z^2 - a_1 z - a_0, \qquad z \in \mathbb{C},$$

is called the *characteristic polynomial* of (1.43). Accordingly, the roots of $P(z)$ are usually referred to as the *characteristic roots* of (1.43). Note that, in practice, $P(D)$ can be obtained from $P(z)$ by exchanging z for D.

Subsequently, we will distinguish two different situations, according to the number of distinct roots of $P(z)$. As the roots of $P(z)$ are

$$\frac{a_1 + \left(a_1^2 + 4a_0\right)^{\frac{1}{2}}}{2},$$

where $(a_1^2 + 4a_0)^{\frac{1}{2}}$ is 0 if $a_1^2 + 4a_0 = 0$, while it indicates the two distinct complex roots of $z^2 = a_1^2 + 4a_0$ if $a_1^2 + 4a_0 \neq 0$, we will distinguish these two cases.

1.8.1   *The case of $a_1^2 + 4a_0 \neq 0$*

In the case of $a_1^2 + 4a_0 \neq 0$, $P(z)$ possesses two complex roots, $\lambda_1 \neq \lambda_2$, and hence,

$$h_1(t) = e^{\lambda_1 t}, \qquad h_2(t) := e^{\lambda_2 t}, \qquad t \in \mathbb{R},$$

provide us with two solutions to the homogeneous equation (1.43). These solutions are linearly independent because if there are

$c_1, c_2 \in \mathbb{C}$ such that

$$c_1 e^{\lambda_1 t} + c_2 e^{\lambda_2 t} = 0 \quad \text{for all } t \in \mathbb{R},$$

then differentiating with respect to t yields

$$c_1 \lambda_1 e^{\lambda_1 t} + c_2 \lambda_2 e^{\lambda_2 t} = 0 \quad \text{for all } t \in \mathbb{R},$$

and hence, by evaluating the previous two expressions at $t = 0$, we find that

$$\begin{cases} c_1 + c_2 = 0 \\ c_1 \lambda_1 + c_2 \lambda_2 = 0 \end{cases} \iff \begin{pmatrix} 1 & 1 \\ \lambda_1 & \lambda_2 \end{pmatrix} \begin{pmatrix} c_1 \\ c_2 \end{pmatrix} = \begin{pmatrix} 0 \\ 0 \end{pmatrix} \iff c_1 = c_2 = 0$$

since the determinant of the matrix is $\lambda_2 - \lambda_1 \neq 0$. Therefore,

$$\Sigma_0^e = \mathrm{span}[e^{\lambda_1 t}, e^{\lambda_2 t}]. \tag{1.45}$$

In other words,

$$h(t) = e^{\lambda_1 t} x_1 + e^{\lambda_2 t} x_2, \qquad x_1, x_2 \in \mathbb{C},$$

is the general solution of the homogeneous equation (1.43).

Setting $v := u'$, it becomes apparent that u solves (1.42) if and only if

$$\begin{pmatrix} u \\ v \end{pmatrix}' = \begin{pmatrix} 0 & 1 \\ a_0 & a_1 \end{pmatrix} \begin{pmatrix} u \\ v \end{pmatrix} + \begin{pmatrix} 0 \\ b(t) \end{pmatrix}, \tag{1.46}$$

which provides us with the linear first-order system associated with (1.42). Although we will always maintain the notations introduced in (1.35) when dealing with general Nth-order equations, in some concrete applications with $N = 2$, we will denote

$$u_1 = u, \qquad u_2 = u' = v,$$

because, if we adopt the perspective of Newtonian mechanics, $v(t) = u'(t)$ is the *velocity*, which is associated with the *displacement* $u(t)$. So, we set

$$(u_1, u_2) = (u, u') = (u, v).$$

From (1.45), thanks to the analysis carried out after Theorem 1.19, we know that the Wronskian matrix of the basis $\{e^{\lambda_1 t}, e^{\lambda_2 t}\}$,

$$W(t) := \begin{pmatrix} e^{\lambda_1 t} & e^{\lambda_2 t} \\ \lambda_1 e^{\lambda_1 t} & \lambda_2 e^{\lambda_2 t} \end{pmatrix}, \qquad t \in \mathbb{R},$$

is a fundamental matrix of solutions of the homogeneous system associated with (1.46). Moreover, by Theorem 1.17, the solutions of (1.42) are the first components of the solutions of (1.46). To solve (1.46), we look for all the possible functions, $C_1(t)$ and $C_2(t)$, such that

$$\begin{pmatrix} u \\ v \end{pmatrix} = W(t) \begin{pmatrix} C_1(t) \\ C_2(t) \end{pmatrix} \tag{1.47}$$

solves (1.46). Actually, performing this change of variables is nothing more than varying coefficients. Since

$$W'(t) = \begin{pmatrix} 0 & 1 \\ a_0 & a_1 \end{pmatrix} W(t), \qquad t \in \mathbb{R},$$

substituting (1.47) in (1.46) and simplifying yield

$$W(t) \begin{pmatrix} C_1'(t) \\ C_2'(t) \end{pmatrix} = \begin{pmatrix} 0 \\ b(t) \end{pmatrix}.$$

Therefore, thanks to Theorem 1.13,

$$\begin{pmatrix} C_1'(t) \\ C_2'(t) \end{pmatrix} = W^{-1}(t) \begin{pmatrix} 0 \\ b(t) \end{pmatrix},$$

and hence, it follows from Cramer's rule that

$$C_1'(t) = \frac{\begin{vmatrix} 0 & e^{\lambda_2 t} \\ b(t) & \lambda_2 e^{\lambda_2 t} \end{vmatrix}}{\begin{vmatrix} e^{\lambda_1 t} & e^{\lambda_2 t} \\ \lambda_1 e^{\lambda_1 t} & \lambda_2 e^{\lambda_2 t} \end{vmatrix}} = \frac{-e^{-\lambda_1 t} b(t)}{\lambda_2 - \lambda_1},$$

$$C_2'(t) = \frac{\begin{vmatrix} e^{\lambda_1 t} & 0 \\ \lambda_1 e^{\lambda_1 t} & b(t) \end{vmatrix}}{\begin{vmatrix} e^{\lambda_1 t} & e^{\lambda_2 t} \\ \lambda_1 e^{\lambda_1 t} & \lambda_2 e^{\lambda_2 t} \end{vmatrix}} = \frac{e^{-\lambda_2 t} b(t)}{\lambda_2 - \lambda_1}.$$

Thus, for any given $t_0 \in [\alpha, \beta]$, we find that

$$C_1(t) = \frac{-1}{\lambda_2 - \lambda_1} \int_{t_0}^{t} b(s) e^{-\lambda_1 s}\, ds + x_1,$$

$$C_2(t) = \frac{1}{\lambda_2 - \lambda_1} \int_{t_0}^{t} b(s) e^{-\lambda_2 s}\, ds + x_2,$$

where x_1 and x_2 are arbitrary complex numbers. Consequently, going back to (1.47) and taking the first component u, we find that

$$u(t) = e^{\lambda_1 t} C_1(t) + e^{\lambda_2 t} C_2(t)$$

$$= e^{\lambda_1 t} x_1 + e^{\lambda_2 t} x_2 + \int_{t_0}^{t} \frac{e^{\lambda_2(t-s)} - e^{\lambda_1(t-s)}}{\lambda_2 - \lambda_1} b(s)\, ds, \qquad (1.48)$$

with arbitrary $x_1, x_2 \in \mathbb{C}$, represents the parametric equation of the affine plane Σ_b^e. In other words, (1.48) gives the general solution of (1.42). According to Corollary 1.18, for every $t_0 \in [\alpha, \beta]$ and $u_0, v_0 \in \mathbb{C}$, there exist two unique $x_1, x_2 \in \mathbb{C}$ for which (1.48) fits the initial conditions

$$u(t_0) = u_0, \qquad u'(t_0) = v_0.$$

In the special case when $a_0, a_1 \in \mathbb{R}$ and $b \in \mathcal{C}([\alpha, \beta]; \mathbb{R})$, the characteristic roots λ_1 and λ_2 are real if $a_1^2 + 4a_0 > 0$ and non-real if $a_1^2 + 4a_0 < 0$. More precisely,

$$\lambda_1 = \frac{a_1}{2} - \frac{\sqrt{a_1^2 + 4a_0}}{2} < \lambda_2 = \frac{a_1}{2} + \frac{\sqrt{a_1^2 + 4a_0}}{2} \qquad \text{if } a_1^2 + 4a_0 > 0,$$

whereas, when $a_1^2 + 4a_0 < 0$, setting

$$r := \frac{a_1}{2}, \qquad \omega := \frac{\sqrt{|a_1^2 + 4a_0|}}{2} > 0,$$

the characteristic roots become

$$\lambda_1 = r - i\omega, \qquad \lambda_2 = r + i\omega.$$

In the first case, (1.48) still provides us with the set of real solutions of (1.42) by restricting ourselves to considering $x_1, x_2 \in \mathbb{R}$. In the second case, in order to find the set of real solutions directly, a possibility is repeating the previous process starting from the very beginning with the construction of a real basis in Σ_0^e, as, for example,

$$h_1(t) := \frac{e^{(r+i\omega)t} + e^{(r-i\omega)t}}{2} = e^{rt}\frac{e^{i\omega t} + e^{-i\omega t}}{2} = e^{rt}\cos(\omega t),$$

$$h_2(t) := \frac{e^{(r+i\omega)t} - e^{(r-i\omega)t}}{2i} = e^{rt}\frac{e^{i\omega t} - e^{-i\omega t}}{2i} = e^{rt}\sin(\omega t).$$

Since the functions $h_1(t)$ and $h_2(t)$ are linearly independent,

$$\Sigma_0^e = \operatorname{span}\left[e^{rt}\cos(\omega t), e^{rt}\sin(\omega t)\right].$$

Indeed, if there exists $c_1, c_2 \in \mathbb{C}$ such that, for every $t \in \mathbb{R}$,

$$c_1 e^{rt}\cos(\omega t) + c_2 e^{rt}\sin(\omega t) = 0,$$

then

$$c_1 \cos(\omega t) + c_2 \sin(\omega t) = 0,$$

and evaluating at $t = 0$ shows that $c_1 = 0$. Thus, $c_2 \sin(\omega t) = 0$ for all $t \in \mathbb{R}$ and, therefore, $c_2 = 0$. The associated Wronskian matrix is given by

$$W(t) = \begin{pmatrix} e^{rt}\cos(\omega t) & e^{rt}\sin(\omega t) \\ e^{rt}\left[r\cos(\omega t) - \omega\sin(\omega t)\right] & e^{rt}\left[r\sin(\omega t) + \omega\cos(\omega t)\right] \end{pmatrix}.$$

Thus,

$$\det W(t) = \omega e^{2rt}$$

and

$$C_1'(t) = \frac{\begin{vmatrix} 0 & e^{rt}\sin(\omega t) \\ b(t) & e^{rt}\left[r\sin(\omega t) + \omega\cos(\omega t)\right] \end{vmatrix}}{\omega e^{2rt}} = \frac{-e^{-rt}\sin(\omega t)b(t)}{\omega},$$

$$C_2'(t) = \frac{\begin{vmatrix} e^{rt}\cos(\omega t) & 0 \\ e^{rt}\left[r\cos(\omega t) - \omega\sin(\omega t)\right] & b(t) \end{vmatrix}}{\omega e^{2rt}} = \frac{e^{-rt}\cos(\omega t)b(t)}{\omega}.$$

Hence,

$$C_1(t) = -\frac{1}{\omega}\int_{t_0}^{t} e^{-rs}\sin(\omega s)b(s)\,ds + x_1,$$

$$C_2(t) = \frac{1}{\omega}\int_{t_0}^{t} e^{-rs}\cos(\omega s)b(s)\,ds + x_2,$$

where $x_1, x_2 \in \mathbb{R}$ are two arbitrary integration constants. Therefore, going back to (1.47), it becomes apparent that

$$u(t) = e^{rt}\cos(\omega t)C_1(t) + e^{rt}\sin(\omega t)C_2(t)$$

$$= e^{rt}\cos(\omega t)x_1 + e^{rt}\sin(\omega t)x_2$$

$$+ \frac{1}{\omega}\int_{t_0}^{t} e^{r(t-s)}\left[\sin(\omega t)\cos(\omega s) - \cos(\omega t)\sin(\omega s)\right]b(s)\,ds$$

$$= e^{rt}\cos(\omega t)x_1 + e^{rt}\sin(\omega t)x_2 + \frac{1}{\omega}\int_{t_0}^{t} e^{r(t-s)}\sin[\omega(t-s)]b(s)\,ds$$

provides us with the set of real solutions of (1.42) when $x_1, x_2 \in \mathbb{R}$.

1.8.2   *The case of $a_1^2 + 4a_0 = 0$*

In this case, $h(t) = e^{\lambda t}$, with $\lambda := a_1/2$, is the unique solution of (1.43) of the form e^{zt} for some $z \in \mathbb{C}$ because

$$P(z) = z^2 - a_1 z - a_0 = (z - \lambda)^2, \qquad z \in \mathbb{C}.$$

Indeed, $2\lambda = a_1$ and $\lambda^2 = a_1^2/4 = -a_0$. To construct another function in a basis of Σ_0^e, one can perform the change of variable

$$u(t) = e^{\lambda t}w(t) \tag{1.49}$$

to solve the homogeneous differential equation

$$u'' - 2\lambda u' + \lambda^2 u = 0. \tag{1.50}$$

By doing so and substituting (1.49) in (1.50), we are led to

$$\lambda^2 e^{\lambda t}w(t) + 2\lambda e^{\lambda t}w'(t) + e^{\lambda t}w''(t)$$

$$- 2\lambda[\lambda e^{\lambda t}w(t) + e^{\lambda t}w'(t)] + \lambda^2 e^{\lambda t}w(t) = 0,$$

which simplifies to $w'' = 0$. Thus,

$$w(t) = x_1 + x_2 t, \qquad x_1, x_2 \in \mathbb{C};$$

therefore, the general solution of (1.50) is given by

$$u(t) = e^{\lambda t}x_1 + te^{\lambda t}x_2, \qquad x_1, x_2 \in \mathbb{C}.$$

In particular, by choosing $(x_1, x_2) = (0, 1)$, we find that $te^{\lambda t}$ solves (1.50). Note that the functions $e^{\lambda t}$ and $te^{\lambda t}$ are linearly independent.

Indeed, if there are $c_1, c_2 \in \mathbb{C}$ such that

$$c_1 e^{\lambda t} + c_2 t e^{\lambda t} = 0$$

for all $t \in \mathbb{R}$, then $c_1 + c_2 t = 0$ for all $t \in \mathbb{R}$ and, hence, $c_1 = c_2 = 0$. Consequently, in this case,

$$\Sigma_0^e = \mathrm{span}[e^{\lambda t}, t e^{\lambda t}],$$

and the Wronskian matrix of this basis,

$$W(t) := \begin{pmatrix} e^{\lambda t} & t e^{\lambda t} \\ \lambda e^{\lambda t} & (\lambda t + 1) e^{\lambda t} \end{pmatrix},$$

is a fundamental matrix of solutions of the homogeneous system associated with (1.46). Since $\det W(t) = e^{2\lambda t}$, arguing as in the previous cases, we find that

$$C_1'(t) = \frac{\begin{vmatrix} 0 & t e^{\lambda t} \\ b(t) & (\lambda t + 1) e^{\lambda t} \end{vmatrix}}{e^{2\lambda t}} = -b(t) t e^{-\lambda t},$$

$$C_2'(t) = \frac{\begin{vmatrix} e^{\lambda t} & 0 \\ \lambda e^{\lambda t} & b(t) \end{vmatrix}}{e^{2\lambda t}} = b(t) e^{-\lambda t}.$$

Hence, for every $t_0 \in \mathbb{R}$ and $x_1, x_2 \in \mathbb{C}$,

$$C_1(t) = -\int_{t_0}^t b(s) s e^{-\lambda s}\, ds + x_1, \qquad C_2(t) = \int_{t_0}^t b(s) e^{-\lambda s}\, ds + x_2.$$

Therefore, by going back to (1.47) and getting the first component, it becomes apparent that

$$u(t) = e^{\lambda t} x_1 + t e^{\lambda t} x_2 + \int_{t_0}^t (t - s) e^{\lambda(t-s)} b(s)\, ds, \qquad (1.51)$$

with $x_1, x_2 \in \mathbb{C}$, provides us with the general (complex) solution of (1.42).

58   *Ordinary Differential Equations*

Since

$$\lim_{\lambda_2 \to \lambda_1} \frac{e^{\lambda_2(t-s)} - e^{\lambda_1(t-s)}}{\lambda_2 - \lambda_1} = (t-s)e^{\lambda_1(t-s)},$$

we note that, when $x_1 = x_2 = 0$, the particular solution of (1.51) with $\lambda = \lambda_1$ can be obtained, at least formally, from the corresponding solution of (1.48) by simply letting $\lambda_2 \to \lambda_1$.

When $a_0, a_1 \in \mathbb{R}$ and $b \in \mathcal{C}([\alpha, \beta]; \mathbb{R})$, it is obvious that (1.51) also provides us with the general solution of (1.42) in $\mathbb{R}$ by simply restricting ourselves to consider $x_1, x_2 \in \mathbb{R}$.

Quite astonishingly, when $a_0(t)$ and $a_1(t)$ are general continuous functions in some compact interval, $[\alpha, \beta]$, no general method is available to construct a basis of Σ_0^e. Nevertheless, in some special cases of interest from the point of view of applications—where, very often, $a_0(t)$ and $a_1(t)$ are real analytic—one can solve the associated homogeneous equation through the *Taylor expansion method*, which will be described in Chapter 3.

1.9   Order Reduction

In the discussion carried out in Section 1.8.2, we have already seen how a second-order linear differential equation can be integrated once a particular solution is known. In that case, where the coefficients of the associated homogeneous equation were constant, we started from the exponential function $e^{\lambda t}$. More generally, let $h \neq 0$ be an arbitrary solution of the homogeneous equation

$$h'' - a_1(t)h' - a_0(t)h = 0. \tag{1.52}$$

Then, in order to obtain the general solution of

$$u'' - a_1(t)u' - a_0(t)u = b(t), \tag{1.53}$$

a good strategy is performing the change of variable

$$u(t) = h(t)v(t),$$

which transforms (1.53) into

$$h''v + 2h'v' + hv'' - a_1(h'v + hv') - a_0hv = b,$$

or, equivalently,

$$hv'' + (2h' - a_1 h)v' + (h'' - a_1 h' - a_0 h)v = b.$$

Thus, since h satisfies (1.52), the previous equation reduces to

$$hv'' + (2h' - a_1 h)v' = b.$$

As the change of variable $w = v'$ transforms this second-order linear equation into the first-order linear equation

$$w' = \left(a_1 - 2\frac{h'}{h} \right) w + \frac{b}{h},$$

the variation of constants formula (1.4) shows that

$$v'(t) = w(t) = e^{\int_{t_0}^{t} \left(a_1(\sigma) - 2\frac{h'(\sigma)}{h(\sigma)} \right) d\sigma} x_1 + \int_{t_0}^{t} e^{\int_{s}^{t} \left(a_1(\sigma) - 2\frac{h'(\sigma)}{h(\sigma)} \right) d\sigma} \frac{b(s)}{h(s)} \, ds,$$

where $x_1 \in \mathbb{K}$ is an integration constant. Consequently,

$$v(t) = \int_{t_0}^{t} e^{\int_{t_0}^{\tau} \left(a_1(\sigma) - 2\frac{h'(\sigma)}{h(\sigma)} \right) d\sigma} d\tau \, x_1 + x_2$$

$$+ \int_{t_0}^{t} \int_{t_0}^{\tau} e^{\int_{s}^{\tau} \left(a_1(\sigma) - 2\frac{h'(\sigma)}{h(\sigma)} \right) d\sigma} \frac{b(s)}{h(s)} \, ds \, d\tau,$$

where $x_2 \in \mathbb{K}$ is another integration constant, and we conclude that

$$u(t) = h(t) \int_{t_0}^{t} e^{\int_{t_0}^{\tau} \left(a_1(\sigma) - 2\frac{h'(\sigma)}{h(\sigma)} \right) d\sigma} d\tau \, x_1 + h(t)x_2$$

$$+ h(t) \int_{t_0}^{t} \int_{t_0}^{\tau} e^{\int_{s}^{\tau} \left(a_1(\sigma) - 2\frac{h'(\sigma)}{h(\sigma)} \right) d\sigma} \frac{b(s)}{h(s)} \, ds \, d\tau$$

provides us with the general solution of (1.53). By making the special choice of $x_1 = x_2 = 0$, it becomes apparent that

$$p(t) := h(t) \int_{t_0}^{t} \int_{t_0}^{\tau} e^{\int_{s}^{\tau} \left(a_1(\sigma) - 2\frac{h'(\sigma)}{h(\sigma)} \right) d\sigma} \frac{b(s)}{h(s)} \, ds \, d\tau$$

is a solution of (1.53). Thus, by the structural properties of linear equations, presented in Section 1.7,

$$h(t) \int_{t_0}^{t} e^{\int_{t_0}^{\tau} \left(a_1(\sigma) - 2\frac{h'(\sigma)}{h(\sigma)} \right) d\sigma} d\tau \, x_1 + h(t)x_2, \qquad x_1, x_2 \in \mathbb{K},$$

is the general solution of the associated homogeneous equation (1.52).

In the previous calculations, we have assumed that $h \neq 0$ in order to divide by $h(t)$ in the first-order differential equation satisfied by w. The following result establishes that $h(t)$ admits, at most, finitely many zeros in $[\alpha, \beta]$ provided a_0 and a_1 are continuous functions on $[\alpha, \beta]$. Consequently, the previous calculations are perfectly justified in each of the intervals where $h(t)$ has a constant sign.

Lemma 1.20. *Suppose* $-\infty < \alpha < \beta < +\infty$, $a_0, a_1 \in \mathcal{C}([\alpha, \beta]; \mathbb{K})$ *and* $h \in \mathcal{C}^2([\alpha, \beta]; \mathbb{K})$ *is a non-zero solution of* (1.52). *Then,* h *possesses, at most, finitely many zeros in* $[\alpha, \beta]$.

Proof. Let $t_0 \in [\alpha, \beta]$ such that $h(t_0) = 0$. We now show that $h'(t_0) \neq 0$ and, hence, t_0 is an isolated zero of $h(t)$. Indeed, if, by contradiction, $h(t_0) = h'(t_0) = 0$, then it follows from Corollary 1.18 that $h = 0$ in $[\alpha, \beta]$, which contradicts our assumption that $h \neq 0$.

Suppose, again by contradiction, that there exists a sequence $\{t_n\}_{n \geq 1}$ of distinct zeros of h in $[\alpha, \beta]$. Then, by the Bolzano–Weierstrass theorem, there exists a subsequence of $\{t_n\}_{n \geq 1}$, $\{t_{n_m}\}_{m \geq 1}$, and $t_0 \in [\alpha, \beta]$ such that

$$\lim_{m \to \infty} t_{n_m} = t_0.$$

Since h is continuous in $[\alpha, \beta]$ and $h(t_{n_m}) = 0$ for all $m \geq 1$, it becomes apparent that $h(t_0) = 0$. Thus, t_0 is a non-isolated zero of $h(t)$, in contrast what we have proved in the first part. This contradiction shows that h admits, at most, finitely many zeros in $[\alpha, \beta]$. $\qquad\square$

More generally, we can extend the procedure of order reduction to equations of higher orders. Indeed, suppose that a particular solution, $h(t)$, of the linear homogeneous equation

$$h^{N)} = \sum_{j=0}^{N-1} a_j(t) h^{j)} \tag{1.54}$$

is known, where we denote $h^{0)} := h$. Then, owing to Leibniz's rule, the change of variables

$$u(t) = h(t) v(t)$$

transforms the linear inhomogeneous equation

$$u^{N)} = \sum_{j=0}^{N-1} a_j(t)u^{j)} + b(t)$$

into

$$\sum_{j=0}^{N} \binom{N}{j} h^{j)} v^{N-j)} = \sum_{j=0}^{N-1} a_j \sum_{i=0}^{j} \binom{j}{i} h^{i)} v^{j-i)} + b.$$

Equivalently,

$$\sum_{j=0}^{N-1} \binom{N}{j} h^{j)} v^{N-j)} + h^{N)} v = \sum_{j=0}^{N-1} a_j h^{j)} v + \sum_{j=0}^{N-1} a_j \sum_{i=0}^{j-1} \binom{j}{i} h^{i)} v^{j-i)} + b.$$

Thus, thanks to (1.54), we find that

$$\sum_{j=0}^{N-1} \binom{N}{j} h^{j)} v^{N-j)} = \sum_{j=0}^{N-1} a_j \sum_{i=0}^{j-1} \binom{j}{i} h^{i)} v^{j-i)} + b. \qquad (1.55)$$

As the change of variable $v' = w$ transforms (1.55) into the linear equation of order $N - 1$,

$$\sum_{j=0}^{N-1} \binom{N}{j} h^{j)} w^{N-1-j)} = \sum_{j=0}^{N-1} a_j \sum_{i=0}^{j-1} \binom{j}{i} h^{i)} w^{j-1-i)} + b, \qquad (1.56)$$

we have reduced the complexity of the original problem by passing from a linear equation of order N to a different one of order $N - 1$. By reasoning as in the proof of Lemma 1.20, one can show that, if the coefficients of the homogeneous equation (1.54) are continuous, then any solution $h \neq 0$ of such an equation has at most a finite number of vanishing points in $[\alpha, \beta]$. Thus, one can work in any interval where h does not vanish and consider the following equation, which is equivalent to (1.56):

$$w^{N-1)} = -\frac{1}{h}\sum_{j=1}^{N-1}\binom{N}{j} h^{j)} w^{N-1-j)} + \frac{1}{h}\sum_{j=0}^{N-1} a_j \sum_{i=0}^{j-1}\binom{j}{i} h^{i)} w^{j-1-i)} + \frac{b}{h}.$$

We conclude this section by analyzing a simple example, in which we determine the general solution of the differential equation

$$u'' - tu' + u = e^{\frac{t^2}{2}} \tag{1.57}$$

by reducing its order. Since $h(t) = t$ satisfies the associated homogeneous equation,

$$h'' - th' + h = 0,$$

the change of variable

$$u = tv \tag{1.58}$$

allows us to reduce the order of (1.57). Indeed, substituting (1.58) into (1.57) leads to

$$2v' + tv'' - t(v + tv') + tv = e^{\frac{t^2}{2}},$$

or, equivalently, after simplifying and dividing by t,

$$v'' + \frac{2 - t^2}{t}v' = t^{-1}e^{\frac{t^2}{2}}. \tag{1.59}$$

To integrate (1.59), we make the additional change of variable

$$w := v', \tag{1.60}$$

which transforms (1.59) into the first-order linear equation

$$w' = \left(t - \frac{2}{t}\right) w + t^{-1}e^{\frac{t^2}{2}}, \tag{1.61}$$

whose integration can be carried out as described in Section 1.2. The linear homogeneous equation associated with (1.61) is

$$h' = \left(t - \frac{2}{t}\right) h, \tag{1.62}$$

or, equivalently,

$$\frac{d}{dt}\operatorname{Log} h(t) = \frac{h'(t)}{h(t)} = t - \frac{2}{t}.$$

Thus,

$$\operatorname{Log} h(t) = \frac{t^2}{2} - 2\operatorname{Log} t + A$$

for some constant $A \in \mathbb{K}$. Hence, the general solution of (1.62) is given by

$$h(t) = e^{\frac{t^2}{2} - 2\operatorname{Log} t}x = t^{-2}e^{\frac{t^2}{2}}x, \qquad x \in \mathbb{K}.$$

Consequently, the change of variable

$$w(t) = t^{-2}e^{\frac{t^2}{2}}z(t) \tag{1.63}$$

transforms (1.61) into

$$t^{-2}e^{\frac{t^2}{2}}z'(t) = t^{-1}e^{\frac{t^2}{2}},$$

which can be written as $z'(t) = t$. So,

$$z(t) = \frac{t^2}{2} + x$$

for some integration constant $x \in \mathbb{K}$. By substitution in (1.63), it becomes apparent that

$$w(t) = t^{-2}e^{\frac{t^2}{2}}\left(\frac{t^2}{2} + x\right) = t^{-2}e^{\frac{t^2}{2}}x + \frac{1}{2}e^{\frac{t^2}{2}}$$

provides us with the general solution of (1.61). By integrating (1.60), we find that

$$v(t) = \int t^{-2}e^{\frac{t^2}{2}}\,dt\,x + y + \frac{1}{2}\int e^{\frac{t^2}{2}}\,dt,$$

where $y \in \mathbb{K}$ is another integration constant, provides us with the general solution of (1.59). Therefore, by (1.58),

$$u(t) = t\int t^{-2}e^{\frac{t^2}{2}}\,dt\,x + ty + \frac{t}{2}\int e^{\frac{t^2}{2}}\,dt \tag{1.64}$$

provides us with the general solution of (1.57). Thanks to the structural properties of (1.57),

$$h(t) = t \int t^{-2} e^{\frac{t^2}{2}} \, dt \, x + ty, \qquad x, y \in \mathbb{K},$$

is the general solution of the associated homogeneous equation, and

$$p(t) = \frac{t}{2} \int e^{\frac{t^2}{2}} \, dt$$

is a particular solution of (1.57), the one obtained by choosing $x = y = 0$ in (1.64). We point out that, since $e^{\frac{t^2}{2}}$ does not admit any primitives in terms of elementary functions, it is not possible to give a closed formula for the general solution of (1.57). Actually, this happens in the vast majority of differential equations, and it is one of the main motivations for their *qualitative study*, which we will present later in this book.

1.10   The Method of Indeterminate Coefficients

Varying the coefficients, as described in Sections 1.6 and 1.8, is the most versatile method to get the general solution of linear inhomogeneous equations, in the sense that it allows one to deal with inhomogeneities, $b(t)$, of the general type. Nonetheless, it requires a number of calculations. We will see in this section that, in certain situations when $b(t)$ has some specific form, getting a particular solution for the inhomogeneous equation can be a rather direct task by using the method of indeterminate coefficients. We discuss here three of these particular, but important, cases.

Example 1. First, consider the second-order linear equation

$$u'' = a_0 u + a_1 u' + b, \tag{1.65}$$

with $a_0, a_1, b \in \mathbb{C}$ and $b \neq 0$. We already know from Section 1.8 how to construct the general solution of the associated homogeneous equation; therefore, here, we only focus on determining a particular solution, $p(t)$.

If $a_0 \neq 0$, since b is constant, we look for a particular solution of the same kind, i.e. $p(t) = c$, where $c \in \mathbb{C}$ is an indeterminate coefficient (this is where the name of the method comes from) that has to be adjusted in order to obtain a solution for the differential equation.

If we substitute this *Ansatz* in the differential equation, we easily see that c has to satisfy

$$0 = a_0 c + b.$$

Hence, $p(t) = -b/a_0$ is the desired particular solution of (1.65) in this case.

When $a_0 = 0$ and $a_1 \neq 0$, instead, (1.65) reduces to

$$u'' = a_1 u' + b,$$

which cannot admit any constant solution (indeed constant functions, in this case, are solutions of the associated homogeneous equation, and we are assuming $b \neq 0$). The change of variable $v = u'$ reduces the order of the equation and transforms it into

$$v' = a_1 v + b. \tag{1.66}$$

Now, we can reason as above and obtain that $v = -b/a_1$ is a constant solution of (1.66). Hence, $u(t) = -\frac{b}{a_1}t$ provides us with a particular solution of (1.65); therefore,

$$\Sigma_b^e = \Sigma_0^e - \frac{b}{a_1}t.$$

Finally, when $a_0 = a_1 = 0$, then (1.65) becomes $u'' = b$, and a direct integration shows that $p(t) = \frac{b}{2}t^2$ provides us with the desired particular solution of (1.65).

More generally, when $a_0 = a_1 = 0$, by two direct integrations, it becomes apparent that the general solution of the inhomogeneous equation $u'' = b(t)$, where $b \in \mathcal{C}([\alpha, \beta]; \mathbb{C})$ is arbitrary, is given, for all $t_0, t \in [\alpha, \beta]$, by

$$u(t) = x_1 + tx_2 + \int_{t_0}^{t} \int_{t_0}^{s} b(\sigma)\, d\sigma\, ds, \qquad x_1, x_2 \in \mathbb{C}.$$

Example 2. As a second simple example, consider the differential equation

$$u'' = a_0 u + a_1 u' + m e^{\lambda t}, \tag{1.67}$$

where $a_0, a_1, m, \lambda \in \mathbb{C}$, with $m \neq 0$. By using the notation and terminology introduced in Section 1.8, we distinguish the following three cases:

(a) $P(\lambda) \neq 0$, i.e. λ is not a characteristic root of the homogeneous differential equation associated with (1.67);
(b) $P(\lambda) = 0$ but $P'(\lambda) \neq 0$, i.e. λ is a simple characteristic root;
(c) $P(\lambda) = P'(\lambda) = 0$, i.e. λ is a double characteristic root.

In case (a), i.e. when $P(\lambda) \neq 0$, (1.67) possesses a solution of the form $ce^{\lambda t}$ for some *indeterminate constant* $c \in \mathbb{C}$. Indeed, since

$$P(D)(ce^{\lambda t}) = cP(\lambda)e^{\lambda t},$$

it becomes apparent that

$$u(t) = \frac{m}{P(\lambda)} e^{\lambda t}$$

solves (1.67). Thus, in case (a),

$$\Sigma_{me^{\lambda t}}^{e} = \Sigma_{0}^{e} + \frac{m}{P(\lambda)} e^{\lambda t}.$$

The previous construction fails if $P(\lambda) = 0$ because, in such a case,

$$P(D)(ce^{\lambda t}) = cP(\lambda)e^{\lambda t} = 0 \neq me^{\lambda t}$$

since $m \neq 0$. Suppose we consider case (b), i.e.

$$P(\lambda) = \lambda^2 - a_1 \lambda - a_0 = 0, \qquad P'(\lambda) = 2\lambda - a_1 \neq 0. \tag{1.68}$$

Then, $e^{\lambda t}$ solves the associated homogeneous equation, and hence, arguing as in Section 1.9, the change of variable $u = e^{\lambda t} v$ transforms

(1.67) into

$$\lambda^2 e^{\lambda t}v + 2\lambda e^{\lambda t}v' + e^{\lambda t}v'' = a_0 e^{\lambda t}v + a_1(\lambda e^{\lambda t}v + e^{\lambda t}v') + me^{\lambda t}.$$

Thus, dividing by $e^{\lambda t}$, rearranging terms, and using (1.68) give

$$v'' = (a_1 - 2\lambda)v' + m. \tag{1.69}$$

The change of variable $w = v'$ reduces (1.69) to

$$w' = (a_1 - 2\lambda)w + m,$$

and, due to (1.68), since $a_1 - 2\lambda \neq 0$, the constant $w = \frac{m}{2\lambda - a_1}$ solves this equation. Hence, by integrating, it becomes apparent that

$$v(t) = \frac{m}{2\lambda - a_1}t$$

solves (1.69). Therefore,

$$u(t) = \frac{m}{2\lambda - a_1}te^{\lambda t} = \frac{m}{P'(\lambda)}te^{\lambda t}$$

solves (1.67), and consequently, in case (b), we have that

$$\Sigma^e_{me^{\lambda t}} = \Sigma^e_0 + \frac{m}{P'(\lambda)}te^{\lambda t}.$$

Finally, suppose that

$$P(\lambda) = \lambda^2 - a_1\lambda - a_0 = 0, \qquad P'(\lambda) = 2\lambda - a_1 = 0.$$

Then, (1.69) reduces to $v'' = m$, and obviously, $v(t) = \frac{m}{2}t^2$ solves this equation. Therefore, the function

$$u(t) = \frac{m}{2}t^2 e^{\lambda t} = \frac{m}{P''(\lambda)}t^2 e^{\lambda t}$$

solves (1.67), and hence, in case (c), it is apparent that

$$\Sigma^e_{me^{\lambda t}} = \Sigma^e_0 + \frac{m}{P''(\lambda)}t^2 e^{\lambda t}.$$

We point out that the equation treated in Example 1 is a particular instance of (1.67), with $\lambda = 0$ and $m = b$. Actually, one can recognize the cases analyzed in Example 1 as particular instances of the cases (a), (b), and (c) discussed here and also understand the deep relation with the characteristic roots in such a situation.

Example 3. To conclude this section with a slightly more sophisticated example, we consider the differential equation

$$P(D)u = b + me^{\lambda t}, \tag{1.70}$$

where

$$P(z) = (z - \lambda_1)(z - \lambda_2) = z^2 - (\lambda_1 + \lambda_2)z + \lambda_1\lambda_2 = z^2 - a_1 z - a_0$$

for some $\lambda_1, \lambda_2, \lambda \in \mathbb{C}$ and $b, m \in \mathbb{C} \setminus \{0\}$. To determine a particular solution of (1.70), it is enough to look for a particular solution of each of the differential equations

$$P(D)u_1 = b, \qquad P(D)u_2 = me^{\lambda t},$$

which have been analyzed in the previous examples. Indeed, thanks to the generalized superposition principle in Exercise 1 of Chapter 1, their sum

$$u := u_1 + u_2$$

solves (1.70). In particular, if we assume, in addition, that $\lambda_1, \lambda_2, \lambda \in \mathbb{C} \setminus \{0\}$ and

$$\lambda_1 \neq \lambda_2 \neq \lambda \neq \lambda_1,$$

since $a_0 = -\lambda_1\lambda_2 \neq 0$ and $P(\lambda) \neq 0$, according to the previous analysis, we can take

$$u_1(t) = -\frac{b}{a_0}, \qquad u_2(t) = \frac{m}{P(\lambda)}e^{\lambda t}.$$

Therefore,

$$\Sigma^e_{b+me^{\lambda t}} = \text{span}[e^{\lambda_1 t}, e^{\lambda_2 t}] - \frac{b}{a_0} + \frac{m}{P(\lambda)}e^{\lambda t}.$$

In other words,

$$u(t) = e^{\lambda_1 t}x_1 + e^{\lambda_2 t}x_2 - \frac{b}{a_0} + \frac{m}{P(\lambda)}e^{\lambda t}, \qquad x_1, x_2 \in \mathbb{C},$$

provides us with the general solution of (1.70).

We conclude this section with the observation that the method of indeterminate coefficients relies on the prediction of a particular solution for the considered equation, which can also work for equations of higher orders. Unless the equation is very easy or one has gained some experience and hence a reasonable *Ansatz* can be quickly found, one should not spend too much time trying to figure out the form of a particular solution. If it is not easy to find a particular solution, instead, as a general recommendation, it is preferable to vary the coefficients, as described in the previous sections.

1.11 Harmonic Motion

1.11.1 *Simple harmonic motion*

Harmonic motion is typified by the motion of a diapason or the motion of a mass hanging from a spring. When the unique force acting on the system is a linear elastic restoring force given by Hooke's law,

$$F := -\kappa u(t),$$

where $\kappa > 0$ is Hooke's constant, the motion is known as a *simple harmonic motion*. If $u(t)$ measures the separation of the mass, $m > 0$, with respect to the equilibrium position, Newton's second law establishes that

$$F = m\, u''(t),$$

and it becomes apparent that $u(t)$ obeys the linear second-order equation

$$u'' + \omega^2 u = 0, \qquad \omega := \sqrt{\frac{\kappa}{m}}. \tag{1.71}$$

Thanks to Corollary 1.18, the motion of the hanging mass is characterized by the initial conditions

$$u(0) = u_0, \qquad u'(0) = v_0. \tag{1.72}$$

Naturally, $u = 0$ if $u_0 = v_0 = 0$. So, suppose $(u_0, v_0) \neq (0, 0)$. As observed in Section 1.8.1,

$$h_1(t) = \sin(\omega t) \quad \text{and} \quad h_2(t) = \cos(\omega t), \qquad t \in \mathbb{R},$$

are two (real) independent solutions of (1.71). Thus, the general (real) solution of (1.71) can be expressed as

$$u(t) = x_1 \sin(\omega t) + x_2 \cos(\omega t),$$

where $x_1, x_2 \in \mathbb{R}$. In particular, all the solutions of (1.71) are $\frac{2\pi}{\omega}$-periodic. Moreover, since

$$v(t) := u'(t) = x_1 \omega \cos(\omega t) - x_2 \omega \sin(\omega t),$$

the initial conditions (1.72) hold if and only if

$$u(0) = x_2 = u_0, \qquad u'(0) = \omega x_1 = v_0.$$

Thus, the function

$$u(t) = \frac{v_0}{\omega} \sin(\omega t) + u_0 \cos(\omega t), \qquad t \in \mathbb{R}, \tag{1.73}$$

provides us with the unique solution of (1.71) satisfying the initial conditions (1.72). Since $(u_0, v_0) \neq (0,0)$, setting

$$A := \sqrt{\left(\frac{v_0}{\omega}\right)^2 + u_0^2} > 0$$

and denoting by φ the unique $\varphi \in [-\pi, \pi)$ such that

$$\cos \varphi = \frac{\frac{v_0}{\omega}}{\sqrt{\left(\frac{v_0}{\omega}\right)^2 + u_0^2}} \qquad \text{and} \qquad \sin \varphi = \frac{u_0}{\sqrt{\left(\frac{v_0}{\omega}\right)^2 + u_0^2}},$$

it follows that (1.73) can be rewritten as

$$u(t) = \sqrt{\left(\frac{v_0}{\omega}\right)^2 + u_0^2} \left[\frac{\frac{v_0}{\omega}}{\sqrt{\left(\frac{v_0}{\omega}\right)^2 + u_0^2}} \sin(\omega t) + \frac{u_0}{\sqrt{\left(\frac{v_0}{\omega}\right)^2 + u_0^2}} \cos(\omega t) \right]$$

$$= A[\cos \varphi \sin(\omega t) + \sin \varphi \cos(\omega t)].$$

Therefore,

$$u(t) = A \sin(\omega t + \varphi) \qquad \text{for all } t \in \mathbb{R}. \tag{1.74}$$

Since A equals the maximal separation from the equilibrium position, $u = 0$, of the mass m, it is the *amplitude* of the simple harmonic motion. The *phase*, φ, measures the separation of the motion with respect to the "pure" sinusoidal function, $A\sin(\omega t)$. Naturally, the period of the oscillations is given by $T := \frac{2\pi}{\omega}$, and their *frequency* is

$$\nu := \frac{1}{T} = \frac{\omega}{2\pi}.$$

The parameter ω is usually referred to as the *angular frequency*, or *pulse*, of the oscillator.

Next, we deduce a fundamental physical property of the simple harmonic motion. Multiplying

$$mu'' + \kappa u = 0$$

by $v(t) := u'(t)$ yields

$$mu'(t)u''(t) + \kappa u(t)u'(t) = 0,$$

or, equivalently,

$$\frac{d}{dt}\left(\frac{m}{2}v^2(t) + \frac{\kappa}{2}u^2(t)\right) = 0 \quad \text{for all } t \in \mathbb{R}.$$

Therefore,

$$\frac{m}{2}v^2(t) + \frac{\kappa}{2}u^2(t) = \frac{m}{2}v_0^2 + \frac{\kappa}{2}u_0^2 \quad \text{for all } t \in \mathbb{R}. \tag{1.75}$$

As the *kinetic, potential,* and *total* energies of the oscillator are defined by

$$E_{\mathrm{k}} := \frac{m}{2}v^2, \qquad E_{\mathrm{p}} := \frac{\kappa}{2}u^2, \qquad \text{and} \qquad E := E_{\mathrm{k}} + E_{\mathrm{p}},$$

respectively, (1.75) establishes the conservation of the total energy of the simple harmonic oscillator, i.e.

$$E(t) = E(0), \qquad \text{for all } t \in \mathbb{R}.$$

Since

$$E(0) = \frac{m}{2}v_0^2 + \frac{\kappa}{2}u_0^2 = \frac{m}{2}v_0^2 + \frac{\kappa}{2}\left(A^2 - \frac{v_0^2}{\omega^2}\right) = \frac{\kappa}{2}A^2,$$

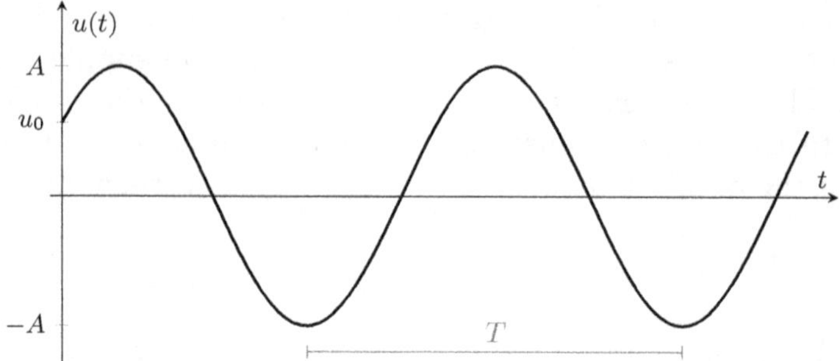

Fig. 1.2. Displacement $u(t)$ of a simple harmonic oscillator.

the total energy of the simple oscillator equals $\frac{\kappa}{2}A^2$ and, hence,

$$\frac{m}{2}v^2(t) + \frac{\kappa}{2}u^2(t) = \frac{\kappa}{2}A^2 \quad \text{for all } t \in \mathbb{R}.$$

In particular, the velocity of the oscillator, $v(t)$, can be expressed in terms of its position, $u(t)$, through the fundamental relation

$$v(t) = \pm\omega\sqrt{A^2 - u^2(t)}, \qquad t \in \mathbb{R}, \tag{1.76}$$

which can be also derived directly from (1.74). According to (1.76), the velocity of the oscillator vanishes whenever it reaches its maximal separation with respect to the equilibrium position at $u = \pm A$, whereas it is maximal, $v = \pm\omega A$, whenever the oscillator crosses the equilibrium position at $u = 0$. Figure 1.2 illustrates the displacement $u(t)$ of a simple harmonic oscillator.

1.11.2 *Damped harmonic motion*

In the previous analysis of simple harmonic oscillators, the amplitude of the oscillations, A, remains unchanged in time. However, in real oscillations, frictional forces caused by the fluid viscosity of air diminish the amplitude of the oscillations and give rise to *damped harmonic oscillations*. Actually, experience dictates that the amplitude A is a function of time, $A(t)$, satisfying

$$\lim_{t\to\infty} A(t) = 0.$$

As a result, in practice, the ideal simple oscillations around the equilibrium position are damped as time goes by at a rate regulated by the viscosity of the medium where the oscillations occur. To quantify these effects mathematically, one can add a damping term, $-ru'(t)$, to the elastic restoring force. In this way, the total force becomes

$$F = -\kappa u(t) - ru'(t),$$

where $r > 0$ measures the resistance, or viscosity, of the medium, which is assumed to be constant. Under this assumption, the differential equation modeling the free damped oscillations becomes

$$mu'' + ru' + \kappa u = 0, \tag{1.77}$$

whose characteristic roots are

$$\lambda_1 := \frac{-r - \sqrt{r^2 - 4\kappa m}}{2m} \quad \text{and} \quad \lambda_2 := \frac{-r + \sqrt{r^2 - 4\kappa m}}{2m}.$$

If the viscosity r is sufficiently strong so that $r^2 > 4\kappa m$, then, λ_1 and λ_2 are real and satisfy $\lambda_1 < \lambda_2 < 0$. Thus, since the general solution of (1.77) is

$$u(t) = xe^{\lambda_1 t} + ye^{\lambda_2 t}, \qquad x, y \in \mathbb{R},$$

the solutions of (1.77) exponentially converge to zero as $t \uparrow \infty$, not being able to oscillate around the equilibrium position $u = 0$. This system is referred to as an *overdamped oscillator* and one of its solutions is represented in the left-hand plot of Figure 1.3.

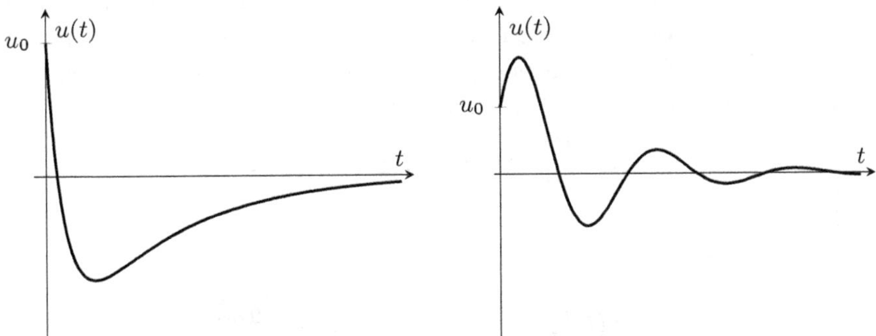

Fig. 1.3. Displacement $u(t)$ of a damped harmonic oscillator: overdamped case (left) and underdamped case (right).

When the viscosity r is sufficiently small so that $r^2 < 4\kappa m$, the system is referred to as an *underdamped oscillator*. In this case,

$$\lambda_1 = \frac{-r}{2m} - \frac{\sqrt{4\kappa m - r^2}}{2m}i \qquad \text{and} \qquad \lambda_2 = \frac{-r}{2m} + \frac{\sqrt{4\kappa m - r^2}}{2m}i,$$

are the complex conjugate characteristic roots of (1.77) with a negative real part, $\frac{-r}{2m} < 0$. Thus, as shown in Section 1.8.1, the general solution of (1.77) in this case can be expressed as

$$u(t) = e^{-\frac{r}{2m}t}\sin(\Omega t)x_1 + e^{-\frac{r}{2m}t}\cos(\Omega t)x_2,$$

where $x_1, x_2 \in \mathbb{R}$ and

$$\Omega := \sqrt{\frac{\kappa}{m} - \frac{r^2}{4m^2}} < \sqrt{\frac{\kappa}{m}} = \omega. \tag{1.78}$$

By reasoning as in the case of simple oscillations, which corresponds to $r = 0$, it follows that there exist two constants, $A_d \geq 0$ and $\varphi_d \in [-\pi, \pi)$, such that

$$u(t) = e^{-\frac{r}{2m}t}A_d\sin(\Omega t + \varphi_d), \qquad t \in \mathbb{R}.$$

Therefore, the amplitude of the underdamped oscillations at time t,

$$A(t) := e^{-\frac{r}{2m}t}A_d, \qquad t \in \mathbb{R},$$

decays exponentially to zero as $t \uparrow \infty$ at the rate $\frac{-r}{2m}$. Another effect of the damping term $-ru'$ is the reduction in the angular frequency since $\Omega < \omega$. Hence, the period of the underdamped oscillations, $\frac{2\pi}{\Omega}$, is greater than the period of the simple oscillations, $\frac{2\pi}{\omega}$, which is very natural from a physical point of view. Note that both periods, $\frac{2\pi}{\omega}$ and $\frac{2\pi}{\Omega}$, are independent of the initial conditions u_0 and v_0. A remarkable effect of the viscosity, $-ru'$, closely related to the previous ones, is the dissipation of the total energy of the oscillator as time increases. Indeed, differentiating with respect to t the total energy

$$E(t) := \frac{m}{2}v^2(t) + \frac{\kappa}{2}u^2(t), \qquad t \in \mathbb{R},$$

yields

$$E'(t) = mv(t)v'(t) + \kappa u(t)u'(t)$$

$$= v(t)[mu''(t) + \kappa u(t)] = -rv^2(t) \leq 0$$

for all $t > 0$. Thus, the total energy decreases as an effect of the viscosity, in strong contrast to the case of $r = 0$, when it remains constant for all time. Thus, the damping term in the equation makes the mechanical system *dissipative*, in the sense that the total energy is *dissipated by friction*. For analogous reasons, the system for $r = 0$ is said to be of *conservative* type.

1.11.3 *Forced oscillations*

Although in many circumstances the dissipation of energy given by damping is desirable since it helps to avoid unpleasant destabilizing effects, in many others it is convenient to add some external energy to the system in order to keep the oscillations active all the time. This can be easily achieved by including an additional *forcing term* to the differential equation. The resulting oscillations are then referred to as *forced oscillations*, in contrast to the previous cases, which are known as *free oscillations*. We consider the following equation as a paradigm for modeling forced mechanical oscillations:

$$mu'' + ru' + \kappa u = a\sin(\gamma t), \tag{1.79}$$

where γ is the angular frequency of the external force, whose *amplitude*, denoted by $a \in \mathbb{R} \setminus \{0\}$, is assumed to be constant. To get a particular solution for (1.79), it is natural to search for values of c_1 and c_2 such that

$$u(t) = c_1 \sin(\gamma t) + c_2 \cos(\gamma t) \tag{1.80}$$

solves (1.79). Substituting (1.80) in (1.79), it is apparent that c_1 and c_2 have to be chosen so that, for all $t \in \mathbb{R}$,

$$\begin{aligned}
a\sin(\gamma t) &= m\left[-c_1\gamma^2 \sin(\gamma t) - c_2\gamma^2 \cos(\gamma t)\right] \\
&\quad + r\left[c_1\gamma \cos(\gamma t) - c_2\gamma \sin(\gamma t)\right] + \kappa\left[c_1 \sin(\gamma t) + c_2 \cos(\gamma t)\right] \\
&= \left[c_1(\kappa - m\gamma^2) - c_2 r\gamma\right]\sin(\gamma t) \\
&\quad + \left[c_2(\kappa - m\gamma^2) + c_1 r\gamma\right]\cos(\gamma t).
\end{aligned}$$

Equivalently,

$$\begin{pmatrix} \kappa - m\gamma^2 & -r\gamma \\ r\gamma & \kappa - m\gamma^2 \end{pmatrix} \begin{pmatrix} c_1 \\ c_2 \end{pmatrix} = \begin{pmatrix} a \\ 0 \end{pmatrix}.$$

Thus, Cramer's rule yields

$$c_1 = \frac{1}{(\kappa - m\gamma^2)^2 + r^2\gamma^2} \begin{vmatrix} a & -r\gamma \\ 0 & \kappa - m\gamma^2 \end{vmatrix} = \frac{\kappa - m\gamma^2}{(\kappa - m\gamma^2)^2 + r^2\gamma^2} a,$$

$$c_2 = \frac{1}{(\kappa - m\gamma^2)^2 + r^2\gamma^2} \begin{vmatrix} \kappa - m\gamma^2 & a \\ r\gamma & 0 \end{vmatrix} = \frac{-r\gamma}{(\kappa - m\gamma^2)^2 + r^2\gamma^2} a,$$

and, therefore, substituting in (1.80) and taking $\psi \in [-\pi, \pi)$ such that

$$\cos \psi = \frac{\kappa - m\gamma^2}{\sqrt{(\kappa - m\gamma^2)^2 + r^2\gamma^2}}, \qquad \sin \psi = \frac{-r\gamma}{\sqrt{(\kappa - m\gamma^2)^2 + r^2\gamma^2}},$$

we find that

$$u(t) = \frac{a}{\sqrt{(\kappa - m\gamma^2)^2 + r^2\gamma^2}} [\cos \psi \sin(\gamma t) + \sin \psi \cos(\gamma t)]$$

$$= \frac{a}{\sqrt{(\kappa - m\gamma^2)^2 + r^2\gamma^2}} \sin(\gamma t + \psi)$$

solves the forced equation (1.79). Thus, if we further assume that the system is in the underdamped case $r^2 - 4m\kappa < 0$, the general solution of (1.79) is

$$u(t) = e^{-\frac{r}{2m}t} \sin(\Omega t)x_1 + e^{-\frac{r}{2m}t} \cos(\Omega t)x_2$$

$$+ \frac{a}{\sqrt{(\kappa - m\gamma^2)^2 + r^2\gamma^2}} \sin(\gamma t + \psi),$$

where $x_1, x_2 \in \mathbb{R}$ and Ω is the constant defined in (1.78). Since the first two terms of $u(t)$ exponentially approach zero as $t \uparrow \infty$, it becomes apparent that, in fact, the solution

$$u(t) = \frac{a}{\sqrt{(\kappa - m\gamma^2)^2 + r^2\gamma^2}} \sin(\gamma t + \psi) \tag{1.81}$$

provides us with the *permanent term of the forced oscillation*. Note that the permanent term (1.81) has the same form as the forcing term, with two exceptions: the change of phase ψ and the *amplification*, or *reduction*, factor

$$f(\gamma) := \frac{1}{\sqrt{(\kappa - m\gamma^2)^2 + r^2\gamma^2}}.$$

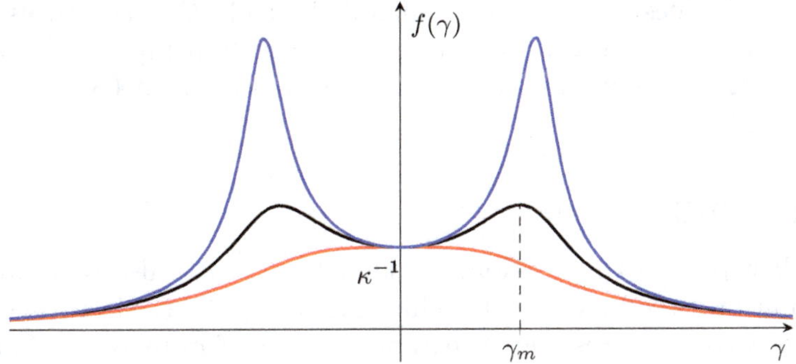

Fig. 1.4. Graph of the amplification, or reduction, factor $f(\gamma)$, for decreasing values of r (given by the red, black, and blue plots in that order).

Figure 1.4 shows the graph of $f(\gamma)$ for three values of r. The function $f(\gamma)$ is positive, even, and satisfies

$$f(0) = \kappa^{-1} > 0, \qquad \lim_{\gamma \to \infty} f(\gamma) = 0.$$

Moreover, for $r \in \left(0, \sqrt{2m\kappa}\right)$, it has a unique critical point in $(0, \infty)$ at

$$\gamma_m := \sqrt{\frac{\kappa}{m} - \frac{r^2}{2m^2}} < \Omega = \sqrt{\frac{\kappa}{m} - \frac{r^2}{4m^2}} < \omega = \sqrt{\frac{\kappa}{m}},$$

where f necessarily has a global maximum and takes the value

$$f(\gamma_m) = \frac{1}{r\sqrt{\frac{\kappa}{m} - \frac{r^2}{4m^2}}} = \frac{1}{r\Omega}.$$

In Physics, the constant γ_m is known as the *resonant frequency*. For small values of r, it is very close to the natural frequency of the free simple system, ω. Therefore, as the resistance r goes to 0, the forced oscillations attain their maximum amplitude for a frequency of the external force close to the frequency of the free undamped system. Moreover, since

$$\lim_{r \to 0^+} f(\gamma_m) = +\infty,$$

the amplitude of the resonant forced oscillations can be arbitrarily large. Naturally, though this phenomenon might be desired in many electrical devices, it must be controlled in elastic structures, such as bridges or buildings. The analysis of the limiting case of $r = 0$ and $\gamma = \omega$ is left as an exercise for the reader (see Exercise 21 of Chapter 1).

1.11.4   RLC *circuits*

In the last part of this section, we show how (1.79) also regulates the evolution of forced electrical oscillations in an RLC circuit, consisting of a resistor of resistance R, an inductor of inductance L, and a capacitor of capacitance C connected either in series or in parallel. Some resistance is unavoidable in real circuits even if a resistor is not included as a component: pure LC circuits exist only in the field of superconductivity. Suppose the capacitor is initially charged, and let $q(t)$, $i(t)$, and $u(t)$, respectively, denote the charge, the current intensity, and the potential of the circuit. Then, since $q = Cu$, we have that

$$u = \frac{q}{C} \qquad \text{and} \qquad i = -q'(t) \tag{1.82}$$

because i is assumed to be positive, by convention, representing discharge, i.e. when $q'(t) < 0$. On the other hand, any variation in the intensity $i(t)$ induces a change in the magnetic field and, hence, causes a counter-electromotive force (CEMF) by magnetic induction, called self-induction, which is given by $-Li'(t) = Lq''(t)$. Thus, if the elements of the circuit are connected in series, the total electromotive force at our disposal is given by $u(t) + Lq''(t)$, and hence, Ohm's law gives

$$u(t) + Lq''(t) = Ri(t).$$

Consequently, by using (1.82), it is easily seen that

$$Lq''(t) + Rq'(t) + \frac{1}{C}q(t) = 0,$$

which is the same differential equation as the free damped oscillator (1.77), with m exchanged for L, r for R, κ for $\frac{1}{C}$, and the displacement for $q(t)$. Therefore, the charge fluctuations, $q(t)$, in an electrical circuit obey the same laws as the oscillations in a mechanical system whose mass, m, equals the self-induction L, whose resistance, r, equals the electrical resistance R, and whose elastic constant, κ, equals the inverse of the capacitance, C^{-1}.

This is an example of the huge versatility and tremendous power of Mathematics, as the same mathematical models and, therefore, the same solutions can describe the behavior of completely different systems. Thus, linear second-order equations establish a hidden unexpected connection between the mechanical and electrical worlds, which are *a priori* different.

The importance of mathematical models lies in the fact that they provide us with an idealized description against which the behavior of reality can be contrasted and measured. Furthermore, they promote the design of illuminating experiments and also enhance interdisciplinary studies. The main paradigm of this is the Newtonian synthesis in Physics: Newton introduced the concept of derivative and, hence, differential equations to obtain a rigorous model for the planetary laws of Kepler. However, Newtonian synthesis, the outcome of Newton's tremendous audacity, innovation, and genius, would have not been possible without the intensive empirical studies by Brahe and his predecessors, as well as the exquisite mathematical taste of Kepler. The idea of combining these methodologies led to one of the most relevant scientific revolutions of all time.

Moreover, in many other circumstances, mathematical models have shown to be imperative for making predictions of highly unexpected phenomena, such as the existence of the Hertzian waves from the Maxwell equations, the curvature of light predicted by Einstein from the equations of gravitational fields, or the existence of black holes, ultimately established by Hawking and Penrose.

1.12 Homogeneous Equations with Constant Coefficients

The main goal of this section is to construct a basis for the set of solutions of the linear homogeneous equation with constant coefficients in the general case when $N \geq 2$:

$$u^{N)} = a_0 u + a_1 u' + \cdots + a_{N-1} u^{N-1)}. \tag{1.83}$$

Let $P_N(z)$ denote the characteristic polynomial of (1.83):

$$P_N(z) = z^n - a_{N-1}z^{N-1} - a_{N-2}z^{N-2} - \cdots - a_1 z - a_0.$$

Setting

$$P_N(D) = D^N - a_{N-1}D^{N-1} - a_{N-2}D^{N-2} - \cdots - a_1 D - a_0, \quad D = \frac{d}{dt},$$

(1.83) can be equivalently written as

$$P_N(D)u = 0.$$

Therefore, since

$$P_N(D)e^{\lambda t} = P_N(\lambda)e^{\lambda t},$$

it is apparent that $e^{\lambda t}$ solves (1.83) if and only if $P_N(\lambda) = 0$. The following result of technical nature will be extremely useful for our study.

Lemma 1.21. *Let $\lambda_j \in \mathbb{C}$, $j \in \{1, \ldots, q\}$, be the (distinct) roots of $P_N(z)$, with respective algebraic multiplicities $m_j \geq 1$, $j \in \{1, \ldots, q\}$, i.e.*

$$P_N(z) = (z - \lambda_1)^{m_1} \cdots (z - \lambda_q)^{m_q} = \prod_{j=1}^{q}(z - \lambda_j)^{m_j}. \qquad (1.84)$$

Then,

$$P_N(D) = (D - \lambda_1)^{m_1} \cdots (D - \lambda_q)^{m_q} = \prod_{j=1}^{q}(D - \lambda_j)^{m_j}, \qquad (1.85)$$

independently of the order of the factors.

Proof. Suppose $N = 2$. Then, there exist $\mu_1, \mu_2 \in \mathbb{C}$, not necessarily distinct, such that

$$P_2(z) = (z - \mu_1)(z - \mu_2) = z^2 - (\mu_1 + \mu_2)z + \mu_1\mu_2$$

and, hence,

$$a_1 = \mu_1 + \mu_2 \quad \text{and} \quad a_0 = -\mu_1\mu_2.$$

Thus,

$$P_2(D) = D^2 - a_1 D - a_0 = D^2 - (\mu_1 + \mu_2)D + \mu_1\mu_2$$
$$= (D - \mu_1)(D - \mu_2),$$

and since the sum and product in $\mathbb{C}$ are commutative, it is apparent that the order of μ_1 and μ_2 is irrelevant, and

$$P_2(D) = (D - \mu_1)(D - \mu_2) = (D - \mu_2)(D - \mu_1).$$

Similarly, in the case of $N = 3$, there exist $\mu_1, \mu_2, \mu_3 \in \mathbb{C}$ such that

$$P_3(z) = (z - \mu_1)(z - \mu_2)(z - \mu_3)$$
$$= z^3 - (\mu_1 + \mu_2 + \mu_3)z^2 + (\mu_1\mu_2 + \mu_1\mu_3 + \mu_2\mu_3)z - \mu_1\mu_2\mu_3$$

and, hence,

$$a_2 = \mu_1 + \mu_2 + \mu_3, \quad a_1 = -(\mu_1\mu_2 + \mu_1\mu_3 + \mu_2\mu_3), \quad \text{and} \quad a_0 = \mu_1\mu_2\mu_3.$$

Therefore,

$$P_3(D) = D^3 - a_2 D^2 - a_1 D - a_0$$
$$= D^3 - (\mu_1 + \mu_2 + \mu_3)D^2 + (\mu_1\mu_2 + \mu_1\mu_3 + \mu_2\mu_3)D - \mu_1\mu_2\mu_3$$
$$= (D - \mu_1)(D - \mu_2)(D - \mu_3),$$

and, once again, we can commute all these factors, $D - \mu_j$, without changing the result.

In the general case, one can reason as above, with the help of the Cardano–Vieta relations between the roots of the polynomial and its coefficients, which are invariant under permutation of the roots. $\square$

According to (1.85) and because of the independence of the order of the factors, we claim that, for every $j \in \{1, \dots, q\}$ and $i \in \{0, \dots, m_j - 1\}$,

$$P_N(D)(t^i e^{\lambda_j t}) = \prod_{\substack{k=1 \\ k \neq j}}^{q} (D - \lambda_k)^{m_k} (D - \lambda_j)^{m_j} (t^i e^{\lambda_j t}) = 0.$$

Indeed, since $m_j - i \geq 1$, we have that

$$
\begin{aligned}
(D - \lambda_j)^{m_j}(t^i e^{\lambda_j t}) &= (D - \lambda_j)^{m_j - 1}(D - \lambda_j)(t^i e^{\lambda_j t}) \\
&= (D - \lambda_j)^{m_j - 1}(it^{i-1} e^{\lambda_j t} + \lambda_j t^i e^{\lambda_j t} - \lambda_j t^i e^{\lambda_j t}) \\
&= i(D - \lambda_j)^{m_j - 1}(t^{i-1} e^{\lambda_j t}) \\
&= i(i-1)(D - \lambda_j)^{m_j - 2}(t^{i-2} e^{\lambda_j t}) = \cdots \\
&= i!(D - \lambda_j)^{m_j - i}(e^{\lambda_j t}) = 0.
\end{aligned}
$$

Therefore, the set

$$
\mathfrak{B} := \left\{ t^i e^{\lambda_j t} : i \in \{0, \ldots, m_j - 1\},\ j \in \{1, \ldots, q\} \right\} \tag{1.86}
$$

provides us with

$$
m_1 + m_2 + \cdots + m_q = N
$$

solutions of (1.83). The following result establishes that $\mathfrak{B}$ is a basis of Σ_0^e.

Theorem 1.22. *Suppose* (1.84), *with* $\lambda_i \neq \lambda_j$, *if* $i \neq j$. *Then, the set* $\mathfrak{B}$ *defined in* (1.86) *is a basis for the set of solutions of* (1.83). *In other words,*

$$
\Sigma_0^e = \mathrm{span} \left[t^i e^{\lambda_j t} : i \in \{0, \ldots, m_j - 1\},\ j \in \{1, \ldots, q\} \right].
$$

Proof. As $\mathfrak{B}$ possesses N functions and, according to Theorem 1.19, the dimension of Σ_0^e is precisely N, it suffices to show that the solutions of $\mathfrak{B}$ are linearly independent. This will be an easy consequence of the fact that, if we take any set of polynomials with complex coefficients, $P_j(t)$, $j \in \{1, \ldots, q\}$, such that

$$
\sum_{j=1}^{q} P_j(t) e^{\lambda_j t} = 0 \quad \text{for all } t \in \mathbb{R}, \tag{1.87}
$$

then, necessarily, $P_j = 0$ for all $j \in \{1, \ldots, q\}$.

We will prove such a result by induction on q. Naturally, it holds for $q = 1$, since $P_1(t)e^{\lambda_1 t} = 0$ for all $t \in \mathbb{R}$ implies $P_1(t) = 0$ for all $t \in \mathbb{R}$, and hence $P_1 = 0$. Suppose the result holds for any "linear" combination of $q - 1 \geq 1$ exponentials with different exponents and where the coefficients of the combination are polynomials in t, and assume (1.87). Then,

$$\sum_{j=1}^{q-1} P_j(t)e^{(\lambda_j - \lambda_q)t} = -P_q(t) \quad \text{for all } t \in \mathbb{R},$$

and hence, differentiating $k := 1 + \deg P_q$ times, this identity yields

$$\sum_{j=1}^{q-1} \frac{d^k}{dt^k}\left(P_j(t)e^{(\lambda_j - \lambda_q)t} \right) = 0 \quad \text{for all } t \in \mathbb{R}.$$

Thus, thanks to Leibniz's rule, we find that, for every $t \in \mathbb{R}$,

$$\sum_{j=1}^{q-1} \left(\sum_{i=0}^{k} \binom{k}{i} (\lambda_j - \lambda_q)^i D^{k-i} P_j(t) \right) e^{(\lambda_j - \lambda_q)t} = 0.$$

Therefore, by the induction hypothesis,

$$\sum_{i=0}^{k} \binom{k}{i} (\lambda_j - \lambda_q)^i D^{k-i} P_j(t) = 0$$

for all $j \in \{1, \ldots, q-1\}$ and $t \in \mathbb{R}$. In other words,

$$(D + \lambda_j - \lambda_q)^k P_j = 0 \qquad \text{for all } j \in \{1, \ldots, q-1\}. \tag{1.88}$$

Fix $j \in \{1, \ldots, q-1\}$. According to (1.88), the polynomial

$$Q_j := (D + \lambda_j - \lambda_q)^{k-1} P_j$$

satisfies

$$(D + \lambda_j - \lambda_q)Q_j = 0.$$

Equivalently,

$$Q_j' = (\lambda_q - \lambda_j)Q_j$$

and, hence,

$$Q_j(t) = e^{(\lambda_q - \lambda_j)t} Q_j(0)$$

for all $t \in \mathbb{R}$. Therefore, since $Q_j(t)$ is a polynomial and $\lambda_q \neq \lambda_j$, by differentiating the previous equality $\tilde{k}$ times, where $\tilde{k} := 1 + \deg Q_j$, we obtain that $Q_j(0) = 0$, which in turn implies $Q_j = 0$. Thus, we have shown that

$$(D + \lambda_j - \lambda_q)^{k-1} P_j = 0 \quad \text{for all } j \in \{1, \ldots, q-1\},$$

i.e. we have decreased by 1 the exponent in (1.88). By iterating this argument until the exponent equals 0, it becomes apparent that $P_j = 0$ for all $j \in \{1, \ldots, q-1\}$. Thus, we have reduced the problem to the situation in which only the summand with $j = q$ is present in (1.87), and the base case of induction allows us to complete the proof. $\qquad\square$

In the special case when all the coefficients of the differential equation (1.83), a_j, $j \in \{0, \ldots, N-1\}$, are real numbers, the characteristic polynomial, $P_N(z)$, has real coefficients, and hence, the complex roots, if any, appear in pairs because $P(\lambda) = 0$ if and only if $P(\bar{\lambda}) = 0$, where $\bar{\lambda}$ indicates the conjugate of $\lambda \in \mathbb{C}$. Moreover, the algebraic multiplicities of λ and $\bar{\lambda}$ coincide.

In such a setting, if there are no complex roots, Theorem 1.22 already provides us with a basis for Σ_0^e, consisting of real functions. Otherwise, i.e. if $P(z)$ admits some complex roots, a basis of Σ_0^e consisting of real functions is constructed as shown in the following result.

Theorem 1.23. *Suppose that $a_j \in \mathbb{R}$ for all $j \in \{0, \ldots, N-1\}$ and that $P_N(z)$ possesses $r \geq 1$ distinct real roots, $\lambda_j \in \mathbb{R}$, $j \in \{1, \ldots, r\}$, with algebraic multiplicities $m_j \geq 1$, respectively, and $2c \geq 2$ distinct complex roots,*

$$\alpha_k \pm i\beta_k, \qquad \beta_k > 0, \qquad k \in \{1, \ldots, c\},$$

with algebraic multiplicities $n_k \geq 1$, respectively. In other words,

$$N = \sum_{j=1}^{r} m_j + 2 \sum_{k=1}^{c} n_k, \tag{1.89}$$

and

$$P_N(z) = z^N - \sum_{j=0}^{N-1} a_j z^j = \prod_{j=1}^{r} (z - \lambda_j)^{m_j} \prod_{k=1}^{c} \left[(z - \alpha_k)^2 + \beta_k^2\right]^{n_k}.$$

Then, the following set, formed by N (distinct) real functions,

$$\mathfrak{B} = \left\{ t^\ell e^{\lambda_j t} : \ell \in \{0, \ldots, m_j - 1\},\ j \in \{1, \ldots, r\} \right\}$$

$$\cup \left\{ t^\ell e^{\alpha_k t} \cos(\beta_k t) : \ell \in \{0, \ldots, n_k - 1\},\ k \in \{1, \ldots, c\} \right\}$$

$$\cup \left\{ t^\ell e^{\alpha_k t} \sin(\beta_k t) : \ell \in \{0, \ldots, n_k - 1\},\ k \in \{1, \ldots, c\} \right\},$$

provides us with a basis for Σ_0^e consisting of real functions. If $r = 0$, i.e. when $P_N(z)$ only possesses $2c \geq 2$ distinct complex roots,

$$\alpha_k \pm i\beta_k, \qquad \beta_k > 0, \qquad k \in \{1, \ldots, c\},$$

with algebraic multiplicities $n_k \geq 1$, respectively, so that $N = 2\sum_{k=1}^{c} n_k$, a similar result holds, in the sense that

$$\mathfrak{B} = \left\{ t^\ell e^{\alpha_k t} \cos(\beta_k t) : \ell \in \{0, \ldots, n_k - 1\},\ k \in \{1, \ldots, c\} \right\}$$

$$\cup \left\{ t^\ell e^{\alpha_k t} \sin(\beta_k t) : \ell \in \{0, \ldots, n_k - 1\},\ k \in \{1, \ldots, c\} \right\}$$

provides us with a basis of Σ_0^e consisting of real functions.

Proof. We detail the case $r > 0$, as the case $r = 0$ follows analogously, with small adaptations. Since

$$t^\ell e^{\alpha_k t} \cos(\beta_k t) = \frac{t^\ell e^{(\alpha_k + i\beta_k)t} + t^\ell e^{(\alpha_k - i\beta_k)t}}{2},$$

$$t^\ell e^{\alpha_k t} \sin(\beta_k t) = \frac{t^\ell e^{(\alpha_k + i\beta_k)t} - t^\ell e^{(\alpha_k - i\beta_k)t}}{2i},$$

it is apparent by linearity that all the functions of $\mathfrak{B}$ solve (1.83). Thus, since $\mathfrak{B}$ consists of N elements owing to (1.89), thanks to Theorem 1.19, it suffices to show that $\mathfrak{B}$ is linearly independent. Indeed, let $\rho_{\ell j} \in \mathbb{R}$, $\ell \in \{0, \ldots, m_j - 1\}$, $j \in \{1, \ldots, r\}$, and $\gamma_{\ell k}, \mu_{\ell k} \in \mathbb{R}$, $\ell \in \{0, \ldots, n_k - 1\}$, $k \in \{1, \ldots, c\}$, be such that, for every $t \in \mathbb{R}$,

$$\sum_{j=1}^{r} \sum_{\ell=0}^{m_j-1} \rho_{\ell j} t^\ell e^{\lambda_j t} + \sum_{k=1}^{c} \sum_{\ell=0}^{n_k-1} \left[\gamma_{\ell k} t^\ell e^{\alpha_k t} \cos(\beta_k t) + \mu_{\ell k} t^\ell e^{\alpha_k t} \sin(\beta_k t) \right]$$
$$= 0.$$

Then, for every $t \in \mathbb{R}$,

$$0 = \sum_{j=1}^{r} \sum_{\ell=0}^{m_j-1} \rho_{\ell j} t^\ell e^{\lambda_j t} + \sum_{k=1}^{c} \sum_{\ell=0}^{n_k-1} t^\ell e^{\alpha_k t}$$
$$\times \left(\gamma_{\ell k} \frac{e^{i\beta_k t} + e^{-i\beta_k t}}{2} + \mu_{\ell k} \frac{e^{i\beta_k t} - e^{-i\beta_k t}}{2i} \right)$$
$$= \sum_{j=1}^{r} \sum_{\ell=0}^{m_j-1} \rho_{\ell j} t^\ell e^{\lambda_j t} + \sum_{k=1}^{c} \sum_{\ell=0}^{n_k-1} t^\ell$$
$$\times \left(e^{(\alpha_k + i\beta_k)t} \frac{\gamma_{\ell k} - i\mu_{\ell k}}{2} + e^{(\alpha_k - i\beta_k)t} \frac{\gamma_{\ell k} + i\mu_{\ell k}}{2} \right).$$

Thus, by reasoning as in the proof of Theorem 1.22, we conclude that $\rho_{\ell j} = 0$ for all $j \in \{1, \ldots, r\}$ and $\ell \in \{0, \ldots, m_j - 1\}$, and

$$\gamma_{\ell k} - i\mu_{\ell k} = \gamma_{\ell k} + i\mu_{\ell k} = 0, \qquad k \in \{1, \ldots, c\}, \ \ell \in \{0, \ldots, n_k - 1\}.$$

Therefore, also,

$$\gamma_{\ell k} = \mu_{\ell k} = 0$$

for all $k \in \{1, \ldots, c\}$ and $\ell \in \{0, \ldots, n_k - 1\}$. $\qquad\qquad \square$

1.13 Systems of Linear Equations of Arbitrary Order

The theory developed in the previous sections to treat first-order systems and linear equations of arbitrary order can be very easily adapted to deal with general systems of linear equations of arbitrary order. For example, consider the following system with two equations of order $N_1 = 2$ and $N_2 = 3$, respectively:

$$\begin{cases} u_1'' = a_{1,0}^1 u_1 + a_{1,1}^1 u_1' + a_{2,0}^1 u_2 + a_{2,1}^1 u_2' + a_{2,2}^1 u_2'' + b_1, \\ u_2''' = a_{1,0}^2 u_1 + a_{1,1}^2 u_1' + a_{2,0}^2 u_2 + a_{2,1}^2 u_2' + a_{2,2}^2 u_2'' + b_2, \end{cases} \tag{1.90}$$

where

$$a_{j,k}^i, \ b_i \in \mathcal{C}([\alpha, \beta]; \mathbb{K}), \qquad k \in \{0, \ldots, N_j - 1\}, \quad j, i \in \{1, 2\}. \tag{1.91}$$

Naturally, if

$$u := (u_1, u_2) \in \mathcal{C}^2([\alpha, \beta]; \mathbb{K}) \times \mathcal{C}^3([\alpha, \beta]; \mathbb{K})$$

solves (1.90), then

$$U := (u_1, u_1', u_2, u_2', u_2'')^T \in \mathcal{C}^1([\alpha, \beta]; \mathbb{K}^5)$$

solves the first-order system

$$U' = A(t)U + B(t), \tag{1.92}$$

where

$$A(t) := \begin{pmatrix} 0 & 1 & 0 & 0 & 0 \\ a_{1,0}^1(t) & a_{1,1}^1(t) & a_{2,0}^1(t) & a_{2,1}^1(t) & a_{2,2}^1(t) \\ 0 & 0 & 0 & 1 & 0 \\ 0 & 0 & 0 & 0 & 1 \\ a_{1,0}^2(t) & a_{1,1}^2(t) & a_{2,0}^2(t) & a_{2,1}^2(t) & a_{2,2}^2(t) \end{pmatrix}$$

and

$$B(t) = (0, b_1(t), 0, 0, b_2(t))^T.$$

Moreover, within the spirit of Theorem 1.17, there is a bijection between the solutions of (1.90) and the corresponding solutions of

(1.92). Thus, by Theorem 1.4, if we assume (1.91) and take any $t_0 \in [\alpha, \beta]$ and any

$$U_0 = (U_{0,1}, U_{0,2}, U_{0,3}, U_{0,4}, U_{0,5})^T \in \mathbb{K}^5,$$

the linear system (1.90) possesses a unique solution, $(u_1(t), u_2(t))$, such that

$$u_1(t_0) = U_{0,1}, \quad u_1'(t_0) = U_{0,2},$$
$$u_2(t_0) = U_{0,3}, \quad u_2'(t_0) = U_{0,4}, \quad \text{and} \quad u_2''(t_0) = U_{0,5},$$

and it is defined in $[\alpha, \beta]$.

More generally, let $n \geq 1$ be an arbitrary integer and consider, for each $i \in \{1, \ldots, n\}$, another integer $N_i \geq 1$, as well as the system of n linear equations of respective orders N_i:

$$u_i^{N_i)} = \sum_{j=1}^{n} \sum_{k=0}^{N_j-1} a_{j,k}^i(t) u_j^{k)} + b_i(t), \qquad i \in \{1, \ldots, n\}, \qquad (1.93)$$

where, for every $i, j \in \{1, \ldots, n\}$ and $k \in \{0, \ldots, N_j - 1\}$,

$$a_{j,k}^i, \; b_i \in \mathcal{C}([\alpha, \beta]; \mathbb{K}). \qquad (1.94)$$

Its associated homogeneous system is obtained by removing the inhomogeneous terms b_i, $i \in \{1, \ldots, n\}$, in (1.93), i.e.

$$u_i^{N_i)} = \sum_{j=1}^{n} \sum_{k=0}^{N_j-1} a_{j,k}^i(t) u_j^{k)}, \qquad i \in \{1, \ldots, n\}. \qquad (1.95)$$

A solution of (1.93) in the interval $[\alpha, \beta]$ is defined as a vectorial function,

$$(u_1, \ldots, u_n) \in \mathcal{C}^{N_1}([\alpha, \beta]; \mathbb{K}) \times \cdots \times \mathcal{C}^{N_n}([\alpha, \beta]; \mathbb{K}),$$

such that

$$u_i^{N_i)}(t) = \sum_{j=1}^{n} \sum_{k=0}^{N_j-1} a_{j,k}^i(t) u_j^{k)}(t) + b_i(t)$$

is satisfied for all $i \in \{1, \ldots, n\}$ and $t \in [\alpha, \beta]$.

In analogy with Section 1.7, we denote by Σ_b^{se}, $b = (b_1, \ldots, b_n)^T$, the set of solutions of the *system of equations* (1.93) and by Σ_0^{se} the set of solutions of the system of associated homogeneous equations (1.95). At this stage of this chapter, checking that Σ_0^{se} is a vector subspace of

$$\mathcal{C}^{N_1}([\alpha, \beta]; \mathbb{K}) \times \cdots \times \mathcal{C}^{N_n}([\alpha, \beta]; \mathbb{K})$$

should be quite easy, and we leave it as an exercise for the reader. Similarly, arguing as in Lemma 1.1, it is easily seen that Σ_b^{se} is an affine subspace of the same space, parallel to Σ_0^{se}. Next, for every

$$u := (u_1, \ldots, u_n) \in \Sigma_b^{se},$$

we consider

$$\mathcal{D}(u) := \left(u_1, u_1', \ldots, u_1^{N_1-1)}, \ldots, u_j, u_j', \ldots, u_j^{N_j-1)}, \ldots, \right.$$
$$\left. u_n, u_n', \ldots, u_n^{N_n-1)} \right)^T,$$

which is a vectorial function in $\mathcal{C}^1([\alpha, \beta]; \mathbb{K}^N)$, with $N := \sum_{j=1}^n N_j$.

Then, $U := \mathcal{D}(u)$ satisfies a certain first-order linear system with N equations of the same type as (1.92), and much like in Theorem 1.17, the map $\mathcal{D}$ establishes a bijection between Σ_b^{se} and the set of solutions of the associated first-order linear system, Σ_B^{fos}. Actually, $\mathcal{D}$ establishes a linear isomorphism between Σ_0^{se} and Σ_0^{fos}. Thus, by Theorem 1.4, the following result holds.

Corollary 1.24. *Under condition* (1.94), *for every* $t_0 \in [\alpha, \beta]$ *and*

$$U_0 = (U_{0,1}, \ldots, U_{0,N})^T \in \mathbb{K}^N,$$

the linear system (1.93) *possesses a unique solution,*

$$u(t) := (u_1(t), u_2(t), \ldots, u_n(t)),$$

such that $\mathcal{D}u(t_0) = U_0.$

1.14 Exercises

1. Prove the following *generalized superposition principle*. Let $m \geq 2$ be an integer, and

$$A(t) := (a_{ij}(t))_{1 \leq i,j \leq N} \qquad \text{and} \qquad B_k(t) := (b_{i,k}(t))_{\substack{1 \leq i \leq N \\ 1 \leq k \leq m}}.$$

If u_k, $1 \leq k \leq m$, are, respectively, the solutions of the linear systems

$$u' = A(t)u + B_k(t),$$

then

$$u(t) := \sum_{k=1}^{m} u_k(t)$$

solves the linear system

$$u' = A(t)u + \sum_{k=1}^{m} B_k(t).$$

2. Find the general solution of the linear scalar equation

$$u' - 2tu = t.$$

3. Find the unique solution of the initial value problem

$$\begin{cases} u' + 2tu = t, & t \in \mathbb{R}, \\ u(1) = 2. \end{cases}$$

4. Find the unique solution of the initial value problem

$$\begin{cases} u' = \frac{u}{t} + e^t, & t \in (0, \infty), \\ u(1) = 2. \end{cases}$$

5. Find the unique solution of the initial value problem

$$\begin{cases} u' + u = \frac{1}{1+t^2}, & t \in \mathbb{R}, \\ u(2) = 3. \end{cases}$$

6. Find the unique continuous solution of the initial value problem

$$\begin{cases} u' + u = b(t), \\ u(0) = 0 \end{cases}$$

in the interval $[0, \infty)$, where

$$b(t) = \begin{cases} 2 & \text{if } 0 \le t \le 1, \\ 0 & \text{if } t > 1. \end{cases}$$

7. Let $a, m, \omega \in \mathbb{R}$ be such that $a > 0$ and $\omega > 0$. Find the general solution of

$$u' = -au + me^{-\omega t}$$

and show that every solution approaches zero as $t \uparrow \infty$.

8. Let $a, b \in \mathcal{C}(\mathbb{R}; \mathbb{R})$ be such that

$$\sup_{\mathbb{R}} a \le -c < 0 \quad \text{and} \quad \lim_{t \uparrow \infty} b(t) = 0,$$

for some constant c. Show that every solution of $u' = a(t)u + b(t)$ approaches zero as $t \uparrow \infty$.

9. E. Rutherford established that the radioactivity of a substance is proportional to the number of its constituent atoms. Thus, if $N(t)$ is the number of atoms present at time t, then $N'(t)$, the number of atoms that disintegrate per unit of time, is proportional to N, i.e.

$$N'(t) = -\lambda N(t),$$

where $\lambda > 0$ is the decay constant of the substance. Observe that the larger the value of λ, the faster the substance decays. One way to measure the rate of disintegration of a given substance is the *half-life*, defined as the time required for the quantity of the substance to reduce to half of its initial quantity. Determine the half-life in terms of λ.

10. Given $\alpha < \beta$, $t_0 \in [\alpha, \beta]$, $\eta > 0$, and a norm $\|\cdot\|$ in $\mathcal{C}([\alpha, \beta]; \mathbb{R}^N)$, prove that

$$\|u\|_\eta := \max_{t \in [\alpha, \beta]} \{ e^{-\eta|t-t_0|} \|u(t)\| \}$$

defines a norm in $\mathcal{C}([\alpha, \beta]; \mathbb{R}^N)$.

11. For every integer $n \geq 1$, consider the function $u_n(t) = t^n$, $t \in [0, 1]$. Prove that $\{u_n\}_{n \geq 1}$ is point-wise convergent in $[0, 1]$ but cannot converge uniformly in $[0, 1]$ as $n \to \infty$.

12. Given $\mathbb{K} \in \{\mathbb{R}, \mathbb{C}\}$ and $\alpha, \beta \in \mathbb{R}$ with $\alpha < \beta$, let $\mathcal{C}^1([\alpha, \beta]; \mathbb{K}^N)$ denote the vector space of all functions $u : [\alpha, \beta] \to \mathbb{K}^N$ such that $u'(t)$ exists for all $t \in [\alpha, \beta]$ and $u' \in \mathcal{C}([\alpha, \beta]; \mathbb{K}^N)$. Prove that

$$\|u\|_{\mathcal{C}^1([\alpha,\beta];\mathbb{K}^N)} := \|u\|_\infty + \|u'\|_\infty$$

equips $\mathcal{C}^1([\alpha, \beta]; \mathbb{K}^N)$ with a structure of Banach space.

13. Show that the operator norm $\| \cdot \|_{\mathcal{L}(\mathbb{K}^N)}$ induced by a norm $\| \cdot \|$ of $\mathbb{K}^N$, which is defined as

$$\|A\|_{\mathcal{L}(\mathbb{K}^N)} = \max\{\|Ax\| : \|x\| = 1\},$$

is a norm in the space $\mathcal{L}(\mathbb{K}^N)$ of linear operators from $\mathbb{K}^N$ to $\mathbb{K}^N$.

14. Show that, if $\| \cdot \|_{\mathcal{L}(\mathbb{K}^N)}$ is the operator norm induced by the p-norm of $\mathbb{K}^N$, in general, one has $\|A\|_{\mathcal{L}(\mathbb{K}^N)} \neq \|A\|_p$, where in the latter norm, one considers A as a vector of $\mathbb{K}^{N^2}$.

15. Suppose that $\mathbb{K} \in \{\mathbb{R}, \mathbb{C}\}$, $\alpha \in \mathbb{R} \cup \{-\infty\}$, $\beta \in \mathbb{R} \cup \{+\infty\}$, $\alpha < \beta$, and $a_{ij}, b_i \in \mathcal{C}((\alpha, \beta); \mathbb{K}^N)$ for all $i, j \in \{1, \ldots, N\}$. Prove that, for every $t_0 \in (\alpha, \beta)$ and $u_0 \in \mathbb{K}^N$, the problem

$$\begin{cases} u' = A(t)u + B(t), \\ u(t_0) = u_0, \end{cases}$$

where $A(t) = (a_{ij}(t))_{1 \leq i,j \leq N}$ and $B(t) = (b_i(t))_{1 \leq i \leq N}$, possesses a unique solution in (α, β).

16. For each of the following initial value problems, compute the associated Picard–Lindelöf iterates and show that they converge to the corresponding unique solution:

(a) $\begin{cases} u' = u - 1, \\ u(0) = u_0 \in \mathbb{R}. \end{cases}$ (b) $\begin{cases} u'' = u, \\ u(0) = 1, \quad u'(0) = -1. \end{cases}$

(c) $\begin{cases} u'' = -u, \\ u(0) = 1, \quad u'(0) = 0. \end{cases}$ (d) $\begin{cases} u'' = -u, \\ u(0) = 0, \quad u'(0) = 1. \end{cases}$

17. Use induction on N to prove identity (1.23).
18. Find the general solution of the differential equation

$$u'' - 3u' + 2u = \pi + e^{\sqrt{2}t}$$

by using each of the following methods:

(a) varying coefficients, as explained in Section 1.8;
(b) reducing the order of the equation, as explained in Section 1.9;
(c) by using the method of indeterminate coefficients, as explained in Section 1.10.

19. Repeat the previous exercise with the following differential equation

$$u'' - 2u' + u = 1 + e^{2t}.$$

20. Find the unique solution of the problem

$$\begin{cases} u'' + u = \sec^3 t, \\ u(0) = 1, \quad u'(0) = \tfrac{1}{2} \end{cases}$$

in the interval $\left(-\frac{\pi}{2}, \frac{\pi}{2}\right)$.

21. Suppose that γ and ω are (distinct) positive real numbers and that $F_0 \in \mathbb{R}$.

(a) Find the solution, $u_\gamma(t)$, of the problem

$$\begin{cases} u'' + \omega^2 u = F_0 \cos(\gamma t), \\ u(0) = u'(0) = 0. \end{cases}$$

(b) Find the solution, $u_\omega(t)$, of the problem

$$\begin{cases} u'' + \omega^2 u = F_0 \cos(\omega t), \\ u(0) = u'(0) = 0. \end{cases}$$

(c) Prove that

$$u_\omega(t) = \lim_{\gamma \to \omega} u_\gamma(t), \qquad t \in \mathbb{R}.$$

(d) Assume that $\gamma \neq \omega$, $\gamma \sim \omega$, and show that $u_\gamma(t)$, determined in part (a), has an oscillating behavior with an amplitude that itself oscillates in time, having a small frequency. Determine the frequency and represent the graph of the solution. In Physics, this phenomenon is known as *beat*.

(e) Represent the graph of $u_\omega(t)$. The underlying physical phenomenon is known as a *resonance*. Compare the graph of $u_\omega(t)$ with the one of $u_\gamma(t)$, for $\gamma \neq \omega$, $\gamma \sim \omega$.

22. Use the solutions determined in the previous exercise to obtain the solutions of the corresponding differential equation with arbitrary initial conditions: $u(0) = u_0$, $u'(0) = v_0$.

23. Let $\lambda_j \in \mathbb{R}$, $j \in \{1, 2, 3\}$, be three different real numbers, and consider the differential operator

$$P(D) := (D - \lambda_1)(D - \lambda_2)(D - \lambda_3).$$

Find the general solution of

$$P(D)u = b(t),$$

where $b \in \mathcal{C}([\alpha, \beta]; \mathbb{R})$ for some $\alpha < \beta$.

24. Let $r, \kappa \in \mathbb{R}$. Prove that the change of temporal scale $t = e^s$, $s \in \mathbb{R}$, transforms the *Euler equation*

$$t^2 u'' + rtu' + \kappa u = 0, \qquad t > 0,$$

into a differential equation with constant coefficients. Find a basis for the set of solutions of the Euler equation according to all possible values of the parameters r and κ.

25. Let $\lambda, \mu, \omega \in \mathbb{C}$ be three distinct complex numbers.

(a) Find the unique solution, u_ω, of the Cauchy problem

$$\begin{cases} (D - \lambda)(D - \mu)u = e^{\omega t}, & t \in \mathbb{R}, \\ u(0) = u'(0) = 0. \end{cases}$$

(b) Find the unique solution, u_λ, of

$$\begin{cases} (D - \lambda)(D - \mu)u = e^{\lambda t}, & t \in \mathbb{R}, \\ u(0) = u'(0) = 0. \end{cases}$$

(c) Compare u_λ and u_ω.

(d) Find the unique solution of

$$\begin{cases} (D - \lambda)^2 u = e^{\lambda t}, & t \in \mathbb{R}, \\ u(0) = u'(0) = 0. \end{cases}$$

26. Given $L > 0$, let U_d denote the vector space of the functions $u \in \mathcal{C}^2([0, L]; \mathbb{R})$ such that $u(0) = u(L) = 0$, and consider the differential operator

$$-D^2 : U_d \to V := \mathcal{C}([0, L]; \mathbb{R}),$$

defined as

$$-D^2 u = -u'' \quad \text{for all } u \in U_d.$$

A number $\lambda \in \mathbb{R}$ is said to be an *eigenvalue* of the *differential operator* $-D^2$ in U_d if there exists $u \in U_d$, $u \neq 0$, such that $-D^2 u = \lambda u$. In such a case, u is said to be an *eigenfunction* associated with the eigenvalue λ. Find the eigenvalues and eigenfunctions of $-D^2$ in U_d, and represent carefully all the eigenfunctions. (The subindex d in U_d refers to the conditions $u(0) = u(L) = 0$, which are known as the *Dirichlet* boundary conditions.)

27. Given $L > 0$, let U_n denote the vector space of the functions $u \in \mathcal{C}^2([0, L]; \mathbb{R})$ such that $u'(0) = u'(L) = 0$, and consider the differential operator

$$-D^2 : U_n \to V := \mathcal{C}([0, L]; \mathbb{R}),$$

defined as

$$-D^2 u = -u'' \quad \text{for all } u \in U_n.$$

A number $\lambda \in \mathbb{R}$ is said to be an *eigenvalue* of the *differential operator* $-D^2$ in U_n if there exists $u \in U_n$, $u \neq 0$, such that $-D^2 u = \lambda u$. In such a case, u is said to be an *eigenfunction* associated with the eigenvalue λ. Find the eigenvalues and eigenfunctions of $-D^2$ in U_n, and represent carefully all the eigenfunctions. (The subindex n in U_n refers to the conditions $u'(0) = u'(L) = 0$, which are known as the *Neumann* boundary conditions.)

28. Show that the function $u(t) = e^t$ solves the differential equation

$$u'' - tu' + (t - 1)u = 0, \qquad t \in \mathbb{R}.$$

Find the general solution of

$$u'' - tu' + (t - 1)u = e^{\frac{t^2}{2}}.$$

29. Given the differential equation

$$u'' + \frac{1}{t}u' + \left(1 - \frac{1}{4t^2}\right)u = 0, \qquad t > 0, \qquad (1.96)$$

use the change of variable $u(t) = t^\alpha v(t)$, for a certain choice of α, in order to find the general solution of (1.96). Then, find the general solution of the inhomogeneous equation

$$u'' + \frac{1}{t}u' + \left(1 - \frac{1}{4t^2}\right)u = 3\sqrt{t}\sin t, \qquad t > 0.$$

30. Find the general solution of the inhomogeneous differential equation

$$u''' + 3u'' + 3u' + u = \pi + e^t.$$

31. For a given $\omega > 0$, find the general (real and complex) solution of each of the following differential equations:

(a) $u^{6)} - 8u^{5)} + 25u^{4)} - 32u^{3)} - u'' + 40u' - 25u = 1 + e^{2t}$.
(b) $u^{4)} - 4u^{3)} + 12u'' + 4u' - 13u = 1 + e^{2t} + \cos(\omega t)$.
(c) $u^{5)} + 3u^{4)} + 3u^{3)} + u'' = 1 + e^{2t}$.

32. Given an integer $N \geq 2$ and $a_0, \ldots, a_{N-1} \in \mathbb{C}$, consider the Nth-order differential equation

$$u^{N)} = a_{N-1}u^{N-1)} + \cdots + a_1 u' + a_0 u,$$

and assume that $e^{\lambda_j t}$ is a solution for all $j \in \{1, \ldots, N\}$, with $\lambda_j \in \mathbb{C}$ and $\lambda_j \neq \lambda_k$ if $j \neq k$.

(a) Write a first-order system $U' = AU$, with $A \in \mathcal{M}_N(\mathbb{C})$, equivalent to the given differential equation.
(b) Determine the characteristic polynomial, eigenvalues, and eigenvectors of A.

(c) Determine a matrix of solutions of $U' = AU$.

(d) Use the theory in Sections 1.5 and 1.12 to prove that the Vandermonde matrix

$$
\begin{pmatrix}
1 & \cdots & 1 \\
\lambda_1 & \cdots & \lambda_N \\
\vdots & \cdots & \vdots \\
\lambda_1^{N-1} & \cdots & \lambda_N^{N-1}
\end{pmatrix}
$$

is invertible.

33. Consider all the solutions of the differential equations in this chapter that are not defined on the whole $\mathbb{R}$. Analyze the behavior of such solutions at the boundary of their domain of definition.

34. Assume that

$$
A(t) = (a_{ij}(t))_{1 \leq i,j \leq N} \in \mathcal{M}_N(\mathcal{C}([\alpha, \beta]; \mathbb{K})),
$$

with

$$
\operatorname{tr} A(t) = 0, \qquad \text{for all } t \in [\alpha, \beta].
$$

For every $t \in [\alpha, \beta]$, define

$$
\Phi(t) : \mathbb{K}^N \to \mathbb{K}^N,
$$

$$
x \mapsto h(t; \alpha, x),
$$

where $h(t; \alpha, x)$ denotes the unique solution of

$$
\begin{cases}
h' = A(t)h, \\
h(\alpha) = x.
\end{cases}
$$

Show that, for every $t \in [\alpha, \beta]$, $\Phi_t := \Phi(t)$ is a linear isomorphism of $\mathbb{K}^N$ that preserves the volume, in the sense that, for every measurable set $A \subset \mathbb{K}^N$,

$$
\operatorname{vol}(A) := \int_A 1 \, dx = \int_{\Phi_t(A)} 1 \, dy =: \operatorname{vol}(\Phi_t(A)).
$$

1.15 Final Comments

As had already been pointed out at the end of Section 1.8, except when the coefficients are constant, or analytic, in the general case when $N \geq 2$, there is no general way to construct a basis of Σ_0^e in terms of elementary functions, such as polynomial, rational, trigonometric, or hyperbolic functions.

Naturally, this should not discourage the reader, who should plainly accept the reality that only an extremely small portion among all differential equations can be solved in terms of elementary functions. Something similar occurs in the problem of finding primitives of elementary functions in integral calculus! Actually, despite this apparent limitation, the analysis in this chapter has provided us with extensive information about the nature and global structures of the solutions of a wide class of linear ordinary differential equations and systems.

D. Hilbert was the first scientist to rather successfully overcome these shortcomings by changing the old but well-established concept of mathematical proof. In early 1888, at the age of 26, he organized a trip to visit 21 prominent mathematicians, among them was P. Gordan at Erlangen, who had the good fortune to venture into the theory of algebraic invariants just when that theory moved to a new level. The early years of development had been devoted to determining the general laws governing the structure of invariants. Then, most of the efforts focused on their systematic classification, and this was Gordan's specialty.

A typical paper by Gordan contained nothing but formulas for at least 20 pages. "Formulas were the indispensable support for the formation of his thoughts, his conclusions, and his mode of expression", a friend wrote to Hilbert some time later. Gordan's strength and reputation in the algebraic invariant theory were so high that he earned his title of "king of invariants" by making a breakthrough in a famous invariant problem. The general problem of the theory, which remains unsolved, was named in his honor: "Gordan's problem".

Hilbert had learned of this problem from C. Hermite during his visit to Paris two years earlier, in 1886, but on that occasion, the problem did not catch his attention. Gordan's problem was a sophisticated question of pure mathematics posed for its mathematical interest. It had no empirical interest. It concerned the totality of invariants, and the question could be formulated as follows: *Is there a basis, or finite system of invariants, in terms of which all other invariants, although infinite in number, could be rationally and integrally expressed?*

Gordan's great achievement, 20 years before his first meeting with the young D. Hilbert, had been to prove the existence of a basis for binary forms, the simplest of all algebraic forms. His proof was a "crude computation". Hilbert, listening to Gordan himself, was attracted to the problem in an extraordinary way. Back home in Könisgsberg, the problem accompanied him day and night, even at dances, which Hilbert loved to attend. In August, he went to Rauschen, as was his custom in the summer, and from Ruschen, on September 6, 1888, he submitted a short note to *Nachrichten* of the Göttingen Scientific Society. In his note, he showed, in a totally unexpected and original way, how Gordan's theorem could be established for forms with an arbitrary number of variables. No expert was prepared for the announcement of the solution of the famous old problem, and the first reaction from the mathematical community was almost pure disbelief. Instead of attempting to construct the basis of invariant forms, as all those attacking the open problem had previously attempted, Hilbert simply established their existence.

Going back to the material presented in this chapter, the reader should realize that, instead of deriving all global properties of Σ_0 from the contractive mapping theorem, as we have done in this chapter, during a long period of time, most experts in differential equations had tried to construct a basis of Σ_0, which is an impossible task except in some special cases. Here lied the tremendous conceptual difficulty of solving Gordan's problem.

To demonstrate the finiteness of the basis of the invariant system, one did not have to construct it, as Gordan and everyone else, including A. Cayley, had tried to do. One did not even have to show how it could be constructed. All that had to be done was to show that a finite basis must exist, and this is what Hilbert did. Gordan himself announced, "Das ist nicht Mathematik. Das ist Theologie". It took almost two years before the exceptionality of Hilbert's work was widely recognized and accepted. Even Gordan finally acknowledged that "Herr Hilbert's proof is completely correct".

However, 130 years after Gordan's problem was solved, Hilbert's way of thinking seems to be far from being adopted in the context of differential equations since most introductory textbooks in this field mainly focus on the construction of a basis for Σ_0 in some specific cases, without stressing that, in general, this is an impossible task. Actually, even when the coefficient functions $a_0(t)$ and $a_1(t)$ are analytic, the basis of the solution set of the second-order linear differential equation

$$u'' = a_0(t)u + a_1(t)u'$$

cannot be expressed in terms of elementary functions. In particular, as will be apparent in Chapter 3, in many circumstances where the differential equation is of great empirical interest, the Taylor series defining its solutions deserve to be named and included in encyclopedias of *special functions*. Even in the case of constant coefficients, deriving the basis of Σ_0 for large N might be a huge task, as it requires obtaining the roots of polynomials of high degrees.

A. L. Cauchy was the first scientist to realize that, even for analytic scalar $f(t,u)$, the problem of finding all the solutions of $u' = f(t,u)$ was not, in general, solvable. Instead, he proposed the initial value problem

$$\begin{cases} u' = f(t,u), \\ u(t_0) = u_0, \end{cases} \tag{1.97}$$

and got the first existence and uniqueness result in the spirit of Theorems 1.2 and 1.4. This explains why initial value problems such as (1.97) are named *Cauchy problems* after him. His pioneering results appeared in print as early as 1840 in a series of lecture notes taken and edited by one of his former students, Moigno (1840). Actually, the interest of Cauchy in these problems pushed him to develop the theory of analytic functions since, once a simple scalar prototype of (1.97) with an analytic $f(t,u)$ was solved through a formal power series, it was very natural to ask whether or not such a power series was convergent.

The proof of Theorem 1.4 is based on the convergence of the iterates of the integral operator, $\mathcal{K}$ associated with (1.2). Although this idea is usually attributed to Picard (1890) and Lindelöf (1894), who gave their proofs of Theorem 1.4 in 1890 and 1894, respectively, by establishing the convergence of the iterative $\mathcal{K}$-scheme,

$$h_n = \mathcal{K}h_{n-1}, \qquad n \geq 1,$$

to the solution, according to Hartman (2002), the method of successive approximations had already been used by A. L. Cauchy and J. Liouville in some special cases.

In its greatest generality, the contractive mapping theorem (see Theorem 1.8) goes back to S. Banach's Ph.D. thesis (cf. Banach, 1922), under the supervision of H. D. Steinhaus, which is considered a cornerstone for the construction of functional analysis. In it, Banach realized the transcendence of the notion of completeness in abstract normed vector spaces and introduced the notion of a linear transformation between them. As a consequence, a number of specific examples that had hitherto been considered independently then fitted within the framework of the unifying Banach theory. For this reason, such spaces are called "Banach spaces" in his honor. In addition, Banach discussed two other seminal findings in his thesis. The first one asserted that the point-wise limit of any bounded sequence of linear and continuous operators is necessarily linear and continuous. It was a joint theorem with Steinhaus, referred to as the *uniform boundedness principle*. The second one was Theorem 1.8, which is also referred to as the *Banach fixed point theorem*. Theorem 1.8 is an abstract, highly versatile version of the already classical method of successive approximations, which had been developed by Picard (1890) and Lindelöf (1894) about 30 years before Banach defended his Ph.D. thesis. As is apparent from this chapter, most of the developed abstract theory is based on the contractive mapping theorem.

The norm $\| \cdot \|_\eta$ introduced in (1.16) goes back to Bielecki (1956). It is an extremely useful device in the context of integral equations. Indeed, if we had not introduced it, the integral operator $\mathcal{K}$ associated with the Cauchy problem (1.2) would have been a contraction only for sufficiently small $\beta - \alpha$. Indeed, as there is a constant $L > 0$ such that, for every $u, v \in \mathcal{C}([\alpha, \beta]; \mathbb{K}^N)$ and $t \in [\alpha, \beta]$,

$$\|\mathcal{K}u(t) - \mathcal{K}v(t)\| \leq L \int_{\min\{t_0, t\}}^{\max\{t_0, t\}} \|u(s) - v(s)\| \, ds \leq L(\beta - \alpha)\|u - v\|_\infty,$$

we have that

$$\|\mathcal{K}u - \mathcal{K}v\|_\infty \leq L(\beta - \alpha)\|u - v\|_\infty.$$

Therefore, $\mathcal{K}$ is a contraction with respect to the $\| \cdot \|_\infty$ norm if and only if

$$L(\beta - \alpha) < 1.$$

Although this is sufficient to obtain the existence and uniqueness of a solution of (1.2) in the whole interval $[\alpha, \beta]$, by means of an additional global continuation argument of the solution, the reader would agree that using $\| \cdot \|_\eta$ is more elegant since it directly allows us to obtain the result on the whole interval.

In the light of the example with discontinuous $b(t)$, presented in the last part of Section 1.2, it is clear that Theorem 1.2 can be valid even if $a(t)$ and $b(t)$ are not continuous in $[\alpha, \beta]$. Actually, Carathéodory's existence theorem only requires some milder integrability conditions on $a(t)$ and $b(t)$, though, in such a case, one has to consider a more general solution concept, the so-called Carathéodory solutions. A function $u(t)$ is said to be a Carathéodory solution of

$$\begin{cases} u'(t) = a(t)u + b(t), \\ u(t_0) = u_0, \end{cases}$$

if it is absolutely continuous, it satisfies the differential equation almost everywhere, and $u(t)$ satisfies the initial condition (see, e.g. Coddington and Levinson, 1955, p. 42). A function $u(t)$ is absolutely continuous in $[\alpha, \beta]$ if and only if u' exists almost everywhere, it is Lebesgue integrable, and, for every $t, t_0 \in [\alpha, \beta]$,

$$u(t) = u(t_0) + \int_{t_0}^{t} u'(s)\, ds.$$

The integrability of the coefficient $a(t)$ in a neighborhood of t_0 is imperative for getting the uniqueness result established by Theorem 1.2. Indeed, $u = 0$ and $u = t$ provide us with two (real) solutions of the initial value problem

$$\begin{cases} u' = \frac{u}{t}, & t > 0, \\ u(0) = 0, \end{cases}$$

where the initial condition has to be understood in a limiting sense, i.e. $\lim_{t \to 0^+} u(t) = 0$. However, it turns out that, for every $q \in (0, 1)$, $u = 0$ is the unique solution of

$$\begin{cases} u' = \frac{u}{t^q}, & t > 0, \\ u(0) = 0. \end{cases}$$

Indeed, the change of variable

$$u(t) := e^{\frac{t^{1-q}}{1-q}} \, v(t)$$

transforms this problem into $v' = 0$, with $v(0) = 0$, whose unique solution is $v = 0$. Thus, $u = 0$. Note that

$$\int_0^t \frac{ds}{s} = \infty, \quad \text{while} \quad \int_0^t \frac{ds}{s^q} = \frac{t^{1-q}}{1-q},$$

for all $t > 0$.

The content of Section 1.11 has been inspired by Chapter 11 in the work of Puig-Adam (1978). The proof of Theorem 1.22 seems to go back to López-Gómez (2001a).

Chapter 2

First-Order Linear Systems with Constant Coefficients

The main purpose of this chapter is to construct a fundamental matrix of solutions for the homogeneous first-order linear system with constant coefficients

$$u' = Au, \tag{2.1}$$

where $A \in \mathcal{M}_N(\mathbb{C})$ is a matrix of order $N \geq 2$ with constant complex entries. In Chapter 1 (see Section 1.12), we have treated the case in which the matrix A has the special form (1.39), with $a_j(t) = a_j \in \mathbb{C}$ for all $j \in \{0, \ldots, N-1\}$. Here, we consider arbitrary matrices in $\mathcal{M}_N(\mathbb{C})$.

By analogy with the theory developed in Section 1.12 for linear equations of arbitrary order, the most natural strategy to achieve our goal consists of looking for solutions to (2.1) of the form

$$u(t) = e^{\lambda t} v, \qquad t \in \mathbb{R}, \tag{2.2}$$

for some $\lambda \in \mathbb{C}$ and $v \in \mathbb{C}^N \setminus \{0\}$. Indeed, by substituting (2.2) in (2.1), it is apparent that (2.2) solves (2.1) if and only if

$$\lambda e^{\lambda t} v = e^{\lambda t} A v$$

for all $t \in \mathbb{R}$, or, in other words, if

$$Av = \lambda v. \tag{2.3}$$

Therefore, (2.2) solves (2.1) if and only if λ is an *eigenvalue* of the matrix A, and v is an *eigenvector* associated with λ.

105

Let us briefly introduce some notations from linear algebra: throughout the rest of this book, for any given matrix $M \in \mathcal{M}_N(\mathbb{C})$, we denote by $N[M]$ the null space, or kernel, of M, and by $R[M]$ the range, or image, of M. The identity map in $\mathbb{C}^N$ is denoted by I.

As we have shown above, the following concepts are pivotal to our analysis.

Definition 2.1 (Eigenvalue and eigenvector). A complex number $\lambda \in \mathbb{C}$ is said to be an *eigenvalue* of the matrix $A \in \mathcal{M}_N(\mathbb{C})$ if

$$Av = \lambda v$$

for some $v \in \mathbb{C}^N \setminus \{0\}$. In such a case, every $v \in N[A - \lambda I] \setminus \{0\}$ is called an *eigenvector* of A associated with the eigenvalue λ, and the null space $N[A - \lambda I]$ is called the *eigenspace* of A associated with the eigenvalue λ.

Obviously, the homogeneous linear system (2.3) admits a solution $v \neq 0$ if and only if

$$\det(A - \lambda I) = 0.$$

Thus, the eigenvalues of the matrix A are the roots of the polynomial

$$P_A(z) := \det(A - zI), \qquad z \in \mathbb{C},$$

which is referred to as the *characteristic polynomial* of the matrix A. Such a polynomial is invariant under changes of basis, and hence, so are the eigenvalues. Indeed, for every $Q, A \in \mathcal{M}_N(\mathbb{C})$ with $\det Q \neq 0$, and every $z \in \mathbb{C}$,

$$P_{Q^{-1}AQ}(z) = \det(Q^{-1}AQ - zI) = \det(Q^{-1}(A - zI)Q)$$

$$= \det Q^{-1} \det(A - zI) \det Q = \det(A - zI) = P_A(z).$$

This shows the consistency of referring to the characteristic polynomial and the eigenvalues of a linear operator in $\mathbb{C}^N$ since they do not depend on the choice of the bases used to represent the operator. However, it is important to note that the eigenvectors are not invariant under changes of basis, instead. The equation $P_A(z) = 0$ is usually referred to as the *characteristic equation* of the matrix A.

It is easily seen that $P_A(z)$ is a polynomial of degree N, with the leading coefficient $(-1)^N$. Therefore, it possesses N complex roots, possibly repeated. So, it can be expressed as

$$P_A(z) = (-1)^N \prod_{j=1}^{q} (z - \lambda_j)^{m_a(\lambda_j)}, \qquad z \in \mathbb{C},$$

for some distinct complex numbers $\lambda_1, \ldots, \lambda_q$ and integers

$$m_a(\lambda_1), \ldots, m_a(\lambda_q) \geq 1,$$

which are the respective algebraic multiplicities of the roots $\lambda_1, \ldots, \lambda_q$. Then, the eigenvalues of the matrix A are $\lambda_1, \ldots, \lambda_q$. The set

$$\sigma(A) := \{\lambda_1, \ldots, \lambda_q\}$$

is referred to as the *spectrum* of A. Naturally,

$$N = m_a(\lambda_1) + \cdots + m_a(\lambda_q).$$

These notations will be maintained throughout this chapter; in particular, the following convention is used:

$$\lambda_i \neq \lambda_j \qquad \text{if } 1 \leq i < j \leq q.$$

For each $j \in \{1, \ldots, q\}$, the *geometric multiplicity* of the eigenvalue λ_j, denoted by $m_g(\lambda_j)$, is defined as

$$m_g(\lambda_j) := \dim N[A - \lambda_j I] \geq 1$$

and satisfies

$$m_g(\lambda_j) \leq m_a(\lambda_j).$$

Naturally, when $m_a(\lambda_j) = 1$ for all $j \in \{1, \ldots, q\}$, i.e. all the eigenvalues of A are *algebraically simple*, necessarily, $q = N$ and

$$m_a(\lambda_j) = m_g(\lambda_j) = \dim N[A - \lambda_j I] = 1.$$

Moreover, in such a case, there exist $v_j \in \mathbb{C}^N \setminus \{0\}$ such that $A v_j = \lambda_j v_j$ for all $j \in \{1, \ldots, N\}$; therefore, we have at our disposal the following set of N solutions of $u' = Au$:

$$\mathfrak{B} = \{e^{\lambda_j t} v_j : 1 \leq j \leq N\},$$

which provides us with a basis of Σ_0, the linear subspace of solutions of the homogeneous first-order system $u' = Au$. By Theorem 1.11, to prove that $\mathfrak{B}$ is a basis of Σ_0, it suffices to show that it is linearly independent. This follows from the following result of technical nature.

Lemma 2.2. *For every integer $N \geq 1$, N distinct complex numbers, $\lambda_1, \ldots, \lambda_N \in \mathbb{C}$, and N vectors $v_j \in \mathbb{C}^N \setminus \{0\}$, $1 \leq j \leq N$, the set of vectorial functions*

$$\{e^{\lambda_j t} v_j : 1 \leq j \leq N\}$$

is linearly independent.

Proof. We argue by induction on N. If $N = 1$ and

$$c_1 e^{\lambda_1 t} v_1 = 0 \quad \text{for all } t \in \mathbb{R},$$

then $c_1 = 0$ because $v_1 \neq 0$. Suppose the result holds for some integer $N - 1$, with $N \geq 2$, and let $c_1, \ldots, c_N \in \mathbb{C}$ be such that

$$c_1 e^{\lambda_1 t} v_1 + c_2 e^{\lambda_2 t} v_2 + \cdots + c_N e^{\lambda_N t} v_N = 0 \quad \text{for all } t \in \mathbb{R}. \qquad (2.4)$$

Then, multiplying by $e^{-\lambda_1 t}$, this identity yields

$$c_1 v_1 + c_2 e^{(\lambda_2 - \lambda_1)t} v_2 + \cdots + c_N e^{(\lambda_N - \lambda_1)t} v_N = 0 \quad \text{for all } t \in \mathbb{R},$$

and hence, differentiating with respect to t shows that

$$(\lambda_2 - \lambda_1) c_2 e^{(\lambda_2 - \lambda_1)t} v_2 + \cdots + (\lambda_N - \lambda_1) c_N e^{(\lambda_N - \lambda_1)t} v_N = 0$$

for all $t \in \mathbb{R}$. Thus, since $\lambda_j - \lambda_1 \neq \lambda_i - \lambda_1$ if $i \neq j$, we find from the induction hypothesis that

$$(\lambda_2 - \lambda_1) c_2 = \cdots = (\lambda_N - \lambda_1) c_N = 0$$

and, hence,

$$c_2 = \cdots = c_N = 0.$$

Therefore, going back to (2.4), we have $c_1 e^{\lambda_1 t} v_1 = 0$ for all $t \in \mathbb{R}$, and we also get that $c_1 = 0$, which completes the proof. $\qquad \square$

The previous result implies that, when all the eigenvalues of A are algebraically simple,

$$\Phi(t) = \left(e^{\lambda_1 t} v_1 \cdots e^{\lambda_N t} v_N \right), \qquad t \in \mathbb{R},$$

is a fundamental matrix of solutions of $u' = Au$. Thus, by Theorem 1.13, $\Phi(t)$ is invertible for all $t \in \mathbb{R}$ and, in particular,

$$\det \Phi(0) = \det \left(v_1 \cdots v_N \right) \neq 0.$$

Therefore, the set of eigenvectors provides us with a basis of $\mathbb{C}^N$, i.e.

$$\mathbb{C}^N = \mathrm{span}[v_1, \ldots, v_N].$$

Naturally, this property entails that

$$\mathbb{C}^N = N[A - \lambda_1 I] + \cdots + N[A - \lambda_N I],$$

and since

$$N[A - \lambda_i I] \cap N[A - \lambda_j I] = \mathrm{span}[0] \quad \text{if } i \neq j,$$

we actually have that

$$\mathbb{C}^N = N[A - \lambda_1 I] \oplus \cdots \oplus N[A - \lambda_N I]. \tag{2.5}$$

Indeed, if

$$Au = \lambda_j u = \lambda_i u$$

for some $u \neq 0$, necessarily $\lambda_i = \lambda_j$, which implies $i = j$. More generally, when $q < N$ but still

$$\mathbb{C}^N = N[A - \lambda_1 I] \oplus \cdots \oplus N[A - \lambda_q I], \tag{2.6}$$

the following result holds.

Theorem 2.3. *Suppose* (2.6) *and, for every* $j \in \{1, \ldots, q\}$,

$$N[A - \lambda_j I] = \mathrm{span}\left[v_{j,i} \colon i \in \{1, \ldots, m_g(\lambda_j)\} \right]. \tag{2.7}$$

Then,

$$\Sigma_0 = \mathrm{span}\left[e^{\lambda_j t} v_{j,i} \colon j \in \{1, \ldots, q\}, \ i \in \{1, \ldots, m_g(\lambda_j)\} \right].$$

Proof. As, according to (2.6), we have that

$$N = m_g(\lambda_1) + \cdots + m_g(\lambda_q),$$

it suffices to establish the linear independence of the solutions $e^{\lambda_j t} v_{j,i}$. Let $c_{j,i} \in \mathbb{C}$ be such that, for every $t \in \mathbb{R}$,

$$\sum_{j=1}^{q} \left(c_{j,1} e^{\lambda_j t} v_{j,1} + \cdots + c_{j,m_g(\lambda_j)} e^{\lambda_j t} v_{j,m_g(\lambda_j)} \right) = 0.$$

Then,

$$\sum_{j=1}^{q} \left(c_{j,1} v_{j,1} + \cdots + c_{j,m_g(\lambda_j)} v_{j,m_g(\lambda_j)} \right) e^{\lambda_j t} = 0$$

for all $t \in \mathbb{R}$ and, hence, owing to Lemma 2.2,

$$c_{j,1} v_{j,1} + \cdots + c_{j,m_g(\lambda_j)} v_{j,m_g(\lambda_j)} = 0$$

for all $j \in \{1, \ldots, q\}$. Therefore, we can infer from (2.7) that $c_{j,i} = 0$ for all $i \in \{1, \ldots, m_g(\lambda_j)\}$ and $j \in \{1, \ldots, q\}$, which completes the proof. $\qquad\square$

To construct a fundamental matrix of solutions in the general case when (2.6) fails, we need to resort to the Jordan decomposition theorem and the concept of exponential matrix e^{tA}, which will provide us with the desired fundamental matrix of solutions of $u' = Au$. The scheme is the following: the Jordan decomposition theorem will allow us to construct the Jordan canonical form, which is imperative to compute e^{tA}.

2.1   Jordan Decomposition Theorem

As already mentioned, for every $\lambda \in \sigma(A)$, $m_g(\lambda) \leq m_a(\lambda)$. Nevertheless, in general, the algebraic and geometric multiplicities do not always coincide. For example, for every $\lambda \in \mathbb{C}$, λ is the unique eigenvalue of the matrix

$$A = \begin{pmatrix} \lambda & 0 \\ 1 & \lambda \end{pmatrix},$$

and, hence, $m_a(\lambda) = 2$. But $m_g(\lambda) = 1$ because

$$N[A - \lambda I] = \text{span}\left[\begin{pmatrix} 0 \\ 1 \end{pmatrix}\right].$$

The eigenvalues $\lambda \in \sigma(A)$ for which

$$m_g(\lambda) = m_a(\lambda)$$

holds are called *semisimple*. It turns out that a matrix A is diagonalizable if and only if $\sigma(A)$ consists of semisimple eigenvalues. Should it be the case, then (2.6) holds true. Observe that an (algebraically) simple eigenvalue $\lambda \in \sigma(A)$ satisfies $m_a(\lambda) = 1$. Then, necessarily, $m_g(\lambda) = 1$, and hence, it is semisimple. All the eigenvalues of the matrix A are simple if and only if (2.5) holds.

Besides $N[A - \lambda I]$, associated with each eigenvalue $\lambda \in \sigma(A)$, we consider the following chain of *generalized eigenspaces*:

$$N[A - \lambda I] \subset N[(A - \lambda I)^2] \subset N[(A - \lambda I)^3] \subset \cdots, \tag{2.8}$$

which is non-decreasing because $N[C] \subset N[BC]$ for all $B, C \in \mathcal{M}_N(\mathbb{C})$. For every integer $j \geq 1$, the vectors in $N[(A - \lambda I)^j]$ are called the *generalized eigenvectors* of A associated with λ. Naturally, since $N = \dim \mathbb{C}^N$, the chain (2.8) must stabilize. Thus, there exists a minimal integer $k \geq 1$ such that

$$N[(A - \lambda I)^{k+1}] = N[(A - \lambda I)^k]. \tag{2.9}$$

In such a case,

$$N[(A - \lambda I)^j] = N[(A - \lambda I)^k] \quad \text{for all } j > k.$$

We prove this by induction on j. The base case of $j = k + 1$ is guaranteed by (2.9). Next, suppose

$$N[(A - \lambda I)^j] = N[(A - \lambda I)^k]$$

for some $j > k$, and consider $v \in N[(A - \lambda I)^{j+1}]$. Then,

$$(A - \lambda I)v \in N[(A - \lambda I)^j] = N[(A - \lambda I)^k],$$

and hence,

$$v \in N[(A - \lambda I)^{k+1}] = N[(A - \lambda I)^k].$$

Therefore,

$$N[(A - \lambda I)^{j+1}] \subset N[(A - \lambda I)^k],$$

and since $j+1 > k$, these kernels must coincide, which concludes the proof by induction.

The minimal integer for which the generalized eigenspaces stabilize plays a fundamental role in the subsequent analysis; thus, it deserves the following definition.

Definition 2.4 (Algebraic ascent and ascent generalized eigenspace). For every $\lambda \in \sigma(A)$, the *algebraic ascent* of λ, $\nu(\lambda)$, is the minimum integer $k \geq 1$ for which (2.9) holds.

Moreover, the subspace $N[(A-\lambda I)^{\nu(\lambda)}]$ is referred to as the *ascent generalized eigenspace* of A associated with the eigenvalue λ.

The previous analysis shows that, for every $\lambda \in \sigma(A)$ with $\nu(\lambda) = 1$, one has that

$$N[A - \lambda I] = N[(A - \lambda I)^j]$$

for all $j > 1$, while, if $\nu(\lambda) > 1$,

$$N[A - \lambda I] \subsetneq N[(A - \lambda I)^i] \subsetneq N[(A - \lambda I)^{\nu(\lambda)}] = N[(A - \lambda I)^j]$$

whenever $1 < i < \nu(\lambda) < j$.

We now state the Jordan decomposition theorem. We refer the reader to López-Gómez and Mora-Corral (2007, Theorem 1.2.1) or López-Gómez (2001a, Theorem 6.1.3) for a self-contained proof of this theorem.

Theorem 2.5 (Jordan decomposition theorem). *Let $A \in \mathcal{M}_N(\mathbb{C})$ and $\sigma(A) = \{\lambda_1, \ldots, \lambda_q\}$, with $\lambda_i \neq \lambda_j$ whenever $i \neq j$. Then,*

$$A\left(N[(A - \lambda_j I)^{\nu(\lambda_j)}]\right) \subset N[(A - \lambda_j I)^{\nu(\lambda_j)}] \quad \text{for all } j \in \{1, \ldots, q\}.$$

In addition, if we denote by A_j the restriction of A to the ascent generalized eigenspace $N[(A-\lambda_j I)^{\nu(\lambda_j)}]$, $j \in \{1, \ldots, q\}$, then $\sigma(A_j) = \{\lambda_j\}$. Moreover,

$$\mathbb{C}^N = \bigoplus_{j=1}^{q} N[(A - \lambda_j I)^{\nu(\lambda_j)}] \tag{2.10}$$

and

$$m_a(\lambda_j) = \dim N[(A - \lambda_j I)^{\nu(\lambda_j)}] \quad \text{for all } j \in \{1, \ldots, q\}.$$

In the following section, we use this theorem to obtain the Jordan canonical form of a matrix $A \in \mathcal{M}_N(\mathbb{C})$.

2.2 Jordan Canonical Form

Given a matrix $A \in \mathcal{M}_N(\mathbb{C})$ and an eigenvalue $\lambda \in \sigma(A)$, the main goal of this section consists of constructing a basis,

$$\mathcal{B}_\lambda := \{v_{\lambda,1}, \ldots, v_{\lambda,m_a(\lambda)}\},$$

of the ascent generalized eigenspace $N[(A - \lambda I)^{\nu(\lambda)}]$ so that the operator

$$A_\lambda := A|_{N[(A-\lambda I)^{\nu(\lambda)}]} : N[(A - \lambda I)^{\nu(\lambda)}] \to N[(A - \lambda I)^{\nu(\lambda)}]$$

can be expressed, with respect to the basis $\mathcal{B}_\lambda$, as

$$A_{\lambda,\mathcal{B}_\lambda} = \operatorname{diag}\{J_{\lambda,i_1}, J_{\lambda,i_2}, \ldots, J_{\lambda,i_{n_\lambda}}\} \tag{2.11}$$

for some integer number $n_\lambda \geq 1$, where $i_h \in \{1, \ldots, \nu(\lambda)\}$ for every $h \in \{1, \ldots, n_\lambda\}$. Here, and in the following, for any given natural number $\ell \geq 1$, we define

$$\operatorname{diag}\{B_1, \ldots, B_\ell\} := \begin{pmatrix} B_1 & 0 & \cdots & 0 \\ 0 & B_2 & \cdots & 0 \\ \vdots & \vdots & \ddots & \vdots \\ 0 & 0 & \cdots & B_\ell \end{pmatrix} \in \mathcal{M}_N(\mathbb{C}),$$

where for every integer j, $1 \leq j \leq \ell$, $B_j \in \mathcal{M}_{\kappa_j}(\mathbb{C})$ and

$$N = \kappa_1 + \cdots + \kappa_\ell.$$

Moreover the diagonal blocks in (2.11) are defined as follows: for each natural number $n \geq 1$, the matrix $J_{\lambda,n} \in \mathcal{M}_n(\mathbb{C})$ is the *Jordan block* of order n associated with the eigenvalue λ, i.e.

$$J_{\lambda,n} := \begin{pmatrix} \lambda & 0 & 0 & \cdots & 0 & 0 \\ 1 & \lambda & 0 & \cdots & 0 & 0 \\ 0 & 1 & \lambda & \cdots & 0 & 0 \\ \vdots & \vdots & \vdots & \ddots & \vdots & \vdots \\ 0 & 0 & 0 & \cdots & \lambda & 0 \\ 0 & 0 & 0 & \cdots & 1 & \lambda \end{pmatrix}_{n}.$$

Thus, the first Jordan blocks, for $1 \leq n \leq 5$, are

$$
J_{\lambda,1} := \lambda, \quad J_{\lambda,2} := \begin{pmatrix} \lambda & 0 \\ 1 & \lambda \end{pmatrix}, \quad J_{\lambda,3} := \begin{pmatrix} \lambda & 0 & 0 \\ 1 & \lambda & 0 \\ 0 & 1 & \lambda \end{pmatrix},
$$

$$
J_{\lambda,4} := \begin{pmatrix} \lambda & 0 & 0 & 0 \\ 1 & \lambda & 0 & 0 \\ 0 & 1 & \lambda & 0 \\ 0 & 0 & 1 & \lambda \end{pmatrix}, \quad J_{\lambda,5} := \begin{pmatrix} \lambda & 0 & 0 & 0 & 0 \\ 1 & \lambda & 0 & 0 & 0 \\ 0 & 1 & \lambda & 0 & 0 \\ 0 & 0 & 1 & \lambda & 0 \\ 0 & 0 & 0 & 1 & \lambda \end{pmatrix}.
$$

When $\nu(\lambda) = 1$, then $m_a(\lambda) = m_g(\lambda)$, and $\mathcal{B}_\lambda = \{v_{\lambda,1}, \ldots, v_{\lambda,m_a(\lambda)}\}$ consists of eigenvectors, i.e.

$$
A_\lambda v_{\lambda,k} = \lambda v_{\lambda,k} \qquad \text{for all } 1 \leq k \leq m_a(\lambda) = m_g(\lambda).
$$

In such a case, $A_{\lambda,\mathcal{B}_\lambda}$ is diagonal and consists of $m_a(\lambda)$ Jordan blocks $J_{\lambda,1}$.

Suppose $\nu(\lambda) \geq 2$, and, for each $k \in \{1, \ldots, \nu(\lambda) - 1\}$, let $C_{\lambda,k}$ be a linear supplement of $N[(A - \lambda I)^k]$ in $N[(A - \lambda I)^{k+1}]$, i.e.

$$
N[(A - \lambda I)^{k+1}] = N[(A - \lambda I)^k] \oplus C_{\lambda,k}. \tag{2.12}
$$

Then, setting

$$
C_{\lambda,0} := N[A - \lambda I],
$$

we have that

$$
N[(A - \lambda I)^{\nu(\lambda)}] = \bigoplus_{k=0}^{\nu(\lambda)-1} C_{\lambda,k}.
$$

So, in order to construct a basis for $N[(A - \lambda I)^{\nu(\lambda)}]$, it suffices to construct a basis for $C_{\lambda,k}$, for each $k \in \{0, \ldots, \nu(\lambda) - 1\}$. Subsequently, we set

$$
h_k := \dim C_{\lambda,k}, \qquad k \in \{0, \ldots, \nu(\lambda) - 1\}.
$$

Then,

$$h_k = \dim N[(A - \lambda I)^{k+1}] - \dim N[(A - \lambda I)^k] \geq 1.$$

Let $\{v_{\lambda,1}, \ldots, v_{\lambda,h_{\nu(\lambda)-1}}\}$ be a basis of $C_{\lambda,\nu(\lambda)-1}$, and consider the images of the elements of this basis under the action of the linear operator $A - \lambda I$:

$$\left\{ (A - \lambda I)v_{\lambda,i} : i \in \{1, \ldots, h_{\nu(\lambda)-1}\} \right\}. \tag{2.13}$$

These images provide us with a linearly independent system of $C_{\lambda,\nu(\lambda)-2}$. Indeed, it is easy to see that such vectors belong to $N[(A-\lambda I)^{\nu(\lambda)-1}]$ but do not belong to $N[(A-\lambda I)^{\nu(\lambda)-2}]$ (this follows since the vectors $v_{\lambda,i}$ belong to $N[(A - \lambda I)^{\nu(\lambda)}]$ but do not belong to $N[(A - \lambda I)^{\nu(\lambda)-1}]$). Moreover, suppose that there exist $c_i \in \mathbb{C}$, $1 \leq i \leq h_{\nu(\lambda)-1}$, such that

$$\sum_{i=1}^{h_{\nu(\lambda)-1}} c_i(A - \lambda I)v_{\lambda,i} = 0.$$

Then, since $\nu(\lambda) - 1 \geq 1$,

$$\sum_{i=1}^{h_{\nu(\lambda)-1}} c_i v_{\lambda,i} \in N[A - \lambda I] \subset N[(A - \lambda I)^{\nu(\lambda)-1}]$$

and hence, by construction,

$$\sum_{i=1}^{h_{\nu(\lambda)-1}} c_i v_{\lambda,i} \in N[(A - \lambda I)^{\nu(\lambda)-1}] \cap C_{\lambda,\nu(\lambda)-1}.$$

Thus, it follows from (2.12) that

$$\sum_{i=1}^{h_{\nu(\lambda)-1}} c_i v_{\lambda,i} = 0.$$

As $\{v_{\lambda,i}\colon i \in \{1,\ldots,h_{\nu(\lambda)-1}\}\}$ is a linearly independent system, $c_i = 0$ for all $i \in \{1,\ldots,h_{\nu(\lambda)-1}\}$. Therefore, (2.13) is a linearly independent subset of $C_{\lambda,\nu(\lambda)-2}$, as claimed above. Consequently, if $h_{\nu(\lambda)-2} = h_{\nu(\lambda)-1}$, then

$$C_{\lambda,\nu(\lambda)-2} = \mathrm{span}\left[(A - \lambda I)v_{\lambda,1},\ldots,(A - \lambda I)v_{\lambda,h_{\nu(\lambda)-1}}\right],$$

while, if $h_{\nu(\lambda)-2} > h_{\nu(\lambda)-1}$, we can extend (2.13) to complete a basis of $C_{\lambda,\nu(\lambda)-2}$, say,

$$\{(A - \lambda I)v_{\lambda,1},\ldots,(A - \lambda I)v_{\lambda,h_{\nu(\lambda)-1}},v_{\lambda,h_{\nu(\lambda)-1}+1},\ldots,v_{\lambda,h_{\nu(\lambda)-2}}\}.$$

Suppose $\nu(\lambda) = 2$. Then,

$$C_{\lambda,\nu(\lambda)-2} = N[A - \lambda I]$$

and the set

$$\mathcal{B}_\lambda := \{v_{\lambda,1},(A - \lambda I)v_{\lambda,1},v_{\lambda,2},(A - \lambda I)v_{\lambda,2},\ldots,$$
$$v_{\lambda,h_{\nu(\lambda)-1}},(A - \lambda I)v_{\lambda,h_{\nu(\lambda)-1}},v_{\lambda,h_{\nu(\lambda)-1}+1},\ldots,v_{\lambda,h_{\nu(\lambda)-2}}\}$$

provides us with a basis of $N[(A - \lambda I)^2]$. Moreover, since

$$Av_{\lambda,i} = \lambda v_{\lambda,i} + (A - \lambda I)v_{\lambda,i}, \qquad 1 \leq i \leq h_{\nu(\lambda)-1},$$
$$A(A - \lambda I)v_{\lambda,i} = \lambda(A - \lambda I)v_{\lambda,i}, \qquad 1 \leq i \leq h_{\nu(\lambda)-1},$$
$$Av_{\lambda,i} = \lambda v_{\lambda,i}, \qquad h_{\nu(\lambda)-1} + 1 \leq i \leq h_{\nu(\lambda)-2},$$

the matrix of the operator A_λ with respect to the basis $\mathcal{B}_\lambda$ is the following:

$$A_{\lambda,\mathcal{B}_\lambda} = \mathrm{diag}\{\underbrace{J_{\lambda,2},J_{\lambda,2},\ldots,J_{\lambda,2}}_{h_{\nu(\lambda)-1}},\underbrace{J_{\lambda,1},J_{\lambda,1},\ldots,J_{\lambda,1}}_{h_{\nu(\lambda)-2}-h_{\nu(\lambda)-1}}\}.$$

When $\nu(\lambda) \geq 3$, as in the previous step, we consider the system formed by all images under the action of $A - \lambda I$ of the vectors of the basis of $C_{\lambda,\nu(\lambda)-2}$:

$$\begin{aligned}
&\left\{(A-\lambda I)^2 v_{\lambda,i}: i \in \{1,\ldots,h_{\nu(\lambda)-1}\}\right\} \\
&\cup \left\{(A-\lambda I)v_{\lambda,i}: i \in \{h_{\nu(\lambda)-1}+1,\ldots,h_{\nu(\lambda)-2}\}\right\}.
\end{aligned} \tag{2.14}$$

This system is linearly independent in $C_{\lambda,\nu(\lambda)-3}$. Indeed, let $c_i \in \mathbb{C}$, $1 \le i \le h_{\nu(\lambda)-2}$, satisfy

$$\sum_{i=1}^{h_{\nu(\lambda)-1}} c_i(A-\lambda I)^2 v_{\lambda,i} + \sum_{i=h_{\nu(\lambda)-1}+1}^{h_{\nu(\lambda)-2}} c_i(A-\lambda I)v_{\lambda,i} = 0.$$

Then, arguing as above, we find from $\nu(\lambda)-2 \ge 1$ that

$$\sum_{i=1}^{h_{\nu(\lambda)-1}} c_i(A-\lambda I)v_{\lambda,i} + \sum_{i=h_{\nu(\lambda)-1}+1}^{h_{\nu(\lambda)-2}} c_i v_{\lambda,i} \in N[(A-\lambda I)^{\nu(\lambda)-2}] \cap C_{\lambda,\nu(\lambda)-2}.$$

Thus, by (2.12), $c_i = 0$ for each $1 \le i \le h_{\nu(\lambda)-2}$. Similarly, (2.14) is a basis of $C_{\lambda,\nu(\lambda)-3}$ if $h_{\nu(\lambda)-3} = h_{\nu(\lambda)-2}$, while it can be extended to complete a basis of $C_{\lambda,\nu(\lambda)-3}$ if $h_{\nu(\lambda)-3} > h_{\nu(\lambda)-2}$, by adding the appropriate system

$$\left\{v_{\lambda,i}: i \in \{h_{\nu(\lambda)-2}+1,\ldots,h_{\nu(\lambda)-3}\}\right\}.$$

If $\nu(\lambda) = 3$, we reason as above, and we are done. In the general case, after $\nu(\lambda)$ steps, we will have constructed an ordered basis of the ascent generalized eigenspace of the form

$$\mathcal{B}_\lambda := \bigcup_{i=1}^{\nu(\lambda)} \bigcup_{\ell=h_{\nu(\lambda)-i+1}+1}^{h_{\nu(\lambda)-i}} \bigcup_{k=0}^{\nu(\lambda)-i} \left\{(A-\lambda I)^k v_{\lambda,\ell}\right\},$$

where we set, for notational convenience, $h_{\nu(\lambda)} = 0$. The construction of such a basis is summarized in the following scheme, where, with respect to the notation used above, we use the abbreviations $v_i := v_{\lambda,i}$ and $\nu := \nu(\lambda)$. We remark that the vectors in $\mathcal{B}_\lambda$ have to be chosen according to the following scheme from left to right and from up to down:

$$
\begin{array}{ccccc}
v_1 & (A-\lambda I)v_1 & (A-\lambda I)^2 v_1 & \cdots & (A-\lambda I)^{\nu-1}v_1 \\
\vdots & \vdots & \vdots & \vdots & \vdots \\
v_{h_{\nu-1}} & (A-\lambda I)v_{h_{\nu-1}} & (A-\lambda I)^2 v_{h_{\nu-1}} & \cdots & (A-\lambda I)^{\nu-1}v_{h_{\nu-1}} \\
& v_{h_{\nu-1}+1} & (A-\lambda I)v_{h_{\nu-1}+1} & \cdots & (A-\lambda I)^{\nu-2}v_{h_{\nu-1}+1} \\
& \vdots & \vdots & \vdots & \vdots \\
& v_{h_{\nu-2}} & (A-\lambda I)v_{h_{\nu-2}} & \cdots & (A-\lambda I)^{\nu-2}v_{h_{\nu-2}} \\
& & v_{h_{\nu-2}+1} & \cdots & (A-\lambda I)^{\nu-3}v_{h_{\nu-2}+1} \\
& & \vdots & & \vdots \\
& & v_{h_{\nu-3}} & \cdots & (A-\lambda I)^{\nu-3}v_{h_{\nu-3}} \\
& & & \cdots & (A-\lambda I)^{\nu-4}v_{h_{\nu-3}+1} \\
& & & & \vdots \\
& & & & (A-\lambda I)^{\nu-4}v_{h_{\nu-4}} \\
& & & & \vdots \\
& & & & (A-\lambda I)v_{h_1} \\
& & & & v_{h_1+1} \\
& & & & \vdots \\
& & & & v_{h_0}
\end{array}
$$

The matrix of A_λ (recall that A_λ is defined as the restriction of A to $N[(A-\lambda I)^{\nu(\lambda)}]$) with respect to $\mathcal{B}_\lambda$, $A_{\lambda,\mathcal{B}_\lambda}$, has $h_{\nu(\lambda)-1}$ Jordan blocks $J_{\lambda,\nu(\lambda)}$, $h_{\nu(\lambda)-2} - h_{\nu(\lambda)-1}$ Jordan blocks $J_{\lambda,\nu(\lambda)-1}$ and, in general, for each $k \in \{2,\ldots,\nu(\lambda)\}$, $h_{\nu(\lambda)-k} - h_{\nu(\lambda)-k+1}$ Jordan blocks $J_{\lambda,\nu(\lambda)-k+1}$. The ordered set of vectors of $\mathcal{B}_\lambda$ is often referred to as a *chain of generalized eigenvectors*.

Since A_λ can be expressed as

$$A_\lambda = \lambda I + (A_\lambda - \lambda I)$$

and

$$(A-\lambda I)^{\nu(\lambda)}\left(N[(A-\lambda I)^{\nu(\lambda)}]\right) = \{0\},$$

it becomes apparent that A_λ is a sum of the diagonal operator λI plus the nilpotent operator $A_\lambda - \lambda I$. The basis $\mathcal{B}_\lambda$ that we have just constructed provides us with a matrix of $(A - \lambda I)|_{N[(A-\lambda I)^{\nu(\lambda)}]}$ with the maximum number of null entries. Suppose that

$$\sigma(A) = \{\lambda_1, \ldots, \lambda_q\},$$

and let P denote the matrix whose columns consist of the coordinates of the vectors of the bases $\mathcal{B}_{\lambda_1}, \ldots, \mathcal{B}_{\lambda_q}$ that we have just constructed. Then, the matrix

$$J := P^{-1}AP$$

is called the *Jordan canonical form* of the matrix A. We have proven that

$$J = \mathrm{diag}\{J_{\lambda_1}, J_{\lambda_2}, \ldots, J_{\lambda_q}\},$$

where, for each $j \in \{1, \ldots, q\}$, the matrix J_{λ_j} consists of a finite diagonal union of Jordan blocks of the type $J_{\lambda_j, k}$, for $k \in \{1, \ldots, \nu(\lambda_j)\}$.

2.2.1 *An example*

Let A be a square matrix such that $\sigma(A) = \{\alpha, \beta, \gamma\}$ and

$$\dim N[A - \alpha I] = \dim N[(A - \alpha I)^2] = 3, \quad \dim N[A - \beta I] = 3,$$

$$\dim N[(A - \beta I)^2] = 4, \quad \dim N[(A - \beta I)^3] = \dim N[(A - \beta I)^4] = 5,$$

$$\dim N[A - \gamma I] = 2, \quad \dim N[(A - \gamma I)^2] = \dim N[(A - \gamma I)^3] = 3.$$
$$\tag{2.15}$$

Then, necessarily, $\alpha \neq \beta \neq \gamma \neq \alpha$ and, by definition,

$$\nu(\alpha) = 1, \quad \nu(\beta) = 3, \quad \text{and} \quad \nu(\gamma) = 2.$$

Moreover, by Theorem 2.5,

$$m_a(\alpha) = \dim N[A - \alpha I] = 3,$$

$$m_a(\beta) = \dim N[(A - \beta I)^3] = 5,$$

$$m_a(\gamma) = \dim N[(A - \gamma I)^2] = 3,$$

and hence, the order of the matrix, N, is

$$N = m_a(\alpha) + m_a(\beta) + m_a(\gamma) = 3 + 5 + 3 = 11.$$

Now, we construct a Jordan basis in each of the ascent generalized eigenspaces $N[A - \alpha I]$, $N[(A - \beta I)^3]$, and $N[(A - \gamma I)^2]$.

By choosing three independent vectors $v_{\alpha,j} \in N[A - \alpha I]$, $j \in \{1, 2, 3\}$, we have that $Av_{\alpha,j} = \alpha v_{\alpha,j}$ for each $j \in \{1, 2, 3\}$, and hence, the matrix of $A|_{N[A-\alpha I]}$ with respect to the basis

$$\mathcal{B}_\alpha = \{v_{\alpha,1}, v_{\alpha,2}, v_{\alpha,3}\}$$

is given by

$$A_\alpha := \begin{pmatrix} \alpha & 0 & 0 \\ 0 & \alpha & 0 \\ 0 & 0 & \alpha \end{pmatrix}.$$

So, it consists of three Jordan blocks of order one associated with the eigenvalue α.

To construct a Jordan basis in $N[(A - \beta I)^3]$, we first consider a supplement, $C_{\beta,1}$, of $N[A - \beta I]$ in $N[(A - \beta I)^2]$ and a supplement, $C_{\beta,2}$, of $N[(A - \beta I)^2]$ in $N[(A - \beta I)^3]$. According to (2.15), we have that

$$\dim C_{\beta,2} = \dim C_{\beta,1} = 1.$$

For any given $v_{\beta,1} \in C_{\beta,2} \setminus \{0\}$, the three vectors

$$v_{\beta,1} \in C_{\beta,2}, \quad v_{\beta,2} := (A - \beta I)v_{\beta,1} \in C_{\beta,1}, \quad \text{and}$$

$$v_{\beta,3} := (A - \beta I)v_{\beta,2} \in N[A - \beta I]$$

satisfy, by definition,

$$Av_{\beta,1} = \beta v_{\beta,1} + v_{\beta,2}, \quad Av_{\beta,2} = \beta v_{\beta,2} + v_{\beta,3}, \quad Av_{\beta,3} = \beta v_{\beta,3}.$$

Thus, they provide us with a Jordan block of order three:

$$J_{\beta,3} = \begin{pmatrix} \beta & 0 & 0 \\ 1 & \beta & 0 \\ 0 & 1 & \beta \end{pmatrix}.$$

Next, consider two additional eigenvectors, $v_{\beta,4}$ and $v_{\beta,5}$, such that

$$N[A - \beta I] = \operatorname{span}[v_{\beta,3}, v_{\beta,4}, v_{\beta,5}].$$

Then, the matrix of $A|_{N[(A-\beta I)^3]}$ with respect to the basis

$$\mathcal{B}_\beta := \{v_{\beta,1}, v_{\beta,2}, v_{\beta,3}, v_{\beta,4}, v_{\beta,5}\}$$

is given by

$$A_\beta := \begin{pmatrix} \beta & 0 & 0 & 0 & 0 \\ 1 & \beta & 0 & 0 & 0 \\ 0 & 1 & \beta & 0 & 0 \\ 0 & 0 & 0 & \beta & 0 \\ 0 & 0 & 0 & 0 & \beta \end{pmatrix}.$$

Finally, consider a supplement, $C_{\gamma,1}$, of $N[A - \gamma I]$ in $N[(A - \gamma I)^2]$. Then, by (2.15),

$$\dim C_{\gamma,1} = 1,$$

and for any given $v_{\gamma,1} \in C_{\gamma,1} \setminus \{0\}$, the two vectors

$$v_{\gamma,1} \in C_{\gamma,1}, \quad v_{\gamma,2} := (A - \gamma I)v_{\gamma,1} \in N[A - \gamma I],$$

satisfy

$$Av_{\gamma,1} = \gamma v_{\gamma,1} + v_{\gamma,2}, \quad Av_{\gamma,2} = \gamma v_{\gamma,2}.$$

So, they provide us with a Jordan block of order two:

$$J_{\gamma,2} = \begin{pmatrix} \gamma & 0 \\ 1 & \gamma \end{pmatrix}.$$

Thus, by considering an additional eigenvector, $v_{\gamma,3}$, such that

$$N[A - \gamma I] = \mathrm{span}[v_{\gamma,2}, v_{\gamma,3}],$$

it becomes apparent that the matrix of $A|_{N[(A-\gamma I)^2]}$ with respect to the basis

$$\mathcal{B}_\gamma := \{v_{\gamma,1}, v_{\gamma,2}, v_{\gamma,3}\}$$

is given by

$$A_\gamma := \begin{pmatrix} \gamma & 0 & 0 \\ 1 & \gamma & 0 \\ 0 & 0 & \gamma \end{pmatrix}.$$

Therefore, the canonical form of A, by blocks, is

$$J = \begin{pmatrix} A_\alpha & 0_{3,5} & 0_{3,3} \\ 0_{5,3} & A_\beta & 0_{5,3} \\ 0_{3,3} & 0_{3,5} & A_\gamma \end{pmatrix},$$

where $0_{i,j}$ denotes the null matrix with i rows and j columns. The matrix J represents A with respect to the basis

$$\mathcal{B} = \{v_{\alpha,1}, v_{\alpha,2}, v_{\alpha,3}, v_{\beta,1}, v_{\beta,2}, v_{\beta,3}, v_{\beta,4}, v_{\beta,5}, v_{\gamma,1}, v_{\gamma,2}, v_{\gamma,3}\}.$$

2.3　Real Jordan Canonical Form

Throughout this section, we assume that $A \in \mathcal{M}_N(\mathbb{R})$. Then, the characteristic polynomial of A has real coefficients and, hence,

$$\sigma(A) = \{\bar{\lambda} : \lambda \in \sigma(A)\},$$

where $\bar{z}$ indicates the complex conjugate of $z \in \mathbb{C}$. Thus, if $\lambda \in \sigma(A) \cap (\mathbb{C} \setminus \mathbb{R})$, both ascent generalized eigenspaces $N[(A - \lambda I)^{\nu(\lambda)}]$ and $N[(A - \bar{\lambda} I)^{\nu(\bar{\lambda})}]$ appear in the Jordan decomposition given by (2.10). Suppose

$$\sigma(A) = \{\lambda_1, \ldots, \lambda_q\},$$

fix $1 \leq j \leq q$, and set $\lambda := \lambda_j$. It is easily seen that the vectors of $\mathcal{B}_\lambda$ constructed in Section 2.2 all have real components if $\lambda \in \mathbb{R}$, whereas, if

$$\lambda = \alpha + i\beta, \qquad \alpha, \beta \in \mathbb{R}, \quad \beta \neq 0,$$

at least one component of such vectors is not real. Suppose this occurs. Then, for every $k \geq 0$ and $v \in \mathbb{C}^N$, we have that

$$(A - \lambda I)^k v = 0 \ \text{ if and only if } \ (A - \bar{\lambda} I)^k \bar{v} = 0, \tag{2.16}$$

where $\bar{v}$ denotes the complex conjugate vector of v, i.e. the vector whose components are the complex conjugate components of v.

Therefore,

$$\dim N[(A - \lambda I)^k] = \dim N[(A - \bar{\lambda} I)^k]$$

for all $k \geq 0$ and, in particular,

$$\nu(\lambda) = \nu(\bar{\lambda}) \quad \text{and} \quad m_a(\lambda) = m_a(\bar{\lambda}).$$

Let

$$\mathcal{B}_\lambda := \{v_{\lambda,1}, \ldots, v_{\lambda,m_a(\lambda)}\}$$

be the basis of $N[(A - \lambda I)^{\nu(\lambda)}]$ constructed in Section 2.2. By (2.16), we have that

$$\mathcal{B}_{\bar{\lambda}} := \{\bar{v}_{\lambda,1}, \ldots, \bar{v}_{\lambda,m_a(\lambda)}\}$$

provides us with the Jordan basis of $N[(A - \bar{\lambda} I)^{\nu(\bar{\lambda})}]$ constructed in Section 2.2. In other words, we take

$$v_{\bar{\lambda},k} := \bar{v}_{\lambda,k}, \qquad 1 \leq k \leq m_a(\lambda) = m_a(\bar{\lambda}).$$

Indeed, by construction, if a set,

$$\{v_{\lambda,k+i} : i \in \{0, \ldots, h-1\}\},$$

of $h \geq 1$ vectors of the basis $\mathcal{B}_\lambda$, where $1 \leq k \leq m_a(\lambda)$, provides us with a Jordan block $J_{\lambda,h}$ of order h, then their conjugate vectors,

$$v_{\bar{\lambda},k+i} = \bar{v}_{\lambda,k+i}, \qquad 0 \leq i \leq h-1,$$

provide us with a Jordan block of type $J_{\bar{\lambda},h}$.

Subsequently, for any integer $N \geq 1$ and $z \in \mathbb{C}^N$, the vectors $\operatorname{Re} z \in \mathbb{R}^N$ and $\operatorname{Im} z \in \mathbb{R}^N$ will denote the real and imaginary parts of z, respectively. According to this notation, when $h = 1$, we have that

$$A\,(\operatorname{Re} v_{\lambda,k} + i \operatorname{Im} v_{\lambda,k}) = A v_{\lambda,k} = \lambda v_{\lambda,k} = (\alpha + i\beta)\,(\operatorname{Re} v_{\lambda,k} + i \operatorname{Im} v_{\lambda,k}),$$

and by identifying the real and imaginary parts,

$$A \operatorname{Re} v_{\lambda,k} = \alpha \operatorname{Re} v_{\lambda,k} - \beta \operatorname{Im} v_{\lambda,k},$$

$$A \operatorname{Im} v_{\lambda,k} = \alpha \operatorname{Im} v_{\lambda,k} + \beta \operatorname{Re} v_{\lambda,k}.$$

Consequently, A leaves the real linear subspace generated by $\operatorname{Re} v_{\lambda,k}$ and $\operatorname{Im} v_{\lambda,k}$ invariant, and the matrix of the restriction of A to this subspace with respect to the basis $\{\operatorname{Re} v_{\lambda,k}, \operatorname{Im} v_{\lambda,k}\}$ is

$$
J^{\mathbb{R}}_{\lambda,2} := \begin{pmatrix} \alpha & \beta \\ -\beta & \alpha \end{pmatrix}.
\tag{2.17}
$$

Note that $\{\operatorname{Re} v_{\lambda,k}, \operatorname{Im} v_{\lambda,k}\}$ is linearly independent because $\det J^{\mathbb{R}}_{\lambda,2} \neq 0$. Similarly, when $h \geq 2$, the following identities hold:

$$
A v_{\lambda,k+i} = \lambda v_{\lambda,k+i} + v_{\lambda,k+i+1}, \qquad 0 \leq i \leq h-2,
$$
$$
A v_{\lambda,k+h-1} = \lambda v_{\lambda,k+h-1},
$$

and identifying the real and imaginary parts in these identities gives

$$
A \operatorname{Re} v_{\lambda,k+i} = \alpha \operatorname{Re} v_{\lambda,k+i} - \beta \operatorname{Im} v_{\lambda,k+i} + \operatorname{Re} v_{\lambda,k+i+1},
$$
$$
A \operatorname{Im} v_{\lambda,k+i} = \alpha \operatorname{Im} v_{\lambda,k+i} + \beta \operatorname{Re} v_{\lambda,k+i} + \operatorname{Im} v_{\lambda,k+i+1},
$$

for all $0 \leq i \leq h-2$, and

$$
A \operatorname{Re} v_{\lambda,k+h-1} = \alpha \operatorname{Re} v_{\lambda,k+h-1} - \beta \operatorname{Im} v_{\lambda,k+h-1},
$$
$$
A \operatorname{Im} v_{\lambda,k+h-1} = \alpha \operatorname{Im} v_{\lambda,k+h-1} + \beta \operatorname{Re} v_{\lambda,k+h-1}.
$$

Therefore, A leaves invariant the real linear subspace

$$
\operatorname{span}\left[\operatorname{Re} v_{\lambda,k}, \operatorname{Im} v_{\lambda,k}, \ldots, \operatorname{Re} v_{\lambda,k+h-1}, \operatorname{Im} v_{\lambda,k+h-1}\right],
\tag{2.18}
$$

and the matrix of A restricted to (2.18) is

$$
J^{\mathbb{R}}_{\lambda,2h} = \begin{pmatrix}
J^{\mathbb{R}}_{\lambda,2} & 0_2 & 0_2 & \cdots & 0_2 & 0_2 \\
I_2 & J^{\mathbb{R}}_{\lambda,2} & 0_2 & \cdots & 0_2 & 0_2 \\
0_2 & I_2 & J^{\mathbb{R}}_{\lambda,2} & \cdots & 0_2 & 0_2 \\
\vdots & \vdots & \vdots & \ddots & \vdots & \vdots \\
0_2 & 0_2 & 0_2 & \cdots & J^{\mathbb{R}}_{\lambda,2} & 0_2 \\
0_2 & 0_2 & 0_2 & \cdots & I_2 & J^{\mathbb{R}}_{\lambda,2}
\end{pmatrix},
\tag{2.19}
$$

where $J^{\mathbb{R}}_{\lambda,2}$ is the Jordan block (2.17), I_2 is the identity matrix of order 2, and $0_2 = 0_{2,2}$ is the null square matrix of order 2. Since

$\det J_{\lambda,2h}^{\mathbb{R}} \neq 0$, the subspace (2.18) has dimension $2h$. To summarize, these properties show that, by taking

$$\mathcal{B}_\lambda^{\mathbb{R}} := \{\operatorname{Re} v_{\lambda,1}, \operatorname{Im} v_{\lambda,1}, \ldots, \operatorname{Re} v_{\lambda,m_a(\lambda)}, \operatorname{Im} v_{\lambda,m_a(\lambda)}\}$$

instead of $\mathcal{B}_\lambda \cup \mathcal{B}_{\bar\lambda}$, the endomorphism A leaves the real linear subspace generated by $\mathcal{B}_\lambda^{\mathbb{R}}$ invariant and that its associated matrix with respect to this basis is real and block-diagonal, with each of its blocks being

$$J_{\lambda,2k}^{\mathbb{R}}, \qquad 1 \leq k \leq \nu(\lambda).$$

By repeating this process for each pair of complex conjugate eigenvalues, one is naturally driven toward the construction of the so-called *real Jordan canonical form* of the matrix A. The block $J_{\lambda,2k}^{\mathbb{R}}$ is often called the *real Jordan block* of order $2k$, associated with the complex eigenvalue λ.

2.3.1 *An example*

Let $A \in \mathcal{M}_N(\mathbb{R})$ be such that $\sigma(A) = \{\lambda, \bar\lambda\}$ and

$$\dim N[A - \lambda I] = 1, \quad \dim N[(A - \lambda I)^2] = \dim N[(A - \lambda I)^3] = 2,$$

where $\lambda = \alpha + i\beta$ with $\beta > 0$. Then, also,

$$\dim N[A - \bar\lambda I] = 1, \quad \dim N[(A - \bar\lambda I)^2] = \dim N[(A - \bar\lambda I)^3] = 2,$$

and hence,

$$\nu(\lambda) = \nu(\bar\lambda) = 2, \qquad m_a(\lambda) = m_a(\bar\lambda) = 2.$$

Thus, $N = 4$. Let $v_{\lambda,1} \in N[(A - \lambda I)^2] \setminus N[A - \lambda I]$. Then,

$$v_{\lambda,1}, \quad v_{\lambda,2} := (A - \lambda I)v_{\lambda,1} \in N[A - \lambda I]$$

provide us with a basis of $N[(A - \lambda I)^2]$. Moreover, since $A \in \mathcal{M}_4(\mathbb{R})$, we also have that $\bar v_{\lambda,1} \in N[(A - \bar\lambda I)^2] \setminus N[A - \bar\lambda I]$ and that

$$\bar v_{\lambda,1}, \quad \bar v_{\lambda,2} := (A - \bar\lambda I)\bar v_{\lambda,1} \in N[A - \bar\lambda I]$$

provide us with a basis of $N[(A - \bar\lambda I)^2]$. Therefore, since

$$\begin{aligned}
A v_{\lambda,1} &= \lambda v_{\lambda,1} + v_{\lambda,2}, \quad & A v_{\lambda,2} &= \lambda v_{\lambda,2}, \\
A \bar v_{\lambda,1} &= \bar\lambda \bar v_{\lambda,1} + \bar v_{\lambda,2}, \quad & A \bar v_{\lambda,2} &= \bar\lambda \bar v_{\lambda,2},
\end{aligned} \tag{2.20}$$

the expression of A with respect to the basis

$$\mathcal{B} := \{v_{\lambda,1}, v_{\lambda,2}, \bar{v}_{\lambda,1}, \bar{v}_{\lambda,2}\}$$

provides us with the complex Jordan form of A:

$$J^{\mathbb{C}} := \begin{pmatrix} \lambda & 0 & 0 & 0 \\ 1 & \lambda & 0 & 0 \\ 0 & 0 & \bar{\lambda} & 0 \\ 0 & 0 & 1 & \bar{\lambda} \end{pmatrix}.$$

If, instead of the basis $\mathcal{B}$, we consider the basis

$$\mathcal{B}^{\mathbb{R}} := \{\operatorname{Re} v_{\lambda,1}, \operatorname{Im} v_{\lambda,1}, \operatorname{Re} v_{\lambda,2}, \operatorname{Im} v_{\lambda,2}\},$$

then, taking the real and imaginary parts in (2.20) leads to

$$A \operatorname{Re} v_{\lambda,1} = \alpha \operatorname{Re} v_{\lambda,1} - \beta \operatorname{Im} v_{\lambda,1} + \operatorname{Re} v_{\lambda,2},$$
$$A \operatorname{Im} v_{\lambda,1} = \alpha \operatorname{Im} v_{\lambda,1} + \beta \operatorname{Re} v_{\lambda,1} + \operatorname{Im} v_{\lambda,2},$$
$$A \operatorname{Re} v_{\lambda,2} = \alpha \operatorname{Re} v_{\lambda,2} - \beta \operatorname{Im} v_{\lambda,2},$$
$$A \operatorname{Im} v_{\lambda,2} = \alpha \operatorname{Im} v_{\lambda,2} + \beta \operatorname{Re} v_{\lambda,2}.$$

Therefore, the matrix of the operator A with respect to the basis $\mathcal{B}^{\mathbb{R}}$ provides us with the real canonical form of A:

$$J^{\mathbb{R}} = \begin{pmatrix} \alpha & \beta & 0 & 0 \\ -\beta & \alpha & 0 & 0 \\ 1 & 0 & \alpha & \beta \\ 0 & 1 & -\beta & \alpha \end{pmatrix}.$$

2.4 Exponential Matrix

The following result allows us to introduce the concept of exponential matrix.

Theorem 2.6 (Definition of exponential matrix). *For every* $A \in \mathcal{M}_N(\mathbb{C})$, *the sequence of matrices*

$$S_n(A) := \sum_{j=0}^{n} \frac{A^j}{j!}, \qquad n \geq 1,$$

converges, absolutely in $\mathcal{L}(\mathbb{C}^N)$, to a matrix denoted by e^A and called the exponential matrix of A. Moreover, for every invertible matrix $P \in \mathcal{M}_N(\mathbb{C})$,

$$e^{P^{-1}AP} = P^{-1}e^A P.$$

Proof. Throughout this proof, we fix a norm, $\| \cdot \|$, in $\mathbb{C}^N$ and consider the induced operator norm in $\mathcal{L}(\mathbb{C}^N)$, i.e.

$$\|A\| := \|A\|_{\mathcal{L}(\mathbb{C}^N)} = \max \left\{ \|Ax\| : \|x\| = 1 \right\}.$$

Then, for every $m, n \geq 1$,

$$\|S_{n+m}(A) - S_n(A)\| = \left\| \sum_{j=n+1}^{n+m} \frac{A^j}{j!} \right\| \leq \sum_{j=n+1}^{n+m} \frac{\|A^j\|}{j!} \leq \sum_{j=n+1}^{n+m} \frac{\|A\|^j}{j!}.$$

Thus, since

$$\sum_{j=0}^{\infty} \frac{\|A\|^j}{j!} = e^{\|A\|} < +\infty,$$

for every $\varepsilon > 0$, there exists $n = n(\varepsilon) \geq 1$ such that

$$\|S_{n+m}(A) - S_n(A)\| \leq \sum_{j=n+1}^{n+m} \frac{\|A\|^j}{j!} < \varepsilon$$

for all $m \geq 1$ and $n \geq n(\varepsilon)$. Therefore, $\{S_n(A)\}_{n \geq 1}$ is a Cauchy sequence in $\mathcal{L}(\mathbb{C}^N)$ and, by completeness, there exists a matrix, denoted by e^A, such that

$$\lim_{n \to \infty} S_n(A) = e^A \quad \text{in } \mathcal{L}(\mathbb{C}^N).$$

Naturally, as usual, the following notation can be used:

$$e^A = \sum_{j=0}^{\infty} \frac{A^j}{j!}.$$

Lastly, let $P \in \mathcal{L}(\mathbb{C}^N)$ be an invertible matrix. Then, for every $n \geq 1$,

$$S_n(P^{-1}AP) = \sum_{j=0}^{n} \frac{(P^{-1}AP)^j}{j!} = \sum_{j=0}^{n} \frac{P^{-1}A^j P}{j!} = P^{-1}S_n(A)P.$$

Thus, letting $n \to \infty$ yields

$$e^{P^{-1}AP} = P^{-1}e^A P,$$

which completes the proof. $\qquad\qquad\qquad\qquad\qquad\qquad\qquad\qquad\quad \square$

The following result collects the main algebraic properties of e^A.

Theorem 2.7. *Let $A, B \in \mathcal{M}_N(\mathbb{C})$. Then:*

(a) $\|e^A\| \le e^{\|A\|}$.
(b) $e^{A+B} = e^A e^B = e^B e^A$ *if $AB = BA$.*
(c) e^A *is invertible, $e^0 = I$, and $(e^A)^{-1} = e^{-A}$.*

Proof. We retain the notations introduced in Theorem 2.6. Then, for every $n \ge 1$,

$$\|S_n(A)\| \le \sum_{j=0}^{n} \frac{\|A\|^j}{j!} \le \sum_{j=0}^{\infty} \frac{\|A\|^j}{j!} = e^{\|A\|}.$$

Thus, letting $n \to \infty$ gives that $\|e^A\| \le e^{\|A\|}$, which completes the proof of Part (a). In order to prove Part (b), suppose that $AB = BA$. Then,

$$(A+B)^2 = (A+B)(A+B) = A^2 + BA + AB + B^2 = A^2 + 2AB + B^2,$$

and hence, a well-known induction argument gives

$$(A + B)^n = \sum_{k=0}^{n} \binom{n}{k} A^k B^{n-k} \quad \text{for all } n \ge 1.$$

Thus,

$$e^{A+B} = \sum_{n=0}^{\infty} \frac{(A+B)^n}{n!} = \sum_{n=0}^{\infty} \frac{1}{n!} \sum_{k=0}^{n} \binom{n}{k} A^k B^{n-k}$$

$$= \sum_{n=0}^{\infty} \sum_{k=0}^{n} \frac{A^k}{k!} \frac{B^{n-k}}{(n-k)!}.$$

As the last series is the Cauchy product of the series defining e^A and e^B, and these series are absolutely convergent, Part (b) holds (see Apostol, 1974, Section 8.24).

Finally, by definition, it is obvious that $e^0 = I$. Thus, since

$$(-A)A = A(-A) = -A^2,$$

from Part (b), it becomes apparent that

$$I = e^0 = e^{A-A} = e^A e^{-A}.$$

Therefore, $\left(e^A\right)^{-1} = e^{-A}$, which completes the proof. $\qquad\square$

The following concept is fundamental in the theory of functions of one complex variable.

Definition 2.8 (Holomorphy). Consider an open subset, $\Omega \subset \mathbb{C}$, and an integer, $N \geq 1$. A vectorial function $f : \Omega \to \mathbb{C}^N$ is said to be *holomorphic* in Ω when

$$\lim_{h \to 0} \frac{f(z+h) - f(z)}{h}$$

exists in Ω for all $z \in \Omega$. In such a case, this limit is denoted by $f'(z)$ and called the derivative of f at z. The set of holomorphic functions in Ω with values in $\mathbb{C}^N$ is denoted by $\mathcal{H}(\Omega; \mathbb{C}^N)$.

By expressing f through its components, $f = (f_1, \ldots, f_N)$, it becomes apparent that $f \in \mathcal{H}(\Omega; \mathbb{C}^N)$ if and only if $f_i \in \mathcal{H}(\Omega; \mathbb{C})$ for all $i \in \{1, \ldots, N\}$. Moreover, in this case,

$$f'(z) = \left(f_1'(z), \ldots, f_N'(z)\right) \quad \text{for all } z \in \Omega.$$

The following result explains the huge interest of exponential matrices in the theory of differential equations. By *entire function*, we mean an holomorphic function in the whole $\mathbb{C}$.

Theorem 2.9. *For every integer $N \geq 1$ and any matrix $A \in \mathcal{M}_N(\mathbb{C})$, the function $\Phi : \mathbb{C} \to \mathcal{M}_N(\mathbb{C})$, defined by*

$$\Phi(z) := e^{zA}, \qquad z \in \mathbb{C},$$

is entire, and it satisfies

$$\Phi'(z) = A\Phi(z), \qquad z \in \mathbb{C}. \tag{2.21}$$

In particular, for every $t_0 \in \mathbb{R}$, the exponential matrix $e^{(t-t_0)A}$, $t \in \mathbb{R}$, provides us with a fundamental matrix of solutions of $u' = Au$, and *the unique solution of the initial value problem*

$$\begin{cases} u' = Au, \\ u(t_0) = u_0 \end{cases}$$

is given by $u(t) = e^{(t-t_0)A} u_0$.

Proof. To prove (2.21), we shall show that

$$\lim_{h \to 0} \frac{e^{(z+h)A} - e^{zA}}{h} = Ae^{zA} \qquad \text{for all } z \in \mathbb{C}.$$

As, according to Theorem 2.7(b),

$$\frac{e^{(z+h)A} - e^{zA}}{h} - Ae^{zA} = \frac{e^{hA}e^{zA} - e^{zA}}{h} - Ae^{zA} = \left(\frac{e^{hA} - I}{h} - A \right) e^{zA},$$

it suffices to establish that

$$\lim_{h \to 0} \left(\frac{e^{hA} - I}{h} - A \right) = 0. \tag{2.22}$$

Indeed, since

$$\frac{e^{hA} - I}{h} - A = h \sum_{n=2}^{\infty} h^{n-2} \frac{A^n}{n!},$$

and, for every $h \in \mathbb{C}$ with $|h| \leq 1$,

$$\left\| \sum_{n=2}^{\infty} h^{n-2} \frac{A^n}{n!} \right\| \leq \sum_{n=2}^{\infty} \frac{\|A\|^n}{n!} = e^{\|A\|} - 1 - \|A\|,$$

it becomes apparent that, whenever $|h| \leq 1$,

$$\left\| \frac{e^{hA} - I}{h} - A \right\| \leq |h| \left(e^{\|A\|} - 1 - \|A\| \right).$$

Therefore, (2.22) follows.

Moving on to the last part, since $\Phi(t) = e^{tA}$ provides us with a matrix of solutions of $u' = Au$ and $\Phi(0) = I$ is invertible, by Theorem 1.13, e^{tA} is a fundamental matrix of solutions of $u' = Au$, and so is $e^{(t-t_0)A}$ for all $t_0 \in \mathbb{R}$. Finally, by linearity, we observe that

$u(t) = e^{(t-t_0)A}u_0$ is a solution of $u' = Au$, and $u(t_0) = Iu_0 = u_0$, which concludes the proof. $\qquad\square$

As a consequence of this theorem, we can repeat the analysis in Section 1.6 by taking $\Phi(t) = e^{(t-t_0)A}$ in (1.27). This leads us to the following result.

Corollary 2.10. *For every integer $N \geq 1$, $A \in \mathcal{M}_N(\mathbb{C})$, $\alpha, \beta \in \mathbb{R}$, with $\alpha < \beta$, $B \in \mathcal{C}([\alpha, \beta]; \mathbb{C}^N)$, $t_0 \in [\alpha, \beta]$, and $u_0 \in \mathbb{C}^N$, the vectorial function*

$$u(t) := e^{(t-t_0)A}u_0 + \int_{t_0}^{t} e^{(t-s)A}B(s)\,ds, \qquad t \in [\alpha, \beta],$$

provides us with the unique solution of the Cauchy problem

$$\begin{cases} u' = Au + B(t), & t \in [\alpha, \beta], \\ u(t_0) = u_0. \end{cases}$$

2.5 Calculating Exponential Matrices

Given any matrix $A \in \mathcal{M}_N(\mathbb{C})$, we show in this section the steps to determine e^{zA}, $z \in \mathbb{C}$. First, one shall find the spectrum

$$\sigma(A) = \{\lambda_1, \ldots, \lambda_q\}, \qquad \lambda_i \neq \lambda_j \text{ if } i \neq j.$$

Then, one shall compute the Jordan canonical form of the matrix A,

$$J = \begin{pmatrix} J_1 & & & \\ & J_2 & & \\ & & \ddots & \\ & & & J_\ell \end{pmatrix},$$

where J_i, $1 \leq i \leq \ell$, are Jordan blocks associated with some eigenvalue of the matrix A (recall that the order of the blocks is smaller than or equal to the ascent of the considered eigenvalue). If we denote by P the matrix whose columns are the components of the vectors

                    Ordinary Differential Equations

of the Jordan basis of A, then

$$A = PJP^{-1},$$

and, thanks to Theorem 2.6,

$$e^{zA} = Pe^{zJ}P^{-1} \quad \text{for all } z \in \mathbb{C}.$$

Moreover, since

$$(zJ)^n = \begin{pmatrix} (zJ_1)^n & & & \\ & (zJ_2)^n & & \\ & & \ddots & \\ & & & (zJ_\ell)^n \end{pmatrix}$$

for all integers $n \geq 0$ and $z \in \mathbb{C}$, it becomes apparent that, for every $z \in \mathbb{C}$,

$$e^{zJ} = \begin{pmatrix} e^{zJ_1} & & & \\ & e^{zJ_2} & & \\ & & \ddots & \\ & & & e^{zJ_\ell} \end{pmatrix}. \tag{2.23}$$

Therefore, e^{zA} can be easily determined if we can compute the exponentials of each of the Jordan blocks appearing in the Jordan canonical form of the matrix A. We now proceed to compute these exponentials. For every $i \in \{1, \ldots, \ell\}$, there exist $\lambda \in \sigma(A)$ and $n \in \{1, \ldots, \nu(\lambda)\}$, such that

$$J_i = J_{\lambda,n} = \underbrace{\begin{pmatrix} \lambda & 0 & 0 & \cdots & 0 & 0 \\ 1 & \lambda & 0 & \cdots & 0 & 0 \\ 0 & 1 & \lambda & \cdots & 0 & 0 \\ \vdots & \vdots & \vdots & \ddots & \vdots & \vdots \\ 0 & 0 & 0 & \cdots & \lambda & 0 \\ 0 & 0 & 0 & \cdots & 1 & \lambda \end{pmatrix}}_{n} = \mathcal{N}_n + \lambda I_n,$$

where I_n is the identity matrix of order n and $\mathcal{N}_n$ is the nilpotent matrix:

$$
\mathcal{N}_n := \begin{pmatrix}
0 & 0 & 0 & \cdots & 0 & 0 \\
1 & 0 & 0 & \cdots & 0 & 0 \\
0 & 1 & 0 & \cdots & 0 & 0 \\
\vdots & \vdots & \vdots & \ddots & \vdots & \vdots \\
0 & 0 & 0 & \cdots & 0 & 0 \\
0 & 0 & 0 & \cdots & 1 & 0
\end{pmatrix} \underbrace{\qquad\qquad\qquad}_{n} .
$$

Since

$$
z J_i = z \mathcal{N}_n + z \lambda I_n \quad \text{for all } z \in \mathbb{C},
$$

and the identity matrix commutes with any other matrix, it is apparent that

$$
e^{z J_i} = e^{z \lambda I_n} e^{z \mathcal{N}_n} = e^{z \lambda} I_n e^{z \mathcal{N}_n} = e^{z \lambda} e^{z \mathcal{N}_n}
$$

for all $z \in \mathbb{C}$. Finally, since $\mathcal{N}_n^n = 0$, it can be easily seen that

$$
e^{z \mathcal{N}_n} = \sum_{j=0}^{n-1} \frac{z^j}{j!} \mathcal{N}_n^j = \begin{pmatrix}
1 & 0 & 0 & \cdots & 0 & 0 & 0 \\
z & 1 & 0 & \cdots & 0 & 0 & 0 \\
\frac{z^2}{2} & z & 1 & \cdots & 0 & 0 & 0 \\
\vdots & \vdots & \vdots & \ddots & \vdots & \vdots & \vdots \\
\frac{z^{n-3}}{(n-3)!} & \frac{z^{n-4}}{(n-4)!} & \frac{z^{n-5}}{(n-5)!} & \cdots & 1 & 0 & 0 \\
\frac{z^{n-2}}{(n-2)!} & \frac{z^{n-3}}{(n-3)!} & \frac{z^{n-4}}{(n-4)!} & \cdots & z & 1 & 0 \\
\frac{z^{n-1}}{(n-1)!} & \frac{z^{n-2}}{(n-2)!} & \frac{z^{n-3}}{(n-3)!} & \cdots & \frac{z^2}{2} & z & 1
\end{pmatrix} .
$$

Therefore,

$$
e^{zJ_i} = e^{z\lambda}
\begin{pmatrix}
1 & 0 & 0 & \cdots & 0 & 0 & 0 \\
z & 1 & 0 & \cdots & 0 & 0 & 0 \\
\frac{z^2}{2} & z & 1 & \cdots & 0 & 0 & 0 \\
\vdots & \vdots & \vdots & \ddots & \vdots & \vdots & \vdots \\
\frac{z^{n-3}}{(n-3)!} & \frac{z^{n-4}}{(n-4)!} & \frac{z^{n-5}}{(n-5)!} & \cdots & 1 & 0 & 0 \\
\frac{z^{n-2}}{(n-2)!} & \frac{z^{n-3}}{(n-3)!} & \frac{z^{n-4}}{(n-4)!} & \cdots & z & 1 & 0 \\
\frac{z^{n-1}}{(n-1)!} & \frac{z^{n-2}}{(n-2)!} & \frac{z^{n-3}}{(n-3)!} & \cdots & \frac{z^2}{2} & z & 1
\end{pmatrix}.
$$

Observe that these matrices are real if $\lambda, z \in \mathbb{R}$. When A is a matrix with real coefficients and $\lambda = \alpha + i\beta \in \sigma(A)$ for some $\alpha, \beta \in \mathbb{R}$, with $\beta > 0$, then $\bar{\lambda} = \alpha - i\beta \in \sigma(A)$. In such a case, our goal is to express e^{zA} as a real matrix, starting with the real canonical form of A. To achieve this purpose, we have to compute the exponential of the real Jordan blocks

$$
J_{\lambda,2}^{\mathbb{R}} = \begin{pmatrix} \alpha & \beta \\ -\beta & \alpha \end{pmatrix}, \qquad
J_{\lambda,4}^{\mathbb{R}} = \begin{pmatrix}
\alpha & \beta & 0 & 0 \\
-\beta & \alpha & 0 & 0 \\
1 & 0 & \alpha & \beta \\
0 & 1 & -\beta & \alpha
\end{pmatrix}, \qquad \cdots
$$

Since, for every $z \in \mathbb{C}$,

$$
zJ_{\lambda,2}^{\mathbb{R}} = z\begin{pmatrix} \alpha & \beta \\ -\beta & \alpha \end{pmatrix} = z\alpha \begin{pmatrix} 1 & 0 \\ 0 & 1 \end{pmatrix} + z\beta \begin{pmatrix} 0 & 1 \\ -1 & 0 \end{pmatrix}
$$

and the identity matrix commutes with any other matrix, by Theorem 2.7(b), we find that

$$
e^{zJ_{\lambda,2}^{\mathbb{R}}} = e^{z\alpha} \sum_{n=0}^{\infty} \frac{(z\beta)^n}{n!} \begin{pmatrix} 0 & 1 \\ -1 & 0 \end{pmatrix}^n = e^{z\alpha} \begin{pmatrix} \cos(\beta z) & \sin(\beta z) \\ -\sin(\beta z) & \cos(\beta z) \end{pmatrix}.
$$

Similarly, for every $z \in \mathbb{C}$,

$$
zJ_{\lambda,4}^{\mathbb{R}} = z\begin{pmatrix}
\alpha & \beta & 0 & 0 \\
-\beta & \alpha & 0 & 0 \\
1 & 0 & \alpha & \beta \\
0 & 1 & -\beta & \alpha
\end{pmatrix} = z\begin{pmatrix} J_{\lambda,2}^{\mathbb{R}} & 0_2 \\ 0_2 & J_{\lambda,2}^{\mathbb{R}} \end{pmatrix} + z\begin{pmatrix} 0_2 & 0_2 \\ I_2 & 0_2 \end{pmatrix},
$$

where 0_2 and I_2 are the zero and identity matrices of order two, respectively. Since

$$\begin{pmatrix} J^{\mathbb{R}}_{\lambda,2} & 0_2 \\ 0_2 & J^{\mathbb{R}}_{\lambda,2} \end{pmatrix} \begin{pmatrix} 0_2 & 0_2 \\ I_2 & 0_2 \end{pmatrix} = \begin{pmatrix} 0_2 & 0_2 \\ J^{\mathbb{R}}_{\lambda,2} & 0_2 \end{pmatrix} = \begin{pmatrix} 0_2 & 0_2 \\ I_2 & 0_2 \end{pmatrix} \begin{pmatrix} J^{\mathbb{R}}_{\lambda,2} & 0_2 \\ 0_2 & J^{\mathbb{R}}_{\lambda,2} \end{pmatrix},$$

according to Theorem 2.7(b), we find that

$$e^{zJ^{\mathbb{R}}_{\lambda,4}} = \exp\left[z \begin{pmatrix} J^{\mathbb{R}}_{\lambda,2} & 0_2 \\ 0_2 & J^{\mathbb{R}}_{\lambda,2} \end{pmatrix} \right] \exp\left[z \begin{pmatrix} 0_2 & 0_2 \\ I_2 & 0_2 \end{pmatrix} \right]$$

$$= \begin{pmatrix} e^{zJ^{\mathbb{R}}_{\lambda,2}} & 0_2 \\ 0_2 & e^{zJ^{\mathbb{R}}_{\lambda,2}} \end{pmatrix} \exp\left[z \begin{pmatrix} 0_2 & 0_2 \\ I_2 & 0_2 \end{pmatrix} \right],$$

where we have used the notation $\exp[M] := e^M$. On the other hand, by definition,

$$\exp\left[z \begin{pmatrix} 0_2 & 0_2 \\ I_2 & 0_2 \end{pmatrix} \right] = \begin{pmatrix} I_2 & 0_2 \\ 0_2 & I_2 \end{pmatrix} + \begin{pmatrix} 0_2 & 0_2 \\ zI_2 & 0_2 \end{pmatrix} = \begin{pmatrix} I_2 & 0_2 \\ zI_2 & I_2 \end{pmatrix}.$$

Therefore,

$$e^{zJ^{\mathbb{R}}_{\lambda,4}} = \begin{pmatrix} e^{zJ^{\mathbb{R}}_{\lambda,2}} & 0_2 \\ 0_2 & e^{zJ^{\mathbb{R}}_{\lambda,2}} \end{pmatrix} \begin{pmatrix} I_2 & 0_2 \\ zI_2 & I_2 \end{pmatrix} = \begin{pmatrix} e^{zJ^{\mathbb{R}}_{\lambda,2}} & 0_2 \\ ze^{zJ^{\mathbb{R}}_{\lambda,2}} & e^{zJ^{\mathbb{R}}_{\lambda,2}} \end{pmatrix},$$

which is a real matrix if $z \in \mathbb{R}$.

By reasoning in a similar way, one can compute $e^{zJ^{\mathbb{R}}_{\lambda,2h}}$, where $J^{\mathbb{R}}_{\lambda,2h}$ is the general real Jordan block defined in (2.19), obtaining

$$e^{zJ^{\mathbb{R}}_{\lambda,2h}} = \begin{pmatrix} e^{zJ^{\mathbb{R}}_{\lambda,2}} & 0_2 & 0_2 & \cdots & 0_2 & 0_2 & 0_2 \\ ze^{zJ^{\mathbb{R}}_{\lambda,2}} & e^{zJ^{\mathbb{R}}_{\lambda,2}} & 0_2 & \cdots & 0_2 & 0_2 & 0_2 \\ \frac{z^2}{2}e^{zJ^{\mathbb{R}}_{\lambda,2}} & ze^{zJ^{\mathbb{R}}_{\lambda,2}} & e^{zJ^{\mathbb{R}}_{\lambda,2}} & \cdots & 0_2 & 0_2 & 0_2 \\ \vdots & \vdots & \vdots & \ddots & \vdots & \vdots & \vdots \\ \frac{z^{h-3}}{(h-3)!}e^{zJ^{\mathbb{R}}_{\lambda,2}} & \frac{z^{h-4}}{(h-4)!}e^{zJ^{\mathbb{R}}_{\lambda,2}} & \frac{z^{h-5}}{(h-5)!}e^{zJ^{\mathbb{R}}_{\lambda,2}} & \cdots & e^{zJ^{\mathbb{R}}_{\lambda,2}} & 0_2 & 0_2 \\ \frac{z^{h-2}}{(h-2)!}e^{zJ^{\mathbb{R}}_{\lambda,2}} & \frac{z^{h-3}}{(h-3)!}e^{zJ^{\mathbb{R}}_{\lambda,2}} & \frac{z^{h-4}}{(h-4)!}e^{zJ^{\mathbb{R}}_{\lambda,2}} & \cdots & ze^{zJ^{\mathbb{R}}_{\lambda,2}} & e^{zJ^{\mathbb{R}}_{\lambda,2}} & 0_2 \\ \frac{z^{h-1}}{(h-1)!}e^{zJ^{\mathbb{R}}_{\lambda,2}} & \frac{z^{h-2}}{(h-2)!}e^{zJ^{\mathbb{R}}_{\lambda,2}} & \frac{z^{h-3}}{(h-3)!}e^{zJ^{\mathbb{R}}_{\lambda,2}} & \cdots & \frac{z^2}{2}e^{zJ^{\mathbb{R}}_{\lambda,2}} & ze^{zJ^{\mathbb{R}}_{\lambda,2}} & e^{zJ^{\mathbb{R}}_{\lambda,2}} \end{pmatrix},$$

and then combine all the Jordan blocks associated with the different (and non-conjugated) eigenvalues, as in (2.23), to obtain the exponential e^{zJ}, where J is the real Jordan canonical form of the matrix A.

2.6   A Direct Approach for Constructing a Basis of Solutions of $u' = Au$

Let $A \in \mathcal{M}_N(\mathbb{C})$ be such that $\sigma(A) = \{\lambda_1, \ldots, \lambda_q\}$, with $\lambda_i \neq \lambda_j$ if $i \neq j$. Then, according to Theorem 2.5,

$$\mathbb{C}^N = N[(A - \lambda_1 I)^{\nu(\lambda_1)}] \oplus \cdots \oplus N[(A - \lambda_q I)^{\nu(\lambda_q)}], \qquad (2.24)$$

and, for every $u_0 \in \mathbb{C}^N$, there exist a unique

$$u_{0,j} \in N[(A - \lambda_j I)^{\nu(\lambda_j)}], \qquad j \in \{1, \ldots, q\},$$

such that

$$u_0 = \sum_{j=1}^{q} u_{0,j}.$$

Since

$$(A - \lambda_j I)^{\nu(\lambda_j)} u_{0,j} = 0 \quad \text{ for each } j \in \{1, \ldots, q\},$$

we find that

$$e^{zA} u_0 = \sum_{j=1}^{q} e^{zA} u_{0,j} = \sum_{j=1}^{q} e^{z(A - \lambda_j I) + z\lambda_j I} u_{0,j} = \sum_{j=1}^{q} e^{z\lambda_j} e^{z(A - \lambda_j I)} u_{0,j}$$

$$= \sum_{j=1}^{q} e^{z\lambda_j} \left[I + z(A - \lambda_j I) + \cdots + \frac{z^{\nu(\lambda_j)-1}}{(\nu(\lambda_j)-1)!} (A - \lambda_j I)^{\nu(\lambda_j)-1} \right] u_{0,j},$$

which provides us with another method to construct a fundamental matrix of solutions of $u' = Au$. Indeed, suppose

$$\mathcal{B}_j := \{u_{\lambda_j,i} : i \in \{1, \ldots, m_a(\lambda_j)\}\}$$

is a basis of $N[(A - \lambda_j I)^{\nu(\lambda_j)}]$ for every $j \in \{1, \ldots, q\}$. This basis does not have to be of Jordan type but can be arbitrary.

Then, thanks to (2.24),

$$\mathcal{B} = \bigcup_{j=1}^{q} \mathcal{B}_j$$

provides us with a basis of $\mathbb{C}^N$. Thus, thanks to Theorem 1.11, it becomes apparent that

$$\Sigma_0 = \operatorname{span}\left[e^{tA}u_{\lambda_j,i} : 1 \le i \le m_a(\lambda_j),\ 1 \le j \le q\right],$$

where Σ_0 denotes the set of solutions of $u' = Au$. According to the previous calculation, for every $j \in \{1,\dots,q\}$ and $i \in \{1,\dots,m_a(\lambda_j)\}$,

$$e^{tA}u_{\lambda_j,i} = e^{t\lambda_j}\left[I + t(A - \lambda_j I)\right.$$
$$\left. + \cdots + \frac{t^{\nu(\lambda_j)-1}}{(\nu(\lambda_j)-1)!}(A - \lambda_j I)^{\nu(\lambda_j)-1}\right]u_{\lambda_j,i}.$$

Therefore, the solutions of this basis of Σ_0 can be determined in a finite number of steps. Moreover, by Proposition 1.15, setting

$$\Phi(t) := \left(e^{tA}u_{\lambda_1,1} \cdots e^{tA}u_{\lambda_1,m_a(\lambda_1)} \cdots e^{tA}u_{\lambda_q,1} \cdots e^{tA}u_{\lambda_q,m_a(\lambda_q)}\right),$$

it becomes apparent that there exists $C \in \operatorname{Iso}(\mathbb{C}^N)$ such that

$$\Phi(t) = e^{tA}C \quad \text{for all } t \in \mathbb{R}$$

because e^{tA} is also a fundamental matrix of solutions of $u' = Au$. Actually, since $e^0 = I$, necessarily,

$$C = \Phi(0) = \left(u_{\lambda_1,1} \cdots u_{\lambda_1,m_a(\lambda_1)} \cdots u_{\lambda_q,1} \cdots u_{\lambda_q,m_a(\lambda_q)}\right).$$

Equivalently,

$$e^{tA} = \Phi(t)C^{-1}.$$

As a byproduct of this analysis, it becomes apparent that, in order to construct e^{tA}, resorting to Jordan basis is far from necessary. Nevertheless, the reader should be aware that the underlying calculations are facilitated by choosing a Jordan basis, of course.

2.7   Periodic Systems. Floquet Theory

In this section, we focus on the T-periodic homogeneous system

$$u' = A(t)u, \tag{2.25}$$

where $A \in \mathcal{M}_N(\mathcal{C}(\mathbb{R}; \mathbb{C}))$ is a T-periodic matrix with a minimal period, $T > 0$, i.e.

$$A(t + T) = A(t) \quad \text{for all } t \in \mathbb{R}. \tag{2.26}$$

According to Theorem 1.11,

$$\mathbb{C}^N = \text{span}[e_1, \ldots, e_N]$$

implies that, for every $t_0 \in \mathbb{R}$,

$$\Sigma_0 = \text{span}[h(t; t_0, e_1), \ldots, h(t; t_0, e_N)],$$

where, for all $j \in \{1, \ldots, N\}$, $h(t; t_0, e_j)$ is the unique solution of (2.25) such that $u(t_0) = e_j$. Thus,

$$\Phi(t) := (h(t; t_0, e_1) \cdots h(t; t_0, e_N))$$

is a fundamental matrix of solutions of (2.25). In general, $\Phi(t)$ is not T-periodic in time. Actually, according to the following result, modulus some T-periodic invertible matrix, the fundamental matrix $\Phi(t)$ behaves much like e^{tB} for some constant matrix $B \in \mathcal{M}_N(\mathbb{C})$. Subsequently, we denote by $\Sigma_{0,B}$ and $\Sigma_{0,A(t)}$ the set of solutions of the linear homogeneous systems

$$v' = Bv \qquad \text{and} \qquad u' = A(t)u,$$

respectively.

Theorem 2.11 (Floquet factorization). *For every fundamental matrix of solutions, $\Phi(t)$, of (2.25), there exist $B \in \mathcal{M}_N(\mathbb{C})$ and a T-periodic matrix $S \in \mathcal{M}_N(\mathcal{C}^1(\mathbb{R}; \mathbb{C}))$,*

$$S(t + T) = S(t), \qquad t \in \mathbb{R},$$

such that

$$\Phi(t) = S(t)e^{tB} \quad \text{for all } t \in \mathbb{R}. \tag{2.27}$$

In addition, the map

$$S(t) : \Sigma_{0,B} \to \Sigma_{0,A(t)}$$

$$v(t) \mapsto S(t)v(t),$$

is a linear isomorphism. Relation (2.27) *is usually referred to as a* Floquet factorization *of* Φ.

Proof. Consider the auxiliary matrix

$$\Psi(t) := \Phi(t+T), \qquad t \in \mathbb{R}.$$

Since $\Phi(t)$ is a matrix of solutions of (2.25), by (2.26), we find that

$$\Psi'(t) = \Phi'(t+T) = A(t+T)\Phi(t+T) = A(t)\Psi(t).$$

Moreover, since $\Phi(t)$ is a fundamental matrix of solutions, Theorem 1.13 ensures that $\Phi(t)$, and hence $\Psi(t)$, is invertible for all $t \in \mathbb{R}$. Thus, $\Psi(t)$ is also a fundamental matrix of solutions of (2.25). Owing to Proposition 1.15, there exists an invertible matrix $C \in \mathcal{M}_N(\mathbb{C})$, i.e. having constant entries, such that

$$\Phi(t+T) = \Psi(t) = \Phi(t)C \qquad \text{for all } t \in \mathbb{R}. \tag{2.28}$$

On the other hand, according to Exercise 6 of Chapter 2, there exists a constant matrix, $B \in \mathcal{M}_N(\mathbb{C})$, such that

$$C = e^{TB}. \tag{2.29}$$

To complete the proof of the first assertion, it suffices to show that

$$S(t) := \Phi(t)e^{-tB}, \qquad t \in \mathbb{R},$$

satisfies all the requirements of the statement. Indeed, by definition, the *Floquet factorization* (2.27) holds true. Moreover, by (2.28) and (2.29),

$$S(t+T) = \Phi(t+T)e^{-(t+T)B} = \Phi(t)e^{TB}e^{-TB-tB} = \Phi(t)e^{-tB} = S(t)$$

for all $t \in \mathbb{R}$. Thus, $S(t)$ is T-periodic. As $S(t)$ is of class $\mathcal{C}^1$, because it is a product of two matrices of class $\mathcal{C}^1$, the proof of the first assertion is complete.

The fact that $S(t)$ carries the solutions of $\Sigma_{0,B}$ onto the solutions of $\Sigma_{0,A(t)}$ is an immediate consequence of the fact that any solution, $v \in \Sigma_{0,B}$, must satisfy $v(t) = e^{tB}v(0)$, and hence,

$$u(t) := S(t)v(t) = S(t)e^{tB}v(0) = \Phi(t)v(0), \qquad t \in \mathbb{R},$$

lies in $\Sigma_{0,A(t)}$. Finally, $S(t)$ is linear and invertible, as it is a product of two invertible matrices. The proof is complete. $\qquad\square$

Essentially, Theorem 2.11 establishes that the T-periodic change of variable $u = S(t)v$ transforms the periodic system $u' = A(t)u$ into a system with constant coefficients, $v' = Bv$.

Observe that, even for a fixed fundamental matrix of solutions of (2.25), $\Phi(t)$, the Floquet factorization is not unique. Indeed, if (2.27) holds, then, for every integer $n \in \mathbb{Z}$, we also have that

$$\Phi(t) = S(t)e^{-n\frac{2\pi i}{T}t}e^{t(B+n\frac{2\pi i}{T}I)}.$$

Therefore, $S_n(t) := S(t)e^{-n\frac{2\pi i}{T}t}$, which is still T-periodic, and

$$B_n := B + n\frac{2\pi i}{T}I$$

provide us with a sequence of (different) Floquet factorizations of $\Phi(t)$.

Naturally, besides $\Phi(t)$, we can consider any other fundamental matrix of solutions of (2.25), $\Psi(t)$, and any Floquet factorization of it, say,

$$\Psi(t) = S_1(t)e^{tB_1}.$$

In such a case, the matrices e^{TB_1} and e^{TB} must be conjugate. In other words, there exists an invertible matrix, C, such that

$$e^{TB_1} = C^{-1}e^{TB}C. \tag{2.30}$$

Indeed, thanks to Proposition 1.15(b), there exists an invertible matrix, C, such that

$$\Psi(t) = \Phi(t)C.$$

Thus, it follows from (2.27) and (2.28) that

$$\Psi(t+T) = \Phi(t+T)C = S(t+T)e^{(t+T)B}C$$
$$= S(t)e^{tB}e^{TB}C = \Phi(t)e^{TB}C.$$

Similarly,

$$\Psi(t+T) = S_1(t+T)e^{(t+T)B_1} = \Psi(t)e^{TB_1} = \Phi(t)Ce^{TB_1}.$$

Therefore, since $\Phi(t)$ is invertible, it is apparent that

$$Ce^{TB_1} = e^{TB}C,$$

which implies (2.30). This allows us to conclude that, up to conjugacy, the matrix e^{TB} is uniquely determined by the periodic system (2.25).

Next, suppose that $\Phi(t)$ has been chosen so that

$$\Phi(0) = I.$$

Then, since $S(0) = \Phi(0) = I$, we find that

$$\Phi(T) = S(T)e^{TB} = S(0)e^{TB} = e^{TB}.$$

Thus, in such a case, the T-time map of the homogeneous system (2.25), $\Phi(T)$, is given by the exponential matrix e^{TB}. This matrix is often referred to as the *Poincaré map*, or *monodromy operator*, of (2.25).

Based on these features, we introduce the following concepts, which are pivotal in the theory of Dynamical Systems.

Definition 2.12 (Multipliers and Lyapunov exponents). Suppose that $\Phi(0) = I$. Then:

(a) the *characteristic multipliers* of (2.25) are the eigenvalues of the monodromy operator e^{TB};
(b) the *characteristic exponents*, also known as the *Floquet exponents*, are the eigenvalues of the matrix B;
(c) the *Lyapunov exponents* are the real parts of the Floquet exponents.

As, according to Exercise 5 of Chapter 2,

$$\sigma(e^{TB}) = e^{T\sigma(B)},$$

it becomes apparent that the characteristic multipliers of (2.25), m_j, $j \in \{1, \ldots, q\}$, can be constructed from the characteristic exponents of (2.25), ℓ_j, $j \in \{1, \ldots, q\}$, by simply taking

$$m_j := e^{T\ell_j}, \qquad j \in \{1, \ldots, q\}.$$

Since $e^{2\pi ni} = 1$ for all $n \in \mathbb{Z}$, the characteristic exponents are determined only up to integer multiples of $\frac{2\pi i}{T}$. Nevertheless, their real parts, i.e. the Lyapunov exponents, which are uniquely determined by $u' = A(t)u$, can be used to ascertain the orbital stability of periodic solutions through a celebrated theorem by Poincaré (cf. Chicone, 2006, Section 2.4; Hartman, 2002, Theorem 11.1, p. 254), which is a pivotal result in the theory of Dynamical Systems.

2.8   Exercises

1. Determine the general solution of the system $u' = Au$ (both complex and real) for each of the following real matrices A:

$$\begin{pmatrix} -1 & 2 \\ -7 & 8 \end{pmatrix}, \quad \begin{pmatrix} 1 & -1 \\ 5 & -3 \end{pmatrix}, \quad \begin{pmatrix} 0 & 1 & 1 \\ 1 & 0 & 1 \\ 1 & 1 & 0 \end{pmatrix}, \quad \begin{pmatrix} 10 & 4 & 13 \\ 5 & 3 & 7 \\ -9 & -4 & -12 \end{pmatrix}.$$

2. Determine the complex and real canonical forms of the matrix

$$A = \begin{pmatrix} 2 & 2 & -1 & -1 \\ -1 & 3 & -1 & 1 \\ 1 & 1 & 1 & 0 \\ 0 & 1 & -1 & 2 \end{pmatrix},$$

as well as the exponential matrix e^{zA} from the real canonical form of A.

3. Let A be a real square matrix such that $\sigma(A) = \{\alpha \pm i\beta\}$, with $\alpha, \beta \in \mathbb{R}$, $\beta > 0$, such that

$$\dim N[A - (\alpha + i\beta)I] = 1, \quad \dim N[(A - (\alpha + i\beta)I)^2] = 2,$$

$$\dim N[(A - (\alpha + i\beta)I)^3] = \dim N[(A - (\alpha + i\beta)I)^4] = 3.$$

Determine its order, its real canonical form, and e^{zA}.

4. Let A be a real square matrix such that

$$\sigma(A) = \{\lambda, \mu, \alpha \pm i\beta\},$$

for some $\lambda, \mu, \alpha, \beta \in \mathbb{R}$, with $\beta > 0$, for which

$$\dim N[A - \lambda I] = 2, \quad \dim N[(A - \lambda I)^2] = 4,$$

$$\dim N[(A - \lambda I)^3] = \dim N[(A - \lambda I)^4] = 6,$$

$$\dim N[A - \mu I] = 4, \quad \dim N[(A - \mu I)^2] = 8,$$

$$\dim N[(A - \mu I)^3] = \dim N[(A - \mu I)^4] = 10,$$

$$\dim N[A - (\alpha + i\beta)I] = 1,$$

$$\dim N[(A - (\alpha + i\beta)I)^2] = \dim N[(A - (\alpha + i\beta)I)^3] = 2.$$

Determine the order of A, its real canonical form, and e^{zA}.

5. Prove that, for every matrix $A \in \mathcal{M}_N(\mathbb{C})$,

$$\sigma(e^A) = e^{\sigma(A)} := \left\{e^\lambda \colon \lambda \in \sigma(A)\right\}.$$

6. Prove that, for every invertible matrix $A \in \mathcal{M}_N(\mathbb{C})$, there exists a matrix $L \in \mathcal{M}_N(\mathbb{C})$ such that

$$e^L = A.$$

In other words, any invertible matrix, A, admits a logarithm, L.

7. Prove that, for all $A \in \mathcal{M}_N(\mathbb{C})$ and $z \in \mathbb{C}$,

$$\det e^{zA} = e^{z \operatorname{tr} A}.$$

8. For every integer $N \geq 1$, $A \in \mathcal{M}_N(\mathbb{C})$, $\alpha, \beta \in \mathbb{R}$, with $\alpha < \beta$, $B \in \mathcal{C}([\alpha, \beta]; \mathbb{C}^N)$, $t_0 \in [\alpha, \beta]$, and $u_0 \in \mathbb{C}^N$, consider the Cauchy problem

$$\begin{cases} u' = Au + B(t), & t \in [\alpha, \beta], \\ u(t_0) = u_0. \end{cases}$$

Look for a solution of the form $u(t) = e^{(t-t_0)A}v(t)$, with $v(t)$ to be determined, with the aim of showing the validity of Corollary 2.10 directly.

9. Consider $A \in \mathcal{M}_N(\mathbb{C})$, $T > 0$, and assume that $f \in \mathcal{C}(\mathbb{R}; \mathbb{C}^N)$ is T-periodic, i.e.

$$f(t + T) = f(t) \quad \text{for all } t \in \mathbb{R}.$$

(a) Assuming that $u \in \mathcal{C}^1(\mathbb{R}; \mathbb{C}^N)$ solves the linear system

$$u' = Au + f(t), \tag{2.31}$$

prove that $u(t)$ is T-periodic if and only if $u(0) = u(T)$.

(b) Prove that if, in addition,

$$\sigma(A) \cap \left(\frac{2\pi i}{T} \mathbb{Z} \right) = \emptyset,$$

then (2.31) admits a unique T-periodic solution.

2.9   Final Comments

A detailed completely self-contained proof of Theorem 2.5 can be found in the work of López-Gómez and Mora-Corral (2007, Theorem 1.2.1) or López-Gómez (2001a, Theorem 6.1.3). Actually, this chapter has been re-elaborated from some materials of López-Gómez (2001a) and López-Gómez and Mora-Corral (2007), except for the Floquet theory in Section 2.7, which was originally inspired by Pontriaguin's (1975) textbook, though Section 2.7 tidies up considerably the classical treatment by Pontriaguin (1975).

According to Galois theory (see, e.g. Lang, 1994), there is no general way of ascertaining the roots of a polynomial of degree $N \geq 5$.

For instance, a classical example of a non-solvable quintic polynomial, named after E. Artin, is

$$P(z) = z^5 - z - 1, \qquad z \in \mathbb{C},$$

(see Lang, 1994, p. 121). Thus, in practice, the abstract theory developed in this chapter might not explicitly give the solutions of $u' = Au$ if $N \geq 5$. So, once again, we should resort to Hilbert's approach, which has been discussed in Section 1.15, to avoid any kind of discomfort caused by this. Nevertheless, since the roots of a polynomial of any order can be numerically approximated with an arbitrary good precision, such an "inconvenience" is not important from a practical point of view.

The result in Exercise 5 of Chapter 2 is a special and very important case of the *spectral mapping theorem*, according to which

$$\sigma(f(A)) = f(\sigma(A))$$

for all $A \in \mathcal{M}_N(\mathbb{C})$ and any holomorphic function $f : \mathbb{C} \to \mathbb{C}$ (see, e.g. López-Gómez and Mora-Corral, 2007, Theorem 2.4.2). The exercise can be solved by means of the Jordan canonical form, but, in fact, such a proof can be easily generalized to get the general result.

Complementing Exercise 6 of Chapter 2, when the matrix $A \in \mathcal{M}_N(\mathbb{C})$ satisfies

$$\|A - I\|_{\mathcal{L}(\mathbb{C}^N)} < 1,$$

then A is invertible, and one can define a logarithm of the matrix A through the series

$$\operatorname{Log} A := \sum_{n=1}^{\infty} \frac{(-1)^{n+1}}{n}(A - I)^n.$$

The reader should try to show that, in such a case,

$$e^{\operatorname{Log} A} = A.$$

Chapter 3

First-Order Linear Systems with Holomorphic Coefficients

The main goal of this chapter is to analyze the first-order linear system

$$u' = A(z)u + B(z), \quad z \in \Omega, \tag{3.1}$$

where Ω is a convex open subset of $\mathbb{C}$ and

$$A \in \mathcal{M}_N(\mathcal{H}(\Omega)), \quad B \in \mathcal{H}(\Omega; \mathbb{C}^N). \tag{3.2}$$

Here and throughout this chapter, $\mathcal{H}(\Omega)$ denotes the set of holomorphic functions from Ω to $\mathbb{C}$, and $\mathcal{H}(\Omega; \mathbb{C}^N)$ indicates the set of vectorial holomorphic functions from Ω to $\mathbb{C}^N$.

Since we are dealing with complex derivatives, we must work in open subsets of $\mathbb{C}$. This will entail some substantial technical changes in some of the proofs.

Although the convexity hypothesis of Ω could be substantially relaxed, as discussed in Section 3.10, we assume it in order to obtain explicit formulas for the solutions of (3.1).

The main result of this chapter establishes that, under condition (3.2), for every $z_0 \in \Omega$ and $u_0 \in \mathbb{C}^N$, the Cauchy problem

$$\begin{cases} u' = A(z)u + B(z), \\ u(z_0) = u_0, \end{cases}$$

possesses a unique solution, $u(z; z_0, u_0)$, in $\mathcal{H}(\Omega; \mathbb{C}^N)$.

147

3.1   Some Fundamental Results on Holomorphic Functions

The following result is a direct consequence of Theorems 3.7.3 and 3.7.4 given by López-Gómez (2001a).

Theorem 3.1 (of analyticity). *Consider an open subset Ω of $\mathbb{C}$ and a function $f : \Omega \to \mathbb{C}$. Then, f is holomorphic in Ω, i.e. $f \in \mathcal{H}(\Omega)$, if and only if f is analytic in Ω, in the sense that, for every $z_0 \in \Omega$, there exist $\varepsilon = \varepsilon(z_0) > 0$ with*

$$D_\varepsilon(z_0) := \{z \in \mathbb{C} : |z - z_0| < \varepsilon\} \subset \Omega$$

and a sequence of complex numbers, $\{a_n\}_{n \geq 0}$, such that

$$f(z) = \sum_{n=0}^{\infty} a_n(z - z_0)^n \quad \text{for all } z \in D_\varepsilon(z_0).$$

In such a case, f possesses derivatives of all orders, $f^{n)}$, $n \geq 0$, which are holomorphic in Ω, and

$$f(z) = \sum_{n=0}^{\infty} \frac{f^{n)}(z_0)}{n!}(z - z_0)^n, \tag{3.3}$$

uniformly on $\bar{D}_R(z_0)$, as soon as $\bar{D}_R(z_0) \subset \Omega$. Furthermore, the series for $f^{m)}$, $m \geq 0$, can be obtained by differentiating m times, term by term, the series (3.3):

$$f^{m)}(z) = \sum_{n=0}^{\infty} \frac{f^{n+m)}(z_0)}{n!}(z - z_0)^n, \quad |z - z_0| < R.$$

Naturally, since $f : \Omega \to \mathbb{C}^N$ is holomorphic if and only if all its components, f_j, $1 \leq j \leq N$, are holomorphic, $f \in \mathcal{H}(\Omega; \mathbb{C}^N)$ if and only if each of the components, f_j, $1 \leq j \leq N$, is an analytic function in Ω. In such a case, for every $j \in \{1, \ldots, N\}$, it is apparent that

$$f_j(z) = \sum_{n=0}^{\infty} \frac{f_j^{n)}(z_0)}{n!}(z - z_0)^n$$

uniformly on $\bar{D}_R(z_0)$, provided $\bar{D}_R(z_0) \subset \Omega$.

For any given open subset $\Omega \subset \mathbb{C}$, any numbers $a, b \in \mathbb{R}$, with $a < b$, and any curve $\gamma : [a, b] \to \Omega$ piecewise of class $\mathcal{C}^1$ in $[a, b]$, the line integral

$$\int_\gamma : \quad \mathcal{C}(\Omega; \mathbb{C}) \to \mathbb{C}$$

is defined as

$$\int_\gamma f = \int_\gamma f(\zeta)\, d\zeta := \int_a^b f(\gamma(t))\, \gamma'(t)\, dt \quad \text{for all } f \in \mathcal{C}(\Omega; \mathbb{C}),$$

where, setting

$$u(t) + iv(t) := f(\gamma(t))\, \gamma'(t),$$

by definition,

$$\int_a^b (u(t) + iv(t))\, dt := \int_a^b u(t)\, dt + i \int_a^b v(t)\, dt.$$

When $f = (f_1, \ldots, f_N)$, the line integral is defined component-wise, i.e.

$$\int_\gamma f := \left(\int_\gamma f_1, \ldots, \int_\gamma f_N \right).$$

The following well-known result characterizes the analyticity of a function on any convex open subset $\Omega \subset \mathbb{C}$. Naturally, a curve $\gamma : [a, b] \to \mathbb{C}$ is said to be closed if $\gamma(a) = \gamma(b)$.

Theorem 3.2 (Morera). *Let Ω be a convex open subset of $\mathbb{C}$. Then, $f \in \mathcal{H}(\Omega; \mathbb{C}^N)$ if and only if*

$$\int_\gamma f(\zeta)\, d\zeta = 0$$

for all closed curves $\gamma : [a, b] \to \Omega$ piecewise of class $\mathcal{C}^1$.

From Theorem 3.2, we can obtain the following version of the Cauchy–Goursat theorem, where we denote by $[z_0, z]$ the segment from z_0 to z, which can be parameterized through

$$\gamma(t) = z_0 + t\,(z - z_0)\,, \quad 0 \le t \le 1. \tag{3.4}$$

Theorem 3.3 (Cauchy–Goursat). *Let Ω be a convex open subset of $\mathbb{C}$ and $f \in \mathcal{H}(\Omega; \mathbb{C}^N)$. Then, for every $z_0 \in \Omega$, the function*

$$F(z) := \int_{[z_0, z]} f(\zeta)\, d\zeta, \quad z \in \Omega,$$

is holomorphic and satisfies

$$F'(z) = f(z) \quad \text{for all } z \in \Omega.$$

In other words, $F(z)$ is a holomorphic primitive of $f(z)$ in the convex open set Ω.

Proof. Note that $[z_0, z] \subset \Omega$ for all $z_0, z \in \Omega$ because Ω is convex. To prove the theorem, it suffices to show that

$$\lim_{h \to 0} \frac{F(z + h) - F(z)}{h} = f(z) \quad \text{for all } z \in \Omega.$$

To show this, we use Theorem 3.2, obtaining

$$\int_{\partial T} f = 0,$$

where $T = [z_0, z + h, z]$ denotes the triangle with vertices z_0, $z + h$, and z (these three points might of course be aligned). Thus,

$$\int_{[z_0, z+h]} f + \int_{[z+h, z]} f + \int_{[z, z_0]} f = 0,$$

and hence,

$$F(z + h) - F(z) = \int_{[z_0, z+h]} f - \int_{[z_0, z]} f = -\int_{[z+h, z]} f = \int_{[z, z+h]} f.$$

Therefore, using the parameterization (3.4),

$$\lim_{h \to 0} \frac{F(z + h) - F(z)}{h} = \lim_{h \to 0} \left(\frac{1}{h} \int_{[z, z+h]} f \right)$$

$$= \lim_{h \to 0} \int_0^1 f(z + th)\, dt = f(z),$$

which completes the proof. $\qquad\square$

In the proof of Theorem 3.3 we have only used Theorem 3.2 for piecewise C^1 curves, which form the boundaries of the triangles contained in Ω. Thus, to prove Theorem 3.3, it suffices to show that, for every $f \in \mathcal{H}(\Omega; \mathbb{C}^N)$, $\int_{\partial T} f(\zeta)\, d\zeta = 0$ for all triangles $T = [a, b, c]$ with vertices a, b, and c. This is done, e.g. in the solution of Exercise 3.13 in López-Gómez (2001b).

Observe, in addition, that one of the implications of Theorem 3.2 follows easily from Theorem 3.3. Indeed, for every $f \in \mathcal{H}(\Omega; \mathbb{C}^N)$ and any closed curve of class C^1, $\gamma : [a, b] \to \Omega$, from $F' = f$, we obtain that

$$
\int_\gamma f = \int_\gamma F' = \int_a^b F'(\gamma(t))\gamma'(t)\, dt
$$

$$
= \int_a^b \frac{d}{dt} F(\gamma(t))\, dt = F(\gamma(b)) - F(\gamma(a)) = 0.
$$

This argument is also valid for piecewise C^1 curves by working separately in each of the intervals where the curve is C^1.

To conclude this section, we recall the following basic facts related to holomorphic functions: the composition of holomorphic functions is holomorphic; the classical chain rule holds; and the exponential function

$$
e^z := \sum_{n=0}^{\infty} \frac{z^n}{n!}, \quad z \in \mathbb{C},
$$

is entire, i.e. holomorphic in $\mathbb{C}$. The reader is referred, e.g. to López-Gómez (2001a) for any further required details.

3.2   Scalar Linear Equations

The main result of this section can be stated as follows.

Theorem 3.4. *Suppose Ω is a convex open subset of $\mathbb{C}$ and $a, b \in \mathcal{H}(\Omega)$. Then, for every $z_0 \in \Omega$ and $u_0 \in \mathbb{C}$, the Cauchy problem*

$$
\begin{cases} u' = a(z)u + b(z), \\ u(z_0) = u_0, \end{cases} \tag{3.5}
$$

possesses a unique solution $u \in \mathcal{H}(\Omega)$, denoted by $u(z; z_0, u_0)$. Moreover, for all $z \in \Omega$,

$$u(z; z_0, u_0) = e^{\int_{[z_0,z]} a(\mathfrak{z})\, d\mathfrak{z}} u_0 + \int_{[z_0,z]} e^{\int_{[\varsigma,z]} a(\mathfrak{z})\, d\mathfrak{z}} b(\varsigma)\, d\varsigma. \tag{3.6}$$

Proof. According to Theorem 3.3, we have that, for every $z \in \Omega$,

$$\frac{d}{dz} \int_{[z_0,z]} a(\varsigma)\, d\varsigma = a(z).$$

Consequently, the function

$$h(z) := e^{\int_{[z_0,z]} a(\varsigma)\, d\varsigma}$$

satisfies $h' = a(z)h$ in Ω and $h(z_0) = 1$. Thus, the change of variable

$$u(z) = h(z)v(z)$$

transforms (3.5) into the problem

$$v'(z) = e^{-\int_{[z_0,z]} a} b(z), \quad v(z_0) = u_0.$$

Hence, for every $z \in \Omega$,

$$v(z) - v(z_0) = \int_{[z_0,z]} v' = \int_{[z_0,z]} e^{-\int_{[z_0,\varsigma]} a(\mathfrak{z})\, d\mathfrak{z}} b(\varsigma)\, d\varsigma.$$

Taking into account the initial condition gives

$$v(z) = u_0 + \int_{[z_0,z]} e^{-\int_{[z_0,\varsigma]} a(\mathfrak{z})\, d\mathfrak{z}} b(\varsigma)\, d\varsigma,$$

and, therefore, the (unique) solution of (3.5) is given by

$$u(z) = e^{\int_{[z_0,z]} a(\mathfrak{z})\, d\mathfrak{z}} \left(u_0 + \int_{[z_0,z]} e^{-\int_{[z_0,\varsigma]} a(\mathfrak{z})\, d\mathfrak{z}} b(\varsigma)\, d\varsigma \right)$$

$$= e^{\int_{[z_0,z]} a(\mathfrak{z})\, d\mathfrak{z}} u_0 + \int_{[z_0,z]} e^{\int_{[z_0,z]} a(\mathfrak{z})\, d\mathfrak{z}} e^{-\int_{[z_0,\varsigma]} a(\mathfrak{z})\, d\mathfrak{z}} b(\varsigma)\, d\varsigma$$

Finally, due to Theorem 3.2, we have that

$$\int_{[z_0,z]} a - \int_{[z_0,\varsigma]} a = \int_{[z_0,z]} a + \int_{[\varsigma,z_0]} a = -\int_{[z,\varsigma]} a = \int_{[\varsigma,z]} a,$$

and it becomes apparent that (3.6) holds. The fact that $u \in \mathcal{H}(\Omega)$ follows from expression (3.6) and Theorem 3.3. $\qquad\square$

Naturally, as in Chapter 1, owing to Theorem 3.4, the set of solutions of the linear equation

$$u' = a(z)u + b(z)$$

in $\mathcal{H}(\Omega)$ is an affine one-dimensional subspace of $\mathcal{H}(\Omega)$ in the direction of $e^{\int_{[z_0,z]} a(\zeta)\, d\zeta}$ passing through the point

$$u(z; z_0, 0) = \int_{[z_0,z]} e^{\int_{[\varsigma,z]} a(3)\, d3} b(\zeta)\, d\zeta.$$

The Taylor coefficients of the solution of (3.5) at z_0 can be determined by differentiating recursively the differential equation. Indeed,

$$u'(z_0) = a(z_0)u_0 + b(z_0), \quad u''(z_0) = a'(z_0)u_0 + a(z_0)u'(z_0) + b'(z_0), \quad \text{etc.}$$

According to Theorem 3.1, the Taylor series of $u(z; z_0, u_0)$ centered at z_0,

$$\sum_{n=0}^{\infty} \frac{u^{n)}(z_0)}{n!}(z - z_0)^n,$$

converges uniformly to $u(z; z_0, u_0)$ in $\bar{D}_R(z_0)$, provided $\bar{D}_R(z_0) \subset \Omega$.

We now consider some examples related to Theorem 3.4. First, we take $a(z) = \frac{1}{z}$, $z \in \mathbb{C} \setminus \{0\}$. Then, the homogeneous equation $u' = a(z)u$ satisfies the assumptions of Theorem 3.4 in

$$\Omega := \{z \in \mathbb{C} : \operatorname{Re} z > 0\}.$$

Thus, for every $z_0 \in \mathbb{C}$, with $\operatorname{Re} z_0 > 0$ and $u_0 \in \mathbb{C}$, the Cauchy problem

$$\begin{cases} u' = a(z)u, \\ u(z_0) = u_0, \end{cases} \tag{3.7}$$

has a unique holomorphic solution in Ω. Necessarily,

$$u(z; z_0, u_0) = \frac{u_0}{z_0} z, \quad z \in \Omega.$$

Actually, this function admits an entire extension. However, if we choose $a(z) = \frac{1}{z^2}$, $z \in \mathbb{C} \setminus \{0\}$, then, by Theorem 3.4, for every $u_0 \in \mathbb{C}$

and every convex open subset $\Omega \subset \mathbb{C}$ such that $0 \notin \Omega$ and $z_0 \in \Omega$, problem (3.7) has a unique solution in $\mathcal{H}(\Omega)$, which is given by

$$u(z; z_0, u_0) = e^{\frac{1}{z_0}} u_0 e^{-\frac{1}{z}}.$$

As 0 is an isolated essential singularity of $u(z; z_0, u_0)$, this solution does not admit any entire extension. Thus, in general, but not always, the solution of (3.7) might develop singularities at the points where $a(z)$ is singular.

3.3   First-Order Linear Systems

This section studies the first-order linear system

$$u' = A(z)u + B(z),$$

where, given a convex open subset Ω of $\mathbb{C}$,

$$A(z) := (a_{ij}(z))_{1 \leq i,j \leq N} = \begin{pmatrix} a_{11}(z) & \cdots & a_{1N}(z) \\ \vdots & \ddots & \vdots \\ a_{N1}(z) & \cdots & a_{NN}(z) \end{pmatrix}, \quad z \in \Omega,$$

is a matrix or order N, with $a_{ij} \in \mathcal{H}(\Omega)$, $i, j \in \{1, \ldots, N\}$, and

$$B(z) := (b_i(z))_{1 \leq i \leq N} = \begin{pmatrix} b_1(z) \\ \vdots \\ b_N(z) \end{pmatrix}, \quad z \in \Omega,$$

with $b_i \in \mathcal{H}(\Omega)$ for all $i \in \{1, \ldots, N\}$. As usual, in this book, u indicates

$$u := \begin{pmatrix} u_1 \\ \vdots \\ u_N \end{pmatrix} = (u_1, \ldots, u_N)^T.$$

The main result of this section is the following N-dimensional counterpart of Theorem 3.4.

Theorem 3.5. *Under the previous general assumptions, for every* $z_0 \in \Omega$ *and* $u_0 \in \mathbb{C}^N$, *the Cauchy problem*

$$\begin{cases} u' = A(z)u + B(z), \\ u(z_0) = u_0, \end{cases} \tag{3.8}$$

possesses a unique solution, $u(z; z_0, u_0)$, *in* $\mathcal{H}(\Omega; \mathbb{C}^N)$.

Proof. The proof of this result is a non-trivial re-elaboration of the proof of Theorem 1.4. Suppose $u \in \mathcal{H}(\Omega; \mathbb{C}^N)$ solves (3.8). Then, for every $z \in \Omega$,

$$\int_{[z_0,z]} u'(\zeta)\, d\zeta = \int_{[z_0,z]} [A(\zeta)u(\zeta) + B(\zeta)]\, d\zeta,$$

where the integrals are taken component-wise. Setting

$$\gamma(t) = z_0 + t(z - z_0), \quad t \in [0,1],$$

it becomes apparent that

$$\int_{[z_0,z]} u' = \int_0^1 u'(\gamma(t))\gamma'(t)\, dt = \int_0^1 \frac{du}{dt}(\gamma(t))\, dt = u(z) - u(z_0).$$

Thus,

$$u(z) = u_0 + \int_{[z_0,z]} [A(\zeta)u(\zeta) + B(\zeta)]\, d\zeta \tag{3.9}$$

for all $z \in \Omega$. Consequently, u must be a fixed point of the integral operator

$$\mathcal{K} : \mathcal{H}(\Omega; \mathbb{C}^N) \to \mathcal{H}(\Omega; \mathbb{C}^N),$$

defined, for every $h \in \mathcal{H}(\Omega; \mathbb{C}^N)$, as

$$\mathcal{K}h(z) := u_0 + \int_{[z_0,z]} [A(\zeta)h(\zeta) + B(\zeta)]\, d\zeta, \quad z \in \Omega.$$

The converse is also true. Indeed, if $u \in \mathcal{H}(\Omega; \mathbb{C}^N)$ is a fixed point of $\mathcal{K}$, then (3.9) holds and, hence, $u(z_0) = u_0$. Moreover, differentiating (3.9) with respect to z, it follows from Theorem 3.3 that

$$u'(z) = A(z)u(z) + B(z)$$

for all $z \in \Omega$. Therefore, $u \in \mathcal{H}(\Omega; \mathbb{C}^N)$ solves (3.8) if and only if $\mathcal{K}u = u$.

Since Ω can be unbounded, the space $\mathcal{H}(\Omega; \mathbb{C}^N)$ does not admit a Banach space structure. Thus, in contrast to the proof of Theorem 1.4, we cannot apply Theorem 1.8 directly to the integral operator $\mathcal{K}$. Instead, the proof of the existence of a unique solution in $\mathcal{H}(\Omega; \mathbb{C}^N)$ is divided into two steps.

Step 1. We begin by establishing Theorem 3.5 on any bounded open and convex subset, D, of Ω, with $\bar{D} \subset \Omega$ and $z_0 \in D$. Then, in Step 2, we prove the result in the general case. Let D be a bounded open and convex subset of Ω such that $z_0 \in D$ and $\bar{D} \subset \Omega$, and consider the Banach space

$$X := \left(\mathcal{C}(\bar{D}) \cap \mathcal{H}(D; \mathbb{C}^N), \|\cdot\|_\infty\right),$$

where $\|\cdot\|_\infty$ denotes the maximum norm

$$\|u\|_\infty := \max_{z \in \bar{D}} \|u(z)\|, \quad u \in \mathcal{C}(\bar{D}) \cap \mathcal{H}(D; \mathbb{C}^N).$$

Naturally, $\|\cdot\|_\infty$ defines a norm in X. Moreover, if $\{u_n\}_{n \geq 1}$ is a Cauchy sequence in X, then, adapting the proof of Theorem 1.6, it is easily seen that there exists $u \in \mathcal{C}(\bar{D})$ such that

$$\lim_{n \to \infty} \|u_n - u\|_\infty = 0.$$

We claim that $u \in \mathcal{H}(D; \mathbb{C}^N)$. Indeed, let $\gamma : [a, b] \to D$ be any closed curve piecewise of class $\mathcal{C}^1$. Then, by Theorem 3.2,

$$0 = \int_\gamma u_n \quad \text{for all } n \geq 1,$$

and, hence, by the uniform convergence in $\bar{D}$,

$$\int_\gamma u = \lim_{n \to \infty} \int_\gamma u_n = 0.$$

So, owing to Theorem 3.2, $u \in \mathcal{H}(D; \mathbb{C}^N)$. Thus, $u \in X$ and, therefore, X is a Banach space. As in the proof of Theorem 1.4 (see Section 1.3.1), for every $\eta > 0$,

$$\|u\|_\eta := \max_{z \in \bar{D}} \left\{ e^{-\eta|z - z_0|} \|u(z)\| \right\}, \quad u \in X,$$

provides us with a norm which is equivalent to $\|\cdot\|_\infty$ in X.

Let $u, v \in X$ be two arbitrary functions in X. Then, for every $z \in \bar{D}$,

$$\|\mathcal{K}u(z) - \mathcal{K}v(z)\| = \left\| \int_{[z_0, z]} A(\zeta)[u(\zeta) - v(\zeta)] \, d\zeta \right\|$$

$$= \left\| \int_0^1 A(\gamma(t))[u(\gamma(t)) - v(\gamma(t))]\gamma'(t) \, dt \right\|,$$

where

$$\gamma(t) = z_0 + t(z - z_0) \quad \text{for all } t \in [0, 1].$$

Consequently, arguing as in the proof of Theorem 1.4 (see Section 1.3.4), there exists a constant $L > 0$ such that

$$\|\mathcal{K}u(z) - \mathcal{K}v(z)\| \le L|z - z_0| \int_0^1 \|u(\gamma(t)) - v(\gamma(t))\| \, dt$$

for all $z \in \bar{D}$. Thus,

$$\|\mathcal{K}u(z) - \mathcal{K}v(z)\|$$

$$\le L|z - z_0| \int_0^1 \|u(\gamma(t)) - v(\gamma(t))\| \, dt$$

$$\le L|z - z_0| \int_0^1 e^{\eta|\gamma(t) - z_0|} e^{-\eta|\gamma(t) - z_0|} \|u(\gamma(t)) - v(\gamma(t))\| \, dt$$

$$\le L|z - z_0| \, \|u - v\|_\eta \int_0^1 e^{\eta|\gamma(t) - z_0|} \, dt$$

$$= L|z - z_0| \, \|u - v\|_\eta \int_0^1 e^{\eta t|z - z_0|} \, dt.$$

By definition,

$$\mathcal{K}u(z_0) - \mathcal{K}v(z_0) = u_0 - u_0 = 0.$$

So, assume that $z \neq z_0$. Then,

$$\int_0^1 e^{\eta t|z - z_0|} \, dt = \frac{e^{\eta|z - z_0|} - 1}{\eta|z - z_0|} < \frac{e^{\eta|z - z_0|}}{\eta|z - z_0|}$$

and, hence, for all $z \in \bar{D}$,

$$\|\mathcal{K}u(z) - \mathcal{K}v(z)\| \leq \frac{L}{\eta} \|u - v\|_\eta \, e^{\eta|z - z_0|}.$$

In other words,

$$e^{-\eta|z - z_0|} \|\mathcal{K}u(z) - \mathcal{K}v(z)\| \leq \frac{L}{\eta} \|u - v\|_\eta, \quad z \in \bar{D}.$$

Therefore, taking the maximum in $z \in \bar{D}$ yields

$$\|\mathcal{K}u - \mathcal{K}v\|_\eta \leq \frac{L}{\eta} \|u - v\|_\eta.$$

By choosing $\eta > L$, it is apparent that $\mathcal{K} : X \to X$ is a contraction; consequently, the fact that $\mathcal{K}$ has a unique fixed point in X is a direct consequence of Theorem 1.8.

Step 2. Subsequently, for sufficiently large $n \geq 1$, say $n \geq n_0$, we consider the open subsets of Ω defined by

$$\Omega_{\frac{1}{n}} := \left\{ z \in \Omega : \text{dist}(z, \partial\Omega) > \tfrac{1}{n} \right\}, \quad n \geq n_0,$$

(n_0 is taken to be sufficiently large to guarantee that $\Omega_{\frac{1}{n_0}} \neq \emptyset$). It is obvious that $\{\Omega_{\frac{1}{n}}\}_{n \geq n_0}$ provides us with a sequence of open subsets of Ω such that $\bar{\Omega}_{\frac{1}{n}} \subset \Omega_{\frac{1}{n+1}}$ for all $n \geq n_0$ and

$$\Omega = \bigcup_{n \geq n_0} \Omega_{\frac{1}{n}}.$$

Moreover, $\Omega_{\frac{1}{n}}$ is convex for all $n \geq n_0$. Indeed, let $z, w \in \Omega_{\frac{1}{n}}$. Then, since, $\text{dist}(z, \partial\Omega) > \frac{1}{n}$ and $\text{dist}(w, \partial\Omega) > \frac{1}{n}$, there exists $\delta > 0$ such that, actually,

$$\text{dist}(z, \partial\Omega) > \tfrac{1}{n} + \delta \quad \text{and} \quad \text{dist}(w, \partial\Omega) > \tfrac{1}{n} + \delta.$$

Thus,

$$\bar{D}_{\frac{1}{n} + \delta}(z) \cup \bar{D}_{\frac{1}{n} + \delta}(w) \subset \Omega.$$

Since Ω is convex, all the segments connecting any two points of these closed disks also lie in Ω. Therefore, for every $\zeta \in [z, w]$, we have that

$\mathrm{dist}(\zeta, \partial\Omega) > \frac{1}{n}$ and, hence, $[z, w] \subset \Omega_{\frac{1}{n}}$, which shows the convexity of $\Omega_{\frac{1}{n}}$.

Now, for every $n \geq n_0$, we consider the open subsets of Ω defined by

$$\mathcal{D}_n := \Omega_{\frac{1}{n}} \cap D_n(z_0), \quad n \geq n_0.$$

For every $n \geq n_0$, $\mathcal{D}_n$ is a convex bounded open subset of Ω such that

$$\mathcal{D}_n \subset \mathcal{D}_{n+1}, \quad \bigcup_{n \geq n_0} \mathcal{D}_n = \Omega.$$

By Step 1, we already know that (3.8) possesses a unique solution, $u_n \in \mathcal{H}(\mathcal{D}_n; \mathbb{C}^N)$, for every $n \geq n_0$. Since $\mathcal{D}_n \subset \mathcal{D}_{n+1}$, by uniqueness, it becomes apparent that $u_{n+1} = u_n$ on $\mathcal{D}_n$. Therefore, the limit

$$u := \lim_{n \to \infty} u_n$$

is well defined, and necessarily, it provides us with the unique solution of (3.8) in Ω. This completes the proof. $\qquad\square$

As in Chapter 1, thanks to Theorem 3.5, when Ω is a convex open subset of $\mathbb{C}$, $A \in \mathcal{M}_N(\mathcal{H}(\Omega))$ and $B \in \mathcal{H}(\Omega; \mathbb{C}^N)$, the set of solutions, Σ_B, of

$$u' = A(z)u + B(z)$$

is a N-dimensional affine subspace of $\mathcal{H}(\Omega; \mathbb{C}^N)$ parallel to Σ_0. Similarly, if

$$\mathbb{C}^N = \mathrm{span}[e_1, \ldots, e_N],$$

then, for every $z_0 \in \mathbb{C}$,

$$\Sigma_0 = \mathrm{span}[h(z; z_0, e_1), \ldots, h(z; z_0, e_N)],$$

where, for every $u_0 \in \mathbb{C}^N$, $h(z; z_0, u_0)$ is the unique solution of the homogeneous problem

$$\begin{cases} h' = A(z)h, \\ h(z_0) = u_0, \end{cases}$$

in $\mathcal{H}(\Omega; \mathbb{C}^N)$. In other words, by extending Definition 1.12 to the case of a complex variable $z \in \Omega$, we have that

$$\Phi(z) := (h(z; z_0, e_1) \cdots h(z; z_0, e_N)), \quad z \in \Omega,$$

is a fundamental matrix of solutions of $u' = A(z)u$ in Ω. Moreover, it is possible to adapt the proof of Theorem 1.13 to the case of a complex variable and show that $\Phi(z)$ is invertible for all $z \in \Omega$.

As usual, to find the solution of (3.8), we can perform the change of variable

$$u(z) = \Phi(z)v(z), \quad z \in \Omega. \tag{3.10}$$

Substituting (3.10) in (3.8) yields

$$u_0 = \Phi(z_0)v(z_0)$$

and

$$\Phi'(z)v(z) + \Phi(z)v'(z) = A(z)\Phi(z)v(z) + B(z), \quad z \in \Omega.$$

Thus, since $\Phi'(z) = A(z)\Phi(z)$, it is apparent that $u(z)$ solves (3.8) if and only if $v(z)$ satisfies

$$\begin{cases} \Phi(z)v'(z) = B(z), \\ \Phi(z_0)v(z_0) = u_0. \end{cases} \tag{3.11}$$

Since $\Phi(z)$ is a fundamental matrix of solutions of $h' = A(z)h$, as remarked above, $\Phi(z)$ is invertible for all $z \in \Omega$ and, hence,

$$v'(z) = \Phi^{-1}(z)B(z) \quad \text{for all } z \in \Omega.$$

Thus,

$$v(z) = v(z_0) + \int_{[z_0, z]} \Phi^{-1}(\zeta)B(\zeta)\, d\zeta$$

$$= \Phi^{-1}(z_0)u_0 + \int_{[z_0, z]} \Phi^{-1}(\zeta)B(\zeta)\, d\zeta$$

is the unique solution of (3.11). Finally, by substituting this expression in (3.10), we infer that

$$u(z; z_0, u_0) = \Phi(z)\Phi^{-1}(z_0)u_0 + \int_{[z_0, z]} \Phi(z)\Phi^{-1}(\zeta)B(\zeta)\, d\zeta \tag{3.12}$$

is the unique solution of (3.8), which is often referred to as the *variation of constants formula* for (3.8). In particular, since

$$u(z; z_0, 0) = \int_{[z_0, z]} \Phi(z)\Phi^{-1}(\zeta)B(\zeta)\, d\zeta,$$

it becomes apparent that

$$\Sigma_B = \Sigma_0 + \int_{[z_0, z]} \Phi(z)\Phi^{-1}(\zeta)B(\zeta)\, d\zeta.$$

Naturally, since $u_0 \in \mathbb{C}^N$ is arbitrary and $\Phi^{-1}(z_0) \in \mathrm{Iso}\,(\mathbb{C}^N)$, the parametric equations of the affine subspace Σ_B can be expressed as

$$\Phi(z)x + \int_{[z_0, z]} \Phi(z)\Phi^{-1}(\zeta)B(\zeta)\, d\zeta, \quad x \in \mathbb{C}^N,$$

or, equivalently, for $x = (x_1, \ldots, x_N) \in \mathbb{C}^N$,

$$x_1 h(z; z_0, e_1) + \cdots + x_N h(z; z_0, e_N) + \int_{[z_0, z]} \Phi(z)\Phi^{-1}(\zeta)B(\zeta)\, d\zeta.$$

In other words,

$$\left\{ h(z; z_0, e_1), \ldots, h(z; z_0, e_N); \int_{[z_0, z]} \Phi(z)\Phi^{-1}(\zeta)B(\zeta)\, d\zeta \right\}$$

is an *affine reference system* of the affine subspace Σ_B of $\mathcal{H}(\Omega; \mathbb{C}^N)$.

In particular, when $A(z) = A$ is a constant matrix and B is entire, i.e. $B \in \mathcal{H}(\mathbb{C}; \mathbb{C}^N)$, by applying Theorem 3.4 with $\Omega = \mathbb{C}$, it follows that Σ_B consists of entire functions.

3.4   Linear Equations of Arbitrary Order

Throughout this section, given a convex open subset Ω of $\mathbb{C}$, a natural number $N \geq 1$, and $N + 1$ holomorphic functions,

$$a_j \in \mathcal{H}(\Omega), \quad 0 \leq j \leq N - 1, \qquad b \in \mathcal{H}(\Omega),$$

we consider the Nth-order holomorphic linear differential equation

$$u^{N)} = \sum_{j=0}^{N-1} a_j(z)u^{j)} + b(z) \tag{3.13}$$

with coefficients $a_j(z)$, $0 \leq j \leq N - 1$, and inhomogeneity $b(z)$. A solution of (3.13) in Ω is a function $u \in \mathcal{H}(\Omega)$ such that

$$u^{N)}(z) = a_0(z)u(z) + a_1(z)u'(z) + \cdots + a_{N-1}(z)u^{N-1)}(z) + b(z)$$

for all $z \in \Omega$. Naturally, as in Chapter 1, the linear equation with $b = 0$,

$$u^{N)} = a_0(z)u + a_1(z)u' + a_2(z)u'' + \cdots + a_{N-1}(z)u^{N-1)},$$

is the *homogeneous equation* associated with (3.13). As in Chapter 1, we denote by Σ_b^e the set of solutions of (3.13) in Ω. It is straightforward to check that Σ_0^e is a linear subspace of $\mathcal{H}(\Omega)$ and that Σ_b^e is an affine subspace of $\mathcal{H}(\Omega)$ parallel to Σ_0^e. Moreover, for every $u \in \Sigma_b^e$, the vectorial function

$$\mathcal{D}(u) := \begin{pmatrix} u \\ u' \\ \vdots \\ u^{N-2)} \\ u^{N-1)} \end{pmatrix} \tag{3.14}$$

is a solution in $\mathcal{H}(\Omega; \mathbb{C}^N)$ of the associated first-order linear system

$$U' = A(z)U + B(z), \tag{3.15}$$

where

$$U = (u_1, u_2, \ldots, u_{N-1}, u_N)^T, \quad B(z) = (0, 0, \ldots, 0, b(z))^T, \tag{3.16}$$

and, for every $z \in \Omega$,

$$A(z) := \begin{pmatrix} 0 & 1 & 0 & \cdots & 0 & 0 \\ 0 & 0 & 1 & 0 & \cdots & 0 \\ \vdots & \ddots & \ddots & \ddots & \ddots & \vdots \\ \vdots & \ddots & \ddots & \ddots & 1 & 0 \\ 0 & \cdots & \cdots & 0 & 0 & 1 \\ a_0(z) & a_1(z) & a_2(z) & \cdots & a_{N-2}(z) & a_{N-1}(z) \end{pmatrix}.$$

$$\tag{3.17}$$

As in Chapter 1, we denote by Σ_B^s the set of solutions of (3.15) in $\mathcal{H}(\Omega; \mathbb{C}^N)$. The following result is the complex-valued counterpart of Theorem 1.17.

Theorem 3.6. *The map* $\mathcal{D} : \Sigma_b^e \to \Sigma_B^s$, *defined by* (3.14), *establishes a bijection between* Σ_b^e *and* Σ_B^s. *Actually,* $\mathcal{D} : \Sigma_0^e \to \Sigma_0^s$ *is a linear isomorphism. Therefore,*

$$N = \dim \Sigma_0^e = \dim \Sigma_b^e,$$

and $\mathcal{D}$ *transforms any basis of* Σ_0^e *into a basis of* Σ_0^s.

Finally, thanks to Theorems 3.5 and 3.6, the following result holds.

Corollary 3.7. *For every* $z_0 \in \Omega$ *and* $U_0 \in \mathbb{C}^N$, *the Cauchy problem*

$$\begin{cases} u^{N)} = a_0(z)u + a_1(z)u' + a_2(z)u'' + \cdots + a_{N-1}(z)u^{N-1)} + b(z), \\ \mathcal{D}(u)(z_0) = \left(u(z_0), u'(z_0), u''(z_0), \ldots, u^{N-1)}(z_0)\right)^T = U_0, \end{cases}$$

possesses a unique solution, $u(z; z_0, U_0)$, *in* $\mathcal{H}(\Omega)$. *Moreover,* $u(z; z_0, U_0)$ *is the first component of the unique solution,* $U(z; z_0, U_0)$, *of the associated problem*

$$\begin{cases} U' = A(z)U + B(z), \\ U(z_0) = U_0, \end{cases}$$

where $B(z)$ *and* $A(z)$ *are given in* (3.16) *and* (3.17), *respectively.*

In particular, when the coefficients $a_j(z)$, $1 \le j \le N - 1$, and $b(z)$ are entire functions, by applying Corollary 3.7 with $\Omega = \mathbb{C}$, it becomes apparent that Σ_b^e consists of entire functions.

3.5   The Airy Equation

The differential equation

$$u'' - zu = 0 \tag{3.18}$$

is named after G. B. Airy. According to Corollary 3.7 applied with $N = 2$, $\Omega = \mathbb{C}$, and $a_0(z) = z$, $a_1(z) = 0$, $b(z) = 0$ for all $z \in \mathbb{C}$, the set of solutions of (3.18) is a vectorial plane of $\mathcal{H}(\mathbb{C})$. Actually, the unique solutions of the two problems

$$\begin{cases} u'' - zu = 0, \\ u(0) = 1, \quad u'(0) = 0, \end{cases} \qquad \begin{cases} u'' - zu = 0, \\ u(0) = 0, \quad u'(0) = 1 \end{cases} \tag{3.19}$$

provide us with a basis of Σ_0^e. As the solutions of (3.18) are entire, owing to Theorem 3.1, their Taylor series at 0,

$$u(z) = \sum_{n=0}^{\infty} u_n z^n, \quad u_n := \frac{u^{n)}(0)}{n!}, \tag{3.20}$$

converge uniformly on compact subsets of $\mathbb{C}$. In particular, these series have an infinite convergence radius and, hence,

$$u'(z) = \sum_{n=1}^{\infty} n u_n z^{n-1}, \quad u''(z) = \sum_{n=2}^{\infty} n(n-1) u_n z^{n-2}, \quad z \in \mathbb{C},$$

uniformly on compact subsets of $\mathbb{C}$. By choosing $(u_0, u_1) = (1, 0)$ and $(u_0, u_1) = (0, 1)$, (3.20) provides us with the solutions of the problems (3.19), respectively. Substituting the above series into (3.18) yields

$$\sum_{n=2}^{\infty} n(n-1) u_n z^{n-2} - \sum_{n=0}^{\infty} u_n z^{n+1} = 0.$$

Equivalently,

$$2u_2 + \sum_{n=3}^{\infty} n(n-1) u_n z^{n-2} - \sum_{n=0}^{\infty} u_n z^{n+1} = 0.$$

Thus, by making the change of index $m = n - 3$ in the first summation, we find that

$$2u_2 + \sum_{m=0}^{\infty} (m+3)(m+2) u_{m+3} z^{m+1} - \sum_{m=0}^{\infty} u_m z^{m+1} = 0.$$

Since these series are uniformly convergent on any compact subset of $\mathbb{C}$, the previous identity can be equivalently expressed as

$$2u_2 + \sum_{n=0}^{\infty} [(n+3)(n+2) u_{n+3} - u_n] z^{n+1} = 0,$$

which entails

$$u_2 = 0, \quad u_{n+3} = \frac{u_n}{(n+3)(n+2)}, \quad n \geq 0.$$

Thus, $u_{3k+2} = 0$ for all $k \geq 0$, while the value of u_0 determines the values of u_{3k} for all $k \geq 1$, and the value of u_1 determines the values of u_{3k+1} for all $k \geq 1$. Indeed, since

$$u_{n+3} = \frac{n+1}{(n+3)(n+2)(n+1)} u_n, \quad n \geq 0,$$

it becomes apparent that, for every $k \geq 1$,

$$u_{3k} = \frac{3(k-1)+1}{(3k)(3k-1)(3k-2)} u_{3(k-1)} = \cdots = \frac{\prod_{j=0}^{k-1}(3j+1)}{(3k)!} u_0.$$

Similarly, for every $k \geq 1$,

$$u_{3k+1} = \frac{3(k-1)+2}{(3k+1)(3k)(3k-1)} u_{3(k-1)+1} = \cdots = \frac{\prod_{j=0}^{k-1}(3j+2)}{(3k+1)!} u_1.$$

Therefore, the functions

$$u_{(1,0)}(z) := \sum_{k=0}^{\infty} \frac{\prod_{j=0}^{k-1}(3j+1)}{(3k)!} z^{3k},$$

$$\tag{3.21}$$

$$u_{(0,1)}(z) := \sum_{k=0}^{\infty} \frac{\prod_{j=0}^{k-1}(3j+2)}{(3k+1)!} z^{3k+1},$$

where we assume, by convention, that the result of the products is 1 if the set of indexes is empty, provide us with the solutions of the problems in (3.19), respectively. Although we already know that the convergence radius of these power series is infinite, considering that not all readers might be completely familiar with the theory of holomorphic functions, we check this directly by using the concept of convergence radius and the root test. Indeed, for every $z \in \mathbb{C}$, we have that

$$\lim_{k \to \infty} \sqrt[k]{\left| \frac{\prod_{j=0}^{k-1}(3j+1)}{(3k)!} z^{3k} \right|} = \lim_{k \to \infty} \sqrt[k]{\frac{\prod_{j=0}^{k-1}(3j+1)}{(3k)!}} |z|^3 = 0$$

because

$$\lim_{k\to\infty} \sqrt[k]{\frac{\prod_{j=0}^{k-1}(3j+1)}{(3k)!}} = \lim_{k\to\infty} \frac{\frac{\prod_{j=0}^{k}(3j+1)}{(3(k+1))!}}{\frac{\prod_{j=0}^{k-1}(3j+1)}{(3k)!}}$$

$$= \lim_{k\to\infty} \frac{1}{(3k+3)(3k+2)} = 0.$$

Here, we are using the fact that, for any sequence of positive real numbers, $\{c_n\}_{n\geq 1}$, such that

$$\lim_{n\to\infty} \frac{c_{n+1}}{c_n} = L \in [0, +\infty],$$

also

$$\lim_{n\to\infty} \sqrt[n]{c_n} = L$$

(see Lemma 2.7.4 in López-Gómez, 2001a, if necessary), and the series $\sum_{n=0}^{\infty} c_n z^n$ converges for all $z \in \mathbb{C}$ with $|z| < \frac{1}{L}$. Similarly,

$$\lim_{k\to\infty} \sqrt[k]{\left|\frac{\prod_{j=0}^{k-1}(3j+2)}{(3k+1)!} z^{3k+1}\right|} = \lim_{k\to\infty} \sqrt[k]{\frac{\prod_{j=0}^{k-1}(3j+2)}{(3k+1)!}} |z|^3 = 0$$

because

$$\lim_{k\to\infty} \sqrt[k]{\frac{\prod_{j=0}^{k-1}(3j+2)}{(3k+1)!}} = \lim_{k\to\infty} \frac{\frac{\prod_{j=0}^{k}(3j+2)}{(3(k+1)+1)!}}{\frac{\prod_{j=0}^{k-1}(3j+2)}{(3k+1)!}}$$

$$= \lim_{k\to\infty} \frac{1}{(3k+4)(3k+3)} = 0.$$

Therefore, the functions defined in (3.21) are indeed entire functions.

The Airy function, $\mathrm{Ai}(z)$, is defined as the unique solution of (3.18) such that

$$\mathrm{Ai}(0) = \frac{1}{3^{\frac{2}{3}}\Gamma(\frac{2}{3})} \quad \text{and} \quad \mathrm{Ai}'(0) = \frac{-1}{3^{\frac{1}{3}}\Gamma(\frac{1}{3})},$$

where

$$\Gamma(z) := \int_0^\infty t^{z-1} e^{-t}\, dt, \quad \operatorname{Re} z > 0,$$

is the Euler gamma function. Thanks to our construction, we conclude that

$$\operatorname{Ai}(z) = \frac{1}{3^{\frac{2}{3}}\Gamma(\frac{2}{3})} \sum_{k=0}^\infty \frac{\prod_{j=0}^{k-1}(3j+1)}{(3k)!} z^{3k}$$
$$- \frac{1}{3^{\frac{1}{3}}\Gamma(\frac{1}{3})} \sum_{k=0}^\infty \frac{\prod_{j=0}^{k-1}(3j+2)}{(3k+1)!} z^{3k+1}.$$

3.6 The Hermite Equation

In this section, we characterize the set of values $\lambda \in \mathbb{R}$ for which the differential equation

$$-\Psi''(x) + x^2 \Psi(x) = \lambda \Psi(x), \quad x \in \mathbb{R}, \tag{3.22}$$

possesses some solution, $\Psi(x)$, such that

$$\int_{-\infty}^\infty \Psi^2(x)\, dx = 1. \tag{3.23}$$

In quantum mechanics, the solutions of (3.22) provide us with the wave functions of the Schrödinger equation for the one-dimensional linear harmonic oscillator (see Chpolski, 1977, p. 487, for further physical details). As their squares are probability densities, (3.23) holds. The values of λ for which a solution of (3.22) and (3.23) exists correspond to the quantized values of the energy of the oscillator and form the spectrum of problem (3.22)–(3.23).

The Gauss function

$$G(x) := \frac{1}{\sqrt[4]{\pi}} e^{-\frac{x^2}{2}}, \quad x \in \mathbb{R},$$

provides us with a solution of (3.22) for $\lambda = 1$ that satisfies (3.23). Indeed, since

$$G'(x) = -\frac{1}{\sqrt[4]{\pi}} x e^{-\frac{x^2}{2}} \quad \text{and} \quad G''(x) = -\frac{1}{\sqrt[4]{\pi}} e^{-\frac{x^2}{2}} + \frac{1}{\sqrt[4]{\pi}} x^2 e^{-\frac{x^2}{2}},$$

it becomes apparent that

$$-G''(x) + x^2 G(x) = \frac{1}{\sqrt[4]{\pi}} e^{-\frac{x^2}{2}} - \frac{1}{\sqrt[4]{\pi}} x^2 e^{-\frac{x^2}{2}} + \frac{1}{\sqrt[4]{\pi}} x^2 e^{-\frac{x^2}{2}} = G(x).$$

Moreover,

$$\int_{-\infty}^{\infty} G^2(x)\, dx = \frac{1}{\sqrt{\pi}} \int_{-\infty}^{\infty} e^{-x^2}\, dx = 1$$

because

$$\left(\int_{-\infty}^{\infty} G^2 \right)^2 = \int_{-\infty}^{\infty} G^2(x)\, dx \int_{-\infty}^{\infty} G^2(y)\, dy$$

$$= \frac{1}{\pi} \int_{\mathbb{R}^2} e^{-x^2 - y^2}\, dx\, dy = \frac{1}{\pi} \int_0^{2\pi} \int_0^{\infty} \rho e^{-\rho^2}\, d\rho\, d\theta$$

$$= -\frac{1}{2\pi} \int_0^{2\pi} \int_0^{\infty} \frac{d}{d\rho} \left(e^{-\rho^2} \right) d\rho\, d\theta = 1.$$

Subsequently, we assume that $\Psi(x)$ is a solution of (3.22)–(3.23), and perform the change of variable

$$\Psi(x) := E(x) H(x), \tag{3.24}$$

where $E(x)$ is the exponential function

$$E(x) := \sqrt[4]{\pi} G(x) = e^{-\frac{x^2}{2}}.$$

Since (3.22) is linear, $E(x)$ solves (3.22) with $\lambda = 1$. Thus,

$$E''(x) + \left(1 - x^2\right) E(x) = 0, \quad x \in \mathbb{R}.$$

Moreover,

$$E'(x) = -x E(x), \quad x \in \mathbb{R}.$$

Thus, substituting (3.24) in (3.22) yields

$$0 = E''(x) H(x) + 2 E'(x) H'(x) + E(x) H''(x) + \left(\lambda - x^2\right) E(x) H(x)$$

$$= \left(x^2 - 1\right) E(x) H(x) - 2x E(x) H'(x) + E(x) H''(x)$$

$$\quad + \left(\lambda - x^2\right) E(x) H(x)$$

$$= E(x) \left[H''(x) - 2x H'(x) + (\lambda - 1) H(x) \right].$$

Hence, $H(x)$ solves the associated differential equation

$$H''(x) - 2xH'(x) + (\lambda - 1)H(x) = 0, \quad x \in \mathbb{R}.$$

To solve this equation, we apply the theory developed in Section 3.4. Thus, we consider the corresponding counterpart with the complex variable:

$$H''(z) - 2zH'(z) + (\lambda - 1)H(z) = 0, \quad z \in \mathbb{C}, \tag{3.25}$$

which is known as the *Hermite equation* of order $\lambda - 1$ and is named after C. Hermite. Moreover, relation (3.23) gives

$$\int_{-\infty}^{\infty} e^{-x^2} H^2(x)\, dx = \int_{-\infty}^{\infty} \Psi^2(x)\, dx = 1. \tag{3.26}$$

According to Corollary 3.7 applied with $N = 2$, $\Omega = \mathbb{C}$, $a_0(z) = 1 - \lambda$, $a_1(z) = 2z$, and $b(z) = 0$ for all $z \in \mathbb{C}$, the set of solutions of (3.25) is a vectorial plane of $\mathcal{H}(\mathbb{C})$. Actually, the unique solutions of the following two problems,

$$\begin{cases} H'' - 2zH' + (\lambda - 1)H = 0, \\ H(0) = 1, \quad H'(0) = 0, \end{cases} \qquad \begin{cases} H'' - 2zH' + (\lambda - 1)H = 0, \\ H(0) = 0, \quad H'(0) = 1, \end{cases}$$

$$\tag{3.27}$$

provide us with a basis of Σ_0^e, where Σ_0^e denotes the set of solutions of (3.25). As the solutions of (3.25) are entire, by Theorem 3.1, their Taylor series at the origin,

$$H(z) = \sum_{n=0}^{\infty} u_n z^n, \quad u_n := \frac{H^{n)}(0)}{n!}, \tag{3.28}$$

must converge in $\mathbb{C}$, uniformly on compact subsets of $\mathbb{C}$. In particular, the convergence radius of this series is infinite and, hence, for every $z \in \mathbb{C}$,

$$H'(z) = \sum_{n=1}^{\infty} n u_n z^{n-1}, \quad H''(z) = \sum_{n=2}^{\infty} n(n-1) u_n z^{n-2}, \tag{3.29}$$

uniformly on compact subsets of $\mathbb{C}$. By choosing $(u_0, u_1) = (1, 0)$ and $(u_0, u_1) = (0, 1)$, (3.28) provides us with the solutions of the problems (3.27), respectively. Substituting the series (3.28) and (3.29) into (3.25) we obtain

$$\sum_{n=2}^{\infty} n(n-1)u_n z^{n-2} - 2\sum_{n=1}^{\infty} nu_n z^n + (\lambda-1)\sum_{n=0}^{\infty} u_n z^n = 0.$$

Equivalently,

$$\sum_{n=0}^{\infty} \left[(n+2)(n+1)u_{n+2} - 2nu_n + (\lambda-1)u_n\right] z^n = 0$$

and, hence,

$$u_{n+2} = \frac{2n+1-\lambda}{(n+2)(n+1)} u_n, \quad n \geq 0.$$

Therefore,

$$u_{2k} = \frac{\prod_{j=0}^{k-1}(4j+1-\lambda)}{(2k)!} u_0, \quad u_{2k+1} = \frac{\prod_{j=0}^{k-1}(4j+3-\lambda)}{(2k+1)!} u_1, \quad k \geq 1,$$

and, consequently,

$$H(z) = \sum_{n=0}^{\infty} u_n z^n = \sum_{k=0}^{\infty} u_{2k} z^{2k} + \sum_{k=0}^{\infty} u_{2k+1} z^{2k+1}$$

$$= \left(1 + \sum_{k=1}^{\infty} \frac{\prod_{j=0}^{k-1}(4j+1-\lambda)}{(2k)!} z^{2k}\right) u_0$$

$$+ \left(z + \sum_{k=1}^{\infty} \frac{\prod_{j=0}^{k-1}(4j+3-\lambda)}{(2k+1)!} z^{2k+1}\right) u_1.$$

In particular, by choosing $(u_0, u_1) = (1,0)$ and $(u_0, u_1) = (0,1)$, we get the solutions of (3.27), respectively. Namely,

$$H_{(1,0)}(z) := 1 + \sum_{k=1}^{\infty} \frac{\prod_{j=0}^{k-1}(4j+1-\lambda)}{(2k)!} z^{2k},$$

$$H_{(0,1)}(z) := z + \sum_{k=1}^{\infty} \frac{\prod_{j=0}^{k-1}(4j+3-\lambda)}{(2k+1)!} z^{2k+1}. \tag{3.30}$$

As already pointed out, by Corollary 3.7, the convergence radii of these series are infinite.

Suppose $\lambda = 4k + 1$ for some integer $k \geq 0$. Then, $H_{(1,0)}(z)$ becomes a polynomial of degree $2k$. Similarly, $H_{(0,1)}(z)$ becomes a polynomial of degree $2k + 1$ if $\lambda = 4k + 3$ for some integer $k \geq 0$. Since these values of λ provide us with all odd natural numbers, we conclude that, when $\lambda = 2n+1$ for some $n \geq 1$, the Hermite equation of order $\lambda - 1 = 2n$ admits a polynomial solution, denoted by $P_n(z)$, which is called the *Hermite polynomial of degree n*. Naturally, since

$$\int_{-\infty}^{\infty} e^{-x^2} P_n^2(x)\, dx < +\infty$$

(see Exercise 4 of Chapter 3, if necessary), there exists a constant, $c_n > 0$, such that

$$\Psi_n(x) := c_n e^{-\frac{x^2}{2}} P_n(x), \quad x \in \mathbb{R},$$

solves (3.22) and (3.23). The previous analysis also shows that every Hermite polynomial is uniquely determined up to a multiplicative constant. Indeed, it must be a multiple of $H_{(1,0)}(z)$ for even n, while it must be a multiple of $H_{(0,1)}(z)$ for odd n. The following table collects the Hermite polynomials of the lowest degree:

$$H_0(z) = 1,$$
$$H_1(z) = 2z,$$
$$H_2(z) = 4z^2 - 2,$$
$$H_3(z) = 8z^3 - 12z, \tag{3.31}$$
$$H_4(z) = 16z^4 - 48z^2 + 12,$$
$$H_5(z) = 32z^5 - 160z^3 + 120z$$
$$H_6(z) = 64z^6 - 480z^4 + 720z^2 - 120.$$

These polynomials have been determined for the choices $\lambda = 2n + 1$, $0 \leq n \leq 6$, by multiplying the corresponding $H_{(1,0)}(z)$ and $H_{(0,1)}(z)$ by appropriate constants (see Exercise 5 of Chapter 3). The multiplying constant is chosen in such a way that the Hermite polynomials satisfy the following formula:

$$H_n(z) = (-1)^n e^{z^2} \frac{d^n}{dz^n}\left(e^{-z^2}\right), \quad n \geq 0$$

(see Exercise 6 of Chapter 3).

To complete our analysis, we show that, when λ is not an odd natural number and $H(z)$ solves the Hermite equation of order $\lambda - 1$, then

$$\lim_{x\to\infty} |H(x)|\, e^{-\frac{x^2}{2}} = +\infty, \tag{3.32}$$

and, as a consequence,

$$\int_{-\infty}^{\infty} e^{-x^2} H^2(x)\, dx = +\infty.$$

This allows us to conclude that the set

$$\lambda = 2n + 1, \quad n \geq 0,$$

provides us with the spectrum of the problem (3.22)–(3.23). Thanks to the previous analysis, in order to prove (3.32), it suffices to check that, for every integer $k \geq 0$,

$$\lim_{x\to\infty} \left|H_{(1,0)}(x)\right| e^{-\frac{x^2}{2}} = \infty \quad \text{if } \lambda \neq 4k + 1, \tag{3.33}$$

and that

$$\lim_{x\to\infty} \left|H_{(0,1)}(x)\right| e^{-\frac{x^2}{2}} = \infty \quad \text{if } \lambda \neq 4k + 3. \tag{3.34}$$

Next, we prove (3.33). As the proof of (3.34) can be easily adapted from the proof of (3.33), it has been proposed in Exercise 7 of Chapter 3. Since $\lambda \neq 4k + 1$ for all integers $k \geq 0$, no coefficient of the series of $H_{(1,0)}$ in (3.30) can vanish. Moreover, there exists $k_0 \geq 1$ such that all the coefficients of such a series share the same sign for all $k \geq k_0$. Thus,

$$\left| H_{(1,0)}(x) - \sum_{k=0}^{k_0-1} u_{2k} x^{2k} \right| = \sum_{k=k_0}^{\infty} b_{2k} x^{2k}, \quad x \in \mathbb{R},$$

where $b_{2k} > 0$ for all $k \geq k_0$, and either $b_{2k} = u_{2k}$, for each $k \geq k_0$, or $b_{2k} = -u_{2k}$, for each $k \geq k_0$. In particular,

$$\frac{b_{2(k+1)}}{b_{2k}} = \frac{4k + 1 - \lambda}{(2k + 2)(2k + 1)}, \quad k \geq k_0.$$

On the other hand,

$$e^{x^2} = \sum_{k=0}^{\infty} c_{2k} x^{2k}, \quad \text{where } c_{2k} := \frac{1}{k!} \text{ for all } k \geq 0,$$

and

$$\lim_{k \to \infty} \frac{b_{2(k+1)}/b_{2k}}{c_{2(k+1)}/c_{2k}} = \lim_{k \to \infty} \frac{(4k + 1 - \lambda)(k + 1)}{(2k + 2)(2k + 1)} = 1.$$

Thus, for any given $\varepsilon \in \left(0, \frac{1}{2}\right)$, there exists $k_\varepsilon \geq k_0$ such that

$$\left| \frac{b_{2(k+1)}/b_{2k}}{c_{2(k+1)}/c_{2k}} - 1 \right| < \varepsilon \quad \text{for all } k \geq k_\varepsilon.$$

In particular,

$$b_{2(k+1)} > (1 - \varepsilon) \frac{c_{2(k+1)}}{c_{2k}} b_{2k} = \frac{1 - \varepsilon}{k + 1} b_{2k} \quad \text{for all } k \geq k_\varepsilon. \tag{3.35}$$

Hence, a recursive application of (3.35) shows that, for every $j \geq 1$,

$$b_{2(k_\varepsilon+j)} > \frac{1 - \varepsilon}{k_\varepsilon + j} b_{2(k_\varepsilon+j-1)} > \frac{(1 - \varepsilon)^2}{(k_\varepsilon + j)(k_\varepsilon + j - 1)} b_{2(k_\varepsilon+j-2)} > \cdots$$

$$> \frac{(1 - \varepsilon)^j}{(k_\varepsilon + j)(k_\varepsilon + j - 1) \cdots (k_\varepsilon + 1)} b_{2k_\varepsilon}$$

$$= \frac{b_{2k_\varepsilon} k_\varepsilon!}{(1 - \varepsilon)^{k_\varepsilon}} \frac{(1 - \varepsilon)^{k_\varepsilon+j}}{(k_\varepsilon + j)!}.$$

Consequently, for every $j \geq 1$ and $x > 0$,

$$b_{2(k_\varepsilon+j)} x^{2(k_\varepsilon+j)} > \frac{b_{2k_\varepsilon} k_\varepsilon!}{(1 - \varepsilon)^{k_\varepsilon}} \frac{\left(x\sqrt{1 - \varepsilon}\right)^{2(k_\varepsilon+j)}}{(k_\varepsilon + j)!}.$$

So,

$$\left| H_{(1,0)}(x) - \sum_{k=0}^{k_0-1} u_{2k} x^{2k} \right| - \sum_{k=k_0}^{k_\varepsilon} b_{2k} x^{2k}$$

$$= \sum_{j=1}^{\infty} b_{2(k_\varepsilon+j)} x^{2(k_\varepsilon+j)} > \frac{b_{2k_\varepsilon} k_\varepsilon!}{(1 - \varepsilon)^{k_\varepsilon}} \sum_{j=1}^{\infty} \frac{\left(x\sqrt{1 - \varepsilon}\right)^{2(k_\varepsilon+j)}}{(k_\varepsilon + j)!}$$

$$= \frac{b_{2k_\varepsilon} k_\varepsilon!}{(1 - \varepsilon)^{k_\varepsilon}} \left(e^{(1-\varepsilon)x^2} - \sum_{n=0}^{k_\varepsilon} \frac{\left(x\sqrt{1 - \varepsilon}\right)^{2n}}{n!} \right).$$

Therefore, if we define the polynomial

$$Q(x) := \sum_{n=k_0}^{k_\varepsilon} b_{2n} x^{2n} - \frac{b_{2k_\varepsilon} k_\varepsilon!}{(1-\varepsilon)^{k_\varepsilon}} \sum_{n=0}^{k_\varepsilon} \frac{(1-\varepsilon)^n x^{2n}}{n!},$$

it becomes apparent that

$$\left| H_{(1,0)}(x) - \sum_{k=0}^{k_0-1} u_{2k} x^{2k} \right| e^{-\frac{x^2}{2}} > Q(x) e^{-\frac{x^2}{2}} + \frac{b_{2k_\varepsilon} k_\varepsilon!}{(1-\varepsilon)^{k_\varepsilon}} e^{\left(\frac{1}{2}-\varepsilon\right)x^2}.$$

Lastly, since $\varepsilon < \frac{1}{2}$, (3.33) readily follows by letting $x \to \infty$.

To conclude this section, we go back to problem (3.22)–(3.23). By making the choice

$$N_n := \sqrt{\frac{1}{2^n \sqrt{\pi} n!}}, \quad n \geq 0,$$

it is easily seen that the solution of the Schrödinger equation

$$-\Psi''(x) + x^2 \Psi(x) = (2n+1)\Psi(x)$$

constructed above,

$$\psi_n(x) := N_n e^{-\frac{x^2}{2}} H_n(x), \quad n \geq 0,$$

also satisfies

$$\int_{-\infty}^{+\infty} \psi_n^2(x)\, dx = 1$$

(see Exercise 8 of Chapter 3). Figure 3.1 shows the graph of the wave function $\psi_0(x)$. It has the shape of the Gauss curve of the distribution of errors. Thus, the most likely position in which an electron can be found is the rest state of the oscillator, corresponding to $x = 0$ (see Chpolski, 1977, p. 495). In Figure 3.2, instead, we have plotted the wave functions of the linear harmonic oscillator for $n = 1, 2, 3, 4, 5, 6$.

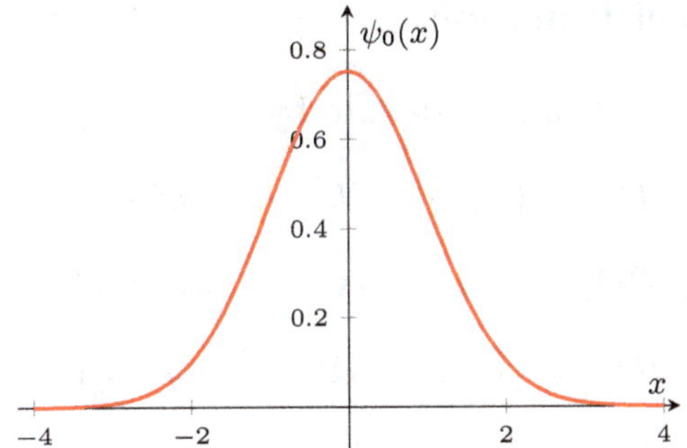

Fig. 3.1. Graph of the wave function $\psi_0(x)$.

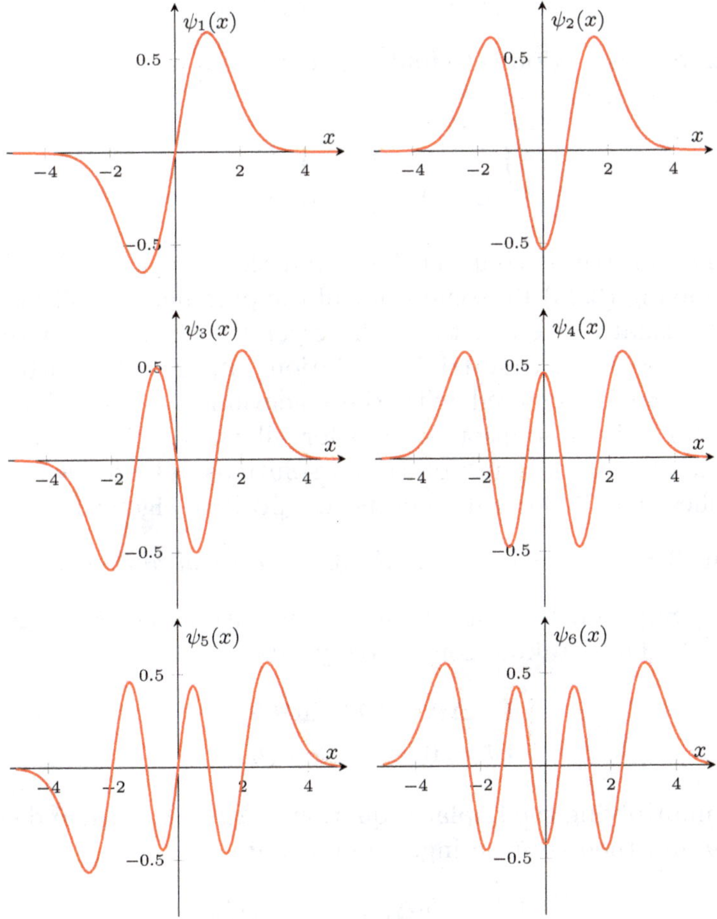

Fig. 3.2. Graph of the wave functions $\psi_n(x)$ for $1 \leq n \leq 6$.

3.7   The Bessel Equation

Throughout this section, we denote by

$$D_1 := \left\{ (x, y) \in \mathbb{R}^2 : x^2 + y^2 < 1 \right\}$$

the unit disk, by ∂D_1 its boundary, the unit circle

$$\partial D_1 := \left\{ (x, y) \in \mathbb{R}^2 : x^2 + y^2 = 1 \right\},$$

and by Δ the bi-dimensional Laplace operator

$$\Delta := \frac{\partial^2}{\partial x^2} + \frac{\partial^2}{\partial y^2}.$$

The aim of this section is to find the values of λ for which the boundary value problem

$$\begin{cases} -\Delta u = \lambda u & \text{in } D_1, \\ u = 0 & \text{on } \partial D_1, \end{cases} \tag{3.36}$$

has some non-trivial solution, i.e. a function $u \in \mathcal{C}^2(D_1) \cap \mathcal{C}^1(\bar{D}_1)$, $u \neq 0$, solving (3.36) for some value of the parameter λ. These values of λ are usually refereed to as the *eigenvalues* of $-\Delta$ in the disk D_1. The associated (non-trivial) solutions, u, are referred to as the *eigenfunctions* associated with the eigenvalue λ. Note that (3.36) admits the trivial solution $u = 0$ for all values of λ; this is why we are interested only in non-trivial solutions. The following result establishes that (3.36) only admits real positive eigenvalues.

Lemma 3.8. *The eigenvalues of* (3.36) *are real and positive.*

Proof. Suppose $\lambda \in \mathbb{C}$ is an eigenvalue with associated eigenfunction $u \neq 0$. Then, taking conjugates yields

$$\begin{cases} -\Delta \bar{u} = \bar{\lambda} \bar{u} & \text{in } D_1, \\ \bar{u} = 0 & \text{on } \partial D_1, \end{cases}$$

and by multiplying the Laplace equation $-\Delta u = \lambda u$ by $\bar{u}$ and $-\Delta \bar{u} = \bar{\lambda} \bar{u}$ by u and then subtracting, we find that

$$-\bar{u}\Delta u + u\Delta \bar{u} = \left(\lambda - \bar{\lambda}\right) |u|^2.$$

Thus, by integrating in D_1, it follows from the divergence theorem that

$$(\lambda - \bar{\lambda}) \int_{D_1} |u|^2 = \int_{D_1} (u\Delta\bar{u} - \bar{u}\Delta u) = \int_{D_1} (u \operatorname{div} \nabla\bar{u} - \bar{u} \operatorname{div} \nabla u)$$

$$= \int_{D_1} \operatorname{div}(u\nabla\bar{u} - \bar{u}\nabla u)$$

$$= \int_{\partial D_1} \left(u\frac{\partial\bar{u}}{\partial n} - \bar{u}\frac{\partial u}{\partial n} \right) dS = 0,$$

where in the last equality, we have used that $u = \bar{u} = 0$ on ∂D_1. Since $u \neq 0$, the integral on the left-hand side is positive, so $\lambda = \bar{\lambda}$. In other words, $\lambda \in \mathbb{R}$. Moreover, if $u = u_1 + iu_2$ is an eigenfunction associated with the eigenvalue $\lambda \in \mathbb{R}$, then identifying the real and imaginary parts in (3.36) shows that both u_1 and u_2 are real solutions of (3.36). Furthermore, since $u = u_1 + iu_2 \neq 0$, at least one of u_1 and u_2 is non-trivial. Thus, we conclude that the problem admits a real, non-trivial eigenfunction, which we denote again by u. Multiplying $-\Delta u = \lambda u$ by u and applying the divergence theorem again yields

$$\lambda \int_{D_1} u^2 = -\int_{D_1} u\Delta u = -\int_{D_1} u \operatorname{div} \nabla u$$

$$= -\int_{D_1} \operatorname{div}(u\nabla u) + \int_{D_1} |\nabla u|^2 = \int_{D_1} |\nabla u|^2.$$

Thus, $\lambda \geq 0$. Suppose $\lambda = 0$. Then,

$$\int_{D_1} |\nabla u|^2 = 0$$

and, hence, $\nabla u = 0$ on D_1. Consequently, u must be constant in D_1 and, since $u = 0$ on ∂D_1, this implies $u = 0$ on $\bar{D}_1$, which has been excluded. Therefore, $\lambda > 0$. The proof is complete. $\qquad\square$

By the radial symmetry of the disk D_1, in order to solve the eigenvalue problem (3.36), it is convenient to change to polar coordinates:

$$x = \rho\cos\theta, \quad y = \rho\sin\theta, \quad v(\rho, \theta) = u(x, y). \qquad (3.37)$$

By the chain rule, for every $(x, y) \in D_1$, $(x, y) \neq (0, 0)$, we have that

$$\frac{\partial u}{\partial x}(x, y) = \frac{\partial v}{\partial \rho}(\rho, \theta)\frac{\partial \rho}{\partial x}(x, y) + \frac{\partial v}{\partial \theta}(\rho, \theta)\frac{\partial \theta}{\partial x}(x, y),$$

$$\frac{\partial u}{\partial y}(x, y) = \frac{\partial v}{\partial \rho}(\rho, \theta)\frac{\partial \rho}{\partial y}(x, y) + \frac{\partial v}{\partial \theta}(\rho, \theta)\frac{\partial \theta}{\partial y}(x, y).$$

Thus, differentiating once more yields

$$\frac{\partial^2 u}{\partial x^2}(x, y) = \left(\frac{\partial^2 v}{\partial \rho^2}\frac{\partial \rho}{\partial x} + \frac{\partial^2 v}{\partial \theta \partial \rho}\frac{\partial \theta}{\partial x}\right)\frac{\partial \rho}{\partial x} + \frac{\partial v}{\partial \rho}\frac{\partial^2 \rho}{\partial x^2}$$
$$+ \left(\frac{\partial^2 v}{\partial \rho \partial \theta}\frac{\partial \rho}{\partial x} + \frac{\partial^2 v}{\partial \theta^2}\frac{\partial \theta}{\partial x}\right)\frac{\partial \theta}{\partial x} + \frac{\partial v}{\partial \theta}\frac{\partial^2 \theta}{\partial x^2}.$$

Similarly,

$$\frac{\partial^2 u}{\partial y^2}(x, y) = \left(\frac{\partial^2 v}{\partial \rho^2}\frac{\partial \rho}{\partial y} + \frac{\partial^2 v}{\partial \theta \partial \rho}\frac{\partial \theta}{\partial y}\right)\frac{\partial \rho}{\partial y} + \frac{\partial v}{\partial \rho}\frac{\partial^2 \rho}{\partial y^2}$$
$$+ \left(\frac{\partial^2 v}{\partial \rho \partial \theta}\frac{\partial \rho}{\partial y} + \frac{\partial^2 v}{\partial \theta^2}\frac{\partial \theta}{\partial y}\right)\frac{\partial \theta}{\partial y} + \frac{\partial v}{\partial \theta}\frac{\partial^2 \theta}{\partial y^2}.$$

Adding the two identities, rearranging terms, and taking into account that we are requiring $u \in \mathcal{C}^2(D_1)$, it becomes apparent that

$$\Delta u = \frac{\partial^2 v}{\partial \rho^2}\left[\left(\frac{\partial \rho}{\partial x}\right)^2 + \left(\frac{\partial \rho}{\partial y}\right)^2\right] + 2\frac{\partial^2 v}{\partial \theta \partial \rho}\left[\frac{\partial \rho}{\partial x}\frac{\partial \theta}{\partial x} + \frac{\partial \rho}{\partial y}\frac{\partial \theta}{\partial y}\right]$$
$$+ \frac{\partial^2 v}{\partial \theta^2}\left[\left(\frac{\partial \theta}{\partial x}\right)^2 + \left(\frac{\partial \theta}{\partial y}\right)^2\right] + \frac{\partial v}{\partial \rho}\Delta \rho + \frac{\partial v}{\partial \theta}\Delta \theta.$$

Equivalently,

$$\Delta u = \frac{\partial^2 v}{\partial \rho^2}|\nabla \rho|^2 + 2\frac{\partial^2 v}{\partial \rho \partial \theta}\langle \nabla \rho, \nabla \theta \rangle + \frac{\partial^2 v}{\partial \theta^2}|\nabla \theta|^2 + \frac{\partial v}{\partial \rho}\Delta \rho + \frac{\partial v}{\partial \theta}\Delta \theta.$$

$$(3.38)$$

Moreover, by implicitly differentiating with respect to x and y in (3.37), it follows that, for every $(x, y) \neq (0, 0)$, i.e. whenever

$$\rho^2 = x^2 + y^2 \neq 0,$$

$$\frac{\partial \rho}{\partial x} = \frac{x}{\rho}, \quad \frac{\partial \rho}{\partial y} = \frac{y}{\rho}, \quad \frac{\partial \theta}{\partial x} = -\frac{y}{\rho^2}, \quad \frac{\partial \theta}{\partial y} = \frac{x}{\rho^2}.$$

Thus,

$$\frac{\partial^2 \rho}{\partial x^2} = \frac{\rho^2 - x^2}{\rho^3}, \quad \frac{\partial^2 \rho}{\partial y^2} = \frac{\rho^2 - y^2}{\rho^3}, \quad \frac{\partial^2 \theta}{\partial x^2} = 2\frac{xy}{\rho^4}, \quad \frac{\partial^2 \theta}{\partial y^2} = -2\frac{xy}{\rho^4},$$

and consequently,

$$|\nabla \rho|^2 = \left(\frac{\partial \rho}{\partial x}\right)^2 + \left(\frac{\partial \rho}{\partial y}\right)^2 = \frac{x^2 + y^2}{\rho^2} = 1,$$

$$|\nabla \theta|^2 = \left(\frac{\partial \theta}{\partial x}\right)^2 + \left(\frac{\partial \theta}{\partial y}\right)^2 = \frac{x^2 + y^2}{\rho^4} = \frac{1}{\rho^2},$$

$$\langle \nabla \rho, \nabla \theta \rangle = \frac{\partial \rho}{\partial x}\frac{\partial \theta}{\partial x} + \frac{\partial \rho}{\partial y}\frac{\partial \theta}{\partial y} = \frac{-xy + xy}{\rho^3} = 0,$$

and

$$\Delta \rho = \frac{1}{\rho}, \quad \Delta \theta = 0.$$

Therefore, substituting these identities in (3.38) gives

$$\Delta u = \frac{\partial^2 v}{\partial \rho^2} + \frac{1}{\rho}\frac{\partial v}{\partial \rho} + \frac{1}{\rho^2}\frac{\partial^2 v}{\partial \theta^2} \quad \text{in } D_1 \setminus \{(0,0)\}, \tag{3.39}$$

which expresses the Laplace operator in polar coordinates.

According to (3.39), if u is a smooth solution of (3.36), then the function $v(\rho, \theta) = u(x, y)$ solves the problem

$$\begin{cases} \frac{\partial^2 v}{\partial \rho^2} + \frac{1}{\rho}\frac{\partial v}{\partial \rho} + \frac{1}{\rho^2}\frac{\partial^2 v}{\partial \theta^2} = -\lambda v(\rho, \theta) & (\rho, \theta) \in (0,1) \times \mathbb{R}, \\ v(1, \theta) = 0 & \theta \in \mathbb{R}, \\ v(\rho, \theta + 2\pi) = v(\rho, \theta) & (\rho, \theta) \in (0,1] \times \mathbb{R}. \end{cases} \tag{3.40}$$

To solve the eigenvalue problem (3.40), we use the method of separation of variables, which consists of searching for elementary solutions of the form

$$v(\rho, \theta) = X(\rho)\varphi(\theta), \qquad (3.41)$$

for some smooth 2π-periodic function, $\varphi(\theta)$, and some smooth $X(\rho)$ such that $X(1) = 0$ and

$$\frac{\partial^2 v}{\partial \rho^2} + \frac{1}{\rho}\frac{\partial v}{\partial \rho} + \frac{1}{\rho^2}\frac{\partial^2 v}{\partial \theta^2} = -\lambda v(\rho, \theta). \qquad (3.42)$$

If either $X = 0$ or $\varphi = 0$, then (3.41) provides us with the *trivial solution* of (3.42), $v = 0$, which is 2π-periodic and obviously satisfies (3.40). Suppose instead that $X(\rho_0) \neq 0$ and $\varphi(\theta_0) \neq 0$ for some $(\rho_0, \theta_0) \in (0, 1) \times \mathbb{R}$. Then, $X(\rho)\varphi(\theta) \neq 0$ in a neighborhood of (ρ_0, θ_0).[1] By substituting (3.41) in (3.42) and dividing by $X(\rho)\varphi(\theta)$, it is apparent that

$$\frac{X''(\rho)}{X(\rho)} + \frac{1}{\rho}\frac{X'(\rho)}{X(\rho)} + \frac{1}{\rho^2}\frac{\ddot{\varphi}(\theta)}{\varphi(\theta)} = -\lambda,$$

where $'$ indicates derivation with respect to ρ and $\cdot$ indicates derivation with respect to θ. Equivalently,

$$\rho^2 \frac{X''(\rho)}{X(\rho)} + \rho\frac{X'(\rho)}{X(\rho)} + \lambda\rho^2 = -\frac{\ddot{\varphi}(\theta)}{\varphi(\theta)}.$$

As the left-hand side of this identity is independent of $\theta \in \mathbb{R}$ and the right-hand side is independent of $\rho \in (0, 1)$, both sides do not depend on ρ or θ. Thus, there exists a constant $\mu \in \mathbb{R}$ such that

$$\rho^2 \frac{X''(\rho)}{X(\rho)} + \rho\frac{X'(\rho)}{X(\rho)} + \lambda\rho^2 = -\frac{\ddot{\varphi}(\theta)}{\varphi(\theta)} = \mu \qquad (3.43)$$

---

[1]Observe that our argument is only rigorous in the neighborhood of (ρ_0, θ_0) where $X(\rho)\varphi(\theta) \neq 0$. Indeed, far from (ρ_0, θ_0) the product may vanish. In such a case, the subsequent argument should be thought as heuristic. In spite of that, it provides us with all the solutions of the eigenvalue problem (3.36), as it will be pointed out at the end of this section.

for all $\theta \in \mathbb{R}$ and $\rho \in (0,1)$. So, we are naturally led to solve the eigenvalue problem

$$\begin{cases} -\ddot{\varphi}(\theta) = \mu\,\varphi(\theta) & \theta \in \mathbb{R}, \\ \varphi(\theta + 2\pi) = \varphi(\theta) & \theta \in \mathbb{R}. \end{cases} \tag{3.44}$$

Multiplying the differential equation of (3.44) by φ and integrating in $(0, 2\pi)$ yields

$$\mu \int_0^{2\pi} \varphi^2 = -\int_0^{2\pi} \ddot{\varphi}(\theta)\varphi(\theta)\, d\theta$$

$$= -\int_0^{2\pi} \frac{d}{d\theta}\left(\dot{\varphi}(\theta)\varphi(\theta)\right)\, d\theta + \int_0^{2\pi} (\dot{\varphi}(\theta))^2 d\theta$$

$$= \dot{\varphi}(0)\varphi(0) - \dot{\varphi}(2\pi)\varphi(2\pi) + \int_0^{2\pi} (\dot{\varphi}(\theta))^2 d\theta$$

$$= \int_0^{2\pi} (\dot{\varphi}(\theta))^2 d\theta$$

(observe that we have used $\dot{\varphi}(0) = \dot{\varphi}(2\pi)$, which follows by differentiating the periodicity condition in (3.44)). Hence, $\mu \geq 0$. Suppose $\mu = 0$. Then, the general solution of $\ddot{\varphi} = 0$ is

$$\varphi(\theta) = A\theta + B$$

for some constants A and B and, since φ must be 2π-periodic,

$$B = \varphi(0) = \varphi(2\pi) = 2\pi A + B.$$

Therefore, $A = 0$. Since any constant provides us with a 2π-periodic function, associated with the lowest eigenvalue, $\mu_0 := 0$, we can choose the eigenfunction $\varphi_0(\theta) = 1$. Any other eigenfunction of (3.44) associated with $\mu_0 = 0$ must be constant. Subsequently, we suppose that $\mu > 0$. Then, the general solution of

$$-\ddot{\varphi} = \mu\varphi$$

is given by

$$\varphi(\theta) = Ae^{i\sqrt{\mu}\theta} + Be^{-i\sqrt{\mu}\theta}, \quad \theta \in \mathbb{R},$$

where $A, B \in \mathbb{C}$ are arbitrary constants. For these functions to solve (3.44), they have to satisfy $\varphi(0) = \varphi(2\pi)$ and $\dot{\varphi}(0) = \dot{\varphi}(2\pi)$. So,

$$A + B = Ae^{i\sqrt{\mu}2\pi} + Be^{-i\sqrt{\mu}2\pi}, \quad A - B = Ae^{i\sqrt{\mu}2\pi} - Be^{-i\sqrt{\mu}2\pi}.$$

Summing and subtracting these identities yields

$$0 = A\left(1 - e^{i\sqrt{\mu}2\pi}\right) = B\left(1 - e^{-i\sqrt{\mu}2\pi}\right);$$

therefore, since $(A, B) \neq (0, 0)$, necessarily,

$$\mu = n^2$$

for some integer $n \geq 1$. Actually, in such a case, the function

$$\varphi(\theta) = Ae^{in\theta} + Be^{-in\theta}, \quad \theta \in \mathbb{R},$$

is 2π-periodic for all $A, B \in \mathbb{C}$. Consequently, for every integer $n \geq 1$, $\mu_n := n^2$ is an eigenvalue of (3.44) with associated eigenfunctions

$$\varphi_n(\theta) = a_{-n}e^{-in\theta} + a_n e^{in\theta}, \quad \theta \in \mathbb{R},$$

for all $a_{-n}, a_n \in \mathbb{C}$. If one looks for real solutions, reasoning as in Section 1.8.1 easily gives that the general (real) solution of (3.44) is

$$\varphi_n(\theta) = b_n \cos(n\theta) + c_n \sin(n\theta), \quad \theta \in \mathbb{R}.$$

Summarizing, (3.44) admits some non-trivial solution if and only if $\mu = n^2$ for some integer $n \geq 0$.

Going back to (3.43), it remains to determine the set of positive λ's for which the boundary value problem

$$\begin{cases} X''(\rho) + \frac{1}{\rho}X'(\rho) + \left(\lambda - \frac{n^2}{\rho^2}\right)X(\rho) = 0 & \rho \in (0, 1), \\ X(1) = 0 \end{cases} \tag{3.45}$$

admits a non-trivial solution for some integer $n \geq 0$. All such values of $\lambda > 0$ provide us with eigenvalues of (3.36). As the change of variable

$$X(\rho) := J(\sqrt{\lambda}\rho), \quad \rho > 0,$$

transforms (3.45) into the problem

$$\begin{cases} \lambda J''(\sqrt{\lambda}\rho) + \dfrac{\lambda}{\sqrt{\lambda}\rho} J'(\sqrt{\lambda}\rho) + \lambda \left(1 - \dfrac{n^2}{(\sqrt{\lambda}\rho)^2}\right) J(\sqrt{\lambda}\rho) = 0 & \rho \in (0,1), \\ J(\sqrt{\lambda}) = 0, \end{cases}$$

it becomes apparent that, if we can find an entire solution $J(z)$ of the differential equation

$$J''(z) + \frac{1}{z} J'(z) + \left(1 - \frac{n^2}{z^2}\right) J(z) = 0, \quad n \geq 0, \tag{3.46}$$

such that $J(x_0) = 0$ for some $x_0 = x_0(n) > 0$, then $\lambda = x_0^2 > 0$ is an eigenvalue of (3.36). The differential equation (3.46) is known as the *Bessel equation of order $n \geq 0$*, named after F. W. Bessel. The change of variable

$$J(z) = z^n \Psi(z), \quad z \in \mathbb{C},$$

transforms (3.46) into

$$n(n-1)z^{n-2}\Psi(z) + 2nz^{n-1}\Psi'(z) + z^n \Psi''(z)$$

$$+ \frac{1}{z}\left(nz^{n-1}\Psi(z) + z^n \Psi'(z)\right) + \left(1 - \frac{n^2}{z^2}\right) z^n \Psi(z) = 0,$$

or, equivalently,

$$\Psi''(z) + \frac{2n+1}{z}\Psi'(z) + \Psi(z) = 0, \quad z \in \mathbb{C} \setminus \{0\}. \tag{3.47}$$

Although this equation does not fit into the setting of Corollary 3.7, due to the singularity of the coefficient $\frac{2n+1}{z}$ at $z = 0$, we will show that, for every $n \geq 0$, it admits an entire solution,

$$\Psi(z) = \sum_{k=0}^{\infty} \psi_k z^k, \tag{3.48}$$

for some sequence of coefficients $\{\psi_k\}_{k\geq 0}$. Indeed, in the convergence disk of the series (3.48), we have that

$$\Psi'(z) = \sum_{k=1}^{\infty} k\psi_k z^{k-1}, \quad \Psi''(z) = \sum_{k=2}^{\infty} k(k-1)\psi_k z^{k-2},$$

and, hence, substituting in (3.47) yields

$$\sum_{k=2}^{\infty} k(k-1)\psi_k z^{k-2} + (2n+1)\sum_{k=1}^{\infty} k\psi_k z^{k-2} + \sum_{k=0}^{\infty} \psi_k z^k = 0.$$

Since the only singular power of z appears in the second series for $k = 1$, we set $\psi_1 = 0$. As a consequence,

$$\sum_{m=0}^{\infty} [(m+2)(m+2+2n)\psi_{m+2} + \psi_m]z^m = 0.$$

In other words,

$$\psi_1 = 0, \quad \psi_{k+2} = \frac{-\psi_k}{(k+2)(k+2+2n)}, \quad k \geq 0.$$

Therefore, $\psi_{2k+1} = 0$ for all $k \geq 0$, while, for all $k \geq 1$,

$$\psi_{2k} = \frac{-\psi_{2(k-1)}}{(2k)(2k+2n)} = \frac{-\psi_{2(k-1)}}{2^2 k(k+n)} = \cdots = \frac{(-1)^k \psi_0}{2^{2k} k! \prod_{j=1}^{k}(n+j)}.$$

Consequently, for every $\psi_0 \in \mathbb{C}$, the power series

$$J(z) = \psi_0 z^n \left(1 + \sum_{k=1}^{\infty} \frac{(-1)^k}{k! \prod_{j=1}^{k}(n+j)} \left(\frac{z}{2}\right)^{2k}\right) \tag{3.49}$$

provides us with a solution of (3.46) if it converges. Since

$$\lim_{k\to\infty} \frac{\dfrac{1}{(k+1)! \prod_{j=1}^{k+1}(n+j)}}{\dfrac{1}{k! \prod_{j=1}^{k}(n+j)}} = \lim_{k\to\infty} \frac{1}{(k+1)(n+k+1)} = 0,$$

the convergence radius of the series defining $J(z)$ is infinite and, hence, $J(z)$ is an entire function solving the Bessel equation (3.46) in $\mathbb{C}$. In particular, by choosing

$$\psi_0 = \frac{1}{2^n n!},$$

the function (3.49) becomes

$$J_n(z) := \left(\frac{z}{2}\right)^n \sum_{k=0}^{\infty} \frac{(-1)^k}{k!(n+k)!} \left(\frac{z}{2}\right)^{2k}, \tag{3.50}$$

which is usually known as the *Bessel function of the first kind and order n*. Figure 3.3 shows the plots of the Bessel functions $J_n(x)$ for $x \in \mathbb{R}$, $x \geq 0$, and $n = 0, 1, 2, 3$.

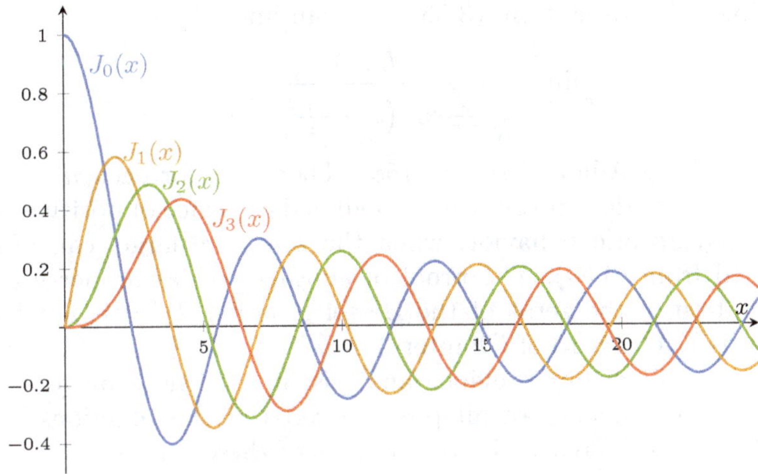

Fig. 3.3.   Graphs of the Bessel functions $J_n(x)$ for $n = 0, 1, 2, 3$.

The following recurrence formula, whose proof is left to the reader as Exercise 9 of Chapter 3,

$$J_{n+1}(z) + J_{n-1}(z) = \frac{2n}{z} J_n(z), \quad z \in \mathbb{C}, \quad n \geq 1, \tag{3.51}$$

provides us with the function $J_{n+1}(z)$ in terms of the two previous Bessel functions $J_{n-1}(z)$ and $J_n(z)$. Thus, if we iterate (3.51), we can express $J_n(z)$ for all $n \geq 2$ starting from $J_0(z)$ and $J_1(z)$. Similarly, by differentiating (3.50), it is easily seen that

$$z J_n'(z) = n J_n(z) - z J_{n+1}(z), \quad z \in \mathbb{C}, \quad n \geq 0. \tag{3.52}$$

Moreover, the Bessel functions $J_n(z)$ satisfy

$$J_n(x) = \frac{1}{\pi} \int_0^\pi \cos(nt - x \sin t)\, dt, \quad x \in \mathbb{R}, \tag{3.53}$$

for all integers $n \geq 0$ (see, e.g. Puig Adam, 1978, p. 160). Although the series defining $J_n(z)$ converge very quickly for small values of $|z|$, it turns out that they converge very slowly for large values of $|z|$, and this explains the huge interest, from the point of view of applications, in representing these functions in the integral form (3.53).

As we are interested in the existence of zeros of $J_n(z)$, we observe first of all that, since $J_n(z)$ is an entire function which is not identically equal to 0, its zeros must be isolated, according to the identity principle. Moreover, from (3.53), one can infer that

$$\lim_{x \to +\infty} \frac{J_n(x)}{\sqrt{\frac{2}{\pi x}} \cos\left(x - \frac{n\pi}{2} - \frac{\pi}{4}\right)} = 1 \qquad (3.54)$$

(see, e.g. Puig Adam, 1978, p. 158). Therefore, for each $n \geq 0$, the function $J_n(x)$ does possess an unbounded sequence of positive zeros, whose asymptotic behavior, when the zeros are large, can also be obtained from (3.54). The proof of another crucial property of the distribution of the zeros of the Bessel function $J_n(x)$ is left to the reader as Exercise 10 of Chapter 3.

As a result of our previous analysis, if we denote by $\varrho_{n,m} > 0$, $m \geq 1$, the sequence of all positive zeros of the functions $J_n(x)$, $n \geq 0$, ordered increasingly, the squares of these zeroes,

$$\lambda = \varrho_{n,m}^2, \quad n \geq 0, \ m \geq 1,$$

provide us with eigenvalues of $-\Delta$ in the unit disk. By looking at Figure 3.3, it can be seen that the first eigenvalues are

$$\lambda_1 = \varrho_{0,1}^2 \sim 5.7832,$$

$$\lambda_2 = \varrho_{1,1}^2 \sim 14.6820,$$

$$\lambda_3 = \varrho_{2,1}^2 \sim 26.3746,$$

$$\lambda_4 = \varrho_{0,2}^2 \sim 30.4713,$$

$$\lambda_5 = \varrho_{3,1}^2 \sim 40.7065,$$

$$\lambda_6 = \varrho_{1,2}^2 \sim 49.2185.$$

Moreover, our construction shows that the eigenvalues corresponding to $\varrho_{0,m}$ are simple, with the corresponding eigenspace generated by

$$u(x,y) = v(\rho,\theta) = J_0(\varrho_{0,m}\rho),$$

while the ones corresponding to $\varrho_{n,m}$, with $n \geq 1$, have a two-dimensional eigenspace generated by

$$u_1(x,y) = v_1(\rho,\theta) = J_n(\varrho_{n,m}\rho)\cos(n\theta),$$

$$u_2(x,y) = v_2(\rho,\theta) = J_n(\varrho_{n,m}\rho)\sin(n\theta).$$

Such eigenfunctions, corresponding to λ_n for $0 \leq n \leq 6$, are represented in Figures 3.4 and 3.5.

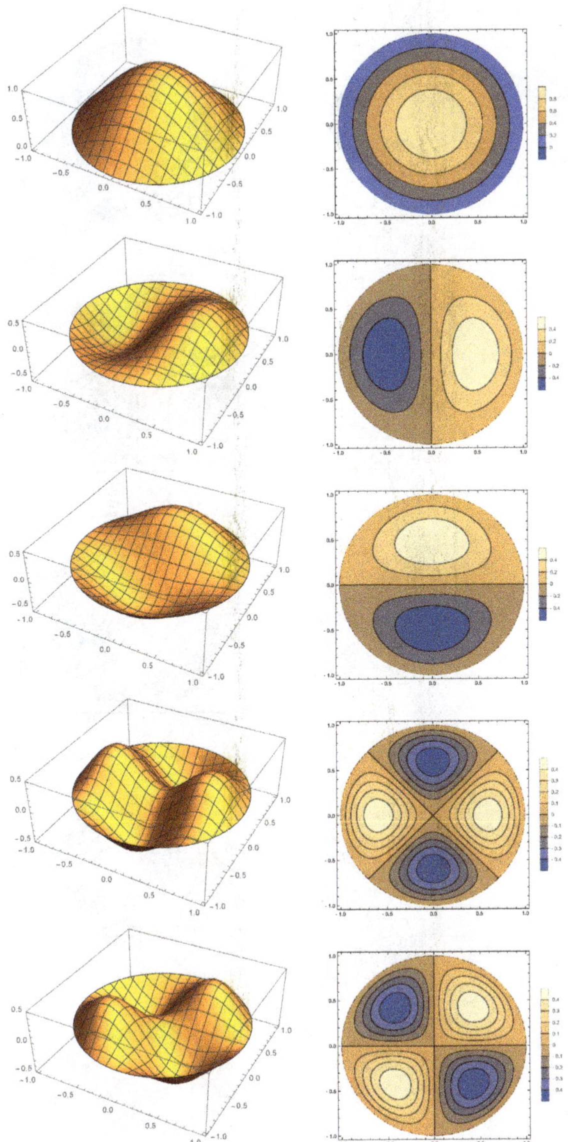

Fig. 3.4. Eigenfunctions of $-\Delta$ in the unit disk (left) and corresponding contour lines (right). The associated eigenvalues are λ_1 (first row), λ_2 (second and third row), λ_3 (fourth and fifth row).

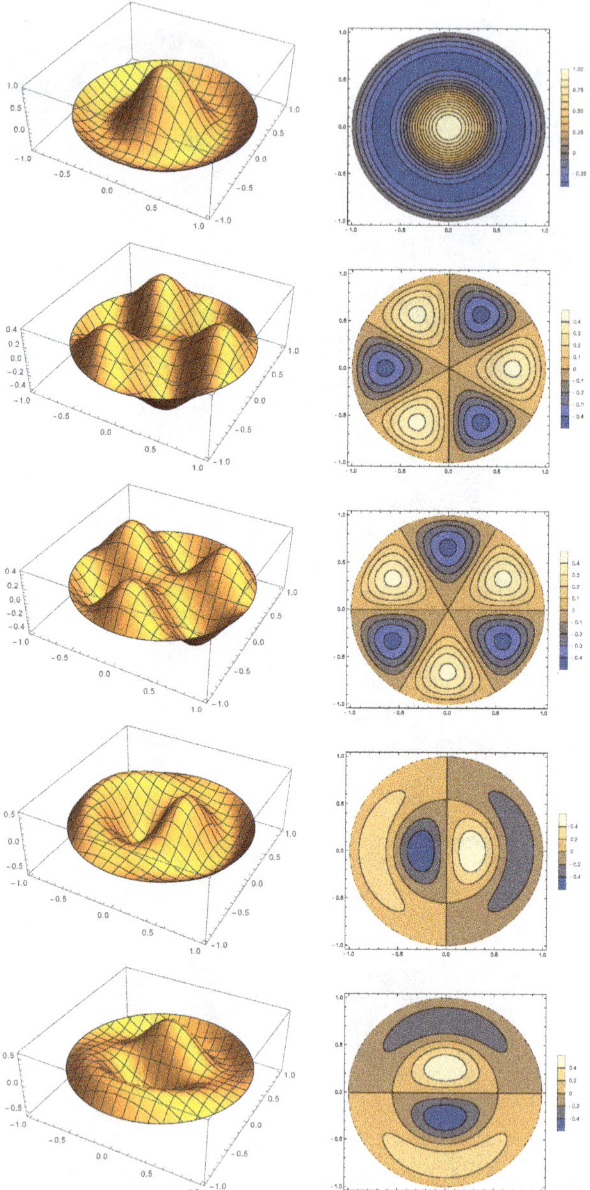

Fig. 3.5. Eigenfunctions of $-\Delta$ in the unit disk (left) and corresponding contour lines (right). The associated eigenvalues are λ_4 (first row), λ_5 (second and third row), λ_6 (fourth and fifth row).

Conversely, with some (substantial) additional work, one might prove that, actually, there are no other eigenvalues for problem (3.36), but this goes beyond the general scope of this textbook. We refer the interested reader, e.g. to Vladimirov (1971, Section 18) for a presentation of the Hilbert–Schmidt theory, which allows one to obtain such a result.

3.8 Frobenius Theorem

The goal of this section is to prove a theorem by Frobenius, which is a general result related to the existence of regular solutions of linear equations with singular coefficients. In particular, it can be applied to show that the Bessel equation (3.46) admits a regular solution in a neighborhood of $z = 0$.

Subsequently, for every $z_0 \in \mathbb{C}$ and $R > 0$, we denote by $D_R(z_0)$ the open disk of radius R centered at z_0, i.e.

$$D_R(z_0) := \{z \in \mathbb{C} : |z - z_0| < R\}.$$

Theorem 3.9 (Frobenius). *Take $z_0 \in \mathbb{C}$, $R > 0$ and $a, b \in \mathcal{H}(D_R(z_0))$. Then, there exist $\varepsilon \in (0, R)$, $r \in \mathbb{C}$, and $v \in \mathcal{H}(D_\varepsilon(z_0))$ such that the function*

$$u(z) := (z - z_0)^r v(z), \quad |z - z_0| < \varepsilon,$$

solves the singular differential equation

$$u''(z) + \frac{a(z)}{z - z_0} u'(z) + \frac{b(z)}{(z - z_0)^2} u(z) = 0 \tag{3.55}$$

in the disk $D_\varepsilon(z_0)$.

Proof. We point out that, here and in the rest of this book, $(z - z_0)^r$ means the principal determination of $(z - z_0)^r$, i.e.

$$(z - z_0)^r := e^{r \, \mathrm{Log} \, (z - z_0)} = e^{r[\ln |z - z_0| + i \mathrm{Arg} \, (z - z_0)]},$$

where Arg denotes the principal determination of the argument (see, e.g. López-Gómez, 2001a, Chapter 1, if necessary). Throughout this proof, for every integer $n \geq 0$, we denote

$$a_n := \frac{a^{n)}(z_0)}{n!}, \quad b_n := \frac{b^{n)}(z_0)}{n!},$$

which are the coefficients of the Taylor series of the functions $a(z)$ and $b(z)$, respectively. Since $a, b \in \mathcal{H}(D_R(z_0))$, we can infer from Theorem 3.1 and the root test for the convergence of series that

$$\limsup_{n \to \infty} \sqrt[n]{|a_n|} \leq \frac{1}{R}, \quad \limsup_{n \to \infty} \sqrt[n]{|b_n|} \leq \frac{1}{R}. \tag{3.56}$$

To prove the theorem, we have to find $r \in \mathbb{C}$ and a power series,

$$v(z) := \sum_{n=0}^{\infty} u_n (z - z_0)^n, \tag{3.57}$$

with positive convergence radius, $0 < \varepsilon \leq R$, so that

$$u(z) := (z - z_0)^r v(z) = \sum_{n=0}^{\infty} u_n (z - z_0)^{n+r} \tag{3.58}$$

solves the differential equation (3.55). Then, for every $z \in D_\varepsilon(z_0)$,

$$u'(z) = \sum_{n=0}^{\infty} (n + r) u_n (z - z_0)^{n+r-1},$$

$$u''(z) = \sum_{n=0}^{\infty} (n + r)(n + r - 1) u_n (z - z_0)^{n+r-2}.$$

Thus, by substituting these derivatives in (3.55) and rearranging terms, it is apparent that $u(z)$ solves (3.55) if and only if

$$\sum_{n=0}^{\infty} (n + r)(n + r - 1) u_n (z - z_0)^{n+r-2}$$

$$+ \sum_{n=0}^{\infty} a_n (z - z_0)^n \sum_{n=0}^{\infty} (n + r) u_n (z - z_0)^{n+r-2}$$

$$+ \sum_{n=0}^{\infty} b_n (z - z_0)^n \sum_{n=0}^{\infty} u_n (z - z_0)^{n+r-2} = 0.$$

Multiplying by $(z - z_0)^{2-r}$ this identity leads to

$$\sum_{n=0}^{\infty} (n+r)(n+r-1)u_n(z-z_0)^n$$

$$+ \sum_{n=0}^{\infty} a_n(z-z_0)^n \sum_{n=0}^{\infty}(n+r)u_n(z-z_0)^n$$

$$+ \sum_{n=0}^{\infty} b_n(z-z_0)^n \sum_{n=0}^{\infty} u_n(z-z_0)^n = 0.$$

Thus, by taking the Cauchy product of the involved series, it becomes apparent that $u(z)$ solves (3.55) provided

$$\sum_{n=0}^{\infty} (n+r)(n+r-1)u_n(z-z_0)^n$$

$$+ \sum_{n=0}^{\infty} \sum_{k=0}^{n} a_k(n+r-k)u_{n-k}(z-z_0)^n$$

$$+ \sum_{n=0}^{\infty} \sum_{k=0}^{n} b_k u_{n-k}(z-z_0)^n = 0.$$

Equivalently,

$$\sum_{n=0}^{\infty} \left\{ (n+r)(n+r-1)u_n \right.$$

$$\left. + \sum_{k=0}^{n} [a_k(n+r-k) + b_k]\, u_{n-k} \right\} (z-z_0)^n = 0.$$

So, isolating the terms with $n = 0$ and $k = 0$ in the two sums of this identity yields

$$[r(r-1) + ra_0 + b_0]\, u_0$$

$$+ \sum_{n=1}^{\infty} \left\{ [(n+r)(n+r-1) + (n+r)a_0 + b_0]\, u_n \right.$$

$$\left. + \sum_{k=1}^{n} [(n+r-k)a_k + b_k]\, u_{n-k} \right\} (z-z_0)^n = 0.$$

Hence, setting

$$P(z) := z(z-1) + a_0 z + b_0 = z^2 + (a_0 - 1)z + b_0 \qquad (3.59)$$

and

$$\varphi_n := \sum_{k=1}^{n} [(n + r - k)a_k + b_k] u_{n-k}, \quad n \geq 1,$$

the previous identity can be written as

$$P(r)u_0 + \sum_{n=1}^{\infty} \left[P(r+n)u_n + \varphi_n \right] (z - z_0)^n = 0. \qquad (3.60)$$

Note that, for every integer $n \geq 1$, φ_n depends on u_j, $0 \leq j \leq n-1$, and a_j, b_j, $1 \leq j \leq n$. The simplest way to achieve (3.60) is by imposing that

$$P(r)u_0 = 0, \quad \text{and} \quad P(r+n)u_n + \varphi_n = 0 \quad \text{for all } n \geq 1. \qquad (3.61)$$

We take $u_0 = 1$ and r as a root of the polynomial $P(z)$, so that the first identity in (3.61) is trivially satisfied. More precisely, we suppose that

$$P(z) = (z - r)(z - s), \quad z \in \mathbb{C}, \qquad (3.62)$$

with

$$\operatorname{Re} r \geq \operatorname{Re} s.$$

Hence, setting

$$d := r - s,$$

we have that

$$\operatorname{Re} d = \operatorname{Re} r - \operatorname{Re} s \geq 0.$$

Moreover, from (3.62), we obtain, for each integer $n \geq 1$,

$$P(r+n) = n(n+d) \neq 0$$

because d has a non-negative real part. Thus, once r has been chosen so that $P(r) = 0$, (3.61) holds if and only if, for every $n \geq 1$,

$$u_n = \frac{-\varphi_n}{P(r+n)} = \frac{-\sum_{k=1}^{n} [(n + r - k)a_k + b_k] u_{n-k}}{P(r+n)}. \qquad (3.63)$$

This scheme provides us with the values of u_n, $n \geq 1$, in such a way that the series (3.58) solves (3.55), as long as it converges. In order to check such a convergence, it is enough to show that the convergence radius of the series (3.57) is positive, i.e. it suffices to prove that

$$\limsup_{n\to\infty} \sqrt[n]{|u_n|} < +\infty. \tag{3.64}$$

Let $\rho \in (0, R)$ be arbitrary. From (3.56), we have

$$\limsup_{n\to\infty} \sqrt[n]{|a_n|} < \rho^{-1}, \quad \limsup_{n\to\infty} \sqrt[n]{|b_n|} < \rho^{-1}.$$

Thus, there exists an integer $n_0 = n_0(\rho) \geq 1$ such that

$$|a_n| < \rho^{-n} \quad \text{and} \quad |b_n| < \rho^{-n} \quad \text{for all } n \geq n_0. \tag{3.65}$$

Moreover, since (3.65) implies that

$$|ra_n + b_n| \leq (|r| + 1)\, \rho^{-n} \quad \text{for all } n \geq n_0,$$

there exists a constant $M > 1$ such that

$$\max\{|a_n|, |b_n|, |ra_n + b_n|\} \leq M\rho^{-n} \quad \text{for all } n \geq 0. \tag{3.66}$$

On the other hand, since

$$P(r + n) = n(n + d), \quad \text{with } \operatorname{Re} d \geq 0,$$

one also has, for all integers $n \geq 1$,

$$|P(r + n)| = |n(n + d)| \geq \operatorname{Re}(n^2 + nd) \geq n^2. \tag{3.67}$$

Thus, thanks to (3.63), (3.66), and (3.67),

$$|u_1| = \frac{|ra_1 + b_1|}{|P(r + 1)|} \leq \frac{M}{\rho}.$$

Reasoning by induction, assume that, for some $n \geq 2$, one has

$$|u_j| \leq \left(\frac{M}{\rho}\right)^j, \quad 1 \leq j \leq n - 1 \tag{3.68}$$

(observe that, as we have taken $u_0 = 1$, (3.68) also holds for $j = 0$). Then, according to (3.63), (3.66), (3.67), and (3.68), we find that

$$|u_n| = \frac{\left|\sum_{k=1}^{n} \left[(n+r-k)a_k + b_k\right] u_{n-k}\right|}{|P(r+n)|}$$

$$\leq \frac{1}{n^2} \left(\sum_{k=1}^{n} |ra_k + b_k||u_{n-k}| + \sum_{k=1}^{n-1} (n-k)|a_k||u_{n-k}| \right)$$

$$\leq \frac{1}{n^2} \left(\sum_{k=1}^{n} \frac{M}{\rho^k} \left(\frac{M}{\rho}\right)^{n-k} + \sum_{k=1}^{n-1} (n-k)\frac{M}{\rho^k} \left(\frac{M}{\rho}\right)^{n-k} \right).$$

Thus, since $M > 1$, we obtain that, for all $n \geq 1$,

$$|u_n| \leq \frac{1}{n^2 \rho^n} \left(\sum_{k=1}^{n} M^{n-k+1} + \sum_{k=1}^{n-1} (n-k)M^{n-k+1} \right)$$

$$\leq \frac{nM^n + (n-1)^2 M^n}{n^2 \rho^n} \leq \left(\frac{M}{\rho}\right)^n.$$

Therefore,

$$\sqrt[n]{|u_n|} \leq \frac{M}{\rho} \quad \text{for all } n \geq 1,$$

and, consequently, (3.64) holds true, concluding the proof. $\square$

As an example, we consider the following differential equation:

$$u''(z) + \frac{1}{z}u'(z) + \left(1 - \frac{1}{4z^2}\right) u(z) = 0, \tag{3.69}$$

which is the Bessel equation of order $\alpha = \frac{1}{2}$ (see Exercise 11 of Chapter 3). Observe that (3.69) is a special case of (3.55), with

$$z_0 = 0, \quad a(z) = 1, \quad b(z) = z^2 - \frac{1}{4},$$

which are entire functions. A straightforward calculation shows that the change of variable

$$u(z) = z^{-1/2}w(z), \tag{3.70}$$

where, as usual, $z^{-1/2}$ denotes the corresponding principal determination, transforms the differential equation (3.69) into

$$w''(z) + w(z) = 0,$$

whose general solution is

$$w(z) = c_1 \sin z + c_2 \cos z, \quad c_1, c_2 \in \mathbb{C}.$$

Thus, the general solution of (3.69) is

$$u(z) = c_1 \frac{\sin z}{\sqrt{z}} + c_2 \frac{\cos z}{\sqrt{z}}, \quad c_1, c_2 \in \mathbb{C}.$$

If we repeat the proof of Theorem 3.9 in the particular case of equation (3.69), we obtain from (3.59) that $P(z) = z^2 - \frac{1}{4}$. Thus, $r = \frac{1}{2}$, and by comparing with the general solution $u(z)$ above, we observe that a function $v \in \mathcal{H}(D_\varepsilon(0))$ whose existence is guaranteed by Theorem 3.9 is, in this case,

$$v(z) = \frac{\sin z}{z}.$$

This example also shows that no solution $u \neq 0$ of equation (3.69) is holomorphic at $z = 0$, though all the solutions are of the form (3.70) with $w(z)$ entire. More generally, the singular equation (3.55) might not admit any nontrivial holomorphic solution at $z = z_0$, though, owing to Theorem 3.9, it admits a solution of the form

$$u(z) = (z - z_0)^r v(z),$$

with $v(z)$ holomorphic at $z_0 = 0$, for some $r \in \mathbb{C}$.

The following refinement of Theorem 3.9 actually shows that the function $v(z)$ can be taken to be holomorphic in the entire disk $D_R(z_0)$. When $R = +\infty$, we identify $D_R(z_0) = \mathbb{C}$.

Theorem 3.10. *Suppose $z_0 \in \mathbb{C}$, $R \in (0, +\infty]$ and $a, b \in \mathcal{H}(D_R(z_0))$. Then, there exist $r \in \mathbb{C}$ and $v \in \mathcal{H}(D_R(z_0))$ such that the function*

$$u(z) := (z - z_0)^r v(z), \quad |z - z_0| < R, \tag{3.71}$$

solves (3.55) in the disk $D_R(z_0)$.

Proof. According to Theorem 3.9, there exist $r \in \mathbb{C}$, $\varepsilon \in (0, R)$, and $v \in \mathcal{H}(D_\varepsilon(z_0))$ such that (3.71) solves (3.55) in $D_\varepsilon(z_0)$. Next, consider the open convex subset of $\mathbb{C}$ defined by

$$D_1 := \{ z \in D_R(z_0) : \operatorname{Re}(z - z_0) > 0 \}.$$

Since $\varepsilon < R$, it is apparent that

$$\zeta_0 := z_0 + \frac{\varepsilon}{2} \in D_\varepsilon(z_0) \cap D_1.$$

Now, consider the Cauchy problem related to (3.55):

$$\begin{cases} w''(z) + \dfrac{a(z)}{z - z_0} w'(z) + \dfrac{b(z)}{(z - z_0)^2} w(z) = 0, \\ w(\zeta_0) = u(\zeta_0), \quad w'(\zeta_0) = u'(\zeta_0). \end{cases} \tag{3.72}$$

According to Corollary 3.7, the initial value problem (3.72) admits a unique solution, $w(z)$, holomorphic in D_1. Moreover, again by Corollary 3.7, $u(z)$ must be the unique holomorphic solution of (3.72) in $D_\varepsilon(z_0) \cap D_1$. Thus, $u = w$ in $D_\varepsilon(z_0) \cap D_1$; therefore, $u(z)$ admits a holomorphic extension to D_1 still solving (3.55).

By repeating the previous argument in each of the remaining quadrants,

$$D_2 := \{ z \in D_R(z_0) : \operatorname{Im}(z - z_0) > 0 \},$$
$$D_3 := \{ z \in D_R(z_0) : \operatorname{Re}(z - z_0) < 0 \},$$
$$D_4 := \{ z \in D_R(z_0) : \operatorname{Im}(z - z_0) < 0 \},$$

it becomes apparent that $u(z)$ admits a holomorphic extension, necessarily unique, to $D_R(z_0) \backslash \{z_0\}$. Thus, $v \in \mathcal{H}(D_R(z_0))$. This completes the proof. $\qquad\square$

Actually, since Corollary 3.7 holds in any convex open subset, Ω, of $\mathbb{C}$ containing z_0, the following result holds.

Theorem 3.11. *Let Ω be a convex open subset of $\mathbb{C}$. Then, for every $a, b \in \mathcal{H}(\Omega)$ and $z_0 \in \Omega$, there exist $r \in \mathbb{C}$ and $v \in \mathcal{H}(\Omega)$ such that*

$$u(z) := (z - z_0)^r v(z), \quad z \in \Omega,$$

solves (3.55) *in* Ω.

The technical details of the proof are left to the reader as Exercise 14 of Chapter 3.

3.9 Exercises

1. Expand in Taylor series centered at zero to find the general solutions of the differential equations
$$u''(z) - \omega^2 u(z) = 0 \quad \text{and} \quad u''(z) + \omega^2 u(z) = 0,$$
where $\omega > 0$ is constant.

2. Expand in Taylor series centered at zero to find the general solutions of the differential equations
$$u''(z) - 2zu(z) = 0 \quad \text{and} \quad u''(z) + 2zu(z) = 0.$$

3. Expand in Taylor series centered at zero to find the general solutions of the differential equations
$$u''(z) - z^2 u(z) = 0 \quad \text{and} \quad u''(z) + z^2 u(z) = 0.$$

4. Let $P(x)$ be an arbitrary polynomial of degree $n \geq 1$. Prove that, for every $\varepsilon > 0$, there exists a constant $C = C(\varepsilon) > 0$ such that
$$e^{-\varepsilon x^2} P(x) \leq C \quad \text{for all } x \in \mathbb{R}.$$
Deduce that
$$\int_{-\infty}^{\infty} e^{-x^2} P(x)\, dx < +\infty.$$

5. Check that the polynomials introduced in (3.31) are indeed Hermite polynomials.

6. Prove that, for every integer $n \geq 0$, the functions
$$H_n(z) = (-1)^n e^{z^2} \frac{d^n}{dz^n} \left(e^{-z^2} \right), \quad n \geq 0,$$
are polynomials of degree n that solve the Hermite equation (3.25) for $\lambda = 2n + 1$. Then, prove the validity of the recurrence formula
$$H_{n+1}(z) = 2z H_n(z) - 2n H_{n-1}(z).$$
This formula can be used to determine H_{n+1} from the two previous polynomials, H_n and H_{n-1}. Thus, starting with $H_0(z) = 1$ and $H_1(z) = 2z$, one can easily determine the remaining Hermite polynomials.
Prove that the leading coefficient of $H_n(x)$ equals 2^n.

7. Adapt the proof of (3.33) to obtain (3.34).
8. Prove that the wave functions

$$\psi_n(x) := \sqrt{\frac{1}{2^n \sqrt{\pi} n!}} \, e^{-\frac{x^2}{2}} H_n(x), \quad n \geq 0,$$

satisfy

$$\int_{-\infty}^{+\infty} \psi_n^2(x)\, dx = 1, \quad n \geq 0,$$

as well as the orthogonality conditions

$$\int_{-\infty}^{+\infty} \psi_n(x)\psi_m(x)\, dx = 0, \quad n \neq m, \quad n, m \geq 0.$$

9. Prove the recurrence formulas (3.51) and (3.52), and deduce from such relations that, for all $z \in \mathbb{C}$ and $n \geq 1$, the following hold:

$$2J_n'(z) = J_{n-1}(z) - J_{n+1}(z),$$

$$\frac{d}{dz}\left(z^n J_n(z)\right) = z^n J_{n-1}(z), \tag{3.73}$$

$$\frac{d}{dz}\left(z^{-n} J_n(z)\right) = -z^{-n} J_{n+1}(z). \tag{3.74}$$

10. Show that, for all $n \geq 0$, between any two (real) positive consecutive zeros of J_n, there exists exactly one zero of J_{n+1}. [Hint: Use (3.73) and (3.74).]
11. Consider, for every $\alpha \in [0, \infty)$, the Bessel equation of order α:

$$u''(z) + \frac{1}{z}u'(z) + \left(1 - \frac{\alpha^2}{z^2}\right)u(z) = 0. \tag{3.75}$$

Use a change of variable of the form

$$u(z) = z^\alpha v(z)$$

and solve the resulting equation in $v(z)$ by means of a Taylor series centered at the origin to find the Bessel functions of non-negative real order α.

12. Given $\alpha \in [0, \infty)$, solve the hyperbolic Bessel equation of order α,

$$u''(z) + \frac{1}{z}u'(z) + \left(1 + \frac{\alpha^2}{z^2}\right)u(z) = 0,$$

by means of the method used in the proof of Theorem 3.9.

13. Adapt the study of equation (3.69) to find the general solution of

$$u''(z) + \frac{1}{z}u'(z) + \left(1 - \frac{1}{4z^2}\right)u(z) = 3\sqrt{z}\sin z, \quad z \in \mathbb{C} \setminus \{0\}.$$

14. Adapt the proof of Theorem 3.10 to complete the technical details of the proof of Theorem 3.11.

15. Consider the eigenvalue problem

$$\begin{cases} -\Delta u = \lambda u & \text{in } Q, \\ u = 0 & \text{on } \partial Q, \end{cases} \tag{3.76}$$

where, given $L, M > 0$, Q denotes the rectangle

$$Q := [0, L] \times [0, M].$$

Proceed as in Section 3.7 to solve the eigenvalue problem (3.76) by searching for solutions of the form

$$u(x, y) = X(x)Y(y) \neq 0$$

and solving the resulting differential equations for the unknowns $X(x)$ and $Y(y)$.

16. Expand in a Taylor series to find a solution of

$$2zu''(z) + u'(z) - u(z) = 0$$

in a neighborhood of zero.

17. Expand in a Taylor series to find a solution of

$$(z - 1)u''(z) + zu'(z) + u(z) = 0$$

in a neighborhood of 1.

18. For every $\lambda \in \mathbb{R}$, the differential equation

$$zu''(z) + (1 - z)u'(z) + \lambda u(z) = 0$$

is known as the *Laguerre equation*, named after E. Laguerre. Solve it using the method introduced in the proof of Theorem 3.9, and characterize the values of the spectral parameter λ for which the equation admits a polynomial solution. Such polynomial solutions are known as the *Laguerre polynomials*.

19. For every $\lambda \in \mathbb{R}$, the differential equation

$$(1 - z^2)u''(z) - 2zu'(z) + \lambda u(z) = 0$$

is known as the *Legendre equation*, named after A. M. Legendre. Solve it using the method introduced in the proof of Theorem 3.9, and characterize the values of the spectral parameter λ for which it admits a polynomial solution. Such polynomial solutions are known as the *Legendre polynomials*.

20. The Schrödinger equation for the hydrogen atom is given by

$$-\frac{\hbar^2}{2m}\Delta\psi - \frac{e^2}{\sqrt{x^2 + y^2 + z^2}}\psi = E\psi, \qquad (3.77)$$

where $\hbar$ is the Planck constant, named after M. Planck; m, e, and E are, respectively, the mass, charge, and energy of the electron; and

$$\Delta = \frac{\partial^2}{\partial x^2} + \frac{\partial^2}{\partial y^2} + \frac{\partial^2}{\partial z^2}.$$

The solutions of (3.77) are the wave functions of the hydrogen atom. Their squares are probability densities. Thus, they must satisfy

$$\int_{\mathbb{R}^3} \psi^2(x, y, z)\, dx\, dy\, dz = 1. \qquad (3.78)$$

(a) Convince yourself that the *Ansatz*

$$\psi(x, y, z) = X(x)Y(y)Z(z) \neq 0$$

does not provide us with three uncoupled differential equations in the variables x, y, and z.

(b) Show that, in spherical coordinates,

$$x = \rho \cos\theta \cos\phi, \quad y = \rho \sin\theta \cos\phi, \quad z = \rho \sin\phi,$$

the wave function

$$\varphi(\rho, \theta, \phi) := \psi(x, y, z)$$

satisfies

$$\frac{1}{\rho^2}\frac{\partial}{\partial\rho}\left(\rho^2\frac{\partial\varphi}{\partial\rho}\right) + \frac{1}{\rho^2\cos\phi}\frac{\partial}{\partial\phi}\left(\cos\phi\frac{\partial\varphi}{\partial\phi}\right)$$
$$+ \frac{1}{\rho^2\cos^2\phi}\frac{\partial^2\varphi}{\partial\theta^2} + \frac{1}{\rho}\frac{2me^2}{\hbar^2}\varphi = -\frac{2mE}{\hbar^2}\varphi. \tag{3.79}$$

(c) Prove that (3.79) admits particular solutions of the form

$$\varphi(\rho, \theta, \phi) = R(\rho)\Theta(\theta)\Phi(\phi) \neq 0,$$

where the functions R, Θ, and Φ satisfy certain uncoupled differential equations in the variables ρ, θ, and ϕ, respectively. In particular, show that $R(\rho)$ solves the Laguerre equation introduced in Exercise 18 of Chapter 3.

(d) Find all the 2π-periodic solutions for the equation involving the function Θ.

(e) Perform the change of variable

$$P(x) = \Phi(\phi), \quad x = \sin\phi,$$

in the equation satisfied by Φ to get the Legendre equation, whose polynomial solutions have been constructed in Exercise 19 of Chapter 3.

(f) Proceed as in Section 3.6 to determine all the solutions of the equation involving the function R that satisfy

$$\int_{-\infty}^{\infty} R^2(\rho)\, d\rho < +\infty.$$

Infer from this analysis that, except for a sequence of values of the energy E, (3.77) cannot admit a solution ψ satisfying (3.78).

   (g) Analyze the behavior of the wave functions of the hydrogen
       atom.

21. Let $a_j \in \mathbb{C}$, for all $j \in \{0, \ldots, n-1\}$, and let $P(z)$ be a polynomial
    of the form

$$P(z) = z^n + \sum_{j=0}^{n-1} a_j z^j$$

and $R > 0$ such that the open disk of radius R centered at 0,
D_R, contains all the roots of $P(z)$.

(a) Show that

$$u(z) = \frac{1}{2\pi i} \int_{|\zeta|=R} \frac{e^{z\zeta}}{P(\zeta)} \, d\zeta, \quad z \in \mathbb{C},$$

   solves $P(D)u = 0$.
(b) Show that $u(z)$ also satisfies

$$D^{n-1}u(0) = 1, \quad D^j u(0) = 0, \quad 0 \leq j \leq n-2.$$

3.10   Final Comments

The contents of Sections 3.3 and 3.5 have been developed from the
lecture notes of the course *"Elements on the theory of functions of
a complex variable and differential equations"*, delivered by López-
Gómez at Complutense University of Madrid from the late 1990s
until the early 2010s. Such notes were originally intended to complement the material on the theory of functions of one complex variable
covered by López-Gómez's (2001a) textbook, with a paradigmatic
application to the fundamental theory of linear systems of ordinary
differential equations. The lectures notes of K. O. Friedrichs, also
cover these topics (see Friedrichs, 1956, Chapter V) from a more
advanced perspective.

   Although in this chapter we have developed the existence theory
of (3.1) in an arbitrary convex subset Ω of $\mathbb{C}$, Theorems 3.4 and 3.5
and Corollary 3.7 still hold true if Ω is a simply connected domain of
$\mathbb{C}$. Indeed, since the Cauchy–Goursat theorem (see Theorem 3.3) is
valid in simply connected subsets of $\mathbb{C}$, the following sharper version
of Theorem 3.4 holds.

Theorem 3.12. *Suppose Ω is a simply connected open subset of $\mathbb{C}$ and $a, b \in \mathcal{H}(\Omega)$. Then, for every $z_0 \in \Omega$ and $u_0 \in \mathbb{C}$, the Cauchy problem*

$$\begin{cases} u' = a(z)u + b(z), \\ u(z_0) = u_0, \end{cases} \tag{3.80}$$

possesses a unique solution $u \in \mathcal{H}(\Omega)$, which is denoted by $u(z; z_0, u_0)$.

Proof. According to the Cauchy–Goursat theorem, the function $a \in \mathcal{H}(\Omega)$ possesses a holomorphic primitive, $A \in \mathcal{H}(\Omega)$, i.e. a function $A(z)$ satisfying

$$A'(z) = a(z) \quad \text{for all } z \in \Omega.$$

Moreover, since $A(z)+c$ provides us with another primitive of $a(z)$ in Ω for every constant $c \in \mathbb{C}$, we can assume without loss of generality that $A(z_0) = 0$ (indeed, it suffices to choose $c = -A(z_0)$). For such a choice, the function $h(z) := e^{A(z)}$ satisfies $h(z_0) = 1$ and

$$h'(z) = A'(z)h(z) = a(z)h(z) \quad \text{for all } z \in \Omega.$$

Thus, the change of variable

$$u(z) = h(z)v(z)$$

transforms (3.80) into the problem

$$v'(z) = e^{-A(z)}b(z), \quad v(z_0) = u_0. \tag{3.81}$$

Since the function $e^{-A(z)}b(z)$ is holomorphic in Ω, thanks again to the Cauchy–Goursat theorem, there exists a function $v \in \mathcal{H}(\Omega)$ that solves (3.81). Therefore, (3.80) has, at least, a holomorphic solution in Ω.

The uniqueness of the solution of (3.80) can be derived, e.g. with the following argument. Suppose (3.80) has two solutions $u_1(z)$ and $u_2(z)$. Then, according to Theorem 3.4, for every $\varepsilon > 0$ such that $\bar{D}_\varepsilon(z_0) \subset \Omega$, one has that $u_1(z) = u_2(z)$ for all $z \in D_\varepsilon(z_0)$. Therefore, $u_1 = u_2$ in Ω by the *identity principle*. $\qquad\square$

Similarly, one can derive the counterparts of Theorem 3.5 and Corollary 3.7 in any simply connected open set, Ω, with $z_0 \in \Omega$, where the coefficients are holomorphic functions. Actually, in Friedrichs (1956, p. 161) it is explained how any solution defined on a disk can be extended to larger simply connected domains. However, the convexity assumption on Ω allows us to obtain explicit formulas for the solutions of (3.5) and (3.8) (see equations (3.6) and (3.12), respectively).

Actually, the assumption that Ω is simply connected is essential for the validity of Theorem 3.12. Indeed, the problem

$$\begin{cases} u' = \frac{1}{z}, \\ u(1) = 0, \end{cases} \tag{3.82}$$

has holomorphic coefficients in $\mathbb{C}\backslash\{0\}$, which is not simply connected. According to Theorem 3.4, for every convex open subset $\Omega \subset \mathbb{C}\backslash\{0\}$ with $1 \in \Omega$,

$$u(z) := \operatorname{Log} z, \quad z \in \Omega,$$

provides us with the unique solution of (3.82) in Ω. Nevertheless, problem (3.82) does not admit any solution in $\mathcal{H}(\mathbb{C}\backslash\{0\})$. Indeed, assume by contradiction that there exists a solution, $u(z)$, which is holomorphic in $\mathbb{C}\backslash\{0\}$. Then, for every positively oriented closed Jordan curve, γ, enclosing the origin in $\mathbb{C}$, one would obtain that

$$2\pi i = \int_\gamma \frac{1}{z}\, dz = \int_\gamma u'(z)\, dz = 0,$$

which is impossible.

The Airy equation (3.18) was introduced by Airy in 1838. Airy functions are solutions of this equation that arise in microscopy and astronomy, though they also have a number of applications in quantum mechanics in connection with semiconductor devices.

The first differential equation of the Bessel type was actually studied by D. Bernoulli in 1732 in the analysis of the oscillations of a chain hanging from some of its ends. Later, L. Euler, J. L. Lagrange, J. B. J. Fourier, and S. D. Poisson, among others, dealt with a number of physical and mathematical questions that gave rise to certain variants of the Bessel equations. As it was not until 1824 that the astronomer F. W. Bessel carried out a systematic analysis of this family of equations in connection with a problem of planetary motion, not only the differential equation but also its solutions were named after him. As illustrated in Section 3.7, the Bessel equation arises in a rather natural way when analyzing the eigenvalue problem of the Laplace operator in the disk. Thus, its scientific impact is huge in several areas of Science and Technology.

The content of Section 3.6 goes back to López-Gómez (2001b, Chapter 9), and it was inspired by Chpolski (1977), though Chpolski did not give a rigorous proof of the fact that the Schrödinger equation (3.22) does not admit any wave function if $\lambda > 0$ is not an odd integer number. Similarly, Theorem 3.9 has been borrowed from López-Gómez (2001b, Chapter 8).

Finally, we mention that the method used in the proof of Theorem 3.9 goes back to Frobenius (1873).

Part 2
Nonlinear Systems

Chapter 4

An Introduction to Nonlinear Differential Equations

This chapter analyzes some pivotal nonlinear initial value problems of the type

$$\begin{cases} u' = f(t, u), \\ u(t_0) = u_0, \end{cases} \tag{4.1}$$

where $f \in \mathcal{C}((a, b) \times \mathbb{R}; \mathbb{R})$, with

$$a, b \in [-\infty, +\infty], \quad a < b, \quad t_0 \in (a, b), \quad u_0 \in \mathbb{R}.$$

These prototypical nonlinear models will be used as paradigmatic examples of the abstract general theory developed in Parts 2 and 3. As in the linear case analyzed in Chapter 1, by a solution of (4.1) in (a, b) we mean any function $u \in \mathcal{C}^1((a, b); \mathbb{R})$ such that

$$u'(t) = f(t, u(t)) \quad \text{for all } t \in (a, b), \quad \text{and} \quad u(t_0) = u_0.$$

4.1 The Logistic Equation

The logistic equation, whose name dates back to Pearl and Reed (1920), is the most paradigmatic differential equation in Population Dynamics and Ecology, and it also arises in many areas of Science and Technology. It describes the evolution of the number of individuals of a population, $u(t)$, at time $t > 0$ in terms of the intrinsic growth rate of the population, λ, and the carrying capacity of the habitat, $\frac{\lambda}{m}$.

209

Here, λ and m are assumed to be positive constants. In the light of this model, the logistic equation reads

$$u' = \lambda u - mu^2 = (\lambda - mu)\,u. \tag{4.2}$$

Since

$$\lambda - mu(t) = \frac{u'(t)}{u(t)}$$

measures the per capita growth rate of the population, the term $-mu(t)$ can be regarded as the intensity of the intraspecific competition among the individuals of the species: the smaller the value of m, the lower the intraspecific competition effects, and vice versa.

The main goal of this section is to obtain as much information as possible on the solutions of (4.2). Simultaneously, we introduce and discuss some of the most important concepts that are analyzed later in full generality, such as *equilibria, maximal interval of definition of solutions, Lyapunov stability, attractivity* or *repulsivity,* and *continuous dependence with respect to initial data.*

The logistic equation is a very special example of a more general family of differential equations of the type

$$u' = A(t)u + B(t)u^p, \tag{4.3}$$

with $p \in \mathbb{R}$ and $A, B \in \mathcal{C}((a,b);\mathbb{R})$, for some $a < b$, which are referred to as the *Bernoulli equations* after Jakob Bernoulli. When $p \in \{0,1\}$, (4.3) becomes a linear equation, whose integration has already been carried out in Section 1.1. When $p \in \mathbb{R}\setminus\{0,1\}$, there exists a constant $\alpha \in \mathbb{R}$ such that the change of variable

$$u = v^\alpha \tag{4.4}$$

transforms (4.3) into a linear equation. Indeed, substituting (4.4) in (4.3) yields

$$\alpha v^{\alpha-1}v' = A(t)v^\alpha + B(t)v^{\alpha p}.$$

Thus, multiplying by $v^{1-\alpha}$ shows that

$$\alpha v' = A(t)v + B(t)v^{\alpha p+1-\alpha}.$$

Consequently, making the choice

$$\alpha = \frac{1}{1-p},$$

it becomes apparent that the change of variable (4.4) transforms (4.3) into the linear equation

$$v' = (1-p)A(t)v + (1-p)B(t).$$

Therefore, by the variation of constants formula (see Theorem 1.2),

$$u(t) = \left[e^{(1-p)\int_{t_0}^t A(\sigma)\,d\sigma}\mu + (1-p)\int_{t_0}^t e^{(1-p)\int_s^t A(\sigma)\,d\sigma}B(s)\,ds \right]^{\frac{1}{1-p}} \tag{4.5}$$

for all $t \in (a,b)$, provides us with a one-parameter family, in terms of $\mu \in \mathbb{R}$, of solutions of (4.3).

It is extremely important to stress that, unless $p \in \{0,1\}$, (4.3) is a nonlinear equation and, hence, the superposition principle established in Lemma 1.1 fails. Indeed, in the special case of $A \equiv 0$, $B \equiv 1$, and $p = 2$, (4.3) becomes

$$u' = u^2, \tag{4.6}$$

and for every solution $u \neq 0$ of (4.6), the product λu cannot solve (4.6) unless either $\lambda = 0$ or $\lambda = 1$. Similarly, if $u \neq 0$ and $v \neq 0$ solve (4.6), then the function $u + v$ cannot satisfy

$$(u+v)' = (u+v)^2$$

because $(u+v)^2 \neq u^2 + v^2$.

Equilibria of the logistic equation.

In order to analyze the logistic equation (4.2), one should begin by finding its *constant solutions*, as they are the main candidates to describe the *dynamics*[1] of (4.2) and, in addition, because they might be *singular solutions* of (4.5) in the sense that they cannot be recovered for any choice of the parameter μ in (4.5). As constant solutions do not vary with time, they are usually referred to as the

---

[1] By "dynamics", here, we mean the asymptotic behavior of the solutions as $t \uparrow +\infty$, or $t \downarrow -\infty$.

steady states, or *equilibria*, or *rest points*, of the differential equation, names which are highly suggestive from the point of view of applications. Observe that the equilibria of (4.2) are the zeroes of its nonlinearity, i.e. the roots of

$$\lambda u - mu^2 = 0.$$

Thus, the set of equilibria of (4.2), $\mathcal{E}$, is given by

$$\mathcal{E} = \left\{ 0, \frac{\lambda}{m} \right\}.$$

The corresponding constant solutions are represented in Figure 4.1, where we plot $u(t)$ versus t, as the two horizontal lines at the levels 0 and $\frac{\lambda}{m}$.

General solution of the logistic equation.

The rigorous proof of the fact that the Cauchy problem

$$\begin{cases} u' = \lambda u - mu^2, \\ u(0) = x, \end{cases} \tag{4.7}$$

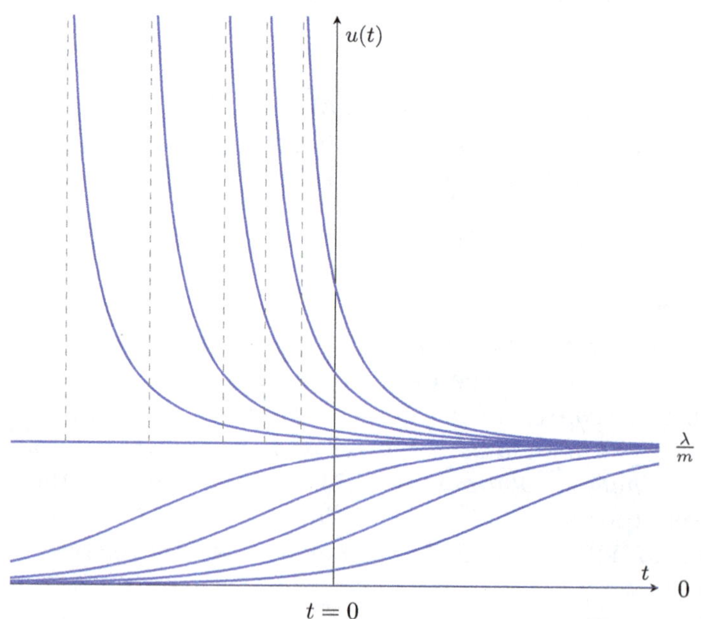

Fig. 4.1. The non-negative solutions of the logistic equation (4.2).

possesses a unique solution for all $x \in \mathbb{R}$ is postponed to the following section when $x \in \mathbb{R} \setminus \mathcal{E}$ and to the abstract general theory developed in Chapter 5 for all $x \in \mathbb{R}$. Next, we find the unique solution of (4.7) for all $x > 0$, which is the most interesting case from an ecological point of view, by performing the change of variable given by (4.4), which, in the case of the logistic equation, reads $u = v^{-1}$ since it is of the Bernoulli type with $p = 2$. The new variable v satisfies the linear equation with constant coefficients

$$\begin{cases} v' = -\lambda v + m, \\ v(0) = 1/x, \end{cases}$$

and hence, for every $t \in \mathbb{R}$,

$$v(t) = e^{-\lambda t} \left(\frac{1}{x} - \frac{m}{\lambda} \right) + \frac{m}{\lambda}.$$

Thus, the (unique) solution of (4.7), denoted by $u(t; x)$, is given by

$$u(t; x) = \frac{1}{e^{-\lambda t} \left(\frac{1}{x} - \frac{m}{\lambda} \right) + \frac{m}{\lambda}}, \tag{4.8}$$

as soon as $v(t) > 0$. It is important to note that, though

$$\lim_{x \downarrow 0} u(t; x) = 0 \quad \text{for all } t \in \mathbb{R},$$

the equilibrium $u = 0$ remains outside the set of solutions represented in (4.8), in the sense that no value of $x \neq 0$ in (4.8) allows us to obtain the zero solution.

This explains why the equilibria of the differential equations should be determined from the very beginning since they might not be obtainable by means of the classical methods of elementary integration. This feature also explains why equilibria are also called *singular solutions* in the classical literature on differential equations.

Qualitative behavior of the solutions.
Since $\lambda > 0$, it follows from (4.8) that

$$\lim_{t \uparrow +\infty} u(t; x) = \frac{\lambda}{m},$$

and this explains why the equilibrium $\frac{\lambda}{m}$ is said to be a *global attractor* as $t \to +\infty$ for all the positive solutions of (4.2). Nevertheless, the nature of all these solutions differs according to the initial size of the population, represented by x. Indeed, if $x < \frac{\lambda}{m}$, then $\frac{1}{x} - \frac{m}{\lambda} > 0$, and hence $u(t; x)$ is increasing in t and globally defined for all $t \in \mathbb{R}$ since the denominator in (4.8) is always positive. Moreover,

$$\lim_{t\downarrow-\infty} u(t; x) = 0 \quad \text{and} \quad \lim_{t\uparrow+\infty} u(t; x) = \frac{\lambda}{m}.$$

Therefore, all these solutions, independently of the size of $x \in (0, \frac{\lambda}{m})$, connect the two equilibria, 0 and $\frac{\lambda}{m}$, when t varies from $-\infty$ to $+\infty$. This is why they are called *heteroclinic connections*, where the Greek-origin word "heteroclinic" refers to *different patterns*. These solutions are represented below the level $\frac{\lambda}{m}$ in Figure 4.1. On the contrary, when $x > \frac{\lambda}{m}$, then $\frac{1}{x} - \frac{m}{\lambda} < 0$, and hence $u(t; x)$ is decreasing in time, t. Actually, in this case, the solutions are only defined in the time interval $(T_{\max}(x), +\infty)$, where $T_{\max}(x) < 0$ denotes the unique value of T such that

$$e^{-\lambda T}\left(\frac{1}{x} - \frac{m}{\lambda}\right) + \frac{m}{\lambda} = 0,$$

which is given by

$$T_{\max}(x) = -\frac{1}{\lambda}\ln\frac{\frac{m}{\lambda}}{\frac{m}{\lambda} - \frac{1}{x}} = \frac{1}{\lambda}\ln\left(1 - \frac{\lambda}{mx}\right).$$

Consequently, $(T_{\max}(x), +\infty)$ is the *maximal existence interval* of these solutions. Moreover,

$$\lim_{t\downarrow T_{\max}(x)} u(t; x) = +\infty \quad \text{and} \quad \lim_{t\uparrow+\infty} u(t; x) = \frac{\lambda}{m}. \tag{4.9}$$

Figure 4.1 also shows the graph of a series of these solutions. It is quite interesting to observe that the blow up times, $T_{\max}(x)$, of these solutions depend on the initial data, x, and that $T_{\max}(x)$ is actually a continuous increasing function of $x > \frac{\lambda}{m}$ such that

$$\lim_{x\downarrow\frac{\lambda}{m}} T_{\max}(x) = -\infty \quad \text{and} \quad \lim_{x\uparrow+\infty} T_{\max}(x) = 0,$$

as illustrated in Figure 4.2. Consequently, the larger the initial value x, the smaller the maximal existence interval of $u(t; x)$, $(T_{\max}(x), +\infty)$, and vice versa.

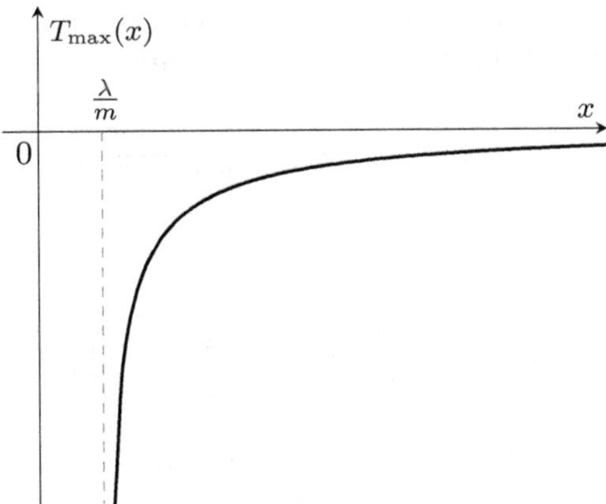

Fig. 4.2. The blow up time, $T_{\max}(x)$, as a function of $x > \frac{\lambda}{m}$.

This is a typical behavior that can only occur in nonlinear equations because in linear equations and systems, the solutions are defined in any interval where the coefficients are continuous, as discussed in Chapter 1. In nonlinear equations, instead, even when $f(t, u)$ is independent of time and analytic, as in the case we are currently analyzing, where

$$f(t, u) = \lambda u - mu^2 \tag{4.10}$$

is a polynomial of degree two, the solutions can blow up in finite time, and the blow up time can vary with the initial datum, x.

Continuous dependence on the initial datum $x \geq 0$.

From (4.8), it is easily seen that, for $x > 0$, $x \neq \frac{\lambda}{m}$, $u(t; x)$ depends continuously on x, in the sense that if we fix any $x_0 > 0$, $x_0 \neq \frac{\lambda}{m}$, a compact interval J contained in the maximal existence interval of $u(t; x_0)$ and an $\varepsilon > 0$, then there exists $\delta = \delta(\varepsilon, J)$ such that, for every $x \in (x_0 - \delta, x_0 + \delta)$, $u(t; x)$ is also defined for all $t \in J$ and

$$u(t; x_0) - \varepsilon < u(t; x) < u(t; x_0) + \varepsilon \quad \text{for all } t \in J,$$

i.e. $u(t; x)$ lies in a tubular ε-neighborhood of $u(t; x_0)$ for $t \in J$.

Proceeding deeper into this analysis, we now show that, although for every $x \in (0, \lambda/m)$, $u(t; x)$ approximates λ/m as $t \uparrow +\infty$, the solution $u(t; x)$ can be maintained as close as we wish to 0 during an arbitrarily large period of time by simply taking x sufficiently small.

Indeed, pick $\varepsilon > 0$ and a time $T > 0$; ε can be as small as we wish, while T is arbitrarily large. Then, the estimate

$$u(T;x) = \frac{1}{e^{-\lambda T}\left(\frac{1}{x} - \frac{m}{\lambda}\right) + \frac{m}{\lambda}} < \varepsilon$$

holds if and only if

$$x < \delta(\varepsilon) := \left[\frac{m}{\lambda} + e^{\lambda T}\left(\frac{1}{\varepsilon} - \frac{m}{\lambda}\right)\right]^{-1}.$$

Therefore, for every $T > 0$ and $\varepsilon > 0$, there exists $\delta = \delta(\varepsilon) \in \left(0, \frac{\lambda}{m}\right)$ such that

$$0 < u(t;x) \le u(T;x) < \varepsilon \quad \text{for all } t \le T \quad \text{if } 0 < x < \delta(\varepsilon)$$

(we are using the fact that the solutions are increasing in t if $0 < x < \frac{\lambda}{m}$). In other words, there is *continuous dependence* of the solution $u(t;x)$ with respect to the initial data x even at $x = 0$, though, inexorably, $u(t;x)$ will approximate to λ/m as $t \uparrow +\infty$. This is why the steady state 0 is said to be a *repeller*. Later, in Section 7.7, it will be shown that the optimal condition to get continuous dependence with respect to the initial data is the uniqueness of solutions of the underlying Cauchy problem. Actually, these are equivalent properties.

Naturally, the continuous dependence with respect to x can also be tested backward in time, not only forward. Indeed arguing as above, it is easily seen that for every $T < 0$ and $\varepsilon > 0$, there exists $\delta = \delta(\varepsilon) > 0$ such that

$$\frac{\lambda}{m} < x < \frac{\lambda}{m} + \delta$$

implies $T_{\max}(x) < T$, and

$$\frac{\lambda}{m} < u(t;x) \le u(T;x) < \frac{\lambda}{m} + \varepsilon \quad \text{for all } t \ge T.$$

Thus, though $u(t;x)$ blows up in finite time at a sufficiently negative time, it can stay as close as we wish to the equilibrium $\frac{\lambda}{m}$ for huge periods of time if x is taken sufficiently close to $\frac{\lambda}{m}$.

Structure of the solutions' set and invariance by time translations.

An important feature of the logistic equation (4.2) is that it is *autonomous*, in the sense that the nonlinearity (4.10) is independent of t. More generally, an equation, or system, $u' = f(t, u)$ is said to be autonomous if

$$f(t, u) = f(u).$$

In such a case, given a solution $u(t)$ defined on some open interval J and an arbitrary $T \in \mathbb{R}$, all shifted functions

$$v(t) := u(t + T), \quad t \in J - \{T\} := \{\tau - T : \tau \in J\},$$

also solve $u' = f(u)$. Indeed,

$$v'(t) = u'(t + T) = f(u(t + T)) = f(v(t)), \quad t \in J - \{T\}.$$

This means that the profiles of the solutions of (4.2) plotted in Figure 4.1 can be obtained by shifting those of the other solutions horizontally. This property extraordinarily facilitates the study of the dynamics of autonomous models. In general, such a study may be much harder for non-autonomous equations and systems.

By adapting the previous analysis (see Exercise 1 of Chapter 4), it is possible to show that, for each $x < 0$, there exists $T_{\max}(x) > 0$ such that the solution $u(t; x)$ of (4.7) is defined in $(-\infty, T_{\max}(x))$, is decreasing, and satisfies

$$\lim_{t \downarrow -\infty} u(t; x) = 0, \qquad \lim_{t \uparrow T_{\max}(x)} = -\infty. \tag{4.11}$$

Stability of equilibria and dynamics.

The most important qualitative features of the logistic equation (4.2) can be synthesized as shown in Figure 4.3, where we plot the graph of the nonlinearity $f(u) = \lambda u - mu^2$ on its upper part, together with the trajectories of the solutions of (4.2) on $\mathbb{R}$ below.

The zeroes of f, 0 and $\frac{\lambda}{m}$, are the steady-state solutions of (4.2), which are marked on the lower part of the figure with two thick points.

In the interval where f is positive, $(0, \frac{\lambda}{m})$, the solutions grow because

$$u'(t) = f(u(t)) > 0,$$

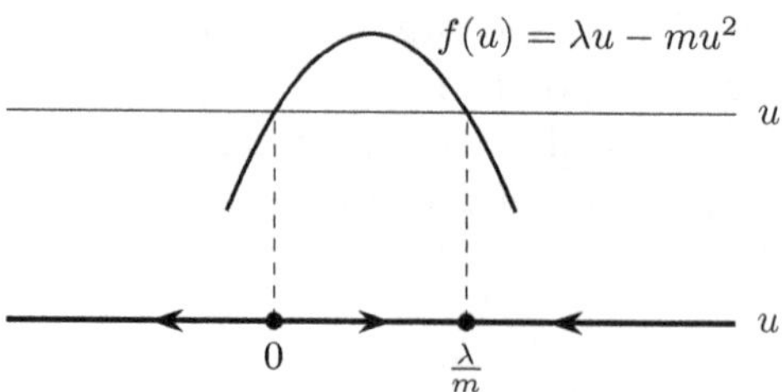

Fig. 4.3.　Nonlinearity (upper plot) and dynamics (lower plot) of the logistic equation.

which is indicated by the right-pointing arrow on the lower line of the figure. In the intervals where f is negative, $(-\infty, 0)$ and $(\frac{\lambda}{m}, +\infty)$, the solutions instead satisfy

$$u'(t) = f(u(t)) < 0$$

and hence decrease, which has been expressed in Figure 4.3 by the two arrows pointing to the left on each of those intervals. Naturally, the trajectories of the steady states are single (rest) points. Figure 4.3 immediately suggests the character of the equilibria: 0 is a repeller, while $\frac{\lambda}{m}$ is a global attractor for all positive solutions of the logistic equation.

　　In many applications, ascertaining this information, i.e. the dynamics of the model, is fairly sufficient for all practical purposes, especially when it is not possible to get explicit solutions.

4.2　Equations with Separated Variables

In this section, we analyze *equations with separated variables*, which are those of the form

$$u' = f(t, u) = F(t)G(u)$$

for some continuous functions $F(t)$ and $G(u)$. Observe that all scalar autonomous equations, $u' = f(u)$ with $f \in \mathcal{C}(\mathbb{R}; \mathbb{R})$, are of this type, particularly the logistic equation analyzed in Section 4.1. We prove that, under certain conditions on F and G, the Cauchy problem

$$\begin{cases} u' = F(t)G(u), \\ u(t_0) = u_0, \end{cases}$$

possesses a unique solution, which is defined (at least) in a neighborhood of $t = t_0$. The proof relies on the implicit function theorem, which we recall here for convenience.

Theorem 4.1 (Implicit function theorem). *Let $(t_0, u_0) \in \mathbb{R}^2$, $T, \varepsilon > 0$, $Q := [t_0 - T, t_0 + T] \times [u_0 - \varepsilon, u_0 + \varepsilon]$, and consider a map $H : Q \to \mathbb{R}$ satisfying*

$$H \in \mathcal{C}^1(Q; \mathbb{R}), \quad H(t_0, u_0) = 0, \quad \frac{\partial H}{\partial u}(t_0, u_0) \neq 0.$$

Then, there exist $\delta \in (0, T)$ and a $\mathcal{C}^1$ function $u : [t_0 - \delta, t_0 + \delta] \to \mathbb{R}$ such that:

- $u(t_0) = u_0$;
- $H(t, u(t)) = 0$ *if* $|t - t_0| \leq \delta$;
- *if $H(t, u) = 0$ with (t, u) sufficiently close to (t_0, u_0), then $u = u(t)$.*

Moreover,

$$u'(t) = -\frac{\partial H}{\partial t}(t, u(t)) \left(\frac{\partial H}{\partial u}(t, u(t)) \right)^{-1}, \quad t \in [t_0 - \delta, t_0 + \delta]. \quad (4.12)$$

Essentially, the implicit function theorem ensures that the set of solutions of the equation $H(t, u) = 0$ can be expressed, in a neighborhood of (t_0, u_0), as a curve $(t, u(t))$. We also point out that (4.12) corresponds to differentiating the relation $H(t, u(t)) = 0$ with respect to t.

To better understand this result, consider the Taylor expansion of $H(t, u)$ centered at (t_0, u_0). Since $H(t_0, u_0) = 0$, there exists a $\mathcal{C}^1$ function $R(t, u)$ such that, for every $(t, u) \in Q$,

$$H(t, u) = \frac{\partial H}{\partial t}(t_0, u_0)\,(t - t_0) + \frac{\partial H}{\partial u}(t_0, u_0)\,(u - u_0) + R(t, u)$$

and

$$\lim_{(t,u) \to (t_0, u_0)} \frac{R(t, u)}{|t - t_0| + |u - u_0|} = 0.$$

Suppose that $R \equiv 0$ in a neighborhood U of (t_0, u_0). Then, since $\frac{\partial H}{\partial u}(t_0, u_0) \neq 0$, it becomes apparent that the straight line

$$u(t) = u_0 - \frac{\frac{\partial H}{\partial t}(t_0, u_0)}{\frac{\partial H}{\partial u}(t_0, u_0)}(t - t_0), \quad t \sim t_0, \quad (4.13)$$

provides us with the set of solutions of $H(t, u) = 0$ in U.

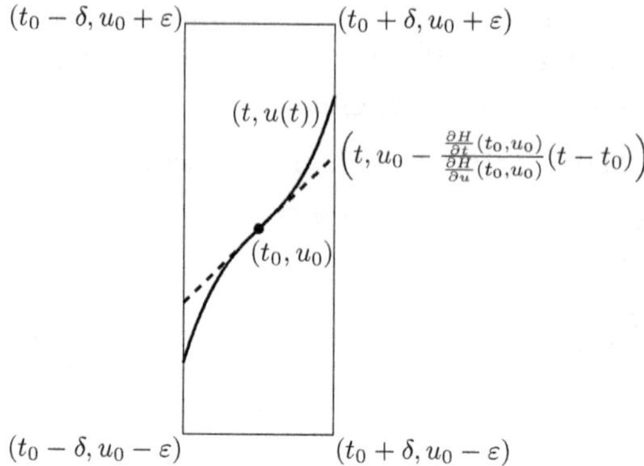

Fig. 4.4. The curve of solutions, $(t, u(t))$, of $H(t, u) = 0$ given by the implicit function theorem, together with its tangent line at (t_0, u_0).

Theorem 4.1 establishes that, even in the general case when $R \not\equiv 0$, there exist $\delta > 0$ and a function of class $\mathcal{C}^1$, $u : [t_0 - \delta, t_0 + \delta] \to \mathbb{R}$, such that

$$u(t) = u_0 - \frac{\frac{\partial H}{\partial t}(t_0, u_0)}{\frac{\partial H}{\partial u}(t_0, u_0)}(t - t_0) + o(|t - t_0|), \quad t \sim t_0,$$

provides us with the set of solutions of $H(t, u) = 0$ in a neighborhood of (t_0, u_0). Here, $o(|t - t_0|)$ means that

$$\lim_{t \to t_0} \frac{u(t) - u_0 + \frac{\frac{\partial H}{\partial t}(t_0, u_0)}{\frac{\partial H}{\partial u}(t_0, u_0)}(t - t_0)}{|t - t_0|} = 0.$$

In other words, the curve of zeroes of H passing through the point (t_0, u_0), $(t, u(t))$, is tangent to the line (4.13). Observe that the last relation implies that

$$u'(t_0) = -\frac{\frac{\partial H}{\partial t}(t_0, u_0)}{\frac{\partial H}{\partial u}(t_0, u_0)},$$

which agrees with (4.12) for $t = t_0$. Figure 4.4 shows the set $H^{-1}(0)$ in a neighborhood of (t_0, u_0).

We can now state and prove the main result for equations with separated variables.

Theorem 4.2. *Suppose* $(t_0, u_0) \in \mathbb{R}^2$, $T > 0$, $R > 0$,

$$F \in \mathcal{C}([t_0 - T, t_0 + T]\,;\mathbb{R}), \quad G \in \mathcal{C}([u_0 - R, u_0 + R]\,;\mathbb{R}),$$

and $G(u_0) \neq 0$. *Then, the Cauchy problem*

$$\begin{cases} u' = F(t)G(u), \\ u(t_0) = u_0, \end{cases} \tag{4.14}$$

possesses a unique local solution, i.e. there exists $\delta \in (0, T]$ *such that* *(4.14) has a unique solution in the interval* $[t_0 - \delta, t_0 + \delta]$.

Remark 4.3. The condition $G(u_0) \neq 0$ is necessary to get uniqueness, unless we impose some additional regularity for $G(u)$ at u_0, such as differentiability (see Corollary 5.3 and Theorem 5.9). This will become clear in Section 4.5. However, such a condition is absolutely irrelevant for existence because if $G(u_0) = 0$, then $u \equiv u_0$ provides us with a constant solution of (4.14).

As a consequence of Theorem 4.2, in equations with separated variables satisfying the assumptions of such a theorem, local non-uniqueness phenomena can only occur at the equilibria. Nevertheless, as will become apparent later (see again Section 4.5), despite the fact that the local solution of (4.14) is unique if $G(u_0) \neq 0$, its *maximal solution* might not be unique. The reader is referred to Section 5.3.2 for the precise definition of a maximal solution.

Proof of Theorem 4.2: Suppose $u(t)$, $t \sim t_0$, is a (local) solution of (4.14). Then, $u(t_0) = u_0$ and, by continuity, $G(u(t)) \neq 0$ for t sufficiently close to t_0. Thus, there exists $\eta > 0$ such that

$$\frac{u'(t)}{G(u(t))} = F(t) \quad \text{for all } t \in [t_0 - \eta, t_0 + \eta].$$

Equivalently,

$$\frac{d}{dt} \int_{u_0}^{u(t)} \frac{1}{G(u)}\, du = F(t) \quad \text{for all } t \in [t_0 - \eta, t_0 + \eta].$$

Thus, there exists a constant C such that, for all $t \in [t_0 - \eta, t_0 + \eta]$,

$$\int_{u_0}^{u(t)} \frac{1}{G(u)}\, du = \int_{t_0}^{t} F(s)\, ds + C.$$

Evaluating at $t = t_0$ gives $C = 0$ and, therefore,

$$\int_{u_0}^{u(t)} \frac{1}{G(u)} \, du = \int_{t_0}^{t} F(s) \, ds \quad \text{for all } t \in [t_0 - \eta, t_0 + \eta]. \qquad (4.15)$$

Therefore, every local solution of the initial value problem (4.14) solves the integral equation (4.15). This motivates the introduction of the function $H \colon Q \to \mathbb{R}$, with $Q := [t_0 - T, t_0 + T] \times [u_0 - \varepsilon, u_0 + \varepsilon]$, defined as

$$H(t, u) = \int_{u_0}^{u} \frac{1}{G(\xi)} \, d\xi - \int_{t_0}^{t} F(s) \, ds,$$

which is well defined for sufficiently small $\varepsilon > 0$, so that $G(u) \neq 0$ for all $|u - u_0| \leq \varepsilon$. Moreover, H is of class C^1 because its partial derivatives,

$$\frac{\partial H}{\partial u} = \frac{1}{G(u)}, \quad \frac{\partial H}{\partial t} = -F(t),$$

are continuous for all $(t, u) \in Q$, and

$$H(t_0, u_0) = 0 \quad \text{and} \quad \frac{\partial H}{\partial u}(t_0, u_0) = \frac{1}{G(u_0)} \neq 0.$$

Therefore, by Theorem 4.1, there exist $\delta \in (0, T]$ and a function of class C^1,

$$u \colon [t_0 - \delta, t_0 + \delta] \to [u_0 - \varepsilon, u_0 + \varepsilon],$$

such that:

  (i) $u(t_0) = u_0$;
 (ii) $H(t, u(t)) = 0$ if $|t - t_0| \leq \delta$;
(iii) if $H(t, u) = 0$ with (t, u) sufficiently close to (t_0, u_0), then $u = u(t)$.

We will now show that $u(t)$ provides us with the (local) unique solution of problem (4.14), which we are looking for. Indeed, the initial condition is satisfied by (i), while differentiating the relation in (ii) with respect to t gives (4.12), which, in this case, reads

$$u'(t) = F(t)G(u(t)) \quad \text{for all } t \in [t_0 - \delta, t_0 + \delta],$$

i.e. $u(t)$ satisfies the differential equation. Finally, (iii) gives the local uniqueness of the solution. $\qquad \square$

According to Theorem 4.2, the initial value problem for the logistic equation (4.7) possesses a unique (local) solution for every $x \in \mathbb{R} \setminus \{0, \frac{\lambda}{m}\}$. In Chapter 5, as a result of the differentiability of

$$f(u) = \lambda u - mu^2,$$

it will be shown that uniqueness also holds at each of the two equilibria. Observe, though, that this property might also be inferred from a careful analysis of the global structure of the set of solutions of (4.7), which is plotted in Figure 4.1.

4.3   A Paradigmatic Example of Blow up in Finite Time

In this section, we study the initial value problem

$$\begin{cases} u' = u^p, \\ u(t_0) = u_0 > 0, \end{cases} \tag{4.16}$$

for some exponent $p > 1$. As it is an autonomous problem, it has separated variables. Thus, by Theorem 4.2, (4.16) possesses a unique local solution for each $u_0 > 0$. The uniqueness at $u_0 = 0$ cannot be guaranteed yet because 0 is a constant solution of $u' = u^p$. Let

$$u(t) = u(t; t_0, u_0)$$

denote the unique local solution, for $|t - t_0| \leq \delta$. Then, as soon as $u(t) > 0$,

$$u^{-p}(t)u'(t) = \frac{u'(t)}{u^p(t)} = 1$$

and, hence,

$$\frac{d}{dt}\left(\frac{u^{-p+1}(t)}{-p+1}\right) = 1.$$

Thus,

$$\frac{u^{-p+1}(t)}{-p+1} = t + C$$

for some constant C such that

$$\frac{u_0^{-p+1}}{-p+1} = t_0 + C.$$

Therefore,

$$\frac{u^{-p+1}(t)}{-p+1} = \frac{u_0^{-p+1}}{-p+1} + t - t_0$$

in the maximal interval of definition of the solution, say I_{u_0}. Consequently,

$$u(t; t_0, u_0) = \left[u_0^{1-p} + (1-p)(t-t_0) \right]^{\frac{-1}{p-1}} \tag{4.17}$$

for all $t \in I_{u_0}$. From (4.17), recalling that $p > 1$, it is easily seen that $u(t; t_0, u_0)$ is increasing in time, which also follows from

$$u'(t) = u^p(t) > 0.$$

Moreover, we observe that the solution $u(t; t_0, u_0)$ given by (4.17) *blows up in a finite time*, $T_{\max} = T_{\max}(u_0)$, given by

$$T_{\max}(u_0) = t_0 + \frac{1}{p-1} \frac{1}{u_0^{p-1}}, \tag{4.18}$$

in the sense that

$$\lim_{t \uparrow T_{\max}(u_0)} u(t; t_0, u_0) = +\infty.$$

Figure 4.5 shows the graph of the maximal existence time of the solution, $T_{\max}(u_0)$, as a function of u_0. We have that this function is decreasing, and

$$\lim_{u_0 \downarrow 0} T_{\max}(u_0) = +\infty, \qquad \lim_{u_0 \uparrow +\infty} T_{\max}(u_0) = t_0.$$

Consequently, the larger the value of u_0, the smaller the blow up time of the solution and, thus, the shorter the maximal existence interval of the solution I_{u_0}, which is given by

$$I_{u_0} = (-\infty, T_{\max}(u_0)).$$

Figure 4.6 plots the graphs of some of the solutions of (4.16) for a series of values $u_0 > 0$. Note that

$$\lim_{t \downarrow -\infty} u(t; t_0, u_0) = 0 \quad \text{for all } u_0 > 0.$$

Therefore, the steady-state solution 0 is a repeller as time grows.

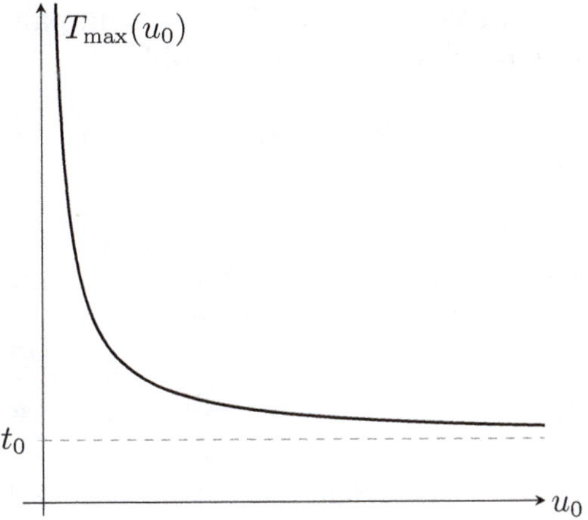

Fig. 4.5. The blow up time of the solutions of (4.16), given by (4.18).

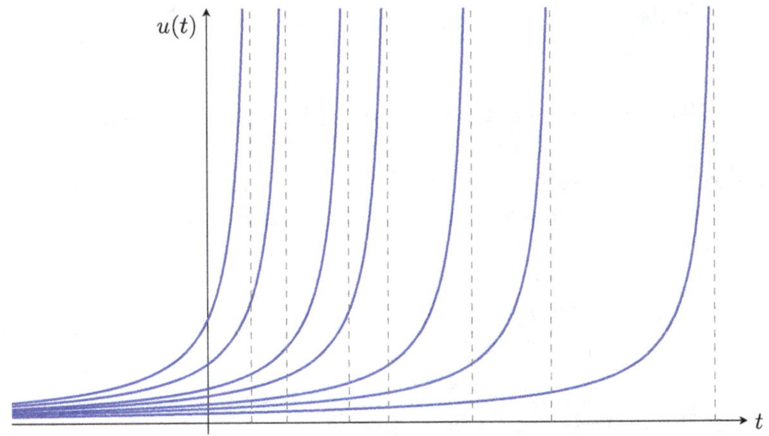

Fig. 4.6. Plots of a series of solutions of (4.16).

Note that the function

$$f(u) = |u|^{p-1}u, \quad u \in \mathbb{R}, \quad p > 1,$$

which coincides with the nonlinearity of problem (4.16) for $u > 0$, is continuous in $\mathbb{R}$, and it is odd because

$$f(-u) = -f(u) \quad \text{for all } u \in \mathbb{R}.$$

Consequently, the solutions of $u' = f(u)$ arise in pairs: u and $-u$. Of course, it makes sense to consider the problem

$$\begin{cases} u' = f(u), \\ u(t_0) = u_0, \end{cases}$$

for all $u_0 \in \mathbb{R}$, and with the usual notation, its unique local solution is denoted as $u(t; t_0, u_0)$. Moreover, for $u_0 > 0$,

$$u(t; t_0, u_0) = -u(t; t_0, -u_0) \quad \text{for all } t \in (-\infty, T_{\max}(u_0)),$$

where $u(t; t_0, u_0)$ and $T_{\max}(u_0)$ are given by (4.17) and (4.18), respectively.

As a result of the abstract general theory developed in Chapter 5, it will become apparent that, since $f \in \mathcal{C}^1(\mathbb{R}; \mathbb{R})$, this problem admits a unique solution for each $(t_0, u_0) \in \mathbb{R}^2$. In particular, $u(\cdot; t_0, 0) \equiv 0$.

4.4 A Simple Non-Autonomous Nonlinear Equation

Throughout the rest of this book, we set $\mathbb{R}_+ := [0, +\infty)$. In this section, we consider a continuous function, $a \in \mathcal{C}(\mathbb{R}_+; \mathbb{R}_+)$, $a \not\equiv 0$, such that

$$I := \int_0^\infty a(s)\,ds = \lim_{t\uparrow +\infty} \int_0^t a(s)\,ds < +\infty, \tag{4.19}$$

as well as the associated Cauchy problem

$$\begin{cases} u' = a(t)u^p, \\ u(0) = u_0 > 0, \end{cases} \tag{4.20}$$

for some constant $p > 1$. Observe, for example, that the function $a(t) = e^{-t}$, as well as any function $a(t)$ with compact support, satisfies (4.19), while no positive constant function does. The differential equation is non-autonomous, as the nonlinearity

$$f(t, u) = a(t)u^p, \quad t \in \mathbb{R}_+, \ u \geq 0,$$

explicitly depends on t. According to Theorem 4.2, (4.20) has a unique local solution,

$$u(t) = u(t; 0, u_0) =: u(t; u_0).$$

Moreover, the solution is non-decreasing in its maximal existence interval, denoted by I_{u_0}, and it satisfies

$$\frac{d}{dt}\left(\frac{u^{-p+1}(t)}{-p+1}\right) = u^{-p}(t)u'(t) = a(t).$$

Thus,

$$\frac{u^{-p+1}(t)}{-p+1} = \frac{u_0^{-p+1}}{-p+1} + \int_0^t a(s)\, ds$$

and, hence,

$$u(t; u_0) = \left[\frac{1}{u_0^{p-1}} - (p-1)\int_0^t a(s)\, ds\right]^{\frac{-1}{p-1}}, \quad t \in I_{u_0}.$$

Let us denote by $u_0^* > 0$ the unique value of u_0 such that

$$0 = \frac{1}{(u_0^*)^{p-1}} - (p-1)\int_0^\infty a(s)\, ds,$$

i.e.

$$u_0^* = \left[\frac{1}{(p-1)I}\right]^{\frac{1}{p-1}}.$$

In strong contrast to the case analyzed in Section 4.1, in the non-autonomous case considered in this section, the asymptotic behavior of $u(t; u_0)$ depends on the size of u_0 with respect to u_0^*. Indeed, suppose that $u_0 < u_0^*$. Then, for every $t > 0$,

$$\frac{1}{u_0^{p-1}} - (p-1)\int_0^t a(s)\, ds > \frac{1}{(u_0^*)^{p-1}} - (p-1)\int_0^t a(s)\, ds$$

$$\geq \frac{1}{(u_0^*)^{p-1}} - (p-1)\int_0^\infty a(s)\, ds = 0$$

and, hence, the solution $u(t; u_0)$ is defined for all $t \in [0, \infty)$. In particular, $I_{u_0} = [0, +\infty)$. Moreover,

$$\lim_{t \uparrow +\infty} u(t; u_0) = \left[\frac{1}{u_0^{p-1}} - (p-1) \int_0^\infty a(s)\, ds \right]^{\frac{-1}{p-1}},$$

i.e. all these solutions stabilize, as $t \uparrow +\infty$, toward the limit

$$L(u_0) := \left[\frac{1}{u_0^{p-1}} - (p-1)I \right]^{\frac{-1}{p-1}} > 0.$$

Note that, since $u \equiv 0$ is the unique steady-state solution of $u' = a(t)u^p$, the constant $L(u_0)$ cannot be an equilibrium of $u' = a(t)u^p$. This phenomenon cannot occur in autonomous equations, as will be shown in Section 4.7. Since $L(u_0)$ is a continuous function of $u_0 > 0$, and

$$\lim_{u_0 \downarrow 0} L(u_0) = 0, \qquad \lim_{u_0 \uparrow u_0^*} L(u_0) = +\infty,$$

the possible limiting values of the solutions, $L(u_0)$, can take any value between 0 and $+\infty$, as u_0 varies from 0 to u_0^*.

Now, suppose that $u_0 > u_0^*$. Then,

$$\frac{1}{u_0^{p-1}} - (p-1) \int_0^\infty a(s)\, ds < \frac{1}{(u_0^*)^{p-1}} - (p-1) \int_0^\infty a(s)\, ds = 0,$$

and hence, setting

$$\varphi(t) := \frac{1}{u_0^{p-1}} - (p-1) \int_0^t a(s)\, ds, \quad t \geq 0,$$

we have that $\varphi(0) > 0$ and $\lim_{t \uparrow \infty} \varphi(t) < 0$. Thus, by the mean value theorem, since φ is continuous, there exists $T_{\max} = T_{\max}(u_0) > 0$ such that

$$\varphi(t) > 0 \quad \text{for all } t \in [0, T_{\max}), \quad \text{and} \quad \varphi(T_{\max}) = 0.$$

Obviously, in this case, the maximal existence interval is given by

$$I_{u_0} = [0, T_{\max}(u_0)),$$

and

$$\lim_{t\uparrow T_{\max}(u_0)} u(t; u_0) = +\infty.$$

Consequently, $u(t; u_0)$ blows up in a finite time, $T_{\max}(u_0)$, if $u_0 > u_0^*$. Since $T_{\max}(u_0)$ is the minimal time, T, for which

$$\frac{1}{u_0^{p-1}} = (p-1) \int_0^T a(s)\, ds \tag{4.21}$$

and $a \geq 0$, $T_{\max}(u_0)$ is non-increasing with respect to u_0. To continue the analysis, we need to prescribe some assumptions on $a(t)$ since different behaviors of $a(t)$ will give rise to different behaviors of the solutions of our problem. Precisely, we begin by distinguishing two possible behaviors for $a(t)$ close to $t = 0$:

(a) $a(t) > 0$ for all $t \in (0, \delta)$, for some $\delta > 0$;
(b) $a(t) = 0$ for all $t \in (0, \delta)$, for some $\delta > 0$.

From (4.21), we obtain

$$\lim_{u_0\uparrow+\infty} T_{\max}(u_0) = 0 \quad \text{in case (a),}$$
$$\lim_{u_0\uparrow+\infty} T_{\max}(u_0) = \delta_0 \quad \text{in case (b),}$$

where

$$\delta_0 := \sup\{\delta > 0 \colon a(t) = 0 \text{ for all } t \in [0, \delta]\} > 0,$$

which is finite since $a \not\equiv 0$.

Instead, the behavior of $T_{\max}(u_0)$ as $u_0 \downarrow u_0^*$ is related to the behavior of $a(t)$ in a neighborhood of infinity. We consider the following possibilities:

(c) $a(t) > 0$ for all $t > \tau$, for some $\tau > 0$;
(d) $a(t) = 0$ for all $t > \tau$, for some $\tau > 0$.

Relation (4.21) now gives

$$\lim_{u_0\downarrow u_0^*} T_{\max}(u_0) = +\infty \quad \text{in case (c),}$$
$$\lim_{u_0\downarrow u_0^*} T_{\max}(u_0) = \tau_0 \quad \text{in case (d),}$$

where

$$\tau_0 := \inf\{\tau > 0 \colon a(t) = 0 \text{ for all } t \geq \tau\},$$

which is positive since $a \not\equiv 0$.

Similarly, in the special case of $u_0 = u_0^*$, the behavior of $u(t; u_0^*)$ depends on the behavior of $a(t)$ in a neighborhood of infinity. Indeed, in case (c), we have that

$$\frac{1}{(u_0^*)^{p-1}} - (p-1)\int_0^t a(s)\,ds > \frac{1}{(u_0^*)^{p-1}} - (p-1)\int_0^\infty a(s)\,ds = 0$$

for all $t > 0$ and, therefore, $u(t; u_0^*)$ is defined in $[0, +\infty)$. In case (d), instead,

$$\frac{1}{(u_0^*)^{p-1}} - (p-1)\int_0^{\tau_0} a(s)\,ds = \frac{1}{(u_0^*)^{p-1}} - (p-1)\int_0^\infty a(s)\,ds = 0,$$

where τ_0 is the one defined above, and hence, $u(t; u_0^*)$ is defined in $[0, \tau_0)$, and

$$\lim_{t \uparrow \tau_0} u(t; u_0) = +\infty.$$

Finally, we go back to the case of $u_0 > u_0^*$ and observe that $T_{\max}(u_0)$ might not be continuous, for example if there exists an interval $[t_1, t_2] \in (0, +\infty)$, with $a(t) = 0$, for all $t \in [t_1, t_2]$. Nevertheless, thanks to the minimality used in the definition of $T_{\max}(u_0)$, one can show that such a function is lower semicontinuous for all $u_0 > u_0^*$ (see Exercise 6 of Chapter 4).

4.5   A Paradigmatic Example of Non-Uniqueness

In this section, we consider the problem

$$\begin{cases} u' = u^q, \\ u(t_0) = u_0 > 0, \end{cases} \tag{4.22}$$

for $q \in (0, 1)$. By Theorem 4.2, (4.22) possesses a unique (local) solution for each $u_0 > 0$, which we denote as

$$u(t) = u(t; t_0, u_0),$$

for $|t - t_0| \leq \delta$. Then, as soon as $u(t) > 0$, which is the region where $u^q(t)$ is well defined for all $q > 0$, we have that

$$u^{-q}(t)u'(t) = \frac{u'(t)}{u^q(t)} = 1$$

and, hence,

$$\frac{d}{dt}\left(\frac{u^{-q+1}(t)}{-q+1}\right) = 1.$$

Consequently, it becomes apparent that

$$u(t; t_0, u_0) = \left[u_0^{1-q} + (1-q)(t - t_0)\right]^{\frac{1}{1-q}}$$

is a solution of (4.22), which is defined (at least) for all $t \geq t_0$. Moreover, there is a critical time, $t_c < t_0$, such that $u(t_c; t_0, u_0) = 0$ and $u(t; t_0, u_0) > 0$ for all $t > t_c$. Namely,

$$t_c = t_0 - \frac{1}{1-q} u_0^{1-q}. \tag{4.23}$$

Therefore, $u(t; t_0, u_0)$ is globally defined in $[t_c, +\infty)$. In addition, we observe that $u(t; t_0, u_0)$ is differentiable as $t \downarrow t_c$ since

$$\lim_{t \downarrow t_c} u'(t; t_0, u_0) = \lim_{t \downarrow t_c} u^q(t; t_0, u_0) = 0. \tag{4.24}$$

Hence, we conclude that the problem

$$\begin{cases} u' = u^q, \\ u(t_c) = 0, \end{cases}$$

possesses, at least, two globally defined solutions, i.e. solutions that are defined for all $t \in \mathbb{R}$: $u = 0$ and $u = \psi$, where

$$\psi(t) := \begin{cases} 0 & \text{if } t \leq t_c, \\ u(t; t_0, u_0) & \text{if } t > t_c. \end{cases} \tag{4.25}$$

The latter is illustrated in Figure 4.7.

Note that (4.24) ensures that ψ is of class $\mathcal{C}^1$, which is one of the conditions that we have required at the beginning of this chapter for a function to be a solution of a first-order differential equation.

From (4.23), we obtain that the critical time $t_c = t_c(u_0)$ is a continuous decreasing function of u_0, satisfying

$$\lim_{u_0 \uparrow +\infty} t_c(u_0) = -\infty \quad \text{and} \quad \lim_{u_0 \downarrow 0} t_c(u_0) = t_0.$$

Consequently, $t_c(u_0)$ ranges from t_0 to $-\infty$ as u_0 ranges from 0 to $+\infty$.

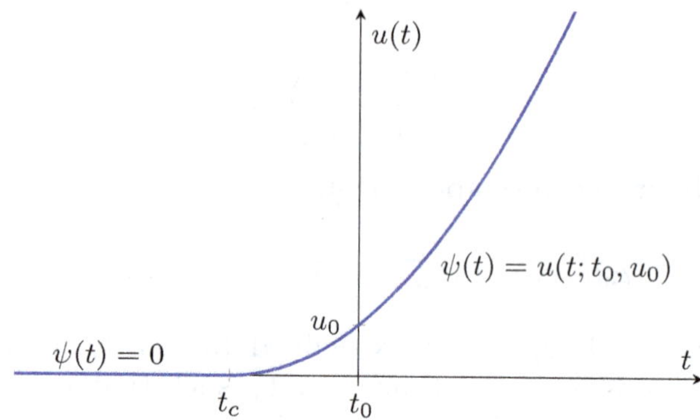

Fig. 4.7.   The global solution $\psi(t)$ of (4.22), defined in (4.25).

As this example exhibits non-uniqueness, not surprisingly, there cannot be continuous dependence of $u(t; t_0, u_0)$ with respect to u_0 as $u_0 \downarrow 0$. Indeed, for $t > t_0$ fixed,

$$\lim_{u_0 \downarrow 0} u(t; t_0, u_0) = [(1 - q)(t - t_0)]^{\frac{1}{1-q}} \neq 0.$$

Actually, the previous analysis can be adapted to show that the function

$$\varphi(t) := \begin{cases} 0 & \text{if } t \leq t_0, \\ [(1 - q)(t - t_0)]^{\frac{1}{1-q}} & \text{if } t > t_0, \end{cases}$$

provides us with a non-negative, non-zero solution of

$$\begin{cases} u' = u^q, \\ u(t_0) = 0, \end{cases} \tag{4.26}$$

which is globally defined in time. Since $u = 0$ is a constant solution of (4.26), such a problem admits at least two (global) solutions. Note that, by construction, $\varphi(t)$ provides us with the *maximal solution* of (4.26) and that 0 must be the *minimal* one.

One of the most significant topological results that we will be presenting in this book, Kneser's theorem (see Theorem 7.17), establishes that whenever an initial value problem possesses two solutions, it must admit a continuum of solutions linking them. By *continuum*, we mean a closed and connected set of solutions. Since (4.26) is

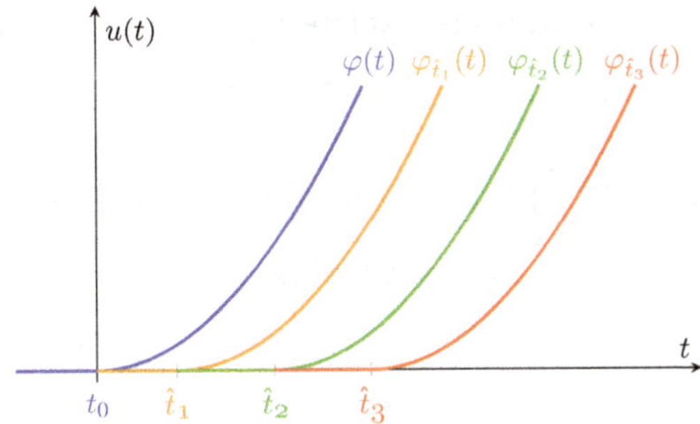

Fig. 4.8. A series of solutions $\varphi_{\hat{t}_j}$ of (4.26) with $\hat{t}_j > t_0$.

autonomous, as observed in Section 4.1, for each $\hat{t} > t_0$, the shifted functions

$$\varphi_{\hat{t}}(t) := \begin{cases} 0 & \text{if } t \leq \hat{t}, \\ \varphi(t_0 + t - \hat{t}) & \text{if } t > \hat{t}, \end{cases}$$

still solve (4.26). Such a one-parameter family of solutions, some of which are illustrated in Figure 4.8, links $\varphi(t)$, as $\hat{t} \downarrow t_0$, with $u = 0$, as $\hat{t} \uparrow \infty$, confirming the validity of Kneser's theorem in this case.

This example also establishes the optimality of Theorem 4.2, in the sense that, in order to have uniqueness, we shall indeed impose $G(u_0) \neq 0$, unless we assume some additional regularity on $G(u)$ at u_0 if $G(u_0) = 0$, such as being of class $\mathcal{C}^1$, as will be shown in Chapter 5.

4.6   A More Sophisticated Example of Non-Uniqueness

In this section, given two arbitrary real numbers, $\omega_1 < \omega_2$, we consider the problem

$$\begin{cases} u' = \sqrt{(u - \omega_1)(\omega_2 - u)}, \\ u(t_0) = u_0 \in (\omega_1, \omega_2). \end{cases} \tag{4.27}$$

Obviously, the nonlinearity

$$f(u) := \sqrt{(u - \omega_1)(\omega_2 - u)}, \quad \omega_1 \leq u \leq \omega_2,$$

is well defined and continuous. Actually, $f \in C^1((\omega_1, \omega_2); \mathbb{R})$ with

$$\lim_{u \downarrow \omega_1} f'(u) = +\infty, \quad \lim_{u \uparrow \omega_2} f'(u) = -\infty.$$

By Theorem 4.2, (4.27) possesses a unique (local) solution,

$$u(t) = u(t; t_0, u_0).$$

Moreover, as soon as $u(t) \in (\omega_1, \omega_2)$, we have that

$$\frac{d}{dt} \int_{u_0}^{u(t)} \frac{dx}{\sqrt{(x - \omega_1)(\omega_2 - x)}} = 1$$

and, hence,

$$\int_{u_0}^{u(t)} \frac{dx}{\sqrt{(x - \omega_1)(\omega_2 - x)}} = t - t_0. \tag{4.28}$$

In order to find a primitive of the integrand in (4.28), straightforward manipulations give

$$(x - \omega_1)(\omega_2 - x) = \left(\frac{\omega_2 - \omega_1}{2} \right)^2 \left[1 - \left(\frac{x - \frac{\omega_1 + \omega_2}{2}}{\frac{\omega_2 - \omega_1}{2}} \right)^2 \right].$$

Thus, by performing the change of variable

$$y = \frac{x - \frac{\omega_1 + \omega_2}{2}}{\frac{\omega_2 - \omega_1}{2}},$$

it is apparent that

$$\int \frac{dx}{\sqrt{(x - \omega_1)(\omega_2 - x)}} = \int \frac{dy}{\sqrt{1 - y^2}} = \arcsin y = \arcsin \frac{x - \frac{\omega_1 + \omega_2}{2}}{\frac{\omega_2 - \omega_1}{2}}.$$

Consequently, (4.28) gives

$$\arcsin \frac{u(t) - \frac{\omega_1 + \omega_2}{2}}{\frac{\omega_2 - \omega_1}{2}} - \arcsin \frac{u_0 - \frac{\omega_1 + \omega_2}{2}}{\frac{\omega_2 - \omega_1}{2}} = t - t_0$$

and, therefore,

$$u(t; t_0, u_0) = \frac{\omega_1 + \omega_2}{2} + \frac{\omega_2 - \omega_1}{2} \sin \left(t - t_0 + \arcsin \frac{u_0 - \frac{\omega_1 + \omega_2}{2}}{\frac{\omega_2 - \omega_1}{2}} \right).$$

From (4.27), it follows that $u'(t) > 0$ if $u(t) \in (\omega_1, \omega_2)$. Thus, $u(t)$ is increasing in such a case. We claim that there exists a minimal time, $t_c^+ > t_0$, such that

$$u(t_c^+; t_0, u_0) = \omega_2. \tag{4.29}$$

Indeed, since (4.29) is equivalent to

$$\frac{\omega_1 + \omega_2}{2} + \frac{\omega_2 - \omega_1}{2} \sin\left(t_c^+ - t_0 + \arcsin \frac{u_0 - \frac{\omega_1 + \omega_2}{2}}{\frac{\omega_2 - \omega_1}{2}} \right) = \omega_2,$$

necessarily,

$$\sin\left(t_c^+ - t_0 + \arcsin \frac{u_0 - \frac{\omega_1 + \omega_2}{2}}{\frac{\omega_2 - \omega_1}{2}} \right) = 1$$

and hence,

$$t_c^+ = t_c^+(u_0) = t_0 + \frac{\pi}{2} - \arcsin \frac{u_0 - \frac{\omega_1 + \omega_2}{2}}{\frac{\omega_2 - \omega_1}{2}}.$$

Similarly, there exists a maximal time, $t_c^- < t_0$, such that

$$u(t_c^-; t_0, u_0) = \omega_1,$$

which is given by

$$t_c^- = t_c^-(u_0) = t_0 - \left(\frac{\pi}{2} + \arcsin \frac{u_0 - \frac{\omega_1 + \omega_2}{2}}{\frac{\omega_2 - \omega_1}{2}} \right).$$

Note that $t_c^- < t_0 < t_c^+$ because

$$-1 = \frac{\omega_1 - \frac{\omega_1 + \omega_2}{2}}{\frac{\omega_2 - \omega_1}{2}} < \frac{u_0 - \frac{\omega_1 + \omega_2}{2}}{\frac{\omega_2 - \omega_1}{2}} < \frac{\omega_2 - \frac{\omega_1 + \omega_2}{2}}{\frac{\omega_2 - \omega_1}{2}} = 1.$$

Moreover, from the differential equation of (4.27) we obtain

$$\lim_{t \downarrow t_c^-} u'(t; t_0, u_0) = \lim_{t \uparrow t_c^+} u'(t; t_0, u_0) = 0.$$

Thus, the $\mathcal{C}^1$ function

$$
\varphi(t) := \begin{cases} \omega_1, & t \leq t_c^-, \\ u(t; t_0, u_0), & t_c^- < t < t_c^+, \\ \omega_2, & t \geq t_c^+ \end{cases}
$$

provides us with a solution of (4.27) that is globally defined in time. In addition, since

$$
\lim_{u_0 \downarrow \omega_1} t_c^-(u_0) = t_0, \qquad \lim_{u_0 \downarrow \omega_1} t_c^+(u_0) = t_0 + \pi
$$

and

$$
\lim_{u_0 \downarrow \omega_1} u(t; t_0, u_0) = \frac{\omega_1 + \omega_2}{2} + \frac{\omega_2 - \omega_1}{2} \sin \left(t - t_0 - \frac{\pi}{2} \right),
$$

it can be easily seen that

$$
\psi(t) := \begin{cases} \omega_1 & \text{if } t \leq t_0, \\ \frac{\omega_1 + \omega_2}{2} + \frac{\omega_2 - \omega_1}{2} \sin \left(t - t_0 - \frac{\pi}{2} \right) & \text{if } t_0 < t < t_0 + \pi, \\ \omega_2 & \text{if } t \geq \pi, \end{cases}
$$

is a global solution of

$$
\begin{cases} u' = \sqrt{(u - \omega_1)(\omega_2 - u)}, \\ u(t_0) = \omega_1. \end{cases} \tag{4.30}
$$

Therefore, this Cauchy problem admits (at least) two global solutions: $\psi(t)$ and the constant ω_1. Actually, since (4.30) is autonomous, for all $t_1 > 0$, the translations

$$
\psi_{t_1}(t) := \psi(t - t_1) \tag{4.31}
$$

are global solutions of (4.30), which connect $\psi(t)$ (obtained by taking the limit $t_1 \downarrow 0$) to the constant ω_1 (obtained as $t_1 \uparrow +\infty$), in agreement with Kneser's theorem 7.17. Some of these solutions are represented in Figure 4.9.

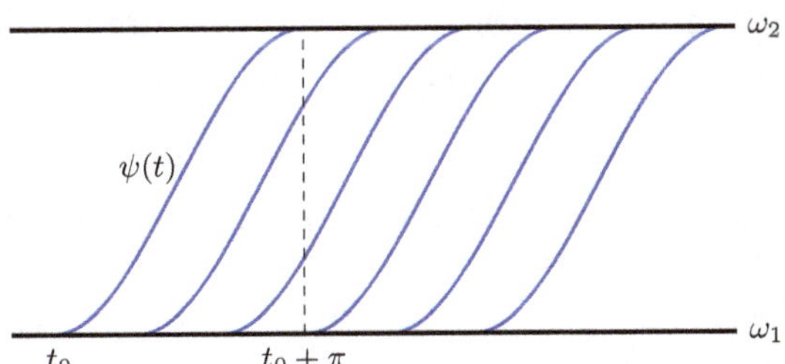

Fig. 4.9. A series of the global solutions of (4.30) given by (4.31).

By adapting the previous analysis, one can readily show that the problem

$$\begin{cases} u' = \sqrt{(u - \omega_1)(\omega_2 - u)}, \\ u(t_c^+) = \omega_2, \end{cases}$$

possesses at least two solutions in $[t_0, t_c^+]$: the constant ω_2 and $u(t; t_0, u_0)$. Actually, proceeding as above, the latter can be extended to obtain a globally defined solution.

Finally, for any given $\omega_1 < \omega_2 < \omega_3$, we consider the function

$$f(u) := \begin{cases} \sqrt{(u - \omega_1)(\omega_2 - u)} & \text{if } u \in [\omega_1, \omega_2], \\ \sqrt{(u - \omega_2)(\omega_3 - u)} & \text{if } u \in (\omega_2, \omega_3], \end{cases}$$

and the autonomous problem

$$\begin{cases} u' = f(u), \\ u(t_0) = u_0 \in (\omega_1, \omega_2). \end{cases} \tag{4.32}$$

We already know that there exists $t_c^+ > t_0$ such that

$$u(t_c^+; t_0, u_0) = \omega_2,$$

where $u(t; t_0, u_0)$ denotes the unique (local) solution of the above initial value problem. By adapting the previous discussion, it is easily seen that the problem

$$\begin{cases} u' = f(u), \\ u(t_c^+) = \omega_2, \end{cases}$$

admits infinitely many global solutions linking the equilibria ω_2 and ω_3. Thus, it becomes apparent that the maximal extension of the local solution $u(t; t_0, u_0)$ of (4.32) is far from being unique.

4.7 Dynamics of Scalar Autonomous Equations

This section describes some general features of the dynamics of the autonomous equation

$$u' = f(u), \tag{4.33}$$

where $f \in \mathcal{C}(\mathbb{R}; \mathbb{R})$, though most of the results are also valid when f is only defined on some subset of $\mathbb{R}$. The set of equilibria of (4.33), which is given by

$$f^{-1}(0) = \{\omega \in \mathbb{R} \colon f(\omega) = 0\},$$

might have a rather intricate structure (it could even be a Cantor subset of $\mathbb{R}$!). For simplicity, here we consider the situation in which the equilibria are separated by open intervals, as is the case for most models which are interesting for applications. Precisely, we assume that $\omega_1 < \omega_2$ are two consecutive zeroes of f and that, to fix ideas,

$$f(x) > 0 \quad \text{for all } x \in (\omega_1, \omega_2).$$

This occurs, for example, in the logistic equation presented in Section 4.1, with $\omega_1 = 0$ and $\omega_2 = \frac{\lambda}{m}$, as well as in the example in Section 4.6.

According to Theorem 4.2, for every $u_0 \in (\omega_1, \omega_2)$, equation (4.33) admits a unique (local) solution, $u(t; t_0, u_0)$, such that $u(t_0) = u_0$, which is defined (at least) for $t \in [t_0 - \delta_0, t_0 + \delta_0]$ for some $\delta_0 > 0$. Moreover, as long as $u(t; t_0, u_0) \in (\omega_1, \omega_2)$, we have

$$u'(t; t_0, u_0) = f(u(t; t_0, u_0)) > 0,$$

and, hence, $u(t; t_0, u_0)$ is increasing in time. In this section, we assume $u(t; t_0, u_0)$ to be defined whenever it is possible, as long as it takes values in (ω_1, ω_2). In Section 5.3.2, we show that this assumption is related to the concept of *maximal solution*, which will be introduced there.

As a direct consequence of these considerations, based on Theorem 4.2 and on the monotonicity of $u(t; t_0, u_0)$, two options may occur:

(a) either there exists a minimal time $t_+ \in (t_0, +\infty)$ such that $u(t; t_0, u_0)$ is defined in $[t_0, t_+)$, and

$$\lim_{t \uparrow t_+} u(t_+; t_0, u_0) = \omega_2,$$

(b) or $u(t; t_0, u_0) < \omega_2$ for all $t \geq t_0$ for which it is defined.

Recall that case (a) occurs in the example in Section 4.6, while (b) occurs in the logistic equation studied in Section 4.1.

Moreover, when (a) holds, from (4.33) and the continuity of f, we have

$$\lim_{t \uparrow t_+} u'(t; t_0, u_0) = f\left(\lim_{t \uparrow t_+} u(t; t_0, u_0) \right) = f(\omega_2) = 0.$$

As a consequence, the function

$$\varphi(t) := \begin{cases} u(t; t_0, u_0) & \text{if } t_0 \leq t < t_+, \\ \omega_2 & \text{if } t \geq t_+, \end{cases}$$

provides us with a solution of

$$\begin{cases} u' = f(u), \\ u(t_0) = u_0, \end{cases} \tag{4.34}$$

which is defined in $[t_0, +\infty)$, i.e. it is globally defined forward in time. The left-hand side plot in Figure 4.10 represents this situation.

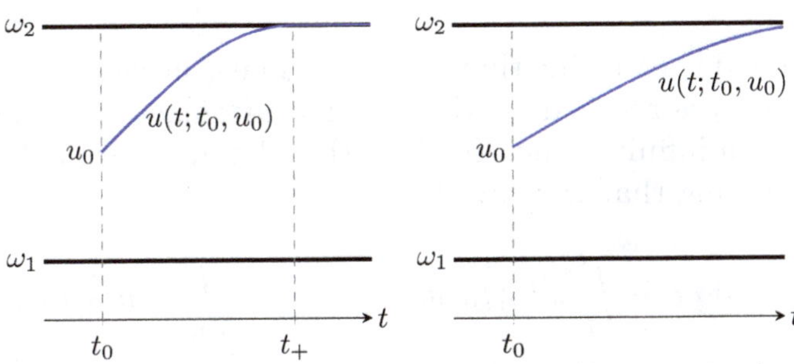

Fig. 4.10. Possible behaviors of the solutions of $u' = f(u)$ with $u_0 \in (\omega_1, \omega_2)$: case (a) on the left, and case (b) on the right.

If option (b) occurs instead, we now show through a global extension argument that the solution is actually defined for $t \in [t_0, +\infty)$. Indeed, if we define the maximal (forward) existence time

$$T_{\max} := \sup\{T > t_0 : u(t; t_0, u_0) \text{ is defined for } t \in [t_0, T]\},$$

we have that $T_{\max} \in [t_0 + \delta_0, +\infty]$. Assume, by contradiction, that $T_{\max} < +\infty$. Then, by the definition of $T_{\max}$ and since we are considering case (b), we have that $\bar{u} := u(T_{\max}; t_0, u_0)$ is well defined, as the limit of $u(t; t_0, u_0)$ for $t \uparrow T_{\max}$, and satisfies $\bar{u} < \omega_2$. Thus, if we consider the Cauchy problem

$$\begin{cases} v' = f(v), \\ v(T_{\max}) = \bar{u}, \end{cases}$$

Theorem 4.2 guarantees that it has a solution which is defined in $[T_{\max}, T_{max} + \delta_1]$ for some $\delta_1 > 0$. As a consequence,

$$w(t) := \begin{cases} u(t; t_0, u_0) & \text{if } t \in [t_0, T_{\max}], \\ v(t) & \text{if } t \in (T_{\max}, T_{\max} + \delta_1], \end{cases}$$

is a solution of (4.34) which properly extends $u(t; t_0, u_0)$. This contradicts the maximality given by the definition of $T_{\max}$ and allows us to conclude that $T_{\max} = +\infty$. In other words, $u(t; t_0, u_0)$ is defined for all $t \geq t_0$.

Since $u(t; t_0, u_0)$ is increasing for all time $t > t_0$, the limit

$$\omega := \lim_{t \uparrow +\infty} u(t; t_0, u_0) \leq \omega_2 \tag{4.35}$$

is well defined. We claim that $\omega = \omega_2$. Thus, in case (b) also, the equilibrium ω_2 is reached by the solution $u(t; t_0, u_0)$, though in this occasion at an infinite time. To show this claim, we argue by contradiction assuming that $\omega < \omega_2$. Then,

$$u(t; t_0, u_0) = u_0 + \int_{t_0}^{t} u'(s; t_0, u_0)\, ds = u_0 + \int_{t_0}^{t} f(u(s; t_0, u_0))\, ds$$

$$\geq u_0 + \int_{t_0}^{t} \min_{u \in [u_0, \omega]} f(u)\, ds = u_0 + (t - t_0) \min_{u \in [u_0, \omega]} f(u).$$

As we assume that $\omega < \omega_2$, we have $\min_{u \in [u_0, \omega]} f(u) > 0$. Thus, the previous relation implies

$$\lim_{t \uparrow +\infty} u(t; t_0, u_0) \geq \lim_{t \uparrow +\infty} \left[u_0 + (t - t_0) \min_{u \in [u_0, \omega]} f(u) \right] = +\infty,$$

which contradicts (4.35). Therefore, $\omega = \omega_2$ and

$$\lim_{t \uparrow +\infty} u(t; t_0, u_0) = \omega_2.$$

So, the solution approximates the equilibrium ω_2 asymptotically as $t \uparrow +\infty$, as is illustrated in the right-hand plot of Figure 4.10. As we have seen in Section 4.1, this situation occurs for the logistic equation when $u_0 \in (0, \frac{\lambda}{m})$. In such a case, $\omega_2 = \frac{\lambda}{m}$ (see, for instance, Figure 4.1).

A similar argument shows that, either $\omega_1 < u(t; t_0, u_0) \leq u_0$ for all $t \leq t_0$, and

$$\lim_{t \downarrow -\infty} u(t; t_0, u_0) = \omega_1,$$

or there exists $t_- < t_0$ such that $\lim_{t \downarrow t_-} u(t; t_0, u_0) = \omega_1$. Consequently, also when we go backward in time, the steady state ω_1 must be reached either in a finite time, at $t = t_-$, or at an infinite time.

Analogously, if we have that

$$f(x) < 0 \quad \text{for all } x \in (\omega_1, \omega_2),$$

then we can prove that $u(t; t_0, u_0)$ is decreasing in time, and the equilibrium ω_1 (respectively, ω_2), must be reached either at a finite, or infinite, forward (respectively, backward) time. As will be stressed in Chapter 5, the forward behavior of the solutions is independent of the backward one.

The previous analysis shows that, to study the dynamics of (4.33), we only need to determine the sign of $f(u)$ between any two consecutive zeroes, as sketched in the example in Figure 4.11. The problem of ascertaining whether each particular equilibrium is reached at a finite or infinite time will be solved in Chapters 5–7.

In the example of Figure 4.11, the equation $u' = f(u)$ has five equilibria since

$$f^{-1}(0) = \{\omega_1, \omega_2, \omega_3, \omega_4, \omega_5\}.$$

The zeroes where f changes sign from positive to negative, ω_1 and ω_3, are *local attractors* for the dynamics, while the zeroes where f

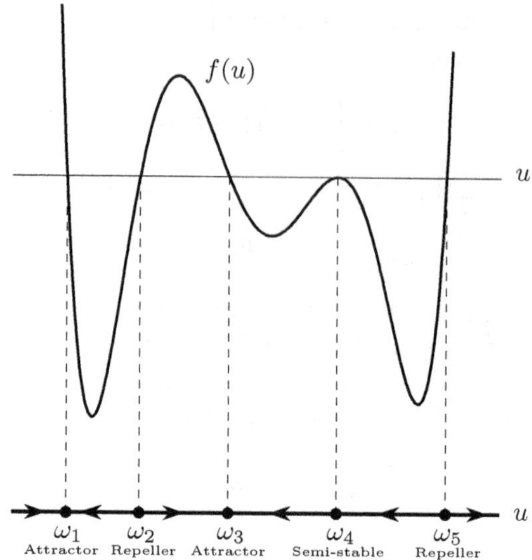

Fig. 4.11. An example of the dynamics of (4.33) (lower plot), corresponding to the nonlinearity $f(u)$ represented in the upper plot.

changes sign from negative to positive, ω_2 and ω_5, are *local repellers*. The zeroes where f does not change of sign (ω_4 in the example in Figure 4.11) are *semi-stable equilibria*, in the sense that they locally attract on one side but locally repel on the other. Of course, in this terminology, we are referring to the forward behavior of the equation since, like most of us, we care about our futures!

To conclude this section, we stress once again the different behaviors of autonomous equations, whose solutions approach (in finite, or infinite, time) some equilibrium, in contrast to the possible behaviors of non-autonomous ones, such as the example analyzed in Section 4.4, where the solutions can stabilize toward constants which are not steady-state solutions of the underlying differential equation.

4.8   Some Special Classes of Differential Equations

In this section, we find the solution of the Cauchy problem

$$\begin{cases} u' = f(t, u), \\ u(t_0) = u_0, \end{cases}$$

for some special classes of continuous functions $f : \mathbb{R} \to \mathbb{R}$ for which the solution can be explicitly determined. In particular, we study the Riccati equation, homogeneous equations, and exact differential equations.

4.8.1 *The Riccati equation*

Suppose that $f(t, u)$ is of class $\mathcal{C}^2$. Then, according to Taylor's theorem, developing f at $u = 0$ gives

$$f(t, u) = f(t, 0) + \frac{\partial f(t, 0)}{\partial u} u + \frac{1}{2} \frac{\partial^2 f(t, 0)}{\partial u^2} u^2 + R(t, u),$$

for some continuous function $R(t, u)$ such that

$$\lim_{u \to 0} \frac{R(t, u)}{|u|^2} = 0.$$

Thus, in a neighborhood of $u = 0$,

$$f(t, u) \approx f(t, 0) + \frac{\partial f(t, 0)}{\partial u} u + \frac{1}{2} \frac{\partial^2 f(t, 0)}{\partial u^2} u^2;$$

therefore, in such a neighborhood, the differential equation

$$u' = f(t, u)$$

can be approximated by the following differential equation with quadratic nonlinearity:

$$u' = a(t) + b(t)u + c(t)u^2, \tag{4.36}$$

where we denote

$$a(t) := f(t, 0), \quad b(t) := \frac{\partial f(t, 0)}{\partial u}, \quad c(t) := \frac{1}{2} \frac{\partial^2 f(t, 0)}{\partial u^2}.$$

The differential equation (4.36), with $a, b, c \in \mathcal{C}(\mathbb{R}; \mathbb{R})$, was studied in 1724 by J. F. Riccati. For this reason, it is known as the *Riccati equation*. Observe that, when $a = 0$, it becomes a Bernoulli equation, which has been considered in Section 4.1. When $c = 0$ instead, it is a linear equation and can be solved using the techniques analyzed in Chapter 1.

In general, when $a \neq 0$ and $c \neq 0$, there is no way to find a particular solution of (4.36). However, from any given solution, $p(t)$, of (4.36), one can construct a one-parameter family of solutions by performing the change of variable

$$u(t) = p(t) + v(t).$$

Indeed, a direct substitution in (4.36) leads to

$$p'(t) + v'(t) = a(t) + b(t)\left[p(t) + v(t)\right] + c(t)\left[p(t) + v(t)\right]^2 ,$$

which can be equivalently rewritten as

$$v'(t) = \left[b(t) + 2c(t)p(t)\right] v(t) + c(t)v^2(t) \tag{4.37}$$

because $p(t)$ solves the equation, i.e.

$$p'(t) = a(t) + b(t)p(t) + c(t)p^2(t).$$

Since (4.37) is a Bernoulli equation with $p = 2$, as detailed in Section 4.1, the change of variable $v = w^{-1}$ transforms (4.37) into

$$w' = -\left(b + 2cp\right) w - c, \tag{4.38}$$

which is a linear equation whose integration has been carried out in Section 1.2. Precisely, the variation of constants formula provides us with the general solution of (4.38):

$$w(t) = e^{-\int_{t_0}^{t} (b(\sigma)+2c(\sigma)p(\sigma))\,d\sigma}\, x$$

$$- \int_{t_0}^{t} e^{-\int_{s}^{t}(b(\sigma)+2c(\sigma)p(\sigma))\,d\sigma} c(s)\,ds, \quad x \in \mathbb{R}.$$

Thus,

$$u(t) = p(t) + \left(e^{-\int_{t_0}^{t}(b(\sigma)+2c(\sigma)p(\sigma))\,d\sigma}\, x \right.$$

$$\left. - \int_{t_0}^{t} e^{-\int_{s}^{t}(b(\sigma)+2c(\sigma)p(\sigma))\,d\sigma} c(s)\,ds \right)^{-1} ,$$

with $x \in \mathbb{R}$, provides us with a one-parameter family of solutions of (4.36).

For example, the differential equation

$$u' = 9 + 6u + u^2 = (u + 3)^2 \tag{4.39}$$

is of Riccati type. It is apparent that $p(t) = -3$ is a constant solution of (4.39). Thus, from the above analysis (anyway, here, it can also be seen directly from (4.39)), the change of variable

$$u = v - 3$$

transforms (4.39) into

$$v' = v^2, \tag{4.40}$$

which is a Bernoulli equation. As the change of variable $v = w^{-1}$ transforms (4.40) into $w' = -1$, we obtain that

$$w(t) = x - t, \quad x \in \mathbb{R};$$

therefore, for every $x \in \mathbb{R}$,

$$u(t) = \frac{1}{x - t} - 3 \tag{4.41}$$

provides us with a solution of (4.39). Observe that, although this method of integration provides us with a one-parameter family of solutions from any given particular solution, $p(t)$, it might not give all the solutions of the equation. For instance, the particular solution $p = -3$ cannot be recovered from (4.41).

More generally, when a, b, and c are constants, there exist $\lambda, \mu \in \mathbb{C}$ such that

$$a + bu + cu^2 = c(u - \lambda)(u - \mu)$$

and, hence, the change of variable

$$u = v + \lambda$$

transforms (4.36) into

$$v' = cv(v + \lambda - \mu) = c(\lambda - \mu)v + cv^2.$$

As above, the change of variable $v = w^{-1}$ transforms this equation into

$$w' = c(\mu - \lambda)w - c.$$

If $\lambda \neq \mu$, then, for every constant $x \in \mathbb{C}$,

$$w(t) = e^{c(\mu - \lambda)t}x + \frac{1}{\mu - \lambda}$$

solves the above equation for w and, therefore,

$$u(t) = \lambda + \left(e^{c(\mu - \lambda)t}x + \frac{1}{\mu - \lambda} \right)^{-1}$$

provides us with a one-parameter family of solutions of (4.36). If $\lambda = \mu$, then w satisfies $w' = -c$, whose general solution is

$$w(t) = -ct + x, \quad x \in \mathbb{C},$$

and, therefore,

$$u(t) = \lambda + \frac{1}{x - ct}$$

solves (4.36) for all $x \in \mathbb{C}$. Observe that, when $\lambda, \mu \in \mathbb{C} \setminus \mathbb{R}$, these solutions are complex, and in contrast to the linear theory, how to obtain a real solution from them is not clear due to a lack of superposition principles for nonlinear equations.

Actually, as already mentioned, there is unfortunately no general method to find a particular solution of the Riccati equation for arbitrary function coefficients $a(t)$, $b(t)$, and $c(t)$. This will also become apparent from Exercise 17 of Chapter 4.

4.8.2   *Homogeneous equations*

Differential equations of the form

$$u' = f\left(\frac{u}{t}\right), \tag{4.42}$$

where $f : \mathbb{R} \to \mathbb{R}$ is a given continuous function, are called *homogeneous equations*. By making the change of variable

$$v(t) = \frac{u(t)}{t}, \tag{4.43}$$

we have that

$$u(t) = tv(t)$$

and, hence, (4.42) is equivalent to

$$v(t) + tv'(t) = f(v(t)),$$

which can be rewritten as

$$v'(t) = \frac{1}{t}\left[f(v(t)) - v(t)\right].$$

As this is a differential equation with separated variables, its integration can be accomplished as described in Section 4.2.

Undoubtedly, the most paradigmatic family of homogeneous equations is the following one:

$$u'(t) = \frac{au(t) + bt}{cu(t) + dt}, \tag{4.44}$$

where $a, b, c, d \in \mathbb{R}$ and $(c, d) \neq (0, 0)$. Indeed, by dividing the numerator and denominator by t, it is apparent that this differential equation can be expressed as

$$u'(t) = \frac{a\frac{u(t)}{t} + b}{c\frac{u(t)}{t} + d}.$$

Thus, it is of the homogeneous type (4.42), with

$$f(x) := \frac{ax + b}{cx + d}.$$

If the coefficients $a, b, c,$ and d satisfy $ad - bc = 0$, then there exists $\lambda \in \mathbb{R}$ such that

$$(a, b) = \lambda(c, d)$$

and, hence,

$$u'(t) = \frac{au(t) + bt}{cu(t) + dt} = \lambda,$$

whose general solution is

$$u(t) = \lambda t + x, \quad x \in \mathbb{R}.$$

Instead, if $ad - bc \neq 0$ but $c = 0$, then $ad \neq 0$ and, hence,

$$u'(t) = \frac{au(t) + bt}{cu(t) + dt} = \frac{a}{d}\frac{1}{t}u(t) + \frac{b}{d},$$

which is a linear equation whose general solution, due to Theorem 1.2, is

$$u(t) = t^{\frac{a}{d}}x + \frac{b}{d-a}t, \quad x \in \mathbb{R},$$

if $a \neq d$, and

$$u(t) = tx + \frac{b}{d}t\ln t, \quad x \in \mathbb{R},$$

if $a = d$.

Thus, it remains to consider the case

$$ad - bc \neq 0, \quad c \neq 0. \tag{4.45}$$

To obtain the set of solutions of (4.44) under conditions (4.45), we determine the solution of the Cauchy problem

$$\begin{cases} u'(t) = \frac{au(t)+bt}{cu(t)+dt}, \\ u(t_0) = u_0, \end{cases} \tag{4.46}$$

for every $(t_0, u_0) \in \mathbb{R}^2$, with $t_0 \neq 0$. By introducing the polynomial

$$P(z) := z^2 + \frac{d-a}{c}z - \frac{b}{c},$$

the change of variable (4.43) transforms (4.46) into

$$\begin{cases} tv'(t) = -\frac{P(v(t))}{v(t)+\frac{d}{c}}, \\ v(t_0) = v_0 := \frac{u_0}{t_0}. \end{cases} \tag{4.47}$$

We observe that, if $P(v_0) = 0$, then $v_0 + \frac{d}{c} \neq 0$ because $P\left(-\frac{d}{c}\right) = 0$ is equivalent to $ad - dc = 0$, which is excluded. Thus, if $P(v_0) = 0$, with $v_0 \in \mathbb{R}$, v_0 is a constant solution of (4.47), and $u(t) = \frac{u_0}{t_0}t$ provides us with the corresponding solution of (4.46).

Next, we assume that

$$P(v_0) \neq 0, \quad v_0 + \frac{d}{c} \neq 0.$$

In this case, (4.47) can be equivalently rewritten as

$$\begin{cases} \frac{v(t)+\frac{d}{c}}{P(v(t))} v'(t) = -\frac{1}{t}, \\ v(t_0) = v_0 := \frac{u_0}{t_0}, \end{cases} \tag{4.48}$$

and, thanks to Theorem 4.2, problem (4.48) possesses a unique local solution, which is the solution of the integral equation

$$\int_{v_0}^{v(t)} \frac{s + \frac{d}{c}}{P(s)} \, ds = -\ln \left| \frac{t}{t_0} \right|. \tag{4.49}$$

To determine a primitive of the integrand on the left-hand side of (4.49), we distinguish three different cases according to the nature of the roots of the polynomial $P(z)$.

Case 1. $P(z)$ has two real roots. This occurs if and only if

$$\left(\frac{a-d}{2c} \right)^2 + \frac{b}{c} > 0.$$

In this case,

$$P(z) = (z - \alpha)(z - \beta),$$

with

$$\alpha := \frac{a-d}{2c} - \sqrt{\left(\frac{a-d}{2c} \right)^2 + \frac{b}{c}} < \frac{a-d}{2c} + \sqrt{\left(\frac{a-d}{2c} \right)^2 + \frac{b}{c}} =: \beta. \tag{4.50}$$

Thus,

$$\frac{s + \frac{d}{c}}{P(s)} = \frac{s + \frac{d}{c}}{(s - \alpha)(s - \beta)} = \frac{A}{s - \alpha} + \frac{B}{s - \beta},$$

where

$$A := \frac{\alpha c + d}{c(\alpha - \beta)}, \quad B := \frac{\beta c + d}{c(\beta - \alpha)}. \tag{4.51}$$

Therefore, integrating and rearranging terms in (4.49) yield

$$\ln\left(\left|\frac{v(t)-\alpha}{v_0-\alpha}\right|^{-A}\left|\frac{v(t)-\beta}{v_0-\beta}\right|^{-B}\right) = \ln\left|\frac{t}{t_0}\right|.$$

Equivalently, taking exponentials and going back to the original variables, t and $u(t)$,

$$\left|\frac{u(t)-\alpha t}{u_0-\alpha t_0}\right|^{-A}\left|\frac{u(t)-\beta t}{u_0-\beta t_0}\right|^{-B} = \left|\frac{t}{t_0}\right|^{1-A-B}. \tag{4.52}$$

On the other hand, from (4.51), it is easily seen that

$$A+B=1 \tag{4.53}$$

and, hence, (4.52) can be equivalently expressed as

$$|u(t)-\alpha t|^A|u(t)-\beta t|^B = |u_0-\alpha t_0|^A|u_0-\beta t_0|^B,$$

which is an implicit equation satisfied by the solution of the Cauchy problem (4.46). The curves in the (u,t)-plane containing the graphs of these solutions are referred to as the *integral curves* of (4.44). Thus, the integral curve of (4.44) through (u_0,t_0) is given, implicitly, by

$$|u-\alpha t|^A|u-\beta t|^B = |u_0-\alpha t_0|^A|u_0-\beta t_0|^B. \tag{4.54}$$

To represent them, it is convenient to introduce the new variables (x,y), defined as

$$\begin{cases} x = u - \alpha t \\ y = u - \beta t \end{cases} \Longleftrightarrow \begin{pmatrix} x \\ y \end{pmatrix} = \begin{pmatrix} 1 & -\alpha \\ 1 & -\beta \end{pmatrix}\begin{pmatrix} u \\ t \end{pmatrix},$$

which gives an admissible (linear) change of coordinates in the plane since $\alpha \neq \beta$. By setting

$$x_0 := u_0 - \alpha t_0, \quad y_0 := u_0 - \beta t_0,$$

in the new variables, the integral curves (4.54) become

$$|x|^A|y|^B = |x_0|^A|y_0|^B. \tag{4.55}$$

The behavior of these curves differs according to the sign of AB. If $AB \geq 0$, then (4.53) implies that $A \geq 0$ and $B \geq 0$. Actually, $A > 0$

and $B > 0$. Indeed, relating the coefficients of $P(z)$ with its roots, it is apparent that

$$\alpha\beta = -\frac{b}{c}, \quad \alpha + \beta = \frac{a-d}{c}.$$

Thus, it follows from (4.51) and (4.45) that

$$AB = \frac{\alpha c + d}{c(\alpha - \beta)}\frac{\beta c + d}{c(\beta - \alpha)} = \frac{bc - ad}{c^2(\beta - \alpha)^2} \neq 0,$$

which entails $A > 0$ and $B > 0$. In this case, (4.55) describes a hyperbolic curve, with axes $x = 0$ and $y = 0$ passing through the point (x_0, y_0). Observe that the axes are also integral curves: they are obtained by choosing $x_0 = 0$ or $y_0 = 0$, and in terms of the variable $v(t)$, they correspond to the constant solutions $v(t) = \alpha$ and $v(t) = \beta$. Figure 4.12 represents this set of integral curves: on the left in (x, y)-coordinates (see (4.55)), on the right in (t, u)-coordinates (see (4.54)). This set of integral curves is often said to be of *saddle type*.

Now, suppose that $AB < 0$. Then, one of A and B is negative, and the other is positive. Thus, it follows from (4.55) that

$$|x| = |x_0||y_0|^{\frac{B}{A}}|y|^{-\frac{B}{A}},$$

with $-\frac{B}{A} > 0$. When $x_0 \neq 0$ and $y_0 \neq 0$, this curve is a parabolic curve of the type shown in Figure 4.13, where the special case

$$-\frac{B}{A} > 1$$

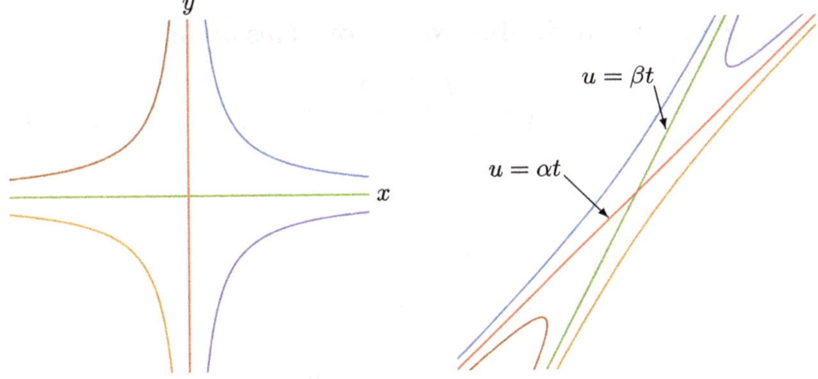

Fig. 4.12. Set of integral curves of saddle type.

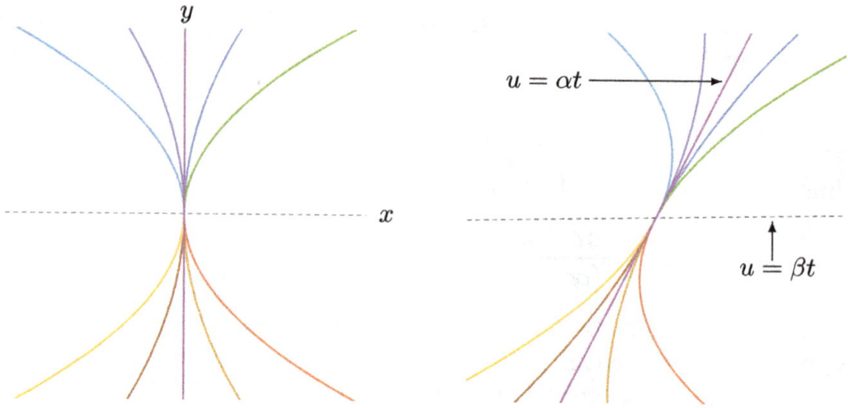

Fig. 4.13. Set of integral curves of node type.

has been represented. When

$$0 < -\frac{B}{A} < 1,$$

the integral curves are of the same type but tangent to the x-axis at the origin, instead to the y-axis. Finally, if

$$-\frac{B}{A} = 1,$$

then the set of integral curves consists of the set of straight lines through $(0,0)$.

In the latter cases, i.e. when $AB < 0$, the set of integral curves is said to be of *node type*.

Case 2. $P(z)$ has a double real root. This occurs if and only if

$$\left(\frac{a-d}{2c}\right)^2 + \frac{b}{c} = 0. \tag{4.56}$$

Then,

$$P(z) = (z - \alpha)^2, \quad \alpha := \frac{a-d}{2c},$$

and hence, the following decomposition holds:

$$\frac{s + \frac{d}{c}}{P(s)} = \frac{s + \frac{d}{c}}{(s - \alpha)^2} = \frac{1}{s - \alpha} + \frac{a+d}{2c}\frac{1}{(s - \alpha)^2}.$$

Thus, by integrating and rearranging terms, (4.49) yields

$$\ln|u(t) - \alpha t| - \frac{a+d}{2c}\frac{t}{u(t) - \alpha t} = \ln|u_0 - \alpha t_0| - \frac{a+d}{2c}\frac{t_0}{u_0 - \alpha t_0}.$$

Therefore, in this case, the solution of (4.46) lies on the integral curve

$$\ln|u - \alpha t| - \frac{a+d}{2c}\frac{t}{u - \alpha t} = \ln|u_0 - \alpha t_0| - \frac{a+d}{2c}\frac{t_0}{u_0 - \alpha t_0}. \quad (4.57)$$

To represent these curves, it is convenient to introduce the new variables

$$\begin{cases} x = u - \alpha t \\ y = \frac{a+d}{2c}t \end{cases} \iff \begin{pmatrix} x \\ y \end{pmatrix} = \begin{pmatrix} 1 & -\alpha \\ 0 & \frac{a+d}{2c} \end{pmatrix}\begin{pmatrix} u \\ t \end{pmatrix}. \quad (4.58)$$

As we are assuming $ad \neq bc$, (4.56) easily gives $a + d \neq 0$; hence, (4.58) defines a linear change of coordinates in the plane. Setting

$$x_0 := u_0 - \alpha t_0, \quad y_0 := \frac{a+d}{2c}t_0,$$

the set of curves (4.57) can be equivalently expressed as

$$\ln|x| - \frac{y}{x} = \ln|x_0| - \frac{y_0}{x_0}.$$

Expressing y as a function of x in this identity gives

$$y = x\ln\left(e^{\frac{y_0}{x_0}}\left|\frac{x}{x_0}\right|\right).$$

These curves are represented in Figure 4.14 for certain values of (x_0, y_0).

Note that, in this case, the unique integral curve which is a straight line is $x = 0$. It corresponds to the constant solution $v(t) = \alpha$, with $v(t) = \frac{u(t)}{t}$. This type of node is usually referred to as a *degenerate node*.

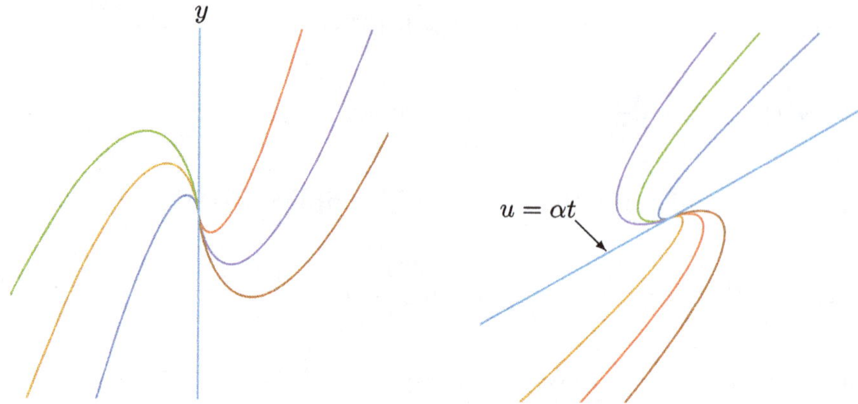

Fig. 4.14.   Set of integral curves of degenerate node type.

Case 3. $P(z)$ has two conjugate complex roots. This occurs if

$$\left(\frac{a-d}{2c}\right)^2 + \frac{b}{c} < 0.$$

Then,

$$P(z) = (z-\alpha)^2 + \beta^2, \quad \alpha := \frac{a-d}{2c}, \quad \beta := \sqrt{-\left(\frac{a-d}{2c}\right)^2 - \frac{b}{c}} > 0,$$

and hence,

$$\frac{s + \frac{d}{c}}{P(s)} = \frac{1}{2}\frac{2(s-\alpha)}{(s-\alpha)^2 + \beta^2} + \frac{\alpha + \frac{d}{c}}{(s-\alpha)^2 + \beta^2}.$$

Thus, integrating (4.49), rearranging terms, and going back to the original variables yield

$$\ln\sqrt{(u(t) - \alpha t)^2 + \beta^2 t^2} + \frac{d + \alpha c}{\beta c}\arctan\frac{u(t) - \alpha t}{\beta t}$$

$$= \ln\sqrt{(u_0 - \alpha t_0)^2 + \beta^2 t_0^2} + \frac{d + \alpha c}{\beta c}\arctan\frac{u_0 - \alpha t_0}{\beta t_0}.$$

Therefore, in this case, the solution of (4.46) lies on the integral curve

$$\ln\sqrt{(u - \alpha t)^2 + \beta^2 t^2} + \frac{d + \alpha c}{\beta c}\arctan\frac{u - \alpha t}{\beta t}$$

$$= \ln\sqrt{(u_0 - \alpha t_0)^2 + \beta^2 t_0^2} + \frac{d + \alpha c}{\beta c}\arctan\frac{u_0 - \alpha t_0}{\beta t_0}. \tag{4.59}$$

In order to represent it, it is convenient to perform the change of variable

$$\begin{cases} x = \beta t \\ y = u - \alpha t \end{cases} \iff \begin{pmatrix} x \\ y \end{pmatrix} = \begin{pmatrix} 0 & \beta \\ 1 & -\alpha \end{pmatrix} \begin{pmatrix} u \\ t \end{pmatrix}. \tag{4.60}$$

Since $\beta > 0$, (4.60) is a linear change of coordinates. As usual, by setting

$$x_0 := \beta t_0, \quad y_0 := u_0 - \alpha t_0,$$

(4.59) can equivalently be expressed as

$$\ln \sqrt{x^2 + y^2} + \frac{d + \alpha c}{\beta c} \arctan \frac{y}{x} = \ln \sqrt{x_0^2 + y_0^2} + \frac{d + \alpha c}{\beta c} \arctan \frac{y_0}{x_0}.$$

In polar coordinates,

$$\rho := \sqrt{x^2 + y^2}, \quad \theta := \arctan \frac{y}{x},$$

after taking exponentials and setting

$$\rho_0 := \sqrt{x_0^2 + y_0^2}, \quad \theta_0 := \arctan \frac{y_0}{x_0},$$

the family of integral curves can be equivalently expressed as

$$\rho = \rho_0 e^{-\frac{d + \alpha c}{\beta c}(\theta - \theta_0)}. \tag{4.61}$$

Since

$$d + \alpha c = \frac{a + d}{2},$$

if $a + d \neq 0$, (4.61) describes a family of logarithmic spirals around $(0, 0)$, and such a family of integral curves is said to be of *focus type*. If $a + d = 0$, instead, (4.61) becomes $\rho = \rho_0$; therefore, in this case, the integral curves consist of the set of all circumferences centered at the origin, forming a so-called *center*.

Figure 4.15 shows two of these sets of integral curves in the original coordinates (u, t). This completes the analysis of the structure of the set of integral curves of (4.44).

Fig. 4.15. Set of integral curves of focus and center types (left and right, respectively).

To conclude this section, we consider a continuous function $F :$ $\mathbb{R} \to \mathbb{R}$ and $a, \tilde{a}, b, \tilde{b}, c, \tilde{c} \in \mathbb{R}$, with $(\tilde{a}, \tilde{b}, \tilde{c}) \neq (0, 0, 0)$. The differential equation

$$u' = F\left(\frac{au(t) + bt + c}{\tilde{a}u(t) + \tilde{b}t + \tilde{c}}\right) \tag{4.62}$$

is homogeneous if and only if $c = \tilde{c} = 0$. Nevertheless, we now show that, even when $(c, \tilde{c}) \neq (0, 0)$, its integration can be reduced to the integration of a homogeneous equation. Indeed, consider the two straight lines in the (t, u)-plane defined by

$$au + bt + c = 0 \quad \text{and} \quad \tilde{a}u + \tilde{b}t + \tilde{c} = 0.$$

If these lines are parallel, there exists a constant $\lambda \in \mathbb{R}$ such that

$$(\tilde{a}, \tilde{b}) = \lambda(a, b).$$

Thus, in such a case, (4.62) can be expressed as

$$u' = F\left(\frac{au(t) + bt + c}{\lambda(au(t) + bt) + \tilde{c}}\right), \tag{4.63}$$

and hence, the change of variable

$$v(t) = au(t) + bt$$

transforms (4.63) into

$$v'(t) = aF\left(\frac{v(t) + c}{\lambda v(t) + \tilde{c}}\right) + b,$$

which is an equation with separated variables, whose integration can be performed as described in Section 4.2.

When the straight lines cross at (t_0, u_0), i.e.

$$au_0 + bt_0 + c = 0 \quad \text{and} \quad \tilde{a}u_0 + \tilde{b}t_0 + \tilde{c} = 0,$$

the change of variable

$$\tau := t - t_0, \quad v(\tau) := u(t) - u_0,$$

transforms (4.62) into

$$\dot{v}(\tau) = u'(t) = F\left(\frac{a\left(v(\tau) + u_0\right) + b\left(\tau + t_0\right) + c}{\tilde{a}\left(v(\tau) + u_0\right) + \tilde{b}\left(\tau + t_0\right) + \tilde{c}}\right)$$
$$= F\left(\frac{av(\tau) + b\tau}{\tilde{a}v(\tau) + \tilde{b}\tau}\right),$$

where u' indicates differentiation of u with respect to t, as usual, while $\dot{v}$ indicates differentiation of v with respect to τ. Thus, we have obtained a homogeneous differential equation in the variables (τ, v), which can be transformed into an equation with separated variables by means of the change of variable $z(\tau) = \frac{v(\tau)}{\tau}$, as described in the first part of this section.

4.8.3 *Exact differential equations*

Given an arbitrary open subset $\Omega \subset \mathbb{R}^2$, a differential equation of the form

$$P(t, u) + Q(t, u)u' = 0, \tag{4.64}$$

where $P, Q : \Omega \to \mathbb{R}$ are functions of class $\mathcal{C}^1$, is said to be an *exact differential equation in* Ω if there exists a function $V : \Omega \to \mathbb{R}$ of class $\mathcal{C}^2(\Omega)$ such that

$$P = \frac{\partial V}{\partial t} \quad \text{and} \quad Q = \frac{\partial V}{\partial u} \quad \text{in } \Omega. \tag{4.65}$$

In such a case, for every $(t_0, u_0) \in \Omega$ and any solution $(t, u(t))$ of the Cauchy problem

$$\begin{cases} P(t, u) + Q(t, u)u' = 0, \\ u(t_0) = u_0, \end{cases} \tag{4.66}$$

such that $(t, u(t)) \in \Omega$ for $t \sim t_0$, the following identity holds true:

$$\frac{\partial V}{\partial t}(t, u(t)) + \frac{\partial V}{\partial u}(t, u(t))u'(t) = 0.$$

Equivalently, thanks to the chain rule,

$$\frac{d}{dt}V(t, u(t)) = 0$$

and, hence,

$$V(t, u(t)) = V(t_0, u_0) \tag{4.67}$$

for all time t such that the solution $(t, u(t))$ of (4.66) lies in Ω. Conversely, by differentiating (4.67) with respect to t and using (4.65), it readily follows that any solution of (4.67) in Ω must be a local solution of (4.66), as proved in the following result, which relies on the implicit function theorem.

Theorem 4.4. *Suppose* (4.65) *and consider* $(t_0, u_0) \in \Omega$ *such that*

$$Q(t_0, u_0) = \frac{\partial V}{\partial u}(t_0, u_0) \neq 0. \tag{4.68}$$

Then, problem (4.66) *has a unique local solution, i.e. there exist* $\delta > 0$ *and a function* $u \in \mathcal{C}^1((t_0 - \delta, t_0 + \delta); \mathbb{R})$ *such that* $(t, u(t)) \in \Omega$ *if* $|t - t_0| < \delta$, *and* $u(t)$ *is the unique solution of* (4.66) *in the interval* $(t_0 - \delta, t_0 + \delta)$.

Proof. As proved above, the solutions of (4.67) are in one-to-one correspondence with the solutions of (4.66). Consider the auxiliary function

$$H(t, u) := V(t, u) - V(t_0, u_0).$$

As V is of class $\mathcal{C}^2(\Omega)$, H is of class $\mathcal{C}^2(\Omega)$ too. Moreover, by (4.68),

$$\frac{\partial H}{\partial u}(t_0, u_0) = \frac{\partial V}{\partial u}(t_0, u_0) \neq 0,$$

and $H(t_0, u_0) = 0$. Thus, the implicit function theorem (see Theorem 4.1) gives the existence of $\delta > 0$ and of a function of class $\mathcal{C}^2$,

$$u : (t_0 - \delta, t_0 + \delta) \to \mathbb{R},$$

such that:

- $u(t_0) = u_0$;
- $H(t, u(t)) = 0$ if $|t - t_0| < \delta$;
- $u = u(t)$ if $H(t, u) = 0$, with (t, u) sufficiently close to (t_0, u_0).

Since $H(t, u(t)) = 0$ if and only if (4.67) holds, the proof is complete. $\square$

The following result characterizes whether or not (4.64) is an exact differential equation in convex domains.

Theorem 4.5. *Suppose Ω is a convex open subset of $\mathbb{R}^2$ and (P, Q) is a planar vector field of class $\mathcal{C}^1(\Omega)$. Then, the following conditions are equivalent:*

(a) *(4.64) is an exact differential equation in Ω.*
(b) *(P, Q) is a gradient-type field, i.e.*

$$(P, Q) = \nabla_{(t,u)} V := \left(\frac{\partial V}{\partial t}, \frac{\partial V}{\partial u} \right) \quad in \ \Omega$$

for some function $V : \Omega \to \mathbb{R}$ of class $\mathcal{C}^2(\Omega)$.
(c) *(P, Q) satisfies*

$$\frac{\partial P}{\partial u} = \frac{\partial Q}{\partial t} \quad in \ \Omega.$$

Proof. Assertions (a) and (b) are equivalent by the definition of exact differential equation. Suppose (b). Then, since V is of class $\mathcal{C}^2$, thanks to Schwarz's theorem on the equality of mixed partial derivatives, we find that

$$\frac{\partial P}{\partial u} = \frac{\partial^2 V}{\partial u \partial t} = \frac{\partial^2 V}{\partial t \partial u} = \frac{\partial Q}{\partial t}.$$

Thus, (c) holds. Finally, suppose (c). To construct a potential V such that, for every $(t, u) \in \Omega$,

$$\frac{\partial V}{\partial t}(t, u) = P(t, u) \quad \text{and} \quad \frac{\partial V}{\partial u}(t, u) = Q(t, u), \tag{4.69}$$

we fix $(t_0, u_0) \in \Omega$, consider any $(t, u) \in \Omega$, and proceed as follows. Since we are assuming that Ω is convex, the segment Γ connecting (t_0, u_0) and (t, u) completely lies in Ω. Such a segment can be parameterized as

$$\gamma_{(t,u)}(s) = \gamma(s) = \begin{pmatrix} \gamma_1(s) \\ \gamma_2(s) \end{pmatrix} := \begin{pmatrix} t_0 \\ u_0 \end{pmatrix} + s \begin{pmatrix} t - t_0 \\ u - u_0 \end{pmatrix}, \quad s \in [0, 1].$$

We set

$$V(t, u) := \int_\Gamma (P, Q) = \int_0^1 \begin{pmatrix} P(\gamma(s)) \\ Q(\gamma(s)) \end{pmatrix} \cdot \dot{\gamma}(s)\, ds$$

$$= \int_0^1 \left(P(\gamma(s))\dot{\gamma}_1(s) + Q(\gamma(s))\dot{\gamma}_2(s) \right) ds, \tag{4.70}$$

with

$$\begin{pmatrix} P(\gamma(s)) \\ Q(\gamma(s)) \end{pmatrix} = \begin{pmatrix} P(t_0 + s(t-t_0), u_0 + s(u-u_0)) \\ Q(t_0 + s(t-t_0), u_0 + s(u-u_0)) \end{pmatrix} \quad \text{and}$$

$$\begin{pmatrix} \dot{\gamma}_1(s) \\ \dot{\gamma}_2(s) \end{pmatrix} = \begin{pmatrix} t - t_0 \\ u - u_0 \end{pmatrix},$$

where $\dot{\gamma}_i(s)$, $i \in \{1, 2\}$, denotes the derivative of $\gamma_i(s)$ with respect to s. We point out that V is well defined since $\Gamma \subset \Omega$. To show the first relation in (4.69), we differentiate (4.70) with respect to t

(it is possible to differentiate under the integral sign because P and Q are $\mathcal{C}^1$), obtaining

$$\frac{\partial V}{\partial t}(t,u) = \int_0^1 \left(\frac{\partial P(\gamma(s))}{\partial t} \frac{\partial \gamma_1(s)}{\partial t} \dot{\gamma}_1(s) + P(\gamma(s)) \frac{\partial \dot{\gamma}_1(s)}{\partial t} \right) ds$$

$$+ \int_0^1 \left(\frac{\partial Q(\gamma(s))}{\partial t} \frac{\partial \gamma_1(s)}{\partial t} \dot{\gamma}_2(s) + Q(\gamma(s)) \frac{\partial \dot{\gamma}_2(s)}{\partial t} \right) ds$$

$$= \int_0^1 \left(s \left[\frac{\partial P(\gamma(s))}{\partial t} \dot{\gamma}_1(s) + \frac{\partial Q(\gamma(s))}{\partial t} \dot{\gamma}_2(s) \right] + P(\gamma(s)) \right) ds,$$

where we have used that

$$\frac{\partial \gamma_1(s)}{\partial t} = s, \quad \frac{\partial \gamma_2(s)}{\partial t} = 0, \quad \frac{\partial \dot{\gamma}_1(s)}{\partial t} = 1, \quad \frac{\partial \dot{\gamma}_2(s)}{\partial t} = 0.$$

Moreover, since $\frac{\partial Q}{\partial t} = \frac{\partial P}{\partial u}$ in Ω, we can rewrite the last expression as

$$\frac{\partial V}{\partial t}(t,u) = \int_0^1 \left(s \left[\frac{\partial P(\gamma(s))}{\partial t} \dot{\gamma}_1(s) + \frac{\partial P(\gamma(s))}{\partial u} \dot{\gamma}_2(s) \right] + P(\gamma(s)) \right) ds$$

$$= \int_0^1 s \left(\nabla P(\gamma(s)) \cdot \dot{\gamma}(s) \right) + P(\gamma(s)) \, ds$$

$$= \int_0^1 s \frac{d}{ds} \left(P(\gamma(s)) \right) + P(\gamma(s)) \, ds,$$

$$= \int_0^1 \frac{d}{ds} \left(s P(\gamma(s)) \right) ds = P(t,u),$$

which is the first relation in (4.69). One can reason similarly to show the second relation in (4.69) by differentiating (4.70) with respect to u. Moreover, since P and Q are of class $\mathcal{C}^1(\Omega)$, it follows from (4.69) that $V \in \mathcal{C}^2(\Omega)$. This concludes the proof. $\qquad\square$

Observe that the same proof works for the case of a star-shaped domain with respect to the point (t_0, u_0).

To illustrate this method, we solve the initial value problem

$$\begin{cases} u'(t) = -\frac{u\cos t + 2te^u}{2+\sin t + t^2 e^u}, \\ u(t_0) = u_0, \end{cases} \qquad (4.71)$$

which can be expressed in the form (4.64) by setting

$$P(t,u) := u\cos t + 2te^u, \quad Q(t,u) := 2 + \sin t + t^2 e^u.$$

Since

$$\frac{\partial P}{\partial u}(t,u) = \cos t + 2te^u = \frac{\partial Q}{\partial t}(t,u),$$

by Theorem 4.5, (P,Q) is a gradient vector field in $\mathbb{R}^2$. In other words,

$$P(t,u) + Q(t,u)u' = 0$$

is an exact differential equation in $\mathbb{R}^2$. To integrate it, we shall find a function $V(t,u)$ such that

$$\begin{cases} \frac{\partial V}{\partial t}(t,u) = P(t,u) = u\cos t + 2te^u, \\ \frac{\partial V}{\partial u}(t,u) = Q(t,u) = 2 + \sin t + t^2 e^u. \end{cases} \qquad (4.72)$$

By integrating the first equation in (4.72) with respect to t, it is clear that all functions of the form

$$V(t,u) = u\sin t + t^2 e^u + \varphi(u),$$

with $\varphi(u)$ arbitrary, satisfy the first identity in (4.72). Thus, imposing the second one, $\varphi(u)$ should satisfy

$$\sin t + t^2 e^u + \varphi'(u) = 2 + \sin t + t^2 e^u.$$

Equivalently, $\varphi'(u) = 2$. Therefore, an admissible potential is

$$V(t,u) = u\sin t + t^2 e^u + 2u, \quad (t,u) \in \mathbb{R}^2,$$

and the solution of (4.71) is implicitly given by

$$u(t)\sin t + t^2 e^{u(t)} + 2u(t) = u_0 \sin t_0 + t_0^2 e^{u_0} + 2u_0. \qquad (4.73)$$

Since

$$\frac{\partial V}{\partial u}(t_0, u_0) = Q(t_0, u_0) = 2 + \sin t_0 + t_0^2 e^{u_0} \geq 1,$$

the existence of a unique local solution of (4.71) readily follows from Theorem 4.4 (and, ultimately, from Theorem 4.1). Note that it is impossible to express $u(t)$ as a function of t from (4.73). Nevertheless, one can consider the Taylor expansion of $u(t)$ at t_0. Indeed, by repeatedly differentiating in (4.71), one can get as many terms as one desires. For example, a direct substitution in (4.71) gives

$$u'(t_0) = -\frac{u_0 \cos t_0 + 2t_0 e^{u_0}}{2 + \sin t_0 + t_0^2 e^{u_0}}$$

and, hence,

$$u(t) = u_0 - \frac{u_0 \cos t_0 + 2t_0 e^{u_0}}{2 + \sin t_0 + t_0^2 e^{u_0}}(t - t_0) + O((t - t_0)^2), \qquad \text{as } t \to t_0.$$

In many circumstances, the differential equation (4.64) is not exact. Nevertheless, one can try to find a function of class $\mathcal{C}^1(\Omega)$, $\mu(t, u)$, such that the *equivalent* equation

$$\mu(t, u)P(t, u) + \mu(t, u)Q(t, u)u'(t) = 0$$

is exact. In such a case, the function $\mu(t, u)$ is said to be an *integrating factor* of (4.64). According to Theorem 4.5, $\mu(t, u)$ is an integrating factor of (4.64) if and only if

$$\frac{\partial}{\partial u}\left[\mu(t, u)P(t, u)\right] = \frac{\partial}{\partial t}\left[\mu(t, u)Q(t, u)\right],$$

which can be equivalently expressed as

$$P(t, u)\frac{\partial \mu}{\partial u}(t, u) - Q(t, u)\frac{\partial \mu}{\partial t}(t, u) = \left(\frac{\partial Q}{\partial t} - \frac{\partial P}{\partial u}\right)\mu(t, u). \qquad (4.74)$$

Although the theory of first-order partial differential equations establishes that, under very general assumptions on P and Q, (4.74) always admits a solution μ, expressing it in terms of elementary functions might not be possible. Moreover, the problem of solving the partial differential equation (4.74) might be substantially more involved than solving the ordinary differential equation (4.64). Nevertheless, in many specific circumstances, (4.74) can be easily solved, and the obtained integrating factor leads to a complete integration of (4.64).

For example, the differential equation

$$u' = -\frac{u^2 + tu}{t^2},$$

which can be rewritten as

$$u^2 + tu + t^2 u' = 0, \qquad (4.75)$$

is not exact because

$$\frac{\partial(u^2 + tu)}{\partial u} = 2u + t \neq 2t = \frac{\partial(t^2)}{\partial t}.$$

However, we will show that it admits an integrating factor of the form

$$\mu(t, u) = t^a u^b$$

for some constants a and b. Indeed, multiplying (4.75) by $t^a u^b$ yields

$$t^a u^{b+2} + t^{a+1} u^{b+1} + t^{a+2} u^b u' = 0, \qquad (4.76)$$

which is an exact differential equation if and only if

$$\frac{\partial(t^a u^{b+2} + t^{a+1} u^{b+1})}{\partial u} = (b + 2)t^a u^{b+1} + (b + 1)t^{a+1} u^b$$

$$= (a + 2)t^{a+1} u^b = \frac{\partial(t^{a+2} u^b)}{\partial t}.$$

On taking a careful look at this identity, it becomes apparent that it can be achieved by taking

$$b + 2 = 0, \quad b + 1 = a + 2.$$

Thus, the choice $b = -2$ and $a = -3$ provides us with the integrating factor

$$\mu(t, u) = t^{-3} u^{-2},$$

and (4.76) becomes the exact equation

$$t^{-3} + t^{-2} u^{-1} + t^{-1} u^{-2} u' = 0.$$

To integrate it, we have to find $V(t, u)$ such that

$$\frac{\partial V}{\partial t} = t^{-3} + t^{-2}u^{-1} \quad \text{and} \quad \frac{\partial V}{\partial u} = t^{-1}u^{-2}.$$

From the second identity, it is apparent that

$$V(t, u) = -t^{-1}u^{-1} + \varphi(t).$$

Thus, by imposing the first one, the function $\varphi(t)$ must satisfy

$$t^{-2}u^{-1} + \varphi'(t) = t^{-3} + t^{-2}u^{-1}.$$

Equivalently, $\varphi'(t) = t^{-3}$, and hence,

$$V(t, u) = -t^{-1}u^{-1} - \frac{1}{2}t^{-2}$$

is an admissible potential. As the solution of (4.75) with $u(t_0) = u_0$ satisfies

$$t^{-1}u^{-1} + \frac{1}{2}t^{-2} = t_0^{-1}u_0^{-1} + \frac{1}{2}t_0^{-2},$$

we obtain that

$$u(t) = \left(t_0^{-1}u_0^{-1}t + \frac{1}{2}(t_0^{-2}t - t^{-1}) \right)^{-1}$$

is the unique solution of (4.75) such that $u(t_0) = u_0$ provided $t_0 \neq 0$ and $u_0 \neq 0$. Thus, the integration in this case is accomplished in the open subset

$$\Omega = \{(t, u) \in \mathbb{R}^2 : t \neq 0, \ u \neq 0\}.$$

4.9 Exercises

1. Generalize the analysis performed in Section 4.1 to study the logistic problem (4.7) for $x < 0$. (Observe that such a case is less interesting for biological models; however, it is completely meaningful from a mathematical point of view.) Specifically:

(a) For all $x < 0$, by analyzing (4.8), which still provides us with a solution of (4.7) also for $x < 0$, show that $u(t; x)$ is decreasing and that there exists $T_{\max}(x) > 0$ such that $u(t; x)$ is defined if and only if $t \in (-\infty, T_{\max}(x))$.

(b) Prove relation (4.11). Deduce that $u(t; x) < 0$ whenever it is defined.

(c) Show that the maximal existence time satisfies

$$T_{\max}(x) = -T_{\max}\left(\frac{\lambda}{m} - x\right) \quad \text{for all } x < 0.$$

Deduce from the previous relation and the analysis after (4.9) that

$$\lim_{x \downarrow -\infty} T_{\max}(x) = 0, \quad \lim_{x \uparrow 0} T_{\max}(x) = +\infty.$$

(d) Show that the solutions of (4.7) depend continuously on x, in the sense discussed in Section 4.1, in a whole neighborhood of $x = 0$.

2. Given the initial value problem

$$\begin{cases} u'(t) = k(a - u)(b + u), \\ u(0) = x \in \mathbb{R}, \end{cases}$$

where k, a, and b are three constants such that $k > 0$ and $a > b > 0$:

(a) determine the equilibria of the differential equation;

(b) find the solution for every $x \in \mathbb{R}$;

(c) analyze the global behavior of the solutions and represent them;

(d) represent the dynamics of the differential equation.

3. Consider the generalized logistic problem

$$\begin{cases} u' = \lambda u - m u^p, \\ u(0) = x, \end{cases} \tag{4.77}$$

with $\lambda > 0$, $m > 0$, $p > 1$, and $x \geq 0$.

(a) Determine the equilibria of the differential equation.

(b) Find the unique solution of (4.77) when $x > 0$ by using the Bernoulli method.

(c) Plot the graphs of the solutions of (4.77) for $x > 0$.

(d) Analyze the dynamics of the generalized logistic equation.

(e) Prove that $u = 0$ is the unique solution of (4.77) for $x = 0$.

4. Given $p > 1$, consider, for every $x > 0$, the initial value problem

$$\begin{cases} u' = e^{-t}u^p, \\ u(0) = x. \end{cases} \tag{4.78}$$

(a) Prove that (4.78) has a unique solution whenever it is defined, and determine it.

(b) Analyze the character of the solution $u(t; x)$ according to all possible values of $x > 0$. In particular, determine

- when $u(t; x)$ is globally defined in $[0, \infty)$;
- when $u(t; x)$ blows up in a finite time.

(c) When $u(t; x)$ blows up in a finite time, compute the blow up time $T_{\max}(x)$ and plot its graph as a function of x.

5. Repeat the previous exercise with the following problem:

$$\begin{cases} u' = a(t)u^p, \\ u(0) = x, \end{cases} \tag{4.79}$$

where $p > 1$ and $a(t)$ is the function defined as

$$a(t) = \begin{cases} \sin t, & t \in [0, \pi], \\ 0, & t > \pi. \end{cases}$$

6. Repeat the previous exercise with problem (4.79), considering $p > 1$ and $a(t)$ defined as

$$a(t) = \begin{cases} \sin t, & t \in [0, \pi] \cup [2\pi, 3\pi], \\ 0, & t \in (\pi, 2\pi) \cup (3\pi, +\infty). \end{cases}$$

7. Given $q \in (0, 1)$, characterize all the functions $a \in \mathcal{C}(\mathbb{R}; \mathbb{R}_+)$ for which $u = 0$ is the unique solution of

$$\begin{cases} u' = a(t)\, u^q, \\ u(0) = 0. \end{cases}$$

8. Find the unique (local) solution of

$$\begin{cases} u' = 2\sqrt{u(\pi - u)}, \\ u(0) = x, \end{cases}$$

for every $x \in (0, \pi)$.

9. Let $f(\theta)$ denote the odd extension of the function

$$F(\theta) := 2\sqrt{\theta(\pi - \theta)}, \quad 0 \le \theta \le \pi,$$

to the interval $[-\pi, \pi]$. Prove that the set of values reached by the solutions of the problem

$$\begin{cases} \theta' = f(\theta), \\ \theta(0) = 0, \end{cases}$$

after an appropriate time, is the interval $[-\pi, \pi]$.

10. Describe the dynamics of the differential equation

$$u' = e^u \sin u.$$

11. Suppose that $\lambda(t)$ and $m(t)$ are continuous and T-periodic, i.e.

$$\lambda(t + T) = \lambda(t) \quad \text{and} \quad m(t + T) = m(t) \quad \text{for all } t \in \mathbb{R}.$$

Suppose, in addition, that

$$\int_0^T \lambda(s)\,ds > 0, \quad \text{and} \quad m(t) > 0 \quad \text{for all } t \in \mathbb{R}.$$

Consider the T-periodic logistic equation

$$u' = \lambda(t)u - m(t)u^2. \tag{4.80}$$

(a) Show, without solving (4.80), that $u \in C^1(\mathbb{R})$ is a T-periodic solution of (4.80) if and only if u solves (4.80) in $\mathbb{R}$ and $u(0) = u(T)$.

(b) Show that, if $u \ge 0$ solves (4.80), then $u(t) > 0$ for all $t \in \mathbb{R}$.

(c) Show that (4.80) admits a unique positive T-periodic solution, and determine it.

12. Solve the initial value problem

$$\begin{cases} u'(t) = \frac{3t^2+4t+2}{2(u-1)}, \\ u(0) = -1. \end{cases}$$

13. Solve the initial value problem

$$\begin{cases} u'(t) = \frac{u\cos t}{1+2u^2}, \\ u(0) = 1. \end{cases}$$

14. Solve the differential equations

$$u'(t) = \frac{u^2 + u}{t^2 - t}, \quad (t^2 + 1)u'(t) = (u + 1)(t^2 + t + 1).$$

15. Solve the differential equation

$$tu^2 u'(t) + u^3 = t\cos t.$$

16. Solve the differential equations

$$u' = 2 - 2tu + u^2, \quad u' + \frac{1}{t}u = tu^2.$$

17. Assume that $c(t) \neq 0$ for all $t \in \mathbb{R}$ and that it is differentiable. Show that the change of variable

$$w(t) = e^{-\int_0^t c(s)u(s)\,ds}$$

transforms the Riccati equation

$$u' = a(t) + b(t)u + c(t)u^2$$

into a second-order linear differential equation of the form

$$w'' + B(t)w' + A(t)w = 0$$

for some appropriate functions $A(t), B(t)$. (As already commented in Part 1, solving this type of equation in general is not possible; thus, finding a particular solution of a Riccati equation might not be possible.)

18. Construct, in full detail, the family of integral curves of the following homogeneous differential equations:

$$u'(t) = \frac{u + 3t}{3u + t}, \qquad u'(t) = \frac{3u + t}{u + 3t}, \qquad u'(t) = \frac{7u + 2t}{-5u + 5t},$$

$$u'(t) = \frac{u + 2t}{-5u - t}, \qquad u'(t) = \frac{u - t}{5u - 3t}, \qquad u'(t) = \frac{-3u - t}{-u - 3t}.$$

19. Construct, in full detail, the family of integral curves of the differential equations

$$u' = \frac{u - 2t - 2}{2t - 2u - 1}, \qquad u' = \frac{u + t + 1}{2u + 2t - 2}.$$

20. Show that

$$u' = \frac{t^2 - u}{t + u^2}$$

is an exact differential, and find a solution such that $u(0) = 1$.

21. Solve the differential equation

$$(ue^u - 2t)u' = u$$

by means of an integrating factor depending on u.

22. Calculate the value of the constant $b \in \mathbb{R}$ so that the following differential equations are exact and solve the associated Cauchy problems:

$$\begin{cases} tu^2 + bt^2u + (t + u)t^2u'(t) = 0, \\ u(1) = -1, \end{cases} \qquad \begin{cases} ue^{2tu} + t + bte^{2tu}u'(t) = 0, \\ u(-1) = 0. \end{cases}$$

In addition, study the existence and uniqueness of the Cauchy problems with the values of b that you have determined.

23. Solve the following differential equations through appropriate integrating factors:

$$t^3 + u^4 + 8tu^3u'(t) = 0, \qquad tu^3 + (t^2u^2 - 1)u'(t) = 0.$$

24. Give general conditions on $f(t, u)$ so that $u' = f(t, u)$ admits smooth integrating factors of the type $\mu(t)$, $\mu(u)$, $\mu(t^2 + u^2)$, $\mu(u + t)$, and $\mu(u - t)$.

4.10   Final Comments

The Bernoulli equation was originally formulated by Jakob Bernoulli. Its resolution by means of the change of variable (4.4), which transforms it into a linear equation, was proposed by G. Leibniz in 1693 and by Johann Bernoulli in 1697.

The logistic equation was introduced by P. F. Verhulst in 1838 to infer a credible law for the growth of a population. Nevertheless, after Verhulst's death, his contributions remained in near-complete oblivion for almost 80 years, until the biologists R. Pearl and L. J. Reed, in 1920, rediscovered it in the same demographical context. They also compared the solutions of the logistic equation with a series of field data. Since then, the logistic equation is also known as the *Verhulst equation*, or the *Verhulst–Pearl equation*. Seventeen years later, in 1937, the diffusive logistic equation, i.e. the following reaction-diffusion equation,

$$\frac{\partial u}{\partial t} - \Delta u = \lambda u - au^2, \tag{4.81}$$

was introduced by A. N. Kolmogorov, I. G. Petrovsky, and N. S. Piskunov in the paper Kolmogorov *et al.* (1937) and independently by R. A. Fisher in 1937 to study some problems of biological nature. Undoubtedly, (4.81) is a pivotal differential equation in many areas of Science and Technology, whose analysis has been a milestone for the development of new techniques in the theory of nonlinear partial differential equations (see, e.g. the monograph by López-Gómez, 2015).

The Riccati equation was introduced by J. F. Riccati in 1724. J. Liouville, on the base of a classification of transcendent numbers, proved in 1841 that the Riccati equation can only be integrated in the cases predicted by D. Bernoulli.

The material in this chapter is based on Chapter 11 in the work of López-Gómez (2001b), though the content here has been substantially polished. The analysis in Section 4.8.2 was inspired by the one by Puig Adam (1978).

Chapter 5

Cauchy–Lipschitz Theory

In the first part of this chapter, given $\alpha, \beta \in \mathbb{R}$, $\alpha < \beta$, and a natural number $N \geq 1$, we consider continuous functions $f \in \mathcal{C}([\alpha, \beta] \times \mathbb{R}^N; \mathbb{R}^N)$ satisfying a *global Lipschitz condition*, uniformly with respect to the time variable, $t \in [\alpha, \beta]$, of the type

$$\|f(t, x) - f(t, y)\| \leq L\|x - y\| \quad \text{for all } x, y \in \mathbb{R}^N, \tag{5.1}$$

where $L \geq 0$ is a constant and, as usual, $\|\cdot\|$ denotes the $\|\cdot\|_1$-norm of $\mathbb{R}^N$.

The main goal is to establish that, under these conditions, for every $t_0 \in [\alpha, \beta]$ and $u_0 \in \mathbb{R}^N$, the Cauchy problem

$$\begin{cases} u' = f(t, u), \\ u(t_0) = u_0, \end{cases} \tag{5.2}$$

possesses a unique solution which is globally defined in $[\alpha, \beta]$. As usual, we use the notation

$$u := (u_1, \ldots, u_N)^T, \quad f = (f_1, \ldots, f_N)^T,$$

where T indicates *transposition*.

The simplest condition ensuring (5.1) is the existence and continuity of all partial derivatives

$$\frac{\partial f_i}{\partial u_j} \in \mathcal{C}([\alpha, \beta] \times \mathbb{R}^N; \mathbb{R}), \quad 1 \leq i, j \leq N,$$

together with the existence of a constant $C > 0$ such that

$$\left| \frac{\partial f_i}{\partial u_j}(t, u) \right| \leq C \quad \text{for all } (t, u) \in [\alpha, \beta] \times \mathbb{R}^N \text{ and } 1 \leq i, j \leq N.$$

Note that, in all the examples treated in Chapter 4 where non-uniqueness occurs, the differentiability of $f(t, u)$ with respect to u fails at the initial datum u_0 (see Sections 4.5 and 4.6).

Later in this chapter, we will replace the global condition (5.1) with a local one and obtain a local existence and uniqueness theorem for the Cauchy problem (5.2). The rest of the chapter is devoted to the study of how such a local solution can be extended as much as possible and to the analysis of its behavior. Finally, we study the dependence of the solution and its existence interval with respect to variations in the initial datum and the vector field of the differential equation.

The chapter is organized as follows. In Section 5.1, we introduce the concepts of globally and locally Lipschitz function (uniformly in t) and show that functions of class $\mathcal{C}^1$ in the u-variables are locally Lipschitz (uniformly in t). In Section 5.2, we state and prove the global Cauchy–Lipschitz theorem, related to the existence and uniqueness of the solution of (5.2). In Section 5.3, we give the local version of this theorem and establish the uniqueness of the solution as long as it exists, as well as the existence of the maximal solution. In Section 5.4, we classify the behaviors of the maximal solutions according to the nature of the domain of definition of $f(t, u)$, and in Section 5.5, we analyze three paradigmatic illustrative examples. Finally, in Section 5.6, the problem of the dependence of the solutions with respect to u_0 and f is addressed and solved in the Lipschitz setting.

5.1   Uniformly Lipschitz Functions

Throughout this section, J denotes an arbitrary interval of $\mathbb{R}$. So, J may be of the form $[\alpha, \beta]$, $(\alpha, \beta]$, $[\alpha, \beta)$, (α, β), $[\alpha, \infty)$, (α, ∞),

$(-\infty, \beta]$, $(-\infty, \beta)$, or $\mathbb{R} = (-\infty, \infty)$, for some $\alpha, \beta \in \mathbb{R}$, with $\alpha < \beta$. The following concept is pivotal in this book.

Definition 5.1 (Uniformly Lipschitz functions). Let J be an interval of $\mathbb{R}$, Ω be an open subset of $\mathbb{R}^N$, and $f : J \times \Omega \to \mathbb{R}^N$ be an arbitrary function.

(a) f is said to be *globally Lipschitz with respect to $u \in \Omega$ uniformly in $t \in J$* if there exists a constant $L \geq 0$ such that

$$\|f(t, x) - f(t, y)\| \leq L\|x - y\| \quad \text{for all } (t, x, y) \in J \times \Omega \times \Omega.$$

Such a constant L will be referred to as a Lipschitz constant for f.

(b) f is said to be *locally Lipschitz with respect to $u \in \Omega$ uniformly in $t \in J$* if, for every $x_0 \in \Omega$, there exists $R = R(x_0) > 0$ such that $B_R(x_0) \subset \Omega$ and $f|_{J \times B_R(x_0)}$ is globally Lipschitz with respect to $u \in B_R(x_0)$ uniformly in $t \in J$, i.e. if there exists $L = L(x_0) \geq 0$ such that, for every $(t, x, y) \in J \times B_R(x_0) \times B_R(x_0)$,

$$\|f(t, x) - f(t, y)\| \leq L\|x - y\|.$$

More generally, for any given subset $S \subset \mathbb{R}^N$ and any interval $J \subset \mathbb{R}$, a function $f : J \times S \to \mathbb{R}^N$ is said to be (globally) Lipschitz with respect to $u \in S$ uniformly in $t \in J$ if there exists a constant $L \geq 0$ such that, for every $x, y \in S$ and $t \in J$,

$$\|f(t, x) - f(t, y)\| \leq L\|x - y\|.$$

Note that, when $f(t, u)$ does not depend on t, i.e. when f is autonomous, the above definition of globally (respectively, locally) Lipschitz function uniformly in t reduces to the usual concept of globally (respectively, locally) Lipschitz function.

We also observe that being globally (or locally) Lipschitz does not depend on the chosen norms: if f is globally (or locally) Lipschitz and other norms are used either in the domain Ω or in the image $\mathbb{R}^N$, an inequality of the Lipschitz type for f will still hold true, possibly with a different Lipschitz constant.

Although more examples will be given later in this section, we already point out that the most paradigmatic example of globally Lipschitz function is

$$f(t, u) = A(t)u + B(t), \quad (t, u) \in [\alpha, \beta] \times \mathbb{R}^N,$$

with

$$A(t) = (a_{ij}(t))_{1 \leq i,j \leq N}, \quad B(t) = (b_j(t))_{1 \leq j \leq N}, \quad t \in [\alpha, \beta],$$

where

$$a_{ij}, b_j \in \mathcal{C}([\alpha, \beta]; \mathbb{R}) \quad \text{for all } 1 \leq i, j \leq N.$$

We claim that such an f is globally Lipschitz with respect to $u \in \mathbb{R}^N$ uniformly in $t \in [\alpha, \beta]$. Indeed, for every $t \in [\alpha, \beta]$ and $x, y \in \mathbb{R}^N$, we have

$$\|f(t, x) - f(t, y)\| = \|A(t)(x - y)\| \leq \|A(t)\|_{\mathcal{L}(\mathbb{R}^N)} \|x - y\|.$$

By the equivalence of the norms in $\mathbb{R}^{N^2}$, there exists a constant $C > 0$ such that

$$\|A(t)\|_{\mathcal{L}(\mathbb{R}^N)} \leq C\|A(t)\|_1 = C \sum_{i,j=1}^{N} |a_{ij}(t)| \leq C \sum_{i,j=1}^{N} \|a_{ij}\|_\infty.$$

So, for every $x, y \in \mathbb{R}^N$ and $t \in [\alpha, \beta]$,

$$\|f(t, x) - f(t, y)\| \leq L\|x - y\|,$$

where

$$L = C \sum_{i,j=1}^{N} \|a_{ij}\|_\infty.$$

This shows that f is globally Lipschitz with respect to $u \in \mathbb{R}^N$ uniformly in $t \in [\alpha, \beta]$.

The following result is a fundamental characterization of uniformly Lipschitz smooth functions. Recall that Ω is convex provided that $x, y \in \Omega$ implies $(1 - s)x + sy \in \Omega$ for all $s \in [0, 1]$.

Proposition 5.2. *Suppose that J is an interval of $\mathbb{R}$, Ω is a convex open subset of $\mathbb{R}^N$, and $f := (f_1, \ldots, f_N)^T \in \mathcal{C}(J \times \Omega; \mathbb{R}^N)$ admits* *continuous partial derivatives in all the u-variables,*

$$\frac{\partial f_i}{\partial u_j} \in \mathcal{C}(J \times \Omega; \mathbb{R}), \quad 1 \le i, j \le N.$$

Then, f is globally Lipschitz with respect to $u \in \Omega$ uniformly in $t \in J$ if and only if

$$M_{ij} := \sup_{(t,u) \in J \times \Omega} \left| \frac{\partial f_i}{\partial u_j}(t, u) \right| < +\infty \quad \text{for all } i, j \in \{1, \ldots, N\}. \tag{5.3}$$

Proof. Assume that (5.3) holds. Then, for every $x, y \in \Omega$ and $t \in J$,

$$f(t, y) - f(t, x) = \int_0^1 \frac{d}{ds} f(t, sy + (1-s)x) \, ds$$

$$= \int_0^1 D_u f(t, sy + (1-s)x)(y - x) \, ds,$$

where

$$D_u f(t, u) = \left(\frac{\partial f_i}{\partial u_j} \right)_{1 \le i, j \le N}$$

indicates the Jacobian matrix of f with respect to the u-variables. By taking norms and using the Minkowski integral inequality (see Section 1.3.3), the previous relation gives

$$\|f(t, x) - f(t, y)\| = \left\| \int_0^1 D_u f(t, sy + (1-s)x)(y - x) \, ds \right\|$$

$$\le \int_0^1 \|D_u f(t, sy + (1-s)x)(y - x)\| \, ds$$

$$\le \int_0^1 \|D_u f(t, sy + (1-s)x)\|_{\mathcal{L}(\mathbb{R}^N)} \, ds \, \|y - x\|.$$

On the other hand, as any two norms in the Euclidean space

$$\mathcal{L}(\mathbb{R}^N) \approx \mathbb{R}^{N^2} \approx \mathcal{M}_N(\mathbb{R})$$

are equivalent, there exists a constant $C > 0$ such that

$$\|D_u f(t, sy + (1-s)x)\|_{\mathcal{L}(\mathbb{R}^N)} \leq C \|D_u f(t, sy + (1-s)x)\|_1$$

$$= C \sum_{i,j=1}^{N} \left| \frac{\partial f_i}{\partial u_j}(t, sy + (1-s)x) \right|$$

for all $s \in [0, 1]$. Hence, it follows from (5.3) that

$$\|D_u f(t, sy + (1-s)x)\|_{\mathcal{L}(\mathbb{R}^N)} \leq C \sum_{i,j=1}^{N} M_{ij}$$

for all $t \in J$, $x, y \in \Omega$ and $s \in [0, 1]$. Therefore, by setting

$$L := C \sum_{i,j=1}^{N} M_{ij},$$

it becomes apparent that

$$\|f(t, x) - f(t, y)\| \leq L\|x - y\| \quad \text{for all } t \in J \text{ and } x, y \in \Omega. \quad (5.4)$$

So, f is globally Lipschitz with respect to $u \in \Omega$ uniformly in $t \in J$.

Conversely, suppose that (5.4) holds for some constant $L \geq 0$. Since $f(t, \cdot)$ is differentiable in Ω for every $t \in J$, we have that

$$D_u f(t, u)v = \lim_{h \to 0} \frac{f(t, u + hv) - f(t, u)}{h}$$

for all $u \in \Omega$ and $v \in \mathbb{R}^N$ with $\|v\| = 1$. According to (5.4), for sufficiently small h so that $u + hv \in \Omega$, we have that

$$\|f(t, u + hv) - f(t, u)\| \leq L\|hv\| = L|h|$$

for all $t \in J$. Thus,

$$\left\| \frac{f(t, u + hv) - f(t, u)}{h} \right\| \leq L,$$

and letting $h \to 0$ in this estimate yields

$$\|D_u f(t, u)v\| \leq L$$

for all $t \in J$, $u \in \Omega$ and $v \in \mathbb{R}^N$ with $\|v\| = 1$. Consequently,

$$\|D_u f(t, u)\|_{\mathcal{L}(\mathbb{R}^N)} = \max_{\|v\|=1} \|D_u f(t, u) v\| \leq L$$

for all $(t, u) \in J \times \Omega$. Therefore, by the equivalence of norms in $\mathcal{L}(\mathbb{R}^N)$, there exists $\tilde{C} > 0$ such that, for every $i, j \in \{1, \ldots, N\}$,

$$\left| \frac{\partial f_i}{\partial u_j}(t, u) \right| \leq \sum_{i,j=1}^{N} \left| \frac{\partial f_i}{\partial u_j}(t, u) \right| = \|D_u f(t, u)\|_1$$

$$\leq \tilde{C} \|D_u f(t, u)\|_{\mathcal{L}(\mathbb{R}^N)} \leq \tilde{C} L$$

for all $(t, u) \in J \times \Omega$. So, taking suprema in $(t, u) \in J \times \Omega$, we find that

$$M_{ij} \leq \tilde{C} L < +\infty \quad \text{for all } i, j \in \{1, \ldots, N\}.$$

This completes the proof. $\qquad\square$

Observe that the convexity of Ω was not required in the previous proof to obtain (5.3) from (5.4). The following consequence of Proposition 5.2 is very useful from the point of view of applications.

Corollary 5.3. *Suppose $N \in \mathbb{N}$, $N \geq 1$, Ω is an (arbitrary) open subset of $\mathbb{R}^N$, $J = [\alpha, \beta]$ for some $\alpha < \beta$, and $f \in \mathcal{C}(J \times \Omega; \mathbb{R}^N)$ satisfies*

$$\frac{\partial f_i}{\partial u_j} \in \mathcal{C}(J \times \Omega; \mathbb{R}), \quad 1 \leq i, j \leq N.$$

Then, f is locally Lipschitz with respect to $u \in \Omega$ uniformly in $t \in J$. Moreover, f is globally Lipschitz, uniformly in $t \in J$, on any open and convex bounded subset D of Ω such that $\bar{D} \subset \Omega$.

Proof. Suppose $D \subset \Omega$ is an open and convex bounded subset of Ω such that $\bar{D} \subset \Omega$. Then, $J \times \bar{D}$ is compact and, since $\frac{\partial f_i}{\partial u_j} \in \mathcal{C}(J \times \bar{D}; \mathbb{R})$ for each $i, j \in \{1, \ldots, N\}$,

$$M_{ij,D} := \max_{(t,u) \in J \times \bar{D}} \left| \frac{\partial f_i}{\partial u_j}(t, u) \right| < +\infty.$$

Therefore, since D is convex, by Proposition 5.2, f is globally Lipschitz with respect to $u \in D$. In particular, choosing $D = B_R(x_0)$ with arbitrary $x_0 \in \Omega$ and sufficiently small $R = R(x_0) > 0$, it becomes apparent that f is locally Lipschitz with respect to $u \in \Omega$ uniformly in $t \in J$. $\qquad\qquad\square$

Example 5.4. We now discuss some examples, with the aim of analyzing whether or not the considered functions are locally or globally Lipschitz, uniformly in t.

Example 5.4.1. Suppose $a \in \mathcal{C}([\alpha, \beta]; \mathbb{R})$ and

$$f(t, u) = a(t) \sin u, \quad (t, u) \in [\alpha, \beta] \times \mathbb{R}.$$

Then, f is continuous in $[\alpha, \beta] \times \mathbb{R}$, as well as

$$\frac{\partial f}{\partial u}(t, u) = a(t) \cos u, \quad (t, u) \in [\alpha, \beta] \times \mathbb{R}.$$

Moreover,

$$\left| \frac{\partial f}{\partial u}(t, u) \right| = |a(t)|\, |\cos u| \leq \|a\|_\infty$$

for all $t \in [\alpha, \beta]$ and $u \in \mathbb{R}$. Therefore, by Proposition 5.2, f is globally Lipschitz with respect to $u \in \mathbb{R}$ uniformly in $t \in [\alpha, \beta]$.

Example 5.4.2. Suppose

$$f(t, u) = \frac{1}{t} \sin u, \quad (t, u) \in (0, \infty) \times \mathbb{R}.$$

Then, f is continuous in $(0, \infty) \times \mathbb{R}$, as well as

$$\frac{\partial f}{\partial u}(t, u) = \frac{1}{t} \cos u, \quad (t, u) \in (0, \infty) \times \mathbb{R}.$$

Moreover,

$$\left| \frac{\partial f}{\partial u}(t, u) \right| = \left| \frac{1}{t} \cos u \right| \leq \frac{1}{t}$$

for all $t > 0$ and $u \in \mathbb{R}$. Thus, for every $\tau > 0$,

$$\max_{t \geq \tau,\, u \in \mathbb{R}} \left| \frac{\partial f}{\partial u}(t, u) \right| = \frac{1}{\tau} < +\infty.$$

Therefore, by Proposition 5.2, f is globally Lipschitz with respect to $u \in \mathbb{R}$ uniformly in $t \in [\tau, \infty)$. Similarly, since

$$\sup_{t > 0,\, u \in \mathbb{R}} \left| \frac{\partial f}{\partial u}(t, u) \right| = \sup_{t > 0} \frac{1}{t} = +\infty,$$

f cannot be globally Lipschitz with respect to $u \in \mathbb{R}$ uniformly in $t \in (0, \infty)$.

Example 5.4.3. Suppose $a \in \mathcal{C}([\alpha, \beta]; \mathbb{R})$, $a \neq 0$, and consider

$$f(t, u) = a(t)|u|^p, \quad t \in [\alpha, \beta], \quad u \in \mathbb{R},$$

for some $p > 1$. Then,

$$\frac{\partial f}{\partial u}(t, u) = p\, a(t)|u|^{p-1}\mathrm{sign}\, u \quad \text{for all } (t, u) \in [\alpha, \beta] \times \mathbb{R}.$$

Since $a \neq 0$, there exists $\tilde{t} \in [\alpha, \beta]$ such that $a(\tilde{t}) \neq 0$, and hence,

$$\lim_{|u| \to +\infty} \left| \frac{\partial f}{\partial u}(\tilde{t}, u) \right| = +\infty.$$

Thus, by Proposition 5.2, f cannot be globally Lipschitz with respect to $u \in \mathbb{R}$ uniformly in $t \in [\alpha, \beta]$. Nevertheless, since $\partial f / \partial u$ is continuous in $[\alpha, \beta] \times \mathbb{R}$, we can infer from Corollary 5.3 that f is locally Lipschitz with respect to $u \in \mathbb{R}$ uniformly in $t \in [\alpha, \beta]$.

In particular, the function $f(t, u) = u^2$ is locally Lipschitz with respect to $u \in \mathbb{R}$ uniformly in $t \in \mathbb{R}$, but it is not globally Lipschitz.

Example 5.4.4. For every $q \in (0, 1)$ and $\varepsilon > 0$, the real function $F(u) = |u|^q$, $u \in \mathbb{R}$, is not globally Lipschitz with respect to $u \in (0, \varepsilon)$. Indeed, suppose F is globally Lipschitz with respect to $u \in (0, \varepsilon)$ for some $\varepsilon > 0$. Then, there exists $L > 0$ such that

$$|F(x) - F(y)| = |x^q - y^q| \leq L\,|x - y| \quad \text{for all } x, y \in (0, \varepsilon).$$

Thus, setting

$$x_n = \frac{1}{2n}, \quad y_n = \frac{1}{n},$$

for sufficiently large $n \in \mathbb{N}$ so that $\frac{1}{n} < \varepsilon$, it follows from the mean value theorem that there exists $\xi_n \in [x_n, y_n]$ such that

$$|x_n^q - y_n^q| = q\, \xi_n^{q-1}\, |x_n - y_n| \leq L\,|x_n - y_n| \text{ for large } n \in \mathbb{N}.$$

Therefore,

$$q\, \xi_n^{q-1} \leq L,$$

which contradicts

$$\lim_{n \to \infty} \xi_n^{q-1} = \infty$$

because $q - 1 < 0$ and $\lim_{n \to \infty} \xi_n = 0$. This contradiction shows that F cannot be globally Lipschitz with respect to u in any interval of the form $(0, \varepsilon)$. As will become apparent later in this chapter, this is the reason why we were able to construct, in Section 4.5, infinitely many solutions of the differential equation $u' = u^q$ with $u(0) = 0$. Observe that, arguing as in Example 5.4.3, $F(u)$ is locally Lipschitz with respect to $u \in (0, \varepsilon)$ for all $\varepsilon > 0$, uniformly in $t \in (-\infty, \infty)$, but it is not locally Lipschitz with respect to $u \in \Omega$ if Ω is an interval containing 0.

Example 5.4.5. Given an integer $N \geq 1$ and a matrix of continuous functions, $a_{ij} \in \mathcal{C}([\alpha, \beta]; \mathbb{R})$, $1 \leq i, j \leq N$,

$$A(t) := (a_{ij}(t))_{1 \leq i,j \leq N},$$

we consider

$$f(t, u) = (f_1(t, u), \ldots, f_N(t, u))^T, \quad u = (u_1, \ldots, u_N)^T,$$

with components defined as

$$f_i(t, u) = \sum_{j=1}^{N} a_{ij}(t) \sin u_j, \quad t \in [\alpha, \beta], \ u \in \mathbb{R}^N, \quad 1 \leq i \leq N.$$

Each of the functions

$$\frac{\partial f_i}{\partial u_j}(t, u) = a_{ij}(t) \cos u_j, \quad t \in [\alpha, \beta], \ u \in \mathbb{R}^N, \quad 1 \leq i, j \leq N,$$

is well defined and continuous in $[\alpha, \beta] \times \mathbb{R}^N$. Moreover,

$$\left| \frac{\partial f_i}{\partial u_j}(t, u) \right| = |a_{ij}(t)| \, |\cos u_j| \leq |a_{ij}(t)| \leq \|a_{ij}\|_\infty$$

for all $t \in [\alpha, \beta]$, $u \in \mathbb{R}^N$ and $1 \leq i, j \leq N$. Hence,

$$\sup_{(t,u)\in[\alpha,\beta]\times\mathbb{R}^N} \left| \frac{\partial f_i}{\partial u_j}(t, u) \right| \leq \|a_{ij}\|_\infty < +\infty \quad \text{for all } 1 \leq i, j \leq N.$$

Therefore, by Proposition 5.2, f is globally Lipschitz with respect to $u \in \mathbb{R}^N$ uniformly in $t \in [\alpha, \beta]$.

Example 5.4.6. Given an integer $N \geq 1$ and a non-zero matrix of continuous functions, $a_{ij} \in \mathcal{C}([\alpha, \beta]; \mathbb{R})$, $1 \leq i, j \leq N$,

$$A(t) := (a_{ij}(t))_{1 \leq i, j \leq N},$$

we consider

$$f(t, u) = (f_1(t, u), \dots, f_N(t, u))^T, \quad u = (u_1, \dots, u_N)^T,$$

with components defined as

$$f_i(t, u) = \sum_{j=1}^N a_{ij}(t) u_j^2, \quad (t, u) \in [\alpha, \beta] \times \mathbb{R}^N, \quad 1 \leq i \leq N.$$

As in the previous example, each of the functions

$$\frac{\partial f_i}{\partial u_j}(t, u) = 2 a_{ij}(t) u_j, \quad (t, u) \in [\alpha, \beta] \times \mathbb{R}^N, \quad 1 \leq i, j \leq N,$$

is well defined and continuous in $[\alpha, \beta] \times \mathbb{R}^N$. Moreover,

$$\left| \frac{\partial f_i}{\partial u_j}(t, u) \right| = 2 \, |a_{ij}(t)| \, |u_j|, \quad (t, u) \in [\alpha, \beta] \times \mathbb{R}^N, \quad 1 \leq i, j \leq N.$$

Since $A \neq 0$, there exist $t_0 \in [\alpha, \beta]$ and $i_0, j_0 \in \{1, \dots, N\}$ such that $a_{i_0 j_0}(t_0) \neq 0$. Thus, if we take

$$u_k := (0, \dots, 0, k, 0, \dots, 0)^T,$$

where the value k is in the j_0th component, we have

$$\lim_{k \to +\infty} \left| \frac{\partial f_{i_0}}{\partial u_{j_0}}(t_0, u_k) \right| = +\infty.$$

Therefore, by Proposition 5.2, f cannot be globally Lipschitz with respect to $u \in \mathbb{R}^N$ uniformly in $t \in [\alpha, \beta]$. However, by Corollary 5.3, f is locally Lipschitz with respect to $u \in \mathbb{R}^N$ uniformly in $t \in [\alpha, \beta]$; actually, it is globally Lipschitz, uniformly in $t \in [\alpha, \beta]$, in any convex bounded subset $D \subset \mathbb{R}^N$.

5.2   Global Cauchy–Lipschitz Theorem

The main result in this section is the following theorem, which is actually the most important of this chapter. Indeed, the remaining results will be derived from it as rather direct consequences.

Theorem 5.5 (Global Cauchy–Lipschitz theorem). *Suppose $\alpha < \beta$, $N \in \mathbb{N}$, $N \geq 1$, and $f \in \mathcal{C}([\alpha, \beta] \times \mathbb{R}^N; \mathbb{R}^N)$ is globally Lipschitz with respect to $u \in \mathbb{R}^N$ uniformly in $t \in [\alpha, \beta]$. In other words, there exists a constant $L \geq 0$ such that, for every $t \in [\alpha, \beta]$ and $x, y \in \mathbb{R}^N$,*

$$\|f(t, x) - f(t, y)\| \leq L\|x - y\|. \tag{5.5}$$

Then, for every $t_0 \in [\alpha, \beta]$ and $u_0 \in \mathbb{R}^N$, the initial value problem

$$\begin{cases} u' = f(t, u), \\ u(t_0) = u_0, \end{cases} \tag{5.6}$$

possesses a unique solution, $u(t; t_0, u_0)$, which is defined in the interval $[\alpha, \beta]$.

This theorem is global in the sense that the solution is *globally defined* in the time interval $[\alpha, \beta]$. The proof of Theorem 5.5 relies on the following lemma.

Lemma 5.6. *Suppose $\alpha < \beta$, $N \in \mathbb{N}$, $N \geq 1$, and $f \in \mathcal{C}([\alpha, \beta] \times \mathbb{R}^N; \mathbb{R}^N)$. Then, the following conditions are equivalent:*

(a) $u \in \mathcal{C}^1([\alpha, \beta]; \mathbb{R}^N)$ is a solution of (5.6) in the interval $[\alpha, \beta]$;
(b) $u \in \mathcal{C}([\alpha, \beta]; \mathbb{R}^N)$ satisfies the integral equation

$$u(t) = u_0 + \int_{t_0}^{t} f(s, u(s))\, ds \quad \text{for all } t \in [\alpha, \beta]. \tag{5.7}$$

Proof.   Assume (a). Then, integrating the differential equation of (5.6) in the interval (t_0, t) yields

$$u(t) - u(t_0) = \int_{t_0}^{t} u'(s)\, ds = \int_{t_0}^{t} f(s, u(s))\, ds$$

for all $t \in [\alpha, \beta]$, which implies (5.7) because $u(t_0) = u_0$.

Conversely, suppose (b). Then, since f is continuous in $[\alpha, \beta] \times \mathbb{R}^N$, by the fundamental theorem of calculus, the function

$$t \mapsto \int_{t_0}^{t} f(s, u(s)) \, ds$$

is differentiable and

$$\frac{d}{dt} \int_{t_0}^{t} f(s, u(s)) \, ds = f(t, u(t)) \quad \text{for all } t \in [\alpha, \beta].$$

Thus, it follows from (5.7) that $u \in \mathcal{C}^1([\alpha, \beta]; \mathbb{R}^N)$ and that

$$u'(t) = f(t, u(t)) \quad \text{for all } t \in [\alpha, \beta].$$

Moreover, evaluating (5.7) at t_0 gives $u(t_0) = u_0$. $\qquad\qquad\square$

Much like in Section 1.3, according to Lemma 5.6, the solutions of the initial value problem (5.6) are the fixed points of the integral operator

$$\mathcal{K} : \mathcal{C}([\alpha, \beta]; \mathbb{R}^N) \to \mathcal{C}([\alpha, \beta]; \mathbb{R}^N)$$

defined, for every $h \in \mathcal{C}([\alpha, \beta]; \mathbb{R}^N)$, as

$$\mathcal{K}h(t) := u_0 + \int_{t_0}^{t} f(s, h(s)) \, ds, \quad t \in [\alpha, \beta]. \tag{5.8}$$

The proof of Theorem 5.5 relies on the fact that $\mathcal{K}$ possesses a unique fixed point. As in the proof of Theorem 1.4, this is a direct consequence of the contractive mapping theorem (see Theorem 1.8).

Proof of Theorem 5.5. As in the proof of Theorem 1.4, to prove that $\mathcal{K}$ establishes a contraction in the Banach space $\mathcal{C}([\alpha, \beta]; \mathbb{R}^N)$, it is appropriate to use the norm

$$\|u\|_\eta := \max_{t \in [\alpha, \beta]} \left\{ e^{-\eta|t-t_0|} \|u(t)\| \right\}, \quad u \in \mathcal{C}([\alpha, \beta]; \mathbb{R}^N),$$

for an appropriate choice of the constant η, which is postponed until the end of the proof. Thanks to (5.8) and (5.5), we have that, for every $u, v \in \mathcal{C}([\alpha, \beta]; \mathbb{R}^N)$ and any $t \in [\alpha, \beta]$,

$$e^{-\eta|t-t_0|} \|\mathcal{K}u(t) - \mathcal{K}v(t)\|$$

$$= e^{-\eta|t-t_0|} \left\| \int_{t_0}^{t} [f(s, u(s)) - f(s, v(s))]\, ds \right\|$$

$$\leq e^{-\eta|t-t_0|} \int_{\min\{t_0,t\}}^{\max\{t_0,t\}} \|f(s, u(s)) - f(s, v(s))\|\, ds$$

$$\leq e^{-\eta|t-t_0|} L \int_{\min\{t_0,t\}}^{\max\{t_0,t\}} \|u(s) - v(s)\|\, ds.$$

Thus,

$$e^{-\eta|t-t_0|} \|\mathcal{K}u(t) - \mathcal{K}v(t)\|$$

$$\leq e^{-\eta|t-t_0|} L \int_{\min\{t_0,t\}}^{\max\{t_0,t\}} e^{\eta|s-t_0|} e^{-\eta|s-t_0|} \|u(s) - v(s)\|\, ds$$

$$\leq e^{-\eta|t-t_0|} L \int_{\min\{t_0,t\}}^{\max\{t_0,t\}} e^{\eta|s-t_0|}\, ds\, \|u - v\|_\eta.$$

On the other hand, as in the proof of Theorem 1.4, we have

$$\int_{\min\{t_0,t\}}^{\max\{t_0,t\}} e^{\eta|s-t_0|}\, ds = \frac{e^{\eta|t-t_0|} - 1}{\eta} < \frac{e^{\eta|t-t_0|}}{\eta}.$$

Consequently, for every $t \in [\alpha, \beta]$, we find that

$$e^{-\eta|t-t_0|} \|\mathcal{K}u(t) - \mathcal{K}v(t)\| \leq \frac{L}{\eta} \|u - v\|_\eta,$$

and hence, taking the maximum for $t \in [\alpha, \beta]$ leads to

$$\|\mathcal{K}u - \mathcal{K}v\|_\eta \leq \frac{L}{\eta} \|u - v\|_\eta.$$

Therefore, by choosing $\eta > L$, it is apparent that $\mathcal{K}$ is a $\frac{L}{\eta}$-contraction, and Theorem 1.8 completes the proof. $\qquad\square$

Actually, according to Theorem 1.8, the following result holds.

Corollary 5.7. *Under the same assumptions of Theorem 5.5, for every $h_0 \in \mathcal{C}([\alpha, \beta]; \mathbb{R}^N)$, the sequence of $\mathcal{K}$-iterates,*

$$h_n := \mathcal{K} h_{n-1} = \mathcal{K}^n h_0, \quad n \geq 1, \tag{5.9}$$

converges, uniformly in $[\alpha, \beta]$, toward the unique solution of problem (5.6), $u(t; t_0, u_0)$, in the interval $[\alpha, \beta]$.

As in Section 1.3.5, in the context of Theorem 5.5, the $\mathcal{K}$-iterates (5.9) are often referred to as the *Picard–Lindelöf iterates* of h_0.

For instance, thanks to the analysis in Example 5.4.5, if $\alpha < \beta$ and $a_{ij} \in \mathcal{C}([\alpha, \beta]; \mathbb{R})$ for every $1 \leq i, j \leq N$, then, according to Theorem 5.5, for every $t_0 \in [\alpha, \beta]$ and $u_0 \in \mathbb{R}^N$, the initial value problem

$$\begin{cases} u_i'(t) = \sum_{j=1}^{N} a_{ij}(t) \sin u_j, & 1 \leq i \leq N, \\ u(t_0) = u_0, \end{cases} \tag{5.10}$$

possesses a unique solution, $u(t; t_0, u_0)$, defined in $[\alpha, \beta]$. Moreover, given an arbitrary $h_0 \in \mathcal{C}([\alpha, \beta]; \mathbb{R}^N)$, the Picard–Lindelöf iterates

$$h_n(t) = \mathcal{K} h_{n-1}(t) = u_0 + \begin{pmatrix} \int_{t_0}^{t} \sum_{j=1}^{N} a_{1j}(s) \sin h_{n-1,j}(s) \, ds \\ \vdots \\ \int_{t_0}^{t} \sum_{j=1}^{N} a_{Nj}(s) \sin h_{n-1,j}(s) \, ds \end{pmatrix}, \quad n \geq 1,$$

where we denote

$$h_i = (h_{i,1}, \ldots, h_{i,N})^T, \quad i \geq 1,$$

converge, thanks to Corollary 5.7, to $u(t; t_0, u_0)$ uniformly in $[\alpha, \beta]$. Nevertheless, observe that it is impossible to explicitly determine the limit of such iterates.

Finally, we reason as in Corollary 1.9 and show how to use the previous result to treat the case in which $f(\cdot, u)$ is defined in an open, possibly unbounded, interval of $\mathbb{R}$.

Theorem 5.8. *Suppose J is an open interval of $\mathbb{R}$, possibly unbounded, and $f \in \mathcal{C}(J \times \mathbb{R}^N; \mathbb{R}^N)$ is globally Lipschitz with respect* to $u \in \mathbb{R}^N$ *uniformly in* $t \in J$, *i.e. there exists a constant* $L \geq 0$ *such that*

$$\|f(t, x) - f(t, y)\| \leq L\|x - y\| \quad \text{for all } t \in J \text{ and } x, y \in \mathbb{R}^N.$$

Then, for every $t_0 \in J$ *and* $u_0 \in \mathbb{R}^N$, *the initial value problem* (5.6) *possesses a unique solution,* $u(t; t_0, u_0)$, *which is (globally) defined in the interval* J.

Proof. Let $\{[\alpha_n, \beta_n]\}_{n \geq 1}$ be any sequence of compact subintervals of J such that

$$\alpha_{n+1} < \alpha_n < \beta_n < \beta_{n+1}$$

for all $n \geq 1$, and

$$\bigcup_{n \geq 1} [\alpha_n, \beta_n] = J.$$

Since $t_0 \in J$, $t_0 \in [\alpha_n, \beta_n]$ for sufficiently large n, say $n \geq n_0$. Thus, thanks to Theorem 5.5, the Cauchy problem (5.6) has a unique solution, u_n, in $[\alpha_n, \beta_n]$. Since $[\alpha_n, \beta_n] \subset (\alpha_{n+1}, \beta_{n+1})$, by uniqueness, it is apparent that

$$u_{n+1}\big|_{[\alpha_n, \beta_n]} = u_n \quad \text{for all } n \geq n_0.$$

Therefore,

$$u := \lim_{n \to \infty} u_n$$

is well defined and provides us with the unique solution of (5.9) in J. This completes the proof. $\qquad\square$

We illustrate this result with an example: suppose that $J = (\alpha, +\infty)$, with $\alpha \in \mathbb{R}$, and that, in (5.10), the functions a_{ij} are defined and continuous in J, for all $1 \leq i, j \leq N$. Then, by Theorem 5.5, for every $n \geq 1$ sufficiently large so that

$$\alpha + \frac{1}{n} < t_0 < t_0 + n,$$

problem (5.10) possesses a unique solution, $u_n(t; t_0, u_0)$, in the compact interval $\left[\alpha + \frac{1}{n}, t_0 + n\right]$. Moreover, by uniqueness, whenever

$m > n$, we have that

$$u_m(t; t_0, u_0) = u_n(t; t_0, u_0) \quad \text{for all } t \in \left[\alpha + \tfrac{1}{n}, t_0 + n\right].$$

Therefore, u_m is an extension of u_n to the larger interval $[\alpha + \tfrac{1}{m}, t_0 + m]$, and the limit

$$u(t; t_0, u_0) := \lim_{n \to \infty} u_n(t; t_0, u_0)$$

is well defined for all $t \in J$ and provides us with the unique solution of (5.10) in J.

It is easily seen that Theorem 5.8 also holds for any interval J, not necessarily open.

5.3 Local Cauchy–Lipschitz Theorem

According to Proposition 5.2, when J is an open interval of $\mathbb{R}$, Ω is an open convex subset of $\mathbb{R}^N$, and $f \in \mathcal{C}(J \times \Omega; \mathbb{R}^N)$ admits continuous partial derivatives with respect to all the u-variables,

$$\frac{\partial f_i}{\partial u_j} \in \mathcal{C}(J \times \Omega; \mathbb{R}), \quad 1 \le i, j \le N,$$

unless the partial derivatives $\frac{\partial f_i}{\partial u_j}$ are uniformly bounded in $J \times \Omega$, f cannot be globally Lipschitz with respect to $u \in \Omega$ uniformly in t on compact subintervals of J. However, it is locally Lipschitz with respect to $u \in \Omega$ uniformly in t on compact subintervals of J. In such a context, we can prove the following local version of the Cauchy–Lipschitz theorem, which is actually optimal, despite its local nature, as discussed in Remark 5.10.

Theorem 5.9 (Local Cauchy–Lipschitz theorem). *Suppose $T > 0$, $t_0 \in \mathbb{R}$, $J = [t_0, t_0 + T]$, Ω is an open subset of $\mathbb{R}^N$, and $f \in \mathcal{C}(J \times \Omega; \mathbb{R}^N)$ is locally Lipschitz with respect to $u \in \Omega$ uniformly in $t \in J$. Then, for every $u_0 \in \Omega$, problem (5.6) has a unique local solution. Precisely:*

(a) *there exists $\delta = \delta(u_0) \in (0, T)$ such that (5.6) possesses a solution, $u(t; t_0, u_0)$, defined in the interval $[t_0, t_0 + \delta]$;*

(b) *if $R > 0$ is such that the Euclidean ball $\bar{B}_R(u_0) \subset \Omega$, $f(t, u)$ is globally Lipschitz with respect to $u \in \bar{B}_R(u_0)$ uniformly in $t \in J$, and we denote*

$$\|f\|_\infty := \max\left\{\|f(t, u)\| : (t, u) \in J \times \bar{B}_R(u_0)\right\}, \tag{5.11}$$

and

$$\delta_0 := \begin{cases} \min\left\{T, \frac{R}{\|f\|_\infty}\right\} & \text{if } \|f\|_\infty > 0, \\ T & \text{if } \|f\|_\infty = 0, \end{cases} \qquad (5.12)$$

then (5.6) possesses a unique solution, $u(t; t_0, u_0)$, in the interval $[t_0, t_0 + \delta_0]$.

Proof. (a) In this part, we work, without loss of generality, with the Euclidean norm of $\mathbb{R}^N$:

$$\|u\| := \|u\|_2 = \sqrt{u_1^2 + \cdots + u_N^2}, \quad u = (u_1, \ldots, u_N) \in \mathbb{R}^N.$$

Fix $u_0 \in \Omega$. Since f is locally Lipschitz with respect to $u \in \Omega$ uniformly in $t \in J$, there exist $R = R(u_0) > 0$ and $L = L(u_0) > 0$ such that $\bar{B}_R(u_0) \subset \Omega$ and, for every $t \in J$ and $x, y \in \bar{B}_R(u_0)$,

$$\|f(t, x) - f(t, y)\| \le L\|x - y\|. \qquad (5.13)$$

Subsequently, we consider the extension of the function $f|_{J \times \bar{B}_R(u_0)}$ to $J \times \mathbb{R}^N$ defined, for every $t \in J = [t_0, t_0 + T]$, as

$$F(t, u) := \begin{cases} f(t, u) & \text{if } \|u - u_0\| \le R, \\ f(t, \tilde{u}) & \text{if } \|u - u_0\| > R, \end{cases} \qquad (5.14)$$

where

$$\tilde{u} := u_0 + \frac{u - u_0}{\|u - u_0\|} R, \quad \|u - u_0\| > R,$$

is the intersection of the ray through u_0 and u with the sphere

$$S(u_0, R) := \{u \in \mathbb{R}^N : \|u - u_0\| = R\},$$

as illustrated in Figure 5.1.

By construction, $F \in \mathcal{C}(J \times \mathbb{R}^N; \mathbb{R}^N)$. Moreover, we claim that F is globally Lipschitz with respect to $u \in \mathbb{R}^N$ uniformly in $t \in J$. To show this, we distinguish three different situations:

Case 1. $u, v \in \bar{B}_R(u_0)$. Then, thanks to (5.13) and (5.14),

$$\|F(t, u) - F(t, v)\| = \|f(t, u) - f(t, v)\| \le L\|u - v\| \quad \text{for all } t \in J.$$

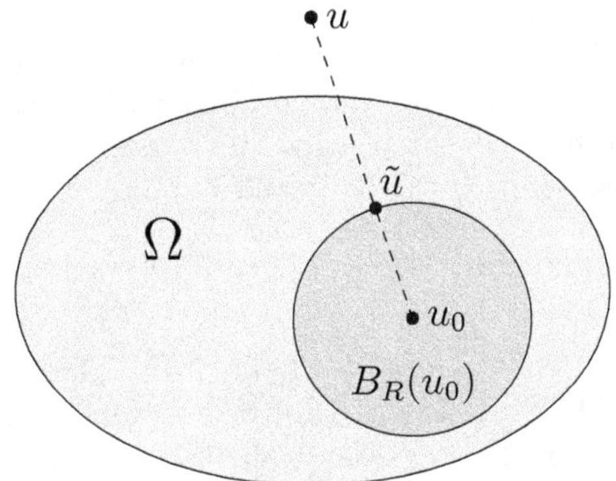

Fig. 5.1. Construction of the point $\tilde{u} = \tilde{u}(u)$.

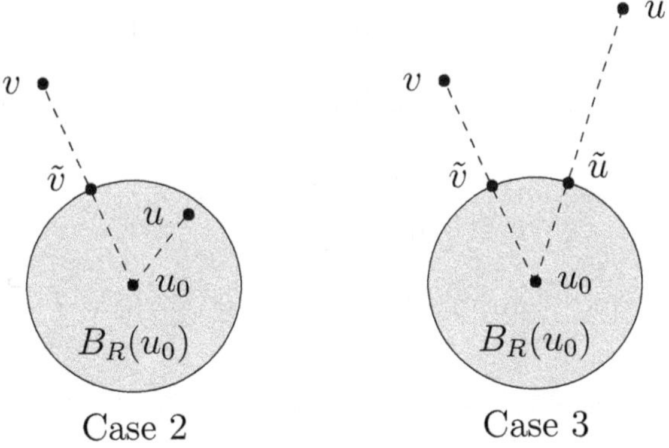

Fig. 5.2. The points $\tilde{u}$ and $\tilde{v}$ in Cases 2 and 3.

Case 2. $u \in \bar{B}_R(u_0)$ and $\|v - u_0\| > R$. This is the case illustrated on the left-hand side of Figure 5.2. Then, again by (5.13) and (5.14),

$$\|F(t, u) - F(t, v)\| = \|f(t, u) - f(t, \tilde{v})\| \leq L\|u - \tilde{v}\|, \qquad (5.15)$$

where

$$\tilde{v} := u_0 + \frac{v - u_0}{\|v - u_0\|} R.$$

By simply looking at the left plot of Figure 5.2, it is easily realized that

$$\|u - \tilde{v}\| \leq \|u - v\|.$$

Indeed, the Euclidean distance of u from the line containing u_0, $\tilde{v}$, and v is obtained by the orthogonal projection of u on such a line, which we denote as $u_\perp$ and lies inside $\bar{B}_R(u_0)$. Moreover, the distance from u to any point P of such a line increases as the distance from P to $u_\perp$ increases. Hence, (5.15) implies that

$$\|F(t,u) - F(t,v)\| \le L\|u - v\|.$$

Case 3. $\|u - u_0\| > R$ and $\|v - u_0\| > R$. This is the case illustrated on the right-hand side of Figure 5.2. One can show by direct computations that, in this case as well, $\|\tilde{u} - \tilde{v}\| \le \|u - v\|$, thus

$$\|F(t,u) - F(t,v)\| = \|f(t,\tilde{u}) - f(t,\tilde{v})\| \le L\|\tilde{u} - \tilde{v}\| \le L\|u - v\|.$$

Consequently, we have shown that F is globally Lipschitz with respect to $u \in \mathbb{R}^N$ uniformly in $t \in J$. The Lipschitz constant of F with respect to the 1-norm might be different from L, but this is irrelevant in our analysis. Therefore, by Theorem 5.5, the problem

$$\begin{cases} u' = F(t,u), \\ u(t_0) = u_0, \end{cases} \tag{5.16}$$

possesses a unique solution, $U(t)$, in $J = [t_0, t_0 + T]$. Moreover, as long as

$$\|U(t) - u_0\| < R, \tag{5.17}$$

it follows from (5.16) and (5.14) that

$$U'(t) = F(t, U(t)) = f(t, U(t)).$$

By continuity, we have that (5.17) holds for $t \in [t_0, t_0 + \delta]$, provided that δ is sufficiently small. Consequently, the restriction $U|_{[t_0, t_0 + \delta]}$ provides us with a local solution of (5.6).

(b) In this second part of the theorem, we work, as usual in this book, with the norm

$$\|u\| := |u_1| + \cdots + |u_N|, \quad u = (u_1, \ldots, u_N) \in \mathbb{R}^N.$$

Observe that the ball of radius R centered at u_0 in this norm is contained in the ball of radius R in the $\|\cdot\|_2$-norm, which has been used in the definition of $F(t,u)$ in (5.14).

If $\|f\|_\infty = 0$, then $f = 0$ in $B_R(u_0)$, and the differential equation in (5.6) becomes $u' = 0$. Hence, $u(t) = u_0$ for all $t \in J$, which completes the proof of uniqueness and the fact that u is defined for all $t \in [t_0, t_0 + \delta_0]$.

Otherwise, i.e. if $\|f\|_\infty > 0$, we prove that any solution, $u(t)$, of (5.6) that is defined in $[t_0, t_0 + \delta]$, with $\delta \le \delta_0$, must satisfy

$$u(t) \in \bar{B}_R(u_0) \quad \text{for all } t \in [t_0, t_0 + \delta]. \tag{5.18}$$

This will entail that $u(t)$ also solves (5.16) in $[t_0, t_0 + \delta]$, which, by Theorem 5.5, has at most one solution. Thus, $u(t)$ must be the unique solution of (5.6) in $[t_0, t_0 + \delta]$, which concludes the proof of the local uniqueness since $\delta \le \delta_0$ is arbitrary.

To prove (5.18), assume by contradiction that there exist $\delta_1 \in (0, \delta_0]$ and $t_1 \in (t_0, t_0 + \delta_1]$ such that $\|u(t_1) - u_0\| > R$. Then, by continuity, there exists $t_2 \in (t_0, t_0 + \delta_1)$ such that

$$\|u(t) - u_0\| < R \text{ if } t_0 \le t < t_2, \quad \text{and} \quad \|u(t_2) - u_0\| = R.$$

Thus,

$$R = \|u(t_2) - u_0\| = \left\| \int_{t_0}^{t_2} f(s, u(s))\, ds \right\| \le \int_{t_0}^{t_2} \|f(s, u(s))\|\, ds$$

$$\le \|f\|_\infty |t_2 - t_0| < \|f\|_\infty \delta_1 \le \|f\|_\infty \delta_0 \le R,$$

where in the last inequality, we have used the definition of δ_0 given in (5.12). This contradiction shows that (5.18) holds true.

The argument of the previous paragraph can be repeated with f and u replaced by F and U defined in Part (a), respectively, to prove that $U(t) \in \bar{B}_R(u_0)$ for all $t \in [t_0, t_0 + \delta_0]$. This shows that the local solution $u(t)$ of (5.6) is indeed defined in $[t_0, t_0 + \delta_0]$, and ends the proof. $\qquad\square$

Remark 5.10. (i) The result obtained in Theorem 5.9 is optimal, in the sense that the existence interval can be arbitrarily small. Indeed, consider the simplest nonlinear autonomous problem

$$\begin{cases} u' = u^2, \\ u(t_0) = u_0 > 0, \end{cases} \tag{5.19}$$

whose unique solution is given by

$$u(t; t_0, u_0) = \frac{u_0}{1 - u_0(t - t_0)}, \quad t \in (-\infty, t_0 + u_0^{-1}).$$

We have seen in Example 5.4.3 that $f(t, u) = u^2$ is only locally Lipschitz with respect to u, uniformly in $t \in \mathbb{R}$. Since

$$\lim_{t \uparrow t_0 + u_0^{-1}} u(t; t_0, u_0) = +\infty,$$

we say that $u(t; t_0, u_0)$ *blows up after a finite time,* u_0^{-1}. As $\delta_0 = \delta_0(u_0) < u_0^{-1}$ can be taken to be arbitrarily small by choosing a sufficiently large $u_0 > 0$, this example shows that, when f is only locally Lipschitz with respect to u, one cannot expect the solution $u(t; t_0, u_0)$ to be globally defined for all $t \in J$ and that, actually, its domain of definition might be arbitrarily small.

(ii) The local Lipschitz character of $f(t, u)$ with respect to $u \in \Omega$ is imperative for the local uniqueness. Indeed, if $q \in (0, 1)$, we know from Example 5.4.4 that $f(t, u) = |u|^q$ for $u \in \Omega := \mathbb{R}$ is not locally Lipschitz with respect to $u \in \Omega$, and the problem

$$\begin{cases} u' = |u|^q, \\ u(t_0) = 0, \end{cases}$$

admits, besides the equilibrium $u = 0$, a one-parameter family of nontrivial solutions in any interval of the form $[t_0, t_0 + \delta]$, for all $\delta > 0$ (see Section 4.5).

(iii) The quantity δ_0 estimated in Theorem 5.9(b) allows us to ensure that the unique local solution of (5.6) is defined at least in the interval $[t_0, t_0 + \delta_0]$. In general, such a quantity depends on u_0, R, and $\|f\|_\infty$.

(iv) Theorem 5.9 allows us to conclude that, in the examples considered in Chapter 4 where $f(t, u)$ is $\mathcal{C}^1$, the local uniqueness of the corresponding Cauchy problems also holds true at equilibria. This situation remained open in Chapter 4.

5.3.1   *Global uniqueness*

It is important to point out that Theorem 5.9 does not guarantee the uniqueness of the solution in any interval of time where it is defined but simply in small intervals of the form $[t_0, t_0 + \delta]$, with $\delta \leq \delta_0$, where δ_0 is defined in (5.12). The following result sharpens this local uniqueness result by establishing that $u(t; t_0, u_0)$ is in fact unique in any interval where it is defined.

Theorem 5.11 (Global uniqueness). *Suppose $T > 0$, $t_0 \in \mathbb{R}$, $J = [t_0, t_0 + T]$, Ω is an open subset of $\mathbb{R}^N$, $u_0 \in \Omega$ and $f \in \mathcal{C}(J \times \Omega; \mathbb{R}^N)$ is* *locally Lipschitz with respect to $u \in \Omega$ uniformly in $t \in J$. Then, for every $0 < h \leq T$ for which (5.6) possesses a solution, $u(t)$, defined in $[t_0, t_0 + h]$, the solution is unique. As a consequence, the uniqueness also holds in any interval of the form $[t_0, t_0 + h)$ where the solution is defined.*

Proof. Suppose by contradiction that (5.6) admits two solutions, $u_1 \neq u_2$, defined in the interval $[t_0, t_0 + h]$ for some $0 < h \leq T$. Note that, since these solutions are defined for all $t \in [t_0, t_0 + h]$, necessarily, $u_1(t) \in \Omega$ and $u_2(t) \in \Omega$ in such a range of t. Consider the set

$$D := \{\delta \in (0, h] : u_1 = u_2 \text{ in } [t_0, t_0 + \delta]\}.$$

According to Theorem 5.9, $D \neq \emptyset$ because $\delta_0 \in D$ ($\delta_0 > 0$ is the one defined in (5.12)). Moreover, $D \subset [0, h]$. Thus, D is bounded and, hence, the supremum

$$\delta^* := \sup D \in (0, h)$$

is well defined (the fact that $\delta^* < h$ follows since we are assuming that $u_1 \neq u_2$ in $[t_0, t_0 + h]$). Let δ_n, $n \geq 1$, be a sequence in D such that

$$\delta_n < \delta_{n+1} < \delta^* \quad \text{for all } n \geq 1, \quad \text{and} \quad \lim_{n\uparrow+\infty} \delta_n = \delta^*.$$

By the definition of D, $u_1 = u_2$ in $[t_0, t_0 + \delta_n]$ for all $n \geq 1$ and, hence, $u_1 = u_2$ in $[t_0, t_0 + \delta^*)$. Moreover, by continuity,

$$u_1(t_0 + \delta^*) = \lim_{n\to+\infty} u_1(t_0 + \delta_n) = \lim_{n\to+\infty} u_2(t_0 + \delta_n) = u_2(t_0 + \delta^*).$$

Thus, $u_1 = u_2$ in $[t_0, t_0 + \delta^*]$. Observe that $u_1(t_0 + \delta^*)$, which coincides with $u_2(t_0 + \delta^*)$, belongs to Ω because $\delta^* < h$ and both u_1 and u_2 are defined up to $t_0 + h$. Now, set

$$t_0^* := t_0 + \delta^* < t_0 + h \leq t_0 + T$$

and consider the Cauchy problem

$$\begin{cases} u' = f(t, u), \\ u(t_0^*) = u_1(t_0^*) = u_2(t_0^*) \in \Omega. \end{cases} \tag{5.20}$$

Thanks again to Theorem 5.9, (5.20) possesses a unique local solution because $t_0^* < t_0 + T$. Therefore, $u_1 = u_2$ in $[t_0, t_0^* + \delta]$ for some $\delta > 0$, which contradicts the definition of δ^* and completes the proof. $\qquad\square$

5.3.2   *Existence of the maximal solution*

Throughout this section, we suppose that $t_0 \in \mathbb{R}$, $T > 0$,

$$J \in \{[t_0, t_0 + T], [t_0, +\infty)\},$$

Ω is an arbitrary open subset of $\mathbb{R}^N$, and $f \in \mathcal{C}(J \times \Omega; \mathbb{R}^N)$ is locally Lipschitz with respect to $u \in \Omega$ uniformly in t on compact subintervals of J. Let Σ denote the set of pairs of the form (I, u_I), where I is an interval of the form $[t_0, t_0 + h]$, or $[t_0, t_0 + h)$, for some $0 < h < T$, or $I = J$, and $u_I \in \mathcal{C}^1(I; \Omega)$ is a solution of (5.6) in I. Subsequently, by a solution of (5.6) we mean an element of Σ. Thus, the solutions of (5.6) are regarded as pairs, (I, u_I), consisting of one of the above-indicated intervals, I, and a function $u_I \in \mathcal{C}^1(I; \Omega)$ such that

$$u_I'(t) = f(t, u_I(t)) \quad \text{for all } t \in I.$$

The following definition establishes a total order in the set of solutions Σ.

Definition 5.12 (Extension of solutions). Let (I_1, u_{I_1}), $(I_2, u_{I_2}) \in \Sigma$ be two solutions of (5.6). Then, the solution (I_2, u_{I_2}) is said to be an *extension to the right* of the solution (I_1, u_{I_1}) if

$$I_1 \subset I_2 \quad \text{and} \quad u_{I_2}\big|_{I_1} = u_{I_1}. \tag{5.21}$$

In such a case, we write

$$(I_1, u_{I_1}) \preccurlyeq (I_2, u_{I_2}).$$

The solution (I_2, u_{I_2}) is said to be a *proper extension* to the right of (I_1, u_{I_1}) if, in addition, $I_1 \subsetneq I_2$.

Note that, in this context, (5.21) is equivalent to $I_1 \subset I_2$ because $u_{I_2}\big|_{I_1}$ provides us with a solution of (5.6) in I_1 and, hence, the identity $u_{I_2}\big|_{I_1} = u_{I_1}$ is guaranteed by the uniqueness established in Theorem 5.11. Consequently, in the context of this chapter, (I_2, u_{I_2}) is an extension to the right of (I_1, u_{I_1}) if $I_1 \subset I_2$, and it is proper if $I_1 \subsetneq I_2$.

Definition 5.12 establishes an order in the solution set $(\Sigma, \preccurlyeq)$. Indeed, it is *reflexive* because $(I, u_I) \preccurlyeq (I, u_I)$ for every solution $(I, u_I) \in \Sigma$, it is *antisymmetric* because if (I_2, u_{I_2}) is an extension of (I_1, u_{I_1}) and, in addition, (I_1, u_{I_1}) is an extension of (I_2, u_{I_2}), then

$I_1 \subset I_2$ and $I_2 \subset I_1$. Thus, $I_1 = I_2$ and, thanks to Theorem 5.11, $u_{I_1} = u_{I_2}$. Consequently,

$$(I_1, u_{I_1}) = (I_2, u_{I_2}).$$

Finally, it is *transitive* because if $(I_2, u_{I_2}) \in \Sigma$ is an extension of $(I_1, u_{I_1}) \in \Sigma$ and $(I_3, u_{I_3}) \in \Sigma$ is an extension of $(I_2, u_{I_2}) \in \Sigma$, then $I_1 \subset I_2 \subset I_3$ and, hence, $I_1 \subset I_3$, i.e. (I_3, u_{I_3}) is an extension of (I_1, u_{I_1}).

Moreover, since $(I_1, u_{I_1}) \preccurlyeq (I_2, u_{I_2})$ holds if and only if $I_1 \subset I_2$, and the order of the intervals of the considered form is total in $\mathbb{R}$ with respect to the set inclusion $\subset$, $(\Sigma, \preccurlyeq)$ is a totally ordered set. Using these preliminaries, we can give the following definition.

Definition 5.13 (Maximal solution). A solution $(I_{\max}, u_{\max}) \in \Sigma$ is said to be maximal when it is a maximal element of the totally ordered set $(\Sigma, \preccurlyeq)$, i.e. when it does not admit any proper extension to the right. In such a case, $I_{\max}$ will be referred to as *maximal existence interval* (to the right) of the solution $u_{\max}$.

The following result establishes the existence and uniqueness of a maximal solution, $(I_{\max}, u_{\max})$, of (5.6) in the general setting of this chapter.

Theorem 5.14. *Suppose $t_0 \in \mathbb{R}$, $T > 0$, $J \in \{[t_0, t_0 + T], [t_0, +\infty)\}$, Ω is an open subset of $\mathbb{R}^N$, $u_0 \in \Omega$ and $f \in \mathcal{C}(J \times \Omega; \mathbb{R}^N)$ is locally Lipschitz with respect to $u \in \Omega$ uniformly in t on compact subintervals of J. Then, problem (5.6) possesses a unique maximal solution, $(I_{\max}, u_{\max})$.*

Proof. We set

$$I_{\max} := \bigcup_{(I, u_I) \in \Sigma} I \tag{5.22}$$

and, for all $t \in I_{\max}$,

$$u_{\max}(t) := u_I(t),$$

where $I \subset I_{\max}$ is any interval of the form specified at the beginning of this section such that $t \in I$. We begin by showing that $u_{\max}$ is well-defined in $I_{\max}$, i.e. that its definition does not depend on the choice of the solution (I, u_I). Indeed, fix $t \in I_{\max}$. If I_1 and I_2 are two intervals of the form considered in this section such that $t \in I_1 \cap I_2$ (we can assume without loss of generality that $I_1 \subset I_2$), we have by Theorem 5.11 that

$$u_{I_2}(t) = u_{I_1}(t) \quad \text{for all } t \in I_1,$$

which shows that $u_{\max}$ is well defined.

We now show that $(I_{\max}, u_{\max})$ is a solution of (5.6). Indeed, $u_{\max}$ is of class $\mathcal{C}^1(I_{\max}; \Omega)$ since, by reasoning as above, we have that

$$u_{\max} = u_I \text{ in } I \text{ whenever } (I, u_I) \in \Sigma.$$

The same argument shows that $(I, u_{\max})$ solves (5.6) in I for all $(I, u_I) \in \Sigma$. Therefore, also, $(I_{\max}, u_{\max})$ solves (5.6).

Due to the construction of $I_{\max}$, $(I_{\max}, u_{\max})$ cannot admit any proper extension. Thus, it is maximal.

Finally, if (I_1, u_{I_1}) and (I_2, u_{I_2}) are two maximal solutions, we have that $I_1 \subset I_2$ due to the maximality of (I_2, u_{I_2}) and that $I_2 \subset I_1$ due to the maximality of (I_1, u_{I_1}). Thus, $I_1 = I_2$. Theorem 5.11 allows us to conclude that $u_{I_1}(t) = u_{I_2}(t)$ for all $t \in I_1 = I_2$, which shows the uniqueness of the maximal solution. $\square$

Hereinafter, for convenience and unless otherwise explicitly stated, we always consider the maximal solution of the Cauchy problem (5.6), which we simply denote as (I, u).

5.4   Extensibility. Fundamental Theorems

The main goal of this section is to analyze the sharp behavior of the maximal solution, (I, u), of (5.6). We will show that, when $f(t, u)$ is defined for $t \in J$ and all $u \in \mathbb{R}^N$, and it is locally Lipschitz with respect to u, uniformly in t on compact subintervals of J, the maximal solution must satisfy one of the following alternatives:

(a) $I = J$;
(b) u blows up in a finite time.

Actually, those are the behaviors that emerged in the simplest prototypical models analyzed in Chapter 4. When $f(t, u)$ is only defined for $u \in \Omega$, where Ω is a proper open subset of $\mathbb{R}^N$, possibly unbounded, we will show that the solution might also approach the boundary of Ω, $\partial\Omega$, at the right endpoint of I.

As the general casuistry can be quite intricate to describe, for the moment, we focus on the most common cases arising in the applications of the abstract theory. Each of these paradigmatic cases will provide us with a "fundamental theorem".

Theorem 5.15 (First fundamental theorem). *Suppose $t_0 \in \mathbb{R}$, $T > 0$, $J = [t_0, t_0 + T]$, and $f \in \mathcal{C}(J \times \mathbb{R}^N; \mathbb{R}^N)$ is locally Lipschitz with respect to $u \in \mathbb{R}^N$ uniformly in $t \in J$. For any given $u_0 \in \mathbb{R}^N$, let (I, u) denote the maximal solution of (5.6). Then, one of the following excluding options occurs:*

(a) $I = J = [t_0, t_0 + T]$;
(b) $I = [t_0, T_{\max})$ *for some* $T_{\max} \in (t_0, t_0 + T]$ *and, in such a case,*

$$\limsup_{t \uparrow T_{\max}} \|u(t)\| = +\infty. \tag{5.23}$$

When (a) *occurs we say that* (I, u) *is **globally defined in time**, while if* (b) *occurs we say that* (I, u) ***blows up in a finite time*** $T_{\max} \le t_0 + T$.

Remark 5.16. (i) The behavior of the maximal solution when option (a) occurs is illustrated in Figure 5.3(a). We have seen in Theorem 5.5 that this is the case, for example, when f is globally Lipschitz in $u \in \mathbb{R}^N$ uniformly in $t \in J$.

(ii) The simple example (5.19) shows how the blow up time, $T_{\max}$, may actually depend on the initial datum u_0.

(iii) As the map $t \mapsto \|u(t)\|$ may oscillate as $t \uparrow T_{\max}$, in general, (5.23) does not imply the stronger property

$$\lim_{t \uparrow T_{\max}} \|u(t)\| = +\infty \tag{5.24}$$

but only guarantees the existence of a sequence, $\{t_k\}_{k \ge 1} \subset I$, such that

$$\lim_{k \to +\infty} t_k = T_{\max} \quad \text{and} \quad \lim_{k \to +\infty} \|u(t_k)\| = +\infty,$$

as illustrated in Figure 5.3(b).

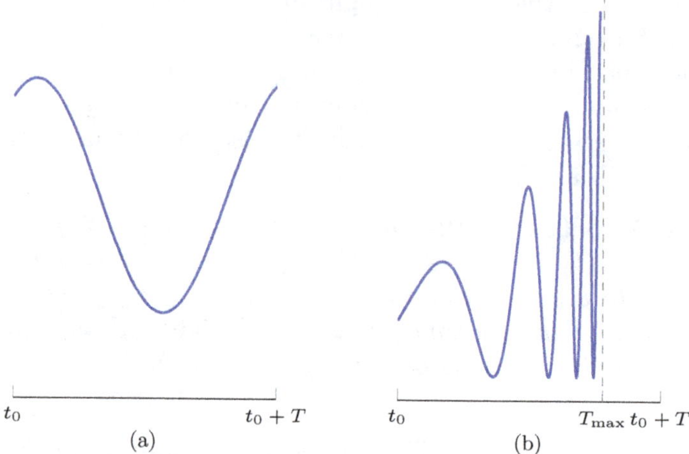

Fig. 5.3. (a) A solution globally defined in J. (b) A solution defined in $I_{\max} = [t_0, T_{\max})$ for some $T_{\max} < t_0 + T$ which is oscillating and blows up at $T_{\max}$.

(iv) When $N \geq 2$, even when (5.24) holds, it does not necessarily mean that

$$\lim_{t \uparrow T_{\max}} |u_j(t)| = +\infty \quad \text{for all } j \in \{1, \ldots, N\}.$$

As an example, consider Exercise 9 of Chapter 10, where $N = 2$, $(u_1, u_2) = (u, u')$ for some $u \in \mathcal{C}^1(I_{\max}; \mathbb{R})$, and it occurs that

$$\lim_{t \uparrow T_{\max}} |u(t)| < +\infty, \quad \text{but} \quad \lim_{t \uparrow T_{\max}} |u'(t)| = +\infty.$$

Proof of Theorem 5.15. Observe that the existence and uniqueness of the maximal solution of (5.6), (I, u), are guaranteed by Theorem 5.14. If option (a) occurs, we are done. So, suppose $I \subsetneq J$, i.e. I is a proper subinterval of J. Then,

- either $I = [t_0, T_{\max}]$ for some $T_{\max} \in (t_0, t_0 + T)$,
- or $I = [t_0, T_{\max})$ for some $T_{\max} \in (t_0, t_0 + T]$.

Suppose the first option occurs, and consider the Cauchy problem

$$\begin{cases} w' = f(t, w), \\ w(T_{\max}) = u(T_{\max}). \end{cases} \tag{5.25}$$

Since $T_{\max} < t_0 + T$, by Theorem 5.9, problem (5.25) possesses a unique solution, $w(t)$, defined in $[T_{\max}, T_{\max} + \delta]$ for sufficiently small $\delta > 0$. Thus, the solution (I^*, u^*) defined by

$$I^* := [t_0, T_{\max} + \delta], \quad u^*(t) := \begin{cases} u(t) & \text{if } t \in [t_0, T_{\max}], \\ w(t) & \text{if } t \in (T_{\max}, T_{\max} + \delta], \end{cases}$$

provides us with a proper extension of (I, u) to the right, which contradicts the maximality of (I, u). Therefore,

$$I = [t_0, T_{\max}) \quad \text{for some } T_{\max} \in (t_0, t_0 + T].$$

To complete the proof of the theorem it suffices to establish (5.23). Consider the (forward) orbit of the solution,

$$\Gamma := \{u(t) \colon t \in I = [t_0, T_{\max})\};$$

we claim that it is unbounded. This will allow us to conclude the proof. Indeed, it will imply that for each natural number $n \geq 1$, there exists $t_n \in I$ such that

$$\|u(t_n)\| \geq n.$$

Naturally, this entails

$$\lim_{n \to +\infty} \|u(t_n)\| = +\infty. \tag{5.26}$$

Since the sequence $\{t_n\}_{n \geq 1}$ is contained in $\bar{I} = [t_0, T_{\max}]$, which is a compact interval, by the Bolzano–Weierstrass theorem, it admits a subsequence, which we still label as $\{t_n\}_{n \geq 1}$, such that

$$\lim_{n \to +\infty} t_n = t_\infty$$

for some $t_\infty \in \bar{I} = [t_0, T_{\max}]$. Suppose $t_\infty < T_{\max}$. Then, since u is continuous at $t_\infty \in I$, it is apparent that

$$\lim_{n \to +\infty} u(t_n) = u(t_\infty),$$

which implies

$$\lim_{n \to +\infty} \|u(t_n)\| = \|u(t_\infty)\| < +\infty$$

and contradicts (5.26). Therefore, $t_\infty = T_{\max}$, and we have

$$\lim_{n \to +\infty} t_n = T_{\max}, \qquad \lim_{n \to +\infty} \|u(t_n)\| = +\infty,$$

which concludes the proof of (5.23).

The fact that Γ is unbounded follows by contradiction: suppose that Γ is bounded, and consider the sequence

$$t_n := T_{\max} - \frac{1}{n} \quad \text{for sufficiently large } n \geq 1.$$

Since $u(t_n) \in \Gamma$ for all $n \geq n_0$ and Γ is bounded, by the Bolzano–Weierstrass theorem, there exists a subsequence, $\{u(t_{n_m})\}_{m \geq 1}$, and some $u^* \in \bar{\Gamma}$ such that

$$\lim_{m \to +\infty} u(t_{n_m}) = u^*. \tag{5.27}$$

Since $J \times \bar{\Gamma}$ is a compact subset of $J \times \mathbb{R}^N$, where f is continuous, the following quantity is well defined:

$$\|f\|_\infty := \max_{(t,\xi) \in J \times \bar{\Gamma}} \|f(t,\xi)\| < +\infty.$$

Thus, for every $t \in I$ and sufficiently large $m \geq 1$,

$$\|u(t_{n_m}) - u(t)\| = \left\| \int_t^{t_{n_m}} u'(s)\, ds \right\| = \left\| \int_t^{t_{n_m}} f(s, u(s))\, ds \right\|$$

$$\leq \int_{\min\{t, t_{n_m}\}}^{\max\{t, t_{n_m}\}} \|f(s, u(s))\|\, ds \leq \|f\|_\infty |t - t_{n_m}|.$$

Hence, by letting $m \to +\infty$, we find from (5.27) that

$$\|u^* - u(t)\| \leq \|f\|_\infty |t - T_{\max}| \quad \text{for all } t \in I.$$

Therefore, by letting $t \uparrow T_{\max}$ in the previous estimate, it becomes apparent that

$$\lim_{t \uparrow T_{\max}} u(t) = u^*.$$

As a consequence, the pair $(\tilde{I}, \tilde{u})$, defined as

$$\tilde{I} := [t_0, T_{\max}], \quad \tilde{u}(t) := \begin{cases} u(t) & \text{if } t \in I, \\ u^* & \text{if } t = T_{\max}, \end{cases}$$

provides us with a proper extension of (I, u), which contradicts its maximality. This contradiction shows that Γ must be unbounded and concludes the proof of the theorem. $\qquad\square$

In the special case when $J = [t_0, +\infty)$, which is always the case, e.g. when dealing with autonomous systems, $u' = f(u)$, where f is globally defined, the following result holds.

Theorem 5.17 (Second fundamental theorem). *Suppose $t_0 \in \mathbb{R}$, $J = [t_0, +\infty)$, and $f \in \mathcal{C}(J \times \mathbb{R}^N; \mathbb{R}^N)$ is locally Lipschitz with respect to $u \in \mathbb{R}^N$ uniformly in t on compact subintervals of J. For any given $u_0 \in \mathbb{R}^N$, let (I, u) denote the maximal solution of (5.6). Then, there exists $t_0 < T_{\max} \le +\infty$ such that $I = [t_0, T_{\max})$. Moreover,*

$$\limsup_{t \uparrow T_{\max}} \|u(t)\| = +\infty \quad \text{if } T_{\max} < +\infty. \tag{5.28}$$

In other words, either u is globally defined in time or it blows up in a finite time, $T_{\max}$.

Proof. Arguing as in the proof of Theorem 5.15, we find that either $I = J = [t_0, +\infty)$, in which case the proof is complete, or I is a proper subinterval of J and then, necessarily, $I = [t_0, T_{\max})$ for some $T_{\max} < +\infty$. Suppose the latter case occurs, and consider the orbit of the solution,

$$\Gamma = \{u(t) \colon t \in I = [t_0, T_{\max})\}.$$

Reasoning as in the proof of Theorem 5.15, Γ must be unbounded. Thus, (5.28) holds, and the proof is complete in this case as well. $\square$

More generally, when $f(t, u)$ is only defined in $[t_0, \infty) \times \Omega$, for some proper open subset $\Omega \subset \mathbb{R}^N$, the following result holds.

Theorem 5.18 (Third fundamental theorem). *Suppose $t_0 \in \mathbb{R}$, $J = [t_0, +\infty)$, Ω is a proper open subset of $\mathbb{R}^N$, and $f \in \mathcal{C}(J \times \Omega; \mathbb{R}^N)$ is locally Lipschitz with respect to $u \in \Omega$ uniformly in t on compact subintervals of J. For any given $u_0 \in \Omega$, let (I, u) denote the maximal solution of (5.6). Then, $I = [t_0, T_{\max})$ for some $t_0 < T_{\max} \le +\infty$. Moreover, when $T_{\max} < +\infty$, one of the following, non-excluding, options occurs: either*

$$\limsup_{t \uparrow T_{\max}} \|u(t)\| = +\infty, \tag{5.29}$$

i.e. **the solution blows up at** $T_{\max}$, *or there exist a sequence of times* $\{t_n\}_{n\geq 1} \subset I$, *with* $\lim_{n\to\infty} t_n = T_{\max}$, *and a point* $u_\omega \in \partial\Omega$ *such that*

$$\lim_{n\to\infty} u(t_n) = u_\omega \in \partial\Omega, \tag{5.30}$$

i.e. **the solution approximates** $\partial\Omega$ **at** $T_{\max}$.

Proof. If $I = J$, then the solution is globally defined in time and the proof is complete. Thus, assume that I is a proper subinterval of J. Then, arguing as in the proof of Theorem 5.15, we get that $I = [t_0, T_{\max})$ for some $T_{\max} \in (t_0, \infty)$. Next, as in the proofs of the previous fundamental theorems, we consider the orbit of the maximal solution

$$\Gamma := \{u(t) \colon t \in I = [t_0, T_{\max})\}.$$

Arguing as in the proof of Theorem 5.15, it becomes apparent that u blows up at $T_{\max}$ if Γ is unbounded. Thus, (5.29) holds in this case.

Subsequently, we suppose that Γ is bounded. Then, $\bar{\Gamma}$ is compact and, since $\Gamma \subset \Omega$, either $\bar{\Gamma} \subset \Omega$ or $\bar{\Gamma} \cap \partial\Omega \neq \emptyset$.

Suppose that $\bar{\Gamma} \subset \Omega$, and consider the sequence of times

$$t_n = T_{\max} - \frac{1}{n}$$

for sufficiently large $n \geq 1$, say, $n \geq n_0$. Then, since $\{u(t_n)\}_{n\geq n_0}$ is bounded, by the Bolzano–Weierstrass theorem, there exist a subsequence, $\{u(t_{n_m})\}_{m\geq 1}$, and a point, $u^* \in \bar{\Gamma} \subset \Omega$, such that

$$\lim_{m\to\infty} u(t_{n_m}) = u^*.$$

Arguing as in the proof of Theorem 5.15, we actually find that

$$\lim_{t\uparrow T_{\max}} u(t) = u^* \in \bar{\Gamma} \subset \Omega.$$

As in the proof of Theorem 5.15, we can extend the solution up to $t = T_{\max}$, which contradicts the maximality of (I, u). This contradiction shows that, necessarily, $\bar{\Gamma} \cap \partial\Omega \neq \emptyset$. Now, pick an element $u_\omega \in \bar{\Gamma} \cap \partial\Omega$, and let $\{t_n\}_{n\geq 1}$ be a sequence of times in I such that

$$\lim_{n\to\infty} u(t_n) = u_\omega \in \partial\Omega.$$

Since I is bounded, up to passing to a subsequence, we can assume that

$$\lim_{n \to \infty} t_n = t_\infty \in [t_0, T_{\max}].$$

Suppose $t_\infty < T_{\max}$. Then, by the continuity of u,

$$u_\omega = \lim_{n \to \infty} u(t_n) = u(t_\infty) \in \Gamma \subset \Omega,$$

which contradicts the choice of u_ω. Therefore, $t_\infty = T_{\max}$. This shows (5.30), which completes the proof. $\square$

Note that, when Γ is unbounded, (5.29) and (5.30) might occur simultaneously. Figure 5.4 illustrates this phenomenology: it shows an unbounded open subset Ω together with the unbounded trajectory of the solution $u(t)$, $t \geq t_0$. Along some sequence of times, $\{s_n\}_{n \geq 1}$, we have that

$$\lim_{n \to \infty} \|u(s_n)\| = +\infty.$$

In addition, there exists another sequence, $\{t_n\}_{n \geq 1}$, with $s_n < t_n < s_{n+1}$, such that

$$\lim_{n \to \infty} u(t_n) = u_\omega$$

for some $u_\omega \in \partial\Omega$. Figure 5.4 also suggests that u_ω may not be necessarily unique. Actually, there are examples for which the set of

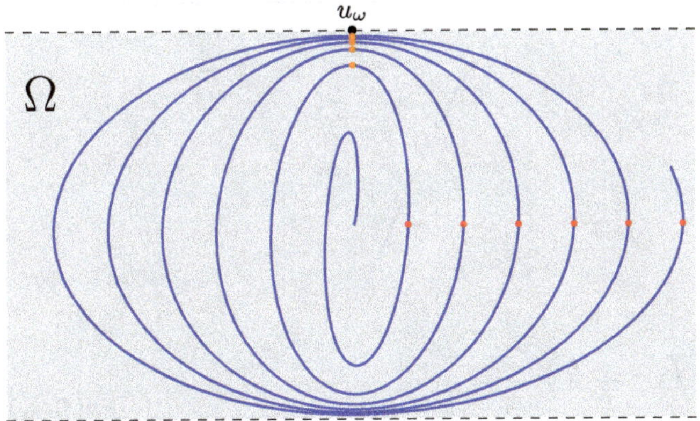

Fig. 5.4. An unbounded trajectory satisfying both (5.29) (e.g. along the red sequence of points) and (5.30) (e.g. along the orange sequence of points).

limit points of Γ as $t \uparrow T_{\max}$ is the whole $\partial\Omega$ (see Exercise 20 of Chapter 5).

The previous results drastically change when $f(t, u)$ is defined on a set of the form $J \times K$ for some compact subset $K \subset \mathbb{R}^N$. In such cases, the following result establishes that the maximal solutions of (5.6) with $u_0 \in \operatorname{int} K$ are globally defined in time, unless they reach the boundary of K, ∂K. Precisely, the following result holds.

Theorem 5.19 (Fourth fundamental theorem). *Suppose $t_0 \in \mathbb{R}$, $T > 0$, $J = [t_0, t_0 + T]$, Ω is an open bounded subset of $\mathbb{R}^N$, and $f \in \mathcal{C}(J \times \bar{\Omega}; \mathbb{R}^N)$ is globally Lipschitz with respect to $u \in \bar{\Omega}$ uniformly in $t \in J$. Then, for any given $u_0 \in \Omega$, (5.6) possesses a unique maximal solution, $(I_{\max}, u_{\max})$. Moreover, either $I_{\max} = J$ or $I_{\max} = [t_0, T_{\max}]$ for some $T_{\max} < t_0 + T$ and, in such a case,*

$$u_{\max}(T_{\max}) \in \partial\Omega.$$

Theorem 5.19 is applicable, for example, when Ω is an open ball, $B_R(u_0)$, for some $u_0 \in \mathbb{R}^N$ and $R > 0$, or a rectangle, as shown in Figure 5.5. Theorem 5.19 establishes that the maximal interval of existence of the solution must be compact, $[t_0, T_{\max}]$, and that, if $T_{\max} < t_0 + T$, then the solution cannot be extended because its trajectory reaches $\partial\Omega$, as illustrated in Figure 5.5. However, it should be noted that $T_{\max}$ is not necessarily the first time when $u_{\max}(t) \in \partial\Omega$.

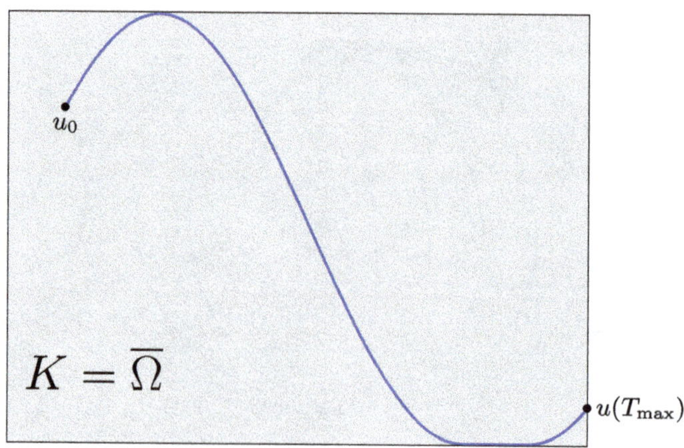

Fig. 5.5. A maximal solution reaching $\partial\Omega$ at $T_{\max} < t_0 + T$.

Proof of Theorem 5.19. The existence of the maximal solution follows the scheme of the proof of Theorem 5.14. Now, the solution pairs (I, u) must satisfy $u(t) \in \bar{\Omega}$ for all $t \in I$. The uniqueness in any interval where a solution is defined is guaranteed by the fact that, thanks to the Tietze-Urysohn theorems, $f(t, u)$ can be extended to a function $F \in \mathcal{C}(J \times \mathbb{R}^N; \mathbb{R}^N)$ globally Lipschitz with respect to $u \in \mathbb{R}^N$ uniformly in J. When Ω is a ball, such an F has been already constructed in the proof of Theorem 5.9. Note that, if $u(t_1) \in \Omega$ for some $t_1 \in [t_0, t_0 + T)$, then, owing to Theorem 5.9, $u(t)$ can be extended to $[t_1, t_1 + \delta]$ for some $\delta > 0$. Instead, if $u(t_1) \in \partial\Omega$ for some $t_1 \in J$, then one of the next two options can occur:

(i) The solution can be extended to $[t_1, t_1 + \delta]$ for some $\delta > 0$, in which case the solution is not maximal since we can add the interval $[t_0, t_1+\delta]$ in the definition of $I_{\max}$ (cf. (5.22)). Note that, since $u(t_1) \in \partial\Omega$, Theorem 5.9 is not directly applicable to obtain the existence of a solution with initial values $(t_1, u(t_1))$ at the right of t_1, which shall be determined using other tools. A necessary condition for this to occur is that $f(t, u(t))$ points in such a way that the solution $u(t)$ remains in $\bar{\Omega}$ in a right neighborhood of t_1.

(ii) The solution cannot be extended to the right of t_1, in which case the solution is maximal, by definition.

Now, we proceed to analyze the global behavior of the maximal solution, $(I, u) \equiv (I_{\max}, u_{\max})$. If $I = J$, the theorem is proved. So, suppose I is a proper subinterval of J. Then, either $I = [t_0, T_{\max})$ for some $T_{\max} \leq t_0 + T$ or $I = [t_0, T_{\max}]$ for some $T_{\max} < t_0 + T$. By reasoning as in the last part of the proof of Theorem 5.15, we can show that the first situation is impossible. Otherwise, we could properly extend the maximal solution based on the fact that, since $\bar{\Omega}$ is compact, the orbit is bounded.

Therefore, $I = [t_0, T_{\max}]$ for some $T_{\max} < t_0 + T$. Suppose

$$u(T_{\max}) \in \Omega.$$

Then, by Theorem 5.9, for sufficiently small $\delta > 0$, the problem

$$\begin{cases} w' = f(t, w), \\ w(T_{\max}) = u(T_{\max}), \end{cases}$$

possesses a unique solution, $w(t)$, defined in $[T_{\max}, T_{\max} + \delta]$. Consequently, the solution

$$\hat{I} = [t_0, T_{\max} + \delta], \quad \hat{u}(t) := \begin{cases} u(t) & \text{if } t \in [t_0, T_{\max}], \\ w(t) & \text{if } t \in (T_{\max}, T_{\max} + \delta], \end{cases}$$

provides us with a proper extension to the right of (I, u), which contradicts the maximality of (I, u). Therefore, $u(T_{\max}) \in \partial\Omega$, and the proof is complete. $\qquad\square$

Remark 5.20. With exactly the same ingredients used in the proof of Theorem 5.19, one can show that, if $J = [t_0, +\infty)$ and $f \in \mathcal{C}(J \times \bar{\Omega}; \mathbb{R}^N)$ is globally Lipschitz with respect to $u \in \bar{\Omega}$ uniformly in t on compact subsets of J, where Ω is an open bounded subset of $\mathbb{R}^N$ containing u_0, then (5.6) has a unique maximal solution, (I, u). Moreover, either $I = [t_0, +\infty)$ or $I = [t_0, T_{\max}]$ for some $T_{\max} \in (t_0, +\infty)$ and, in such a case, $u(T_{\max}) \in \partial\Omega$.

5.4.1　*Going backward in time versus going forward*

All the results in Sections 5.3 and 5.4 can be easily adapted to obtain their corresponding counterparts when f is defined in $J \times \mathbb{R}^N$, with $J = (-\infty, t_0]$, or $J = [t_0 - T, t_0]$ for some $T > 0$. Indeed, the change of variable

$$s = 2t_0 - t, \quad v(s) = u(t), \tag{5.31}$$

transforms the differential equation

$$u'(t) = f(t, u(t))$$

into the new one

$$\dot{v}(s) := \frac{dv}{ds} = u'(t)\frac{dt}{ds} = -f(2t_0 - s, v(s)).$$

Thus, (5.6) becomes

$$\begin{cases} \dot{v} = g(s, v), \\ v(t_0) = u_0, \end{cases}$$

where

$$g(s, v) := -f(2t_0 - s, v), \quad (s, v) \in \tilde{J} \times \mathbb{R}^N,$$

with

$$\tilde{J} := \begin{cases} [t_0, t_0 + T] & \text{if } J = [t_0 - T, t_0], \\ [t_0, +\infty) & \text{if } J = (-\infty, t_0]. \end{cases}$$

As g is locally Lipschitz with respect to $v \in \mathbb{R}^N$ uniformly in s on compact subsets of $\tilde{J}$ if and only if f is locally Lipschitz with respect to $u \in \mathbb{R}^N$ uniformly in t on compact subsets of J, it turns out that, from a mathematical point of view, looking forward in time, i.e. for $t \geq t_0$, or backward, i.e. for $t \leq t_0$, is equivalent, thanks to the change of variables (5.31).

More generally, when J is any subinterval of $\mathbb{R}$ and $t_0 \in \text{int } J$, then the solutions of problem (5.6) on the right and left of t_0 can be glued together to construct a unique solution of (5.6), which is defined (at least) on an open neighborhood of t_0, $(t_0 - \delta, t_0 + \delta)$, because

$$u'(t_0^+) = u'(t_0^-)$$

by the continuity of $t \mapsto f(t, u(t))$. Thus, all the theory developed in this chapter can be very easily adapted to look backward in time, or even forward and backward in time simultaneously. In particular, the concept of maximal solution can be extended to work for $t \leq t_0$, and Theorem 5.14 possesses its backward counterpart, as follows.

Theorem 5.21. *Suppose $t_0 \in \mathbb{R}$, $T > 0$, $J \in \{[t_0 - T, t_0], (-\infty, t_0]\}$, Ω is an open subset of $\mathbb{R}^N$, $u_0 \in \Omega$ and $f \in C(J \times \Omega; \mathbb{R}^N)$ is locally Lipschitz with respect to $u \in \Omega$ uniformly in t on compact subintervals of J. Then, problem (5.6) possesses a unique maximal solution toward the left, $(I_{\max}, u_{\max})$.*

Naturally, by gluing together the maximal solution of (5.6) to the right provided by Theorem 5.14 with the maximal solution to the left given by Theorem 5.21, the following result holds.

Theorem 5.22. *Suppose J is any subinterval of $\mathbb{R}$ with $t_0 \in \text{int } J$, Ω is an open subset of $\mathbb{R}^N$, $u_0 \in \Omega$ and $f \in C(J \times \Omega; \mathbb{R}^N)$ is locally*

Lipschitz with respect to $u \in \Omega$ uniformly in t on compact subintervals of J. Then, problem (5.6) possesses a unique bilateral maximal solution, $(I_{\max}, u_{\max})$.

In the rest of the book, we will denote $T_{\min} := \inf I_{\max}$ and $T_{\max} := \sup I_{\max}$. Nevertheless, unless otherwise explicitly said, we will mostly look forward in time, as this is more natural from the point of view of applications.

To conclude this section, we remark that, obviously, thanks to the change of variable (5.31), the four fundamental theorems in this section can be adapted to obtain their corresponding counterparts backward in time. It is important to note that, as evidenced by the logistic equation analyzed in Section 4.1, the forward and backward behaviors of the maximal solutions are independent of each other.

5.5   Some Illustrative Examples

In this section, we analyze three paradigmatic applications of the abstract theory developed in this chapter. As in other parts of this book, in specific applications with $N = 1, 2$, we denote, instead of $u = (u_1, \ldots, u_N)$,

$$u = u_1, \quad (u, v) = (u_1, u_2),$$

respectively. Similarly, instead of $f = (f_1, \ldots, f_N)$, we respectively set

$$f = f_1, \quad (f, g) = (f_1, f_2).$$

5.5.1   *A first example*

The first example we consider is the Cauchy problem

$$\begin{cases} u' = \lambda(t)u - a(t)u^2 - b(t)uv, \\ v' = \mu(t)v - d(t)v^2 - c(t)uv, \\ (u(0), v(0)) = (u_0, v_0), \end{cases} \tag{5.32}$$

where $\lambda, \mu, a, b, c, d \in \mathcal{C}([0, \infty); \mathbb{R})$ and $(u_0, v_0) \in \mathbb{R}^2$. Setting

$$f(t, u, v) := \lambda(t)u - a(t)u^2 - b(t)uv,$$

$$g(t, u, v) := \mu(t)v - d(t)v^2 - c(t)uv,$$

and differentiating gives

$$\frac{\partial f}{\partial u} = \lambda(t) - 2a(t)u - b(t)v, \qquad \frac{\partial f}{\partial v} = -b(t)u,$$

$$\frac{\partial g}{\partial u} = -c(t)v, \qquad \frac{\partial g}{\partial v} = \mu(t) - 2d(t)v - c(t)u,$$

for every $t \geq 0$ and $(u, v) \in \mathbb{R}^2$. Since all these derivatives are continuous for $(t, u, v) \in [0, +\infty) \times \mathbb{R} \times \mathbb{R}$, it follows from Corollary 5.3 that (f, g) is locally Lipschitz with respect to $(u, v) \in \mathbb{R}^2$ uniformly in t on compact subintervals of $[0, +\infty)$. Therefore, thanks to Theorem 5.14, for every $(u_0, v_0) \in \mathbb{R}^2$, problem (5.32) possesses a unique maximal solution (I, u, v). Moreover, by the second fundamental theorem (see Theorem 5.17), either $I = [0, +\infty)$ or $I = [0, T_{\max})$ for some $T_{\max} < +\infty$ and, in such a case,

$$\limsup_{t \uparrow T_{\max}} (|u(t)| + |v(t)|) = +\infty. \qquad (5.33)$$

Setting

$$A(t) := \lambda(t) - a(t)u(t) - b(t)v(t),$$

$$B(t) := \mu(t) - d(t)v(t) - c(t)u(t),$$

it is apparent from (5.32) that, for every $t \in I$,

$$u'(t) = A(t)u(t), \quad v'(t) = B(t)v(t);$$

hence, by Theorem 1.2, for every $t \in I$,

$$u(t) = e^{\int_0^t A(s)\,ds} u_0, \quad v(t) = e^{\int_0^t B(s)\,ds} v_0,$$

and

$$\operatorname{sign} u(t) = \operatorname{sign} u_0, \quad \operatorname{sign} v(t) = \operatorname{sign} v_0.$$

Therefore, for all $t \in I$, $u(t) > 0$, $u(t) < 0$, and $u(t) = 0$, if $u_0 > 0$, $u_0 < 0$, and $u_0 = 0$, respectively. Similarly, for all $t \in I$, $v(t) > 0$, $v(t) < 0$, and $v(t) = 0$ if $v_0 > 0$, $v_0 < 0$, and $v_0 = 0$, respectively.

Whether or not $T_{\max} < +\infty$ depends on the nature of the weight functions $a(t)$, $b(t)$, $c(t)$, and $d(t)$ as well as on the sign of u_0 and v_0. For example, if the weights satisfy

$$a \geq 0, \quad b \geq 0, \quad c \geq 0, \quad d \geq 0, \tag{5.34}$$

and $u_0 > 0$ and $v_0 > 0$, then $u(t) > 0$ and $v(t) > 0$ for all $t \in I$ and, hence, $A(t) \leq \lambda(t)$ and $B(t) \leq \mu(t)$ for all $t \in I$. Thus,

$$0 < u(t) \leq e^{\int_0^t \lambda(s)\,ds} u_0, \quad 0 < v(t) \leq e^{\int_0^t \mu(s)\,ds} v_0,$$

for all $t \in I$. So, $u(t)$ and $v(t)$ must be bounded if $T_{\max} < +\infty$, which excludes (5.33). Therefore, under assumption (5.34), $I = [0, +\infty)$, i.e. (u, v) is globally defined in time. However, if we take $v_0 = 0$, then $v = 0$ in I. Thus, (I, u) provides us with the maximal solution of

$$\begin{cases} u' = \lambda(t)u - a(t)u^2, \\ u(0) = u_0. \end{cases}$$

By choosing $\lambda = 0$ and $a = -1$, we are led to the problem

$$\begin{cases} u' = u^2, \\ u(0) = u_0, \end{cases}$$

whose unique solution, if $u_0 > 0$, is

$$u(t) = \frac{u_0}{1 - u_0 t}, \quad t \in [0, u_0^{-1}).$$

Therefore, for this choice, $T_{\max} = u_0^{-1} < +\infty$.

We point out that systems of the form (5.32) are extremely relevant from the point of view of applications in Population Dynamics. Some other specific cases of this class of systems will be considered in the exercises of Section 5.7, as well as in Part 3.

5.5.2   *A second example*

The second example we study is the initial value problem

$$\begin{cases} u' = a(t)u + b(t)\frac{u^n}{1+u}, \\ u(0) = 1, \end{cases} \tag{5.35}$$

where $a, b \in \mathcal{C}([0, \infty); \mathbb{R})$ satisfy $a(t) \geq 0$ and $b(t) > 0$ for all $t \geq 0$, and $n \geq 1$ is a natural number. If we set

$$f(t, u) := a(t)u + b(t)\frac{u^n}{1 + u},$$

we have that $f \in \mathcal{C}(J \times \Omega; \mathbb{R})$, where $J := [0, \infty)$ and $\Omega := (-1, \infty)$. Moreover, since

$$\frac{\partial f}{\partial u}(t, u) = a(t) + b(t)\frac{nu^{n-1} + (n-1)u^n}{(1 + u)^2}$$

for all $t \in J$ and $u \in \Omega$, it is apparent that $\frac{\partial f}{\partial u}$ is continuous in $J \times \Omega$. Thus, by Corollary 5.3, f is locally Lipschitz with respect to $u \in \Omega$ uniformly in t on compact subsets of J. Therefore, by Theorem 5.14, (5.35) has a unique maximal solution, (I, u). Moreover, by the third fundamental theorem (see Theorem 5.18), $I = [0, T_{\max})$ for some $0 < T_{\max} \leq +\infty$, and if $T_{\max} < \infty$, then either

$$\limsup_{t \uparrow T_{\max}} |u(t)| = +\infty$$

or, since $\partial\Omega = \{-1\}$, there exists a sequence of times $\{t_n\}_{n \geq 1} \subset I$, with $\lim_{n \to \infty} t_n = T_{\max}$, such that

$$\lim_{n \to \infty} u(t_n) = -1.$$

Setting

$$A(t) := a(t) + b(t)\frac{u^{n-1}(t)}{1 + u(t)}, \quad t \in I,$$

it is apparent that $u'(t) = A(t)u(t)$ for all $t \in I$ and, hence, by Theorem 1.2,

$$u(t) = e^{\int_0^t A(s)\,ds} \quad \text{for all } t \in I.$$

In particular, $u(t) > 0$ for all $t \in I$. Thus, u cannot approach $\partial\Omega = \{-1\}$ and, hence, it blows up at $T_{\max}$ if $T_{\max} < +\infty$. Moreover,

$$u' = a(t)u(t) + b(t)\frac{u^n(t)}{1 + u(t)} > 0$$

for all $t \in I$. Therefore, u is increasing in I. To proceed with the analysis, we subsequently distinguish two different cases.

Suppose $n = 1, 2$. Then,

$$A(t) = a(t) + b(t)\frac{u^{n-1}(t)}{1 + u(t)} \leq a(t) + b(t), \quad t \in I,$$

and, hence,

$$u(t) = e^{\int_0^t A(s)\, ds} \leq e^{\int_0^t [a(s)+b(s)]\, ds}$$

for all $t \in I$. Consequently, u cannot blow up at $T_{\max}$ if $T_{\max} < +\infty$ because, if $T_{\max} < +\infty$, the previous estimate implies

$$u(t) \leq e^{T_{\max}(\sup_I a + \sup_I b)} < +\infty$$

for all $t \in I$. Therefore, $I = [0, +\infty)$. In other words, the solution of (5.35) is globally defined in time for $n = 1, 2$.

Suppose $n = 3$. Then,

$$f(t, u) = a(t)u + b(t)\frac{u^3}{1 + u} = a(t)u + b(t)\frac{u}{1 + u}u^2,$$

and hence, since $u \geq 1$ and the function $u \mapsto \frac{u}{1+u}$ is increasing in $[1, +\infty)$,

$$f(t, u) \geq a(t)u + \frac{b(t)}{2}u^2 \geq \frac{b(t)}{2}u^2, \quad t \geq 0, \ \ u \in [1, +\infty).$$

Therefore, if we further assume that

$$\inf_{[0,+\infty)} b = \omega > 0,$$

then, for every $t \in I$,

$$u'(t) \geq \frac{\omega}{2}u^2(t).$$

We will show that $T_{\max} < +\infty$, and

$$\lim_{t \uparrow T_{\max}} u(t) = +\infty.$$

Indeed, multiplying this differential inequality by $u^{-2}(t)$ gives

$$-(u^{-1}(t))' \geq \frac{\omega}{2}$$

for all $t \in I$. Thus, by integrating in $(0, t)$, with $t < T_{\max}$, it follows that

$$\frac{1}{u(t)} - 1 \leq -\frac{\omega}{2}t$$

and, hence,

$$u(t) \geq \frac{1}{1 - \frac{\omega}{2}t}.$$

Since the denominator on the right-hand side vanishes for $t = \frac{2}{\omega}$, we conclude that, under these conditions, the solution of (5.35) blows up in a finite time, $T_{\max} \leq \frac{2}{\omega}$.

5.5.3 *A third example*

The third example we analyze is the initial value problem

$$\begin{cases} u' = |v|^p, \\ v' = |u|^q, \\ u(0) = v(0) = 1, \end{cases} \tag{5.36}$$

where $p, q \in \mathbb{R}$ satisfy

$$p \geq q > 1.$$

The nonlinearity associated with (5.36) is the autonomous map

$$f(t, u, v) = \begin{pmatrix} |v|^p \\ |u|^q \end{pmatrix}, \quad (t, u, v) \in \mathbb{R} \times \mathbb{R} \times \mathbb{R}.$$

Since $p, q > 1$, f is differentiable at any $(u, v) \in \mathbb{R}^2$ and its Jacobian is given by

$$D_{(u,v)} f(t, u, v) = \begin{pmatrix} 0 & p|v|^{p-1}\mathrm{sign}\, v \\ q|u|^{q-1}\mathrm{sign}\, u & 0 \end{pmatrix}, \quad (u, v) \in \mathbb{R}^2.$$

As the coefficients of this matrix are unbounded continuous functions, by Proposition 5.2, f is not globally Lipschitz with respect to $(u, v) \in \mathbb{R}^2$. Though, by Corollary 5.3, it is locally Lipschitz with respect to $(u, v) \in \mathbb{R}^2$ uniformly in $t \in \mathbb{R}$. Thus, thanks to

Theorem 5.14 and the second fundamental theorem (see Theorem 5.17), (5.36) has a maximal (forward) solution, (I, u, v), $I = [0, T_{\max})$ for some $T_{\max} \leq +\infty$, and

$$\limsup_{t \uparrow T_{\max}} (|u(t)| + |v(t)|) = +\infty \quad \text{if } T_{\max} < +\infty. \tag{5.37}$$

On the other hand, since $u'(t) \geq 0$ and $v'(t) \geq 0$ for all $t \in I$, we have that

$$u(t) \geq u(0) = 1 \quad \text{and} \quad v(t) \geq v(0) = 1 \quad \text{for all } t \in I.$$

Hence,

$$u'(t) = |v(t)|^p \geq 1, \quad v'(t) = |u(t)|^q \geq 1,$$

for all $t \in I$. Therefore, $u(t)$ and $v(t)$ are actually increasing in I, and (u, v) solve the following Cauchy problem:

$$\begin{cases} u'(t) = v^p(t) \\ v'(t) = u^q(t) \\ u(0) = v(0) = 1. \end{cases} \quad \text{for all } t \in I, \tag{5.38}$$

Consequently,

$$u'(t)u^q(t) = v'(t)v^p(t) \quad \text{for all } t \in I,$$

which can be equivalently expressed as

$$\frac{d}{dt}\left(\frac{u^{q+1}(t)}{q+1} - \frac{v^{p+1}(t)}{p+1} \right) = 0.$$

Thus, taking into account the initial conditions, we have

$$\frac{u^{q+1}(t)}{q+1} - \frac{v^{p+1}(t)}{p+1} = \frac{1}{q+1} - \frac{1}{p+1} = \frac{p-q}{(p+1)(q+1)} \geq 0$$

for all $t \in I$. Therefore,

$$u(t) = \left(\frac{q+1}{p+1} v^{p+1}(t) + \frac{p-q}{p+1} \right)^{\frac{1}{q+1}} \quad \text{for all } t \in I. \tag{5.39}$$

In particular, since u and v are increasing, (5.37) and (5.39) imply that

$$\lim_{t\uparrow T_{\max}} u(t) = \lim_{t\uparrow T_{\max}} v(t) = +\infty \quad \text{if } T_{\max} < +\infty. \tag{5.40}$$

By substituting (5.39) in the second equation of (5.38), it is apparent that

$$v'(t) = \left(\frac{q+1}{p+1}v^{p+1}(t) + \frac{p-q}{p+1}\right)^{\frac{q}{q+1}} \geq \left(\frac{q+1}{p+1}\right)^{\frac{q}{q+1}} v^{\frac{q+pq}{q+1}}(t) \tag{5.41}$$

for all $t \in I$. Thus, by setting

$$C := \left(\frac{q+1}{p+1}\right)^{\frac{q}{q+1}}, \quad \varrho := \frac{q+pq}{q+1} > 1,$$

(5.41) can be rewritten as

$$v'(t) \geq Cv^{\varrho}(t) \quad \text{for all } t \in I. \tag{5.42}$$

We will show that this entails $T_{\max} < +\infty$. Indeed, since $v(t) > 0$ for all $t \in I$, (5.42) yields

$$\left(\frac{v^{1-\varrho}(t)}{1-\varrho}\right)' = v'(t)v^{-\varrho}(t) \geq C.$$

Hence, by integrating in $[0, t]$, $t \in I$, it follows that

$$\frac{v^{1-\varrho}(t)}{1-\varrho} - \frac{1}{1-\varrho} \geq Ct \quad \text{for all } t \in I,$$

which implies

$$v^{1-\varrho}(t) \leq 1 - (\varrho-1)Ct \quad \text{for all } t \in I$$

because $1 - \varrho < 0$. Therefore,

$$v(t) \geq \frac{1}{[1 - (\varrho-1)Ct]^{\frac{1}{\varrho-1}}} \quad \text{for all } t \in I$$

and, hence,

$$T_{\max} \leq \frac{1}{(\varrho-1)C} = \left(\frac{p+1}{q+1}\right)^{\frac{q}{q+1}} \frac{q+1}{pq-1} < +\infty. \tag{5.43}$$

In the special case of $p = q$, the equation in (5.41) becomes

$$v'(t) = v^p(t);$$

therefore,

$$v(t) = \frac{1}{[1 - (p-1)t]^{\frac{1}{p-1}}} \quad \text{for all } t \in I.$$

As a byproduct,

$$T_{\max} = \frac{1}{p-1}.$$

Consequently, (5.43) gives a good estimate for $T_{\max}$ in the general case, at least when p is close to q. Actually, in this particular example, the blow up time can be explicitly determined using the following argument. First, note that (5.41) implies

$$t = \int_0^t ds = \int_0^t \frac{v'(s)}{\left(\frac{q+1}{p+1}v^{p+1}(s) + \frac{p-q}{p+1}\right)^{\frac{q}{q+1}}} \, ds$$

for every $t \in I$. Indeed the denominator in the previous expression never vanishes since v is increasing. Thus, the change of variable $\xi = v(t)$ yields

$$t = \int_1^{v(t)} \frac{d\xi}{\left(\frac{q+1}{p+1}\xi^{p+1} + \frac{p-q}{p+1}\right)^{\frac{q}{q+1}}} \quad \text{for all } t \in I.$$

Lastly, according to (5.40), letting $t \uparrow T_{\max}$ in this identity shows that

$$T_{\max} = T_{\max}(p, q) = \int_1^{+\infty} \frac{d\xi}{\left(\frac{q+1}{p+1}\xi^{p+1} + \frac{p-q}{p+1}\right)^{\frac{q}{q+1}}}.$$

Observe that this improper integral is finite since

$$\frac{q(p+1)}{q+1} - 1 = \frac{pq-1}{q+1} > 0.$$

It is easy to see that $T_{\max}(p, p) = \frac{1}{p-1}$, which coincides with the value determined above in the case of $p = q$.

5.6   Lipschitz Dependence on u_0 and f

In the context of this chapter, where we have established the existence of a unique maximal solution (I, u) for the Cauchy problem (5.6), the solution can be regarded as a function of the initial value u_0 and the nonlinearity f. In this section, we analyze the regularity of such a function; for this reason, we denote

$$u(t) := u(t; t_0, u_0, f), \quad t \in [t_0, T_{\max}(t_0, u_0, f)).$$

From the point of view of the applications of differential equations to Science and Engineering, studying the dependence of the solution u with respect to small variations of the initial value u_0 and the nonlinearity f is imperative. In practice, changes in the nonlinearity f can be due to the variation in one, or several, of the parameters involved in the problem. For instance, in the logistic equation (4.2), the function

$$f(t, u) = \lambda u - mu^2$$

increases, for $u > 0$, if the intrinsic growth rate of the species, λ, increases and the other quantities do not vary; similarly, it decreases if the intraspecific competition coefficient, m, increases.

In sum, there is a huge interest in analyzing the continuity properties of the solution operator $\mathcal{S}$, defined as

$$(u_0, f) \mapsto \mathcal{S}(u_0, f) := u(\cdot; t_0, u_0, f). \tag{5.44}$$

The main result in this section establishes that $\mathcal{S}$ is globally Lipschitz with respect to u_0 and f in any compact interval $J = [t_0, t_0+T]$ where the underlying solutions are defined.[1] In the proof of this result, we use an important classical result, which is named after Grönwall and Bellman.

5.6.1   *The Grönwall–Bellman lemma*

The main result of this section is the following lemma, which gives an estimate for a function satisfying an integral inequality. It is extremely useful in several contexts involving differential equations.

---

[1]In the space of continuous functions, we consider the maximum norm.

Lemma 5.23 (Grönwall–Bellman). *Suppose $t_0 \in \mathbb{R}$, $T > 0$, $J = [t_0, t_0 + T]$ and the functions $x, y \in C(J; \mathbb{R})$ and $p \in C(J; [0, \infty))$ satisfy the integral inequality*

$$x(t) \leq y(t) + \int_{t_0}^{t} p(s)x(s)\, ds \qquad (5.45)$$

for all $t \in J$. Then,

$$x(t) \leq y(t) + \int_{t_0}^{t} y(s)p(s)e^{\int_s^t p(\sigma)\, d\sigma}\, ds \quad \text{for all } t \in J. \qquad (5.46)$$

Proof.   Consider the auxiliary function defined as

$$\varphi(t) := \int_{t_0}^{t} p(s)x(s)\, ds, \quad t \in J.$$

By the fundamental theorem of calculus, $\varphi \in C^1(J; \mathbb{R})$ and

$$\varphi'(t) = p(t)x(t) \quad \text{for all } t \in J.$$

Moreover, by the definition of φ, (5.45) can be expressed as

$$x(t) \leq y(t) + \varphi(t), \quad t \in J.$$

Thus, since $p \geq 0$, we find that

$$\varphi'(t) \leq p(t)\varphi(t) + p(t)y(t) \qquad (5.47)$$

for all $t \in J$. The change of variable

$$\varphi(t) = e^{\int_{t_0}^{t} p(s)\, ds}\psi(t), \quad t \in J,$$

transforms (5.47) into

$$p(t)e^{\int_{t_0}^{t} p(s)\, ds}\psi(t) + e^{\int_{t_0}^{t} p(s)\, ds}\psi'(t)$$

$$\leq p(t)e^{\int_{t_0}^{t} p(s)\, ds}\psi(t) + p(t)y(t), \quad t \in J,$$

which leads to

$$e^{\int_{t_0}^{t} p(s)\,ds}\,\psi'(t) \le p(t)y(t), \quad t \in J.$$

Thus, since $0 = \varphi(t_0) = \psi(t_0)$, we find that

$$\psi(t) = \psi(t) - \psi(t_0) = \int_{t_0}^{t} \psi'(s)\,ds \le \int_{t_0}^{t} y(s)p(s)e^{-\int_{t_0}^{s} p(\sigma)\,d\sigma}\,ds$$

for every $t \in J$. Therefore,

$$\varphi(t) = e^{\int_{t_0}^{t} p(\sigma)\,d\sigma}\,\psi(t) \le \int_{t_0}^{t} y(s)p(s)e^{\int_{s}^{t} p(\sigma)\,d\sigma}\,ds \quad \text{for all } t \in J.$$

Thanks to the definition of φ and (5.45), this implies (5.46). $\qquad\square$

5.6.2 *The abstract result*

The following result establishes the Lipschitz dependence of the solution $u(t; t_0, u_0, f)$ of (5.6) with respect to u_0 and f.

Theorem 5.24 (Lipschitz dependence). *Suppose $t_0 \in \mathbb{R}$, $T > 0$, $J := [t_0, t_0 + T]$, $R > 0$, $L > 0$, $u_0 \in \mathbb{R}^N$ and $f \in \mathcal{C}(J \times \bar{B}_R(u_0); \mathbb{R}^N)$ satisfies*

$$\|f(t, x) - f(t, y)\| \le L\|x - y\| \tag{5.48}$$

for all $t \in J$ and $x, y \in \bar{B}_R(u_0)$. Then, for every $u_1,\ u_2 \in \mathcal{C}^1(J; \bar{B}_R(u_0))$, if we set

$$\delta_0 := \|u_1(t_0) - u_2(t_0)\|, \quad \delta_j := \max_{t \in J} \|u_j'(t) - f(t, u_j(t))\|, \quad j \in \{1, 2\},$$

the following estimate holds:

$$\|u_1(t) - u_2(t)\| \le e^{L(t-t_0)}\delta_0 + \frac{e^{L(t-t_0)} - 1}{L}(\delta_1 + \delta_2) \tag{5.49}$$

for all $t \in J$. As a consequence,

$$\max_{t \in J} \|u_1(t) - u_2(t)\| \le e^{LT}\delta_0 + \frac{e^{LT} - 1}{L}(\delta_1 + \delta_2). \tag{5.50}$$

Before proving this result, we give some remarks on its statement and analyze some of its consequences. Naturally, the assumptions of Theorem 5.24 are considered to be satisfied throughout this discussion.

Remark 5.25. (a) We point out that, in order to apply Theorem 5.24, it is crucial that both functions u_1 and u_2 are defined in J.

(b) If we assume that u_1 and u_2 are two solutions of problem (5.6), both defined in J, then we have that

$$\delta_0 = \delta_1 = \delta_2 = 0.$$

Therefore, (5.50) entails $u_1 = u_2$ in J, i.e. we obtain again the global uniqueness of the solution of (5.6). In Section 7.7, it will be shown that, actually, the uniqueness of the solution of (5.6) is equivalent to the continuous dependence of the solutions with respect to u_0.

(c) If u_1 and u_2 solve the system

$$u' = f(t, u)$$

in the interval J, then $\delta_1 = \delta_2 = 0$. In such a case, (5.50) reduces to

$$\max_{t \in J} \|u_1(t) - u_2(t)\| \le e^{LT} \|u_1(t_0) - u_2(t_0)\|, \qquad (5.51)$$

which shows the Lipschitz continuity of $u(t; t_0, u_0, f)$ with respect to u_0 in any compact time interval where it is defined. Indeed, if we introduce the solution operator

$$x \mapsto \mathcal{S}_1(x) := u(\cdot; t_0, x, f) \in \mathcal{C}(J; \bar{B}_R(u_0)),$$

(5.51) implies that, for every $x,\, y \in \bar{B}_R(u_0)$ for which $u(t; t_0, x)$ and $u(t; t_0, y)$ are globally defined in J, the following estimate holds:

$$\|\mathcal{S}_1(x) - \mathcal{S}_1(y)\|_{\mathcal{C}(J;\mathbb{R}^N)} \le e^{LT} \|x - y\|, \qquad (5.52)$$

where, for every $u \in \mathcal{C}(J; \mathbb{R}^N)$, we denote

$$\|u\|_{\mathcal{C}(J;\mathbb{R}^N)} := \max_{t \in J} \|u(t)\|.$$

This also shows that a Lipschitz constant for the solution operator $\mathcal{S}_1$ is given by e^{LT}. Actually, we will see in Exercise 17 of Chapter 5 that this is the optimal admissible Lipschitz constant.

(d) For every pair of functions $f, g \in \mathcal{C}(J \times \bar{B}_R(u_0); \mathbb{R}^N)$ satisfying (5.48), denote by $u(t; f)$ and $u(t; g)$ the solutions of

$$\begin{cases} u' = f(t, u), \\ u(t_0) = u_0, \end{cases} \qquad \begin{cases} u' = g(t, u), \\ u(t_0) = u_0, \end{cases}$$

respectively, and suppose that both are defined in J. Then, $\delta_0 = 0$, and Theorem 5.24 provides us with the following estimate:

$$\|u(\cdot; f) - u(\cdot; g)\|_{\mathcal{C}(J;\mathbb{R}^N)} \leq \frac{e^{LT} - 1}{L} \|f - g\|_{\mathcal{C}(J \times \bar{B}_R(u_0);\mathbb{R}^N)}.$$

Consequently, the solution operator

$$h \mapsto \mathcal{S}_2(h) := u(\cdot; t_0, u_0, h) \in \mathcal{C}(J; \bar{B}_R(u_0))$$

is Lipschitz continuous as well. Indeed,

$$\|\mathcal{S}_2(f) - \mathcal{S}_2(g)\|_{\mathcal{C}(J;\mathbb{R}^N)} \leq \frac{e^{LT} - 1}{L} \|f - g\|_{\mathcal{C}(J \times \bar{B}_R(u_0);\mathbb{R}^N)}. \tag{5.53}$$

We will see in Exercise 17 of Chapter 5 that the Lipschitz constant is optimal in this case as well.

Proof of Theorem 5.24. For every $t \in J$, we have that

$$u_1'(t) - u_2'(t) = u_1'(t) - f(t, u_1(t)) + f(t, u_1(t)) - f(t, u_2(t))$$
$$+ f(t, u_2(t)) - u_2'(t).$$

Thus, thanks to (5.48),

$$\|u_1'(t) - u_2'(t)\| \leq \delta_1 + \delta_2 + L\|u_1(t) - u_2(t)\| \quad \text{for all } t \in J. \tag{5.54}$$

On the other hand, for every $t \in J$,

$$u_1(t) - u_2(t) = u_1(t_0) - u_2(t_0) + \int_{t_0}^{t} \left(u_1'(s) - u_2'(s)\right) ds.$$

So, it follows from (5.54) that

$$\|u_1(t) - u_2(t)\| \leq \delta_0 + \int_{t_0}^{t} \|u_1'(s) - u_2'(s)\| \, ds$$

$$\leq \delta_0 + \int_{t_0}^{t} \left(\delta_1 + \delta_2 + L\|u_1(s) - u_2(s)\|\right) ds$$

$$= \delta_0 + (\delta_1 + \delta_2)(t - t_0) + \int_{t_0}^{t} L\|u_1(s) - u_2(s)\| \, ds.$$

In particular, the continuous function

$$x(t) := \|u_1(t) - u_2(t)\|, \quad t \in J,$$

satisfies the integral inequality (5.45), with

$$y(t) := \delta_0 + (\delta_1 + \delta_2)(t - t_0), \quad p(t) := L, \quad t \in J.$$

Consequently, thanks to Lemma 5.23, we find that

$$x(t) \le \delta_0 + (\delta_1 + \delta_2)(t - t_0) + \int_{t_0}^{t} L\left[\delta_0 + (\delta_1 + \delta_2)(s - t_0)\right] e^{L(t-s)}\, ds.$$

$$(5.55)$$

Integrating by parts yields

$$\int_{t_0}^{t} L\left[\delta_0 + (\delta_1 + \delta_2)(s - t_0)\right] e^{L(t-s)}\, ds$$

$$= -\int_{t_0}^{t} \left[\delta_0 + (\delta_1 + \delta_2)(s - t_0)\right] \frac{d}{ds} e^{L(t-s)}\, ds$$

$$= -\int_{t_0}^{t} \frac{d}{ds} \left\{\left[\delta_0 + (\delta_1 + \delta_2)(s - t_0)\right] e^{L(t-s)}\right\}\, ds$$

$$\quad + (\delta_1 + \delta_2) \int_{t_0}^{t} e^{L(t-s)}\, ds$$

$$= -\left[(\delta_0 + (\delta_1 + \delta_2)(s - t_0))e^{L(t-s)}\right]_{s=t_0}^{s=t} - \frac{\delta_1 + \delta_2}{L} \left[e^{L(t-s)}\right]_{s=t_0}^{s=t}$$

$$= -\delta_0 - (\delta_1 + \delta_2)(t - t_0) + \delta_0 e^{L(t-t_0)} + \frac{e^{L(t-t_0)} - 1}{L}(\delta_1 + \delta_2).$$

Therefore, substituting this identity in (5.55) gives

$$x(t) \le e^{L(t-t_0)}\delta_0 + \frac{e^{L(t-t_0)} - 1}{L}(\delta_1 + \delta_2)$$

for all $t \in J$, which is (5.49). Finally, taking maxima in this relation, we obtain (5.50). The proof is complete. $\square$

5.6.3 *Dependence of the existence interval*

As already pointed out, in order to apply Theorem 5.24, $u(t; t_0, x, f)$ and $u(t; t_0, y, g)$ need to be defined in the same interval $[t_0, t_0 + T]$.

The following result gives some sufficient conditions for this property to hold, provided x and y are sufficiently close and $f = g$.

Theorem 5.26. *Suppose $t_0 \in \mathbb{R}$, $T > 0$, $J = [t_0, t_0 + T]$, $f \in C(J \times \mathbb{R}^N; \mathbb{R}^N)$ is globally Lipschitz with respect to u in any closed ball of $\mathbb{R}^N$, uniformly in $t \in J$, and the solution $u(t; x)$ of problem*

$$\begin{cases} u' = f(t, u), \\ u(t_0) = x, \end{cases}$$

is defined in $[t_0, t_0 + \tau]$ for some $0 < \tau \leq T$. Let $R > 0$ be such that the forward orbit

$$\Gamma_x := \{u(t; x) \colon t_0 \leq t \leq t_0 + \tau\}$$

satisfies $\Gamma_x \subset B_R(x)$, and let $L = L(R) > 0$ be a Lipschitz constant for $f(t, u)$ with respect to $u \in \bar{B}_R(x)$, uniformly in $t \in J$. Then, there exists $\varepsilon > 0$ such that, for every $y \in \bar{B}_\varepsilon(x)$, the solution $u(t; y)$ of problem

$$\begin{cases} u' = f(t, u), \\ u(t_0) = y, \end{cases}$$

is also defined in $[t_0, t_0 + \tau]$. Moreover, for every $y \in \bar{B}_\varepsilon(x)$,

$$\max_{t \in [t_0, t_0 + \tau]} \|u(t; x) - u(t; y)\| \leq e^{L\tau} \|x - y\|.$$

Remark 5.27. In order to apply the previous result, one needs to check that $f(t, u)$ is globally Lipschitz with respect to u in any closed ball of $\mathbb{R}^N$, uniformly in $t \in J$. Nevertheless, in most applications, f possesses continuous partial derivatives with respect to all the u variables and

$$\frac{\partial f_i}{\partial u_j} \in C(J \times \mathbb{R}^N; \mathbb{R}), \quad 1 \leq i, j \leq N.$$

In such cases, thanks to Corollary 5.3, the desired assumption is satisfied.

Proof of Theorem 5.26. Since u is of class C^1 and $[t_0, t_0 + \tau]$ is compact, Γ_x is a compact, differentiable arc of a curve. Thus, by the choice of R, $\Gamma_x \cap \partial B_R(x) = \emptyset$. Moreover, setting

$$\eta := \frac{1}{2}\,\mathrm{dist}\,(\Gamma_x, \partial B_R(x)),$$

we have that $\eta > 0$, and the compact η-neighborhood of Γ_x

$$K_\eta := \Gamma_x + \bar{B}_\eta(0) = \left\{ y \in \mathbb{R}^N : \mathrm{dist}(y, \Gamma_x) \leq \eta \right\}$$

also satisfies

$$K_\eta \subset B_R(x).$$

Take $\varepsilon := e^{-L\tau}\eta$ and $y \in \bar{B}_\varepsilon(x)$. We claim that the solution $u(t; y)$ is defined in, at least, in the interval $[t_0, t_0 + \tau]$ and that, actually,

$$u(t; y) \in B_R(x) \quad \text{for all } t \in [t_0, t_0 + \tau]. \tag{5.56}$$

To show this, we can argue by contradiction. If $u(t; y)$ is not defined in $[t_0, t_0 + \tau]$, by the fourth fundamental theorem (see Theorem 5.19), there exists $t_1 \in (t_0, t_0 + \tau)$ such that

$$u(t; y) \in B_R(x) \quad \text{for all } t \in [t_0, t_1) \quad \text{and} \quad u(t_1; y) \in \partial B_R(x). \tag{5.57}$$

The same conclusion holds if $u(t; y)$ is defined in $[t_0, t_0 + \tau]$ but (5.56) does not hold, though, in such a case, t_1 might be $t_0 + \tau$. In both cases, applying Theorem 5.24 in $[t_0, t_1]$ with $u_0 = x$ gives

$$\max_{t \in [t_0, t_1]} \|u(t; x) - u(t; y)\| \leq e^{L\tau}\|x - y\| \leq e^{L\tau}\varepsilon = \eta.$$

Hence, $u(t; y) \in K_\eta \subset B_R(x)$ for all $t \in [t_0, t_1]$, which contradicts (5.57). Therefore, $u(t; y)$ is indeed defined for all $t \in [t_0, t_0 + \tau]$ and satisfies $u(t; y) \in B_R(x)$ for all $t \in [t_0, t_0 + \tau]$. Consequently, thanks again to Theorem 5.24,

$$\max_{t \in [t_0, t_0 + \tau]} \|u(t; x) - u(t; y)\| \leq e^{L\tau}\|x - y\|$$

for all $y \in \bar{B}_\varepsilon(x)$. The proof is complete. $\qquad\square$

Similarly, the following fundamental result establishes some sufficient conditions for the solutions of two Cauchy problems to be defined in the same interval, when they have the same initial datum and the respective nonlinearities are sufficiently close.

Theorem 5.28. *Suppose* $t_0 \in \mathbb{R}$, $T > 0$, $J = [t_0, t_0 + T]$, $u_0 \in \mathbb{R}^N$ *and* $f, g \in \mathcal{C}(J \times \mathbb{R}^N; \mathbb{R}^N)$ *are globally Lipschitz with respect to* u *in any closed ball of* $\mathbb{R}^N$, *uniformly in* $t \in J$. *For* $h \in \{f, g\}$, *let* $u(t; h)$ *denote the unique solution of*

$$\begin{cases} u' = h(t, u), \\ u(t_0) = u_0. \end{cases}$$

Suppose that $u(t; f)$ *is defined in* $[t_0, t_0 + \tau]$ *for some* $0 < \tau \le T$, *let* $R > 0$ *be such that*

$$\Gamma_f := \{u(t; f) \colon t_0 \le t \le t_0 + \tau\} \subset B_R(u_0),$$

and let L *be a Lipschitz constant for* $f(t, u)$ *with respect to* $u \in \bar{B}_R(u_0)$, *uniformly in* $t \in J$. *Then, there exists* $\varepsilon > 0$ *such that, for every* $g \in \bar{B}_\varepsilon(f) \subset \mathcal{C}(J \times \bar{B}_R(u_0); \mathbb{R}^N)$, *i.e. if*

$$\|f - g\|_{\mathcal{C}(J \times \bar{B}_R(u_0))} := \max_{(t,y) \in J \times \bar{B}_R(u_0)} \|f(t, y) - g(t, y)\| \le \varepsilon, \quad (5.58)$$

the solution $u(t; g)$ *is also defined in* $[t_0, t_0 + \tau]$. *Moreover, for every* $g \in \bar{B}_\varepsilon(f)$,

$$\max_{t \in [t_0, t_0 + \tau]} \|u(t; f) - u(t; g)\| \le \frac{e^{L\tau} - 1}{L} \|f - g\|_{\mathcal{C}(J \times \bar{B}_R(u_0); \mathbb{R}^N)}.$$

Proof. The existence of R as in the statement follows as in the proof of Theorem 5.26. Analogously, setting

$$\eta := \frac{1}{2} \operatorname{dist}(\Gamma_f, \partial B_R(u_0)),$$

we have $\eta > 0$ and

$$K_\eta := \Gamma_f + \bar{B}_\eta(0) \subset B_R(u_0).$$

Next, we set

$$\varepsilon := \frac{L}{e^{L\tau} - 1} \eta > 0. \quad (5.59)$$

To show that, for every $g \in \bar{B}_\varepsilon(f)$, $u(t; g)$ is defined in $[t_0, t_0 + \tau]$ and $u(t; g) \in B_R(u_0)$ for all $t \in [t_0, t_0 + \tau]$, we argue as in the proof of Theorem 5.26.

Precisely, if we assume (5.58) and, by contradiction, that $u(t; g) \in B_R(u_0)$ for all $t \in [t_0, t_1)$ and $u(t_1; g) \in \partial B_R(u_0)$ for some $t_1 \in (t_0, t_0 + \tau]$, then applying Theorem 5.24 in the interval $[t_0, t_1]$ yields

$$\max_{t \in [t_0, t_1]} \|u(t; f) - u(t; g)\| \leq \frac{e^{L\tau} - 1}{L} \|f - g\|_{\mathcal{C}([t_0, t_1] \times \bar{B}_R(u_0); \mathbb{R}^N)}$$

$$\leq \frac{e^{L\tau} - 1}{L} \varepsilon = \eta$$

because of the choice (5.59). Hence, $u(t; g) \in K_\eta \subset B_R(u_0)$ for all $t \in [t_0, t_1]$, which gives the desired contradiction. Therefore, we obtain that $u(t; g)$ is defined in $[t_0, t_0 + \tau]$, and the last estimate is a direct consequence of Theorem 5.24. $\qquad\square$

5.7　Exercises

1. Give an alternative proof of Theorem 5.5 without using the Bielecki norm $\| \cdot \|_\eta$.

2. Given an integer $N \in \mathbb{N}$, $N \geq 1$, and N^2 continuous functions, $a_{ij} \in \mathcal{C}(\mathbb{R}; \mathbb{R})$, $i, j \in \{1, \ldots, N\}$, consider the Cauchy problem

$$\begin{cases} u_i' = \sum_{j=1}^N a_{ij}(t) \cos u_j, & i \in \{1, \ldots, N\}, \\ u(t_0) = u_0. \end{cases}$$

   Prove that, for every $t_0 \in \mathbb{R}$ and $u_0 \in \mathbb{R}^N$, it admits a unique solution, $u(t; t_0, u_0)$, globally defined in $t \in \mathbb{R}$.

3. Consider the following Cauchy problem:

$$\begin{cases} u' = a(t)u + b(t)\frac{u^n}{1+u}, \\ u(0) = u_0, \end{cases}$$

   with $a, b \in \mathcal{C}([0, +\infty); \mathbb{R})$, $n \geq 1$ a natural number, and $u_0 \in (-1, 0)$. Analyze the possible behaviors of the maximal solution to the right.

4. Prove that, for every $t_0 \in \mathbb{R}$ and $(u_0, v_0) \in \mathbb{R}^2$, the Cauchy problem

$$\begin{cases} u' = t \cos t \sin v, \\ v' = t \sin t \cos u, \end{cases} \qquad (u(t_0), v(t_0)) = (u_0, v_0),$$

   has a unique solution globally defined in $t \in \mathbb{R}$.

5. Prove that, for every $t_0 > 0$ and $(u_0, v_0) \in \mathbb{R}^2$, the problem

$$\begin{cases} u' = \cos t \ \log t \ \sin^2 v, \\ v' = e^t \sin t \ \cos^2 u, \\ (u(t_0), v(t_0)) = (u_0, v_0), \end{cases}$$

has a unique solution globally defined in $t \in (0, +\infty)$.

6. Show that the problem

$$\begin{cases} u' = u^2 + v, \\ v' = v^2 + u, \\ (u(0), v(0)) = (1, 1), \end{cases}$$

has a unique maximal solution to the right and to the left. Prove that the former blows up after a finite time. Determine the maximal existence time of the solution. What is the global backward behavior of the solution?

[Hint: Observe that u and v play the same role in the system.]

7. Given $\lambda > 0$, $\mu > 0$, $b > 0$, and $c > 0$, consider the competing species model

$$\begin{cases} u' = \lambda u - u^2 - buv, \\ v' = \mu v - v^2 - cuv, \\ (u(0), v(0)) = (u_0, v_0), \end{cases} \tag{5.60}$$

where $u_0 \geq 0$ and $v_0 \geq 0$.

(a) Prove that (5.60) has a unique maximal solution to the right, (I, u, v).
(b) Determine the maximal solution when $u_0 = 0$ or $v_0 = 0$.
(c) Prove that $u(t) > 0$ and $v(t) > 0$ for all $t \in I$ if $u_0 > 0$ and $v_0 > 0$.
(d) Prove that $I = [0, \infty)$ if $u_0 > 0$ and $v_0 > 0$.

8. Repeat the previous exercise for the predator-prey model, which is given by

$$\begin{cases} u' = \lambda u - u^2 - buv, \\ v' = \mu v - v^2 + cuv, \\ (u(0), v(0)) = (u_0, v_0), \end{cases}$$

where $\lambda > 0$, $\mu > 0$, $b > 0$, $c > 0$, $u_0 \geq 0$, and $v_0 \geq 0$.

9. Given $\lambda > 0$ and $b > 0$, consider the symbiotic model

$$\begin{cases} u' = \lambda u - u^2 + buv, \\ v' = \lambda v - v^2 + buv, \\ (u(0), v(0)) = (1, 1). \end{cases}$$

Show that it admits a unique maximal solution to the right. Determine this solution and its maximal existence time in terms of the parameter $b > 0$, which measures the synergy between the individuals of the species u and v.

10. Given the planar system

$$\begin{cases} u' = u + v - u(u^2 + v^2), \\ v' = -u + v - v(u^2 + v^2), \end{cases}$$

obtain an equivalent one using polar coordinates,

$$u = \rho \cos \theta, \quad v = \rho \sin \theta.$$

Solve the resulting system and represent the solutions of the original one in the (u, v)-plane.

11. Given $p > 0$ and $q > 0$ with $pq < 1$, consider the problem

$$\begin{cases} u' = |v|^p, \\ v' = |u|^q, \\ (u(0), v(0)) = (1, 1). \end{cases}$$

Show the existence of a unique maximal solution to the right, (I, u, v), and analyze its global character.

12. Given $h, j, k, \ell \in \mathbb{N} \setminus \{0\}$, prove that, for every $t_0 \in \mathbb{R}$ and $(u_0, v_0) \in \mathbb{R}^2$, the problem

$$\begin{cases} u' = -u^{2h+1} v^{2\ell}, \\ v' = -u^{2j} v^{2k+1}, \\ (u(t_0), v(t_0)) = (u_0, v_0), \end{cases}$$

has a unique solution, which is globally defined in $[t_0, \infty)$.

13. A function $f \in \mathcal{C}^1(\mathbb{R} \times \mathbb{R}^N; \mathbb{R}^N)$ is said to be of the *Kolmogorov type* if, by denoting

$$u = (u_1, \ldots, u_N)^T \quad \text{and} \quad f = (f_1, \ldots, f_N)^T,$$

there exist functions $g_i \in \mathcal{C}(\mathbb{R} \times \mathbb{R}^N; \mathbb{R})$, $1 \le i \le N$, such that

$$f_i(t, u) = g_i(t, u) u_i \quad \text{for all } i \in \{1, \ldots, N\},\ t \in \mathbb{R},\ u \in \mathbb{R}^N.$$

Assuming that f is of the Kolmogorov type, consider, for every $u_0 \in \mathbb{R}^N$, the initial value problem

$$\begin{cases} u' = f(t, u), \\ u(0) = u_0. \end{cases} \tag{5.61}$$

(a) Prove the existence of a unique maximal solution of (5.61) and discuss all its admissible behaviors.

(b) Let $(I, u(\cdot; u_0))$ be the maximal solution of (5.61). Prove that

$$\operatorname{sign} u_i(t; u_0) = \operatorname{sign} u_{0i}$$

for all $t \in I$ and $i \in \{1, \ldots, N\}$, where $u_0 = (u_{01}, \ldots, u_{0N})^T$.

(c) Suppose that, in addition,

$$g_i(t, u) = \lambda_i - \sum_{j=1}^{N} a_{ij} u_j, \quad 1 \le i \le N,$$

where $\lambda_i \in \mathbb{R}$ and $a_{ij} \ge 0$ are constant for each $i, j \in \{1, \ldots, N\}$. Show that, in such a case, $u(t; u_0)$ is globally defined on $[0, \infty)$ provided $u_{0i} \ge 0$ for all $i \in \{1, \ldots, N\}$. Assuming, moreover, that $\lambda_i < 0$ for every $i \in \{1, \ldots, N\}$, show that, for any of those i's,

$$\lim_{t \uparrow \infty} u_i(t) = 0.$$

(d) Give examples, with $N = 2$ and $a_{12} = a_{21} < 0$, for which the solution blows up in a finite time.

14. Given an integer $N \geq 1$ and a non-zero square matrix

$$A = (a_{ij})_{1 \leq i,j \leq N} \in \mathcal{M}_N(\mathcal{C}(\mathbb{R}; \mathbb{R})),$$

consider the function

$$f(t, u) = (f_1(t, u), \ldots, f_N(t, u))^T, \quad u = (u_1, \ldots, u_N)^T,$$

whose components are defined as

$$f_i(t, u) = \sum_{j=1}^{N} a_{ij}(t) |u_j|^{p_{ij}}, \quad (t, u) \in \mathbb{R} \times \mathbb{R}^N, \quad 1 \leq i \leq N,$$

where, for every $i, j \in \{1, \ldots, N\}$, $p_{ij} \geq 1$. Prove the existence and uniqueness of a maximal solution for the problem

$$\begin{cases} u' = f(t, u), \\ u(0) = u_0 \in \mathbb{R}^N. \end{cases}$$

Discuss all its admissible behaviors, and give examples for all of them in the special case of $N = 2$ by choosing appropriate $a_{ij}(t)$ and p_{ij}.

15. Prove that, for every $\varepsilon \geq 0$, the problem

$$\begin{cases} u'(t) = t \left[\sin u(t) + \varepsilon \cos u(t)\right], \\ u(0) = 1, \end{cases}$$

has a unique solution globally defined in $[0, \infty)$, denoted by u_ε, and estimate $|u_\varepsilon(t) - u_0(t)|$ for every $t > 0$ and $\varepsilon > 0$.

16. Prove that, for every $\varepsilon \geq 0$, the problem

$$\begin{cases} u'(t) = t \sin u(t), \\ u(0) = \varepsilon, \end{cases}$$

has a unique solution globally defined in $[0, \infty)$, denoted by u_ε, and estimate $|u_\varepsilon(t) - u_0(t)|$ for every $t > 0$ and $\varepsilon > 0$.

17. Construct some linear Cauchy problems where the Lipschitz constants for the solution operators $\mathcal{S}_1$ and $\mathcal{S}_2$ in (5.52) and (5.53) are achieved. These examples show that such Lipschitz constants cannot be sharpened.

18. Check the validity of the conclusions of Theorem 5.26 in the case of the Cauchy problem in Exercise 6 of Chapter 4, where the domain of the maximal solution does not depend continuously on the initial datum $x > 0$.

19. State and prove analogous results to Theorems 5.26 and 5.28, allowing small perturbations both in the initial data and the nonlinearities.

20. Consider the planar system

$$\begin{cases} x' = x + y, \\ y' = -x + y. \end{cases} \tag{5.62}$$

(a) Change to polar coordinates, $x = \rho \cos\theta$, $y = \rho \sin\theta$, solve the resulting system, and represent the planar trajectories of its solutions.

(b) Find a conformal planar map $T : \mathbb{R}^2 \to \mathbb{R}^2$, $(u, v) = T(x, y)$, such that

$$T(\mathbb{R}^2) = \big\{ (u, v) \in \mathbb{R}^2 \colon \, -1 < v < 1 \big\}.$$

(c) Construct a planar system of the form

$$\begin{cases} u' = f(u, v), \\ v' = g(u, v), \end{cases}$$

whose trajectories look like those represented in Figure 5.4.

5.8 Final Comments

As already mentioned in Section 1.15, Theorem 5.5 goes back to A. L. Cauchy in the special case when $f(t, u)$ is analytic. Such a result appeared as early as 1840 in a series of lecture notes taken and edited by one of his former students, Moigno (1840). Cauchy considered the simplest scalar prototype of (5.2) with $f(t, u)$ analytic, and expressed the solution as a formal power series. Then, he analyzed the very natural question of whether such a series was convergent. This problem led Cauchy to obtain some of the most pioneering results in the theory of analytic functions.

Thirty-six years later, Lipschitz (1876) was able to extend some of Cauchy's previous findings to obtain Theorem 5.5. As he was the first scientist to realize the importance of the uniform continuity condition (5.1) in the theory of differential equations, the functions satisfying (5.1) were named after him as Lipschitz functions. Apart from the area of differential equations, they play a central role in the theory of functions.

Some years after the seminal work of Lipschitz was published, Picard and Lindelöf, in 1890 and 1894, respectively, gave alternative proofs for Theorem 5.5 by establishing the convergence of the iterative $\mathcal{K}$-scheme,

$$h_n = \mathcal{K}h_{n-1}, \quad n \geq 1,$$

where $\mathcal{K}$ denotes the integral operator associated with (5.6), to the solution $u(t; t_0, u_0)$ of such a problem. In the light of these circumstances, Theorem 5.5 should be attributed to Cauchy and Lipschitz; nevertheless, in most of the available texts on Ordinary Differential Equations, it is attributed to Picard and Lindelöf. This might be explained by the fact that, at the beginning of the 20th century, experts in differential equations were fascinated by the versatility and extremely elegant formulation of the existence problem as a fixed point equation for the associated integral operator.

In around 1890, very intensive studies on linear integral equations were conducted by some of the leading mathematicians of the time, including D. Hilbert, E. I. Fredholm, H. Poincaré, and V. Volterra, in connection with some classical problems in potential theory. The works by Picard (1890) and Lindelöf (1894) contained an extremely fruitful idea, which would prove to be a milestone for the generation of Nonlinear Analysis in a huge variety of contexts.

Picard had a huge mathematical influence. A proof of his recognition is that he was one of five European scientists invited by Clark University in the United States of America to celebrate its Decennial Celebration in 1899. Apart from Picard, the Spanish histologist S. Ramón y Cajal, who later received the 1906 Nobel Prize in Physiology or Medicine, also participated and, in fact, traveled on the same ship as Picard.

Although Hartman emphasized some of these circumstances in his final comments on p. 23 of his advanced textbook Hartman (2002), he still titled Section 1 of Chapter II as "The Picard–Lindelöf Theorem". In more recent books, such as the one by Amann (1990), Theorem 5.5

is simply proposed as an exercise on p. 105, again with the name of "Picard–Lindelöf Theorem". In addition, also quite astonishingly, Hartman (2002) observed that the method of successive approximations had already been used in some special cases by J. Liouville and A. L. Cauchy.

Nowadays, more than 180 years after the publication of Moigno's lecture notes, we are able to gather some of the most important existence and uniqueness results in the context of ordinary differential equations and deliver this material at an undergraduate level by using some of the most elegant and fruitful mathematical devices developed in the 19th and 20th centuries, such as the contractive mapping theorem. Actually, as we have seen, most of the existence results established in the first five chapters of this book are direct consequences of Theorem 1.8. This shows the really high level of abstraction reached by Banach in his doctoral dissertation.

In Section 5.1, we have shown how $\mathcal{C}^1$-functions with respect to the u variables are locally Lipschitz with respect to u. Conversely (see, e.g. López-Gómez, 2013, Theorem 4.3), any locally Lipschitz continuous function is differentiable almost everywhere. "Almost everywhere" means that there exists a subset of $Z \subset \Omega$, with zero Lebesgue measure, such that the function is differentiable in $\Omega \setminus Z$. Therefore, locally Lipschitz continuity is closely related to differentiability.

Actually, a sharper result establishes that a function $F : \Omega \to \mathbb{R}^N$ is globally Lipschitz in Ω if and only if $F \in W^{1,\infty}(\Omega; \mathbb{R}^N)$, where $W^{1,\infty}(\Omega; \mathbb{R}^N)$ indicates the Sobolev space of measurable bounded functions, $L^\infty(\Omega; \mathbb{R}^N)$, with distributional, or weak, derivatives in $L^\infty(\Omega; \mathbb{R}^N)$. The Lebesgue space $L^\infty(\Omega; \mathbb{R}^N)$ was introduced by H. Lebesgue in 1910. Similarly, $F \in W^{1,\infty}_{\mathrm{loc}}(\Omega; \mathbb{R}^N)$ if and only if F is locally Lipschitz continuous in Ω. The interested reader is referred to López-Gómez (2013, Chapter 4) for any further details. The number of applications of these spaces in the context of Partial Differential Equations is huge.

In the proof of Theorem 5.15, in order to show that the forward orbit Γ is unbounded if $I = [t_0, T_{\max})$ for some $T_{\max} \leq t_0 + T$, we have actually established that, if $u(t)$ is a solution of $u' = f(t, u)$ in $(T_{\max} - \varepsilon, T_{\max})$ such that

$$\lim_{n \to +\infty} u(t_n) = u^*$$

along some sequence of times

$$T_{\max} - \varepsilon < t_n < T_{\max}, \ \ n \geq 1, \ \ \ \text{with} \ \ \ \lim_{n \to +\infty} t_n = T_{\max},$$

then

$$\lim_{t \uparrow T_{\max}} u(t) = u^*.$$

This result was proved by Wintner in 1946, and it is sometimes referred to as the "Wintner lemma".

A differential counterpart of Lemma 5.23 was given by Grönwall in 1919, while its integral form presented here was proved by Bellman in 1943.

When f is differentiable in the u variables and, for some open subset $\Omega \subset \mathbb{R}^N$,

$$\frac{\partial f_i}{\partial u_j} \in \mathcal{C}(J \times \Omega; \mathbb{R}^N), \quad 1 \leq i, j \leq N,$$

the celebrated *differentiation theorem* by Peano (1897) establishes that the solution operator, $\mathcal{S}$, defined in (5.44) is actually differentiable with respect to u_0 and f. Nevertheless, this result remains outside the general scope of this introductory textbook.

Although the material in Section 5.6 has been originally inspired by the contents of the works by de Guzmán (1975) and Coppel (1977), Theorems 5.26 and 5.28 might be new as stated.

Chapter 6

High-Order Nonlinear Equations

In this chapter, given $t_0 \in \mathbb{R}$, $T > 0$, $J \in \{[t_0, t_0 + T], [t_0, +\infty)\}$, a natural number $N \geq 1$, an open subset $\Omega \subset \mathbb{R}^N$, and a scalar function $g \in \mathcal{C}(J \times \Omega; \mathbb{R})$, we study the Nth-order nonlinear differential equation

$$u^{N)} = g(t, u, u', u'', \dots, u^{N-1)}). \tag{6.1}$$

A function $u \in \mathcal{C}^N(I; \mathbb{R})$ is said to be a solution of (6.1) in $I \subset J$ if, for every $t \in I$,

$$(u(t), u'(t), u''(t), \dots, u^{N-1)}(t)) \in \Omega,$$

and

$$u^{N)}(t) = g(t, u(t), u'(t), u''(t), \dots, u^{N-1)}(t)).$$

Besides (6.1), we also consider the associated Cauchy problem

$$\begin{cases} u^{N)} = g(t, u, u', u'', \dots, u^{N-1)}), \\ (u(t_0), u'(t_0), u''(t_0), \dots, u^{N-1)}(t_0)) = U_0, \end{cases} \tag{6.2}$$

for every $U_0 \in \Omega$. Throughout this chapter, Σ_e indicates the set of solutions, (I, u), of (6.1), with $I = [t_0, t_0 + h]$, or $I = [t_0, t_0 + h)$, for some $h > 0$ such that $I \subset J$.

The chapter is organized as follows. Section 6.1 adapts the strategy used in Section 1.7 to derive the first-order system associated with the Nth-order equation (6.1). Then, it establishes a bijection

337

between the set of solutions Σ_e and the set of solutions Σ_s of the associated first-order system. Section 6.2 ascertains the relationships between the Lipschitz continuity properties of g and those of the nonlinearity of the associated first-order system. Section 6.3 applies the abstract theory developed in Chapter 5 to the associated first-order system in order to obtain an abstract theory for the Cauchy problem (6.2). Section 6.4 applies the abstract theory to the simple gravity pendulum, a model which is fundamental in Classical Mechanics. Section 6.5, instead, applies the theory developed in this chapter to the diffusive logistic equation, which is the basis on which the mathematical theory of populations and Environmental Sciences has been developed. Moreover, Section 6.5 analyzes three examples related to the so-called *large*, or *explosive*, solutions. Then, similarly to Section 1.13, Section 6.6 applies the general existence theory developed in Chapter 5 to a general class of systems of nonlinear differential equations with arbitrary order. Finally, in Section 6.7, we derive the diffusive logistic equation studied in Section 6.5 from the heat equation, a partial differential equation that is pivotal in many fields of Science and Technology.

6.1   The Associated First-Order System

Arguing as in Section 1.7, it is easily seen that if $u \in \mathcal{C}^N(I; \mathbb{R})$ is a solution of (6.1) in some interval $I \subset J$, then the vectorial function formed by u together with its first $N-1$ derivatives,

$$
\begin{pmatrix} u_1 \\ u_2 \\ \vdots \\ u_{N-1} \\ u_N \end{pmatrix} := \mathcal{D}(u) = \begin{pmatrix} u \\ u' \\ \vdots \\ u^{N-2)} \\ u^{N-1)} \end{pmatrix},
\tag{6.3}
$$

is a solution of the first-order system

$$
U' = f(t, U)
\tag{6.4}
$$

in the interval I, where we denote

$$U := \begin{pmatrix} u_1 \\ u_2 \\ \vdots \\ u_{N-1} \\ u_N \end{pmatrix} \quad \text{and} \quad f(t, U) := \begin{pmatrix} u_2 \\ u_3 \\ \vdots \\ u_N \\ g(t, u_1, u_2, \cdots, u_N) \end{pmatrix}. \quad (6.5)$$

Moreover, if we denote by Σ_s the set of solutions of the first-order system (6.4) in any subinterval $I \subset J$, then the following nonlinear counterpart of Theorem 1.17 holds.

Theorem 6.1. *The map* $\mathcal{D} : \Sigma_e \to \Sigma_s$ *defined by* (6.3) *establishes a bijection between* Σ_e *and* Σ_s.

Proof. By construction, $\mathcal{D}(u) \in \Sigma_s$ if $u \in \Sigma_e$. Moreover, it follows from the definition of $\mathcal{D}$ that $u = v$ if $\mathcal{D}(u) = \mathcal{D}(v)$. Thus, $\mathcal{D}$ is injective. It remains to prove that $\mathcal{D}$ is surjective. Indeed, if $U \in \Sigma_s$ is given by

$$U = (u_1, \ldots, u_N)^T,$$

then, from the first $N - 1$ differential equations of (6.4), we obtain that

$$u_1' = u_2, \quad u_2' = u_3, \quad u_3' = u_4, \quad \ldots, \quad u_{N-1}' = u_N,$$

and, hence,

$$u_2 = u_1', \quad u_3 = u_1'', \quad u_4 = u_1''', \quad \ldots, \quad u_N = u_1^{N-1)}.$$

Thus, $U = \mathcal{D}(u_1)$. Furthermore, the last differential equation in (6.4) gives

$$u_1^{N)} = u_N' = g(t, u_1, u_2, \ldots, u_N) = g\big(t, u_1, u_1', \ldots, u_1^{N-1)}\big)$$

in the interval I. Therefore, u_1 solves (6.1), and the proof is complete. $\square$

Note that the initial condition

$$U(t_0) = (U_{0,1}, U_{0,2}, \ldots, U_{0,N-1}, U_{0,N})^T$$

for the first-order system (6.4) becomes

$$\left(u(t_0), u'(t_0), u''(t_0), \ldots, u^{N-1)}(t_0)\right) = \left(U_{0,1}, U_{0,2}, \ldots, U_{0,N-1}, U_{0,N}\right)$$

for the differential equation (6.1).

6.2 Lipschitz Continuity

Naturally, all the continuity and differentiability properties of the scalar function g are inherited by the associated f defined in (6.5). The following concept adapts to this context the one introduced in Definition 5.1.

Definition 6.2. Suppose that J is a subinterval of $\mathbb{R}$, Ω is an open subset of $\mathbb{R}^N$, and $g \in \mathcal{C}(J \times \Omega; \mathbb{R})$. Then:

(a) g is said to be globally Lipschitz with respect to $U \in \Omega$ uniformly in $t \in J$ if there exists a constant $L > 0$, called a Lipschitz constant for g, such that, for all $(t, x, y) \in J \times \Omega \times \Omega$,

$$|g(t, x) - g(t, y)| \leq L\|x - y\|; \tag{6.6}$$

(b) g is said to be locally Lipschitz with respect to $U \in \Omega$ uniformly in $t \in J$ if, for every $x_0 \in \Omega$, there exists $R = R(x_0) > 0$ such that $B_R(x_0) \subset \Omega$ and $g|_{J \times B_R(x_0)}$ is globally Lipschitz with respect to $U \in B_R(x_0)$ uniformly in $t \in J$, i.e. if there exists $L = L(x_0) > 0$ such that, for every $(t, x, y) \in J \times B_R(x_0) \times B_R(x_0)$,

$$|g(t, x) - g(t, y)| \leq L\|x - y\|.$$

Of course, these definitions can be extended to arbitrary sets $S \subset \mathbb{R}^N$, not necessarily open. The following result holds.

Lemma 6.3. *Suppose that J is a subinterval of $\mathbb{R}$, Ω is an open subset of $\mathbb{R}^N$, and $g \in \mathcal{C}(J \times \Omega; \mathbb{R})$. Let $f \in \mathcal{C}(J \times \Omega; \mathbb{R}^N)$ be the function defined in (6.5). Then:*

(a) *g is globally Lipschitz with respect to $U \in \Omega$ uniformly in $t \in J$ (in the sense of Definition 6.2) if and only if f is globally Lipschitz with respect to $U \in \Omega$ uniformly in $t \in J$ (in the sense of Definition 5.1);*

(b) *g is locally Lipschitz with respect to $U \in \Omega$ uniformly in $t \in J$ (in the sense of Definition 6.2) if and only if f is locally Lipschitz with respect to $U \in \Omega$ uniformly in $t \in J$ (in the sense of Definition 5.1);*

(c) *g is of class C^1 with respect to $U \in \Omega$ if and only if f is of class C^1 with respect to $U \in \Omega$.*

Proof. Suppose that g is globally Lipschitz with respect to $U \in \Omega$, uniformly in $t \in J$. Then, there exists a constant $L > 0$ for which (6.6) holds. Thus, for every $t \in J$ and $x, y \in \Omega$,

$$\|f(t, x) - f(t, y)\| = \sum_{j=2}^{N} |x_j - y_j| + |g(t, x) - g(t, y)|$$

$$\leq \sum_{j=2}^{N} |x_j - y_j| + L\|x - y\| \leq (1 + L)\|x - y\|.$$

Conversely, suppose that f is globally Lipschitz with respect to $U \in \Omega$ uniformly in $t \in J$. Then, there exists $M > 0$ such that

$$\|f(t, x) - f(t, y)\| \leq M\|x - y\| \quad \text{for all } t \in J \text{ and } x, y \in \Omega.$$

Thus,

$$|g(t, x) - g(t, y)| = \|f(t, x) - f(t, y)\| - \sum_{j=2}^{N} |x_j - y_j| \leq M\|x - y\|$$

for all $t \in J$ and $x, y \in \Omega$. So, g is globally Lipschitz with respect to $U \in \Omega$ uniformly in $t \in J$. This completes the proof of Part (a).

Naturally, Part (b) is a direct consequence of Part (a). Finally, Part (c) is a consequence of the particular structure of f, defined in (6.5), and the proof is straightforward. $\qquad\square$

By combining Lemma 6.3 with Proposition 5.2, we obtain the following result.

Proposition 6.4. *Suppose that J is a subinterval of $\mathbb{R}$, Ω is a convex open subset of $\mathbb{R}^N$, and $g \in C(J \times \Omega; \mathbb{R})$ admits partial derivatives* *with respect to all the U variables, which satisfy*

$$\frac{\partial g}{\partial u_j} \in \mathcal{C}(J \times \Omega; \mathbb{R}) \quad \text{for all } j \in \{1, \dots, N\}.$$

Then, g is globally Lipschitz with respect to $U \in \Omega$ uniformly in $t \in J$ if and only if

$$M_j := \sup_{(t,U) \in J \times \Omega} \left| \frac{\partial g}{\partial u_j}(t, U) \right| < +\infty \quad \text{for all } j \in \{1, \dots, N\}. \quad (6.7)$$

Proof. By the structure of f, defined in (6.5), it is apparent that (6.7) is equivalent to

$$\sup_{(t,U) \in J \times \Omega} \left| \frac{\partial f_i}{\partial u_j}(t, U) \right| < +\infty \quad \text{for all } i, j \in \{1, \dots, N\}.$$

Therefore, the proof follows from Proposition 5.2 and Lemma 6.3. $\qquad\square$

Naturally, the following counterpart of Corollary 5.3 holds.

Corollary 6.5. *Suppose Ω is an open subset of $\mathbb{R}^N$, $J = [\alpha, \beta]$ for some $\alpha < \beta$, and $g \in \mathcal{C}(J \times \Omega; \mathbb{R})$ satisfies*

$$\frac{\partial g}{\partial u_j} \in \mathcal{C}(J \times \Omega; \mathbb{R}) \quad \text{for all } j \in \{1, \dots, N\}.$$

Then, g is locally Lipschitz with respect to $U \in \Omega$ uniformly in $t \in J$. Moreover, g is globally Lipschitz uniformly in $t \in J$ with respect to U in any convex open bounded subset $D \subset \Omega$, with $\bar{D} \subset \Omega$.

6.3   Cauchy–Lipschitz Theory

According to Lemma 6.3 and Theorems 6.1 and 5.5, the following global result holds.

Theorem 6.6. *Suppose $J = [\alpha, \beta]$ for some $\alpha < \beta$ and $g \in \mathcal{C}(J \times \mathbb{R}^N; \mathbb{R})$ is globally Lipschitz with respect to $U \in \mathbb{R}^N$ uniformly in $t \in J$. Then, for every $t_0 \in J$ and $U_0 \in \mathbb{R}^N$, the Cauchy problem*

$$\begin{cases} u^{N)} = g\big(t, u, u', \dots, u^{N-1)}\big), \\ \big(u(t_0), u'(t_0), u''(t_0), \dots, u^{N-1)}(t_0)\big) = U_0, \end{cases} \quad (6.8)$$

possesses a unique solution, $u(t; t_0, U_0)$, defined in $J = [\alpha, \beta]$.

In addition, by combining Lemma 6.3 with Theorems 6.1, 6.6, and 5.14, the existence of a unique maximal solution also holds true in the context of high-order differential equations with locally Lipschitz nonlinearities.

Theorem 6.7. *Suppose $t_0 \in \mathbb{R}$, $T > 0$, $J \in \{[t_0, t_0+T], [t_0, +\infty)\}$, Ω is an open subset of $\mathbb{R}^N$ and $g \in \mathcal{C}(J \times \Omega; \mathbb{R})$ is locally Lipschitz with respect to $U \in \Omega$ uniformly in t on compact subintervals of J. Then, for every $U_0 \in \Omega$, problem (6.8) possesses a unique maximal solution, (I, u), i.e. a solution that does not admit any proper extension to the right.*

Similarly, thanks to Lemma 6.3 and Theorems 6.1, 6.6 and 6.7, the following series of fundamental results for the Cauchy problem (6.8) hold true. They are the counterparts of Theorems 5.15, 5.17, 5.18, and 5.19, respectively.

Theorem 6.8 (First fundamental theorem). *Suppose $t_0 \in \mathbb{R}$, $T > 0$, $J = [t_0, t_0 + T]$, $U_0 \in \mathbb{R}^N$ and $g \in \mathcal{C}(J \times \mathbb{R}^N; \mathbb{R})$ is locally Lipschitz with respect to $U \in \mathbb{R}^N$ uniformly in $t \in J$. Let (I, u) be the (unique) maximal solution of (6.8). Then, either $I = J$ or $I = [t_0, T_{\max})$ for some $t_0 < T_{\max} \leq t_0 + T$ and, in such a case,*

$$\limsup_{t \uparrow T_{\max}} \left(|u(t)| + |u'(t)| + \cdots + |u^{N-1)}(t)| \right) = +\infty.$$

If the former option in Theorem 6.8 occurs, i.e. if $I = J$, (I, u) is said to be **globally defined in time**. If the latter occurs instead, (I, u) is said to **blow up in a finite time**, $T_{\max}$. Observe that, in the case of blow up, what Theorem 6.8 guarantees is that at least one component of $\mathcal{D}(u)$ blows up at the maximal existence time, $T_{\max}$.

Theorem 6.9 (Second fundamental theorem). *Suppose $t_0 \in \mathbb{R}$, $J = [t_0, +\infty)$, $U_0 \in \mathbb{R}^N$ and $g \in \mathcal{C}(J \times \mathbb{R}^N; \mathbb{R})$ is locally Lipschitz with respect to $U \in \mathbb{R}^N$ uniformly in t on compact subintervals of J. Let (I, u) be the (unique) maximal solution of (6.8). Then, there exists $t_0 < T_{\max} \leq +\infty$ such that $I = [t_0, T_{\max})$. Moreover,*

$$\limsup_{t \uparrow T_{\max}} \left(|u(t)| + |u'(t)| + \cdots + |u^{N-1)}(t)| \right) = +\infty \quad \text{if } T_{\max} < +\infty.$$

In other words, either u is globally defined in time or $\mathcal{D}(u)$ blows up in a finite time, $T_{\max}$.

Theorem 6.10 (Third fundamental theorem). *Suppose $t_0 \in \mathbb{R}$, $J = [t_0, +\infty)$, Ω is a proper open subset of $\mathbb{R}^N$, and $g \in \mathcal{C}(J \times \Omega; \mathbb{R})$ is locally Lipschitz in $U \in \Omega$ uniformly in t on compact subintervals of J. For any given $U_0 \in \Omega$, let (I, u) be the (unique) maximal solution of (6.8). Then, $I = [t_0, T_{\max})$ for some $t_0 < T_{\max} \leq +\infty$. Moreover, if $T_{\max} < +\infty$, one of the following non-excluding options occurs: either*

$$\limsup_{t \uparrow T_{\max}} \left(|u(t)| + |u'(t)| + \cdots + |u^{N-1)}(t)| \right) = +\infty,$$

*i.e. **the solution blows up at** $T_{\max}$, or there exists a sequence of times $\{t_n\}_{n \geq 1} \subset I$, with $\lim_{n \to \infty} t_n = T_{\max}$, and a point $U_\omega \in \partial\Omega$ such that*

$$\lim_{n \to \infty} \mathcal{D}(u)(t_n) = U_\omega \in \partial\Omega,$$

*i.e. $\mathcal{D}(u)$ **approximates** $\partial\Omega$ at $T_{\max}$.*

Theorem 6.11 (Fourth fundamental theorem). *Suppose $t_0 \in \mathbb{R}$, $T > 0$, $J = [t_0, t_0 + T]$, Ω is an open bounded subset of $\mathbb{R}^N$ with $U_0 \in \Omega$, and $g \in \mathcal{C}(J \times \bar{\Omega}; \mathbb{R})$ is globally Lipschitz with respect to $U \in \bar{\Omega}$ uniformly in $t \in J$. Then, problem (6.8) admits a unique maximal solution, (I, u). Moreover, either $I = J$ or $I = [t_0, T_{\max}]$ for some $T_{\max} < t_0 + T$ and, in such a case,*

$$\mathcal{D}(u)(T_{\max}) \in \partial\Omega.$$

Of course, as in Chapter 5, one can obtain the counterparts of Theorems 6.8–6.11 for the behavior of the maximal solution of (6.8) backward in time.

The following two sections analyze some pivotal second-order non-linear differential equations, which have a huge interest from the point of view of applications. They provide us with some paradigmatic examples where we can apply the general theory presented in this chapter.

6.4 The Simple Gravity Pendulum

The simple gravity pendulum is a simplified model for the real pendulum, which is sketched in Figure 6.1. It consists of a bob of mass M,

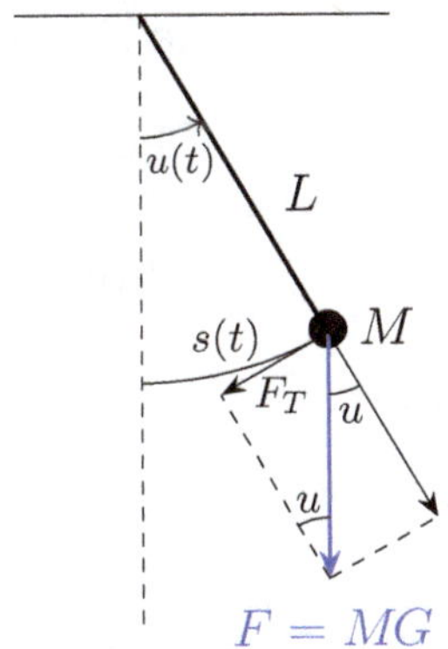

Fig. 6.1. The simple gravity pendulum.

which is assumed to be dimensionless, hanging from an endpoint of
a massless, non-extensible, and always taut rod of length L. The rod
is suspended at a fixed point, and the bob describes an arc of planar
curve as it moves. In the case of the simple pendulum, no energy loss
due to friction or air resistance is taken into account: the motion of
the bob is only due to a uniform gravitational field. In Figure 6.1,
the vertical axis is the equilibrium position, and $u(t)$ indicates the
angle formed by the rod and the vertical axis through the pivot point.
Thus,

$$s(t) = Lu(t)$$

is the (angular) displacement at time t of the bob, measured from its
equilibrium position.

As the only force causing the movement of the mass is the tangential component of a downward gravitational force equal to MG,
where G is the acceleration constant due to gravity, by using Newton's second law, the equation of motion is

$$M\,Lu''(t) = Ms''(t) = -F_T = -M\,G\sin u(t),$$

where F_T denotes the tangential component of $F = MG$ and the
minus sign indicates that such a force is opposite to the direction of
the motion. Therefore,

$$u''(t) = -\frac{G}{L}\sin u(t)$$

provides us with the equation of the simple pendulum. The corresponding Cauchy problem is

$$\begin{cases} u'' = -\frac{G}{L}\sin u, \\ u(0) = u_0, \quad u'(0) = v_0, \end{cases} \tag{6.9}$$

and the associated first-order problem reads

$$\begin{cases} u' = v, \\ v' = -\frac{G}{L}\sin u, \\ u(0) = u_0, \quad v(0) = v_0, \end{cases} \tag{6.10}$$

where $u_0, v_0 \in \mathbb{R}$. The nonlinearity in (6.9),

$$g(t, u, v) := -\frac{G}{L}\sin u, \quad (t, u, v) \in \mathbb{R} \times \mathbb{R} \times \mathbb{R},$$

is autonomous, as it is independent of t, and it is of class $\mathcal{C}^\infty$ in $(u, v) \in \mathbb{R}^2$ with gradient

$$\nabla g(t, u, v) = \left(\frac{\partial g}{\partial u}, \frac{\partial g}{\partial v} \right) = \left(-\frac{G}{L}\cos u, 0 \right).$$

Since all the components of ∇g are globally bounded in $(u, v) \in \mathbb{R}^2$, uniformly in $t \in \mathbb{R}$, it follows from Proposition 6.4 that g is globally Lipschitz with respect to $(u, v) \in \mathbb{R}^2$ uniformly in $t \in \mathbb{R}$. Thus, by Theorem 6.6, for every $T > 0$, the Cauchy problem (6.9), or, equivalently, (6.10), possesses a unique solution, u_T, globally defined in $I_T := [t_0 - T, t_0 + T]$. By uniqueness, $u_T|_{I_S} = u_S$ for all $T > S > 0$. Therefore,

$$u := \lim_{T \uparrow +\infty} u_T$$

provides us with the unique solution of (6.9) in $t \in \mathbb{R}$. This reflects the fact that, from the point of view of Newtonian mechanics, this system is deterministic, in the sense that, once the initial (angular) position and velocity, represented by u_0 and v_0, respectively, are fixed, the motion is completely determined for all times. The precise nature of the solutions of the simple gravity pendulum will be analyzed, in full detail, in Section 10.3.

6.5   The Diffusive Logistic Equation

In this section, for any given $(u_0, v_0) \in \mathbb{R}^2$, we apply the abstract theory developed in Section 6.3 to the Cauchy problem

$$\begin{cases} -u'' = \lambda u - mu^2, \\ u(0) = u_0, \quad u'(0) = v_0, \end{cases} \tag{6.11}$$

where $\lambda, m \in \mathbb{R}$, with $m \neq 0$. This problem represents a second-order counterpart of the first-order logistic equation studied in Section 4.1. As will become clear in Section 6.7, it is referred to as the *(stationary) diffusive logistic equation*, and it arises in numerous applications in different branches of Science and Technology.

The first-order system associated with (6.11) is

$$\begin{cases} u' = v, \\ v' = -\lambda u + mu^2, \\ u(0) = u_0, \quad v(0) = v_0. \end{cases} \tag{6.12}$$

The nonlinearity of (6.11) is the autonomous function

$$g(t, u, v) = -\lambda u + mu^2, \quad (t, u, v) \in \mathbb{R} \times \mathbb{R} \times \mathbb{R},$$

which is of class $\mathcal{C}^\infty$ in $(u, v) \in \mathbb{R}^2$ and whose gradient is

$$\nabla g(t, u, v) = \left(\frac{\partial g}{\partial u}, \frac{\partial g}{\partial v} \right) = (-\lambda + 2mu, 0).$$

Since $2mu$ is unbounded in $u \in \mathbb{R}$, by Proposition 6.4, g is not globally Lipschitz with respect to $(u, v) \in \mathbb{R}^2$, although, thanks to Corollary 6.5, it is locally Lipschitz with respect to $(u, v) \in \mathbb{R}^2$, uniformly in $t \in \mathbb{R}$, because it is autonomous. Therefore, thanks to the second fundamental theorem (see Theorem 6.9) applied to (6.11), or, equivalently, Theorem 5.17 applied to (6.12), problem (6.11) admits a unique maximal solution, (I, u), and $I = [0, T_{\max})$ for some $T_{\max} \leq +\infty$. Moreover,

$$\limsup_{t \uparrow T_{\max}} \left(|u(t)| + |u'(t)| \right) = +\infty \quad \text{if } T_{\max} < +\infty. \tag{6.13}$$

In the following two sections, we analyze two particular examples of the diffusive logistic equation. They provide insight into the so-called *large solutions.* A more general result on large solutions will be presented in Section 6.5.3.

6.5.1   *Large solutions: a simple example*

In this section, we study the Cauchy problem

$$\begin{cases} u'' = u^2, \\ u(0) = x, \quad u'(0) = 0, \end{cases} \tag{6.14}$$

with $x > 0$, whose associated first-order problem is

$$\begin{cases} u' = v, \\ v' = u^2, \\ u(0) = x, \quad v(0) = 0. \end{cases} \tag{6.15}$$

Note that (6.14) is a particular case of (6.11) for the choice $\lambda = 0$ and $m = 1$. Thus, from the discussion above, we already know that (6.14) admits a unique maximal solution, (I, u), and $I = [0, T_{\max})$ for some $T_{\max} \leq +\infty$. Moreover, (6.13) holds. According to (6.14),

$$u''(t) = u^2(t) \geq 0 \quad \text{for all } t \in I.$$

Thus, u' is non-decreasing in I and, since $u'(0) = 0$, we obtain that $u'(t) \geq 0$ for all $t \in I$. So, u is also non-decreasing and, since $u(0) = x > 0$,

$$u(t) \geq x \quad \text{for all } t \in I. \tag{6.16}$$

More precisely, since (6.16) implies

$$u''(t) = u^2(t) \geq x^2 > 0$$

for all $t \in I$, u is convex in I and u' is increasing. Consequently, since $u'(0) = 0$, we find that $u'(t) > 0$ for all $t \in (0, T_{\max})$, and hence, $u(t)$ is also increasing. Therefore, (6.13) becomes

$$\lim_{t \uparrow T_{\max}} \big(u(t) + u'(t)\big) = +\infty \quad \text{if } T_{\max} < +\infty. \tag{6.17}$$

To ascertain the precise behavior of (I, u), we use the special structure of (6.15) as follows: taking the cross products in (6.15) yields

$$v(t)v'(t) = u^2(t)u'(t) \quad \text{for all } t \in I.$$

Equivalently,

$$\frac{d}{dt}\left(\frac{v^2(t)}{2} - \frac{u^3(t)}{3}\right) = 0 \quad \text{for all } t \in I,$$

which implies

$$\frac{v^2(t)}{2} - \frac{u^3(t)}{3} = \frac{v^2(0)}{2} - \frac{u^3(0)}{3} = -\frac{x^3}{3} \quad \text{for all } t \in I.$$

Thus, since $v(t) = u'(t) > 0$ for all $t \in I$, we deduce that

$$v(t) = \sqrt{\frac{2}{3}\left(u^3(t) - x^3\right)} \quad \text{for all } t \in I. \tag{6.18}$$

In particular, from (6.17) and (6.18), it becomes apparent that

$$\lim_{t \uparrow T_{\max}} u(t) = +\infty \quad \text{and} \quad \lim_{t \uparrow T_{\max}} u'(t) = +\infty \tag{6.19}$$

if $T_{\max} < +\infty$. We now show that, starting from (6.18) and using $v(t) = u'(t)$, $T_{\max} < +\infty$. Indeed, for every $t \in I$, we find that

$$t = \int_0^t ds = \int_0^t \frac{u'(s)}{v(s)}\, ds = \sqrt{\frac{3}{2}} \int_0^t \frac{u'(s)}{\sqrt{u^3(s) - x^3}}\, ds.$$

Thus, since $u'(s) > 0$ for all $s \in I \setminus \{0\}$, the change of variable $\xi = u(s)$ yields

$$t = \sqrt{\frac{3}{2}} \int_x^{u(t)} \frac{d\xi}{\sqrt{\xi^3 - x^3}}.$$

Hence, the new change of variable $\xi = x\theta$ gives

$$t = \sqrt{\frac{3}{2x}} \int_1^{u(t)/x} \frac{d\theta}{\sqrt{\theta^3 - 1}}. \tag{6.20}$$

Since we will show later that

$$\imath := \int_1^\infty \frac{d\theta}{\sqrt{\theta^3 - 1}} < +\infty, \tag{6.21}$$

it follows from (6.20) that

$$t < \sqrt{\frac{3}{2x}}\, \imath \quad \text{for all } t \in I$$

and, therefore,

$$T_{\max} \leq \sqrt{\frac{3}{2x}}\, \imath < +\infty.$$

Actually, by letting $t \uparrow T_{\max}$ in (6.20), (6.19) leads to

$$T_{\max}\left(= T_{\max}(x)\right) = \sqrt{\frac{3}{2x}} \int_1^\infty \frac{d\theta}{\sqrt{\theta^3 - 1}}, \tag{6.22}$$

which provides us with the blow up time of the solution of (6.14).

The proof of (6.21) may proceed as follows: for every $M > 1$, we have

$$\int_1^\infty \frac{d\theta}{\sqrt{\theta^3 - 1}} = \int_1^M \frac{d\theta}{\sqrt{\theta^3 - 1}} + \int_M^\infty \frac{d\theta}{\sqrt{\theta^3 - 1}}. \tag{6.23}$$

Since

$$\lim_{\theta \to +\infty} \frac{\sqrt{\theta^3}}{\sqrt{\theta^3 - 1}} = 1,$$

there exists $M > 0$ such that

$$\frac{\sqrt{\theta^3}}{\sqrt{\theta^3 - 1}} \leq 2 \quad \text{for all } \theta \geq M.$$

For this choice of M, it is apparent that

$$\int_M^\infty \frac{d\theta}{\sqrt{\theta^3 - 1}} = \int_M^\infty \frac{\sqrt{\theta^3}}{\sqrt{\theta^3 - 1}} \frac{1}{\sqrt{\theta^3}}\, d\theta \leq 2 \int_M^\infty \theta^{-\frac{3}{2}}\, d\theta = \frac{4}{\sqrt{M}}. \tag{6.24}$$

Moreover, since

$$\theta^3 - 1 = (\theta^2 + \theta + 1)(\theta - 1) \geq 3(\theta - 1) \quad \text{for all } \theta \geq 1,$$

we also find that

$$\int_1^M \frac{d\theta}{\sqrt{\theta^3 - 1}} \leq \frac{1}{\sqrt{3}} \int_1^M \frac{d\theta}{\sqrt{\theta - 1}} = \frac{2}{\sqrt{3}} \sqrt{M - 1}. \tag{6.25}$$

By using (6.24) and (6.25) in (6.23), we finally obtain (6.21).

Going back to the study of (6.14), since the function

$$w(s) := u(-s), \quad s \in (-T_{\max}, 0],$$

satisfies

$$w''(s) = w^2(s) \quad \text{for all } s \in (-T_{\max}, 0],$$

and

$$w(0) = x, \quad w'(0) = 0,$$

by uniqueness, the global bilateral solution of (6.14) is defined in the symmetric interval $(-T_{\max}, T_{\max})$ and satisfies

$$u(-s) = u(s) \quad \text{for all } s \in (-T_{\max}, T_{\max}).$$

Identity (6.22) shows that $T_{\max}(x)$ is decreasing with respect to $x > 0$ and that

$$\lim_{x \uparrow +\infty} T_{\max}(x) = 0, \quad \lim_{x \downarrow 0} T_{\max}(x) = +\infty.$$

Figure 6.2 shows the graph of the maximal solution, $u(t)$, of (6.14) for an arbitrary $x > 0$. In sum, we have shown that its maximal interval of existence is $(-T_{\max}(x), T_{\max}(x))$ and that the solution blows up at both endpoints.

As illustrated in López-Gómez (2015), the following concept has been shown to be pivotal in describing the dynamics of a wide class of nonlinear partial differential equations.

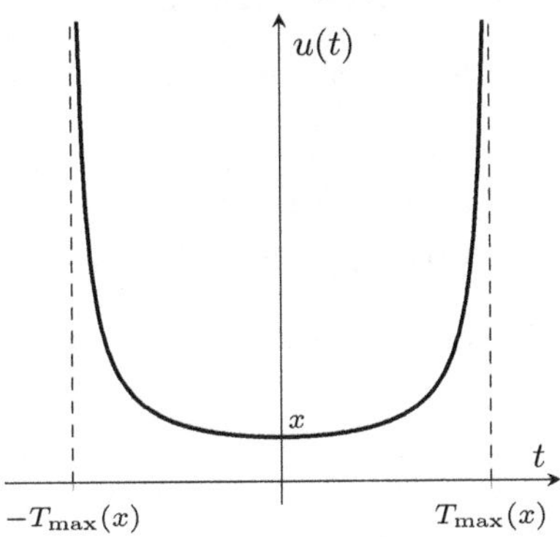

Fig. 6.2. The maximal bilateral solution of problem (6.14).

Definition 6.12. Given $\alpha < \beta$, $J = (\alpha, \beta)$, and $g \in \mathcal{C}(J \times \mathbb{R}^2; \mathbb{R})$, a function $\ell \in \mathcal{C}^2(J; \mathbb{R})$ is called a **large** or **explosive solution** of

$$u'' = g(t, u, u') \tag{6.26}$$

in the interval J if ℓ solves (6.26) in J, and

$$\lim_{t \downarrow \alpha} |\ell(t)| = +\infty, \quad \lim_{t \uparrow \beta} |\ell(t)| = +\infty.$$

According to Definition 6.12, for every $x > 0$, the unique solution of (6.14) provides us with a positive large solution of $u'' = u^2$ in the interval $(-T_{\max}(x), T_{\max}(x))$. Since the function $x \mapsto T_{\max}(x)$ bijectively maps $(0, +\infty)$ onto itself, the following result holds.

Corollary 6.13. *For every $R > 0$, the equation $u'' = u^2$ possesses a unique positive large solution, ℓ, in the interval $(-R, R)$. Moreover,*

$$\ell(0) = \frac{3}{2R^2} \left(\int_1^\infty \frac{d\theta}{\sqrt{\theta^3 - 1}} \right)^2. \tag{6.27}$$

Proof. For every $R > 0$, there exists a unique $x = x(R)$ such that

$$R = T_{\max}(x) = \sqrt{\frac{3}{2x}} \int_1^\infty \frac{d\theta}{\sqrt{\theta^3 - 1}},$$

and x is given by the right-hand side of (6.27). For such an x, the analysis of the Cauchy problem (6.14) performed in this section allows us to conclude the proof. $\qquad \square$

6.5.2   *A more sophisticated example*

In this section, we fix two constants $\lambda, m > 0$ to study the Cauchy problem

$$\begin{cases} u'' = -\lambda u + mu^2, \\ u(0) = x, \quad u'(0) = 0, \end{cases} \tag{6.28}$$

for every $x > 0$. The associated first-order system is given by

$$\begin{cases} u' = v, \\ v' = -\lambda u + mu^2, \\ u(0) = x, \quad v(0) = 0. \end{cases} \tag{6.29}$$

Since it is a particular case of (6.11), we already know that (6.28) admits a unique maximal solution, (I, u), and $I = [0, T_{\max})$ for some $T_{\max} \leq +\infty$. Moreover,

$$\limsup_{t \uparrow T_{\max}} \left(|u(t)| + |u'(t)| \right) = +\infty \quad \text{if } T_{\max} < +\infty. \tag{6.30}$$

From (6.29), we have that

$$u'(t) \left(-\lambda u(t) + m u^2(t) \right) = v(t) v'(t)$$

for all $t \in I$. Thus,

$$\frac{d}{dt} \left(\frac{v^2(t)}{2} + \lambda \frac{u^2(t)}{2} - m \frac{u^3(t)}{3} \right) = 0$$

and, hence,

$$\frac{v^2(t)}{2} + \lambda \frac{u^2(t)}{2} - m \frac{u^3(t)}{3} = \frac{v^2(0)}{2} + \lambda \frac{u^2(0)}{2} - m \frac{u^3(0)}{3} = \lambda \frac{x^2}{2} - m \frac{x^3}{3}$$

for all $t \in I$. In other words, by introducing the auxiliary function

$$\varphi(\xi) := \lambda \frac{\xi^2}{2} - m \frac{\xi^3}{3}, \quad \xi \in \mathbb{R},$$

we have that

$$\frac{v^2(t)}{2} + \varphi(u(t)) = \varphi(x) \quad \text{for all } t \in I. \tag{6.31}$$

From the point of view of applications to Classical Mechanics, φ can be interpreted as the *potential energy* and $v^2/2$ as the *kinetic energy* of a nonlinear oscillator, either mechanical or electrical, governed by (6.28). In such a context, (6.31) establishes the conservation of the total energy of the oscillator.

According to (6.31), we obtain that

$$\varphi(u(t)) \leq \varphi(x) \quad \text{for all } t \in I, \tag{6.32}$$

and hence,

$$v(t) = \pm \sqrt{2 \left[\varphi(x) - \varphi(u(t)) \right]} \quad \text{for all } t \in I. \tag{6.33}$$

The function φ is a cubic polynomial with two roots: $\xi = 0$, which is double, and $\xi = \frac{3}{2} \frac{\lambda}{m} > 0$, which is simple. Moreover, as

$$\varphi'(\xi) = \lambda \xi - m \xi^2 = \xi \left(\lambda - m \xi \right),$$

φ exhibits two critical points at $\xi = 0$ and $\xi = \frac{\lambda}{m}$. Since $\varphi''(0) = \lambda > 0$ and $\varphi''\left(\frac{\lambda}{m}\right) = -\lambda < 0$, 0 is a local minimum, while $\frac{\lambda}{m}$ is a local maximum. Moreover, $\varphi(\xi) > 0$ for all $\xi < 0$ and $\xi \in \left(0, \frac{3}{2}\frac{\lambda}{m}\right)$, $\varphi(\xi) < 0$ for all $\xi > \frac{3}{2}\frac{\lambda}{m}$, and

$$\lim_{\xi\downarrow-\infty} \varphi(\xi) = +\infty, \qquad \lim_{\xi\uparrow+\infty} \varphi(\xi) = -\infty.$$

Figure 6.3 shows a plot of $\varphi(\xi)$.

When $x = \frac{\lambda}{m}$, the (globally defined) constant function $u(t) = \frac{\lambda}{m}$ provides us with the unique solution of (6.28). Subsequently, we distinguish two different cases according to the value of $x > 0$.

Case 1. $x \in \left(0, \frac{\lambda}{m}\right)$. Then, setting

$$\alpha\,(= \alpha(x)) := \varphi^{-1}(\varphi(x)) \cap (-\infty, 0),$$

it becomes apparent from (6.32) that $u(t) \in [\alpha, x]$ for all $t \in I$. Indeed, by simply looking at Figure 6.3, it is obvious that the set of points

$$\{\xi \in \mathbb{R} \colon \varphi(\xi) \le \varphi(x)\}$$

consists of two intervals: one of them containing 0, $[\alpha, x]$, and the other of the form $[\beta, +\infty)$, with

$$\beta\,(= \beta(x)) := \varphi^{-1}(\varphi(x)) \cap \left(\tfrac{\lambda}{m}, +\infty\right).$$

Since $u(0) = x$, by continuity, $u(t) \in [\alpha, x]$ for all $t \in I$. In particular, $u(t)$ is bounded for all $t \in I$ and, as a consequence of (6.33), $v(t)$

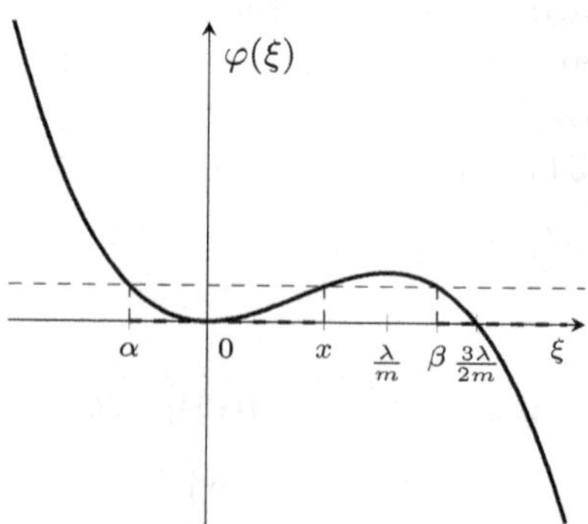

Fig. 6.3. Plot of the potential energy $\varphi(\xi)$, together with the set of points ξ with $\varphi(\xi) \le \varphi(x)$ when $x \in \left(0, \frac{\lambda}{m}\right)$.

is also bounded for all $t \in I$. Therefore, (6.30) cannot occur and, hence, $I = [0, +\infty)$. One can actually establish that, in this case, the maximal bilateral interval where the solution of (6.28) is defined is the entire line $\mathbb{R} = (-\infty, +\infty)$. In Section 10.2, we will show that actually all these solutions are periodic.

Case 2. $x > \frac{\lambda}{m}$. In this case, arguing as above, it is easily seen that

$$u(t) > x > \frac{\lambda}{m} \quad \text{for all } t \in (0, T_{\max}).$$

Indeed, since $u \in \mathcal{C}^2(I)$ and

$$u''(0) = u(0)\left(-\lambda + mu(0)\right) = x\left(-\lambda + mx\right) > 0,$$

there exists $\delta > 0$ such that $u''(t) > 0$ for all $t \in [0, \delta]$. Thus, since $u'(0) = 0$, we find that $u'(t) > 0$ for all $t \in (0, \delta]$ and, in particular, $u(t)$ is increasing in $[0, \delta]$. Thus, $u(t) > x$ for all $x \in (0, \delta]$. Now, arguing by contradiction, suppose that there is $t_1 > \delta$ such that

$$u(t) > x \text{ for all } t \in (0, t_1) \quad \text{and} \quad u(t_1) = x.$$

Then, since $u'(\delta) > 0$ and $u'(t_1) \leq 0$, there exists $t_2 \in (\delta, t_1)$ such that

$$u'(t_2) = 0 \quad \text{and} \quad u''(t_2) = u(t_2)\left(-\lambda + mu(t_2)\right) \leq 0.$$

So, $u(t_2) \leq \lambda/m$, which is impossible because, as $t_2 < t_1$, by construction $u(t_2) > x > \lambda/m$. Consequently, for every $t \in (0, T_{\max})$, $u(t) > x > \lambda/m$, and

$$u''(t) = u(t)\left(-\lambda + mu(t)\right) > 0.$$

Hence, since $u'(0) = 0$, $u'(t) > 0$ for all $t \in I \setminus \{0\}$. Therefore, $u(t)$ is strictly increasing for $t \in (0, T_{\max})$, and (6.33) becomes

$$v(t) = \sqrt{2\left[\varphi(x) - \varphi(u(t))\right]} \quad \text{for all } t \in I.$$

Consequently, for every $t \in (0, T_{\max})$,

$$t = \int_0^t ds = \int_0^t \frac{u'(s)}{v(s)} \, ds = \int_0^t \frac{u'(s)}{\sqrt{2\left[\varphi(x) - \varphi(u(s))\right]}} \, ds.$$

As $s \mapsto u(s)$ is increasing, the change of variable $\xi = u(s)$ gives

$$t = \int_x^{u(t)} \frac{d\xi}{\sqrt{2\left[\varphi(x) - \varphi(\xi)\right]}} = \int_x^{u(t)} \frac{d\xi}{\sqrt{\frac{2}{3}m\left(\xi^3 - x^3\right) - \lambda\left(\xi^2 - x^2\right)}}.$$

Therefore, setting $\xi = x\theta$ yields

$$t = \int_1^{u(t)/x} \frac{d\theta}{\sqrt{\frac{2}{3}mx\left(\theta^3 - 1\right) - \lambda\left(\theta^2 - 1\right)}} \tag{6.34}$$

for all $t \in (0, T_{\max})$. The following result, which is of technical nature, holds.

Lemma 6.14. *For every* $x > \frac{\lambda}{m}$,

$$\int_1^{\infty} \frac{d\theta}{\sqrt{\frac{2}{3}mx\left(\theta^3 - 1\right) - \lambda\left(\theta^2 - 1\right)}} < +\infty.$$

Proof.   Note that the function

$$\varrho(\theta) := \frac{2}{3}mx\left(\theta^3 - 1\right) - \lambda\left(\theta^2 - 1\right), \quad \theta \geq 1,$$

can be rewritten as

$$\varrho(\theta) = \tilde{\varrho}(\theta)\left(\theta - 1\right), \quad \text{with } \tilde{\varrho}(\theta) := \frac{2}{3}mx\left(\theta^2 + \theta + 1\right) - \lambda\left(\theta + 1\right).$$

Moreover,

$$\tilde{\varrho}(1) = 2\left(mx - \lambda\right) > 0$$

and, for every $\theta \geq 1$,

$$\frac{d}{d\theta}\tilde{\varrho}(\theta) = \frac{2}{3}mx\left(2\theta + 1\right) - \lambda \geq 2mx - \lambda > \lambda\,(> 0).$$

Thus, $\tilde{\varrho}(\theta) \geq \tilde{\varrho}(1)$ for all $\theta \geq 1$ and, hence,

$$\varrho(\theta) \geq \tilde{\varrho}(1)\left(\theta - 1\right) = 2\left(mx - \lambda\right)\left(\theta - 1\right) \quad \text{for all } \theta \geq 1.$$

Thus, for every $M > 1$,

$$\int_1^M \frac{d\theta}{\sqrt{\frac{2}{3}mx\left(\theta^3 - 1\right) - \lambda\left(\theta^2 - 1\right)}} \leq \int_1^M \frac{d\theta}{\sqrt{2\left(mx - \lambda\right)\left(\theta - 1\right)}}$$

$$= \sqrt{\frac{2\left(M - 1\right)}{mx - \lambda}}. \qquad (6.35)$$

On the other hand, since

$$\lim_{\theta \uparrow +\infty} \frac{\sqrt{\frac{2}{3}mx\theta^3}}{\sqrt{\frac{2}{3}mx\left(\theta^3 - 1\right) - \lambda\left(\theta^2 - 1\right)}} = 1,$$

there exists $M > 1$ such that

$$\frac{\sqrt{\frac{2}{3}mx\theta^3}}{\sqrt{\frac{2}{3}mx\left(\theta^3 - 1\right) - \lambda\left(\theta^2 - 1\right)}} \leq 2 \quad \text{for all } \theta \geq M.$$

Thus,

$$\int_M^\infty \frac{d\theta}{\sqrt{\frac{2}{3}mx\left(\theta^3 - 1\right) - \lambda\left(\theta^2 - 1\right)}} \leq 2 \int_M^\infty \frac{d\theta}{\sqrt{\frac{2}{3}mx\theta^3}} = \sqrt{\frac{24}{Mmx}}. $$

$$(6.36)$$

By making this choice of M, (6.35) and (6.36) allow us to conclude that

$$\int_1^\infty \frac{d\theta}{\sqrt{\frac{2}{3}mx\left(\theta^3 - 1\right) - \lambda\left(\theta^2 - 1\right)}}$$

$$= \int_1^M \frac{d\theta}{\sqrt{\frac{2}{3}mx\left(\theta^3 - 1\right) - \lambda\left(\theta^2 - 1\right)}}$$

$$+ \int_M^\infty \frac{d\theta}{\sqrt{\frac{2}{3}mx\left(\theta^3 - 1\right) - \lambda\left(\theta^2 - 1\right)}} < +\infty,$$

as claimed. $\qquad \square$

Thanks to Lemma 6.14 and (6.34), it follows that $T_{\max} < +\infty$. Actually, letting $t \uparrow T_{\max}$ in (6.34) yields

$$T_{\max}\left(= T_{\max}(x)\right) = \int_1^\infty \frac{d\theta}{\sqrt{\frac{2}{3}mx\left(\theta^3 - 1\right) - \lambda\left(\theta^2 - 1\right)}} < +\infty.$$

Consequently, the solution $u(t)$ of (6.28) is explosive for every $x > \frac{\lambda}{m}$. Note that $T_{\max}(x)$ is decreasing with respect to $x > \frac{\lambda}{m}$. Moreover, thanks to the estimates (6.35) and (6.36), it becomes apparent that

$$\lim_{x \uparrow +\infty} T_{\max}(x) = 0,$$

whereas, letting $x \downarrow \frac{\lambda}{m}$ leads to

$$\lim_{x \downarrow \frac{\lambda}{m}} T_{\max}(x) = \int_1^\infty \frac{d\theta}{\sqrt{\frac{2}{3}\lambda\left(\theta^3 - 1\right) - \lambda\left(\theta^2 - 1\right)}}$$

$$= \int_1^\infty \frac{\sqrt{3/\lambda}}{\sqrt{2\theta + 1}\,(\theta - 1)}\, d\theta = +\infty. \tag{6.37}$$

Actually, (6.37) can also be obtained from Theorem 5.26. Indeed, since the constant function $u(t) = \frac{\lambda}{m}$, which is (globally) defined for $t \in \mathbb{R}$, is the unique solution of (6.28) for $x = \frac{\lambda}{m}$, it follows from Theorem 5.26 that, for every $T > 0$, there exists $\varepsilon > 0$ for which the unique solution of (6.28), $u(t; x)$, is defined for all $t \in [-T, T]$, provided $\left|x - \frac{\lambda}{m}\right| < \varepsilon$. Thus, for each $x \in \left(\frac{\lambda}{m}, \frac{\lambda}{m} + \varepsilon\right)$, the explosion time $T_{\max}(x)$ satisfies $T_{\max}(x) > T$. Therefore,

$$\lim_{x \downarrow \frac{\lambda}{m}} T_{\max}(x) = +\infty.$$

Summing up, we have obtained the following counterpart of Corollary 6.13.

Corollary 6.15. *For every $R > 0$, the equation*

$$u'' = -\lambda u + mu^2$$

possesses a unique positive large solution, ℓ, in the interval $(-R, R)$. Moreover, $\ell(0)$ is given by the unique $x > \frac{\lambda}{m}$ for which

$$R = \int_1^\infty \frac{d\theta}{\sqrt{\frac{2}{3}mx\left(\theta^3 - 1\right) - \lambda\left(\theta^2 - 1\right)}}.$$

6.5.3 Large solutions: the Keller–Osserman condition

The main purpose of this section is to obtain the counterparts of Corollaries 6.13 and 6.15 for general second-order equations of the form

$$u'' = g(u),$$

where g is locally Lipschitz with respect to $u \in \mathbb{R}$. The key assumption for large solutions to exist is the so-called *Keller–Osserman condition* (6.38), whose name comes from the seminal works by Keller (1957) and Osserman (1957).

Theorem 6.16. *Let $g : [0, +\infty) \to \mathbb{R}$ be a locally Lipschitz function such that there exists $u^* \geq 0$ satisfying $g(u^*) = 0$ and*

$$g(u) > 0 \quad and \quad \frac{g(u)}{u} \text{ is increasing} \quad for\ all\ u > u^*.$$

Assume that the following Keller–Osserman condition *holds true:*

$$H(x) := \int_1^{+\infty} \left(\int_1^\theta g(x\xi)\, d\xi\right)^{-\frac{1}{2}} d\theta < +\infty \quad for\ all\ x > u^*, \quad (6.38)$$

and in addition,

$$\lim_{x\uparrow+\infty} \left(\sqrt{x}H(x)\right) = 0. \tag{6.39}$$

Then, for every $R > 0$, the differential equation

$$u'' = g(u)$$

admits a unique positive large solution in $(-R, R)$ satisfying $u(t) > u^*$ for $t \in (-R, R)$.

Proof. Thanks to the uniqueness of the Cauchy problems associated with $u'' = g(u)$, given by Theorem 6.6, and the fact that, if $u(t)$ is a solution of the differential equation, $w(t) := u(-t)$ also solves the same differential equation with $w(0) = u(0)$ and $w'(0) = -u'(0)$, we have that the positive large solutions of the differential equations in the interval $(-R, R)$ are in one-to-one correspondence with the forward maximal solution of the associated first-order Cauchy problem

$$\begin{cases} u' = v, \\ v' = g(u), \\ u(0) = x, \\ v(0) = 0, \end{cases} \tag{6.40}$$

provided that the forward maximal existence interval is $I = [0, T_{\max})$, with $T_{\max} = T_{\max}(x) = R$. Thus, in the rest of the proof, we show that, once $R > 0$ has been fixed, there exists a unique $x > u^*$ such that the maximal solution (I, u, v) of (6.40) blows up at $T_{\max} = T_{\max}(x) = R$.

Let us fix, for the moment, $x > u^*$, and consider the maximal forward solution (I, u, v) of (6.40), whose existence is guaranteed by Theorem 6.7 and whose global behavior is established in Theorem 6.9. Since $u''(0) = g(x) > 0$ and $u'(0) = 0$, by continuity, there exists $\delta_0 > 0$ such that $u(t) > x$ and $u'(t) > 0$ for all $t \in (0, \delta_0]$. Let $\delta_m \in (\delta_0, T_{\max}]$ denote the maximal $\delta > 0$ for which

$$u(t) > x \quad \text{and} \quad u'(t) > 0 \quad \text{for all } t \in (0, \delta).$$

We claim that $\delta_m = T_{\max}$. Indeed, if $\delta_m < T_{\max}$, then $u(t) > x$ and $u'(t) > 0$ for all $t \in (0, \delta_m)$, though $u'(\delta_m) = 0$. But this is impossible because

$$u'(\delta_m) = \int_0^{\delta_m} u''(s)\, ds = \int_0^{\delta_m} g(u(s))\, ds > 0.$$

Therefore,

$$u(t) > x \quad \text{and} \quad u'(t) > 0 \quad \text{for all } t \in (0, T_{\max}).$$

Next, from (6.40), we obtain

$$v'(t)v(t) - g(u(t))u'(t) = 0$$

for all $t \in I$, which entails

$$\frac{d}{dt}\left(\frac{v^2(t)}{2} - \int_0^{u(t)} g(s)\,ds\right) = 0.$$

Then,

$$\frac{v^2(t)}{2} - \int_0^{u(t)} g(s)\,ds = -\int_0^x g(s)\,ds \quad \text{for all } t \in I,$$

which, since $u'(t) = v(t) > 0$ for all $t \in I$, can be rewritten as

$$v(t) = \left(2\int_x^{u(t)} g(s)\,ds\right)^{\frac{1}{2}} = \left(2\int_1^{u(t)/x} xg(x\xi)\,d\xi\right)^{\frac{1}{2}}, \qquad (6.41)$$

where we have performed the change of variables $s = x\xi$. As a consequence, by performing the changes of variable $u = u(s)$ and $u = x\theta$, we get

$$t = \int_0^t ds = \int_0^t \frac{u'(s)}{v(s)}\,ds = \int_0^t u'(s)\left(2\int_1^{\frac{u(s)}{x}} xg(x\xi)\,d\xi\right)^{-\frac{1}{2}} ds$$

$$= \int_x^{u(t)}\left(2\int_1^{\frac{u}{x}} xg(x\xi)\,d\xi\right)^{-\frac{1}{2}} du = \int_1^{\frac{u(t)}{x}}\left(2\int_1^{\theta} \frac{g(x\xi)}{x}\,d\xi\right)^{-\frac{1}{2}} d\theta,$$

which, thanks to (6.38), gives

$$t < \sqrt{\frac{x}{2}}\int_1^{+\infty}\left(\int_1^{\theta} g(x\xi)\,d\xi\right)^{-\frac{1}{2}} d\theta = \sqrt{\frac{x}{2}}H(x) < +\infty$$

for all $x > u^*$. This relation implies $T_{\max} < +\infty$, and Theorem 6.9, together with (6.41) and the fact that $u(t)$ is increasing for $t \in (0, T_{\max})$, guarantees that

$$\lim_{t \uparrow T_{\max}} u(t) = +\infty.$$

Hence, letting $t \uparrow T_{\max}$ in the above identity gives

$$T_{\max} = T_{\max}(x) = \sqrt{\frac{x}{2}}H(x) = \int_1^{+\infty}\left(2\int_1^{\theta} \frac{g(x\xi)}{x\xi}\xi\,d\xi\right)^{-\frac{1}{2}} d\theta.$$

Since $u \mapsto \frac{g(u)}{u}$ is increasing for all $u > u^*$, it follows that $T_{\max}(x)$ is decreasing. In addition, since the constant function $(u(t), v(t)) = (u^*, 0)$ (globally) solves (6.40) for $x = u^*$, Theorem 5.26 guarantees that

$$\lim_{x \downarrow u^*} T_{\max}(x) = +\infty.$$

Combining this information with (6.39) allows us to conclude that, for every $R > 0$, there exists a unique $x > u^*$ such that $T_{\max}(x) = R$, and the corresponding solution of (6.40) provides us with the unique explosive solution of $u'' = g(u)$ in $(-R, R)$, as claimed. $\qquad\square$

Some specific examples where Theorem 6.16 can be applied are those treated in Section 6.5.1, with $g(u) = u^2$, and in Section 6.5.2, with $g(u) = -\lambda u + m u^2$. We refer the reader to Exercise 3 of Chapter 6 for additional examples of nonlinearities, $g(u)$, for which the existence of large positive solutions can be established.

6.6 Systems of High-Order Nonlinear Equations

The abstract theory developed in Section 1.13 can be adapted to a nonlinear setting in a rather straightforward way. We begin by considering the following nonlinear counterpart of system (1.90):

$$\begin{cases} u_1'' = g_1(t, u_1, u_1', u_2, u_2', u_2''), \\ u_2''' = g_2(t, u_1, u_1', u_2, u_2', u_2''), \end{cases} \tag{6.42}$$

where $J \subset \mathbb{R}$ is an interval, Ω is an open subset of $\mathbb{R}^5$, and $g_1, g_2 \in \mathcal{C}(J \times \Omega; \mathbb{R})$. Naturally, if

$$u := (u_1, u_2) \in \mathcal{C}^2(J; \mathbb{R}) \times \mathcal{C}^3(J; \mathbb{R})$$

solves (6.42) in J, then

$$W := (w_1, w_2, w_3, w_4, w_5)^T = (u_1, u_1', u_2, u_2', u_2'')^T \in \mathcal{C}^1(J; \mathbb{R}^5)$$

solves the first-order system

$$W' = \begin{pmatrix} w_2 \\ g_1(t, w_1, w_2, w_3, w_4, w_5) \\ w_4 \\ w_5 \\ g_2(t, w_1, w_2, w_3, w_4, w_5) \end{pmatrix} =: f(t, W). \tag{6.43}$$

Actually, in the vein of Theorem 6.1, there is a canonical bijection between the solutions of (6.42) and the solutions of (6.43). Moreover, the following counterpart of Lemma 6.3 holds.

Lemma 6.17. *Suppose that J is a subinterval of $\mathbb{R}$, Ω is an open subset of $\mathbb{R}^5$, and $g_1, g_2 \in \mathcal{C}(J \times \Omega; \mathbb{R})$. Let $f \in \mathcal{C}(J \times \Omega; \mathbb{R}^5)$ be the function defined in (6.43). Then:*

(a) *g_1 and g_2 are globally Lipschitz with respect to $W \in \Omega$ uniformly in $t \in J$ if and only if f is globally Lipschitz with respect to $W \in \Omega$ uniformly in $t \in J$;*

(b) *g_1 and g_2 are locally Lipschitz with respect to $W \in \Omega$ uniformly in $t \in J$ if and only if f is locally Lipschitz with respect to $W \in \Omega$ uniformly in $t \in J$;*

(c) *g_1 and g_2 are of class $\mathcal{C}^1$ with respect to $W \in \Omega$ if and only if f is of class $\mathcal{C}^1$ with respect to $W \in \Omega$.*

The following result follows as a direct application of Lemma 6.17 and Theorem 5.8.

Theorem 6.18. *Suppose that J is a subinterval of $\mathbb{R}$ and that $g_1, g_2 \in \mathcal{C}(J \times \mathbb{R}^5; \mathbb{R})$ are globally Lipschitz continuous with respect to $W \in \mathbb{R}^5$ uniformly in t on compact subintervals of J. Then, for every $t_0 \in J$ and $W_0 \in \mathbb{R}^5$, the Cauchy problem*

$$\begin{cases} u_1'' = g_1(t, u_1, u_1', u_2, u_2', u_2''), \\ u_2''' = g_2(t, u_1, u_1', u_2, u_2', u_2''), \\ (u_1(t_0), u_1'(t_0), u_2(t_0), u_2'(t_0), u_2''(t_0))^T = W_0, \end{cases} \tag{6.44}$$

has a unique solution, (u_1, u_2), globally defined in J.

Proof. The result is obvious if $J = [\alpha, \beta]$ is compact. Suppose $J = [\alpha, \beta)$. Then, for sufficiently small $\varepsilon > 0$, the problem (6.44) admits a unique solution, say $(u_{1,\varepsilon}, u_{2,\varepsilon})$, in $[\alpha, \beta - \varepsilon]$. By uniqueness,

$$(u_{1,\varepsilon}, u_{2,\varepsilon})|_{[\alpha, \beta - \delta]} = (u_{1,\delta}, u_{2,\delta})$$

if $\varepsilon < \delta$ are both sufficiently small. Thus, the point-wise limit

$$(u_1, u_2) := \lim_{\varepsilon \downarrow 0}(u_{1,\varepsilon}, u_{2,\varepsilon})$$

is well defined and provides us with the unique solution of (6.44) in $[\alpha, \beta)$.

This argument can be easily adapted to deal with the remaining cases. For instance, if $J = \mathbb{R}$, one can consider the unique solution of (6.44) in $[t_0 - T, t_0 + T]$, say $(u_{1,T}, u_{2,T})$, and consider

$$(u_1, u_2) := \lim_{T \uparrow \infty} (u_{1,T}, u_{2,T}).$$

This completes the proof. $\hfill\square$

Similarly, by combining Lemma 6.17 and Theorem 5.14, the existence of a maximal solution holds true.

Theorem 6.19. *Suppose $t_0 \in \mathbb{R}$, $T > 0$, $J \in \{[t_0, t_0 + T], [t_0, +\infty)\}$, Ω is an open subset of $\mathbb{R}^5$, and $g_1, g_2 \in \mathcal{C}(J \times \Omega; \mathbb{R})$ are locally Lipschitz with respect to $W \in \Omega$ uniformly in t on compact subintervals of J. Then, for every $W_0 \in \Omega$, problem (6.44) possesses a unique maximal solution, (I, u_1, u_2).*

Moreover, the following counterpart of the second fundamental theorem holds. We point out that the other fundamental theorems also have counterparts for (6.44); nevertheless, we do not state them here (see Exercise 1 of Chapter 6).

Theorem 6.20 (Second fundamental theorem for (6.44)). *Suppose $t_0 \in \mathbb{R}$, $J = [t_0, +\infty)$ and $g_1, g_2 \in \mathcal{C}(J \times \mathbb{R}^5; \mathbb{R})$ are locally Lipschitz with respect to $W \in \mathbb{R}^5$ uniformly in t on compact subintervals of J. For any given $W_0 \in \mathbb{R}^5$, let (I, u_1, u_2) denote the (unique) maximal solution of (6.44), whose existence is guaranteed by Theorem 6.19. Then, there exists $t_0 < T_{\max} \le +\infty$ such that $I = [t_0, T_{\max})$. Moreover, if $T_{\max} < +\infty$,*

$$\limsup_{t \uparrow T_{\max}} \left(|u_1(t)| + |u_1'(t)| + |u_2(t)| + |u_2'(t)| + |u_2''(t)| \right) = +\infty.$$

Thus, either (u_1, u_2) is globally defined in time or $(u_1, u_1', u_2, u_2', u_2'')$ blows up at a finite time, $T_{\max}$.

More generally, let $n \ge 1$ be an arbitrary integer. Consider, for every $j \in \{1, \ldots, n\}$, another integer $N_j \ge 1$, as well as the system of n linear equations with respective orders N_j,

$$u_j^{N_j)} = g_j\Big(t, u_1, u_1', \ldots, u_1^{N_1-1)}, \ldots, u_i, u_i', \ldots, u_i^{N_i-1)}, \ldots,$$
$$u_n, u_n', \ldots, u_n^{N_n-1)}\Big),$$

where

$$g_j \in \mathcal{C}(J \times \mathbb{R}^{N_1} \times \cdots \times \mathbb{R}^{N_n}; \mathbb{R}), \quad 1 \le j \le n, \tag{6.45}$$

for some subinterval J of $\mathbb{R}$. By a solution of this system in $I \subset J$, we mean a vectorial function

$$(u_1, \ldots, u_n) \in \mathcal{C}^{N_1}(I; \mathbb{R}) \times \cdots \times \mathcal{C}^{N_n}(I; \mathbb{R})$$

satisfying the system for every $t \in I$. To reduce the complexity of notations, we set

$$\mathcal{D}_i(u_i) := \left(u_i, u_i', \ldots, u_i^{N_i-1)}\right), \quad 1 \le i \le n,$$

and

$$W := (\mathcal{D}_1(u_1), \ldots, \mathcal{D}_i(u_i), \ldots, \mathcal{D}_n(u_n))^T \in \mathbb{R}^{N_1+\cdots+N_n}.$$

Then, the Cauchy problem associated with the previous system can be equivalently expressed as

$$\begin{cases} u_j^{N_j)} = g_j(t, \mathcal{D}_1(u_1), \ldots, \mathcal{D}_i(u_i), \ldots, \mathcal{D}_n(u_n)), \quad 1 \le j \le n, \\ (\mathcal{D}_1(u_1)(t_0), \ldots, \mathcal{D}_i(u_i)(t_0), \ldots, \mathcal{D}_n(u_n)(t_0))^T = W_0 \in \mathbb{R}^N, \end{cases} \tag{6.46}$$

where we have denoted

$$N := N_1 + \cdots + N_n.$$

Then, the following direct consequence of Theorem 5.5 holds.

Theorem 6.21. *Suppose that the functions g_j introduced in (6.45) are globally Lipschitz with respect to $W \in \mathbb{R}^N$ uniformly in t on compact subintervals of J. Then, for every $t_0 \in J$ and $W_0 \in \mathbb{R}^N$, the Cauchy problem (6.46) has a unique solution, which is (globally) defined in J.*

Similarly, the following consequence of Theorems 5.14 and 5.17 holds.

Theorem 6.22 (Second fundamental theorem for (6.46)). *Suppose $J = [t_0, +\infty)$ and the functions g_i introduced in (6.45) are* *locally Lipschitz with respect to $W \in \mathbb{R}^N$ uniformly in t on compact subintervals of J. Then, for every $W_0 \in \mathbb{R}^N$, the Cauchy problem (6.46) has a unique maximal solution, $(I, u_1, \dots, u_n)$. Moreover, there exists $t_0 < T_{\max} \leq +\infty$ such that $I = [t_0, T_{\max})$, and*

$$\limsup_{t \uparrow T_{\max}} \sum_{j=1}^{n} \sum_{i=0}^{N_j - 1} |u_j^{i)}(t)| = +\infty \quad \text{if } T_{\max} < +\infty.$$

6.7　Derivation of the Diffusive Logistic Equation

We derive the diffusive logistic equation already analyzed in Section 6.5 from the celebrated heat equation. In this section, Ω represents a solid body in the space, $\mathbb{R}^3$, with boundary $\partial\Omega$, and $u(x, y, z, t)$ denotes the temperature of the body at an arbitrary location $(x, y, z) \in \Omega$ and time $t \geq 0$. A *heat flux* is generated at every point due to the difference between temperatures with respect to neighboring points. According to the *first Fourier's law* (cf. Fourier, 1822), such a flux is proportional to the gradient of u, i.e.

$$\mathbf{J} := -\mu \nabla u = -\mu \left(\frac{\partial u}{\partial x}, \frac{\partial u}{\partial y}, \frac{\partial u}{\partial z} \right), \tag{6.47}$$

where the coefficient μ measures the *heat conductivity* of the body and the gradient, ∇, involves the spatial variables. Observe that, since the gradient points in the direction of the largest increase in the function, the minus sign in (6.47) takes into account the physical evidence that heat flows from hotter to colder regions.

If the *specific heat* of Ω, c, is assumed to be constant, then, for every $(x, y, z) \in \Omega$ and sufficiently small $\varepsilon > 0$ such that the ball

$$B_\varepsilon := \left\{ (\tilde{x}, \tilde{y}, \tilde{z}) \in \mathbb{R}^3 : (\tilde{x} - x)^2 + (\tilde{y} - y)^2 + (\tilde{z} - z)^2 < \varepsilon^2 \right\},$$

is contained in Ω, the *thermal energy*, or heat content, of B_ε at time $t \geq 0$ is given by

$$Q_\varepsilon(t) := c \int_{B_\varepsilon} u(\tilde{x}, \tilde{y}, \tilde{z}, t) \, dV,$$

where $dV = d\tilde{x} \, d\tilde{y} \, d\tilde{z}$ is the volume element of $\mathbb{R}^3$. Then, the infinitesimal variation of thermal energy in B_ε at time $t > 0$ is given by the time derivative

$$Q'_\varepsilon(t) = c\,\frac{d}{dt}\int_{B_\varepsilon} u(\tilde{x},\tilde{y},\tilde{z},t)\,dV = c\int_{B_\varepsilon}\frac{\partial u}{\partial t}(\tilde{x},\tilde{y},\tilde{z},t)\,dV, \qquad (6.48)$$

where the last equality can be rigorously established, provided u is sufficiently regular.

On the other hand, by the definition of flux, the surface integral

$$\int_{\partial B_\varepsilon} \mathbf{J}\,dS$$

quantifies the total thermal energy that is lost, or gained, through the boundary of B_ε, ∂B_ε. If we further assume that the heat content of B_ε can only vary due to the heat flux through ∂B_ε, then

$$Q'_\varepsilon(t) + \int_{\partial B_\varepsilon} \mathbf{J}\,dS = 0. \qquad (6.49)$$

Consequently, substituting (6.47) and (6.48) in (6.49), together with Gauss' divergence theorem, yields

$$c\int_{B_\varepsilon}\frac{\partial u}{\partial t}\,dV = -\int_{\partial B_\varepsilon}\mathbf{J}\,dS = \mu\int_{\partial B_\varepsilon}\nabla u\,dS$$

$$= \mu\int_{B_\varepsilon}\operatorname{div}\nabla u\,dV = \mu\int_{B_\varepsilon}\Delta u\,dV,$$

where Δ is the Laplace operator

$$\Delta u := \frac{\partial^2 u}{\partial x^2} + \frac{\partial^2 u}{\partial y^2} + \frac{\partial^2 u}{\partial z^2}.$$

Thus, for every $(x,y,z)\in\Omega$ and sufficiently small $\varepsilon > 0$, we find that

$$\int_{B_\varepsilon}\frac{\partial u}{\partial t}\,dV = \frac{\mu}{c}\int_{B_\varepsilon}\Delta u\,dV. \qquad (6.50)$$

So, by setting $\kappa := \frac{\mu}{c}$, dividing (6.50) by the volume of the ball B_ε, $|B_\varepsilon| = \frac{4}{3}\pi\varepsilon^3$, and letting $\varepsilon \downarrow 0$, the integral mean value theorem yields

$$\frac{\partial u}{\partial t}(x,y,z,t) = \kappa\,\Delta u(x,y,z,t)$$

for all $(x, y, z) \in \Omega$ and $t > 0$, which is known as the (linear) *heat equation* and is one of the most pivotal partial differential equations.

In the context of diffusion of chemical substances, the first Fourier's law is usually referred to as the *first Fick's law* (cf. Fick, 1855). According to it, the flux $\mathbf{J}$ of a chemical, whose concentration is measured as $u(x, y, z, t)$, is proportional through some constant $d > 0$—the diffusion coefficient of the chemical in the medium where it is distributed—to the gradient of u, $\nabla u(x, y, z, t)$, and of opposite sign, i.e.

$$\mathbf{J} := -d \, \nabla u,$$

because, as in heat transfer problems, physical experience dictates that the substance diffuses from the regions with higher concentrations toward regions with lower concentrations, much like a milk drop diffuses into a cup of coffee. As the total mass of the chemical in B_ε is given by

$$Q_\varepsilon(t) = \int_{B_\varepsilon} u \, dV,$$

if we further assume that the chemical in B_ε can only vary due to its diffusion through the boundary, ∂B_ε, by reasoning as above, it becomes apparent that the concentration of the chemical in Ω is also described by the *diffusion equation*,

$$\frac{\partial u}{\partial t}(x, y, z, t) = d \, \Delta u(x, y, z, t), \tag{6.51}$$

for all $(x, y, z) \in \Omega$ and $t > 0$.

Similarly, in the context of Population Dynamics, (6.51) models the evolution of a biological species in the habitat Ω under the assumption that the individuals of the species are highly concentrated and randomly disperse through Ω. In this context, $u(x, y, z, t)$ measures the density of the population at the location $(x, y, z) \in \Omega$ and time $t \geq 0$; so, the total population in $B_\varepsilon \subset \Omega$ at time t is given by

$$P(t) = \int_{B_\varepsilon} u \, dV.$$

Surprisingly, random—Brownian—motion provokes regular migration of the species from highly populated to less populated areas, in complete agreement with Fourier's and Fick's laws. The influential essay by Schrödinger (1944) contains an extremely illuminating discussion on the role played by spatial dispersion in population dynamics.

In (6.51), no interaction was considered, only random motion of individuals. Nevertheless, in the context of Population Dynamics, every individual of the species has an interplay with the environment and interacts with other individuals. For example, if the reproduction rate is assumed to be uniform in the whole environment and equals $\Lambda \geq 0$, while, simultaneously, the individuals compete for the available resources at a rate $M > 0$, we are led to the nonlinear diffusive equation

$$\frac{\partial u}{\partial t} = d\Delta u + \Lambda u - M u^2 \tag{6.52}$$

for $(x, y, z) \in \Omega$ and $t > 0$, which is the simplest diffusive counterpart of the logistic equation (4.2).

In applications to Population Dynamics, equation (6.52) should be complemented with appropriate *boundary conditions* on $\partial\Omega$, as well as with the *initial conditions*

$$u(x, y, z, 0) = u_0(x, y, z), \quad (x, y, z) \in \Omega, \tag{6.53}$$

where u_0 is the (given) initial distribution of the population at time $t = 0$. In practice, the most commonly used boundary conditions on $\partial\Omega$ are of three different types:

(a) *Dirichlet boundary conditions*, named after P. G. L. Dirichlet, are of the form

$$u(x, y, z, t) = h(x, y, z, t), \quad (x, y, z) \in \partial\Omega, \quad t > 0,$$

for some given function $h : \partial\Omega \times (0, \infty) \to \mathbb{R}$;

(b) *Neumann boundary conditions*, named after C. G. Neumann, are of type

$$\frac{\partial u}{\partial n}(x, y, z, t) = h(x, y, z, t), \quad (x, y, z) \in \partial\Omega, \quad t > 0, \tag{6.54}$$

for some given function h defined on $\partial\Omega \times (0, \infty)$, where n indicates the outward normal vector field to Ω, along $\partial\Omega$;

(c) *Robin boundary conditions*, named after V. G. Robin, are of the form

$$\frac{\partial u}{\partial n}(x, y, z, t) + \beta(x, y, z, t)u(x, y, z, t) = h(x, y, z, t) \qquad (6.55)$$

for all $(x, y, z) \in \partial\Omega$ and $t > 0$, where $\beta \neq 0$ and h are two arbitrary given functions on $\partial\Omega \times (0, \infty)$. Observe that, in the special case of $\beta = 0$, (6.55) reduces to (6.54).

These conditions are said to be of homogeneous type when $h = 0$. Taking into account the modeling presented above, the Dirichlet boundary conditions prescribe the population density at the boundary of the habitat, while the Neumann boundary conditions prescribe the population flux through the boundary. For this reason, homogeneous Neumann boundary conditions are also referred to as *no-flux (or zero flux) boundary conditions*. The Robin boundary conditions, instead, are used, for example, in transfer processes through cellular membranes, where the flux of the considered density linearly depends on its concentration.

When the diffusive logistic equation (6.52) is completed with homogeneous Dirichlet boundary conditions on $\partial\Omega$, together with the initial condition (6.53), we are led to the following initial boundary value problem

$$\begin{cases} \frac{\partial u}{\partial t} = d\Delta u + \Lambda u - M u^2 & \text{in } \Omega, \quad t > 0, \\ u = 0 & \text{on } \partial\Omega, \quad t > 0, \\ u(\cdot, 0) = u_0 & \text{in } \Omega. \end{cases} \qquad (6.56)$$

As in Section 4.1, we expect the dynamics of (6.56) to be regulated by its steady-state solutions, which are the solutions of the *boundary value problem*

$$\begin{cases} -d\Delta u = \Lambda u - M u^2 & \text{in } \Omega, \\ u = 0 & \text{on } \partial\Omega. \end{cases} \qquad (6.57)$$

Dividing by d and renaming

$$\lambda := \frac{\Lambda}{d}, \quad m := \frac{M}{d}$$

lead us back to the diffusive logistic equation considered in Section 6.5. Nevertheless, observe that the problem analyzed there is not a boundary value problem, like the one presented here. We refer the reader to Section 10.4 and the exercises of Chapter 10 for the treatment of simple one-dimensional prototypes similar to (6.57).

6.8   Exercises

1. State and prove the four fundamental theorems for the Cauchy problem (6.46).
2. Assuming that a frictional force resists the motion of the simple pendulum and that such a force is proportional, with a coefficient $\kappa > 0$, to the speed of the bob, the equation of the simple pendulum becomes

$$ MLu'' + \kappa Lu' + MG\sin u = 0. $$

Analyze the existence, uniqueness, and global behavior of the maximal solutions of the Cauchy problem associated with this differential equation.

3. Given four constants $\lambda \in \mathbb{R}$, $p > 1$, $m > 0$, and $x > 0$, show that the Cauchy problem

$$ \begin{cases} u'' = -\lambda u + mu^p, \\ u(0) = x, \quad u'(0) = 0, \end{cases} $$

admits a unique maximal solution, and prove that it blows up in a finite time. Analyze the dependence on $x > 0$ and $p > 1$ of the blow up time.

4. Characterize the functions $f \in \mathcal{C}^1([0,\infty);\mathbb{R})$ such that $f(0) = 0$ and $f'(u) > 0$ for all $u > 0$ for which the solution of the Cauchy problem

$$ \begin{cases} u'' = f(u), \\ u(0) = x > 0, \quad u'(0) = 0, \end{cases} $$

blows up in a finite time. For any given $R > 0$, analyze the existence of a positive solution for the singular problem

$$ \begin{cases} u'' = f(u), \quad t \in (-R, R), \\ u(-R) = u(R) = +\infty, \end{cases} $$

where the boundary conditions should be understood as

$$ \lim_{t\downarrow -R} u(t) = +\infty, \quad \lim_{t\uparrow R} u(t) = +\infty. $$

5. Analyze the existence, uniqueness, and extensibility of the solutions of the Cauchy problem

$$\begin{cases} -u'' = \lambda u - au^2 - buv, \\ -v'' = \mu v - cuv - dv^2, \\ u(0) = u_0, \ u'(0) = 0, \ v(0) = v_0, \ v'(0) = 0, \end{cases}$$

where λ, μ, a, b, c, d, u_0, and v_0 are positive constants.

6. Analyze the existence, uniqueness, and extensibility of the solutions of the Cauchy problem

$$\begin{cases} -u'' = \lambda u - au^2 - buv, \\ -v'' = \mu v + cuv - dv^2, \\ u(0) = u_0, \ u'(0) = 0, \ v(0) = v_0, \ v'(0) = 0, \end{cases}$$

where $\lambda > 0$, $\mu < 0$, and a, b, c, d, u_0, and v_0 are positive constants.

7. Analyze the existence, uniqueness, and extensibility of the solutions of the Cauchy problem

$$\begin{cases} -u'' = \lambda u - au^2 + buv, \\ -v'' = \mu v + cuv - dv^2, \\ u(0) = u_0, \ u'(0) = 0, \ v(0) = v_0, \ v'(0) = 0, \end{cases}$$

where $\lambda < 0$, $\mu < 0$, and a, b, c, d, u_0, and v_0 are positive constants.

6.9 Final Comments

Traditionally, most classical textbooks on Ordinary Differential Equations focus on second-order equations from Classical Mechanics, in the spirit of 19th century Mathematics. Nevertheless, during the 20th century, diffusive equations were shown to have wide applications in Population Dynamics, Physics, Chemistry, Biology, and Engineering. Although their multidimensional versions consist of Partial Differential Equations, when considered in one dimension, they reduce to Ordinary Differential Equations. In addition, the study of the simplest one-dimensional models has been the pillar of later developments in the design of strategies to tackle the general multidimensional counterparts, leading to significant advancements in Differential Equations during the 20th century. For this reason, we have decided to include their analysis in this introductory book.

A paradigmatic example consists of the large solutions of equation

$$u'' = u^2,$$

which have been constructed in Section 6.5.1. Beginning with the pioneering works of Keller (1957) and Osserman (1957), large solutions have been one of the most active research fields in Partial Differential Equations during the past three decades. Although the uniqueness of large solutions for the general multidimensional problem

$$\Delta u = f(u),$$

with f increasing, remains a major challenge, the analysis of these problems in the one-dimensional context can be carried out by using some elementary techniques of Ordinary Differential Equations, as shown in Section 6.5.3. Actually, some of the most recent advances in this field have been possible thanks to these techniques. The interested reader is referred, e.g. to the astonishing multiplicity results given by López-Gómez *et al.* (2014), López-Gómez and Tellini (2014), and López-Gómez *et al.* (2015), or to the more recent findings by López-Gómez and Maire (2018, 2019) on the Keller–Osserman condition and uniqueness problem of large solutions for equations of the form

$$u'' = f(u)$$

with non-monotone $f(u)$. The Keller–Osserman condition (see (6.38)) is the necessary and sufficient condition that f must satisfy for $u'' = f(u)$ to admit a large solution (see Exercise 4 of Chapter 6).

The huge interest of positive large solutions emerges in the analysis of the dynamics of positive solutions of the one-dimensional diffusive logistic problem

$$\begin{cases} \frac{\partial u}{\partial t} = \frac{\partial^2 u}{\partial x^2} + \lambda u - m(x)u^2 & x \in (-L, L), \quad t > 0, \\ u(-L, t) = u(L, t) = 0 & t > 0, \\ u(x, 0) = u_0(x) \gneq 0 & x \in (-L, L), \end{cases} \tag{6.58}$$

where $m \gneq 0$ satisfies

$$m^{-1}(0) = [-L, -R] \cup [R, L]$$

for some $R \in (0, L)$. So, $m(x) > 0$ if and only if $x \in (-R, R)$. According to the abstract theory of López-Gómez (2015), if we denote by $u(x, t)$ the unique solution of (6.58), then

$$\lim_{t \uparrow +\infty} u(x, t) = \begin{cases} 0 & \text{if } \lambda \leq \left(\frac{\pi}{2L}\right)^2, \\ \theta_\lambda(x) & \text{if } \left(\frac{\pi}{2L}\right)^2 < \lambda < \left(\frac{\pi}{L-R}\right)^2, \end{cases}$$

where $\theta_\lambda(x)$ is the unique positive solution of the problem

$$\begin{cases} -u'' = \lambda u - m(x)u^2 & x \in (-L, L), \\ u(-L) = u(L) = 0. \end{cases} \tag{6.59}$$

The existence of $\theta_\lambda(x)$ for (6.59) was characterized by Fraile *et al.* in 1996 (in Chapter 10, we will show how to construct the positive solutions of (6.59) in the simplest case when m is a positive constant). According to Fraile *et al.* (1996), for the choice of $m(x)$ described above, problem (6.59) cannot admit a positive solution if

$$\lambda \geq \left(\frac{\pi}{L - R}\right)^2.$$

In such a case, according to López-Gómez (2015, Chapter 5), we have that

$$\lim_{t \uparrow +\infty} u(x, t) = \begin{cases} +\infty & \text{if } x \in (-L, -R] \cup [R, L), \\ \ell_\lambda(x) & \text{if } x \in (-R, R), \end{cases}$$

where ℓ_λ is the minimal large positive solution of the differential equation

$$-u'' = \lambda u - m(x)u^2$$

in the interval $(-R, R)$. Therefore, large solutions are imperative to describe the dynamics of spatially heterogeneous diffusive logistic equations. In Section 6.5.2, we have constructed all these large positive solutions in the special case when λ and $m(x)$ are positive constants.

In addition to the heat equation derived in Section 6.7, we point out that another important "dynamical" version of the diffusive logistic equation is the nonlinear counterpart of the *wave equation*

$$\frac{\partial^2 u}{\partial t^2} = c^2 \frac{\partial^2 u}{\partial x^2},$$

where c is a real constant. As its name suggests, this equation can be used to describe oscillatory phenomena. We refer the interested reader to Marsden and Tromba (2013) for details on how this equation arises, e.g. in electromagnetism from Maxwell's equations.

Chapter 7

Peano Theory

In this chapter, given $t_0 \in \mathbb{R}$, $T > 0$, $J \in \{[t_0, t_0 + T], [t_0, +\infty)\}$, an open subset Ω of $\mathbb{R}^N$ with $N \geq 1$, and $f \in \mathcal{C}(J \times \Omega; \mathbb{R}^N)$, the main goal is to establish, for every $u_0 \in \Omega$, the existence of *at least* one maximal solution for the problem

$$\begin{cases} u' = f(t, u), \\ u(t_0) = u_0, \end{cases} \tag{7.1}$$

as well as to ascertain all the admissible behaviors of the maximal solutions of (7.1). When $\Omega = \mathbb{R}^N$, as in the case of locally Lipschitz f with respect to u uniformly in t on compact subintervals of J, considered in Chapter 5, we will see that any maximal solution is either globally defined in time or it blows up in a finite time. When Ω is a proper subset of $\mathbb{R}^N$ instead, a maximal solution might also approximate $\partial\Omega$. The main novelty of this more general context is that, since the problem can have multiple maximal solutions, each of them might exhibit a different behavior.

Even for simple scalar equations (see Sections 4.5 and 4.6), we already know that, if $f(t, u)$ is only assumed to be continuous, (7.1) might exhibit a one-parameter family of solutions. Thus, in the general context of this chapter, (7.1) will not admit, in general, a unique solution, unless some additional conditions are imposed. This reveals that the integral operator $\mathcal{K}$ associated with problem (7.1) is not, in general, a contraction and, hence, the existence of a solution cannot be inferred from the contractive mapping theorem as in the setting of the Cauchy–Lipschitz theory.

377

A very illuminating example of M. Müller, analyzed in Section 7.6, shows that, even in cases where (7.1) has a unique solution, if $f(t, u)$ is not Lipschitz continuous at u_0, one should not expect the iterates of the associated integral operator $\mathcal{K}$ to converge toward the solution of (7.1). This reveals that, in the context of the Cauchy–Lipschitz theory, the local Lipschitz continuity of $f(t, u)$ is not merely a technical assumption but is essential for the validity of the results presented in Chapter 5.

In the setting of this chapter, the existence of a solution is obtained by means of a fundamental result in functional analysis, named after Ascoli and Arzelà, who established a necessary and sufficient condition for a sequence of continuous functions on a compact interval to admit a uniformly converging subsequence. Actually, most of the theory in Chapters 7 and 8 is a rather direct consequence of the Ascoli–Arzelà theorem.

We point out that, as in Chapter 5, we mainly detail the proofs and constructions reasoning "in the future", i.e. at the right of t_0. Thanks to the change of variable discussed in Section 5.4.1, one can deduce the validity of all the results "in the past" as well, i.e. at the left of t_0.

7.1   Ascoli–Arzelà Theorem

Throughout the rest of this book, given two natural numbers, $N \geq 1$ and $M \geq 1$, an open bounded subset $\Omega \subset \mathbb{R}^M$, and the compact set $K := \bar{\Omega}$, we denote by $\mathcal{C}(K; \mathbb{R}^N)$ the vector space of all continuous functions $u : K \to \mathbb{R}^N$ endowed with the maximum norm

$$\|u\|_{\mathcal{C}(K;\mathbb{R}^N)} = \|u\|_\infty := \max_{x \in K} \|u(x)\|, \quad u \in \mathcal{C}(K; \mathbb{R}^N).$$

where, as usual, $\|u(x)\|$ denotes the 1-norm in $\mathbb{R}^N$. The theory in Section 1.3.1 can be easily adapted to show that this space is a real Banach space (see Exercise 1 of Chapter 7).

The Ascoli–Arzelà theorem characterizes the subsets $\mathfrak{F} \subset \mathcal{C}(K; \mathbb{R}^N)$ that are *relatively compact* in $\mathcal{C}(K; \mathbb{R}^N)$. Before continuing, we recall the definition of a *relatively compact* set.

Definition 7.1 (Relatively compact subset). Let E be a real Banach space. A subset $X \subset E$ is said to be relatively compact if the closure of X in E, $\bar{X}$, is a compact subset of E.

The Ascoli–Arzelà theorem is essentially the counterpart in $\mathcal{C}(K; \mathbb{R}^N)$ of the Bolzano–Weierstrass theorem in $\mathbb{R}^d$, $d \geq 1$, which ensures that a subset A of $\mathbb{R}^d$ is relatively compact, i.e. $\bar{A}$ is compact, if and only if A is *bounded*.

It is well known that the following result holds (see, e.g. Dieudonné, 1960).

Proposition 7.2. *With the notation of Definition 7.1, the following conditions are equivalent:*

- *X is a relatively compact subset of E;*
- *for every sequence, $\{x_n\}_{n \geq 1} \subset X$, there exists a subsequence, $\{x_{n_m}\}_{m \geq 1}$, such that, for some $x \in E$,*

$$\lim_{m \to \infty} x_{n_m} = x \quad in \ E.$$

In particular, a subset $\mathfrak{F} \subset \mathcal{C}(K; \mathbb{R}^N)$ is relatively compact in $\mathcal{C}(K; \mathbb{R}^N)$ if and only if from every sequence, $\{f_n\}_{n \geq 1}$, of functions of $\mathfrak{F}$ one can extract a subsequence, $\{f_{n_m}\}_{m \geq 1}$, such that

$$\lim_{m \to \infty} f_{n_m} = f \quad in \ \mathcal{C}(K; \mathbb{R}^N)$$

for some continuous function $f \in \mathcal{C}(K; \mathbb{R}^N)$.

Naturally, as compact subsets in metric spaces are bounded, any relatively compact subset X of a Banach space E must be bounded. In $\mathbb{R}^d$, the converse also holds true, thanks to the Bolzano–Weierstrass theorem. However, in infinite-dimensional Banach spaces the converse fails: as a simple example, consider $E = \mathcal{C}([0, 1]; \mathbb{R})$ and

$$\mathfrak{F} := \{f_n \colon n \geq 1\},$$

where

$$f_n(x) = x^n, \quad x \in [0, 1], \quad n \geq 1, \tag{7.2}$$

(see Figure 7.1). Since

$$|x^n| \leq 1 \quad \text{for all } x \in [0, 1] \text{ and } n \geq 1,$$

which implies

$$\|f_n\|_\infty \leq 1 \quad \text{for all } n \geq 1,$$

we have

$$\mathfrak{F} \subset \bar{B}_1(0) \subset \mathcal{C}([0, 1]; \mathbb{R}),$$

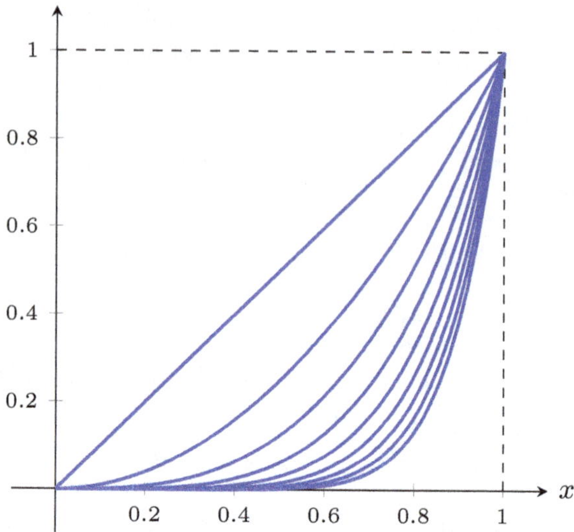

Fig. 7.1. Plots of the graphs of $f_n(x) = x^n$ for $1 \leq n \leq 9$.

where $B_1(0)$ denotes the open unit ball centered at 0 of the Banach space $\mathcal{C}([0,1]; \mathbb{R})$. Thus, $\mathfrak{F} := \{f_n \colon n \geq 1\}$ is bounded. In addition, we have

$$\lim_{n \to \infty} f_n(x) = \lim_{n \to \infty} x^n = \begin{cases} 0 & \text{if } x \in [0,1), \\ 1 & \text{if } x = 1. \end{cases}$$

Therefore, the sequence $\{f_n\}_{n \geq 1}$ is point-wise convergent in $[0,1]$, as $n \to \infty$, to the function $f : [0,1] \to \mathbb{R}$, defined by

$$f(x) := \begin{cases} 0 & \text{if } x \in [0,1), \\ 1 & \text{if } x = 1, \end{cases}$$

which is not continuous at $x = 1$.

As a consequence, no subsequence of $\{f_n\}_{n \geq 1}$ can converge in $\mathcal{C}([0,1]; \mathbb{R})$ because, since $\mathcal{C}([0,1]; \mathbb{R})$ is complete, the limit should be a continuous function, while we have shown that the point-wise limit, f, is discontinuous. So, $\mathfrak{F}$, although bounded, is not relatively compact in $\mathcal{C}([0,1]; \mathbb{R})$.

It turns out that, besides being bounded, the subset $\mathfrak{F}$ must satisfy an additional property for it to be relatively compact in $\mathcal{C}(K; \mathbb{R}^N)$. Such a property is introduced in the following definition.

Definition 7.3 (Equicontinuity). A subset $\mathfrak{F} \subset \mathcal{C}(K; \mathbb{R}^N)$ is said to be *equicontinuous* if for every $\varepsilon > 0$ there exists $\delta = \delta(\varepsilon) > 0$ such that, whenever $x, y \in K$ satisfy $\|x - y\| < \delta$,

$$\|f(x) - f(y)\| < \varepsilon \quad \text{for all } f \in \mathfrak{F}.$$

In other words, "equicontinuity" means that the δ in the definition of uniform continuity of $f \in \mathfrak{F}$ on the compact set K is independent of the function f.

The following result gives a very simple sufficient condition that guarantees equicontinuity.

Proposition 7.4. *Suppose* $\mathfrak{F} = \{f_n \colon n \geq 1\}$ *is a sequence in* $\mathcal{C}(K; \mathbb{R}^N)$ *such that, for every* $n \geq 1$, *the function* f_n *is globally Lipschitz with constant* $L > 0$ *independent of* $n \geq 1$, *i.e.*

$$\|f_n(x) - f_n(y)\| \leq L\|x - y\| \quad \text{for all } x, y \in K \text{ and } n \geq 1.$$

Then, $\mathfrak{F}$ *is equicontinuous.*

Proof. Fix $\varepsilon > 0$, and choose $\delta = \varepsilon/L$. Then, for every $x, y \in K$ such that $\|x - y\| \leq \delta$, we have that

$$\|f_n(x) - f_n(y)\| \leq L\|x - y\| \leq L\delta = \varepsilon.$$

This completes the proof. $\qquad\qquad\qquad\qquad\qquad\qquad\qquad\qquad\square$

The main result in this section is the following theorem. Observe that it immediately implies that the sequence (7.2) is not equicontinuous.

Theorem 7.5 (Ascoli–Arzelà). *Suppose* $\mathfrak{F} \subset \mathcal{C}(K; \mathbb{R}^N)$. *Then, the following assertions are equivalent:*

(a) $\mathfrak{F}$ *is relatively compact in* $\mathcal{C}(K; \mathbb{R}^N)$, *i.e.* $\bar{\mathfrak{F}}$ *is compact in* $\mathcal{C}(K; \mathbb{R}^N)$;

(b) $\mathfrak{F}$ *is bounded and equicontinuous.*

Proof. As is natural, throughout this book, given any Banach space $(E, \|\cdot\|)$, $R > 0$, and $x_0 \in E$, we denote by $B_R(x_0)$ the open ball

$$B_R(x_0) := \{x \in E \colon \|x - x_0\| < R\}.$$

First, we show that (a) implies (b). So, assume that $\mathfrak{F}$ is relatively compact in $\mathcal{C}(K;\mathbb{R}^N)$. Then, $\bar{\mathfrak{F}}$ is compact and, since

$$\bar{\mathfrak{F}} \subset \bigcup_{f \in \mathfrak{F}} B_1(f),$$

there exist $p \geq 1$ and $f_j \in \mathfrak{F}$, $j \in \{1,\ldots,p\}$, such that

$$\bar{\mathfrak{F}} \subset \bigcup_{j=1}^{p} B_1(f_j).$$

Thus, for every $f \in \mathfrak{F}$, there exists $j = j(f) \in \{1,\ldots,p\}$ such that $f \in B_1(f_j)$. Hence, for all $f \in \mathfrak{F}$,

$$\|f\|_\infty = \|f - f_j + f_j\|_\infty \leq \|f - f_j\|_\infty + \|f_j\|_\infty$$
$$< 1 + \|f_j\|_\infty \leq 1 + \max_{1 \leq j \leq p} \|f_j\|_\infty =: R.$$

Therefore, $\mathfrak{F} \subset B_R(0)$, which shows that $\mathfrak{F}$ is bounded. Actually, this argument can be adapted to prove that any compact subset of a Banach space must be bounded.

In order to prove that, in addition, $\mathfrak{F}$ is equicontinuous we argue by contradiction. Assume $\mathfrak{F}$ is not equicontinuous. Then, there is $\varepsilon > 0$ such that, for every natural number $n \geq 1$, there exist two points $x_n, y_n \in K$ and a function $f_n \in \mathfrak{F}$ such that

$$\|x_n - y_n\| \leq \frac{1}{n} \quad \text{and} \quad \|f_n(x_n) - f_n(y_n)\| \geq \varepsilon. \tag{7.3}$$

Since $K \times K \times \bar{\mathfrak{F}}$ is a compact subset of $\mathbb{R}^M \times \mathbb{R}^M \times \mathcal{C}(K;\mathbb{R}^N)$, we can extract a subsequence of (x_n, y_n, f_n), $n \geq 1$, say, $(x_{n_m}, y_{n_m}, f_{n_m})$, $m \geq 1$, such that $n_m \to \infty$ as $m \to \infty$, and

$$\lim_{m \to \infty} (x_{n_m}, y_{n_m}, f_{n_m}) = (x, y, f) \quad \text{in } \mathbb{R}^M \times \mathbb{R}^M \times \mathcal{C}(K;\mathbb{R}^N),$$

for some $x, y \in K$ and $f \in \mathcal{C}(K;\mathbb{R}^N)$. Letting $m \to \infty$ in the estimate

$$\|x_{n_m} - y_{n_m}\| \leq \frac{1}{n_m}$$

yields $x = y$. Moreover, by (7.3), we have that

$$\varepsilon \le \|f_{n_m}(x_{n_m}) - f_{n_m}(y_{n_m})\|$$
$$\le \|f_{n_m}(x_{n_m}) - f(x_{n_m})\| + \|f(x_{n_m}) - f(y_{n_m})\|$$
$$\quad + \|f(y_{n_m}) - f_{n_m}(y_{n_m})\|$$
$$\le 2\|f_{n_m} - f\|_\infty + \|f(x_{n_m}) - f(y_{n_m})\|$$

for all $m \ge 1$. Thus, since

$$\lim_{m \to \infty} \|f_{n_m} - f\|_\infty = 0$$

and, by the continuity of f,

$$\lim_{m \to \infty} \|f(x_{n_m}) - f(y_{n_m})\| = \|f(x) - f(x)\| = 0,$$

letting $m \to \infty$ in the previous chain of inequalities yields $0 < \varepsilon \le 0$, which is impossible. Therefore, $\mathfrak{F}$ is equicontinuous.

In order to prove that (b) implies (a), given any sequence in $\mathfrak{F}$, $\{f_n\}_{n \ge 1}$, we shall extract a subsequence, say $\{g_n\}_{n \ge 1}$, converging in $\mathcal{C}(K; \mathbb{R}^N)$ to some function $g \in \mathcal{C}(K; \mathbb{R}^N)$. The proof of this fact will be divided into three steps:

1. There exists a sequence of points $d_n \in K$, $n \ge 1$, such that the set

$$D := \{d_n : n \ge 1\}$$

   is dense in K, i.e. $\bar{D} = K$.
2. There exists a subsequence of $\{f_n\}_{n \ge 1}$, $\{g_n\}_{n \ge 1}$, point-wise converging in D.
3. The subsequence $\{g_n\}_{n \ge 1}$ is a Cauchy sequence in the Banach space $\mathcal{C}(K; \mathbb{R}^N)$. Therefore, it converges to some function $g \in \mathcal{C}(K; \mathbb{R}^N)$, uniformly in K.

Step 1 holds true in any Banach space, Step 2 is based on the fact that $\mathfrak{F}$ is bounded, and Step 3 follows from the equicontinuity of $\mathfrak{F}$.

Step 1. For each natural number $n \ge 1$,

$$K \subset \bigcup_{x \in K} B_{\frac{1}{n}}(x).$$

Thus, by compactness, for each natural number $n \geq 1$, there exist $p(n)$ points, $x_{n,j} \in K$, $j \in \{1, \ldots, p(n)\}$, such that

$$K \subset \bigcup_{j=1}^{p(n)} B_{\frac{1}{n}}(x_{n,j}). \tag{7.4}$$

By construction, the set

$$D := \{x_{n,j} : n \geq 1, \ 1 \leq j \leq p(n)\}$$

is countable since it is a countable union of finite sets. We show that it is dense in K. Indeed, from (7.4), for every $x \in K$ and $n \geq 1$, there exists $j(n, x) \in \{1, \ldots, p(n)\}$ such that

$$\left\| x - x_{n,j(n,x)} \right\| < \frac{1}{n}.$$

Hence,

$$\lim_{n \to \infty} x_{n,j(n,x)} = x.$$

By renaming its elements, we can set

$$D = \{d_n : n \geq 1\},$$

and this completes the proof of Step 1.

Step 2. Since $\mathfrak{F}$ is bounded in $\mathcal{C}(K; \mathbb{R}^N)$, there exists a constant $M > 0$ such that

$$\|f_n\|_\infty = \max_{x \in K} \|f_n(x)\| \leq M \quad \text{for all } n \geq 1. \tag{7.5}$$

In particular, $\{f_n(d_j)\}_{n \geq 1}$ is a bounded sequence of $\mathbb{R}^N$, for all $j \geq 1$. Thus, thanks to the Bolzano–Weierstrass theorem we can extract a subsequence of $\{f_n\}_{n \geq 1}$, say, $\{f_n^1\}_{n \geq 1}$, such that, for some $\ell_1 \in \mathbb{R}^N$,

$$\lim_{n \to \infty} f_n^1(d_1) = \ell_1. \tag{7.6}$$

Similarly, since $\{f_n^1(d_2)\}_{n \geq 1}$ is bounded due to (7.5), by the Bolzano–Weierstrass theorem we can extract a subsequence of $\{f_n^1\}_{n \geq 1}$, say, $\{f_n^2\}_{n \geq 1}$, such that, for some $\ell_2 \in \mathbb{R}^N$,

$$\lim_{n \to \infty} f_n^2(d_2) = \ell_2. \tag{7.7}$$

Naturally, as $\{f_n^2\}_{n\geq 1}$ is a subsequence of $\{f_n^1\}_{n\geq 1}$, by (7.6), we do not only have (7.7) but also

$$\lim_{n\to\infty} f_n^2(d_j) = \ell_j \quad \text{for } j \in \{1,2\}.$$

By repeating this argument, it becomes apparent that, for every $m \geq 1$, one can construct a subsequence of $\{f_n\}_{n\geq 1}$, denoted by $\{f_n^m\}_{n\geq 1}$, such that:

- $\{f_n^k\}_{n\geq 1}$ is a subsequence of $\{f_n^{k-1}\}_{n\geq 1}$, for all $k \geq 2$;
- for every $j \in \{1,\dots,m\}$, there exists $\ell_j \in \mathbb{R}^N$ such that

$$\lim_{n\to\infty} f_n^m(d_j) = \ell_j.$$

The following table represents all these subsequences, ordered by rows. When the functions of the mth line are evaluated at d_j, $j \in \{1,\dots,m\}$, they converge to ℓ_j. Moreover, each of the sequences is a subsequence of the sequence in the previous line.

$$
\begin{array}{cccccl}
f_1^1 & f_2^1 & f_3^1 & \cdots & f_n^1 & \cdots \quad (\text{converges in } \{d_1\}) \\
f_1^2 & f_2^2 & f_3^2 & \cdots & f_n^2 & \cdots \quad (\text{converges in } \{d_1, d_2\}) \\
f_1^3 & f_2^3 & f_3^3 & \cdots & f_n^3 & \cdots \quad (\text{converges in } \{d_1, d_2, d_3\}) \\
\vdots & \vdots & \vdots & \vdots & \vdots & \vdots \\
f_1^m & f_2^m & f_3^m & \cdots & f_n^m & \cdots \quad (\text{converges in } \{d_1, d_2, d_3, \dots, d_m\}) \\
\vdots & \vdots & \vdots & \vdots & \vdots & \vdots
\end{array}
$$

By construction, the functions on the diagonal of this table,

$$g_n := f_n^n, \quad n \geq 1,$$

provide us with a subsequence of $\{f_n\}_{n\geq 1}$ that converges at d_m for all $m \geq 1$. Indeed, fix an arbitrary $m \geq 1$. Then, since $\{g_n\}_{n\geq m}$ is a subsequence of $\{f_n^m\}_{n\geq 1}$, it converges in $\{d_1, d_2, \dots, d_m\}$. In particular, it converges at d_m. This shows the previous claim and completes the proof of Step 2.

Step 3. We claim that the sequence $\{g_n\}_{n\geq 1} \subset \mathfrak{F}$ constructed in Step 2 is a Cauchy sequence in $\mathcal{C}(K; \mathbb{R}^N)$. Indeed, fix $\varepsilon > 0$. As $\mathfrak{F}$

is equicontinuous, there exists $\delta = \delta(\varepsilon) > 0$ such that, as soon as $x, y \in K$ with $\|x - y\| < \delta$,

$$\|g_n(x) - g_n(y)\| < \frac{\varepsilon}{3} \quad \text{for all } n \geq 1. \tag{7.8}$$

On the other hand, since the set D constructed in Step 1 satisfies $\bar{D} = K$, it is apparent that

$$K \subset \bigcup_{n \geq 1} B_\delta(d_n).$$

Indeed, for every $x \in K$, there exists $n \geq 1$ such that $\|x - d_n\| < \delta$, i.e. $x \in B_\delta(d_n)$. Thus, since K is compact, there exists an integer $q \geq 1$ such that

$$K \subset \bigcup_{j=1}^{q} B_\delta(d_j). \tag{7.9}$$

Furthermore, since

$$\lim_{n \to \infty} g_n(d_j) = \ell_j \quad \text{for all } j \geq 1,$$

$\{g_n(d_j)\}_{n \geq 1}$ is a Cauchy sequence in $\mathbb{R}^N$, for all $j \geq 1$. Thus, for each $j \geq 1$, there exists an integer $n_j = n_j(\varepsilon) \geq 1$ such that

$$\|g_n(d_j) - g_{n+m}(d_j)\| < \frac{\varepsilon}{3} \quad \text{for all } n \geq n_j \text{ and } m \geq 1. \tag{7.10}$$

Set

$$n(\varepsilon) := \max_{1 \leq j \leq q} n_j(\varepsilon)$$

and take any $x \in K$, $n \geq n(\varepsilon)$ and $m \geq 1$. Then, by (7.9), there exists $j = j(x) \in \{1, \ldots, q\}$ such that $x \in B_\delta(d_j)$, i.e. $\|x - d_j\| < \delta$. Thus,

$$\|g_n(x) - g_{n+m}(x)\| \leq \|g_n(x) - g_n(d_j)\| + \|g_n(d_j) - g_{n+m}(d_j)\|$$
$$+ \|g_{n+m}(d_j) - g_{n+m}(x)\|.$$

Since $\|x - d_j\| < \delta$, (7.8) implies that

$$\|g_n(x) - g_n(d_j)\| < \frac{\varepsilon}{3}, \quad \|g_{n+m}(d_j) - g_{n+m}(x)\| < \frac{\varepsilon}{3}.$$

Moreover since $n \geq n(\varepsilon) \geq n_j$, (7.10) holds. Therefore,

$$\|g_n(x) - g_{n+m}(x)\| < 3\frac{\varepsilon}{3} = \varepsilon$$

for all $x \in K$, $n \geq n(\varepsilon)$ and $m \geq 1$. Consequently,

$$\|g_n - g_{n+m}\|_\infty \leq \varepsilon, \quad n \geq n(\varepsilon), \quad m \geq 1.$$

The proof is complete. $\qquad\square$

7.2 Global Peano Theorem

The following result was proven in 1890 by G. Peano.

Theorem 7.6 (Global Peano theorem). *Suppose $J = [t_0, t_0 + T]$ and $f \in \mathcal{C}(J \times \mathbb{R}^N; \mathbb{R}^N)$ is bounded, i.e. there exists a constant $M > 0$ such that*

$$\|f(t, x)\| \leq M \quad \text{for all } (t, x) \in J \times \mathbb{R}^N. \tag{7.11}$$

Then, for every $u_0 \in \mathbb{R}^N$, the Cauchy problem (7.1) possesses, at least, one solution which is globally defined in the interval J.

Observe that the global estimate (7.11) is necessary for the validity of the theorem even if we additionally assume that f is locally Lipschitz with respect to $u \in \mathbb{R}^N$, uniformly in $t \in J$. Indeed, as we have shown in Section 4.3, the unique solution of the problem

$$\begin{cases} u' = u^2, \\ u(0) = u_0 > 0, \end{cases}$$

which is given by

$$u(t; u_0) = \frac{1}{u_0^{-1} - t},$$

blows up at time $T_{\max} = u_0^{-1}$. Thus, condition (7.11) is essential for the validity of the theorem.

Proof of Theorem 7.6. Thanks to Lemma 5.6, we already know that the solutions of (7.1) are the continuous functions $u \in C(J; \mathbb{R}^N)$ such that

$$u(t) = u_0 + \int_{t_0}^t f(s, u(s))\, ds \quad \text{for all } t \in J. \tag{7.12}$$

The basic idea of the proof is to construct an approximating sequence $\{u_n\}_{n \geq 1}$ to a solution of (7.12). This can be done, for every integer $n \geq 1$, by dividing the interval J in n subintervals with the same length, $\frac{T}{n}$,

$$\left[t_0 + \frac{(i-1)T}{n}, t_0 + \frac{iT}{n} \right], \quad 1 \leq i \leq n,$$

and considering, for every $n \geq 1$, the function

$$u_n(t) := \begin{cases} u_0 & \text{if } t \in \left[t_0, t_0 + \frac{T}{n} \right], \\ u_0 + \int_{t_0}^{t - \frac{T}{n}} f(s, u_n(s))\, ds & \text{if } t \in \left(t_0 + \frac{T}{n}, t_0 + T \right]. \end{cases} \tag{7.13}$$

Let us better understand how the function u_n is defined on each of the intervals of the considered partition. In $\left[t_0, t_0 + \frac{T}{n} \right]$, u_n equals the constant u_0. If $t \in \left(t_0 + \frac{T}{n}, t_0 + \frac{2T}{n} \right]$, then

$$t_0 < t - \frac{T}{n} \leq t_0 + \frac{T}{n},$$

and hence, according to (7.13),

$$u_n(t) = u_0 + \int_{t_0}^{t - \frac{T}{n}} f(s, u_0)\, ds.$$

This shows that $u_n(t)$ is continuous in $\left(t_0 + \frac{T}{n}, t_0 + \frac{2T}{n} \right]$ and that

$$\lim_{t \downarrow t_0 + \frac{T}{n}} u_n(t) = u_0 + \int_{t_0}^{t_0} f(s, u_0)\, ds = u_0 = u_n\left(t_0 + \frac{T}{n} \right).$$

Thus, u_n is continuous also at $t_0 + \frac{T}{n}$. Similarly, if $t \in \left(t_0 + \frac{2T}{n}, t_0 + \frac{3T}{n} \right]$, then

$$t_0 + \frac{T}{n} < t - \frac{T}{n} \leq t_0 + \frac{2T}{n},$$

and hence, according to (7.13),

$$u_n(t) = u_0 + \int_{t_0}^{t-\frac{T}{n}} f(s, u_n(s))\, ds$$

$$= u_0 + \int_{t_0}^{t_0+\frac{T}{n}} f(s, u_0)\, ds$$

$$+ \int_{t_0+\frac{T}{n}}^{t-\frac{T}{n}} f\left(s, u_0 + \int_{t_0}^{s-\frac{T}{n}} f(\tau, u_0)\, d\tau\right) ds$$

is also continuous in $\left(t_0 + \frac{2T}{n}, t_0 + \frac{3T}{n}\right]$. Moreover, since

$$\lim_{t\downarrow t_0+\frac{2T}{n}} u_n(t) = u_0 + \int_{t_0}^{t_0+\frac{T}{n}} f(s, u_0)\, ds = u_n\left(t_0 + \frac{2T}{n}\right),$$

u_n is continuous also at $t_0 + \frac{2T}{n}$. By reasoning recursively, it becomes apparent that $u_n \in \mathcal{C}(J; \mathbb{R}^N)$ for all $n \geq 1$. Consequently,

$$\mathfrak{F} := \{u_n : n \geq 1\} \subset \mathcal{C}(J; \mathbb{R}^N).$$

In order to apply the Ascoli–Arzelà theorem (see Theorem 7.5) to the set $\mathfrak{F}$, we shall check that $\mathfrak{F}$ is bounded in $\mathcal{C}(J; \mathbb{R}^N)$ and equicontinuous. Starting with boundedness, for every $n \geq 1$ and $t \in \left[t_0, t_0 + \frac{T}{n}\right]$, it follows from (7.13) that

$$\|u_n(t)\| = \|u_0\|,$$

whereas, for every $n \geq 1$ and $t \in \left[t_0 + \frac{T}{n}, t_0 + T\right]$, (7.13) and (7.11) give

$$\|u_n(t)\| = \left\|u_0 + \int_{t_0}^{t-\frac{T}{n}} f(s, u_n(s))\, ds\right\|$$

$$\leq \|u_0\| + \int_{t_0}^{t-\frac{T}{n}} \|f(s, u_n(s))\|\, ds \leq \|u_0\| + MT.$$

Thus,

$$\|u_n\|_{\mathcal{C}(J;\mathbb{R}^N)} \leq \|u_0\| + MT \quad \text{for all } n \geq 1$$

and, hence, $\mathfrak{F}$ is bounded in $\mathcal{C}(J; \mathbb{R}^N)$.

The equicontinuity of $\mathfrak{F}$ can be established as follows. First, take $t, s \in \left[t_0, t_0 + \frac{T}{n}\right]$. Then, it follows from (7.13) that

$$\|u_n(t) - u_n(s)\| = \|u_0 - u_0\| = 0.$$

Now, take t and s so that

$$t_0 \leq t \leq t_0 + \frac{T}{n} \leq s \leq t_0 + T.$$

For this choice, according to (7.13) and (7.11),

$$\|u_n(t) - u_n(s)\| = \left\| u_0 - u_0 - \int_{t_0}^{s - \frac{T}{n}} f(\tau, u_n(\tau)) \, d\tau \right\|$$

$$\leq M \left| s - \frac{T}{n} - t_0 \right| = M \left| s - \left(t_0 + \frac{T}{n}\right) \right| \leq M \left| t - s \right|.$$

Lastly, suppose that

$$t, s \in \left[t_0 + \frac{T}{n}, t_0 + T\right].$$

In this case, we have

$$\|u_n(t) - u_n(s)\| = \left\| u_0 + \int_{t_0}^{t - \frac{T}{n}} f(\tau, u_n(\tau)) \, d\tau \right.$$

$$\left. - u_0 - \int_{t_0}^{s - \frac{T}{n}} f(\tau, u_n(\tau)) \, d\tau \right\|$$

$$= \left\| \int_{s - \frac{T}{n}}^{t - \frac{T}{n}} f(\tau, u_n(\tau)) \, d\tau \right\| \leq M|t - s|.$$

Thus, for every $t, s \in J$ and $n \geq 1$, we have shown that

$$\|u_n(t) - u_n(s)\| \leq M|t - s|.$$

So, thanks to Proposition 7.4, the set $\mathfrak{F}$ is equicontinuous. Therefore, by Theorem 7.5, there exist a continuous function $u \in \mathcal{C}(J; \mathbb{R}^N)$ and a subsequence, $\{u_{n_m}\}_{m \geq 1}$, of $\{u_n\}_{n \geq 1}$ such that

$$\lim_{m \to \infty} \|u_{n_m} - u\|_{C(J;\mathbb{R}^N)} = 0. \tag{7.14}$$

It remains to check that u solves the integral equation (7.12). Indeed, since

$$\lim_{m \to \infty} \frac{T}{n_m} = 0,$$

for every $t > t_0$, there exists $m_0 \geq 1$ such that

$$t > t_0 + \frac{T}{n_m} \quad \text{for all } m \geq m_0.$$

Thus, thanks to (7.13), for all $m \geq m_0$,

$$\begin{aligned}
u_{n_m}(t) &= u_0 + \int_{t_0}^{t - \frac{T}{n_m}} f(s, u_{n_m}(s))\, ds \\
&= u_0 + \int_{t_0}^{t} f(s, u_{n_m}(s))\, ds - \int_{t - \frac{T}{n_m}}^{t} f(s, u_{n_m}(s))\, ds.
\end{aligned} \tag{7.15}$$

On the other hand, by (7.11),

$$\left\| \int_{t - \frac{T}{n_m}}^{t} f(s, u_{n_m}(s))\, ds \right\| \leq M \frac{T}{n_m}$$

and, hence,

$$\lim_{m \to \infty} \int_{t - \frac{T}{n_m}}^{t} f(s, u_{n_m}(s))\, ds = 0. \tag{7.16}$$

Moreover, as we will show in the last part of the proof,

$$\lim_{m \to \infty} \int_{t_0}^{t} f(s, u_{n_m}(s))\, ds = \int_{t_0}^{t} f(s, u(s))\, ds. \tag{7.17}$$

Thus, letting $m \to \infty$ in (7.15), it follows from (7.16) and (7.17) that

$$u(t) = u_0 + \int_{t_0}^{t} f(s, u(s))\, ds$$

for all $t > t_0$. Moreover, letting $t \to t_0$, we also find that $u(t_0) = u_0$. Therefore, u solves the integral equation (7.12). Equivalently, $u(t)$ solves the Cauchy problem (7.1).

To prove (7.17), for every $R > 0$, consider the tubular neighborhood of the graph of u, $(t, u(t))$, $t \in J$,

$$\mathcal{Q}_R := \left\{ (t, x) \in J \times \mathbb{R}^N : \|x - u(t)\| \le R \right\}.$$

As $\mathcal{Q}_R$ is a bounded and closed subset of $J \times \mathbb{R}^N$, it is compact. Thus, since f is continuous in $J \times \mathbb{R}^N$, it is uniformly continuous in $\mathcal{Q}_R$. Hence, for any given $\varepsilon > 0$ there exists $\delta = \delta(\varepsilon) > 0$ such that, as soon as $(t, x), (t, y) \in \mathcal{Q}_R$ with $\|x - y\| \le \delta$,

$$\|f(t, x) - f(t, y)\| \le \varepsilon. \tag{7.18}$$

On the other hand, owing to (7.14), there exists $m_1 \ge m_0$ such that

$$\|u_{n_m}(t) - u(t)\| \le \delta \quad \text{for all } m \ge m_1 \text{ and } t \in J. \tag{7.19}$$

Thus, for every $m \ge m_1$, from (7.19) and (7.18), we obtain

$$\left\| \int_{t_0}^{t} f(s, u_{n_m}(s)) \, ds - \int_{t_0}^{t} f(s, u(s)) \, ds \right\|$$

$$= \left\| \int_{t_0}^{t} [f(s, u_{n_m}(s)) - f(s, u(s))] \, ds \right\|$$

$$\le \int_{t_0}^{t} \|f(s, u_{n_m}(s)) - f(s, u(s))\| \, ds \le \varepsilon |t - t_0| \le \varepsilon T.$$

The proof is complete. $\qquad\qquad\qquad\qquad\qquad\qquad\qquad\qquad\qquad\square$

We now show that, under the general assumptions of the global Peano theorem (see Theorem 7.6), the Cauchy problem (7.1) can admit infinitely many solutions. Indeed, consider the autonomous function $f(u)$ defined by

$$f(u) := \begin{cases} 0 & \text{if } u \le 0, \\ u^q & \text{if } 0 < u < 1, \\ 1 & \text{if } u \ge 1, \end{cases} \tag{7.20}$$

for some $q \in (0, 1)$, whose graph is plotted in Figure 7.2.

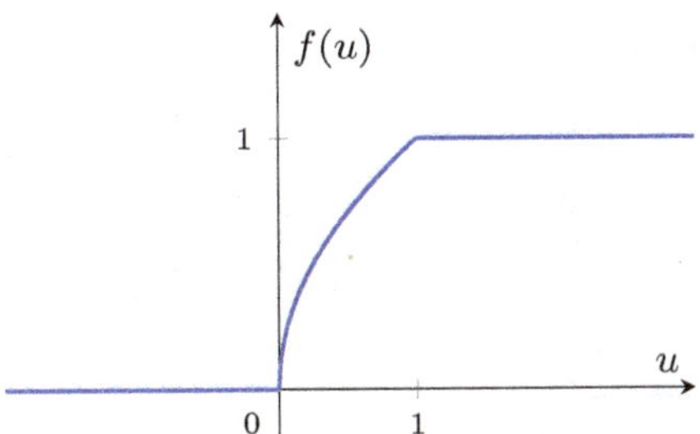

Fig. 7.2. Plot of the function $f(u)$ defined in (7.20).

Obviously, $f \in \mathcal{C}(\mathbb{R}; \mathbb{R})$ and $\|f\|_\infty = 1$. Thus, according to Theorem 7.6, for every $u_0 \geq 0$ and $T > 0$, the Cauchy problem

$$\begin{cases} u' = f(u), \\ u(0) = u_0, \end{cases}$$

has, at least, one solution which is globally defined in $[0, T]$. Observe that, in this case, Proposition 5.2 ensures that f is not Lipschitz continuous at $u = 0$ because

$$\lim_{u \downarrow 0} f'(u) = \lim_{u \downarrow 0} \left(q\, u^{q-1} \right) = +\infty.$$

Thus, neither the Cauchy–Lipschitz theory nor Theorem 4.2 can give, without additional work, the existence of a global solution defined in $[0, T]$. Nevertheless, these results suggest that the natural candidate for getting multiplicity of solutions is the Cauchy problem

$$\begin{cases} u' = f(u), \\ u(0) = 0. \end{cases} \tag{7.21}$$

Thanks to the analysis in Section 4.5, we already know that

$$\varphi(t) := [(1 - q)\, t]^{\frac{1}{1-q}}, \quad t \geq 0,$$

solves problem (7.21) as soon as $0 \leq \varphi(t) \leq 1$, i.e. in the interval $\left[0, \frac{1}{1-q}\right]$. Naturally, if we consider, for $t > \frac{1}{1-q}$, the unique solution of

$$\begin{cases} u' = 1, \\ u\left(\frac{1}{1-q}\right) = 1, \end{cases}$$

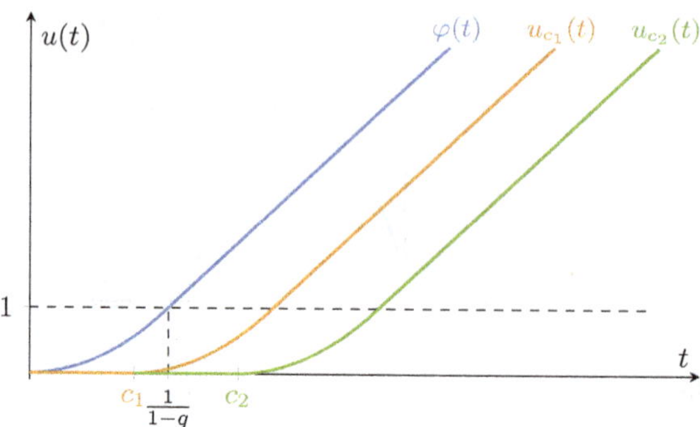

Fig. 7.3. Plots of some solutions $u_c(t) = \varphi(t - c)$ of (7.21), with f defined in (7.20).

which is

$$u(t) = t - \frac{q}{1-q},$$

we can extend $\varphi(t)$ as follows,

$$\varphi(t) := \begin{cases} [(1-q)\,t]^{\frac{1}{1-q}} & \text{if } 0 \le t \le \frac{1}{1-q}, \\ t - \frac{q}{1-q} & \text{if } t > \frac{1}{1-q}, \end{cases}$$

and we get a non-negative solution of (7.21) that is globally defined in $[0, \infty)$. Moreover, arguing as in Section 4.5, for every constant $c > 0$, each of the functions

$$u_c(t) := \begin{cases} 0 & \text{if } 0 \le t \le c, \\ \varphi(t - c) & \text{if } t > c, \end{cases}$$

provides us with another solution to (7.21) because 0 is the equilibrium solution of $u' = f(u)$. Therefore, this example, which fulfills the hypothesis of Theorem 7.6, possesses a one-parameter family of solutions. Figure 7.3 shows the plots of some of them.

7.3   Local Peano Theorem

In this section and the following one, the Peano theory follows very similar patterns to those of the Cauchy–Lipschitz theory, except for

the proof of the existence of a maximal solution, which will rely on
the use of the Kuratowski–Zorn lemma.

The main result in this section is the following local version of the
Peano theorem. The proof is an adaptation of the proof of the local
Cauchy–Lipschitz theorem (see Theorem 5.9). Actually, it is simpler
since checking the assumptions of the global Peano theorem for a
suitable extension of f is easier than checking its global Lipschitz
character.

Theorem 7.7 (Local Peano theorem). *Suppose $J = [t_0, t_0 + T]$, Ω is an open subset of $\mathbb{R}^N$, and $f \in \mathcal{C}(J \times \Omega; \mathbb{R}^N)$. Then, for every $u_0 \in \Omega$, there exists $\delta = \delta(u_0) > 0$ such that the Cauchy problem (7.1) admits, at least, one solution in the interval $[t_0, t_0 + \delta]$.*

Proof. Let $R > 0$ be such that $\bar{B}_R(u_0) \subset \Omega$, and consider the extension of f outside $\bar{B}_R(u_0)$, $F : J \times \mathbb{R}^N \to \mathbb{R}^N$, defined as

$$F(t, u) := \begin{cases} f(t, u) & \text{if } \|u - u_0\| \le R, \\ f(t, \tilde{u}) & \text{if } \|u - u_0\| > R, \end{cases}$$

where, for every $u \in \mathbb{R}^N \setminus \bar{B}_R(u_0)$,

$$\tilde{u} = \tilde{u}(u) := u_0 + \frac{u - u_0}{\|u - u_0\|} R.$$

Observe that $\|\tilde{u} - u_0\| = R$ and that $u \mapsto \tilde{u}(u)$ is continuous. As
a consequence, F is continuous, and the values it attains are those
attained by f on $J \times \bar{B}_R(u_0)$, which is compact. Thus, F is bounded
in $J \times \mathbb{R}^N$, and Theorem 7.6 guarantees that the problem

$$\begin{cases} u' = F(t, u), \\ u(t_0) = u_0, \end{cases}$$

has, at least, one solution, $u(t)$, that is globally defined in J. By
continuity, for sufficiently small $\delta > 0$, $u(t) \in B_R(u_0)$ if $t \in [t_0, t_0 + \delta]$.
Hence, according to the definition of F, $u(t)$ also provides us with a
solution of (7.1) on $[t_0, t_0 + \delta]$. The proof is complete. $\qquad \square$

The following example shows that (7.1) might actually have infinitely many solutions exhibiting a great variety of different behaviors. As in the previous section, here we consider (7.21) but with

$$f(u) := \begin{cases} 0 & \text{if } u \le 0, \\ u^q & \text{if } 0 < u < 1, \\ u^p & \text{if } u \ge 1, \end{cases} \tag{7.22}$$

where $q \in (0,1)$, in order to get multiple solutions, and $p > 1$, to provoke their explosion in a finite time. Figure 7.4 shows the plot of this function.

From the analysis in the previous section, we know that

$$\varphi(t) = [(1-q)\,t]^{\frac{1}{1-q}}, \quad 0 \le t \le \frac{1}{1-q},$$

solves (7.21). Since $\varphi\left(\frac{1}{1-q}\right) = 1$, in order to extend this solution for $t > \frac{1}{1-q}$, we must solve the problem

$$\begin{cases} u' = u^p, \\ u\left(\frac{1}{1-q}\right) = 1, \end{cases} \tag{7.23}$$

According to Theorem 4.2, the Cauchy problem (7.23) has a unique solution, $u(t)$, which is given by

$$\int_{\frac{1}{1-q}}^{t} \frac{d}{ds}\left(\frac{u^{1-p}(s)}{1-p}\right) ds = \int_{\frac{1}{1-q}}^{t} ds = t - \frac{1}{1-q}.$$

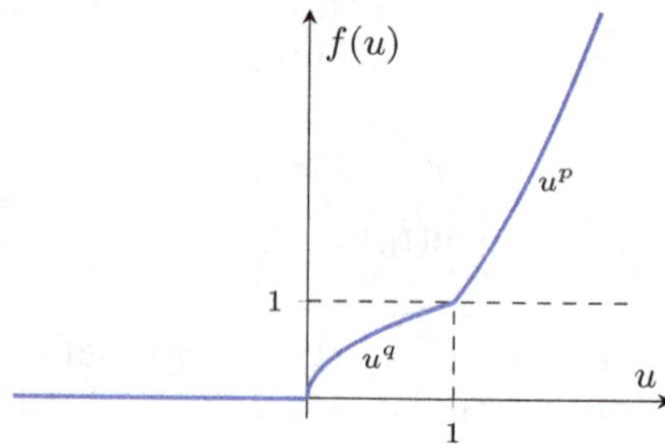

Fig. 7.4. Plot of the function $f(u)$ defined in (7.22).

Thus,

$$\frac{u^{1-p}(t)}{1-p} - \frac{1}{1-p} = t - \frac{1}{1-q},$$

and hence,

$$u(t) = \left[1 + \frac{p-1}{1-q} - (p-1)\,t\right]^{\frac{-1}{p-1}}, \qquad t \geq \frac{1}{1-q}.$$

In particular, this solution blows up in a finite time

$$\tau := \frac{1}{1-q} + \frac{1}{p-1}.$$

Therefore, besides the constant solution $u = 0$, problem (7.21) with the choice (7.22) admits the positive solution

$$\varphi(t) = \begin{cases} \left[(1-q)\,t\right]^{\frac{1}{1-q}} & \text{if } 0 \leq t \leq \frac{1}{1-q}, \\[2mm] \left[1 + \frac{p-1}{1-q} - (p-1)\,t\right]^{\frac{-1}{p-1}} & \text{if } \frac{1}{1-q} < t < \frac{1}{1-q} + \frac{1}{p-1} = \tau. \end{cases}$$

Moreover, since the problem is autonomous, the associated shifted solutions, for every $c > 0$,

$$u_c(t) := \begin{cases} 0 & \text{if } 0 \leq t \leq c, \\[2mm] \varphi(t-c) & \text{if } c < t < \tau + c, \end{cases}$$

are solutions of (7.21). Some of these solutions are plotted in Figure 7.5. Observe that, for every $c > 0$, the solution $([0, \tau + c), u_c(t))$ is maximal, in the sense that it cannot be extended to the right.

Subsequently, for a fixed $T > \tau$, we analyze the nature of the solutions of (7.21) in the interval $J = [0, T]$. The previous construction shows that, for every $T_{\max} \in [\tau, T]$, problem (7.21) admits a positive solution that blows up in $T_{\max}$. Moreover, it admits a family of solutions which are globally defined in $[0, T]$, those lying below the solution that blows up at time T, which is given by $u_{T-\tau}(t)$. Some of these solutions are represented in Figure 7.6.

Summarizing, in the interval $[0, T]$, the Cauchy problem (7.21), with $f(u)$ given by (7.22), possesses infinitely many explosive

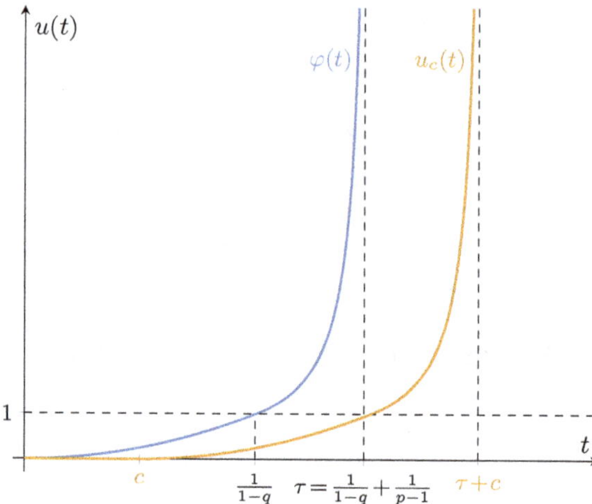

Fig. 7.5. Plots of some solutions of (7.21) with the choice (7.22).

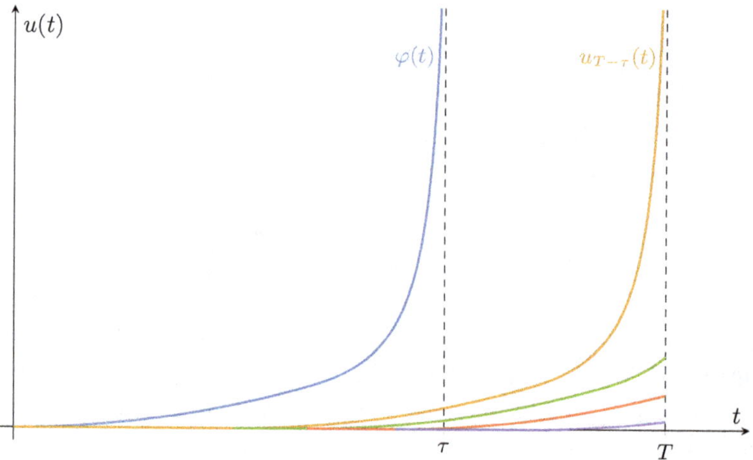

Fig. 7.6. A series of solutions of (7.21) with f given by (7.22), exhibiting different nature.

solutions, which can blow up at any time between τ and T, and simultaneously, it possesses a one-parameter family of positive solutions which are globally defined in $[0, T]$. Moreover, the set of values reached at time T by all these solutions is the interval $[0, +\infty)$. This aspect will be analyzed, in general, in Section 7.8.

7.4 Existence of Maximal Solutions

In this section, we appreciate one of the main differences between the Cauchy–Lipschitz theory and the Peano theory. According to the former, as explained in Chapter 5, problem (7.1) has a unique solution if $f(t, u)$ is locally Lipschitz in u uniformly on compact subintervals of the time interval. This entails that, if (I_1, u_1) and (I_2, u_2) are two solutions of (7.1) with $I_1 \subset I_2$, then $u_2|_{I_1} = u_1$. Nevertheless, the example analyzed at the end of Section 7.3 reveals that this property is not true if the Lipschitz continuity of $f(u)$ fails at u_0 and f is merely continuous. Indeed, none of the global solutions of (7.21) constructed for the special choice (7.22) is an extension of any other of them. Thus, in general, the relation *"being an extension of"* may not be total if the Lipschitz continuity of $f(t, u)$ fails at u_0. This forces us to use the Kuratowski–Zorn lemma to prove the existence of a maximal solution for (7.1) in the context of the Peano theory. This is the content of the remainder of this section.

7.4.1 *Partial inductive orders*

Suppose that $(X, \preccurlyeq)$ is an ordered set. Recall that $(X, \preccurlyeq)$ is said to be *totally ordered* (or that $\preccurlyeq$ is a *total order* in X) if, for every $x, y \in X$, either $x \preccurlyeq y$ or $y \preccurlyeq x$. If the order is not total, it is said that $\preccurlyeq$ is a *partial order* in X, or, equivalently, that $(X, \preccurlyeq)$ is *partially ordered*.

Suppose that $(X, \preccurlyeq)$ is a partially ordered set. A subset $C \subset X$ is said to be a *chain* when $(C, \preccurlyeq)$ is totally ordered. In other words, C is a chain of X if, for every $c_1, c_2 \in C$ either $c_1 \preccurlyeq c_2$ or $c_2 \preccurlyeq c_1$.

A partially ordered set $(X, \preccurlyeq)$ is said to be *inductive* if every chain C has an *upper bound* in X, i.e. an element $s \in X$ such that $c \preccurlyeq s$ for all $c \in C$.

The following result is pivotal in set theory.

Theorem 7.8 (Kuratowski–Zorn Lemma). *Suppose that $(X, \preccurlyeq)$ is an inductive partially ordered set. Then, every chain $C \subset X$ possesses a maximal element, i.e. there exists an element $m \in X$ such that:*

- $c \preccurlyeq m$ *for every* $c \in C$;
- *if* $m \preccurlyeq x$ *for some* $x \in X$, *then* $m = x$.

This result was proved by Kuratowski in 1922 and independently by Zorn in 1935. Moreover, within the Zermelo–Fraenkel axiomatic of set theory, it is equivalent to the *axiom of choice* and to the *well-ordering theorem*. The axiom of choice establishes that, for any given arbitrary family of sets, $\{S_\alpha\}_{\alpha \in A}$, one can construct another set by taking an element $x_\alpha \in S_\alpha$ for every $\alpha \in A$. The well-ordering theorem, instead, guarantees that every set can be well-ordered. A set X is said to be well-ordered by a strict total order if every non-empty subset of X has a least element, i.e. a minimum, with respect to the considered order.

Theorem 7.8 is necessary in the proofs of several fundamental results that cover all areas of Mathematics, such as:

- the theorem establishing that every vector space has a basis;
- the theorem establishing that any product of compact subsets is compact, going back to Tychonoff (1930);
- the theorem establishing that, in a normed vector space, any closed set A and any compact set K that are disjoint and convex can be separated by a closed hyperplane, which was proven, independently, by H. Hahn and S. Banach in the late 1920s;
- the theorem establishing that, in any nonzero unitary ring, every proper ideal is contained in a maximal ideal, which was proven by W. Krull in 1929.

In this book, Theorem 7.8 will be used to establish the existence of a *maximal solution* for the Cauchy problem (7.1), as detailed in the following section.

7.4.2 *Existence of maximal solutions*

Throughout this section, we assume $t_0 \in \mathbb{R}$, $T > 0$, $J \in \{[t_0, t_0 + T],$ $[t_0, +\infty)\}$, Ω is an open subset of $\mathbb{R}^N$, $u_0 \in \Omega$, $f \in \mathcal{C}(J \times \Omega; \mathbb{R}^N)$, and we consider the Cauchy problem (7.1). A pair (I, u) formed by a subinterval I of J of the form

$$I \in \{[t_0, t_0 + h), [t_0, t_0 + h], [t_0, +\infty)\},$$

for some $h > 0$, and a function $u \in \mathcal{C}^1(I; \mathbb{R}^N)$ is said to be a solution of (7.1) if $u(t) \in \Omega$ for all $t \in I$, $u(t_0) = u_0$ and $u'(t) = f(t, u(t))$ for all $t \in I$. In such a case, we indicate

$$(I, u) \in \Sigma,$$

i.e. Σ denotes the set of solutions of (7.1). According to the local Peano theorem (see Theorem 7.7), $\Sigma \neq \emptyset$. Naturally, the concept of (right) extensibility introduced in Definition 5.12 is also valid in the more general setting of this chapter and induces an order relation, $\preccurlyeq$, in the set Σ by setting

$$(I_1, u_1) \preccurlyeq (I_2, u_2)$$

if and only if (I_2, u_2) is an extension (to the right) of (I_1, u_1). Nevertheless, since (7.1) might have several solutions in this setting, the set $(\Sigma, \preccurlyeq)$ is, in general, only partially ordered, as illustrated by the example presented in the previous section.

Obviously, the ordered set $(\Sigma, \preccurlyeq)$ possesses chains. Indeed, by the local Peano theorem (see Theorem 7.7), problem (7.1) possesses a solution of the form (I, u), with $I = [t_0, t_0 + \delta]$, for some $\delta > 0$, and hence, if $n_0 \in \mathbb{N}$ satisfies $\frac{1}{n_0} < \delta$, by setting

$$I_n := \left[t_0, t_0 + \delta - \tfrac{1}{n}\right], \quad n \geq n_0,$$

and $u_n := u|_{I_n}$, it is apparent that

$$C := \{(I_n, u_n) \colon n \geq n_0\}$$

provides us with a chain of $(\Sigma, \preccurlyeq)$. The following result establishes that $(\Sigma, \preccurlyeq)$ is inductive.

Lemma 7.9. *Every chain $C := \{(I_\alpha, u_\alpha) \colon \alpha \in A\}$ has an upper bound in Σ, i.e. the order $\preccurlyeq$ is inductive.*

Proof. Since C is a chain, the pair (I_s, u_s), defined by

$$I_s := \bigcup_{\alpha \in A} I_\alpha, \quad u_s(t) := u_\alpha(t), \quad \text{for } t \in I_\alpha,$$

is well defined and provides us with a solution of (7.1), which is an upper bound for C, i.e. $(I_\alpha, u_\alpha) \preccurlyeq (I_s, u_s)$ for all $\alpha \in A$. $\qquad\square$

Considering these preliminaries, the following result is a direct consequence of Theorem 7.8 and Lemma 7.9.

Theorem 7.10 (Existence of a maximal solution). *Suppose $t_0 \in \mathbb{R}$, $T > 0$, $J \in \{[t_0, t_0+T], [t_0, +\infty)\}$, Ω is an open subset of $\mathbb{R}^N$, $u_0 \in \Omega$, and $f \in C(J \times \Omega; \mathbb{R}^N)$. Then, problem (7.1) admits a maximal solution, i.e. a solution (I, u) that cannot be properly extended to the right.*

To summarize the results of the last two sections, we have shown that, in the framework of the Peano theory, any local solution of (7.1), whose existence is guaranteed by the local Peano theorem (see Theorem 7.7) can be extended, possibly in several ways, to a maximal solution.

7.5   Fundamental Theorems for the Peano Theory

This section collects the most important properties of the maximal solutions (to the right) of problem (7.1). As the proofs are exactly the same as those in Section 5.4, we do not provide them here. We only point out that, in this context, where there might be several maximal solutions of (7.1), the global behavior of each of them is independent of the global behavior of the others. Moreover, we recall that, by performing the change of variable (5.31), presented in Section 5.4.1, all the fundamental theorems in this section have their backward counterparts (see Exercise 2 of Chapter 7).

The first result we state is the following counterpart of Theorem 5.15.

Theorem 7.11 (First fundamental theorem). *Suppose $t_0 \in \mathbb{R}$, $T > 0$, $J = [t_0, t_0 + T]$ and $f \in C(J \times \mathbb{R}^N; \mathbb{R}^N)$. For every $u_0 \in \mathbb{R}^N$, let (I, u) be a maximal solution of (7.1). Then, either $I = J$, i.e. u is globally defined in J, or there exists $T_{\max} \in (t_0, t_0 + T]$ such that $I = [t_0, T_{\max})$ and, in such a case,*

$$\limsup_{t \uparrow T_{\max}} \|u(t)\| = +\infty,$$

i.e. $u(t)$ blows up at $T_{\max}$.

As shown in Section 7.3 for problem (7.21) with f given by (7.22), both behaviors described by Theorem 7.11 can occur in the same

Cauchy problem. Moreover, there can be maximal solutions that blow up at different times.

The following theorem is a natural counterpart of Theorem 5.17.

Theorem 7.12 (Second fundamental theorem). *Suppose $t_0 \in \mathbb{R}$, $J = [t_0, +\infty)$ and $f \in C(J \times \mathbb{R}^N; \mathbb{R}^N)$. For every $u_0 \in \mathbb{R}^N$, let (I, u) be a maximal solution of (7.1). Then, there exists $T_{\max} \in (t_0, +\infty]$ such that $I = [t_0, T_{\max})$ and*

$$\limsup_{t \uparrow T_{\max}} \|u(t)\| = +\infty \quad \text{if } T_{\max} < +\infty.$$

Similarly, the following counterpart of Theorem 5.18 holds.

Theorem 7.13 (Third fundamental theorem). *Suppose $t_0 \in \mathbb{R}$, $J = [t_0, +\infty)$, Ω is a proper open subset of $\mathbb{R}^N$, and $f \in C(J \times \Omega; \mathbb{R}^N)$. For every $u_0 \in \Omega$, let (I, u) be a maximal solution of (7.1). Then, $I = [t_0, T_{\max})$ for some $t_0 < T_{\max} \leq +\infty$. Moreover, when $T_{\max} < +\infty$ one of the following non-excluding options occurs: either*

$$\limsup_{t \uparrow T_{\max}} \|u(t)\| = +\infty,$$

i.e. the solution blows up at $T_{\max}$, or there exist a sequence $\{t_n\}_{n \geq 1} \subset I$, with $\lim_{n \to \infty} t_n = T_{\max}$, and a point $u_\omega \in \partial\Omega$ such that

$$\lim_{n \to \infty} u(t_n) = u_\omega \in \partial\Omega,$$

i.e. the solution approximates $\partial\Omega$ at $T_{\max}$.

Finally, the following theorem is the counterpart of Theorem 5.19.

Theorem 7.14 (Fourth fundamental theorem). *Suppose $J = [t_0, t_0 + T]$ for some $t_0 \in \mathbb{R}$ and $T > 0$, Ω is an open bounded subset of $\mathbb{R}^N$, and $f \in C(J \times \bar{\Omega}; \mathbb{R}^N)$. Then, for every $u_0 \in \Omega$, problem (7.1) possesses (at least) a maximal solution. Moreover, for any of these maximal solutions, (I, u), either $I = J$ or $I = [t_0, T_{\max}]$, with $T_{\max} < t_0 + T$, and, in such a case, $u(T_{\max}) \in \partial\Omega$.*

Naturally there are several cases that do not exactly adjust to any of the previous four fundamental theorems but where the techniques of the proofs can be adapted to establish the existence of maximal solutions for problem (7.1), together with their global behavior.

To conclude this section, we estimate the existence time of any maximal solution, (I, u), of problem (7.1). This result is the counterpart of Theorem 5.9(b) in the context of the Peano theory.

Theorem 7.15. *Suppose $J = [t_0, t_0 + T]$ for some $T > 0$, Ω is an open subset of $\mathbb{R}^N$, and $f \in \mathcal{C}(J \times \Omega; \mathbb{R}^N)$. Fix $u_0 \in \Omega$ and let (I, u) be a maximal solution of (7.1). For every $R > 0$ such that $\bar{B}_R(u_0) \subset \Omega$, consider*

$$\|f\|_\infty := \max_{(t,u) \in J \times \bar{B}_R(u_0)} \|f(t, u)\|$$

and

$$\delta_0 := \begin{cases} \min\left\{T, \frac{R}{\|f\|_\infty}\right\} & \text{if } \|f\|_\infty > 0, \\ T & \text{if } \|f\|_\infty = 0. \end{cases} \qquad (7.24)$$

Then,

$$[t_0, t_0 + \delta_0] \subset I.$$

Proof. If $I = J$, we are done. Observe that this occurs, in particular, if $\|f\|_\infty = 0$ because in such a case, $u(t) = u_0$ for all $t \in I$. So, suppose $I \subsetneq J$ and $\|f\|_\infty > 0$. Then, $I = [t_0, T_{\max})$ for some $T_{\max} \leq t_0 + T$. Indeed, if $I = [t_0, T_{\max}]$ for some $T_{\max} < t_0 + T$, then $u(T_{\max}) \in \Omega$, and arguing as at the beginning of the proof of Theorem 5.15, we could extend this solution, contradicting the maximality of (I, u). Subsequently, we consider the orbit of the solution

$$\Gamma := \{(t, u(t)) : t \in I\}.$$

Suppose $\Gamma \subset \bar{B}_R(u_0)$. Then, $\bar{\Gamma}$ is compact, and we can reason as in the proof of Theorem 5.15 to infer that

$$\lim_{t \uparrow T_{\max}} u(t) = u^*$$

for some $u^* \in \bar{B}_R(u_0) \subset \Omega$, which again contradicts the maximality of (I, u). Thus, there exists $\tilde{t} \in (t_0, T_{\max})$ such that

$$\|u(\tilde{t}) - u_0\| > R.$$

By continuity, there exists $t_1 \in (t_0, \tilde{t})$ such that

$$\|u(t_1) - u_0\| = R, \quad \text{and} \quad \|u(t) - u_0\| < R \quad \text{for all } t \in [t_0, t_1].$$

Therefore,

$$R = \|u(t_1) - u_0\| = \left\| \int_{t_0}^{t_1} u'(s)\, ds \right\| = \left\| \int_{t_0}^{t_1} f(s, u(s))\, ds \right\|$$

$$\leq \int_{t_0}^{t_1} \|f(s, u(s))\|\, ds \leq \|f\|_\infty\, (t_1 - t_0),$$

which implies

$$t_1 - t_0 \geq \frac{R}{\|f\|_\infty}.$$

Consequently,

$$\left[t_0, t_0 + \tfrac{R}{\|f\|_\infty} \right] \subset I,$$

and the proof is complete in this case as well. $\qquad\square$

Observe that, in practice, if we set

$$M(R) := \max_{(t,u)\in J \times \bar{B}_R(u_0)} \|f(t, u)\|,$$

then the function $R \mapsto \frac{R}{M(R)}$ can be maximized to get the sharpest lower estimate for the existence interval of maximal solutions. Moreover, the argument of the proof of Theorem 7.15 can be adapted to show that, if $I \subsetneq J$, then, for any compact subset $K \subset \Omega$, there exists $\tilde{t} \in I$ such that $u(\tilde{t}) \notin K$. Thus, when $I \subsetneq J$, any maximal solution (I, u) must eventually leave any compact subset $K \subset \Omega$. Therefore, if Ω is a proper (possibly unbounded) open subset of $\mathbb{R}^N$, and we take $K = K_n$, where

$$K_n := \left\{ x \in \Omega :\ \mathrm{dist}(x, \partial\Omega) \geq \frac{1}{n} \right\} \cap \bar{B}_n(u_0)$$

for sufficiently large $n \geq 1$, we can conclude that any maximal solution (I, u) with $I \subsetneq J$ is either unbounded or it approximates $\partial\Omega$ as $t \uparrow T_{\max}$. Of course, these two options are not exclusive, and the solution might as well oscillate as it approaches $\partial\Omega$ or ∞ (see Figure 5.4).

Naturally, the results in this section have their counterparts for general nonlinear equations of arbitrary order in the vein of

Chapter 6. We do not state these results here, as they are simple applications of the general theory developed in this chapter. For example, to apply the Peano theorems to the Cauchy problems (6.8), (6.44), or (6.46), it suffices to assume, respectively, that g, g_1 and g_2, or g_j, $1 \le j \le n$, are continuous (see Exercise 3 of Chapter 7).

7.6   Müller's Example

In 1927, Müller gave an example revealing that in the context of the Peano theory one cannot expect the iterates of the integral operator associated with (7.1),

$$\mathcal{K}h(t) = u_0 + \int_{t_0}^{t} f(s, h(s))\, ds, \quad h \in \mathcal{C}(J; \mathbb{R}^N),$$

to converge to a solution of (7.1). Specifically, he considered the function

$$f \colon [0, 1] \times \mathbb{R} \to \mathbb{R},$$

defined, for every $t \in (0, 1]$ and $u \in \mathbb{R}$, as

$$f(t, u) := \begin{cases} 2t & \text{if } u \le 0, \\ 2t - \frac{4}{t}u & \text{if } 0 < u < t^2, \\ -2t & \text{if } u \ge t^2, \end{cases} \tag{7.25}$$

and as $f(0, u) = 0$ for all $u \in \mathbb{R}$. Figure 7.7 shows a plot of $u \mapsto f(t, u)$ for a generic $t \in (0, 1]$.

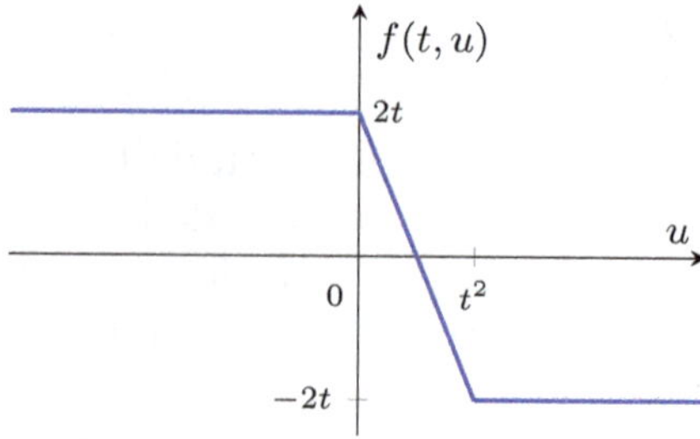

Fig. 7.7.   Plot of $u \mapsto f(t, u)$ defined in (7.25) for $t \in (0, 1]$.

Thus, f is continuous in $[0,1] \times \mathbb{R}$, and it is bounded since

$$|f(t,u)| \le 2 \quad \text{for all } (t,u) \in [0,1] \times \mathbb{R}.$$

Therefore, thanks to the global Peano theorem (see Theorem 7.6), the Cauchy problem

$$\begin{cases} u' = f(t,u), \\ u(0) = 0, \end{cases} \tag{7.26}$$

admits at least a solution, $u(t)$, globally defined in $[0,1]$. Actually, since $u \mapsto f(t,u)$ is non-increasing, this solution is unique. Indeed, suppose $v(t)$ is another solution of (7.26), and set

$$x(t) := (u(t) - v(t))^2 \ge 0, \quad t \in [0,1].$$

Then,

$$x(0) = (u(0) - v(0))^2 = 0.$$

Moreover,

$$\begin{aligned} x'(t) &= 2\left(u(t) - v(t)\right)\left(u'(t) - v'(t)\right) \\ &= 2\left(u(t) - v(t)\right)\left(f(t,u(t)) - f(t,v(t))\right) \le 0 \end{aligned}$$

because

$$f(t,u) \ge f(t,v) \quad \text{if } u \le v.$$

Consequently,

$$0 \le x(t) = x(0) + \int_0^t x'(s)\, ds \le x(0) = 0,$$

and hence $x = 0$, which entails $u(t) = v(t)$ for all $t \in [0,1]$ and that (7.26) has a unique solution. Observe that $f(t,u)$ is not locally Lipschitz continuous at $u = 0$ uniformly in t. Indeed, if there exists a constant $L > 0$ such that

$$\left| f(t,t^2) - f(t,0) \right| \le Lt^2$$

for sufficiently small $t > 0$, then

$$4t = \left| -2t - 2t \right| \le Lt^2,$$

which is impossible.

The integral operator associated with (7.26) is given by

$$\mathcal{K}h(t) = \int_0^t f(s, h(s))\, ds, \quad h \in \mathcal{C}(J; \mathbb{R}^N).$$

Consider the iterates from $h_0 := 0$,

$$h_n := \mathcal{K}h_{n-1}, \quad n \geq 1.$$

By definition,

$$h_1(t) = \mathcal{K}h_0(t) = \int_0^t f(s, h_0(s))\, ds = \int_0^t f(s, 0)\, ds = \int_0^t 2s\, ds = t^2.$$

Similarly,

$$h_2(t) = \mathcal{K}h_1(t) = \int_0^t f(s, h_1(s))\, ds = \int_0^t f(s, s^2)\, ds$$

$$= \int_0^t (-2s)\, ds = -t^2,$$

and

$$h_3(t) = \mathcal{K}h_2(t) = \int_0^t f(s, h_2(s))\, ds = \int_0^t f(s, -s^2)\, ds$$

$$= \int_0^t 2s\, ds = t^2.$$

Therefore, for every integer $n \geq 0$,

$$h_{2n+1}(t) = t^2, \quad t \in [0, 1],$$

and

$$h_{2n+2}(t) = -t^2, \quad t \in [0, 1].$$

However, rather astonishingly, the functions t^2 and $-t^2$ do not solve (7.26) because

$$f(t, t^2) = -2t \neq 2t = \left(t^2\right)' \quad \text{and} \quad f(t, -t^2) = 2t \neq -2t = \left(-t^2\right)'.$$

Consequently, this example shows that the Lipschitz continuity of $f(t, u)$ at u_0, in $t = t_0$, is imperative for the convergence of the $\mathcal{K}$-iterates to some solution of the Cauchy problem rather than for its uniqueness.

7.7 Uniqueness and Continuous Dependence

Obviously, when a Cauchy problem admits multiple solutions, the continuous dependence of solutions with respect to the initial datum fails. In other words, continuous dependence with respect to u_0 entails the uniqueness of the solution of problem (7.1). The following result establishes that, actually, uniqueness is the minimal requirement to obtain the continuous dependence with respect to the initial conditions.

Theorem 7.16. *Let $t_0 \in \mathbb{R}$, $T > 0$, $J = [t_0, t_0 + T]$, $R > 0$, $u_0 \in \mathbb{R}^N$, and $f \in \mathcal{C}(J \times \bar{B}_R(u_0); \mathbb{R}^N)$. Assume that there exist $\eta \in (0, R)$ and $h \in (0, T]$ such that, for every $x \in \bar{B}_\eta(u_0)$, the Cauchy problem*

$$\begin{cases} u' = f(t, u), \\ u(t_0) = x, \end{cases} \tag{7.27}$$

has a unique solution, $u(\cdot; x)$, in $J_h := [t_0, t_0 + h]$; in particular, $u(t; x) \in \bar{B}_R(u_0)$ for all $t \in J_h$. Then, the solution operator

$$\bar{B}_\eta(u_0) \xrightarrow{\;\mathcal{S}\;} \mathcal{C}(J_h; \bar{B}_R(u_0))$$
$$x \longmapsto \mathcal{S}(x) := u(\cdot; x)$$

is continuous.

Proof. Naturally, $\mathcal{C}(J_h; \bar{B}_R(u_0))$ is viewed as a Banach space with the uniform convergence norm.

Let $\{x_n\}_{n \geq 1}$ be a sequence in $\bar{B}_\eta(u_0)$ converging to some x,

$$\lim_{n \to \infty} x_n = x.$$

Necessarily, $x \in \bar{B}_\eta(u_0)$. Our goal is to show that

$$\lim_{n \to \infty} \mathcal{S}(x_n) = \mathcal{S}(x).$$

To prove it, we will show that, for any given subsequence $\{\mathcal{S}(x_{n_m})\}_{m \geq 1}$, one can further extract from it a subsequence converging to $\mathcal{S}(x)$. Indeed, for every $m \geq 1$, consider

$$u_{n_m}(t) := u(t; x_{n_m}), \quad t \in J_h = [t_0, t_0 + h],$$

and define the set of functions

$$\mathfrak{F} := \{ u_{n_m} : m \geq 1 \} .$$

Since $u_{n_m}(t) \in \bar{B}_R(u_0)$ for all $t \in J_h$,

$$\| u_{n_m}(t) \| \leq \| u_0 \| + \| u_{n_m}(t) - u_0 \| \leq \| u_0 \| + R$$

for all $t \in J_h$ and $m \geq 1$, and, hence,

$$\| u_{n_m} \|_{\mathcal{C}(J_h; \mathbb{R}^N)} \leq \| u_0 \| + R, \quad m \geq 1.$$

Thus, $\mathfrak{F}$ is a bounded subset of $\mathcal{C}(J_h; \mathbb{R}^N)$. We now consider the constant

$$M := \| f \|_{\mathcal{C}(J_h \times \bar{B}_R(u_0); \mathbb{R}^N)} = \max_{(t,u) \in J_h \times \bar{B}_R(u_0)} \| f(t, u) \| .$$

Then, for every $m \geq 1$ and $t, s \in J_h$,

$$u_{n_m}(t) - u_{n_m}(s) = \int_s^t u'_{n_m}(\tau) \, d\tau = \int_s^t f(\tau, u_{n_m}(\tau)) \, d\tau$$

and, hence,

$$\| u_{n_m}(t) - u_{n_m}(s) \| \leq M \, |t - s| .$$

Consequently, by Proposition 7.4, $\mathfrak{F}$ is equicontinuous, and due to the Ascoli–Arzelà theorem (see Theorem 7.5), we can extract a subsequence of $\mathfrak{F}$, $\{ u_{n_{m_k}} \}_{k \geq 1}$, such that, for some $u \in \mathcal{C}(J_h; \mathbb{R}^N)$,

$$\lim_{k \to \infty} \| u_{n_{m_k}} - u \|_{\mathcal{C}(J_h; \mathbb{R}^N)} = 0. \tag{7.28}$$

On the other hand, for every $t \in J_h$ and $k \geq 1$, we have that

$$u_{n_{m_k}}(t) = x_{n_{m_k}} + \int_{t_0}^t f(s, u_{n_{m_k}}(s)) \, ds.$$

Thus, letting $k \to \infty$ in this identity, (7.28) implies that

$$u(t) = x + \int_{t_0}^t f(s, u(s)) \, ds.$$

Therefore, $u(t) = u(t; x)$ must be the unique solution of (7.27). In terms of the solution operator $\mathcal{S}$, we have just proved that

$$\lim_{k \to \infty} \mathcal{S}(x_{n_{m_k}}) = \mathcal{S}(x),$$

as we wanted. $\qquad\square$

7.8 Topological Structure of the Set of Solutions

When uniqueness for the Cauchy problem (7.1) fails, the set of solutions of (7.1) can reach an extremely high degree of complexity. Specific examples in this direction were first given by M. Laurentiev in 1925 and P. Hartman in 1963a (see, e.g. the discussion in Section 7.10 for more details).

As a consequence, when $f(t, u)$ is continuous, it is important to obtain as much information as possible about the topological structure of the set of points $\mathcal{A}(t_1) \subset \mathbb{R}^N$ that can be attained by the solutions of (7.1) at a time $t_1 > t_0$. In all of the examples of first-order equations considered up to this point, when multiplicity occurred, $\mathcal{A}(t_1)$ was a non-trivial closed interval of $\mathbb{R}$ for all $t_1 > t_0$ (see, e.g. the construction at the end of Section 4.5).

The main goal of this section is to show that, in the general case of first-order systems, $\mathcal{A}(t_1)$ is a *topological continuum*, i.e. a *closed and connected* subset of $\mathbb{R}^N$. Thus, in the special case when $N = 1$, $\mathcal{A}(t_1)$ must be a closed interval, as it occurred in all the examples previously analyzed in this book. This general result, presented in Theorem 7.17, is due to Kneser (1923). The proof presented here is based on Müller's original proof given in Müller (1928). Two alternative proofs in the case of $N = 1$ are proposed in Exercises 9 and 10 of Chapter 7.

Some remarks are in order. On the one hand, as intervals are convex, one might be tempted to think that $\mathcal{A}(t_1)$ is also convex in the general multidimensional setting, but Exercise 6 of Chapter 7 gives a simple example in the plane where this is not true. On the other hand, observe that, when uniqueness for the Cauchy problem (7.1) holds true, $\mathcal{A}(t_1)$ reduces to one point, which is again a topological continuum, for all $t_1 > t_0$.

Theorem 7.17 (Kneser). *Consider* $J = [t_0, t_0 + T]$, $u_0 \in \mathbb{R}^N$, $R > 0$, $f \in \mathcal{C}(J \times \bar{B}_R(u_0); \mathbb{R}^N)$, *and let* Σ *denote the set of maximal solutions* (I, u) *of* (7.1). *Take* δ_0 *given by* (7.24), *so that* $\tilde{J} := [t_0, t_0 + \delta_0] \subset I$ *by Theorem 7.15. Then, for every*

$$t_1 \in [t_0, t_0 + \delta_0),$$

the set of points of $\mathbb{R}^N$ *attained by the solutions of* (7.1),

$$\mathcal{A}(t_1) := \{u(t_1) : (I, u) \in \Sigma\},$$

is a continuum, i.e. a compact connected subset of $\bar{B}_R(u_0)$.

Proof. If $t_1 = t_0$, then $\mathcal{A}(t_1) = \{u_0\}$. Similarly, if $f = 0$ in $J \times \bar{B}_R(u_0)$, $\mathcal{A}(t_1) = \{u_0\}$ for all $t_1 \in [t_0, t_0 + T]$. Thus, the conclusion is trivial in these cases. So, suppose that $f \neq 0$ in $J \times \bar{B}_R(u_0)$ and that

$$t_0 < t_1 < t_0 + \delta_0 = t_0 + \min\left\{T, \frac{R}{\|f\|_\infty}\right\}. \tag{7.29}$$

By reasoning as in the proof of Theorem 7.15, $\mathcal{A}(t_1) \subset \bar{B}_R(u_0)$. Thus, to prove that $\mathcal{A}(t_1)$ is compact, it suffices to show that it is closed. Let $\{x_n\}_{n\geq 1}$ be a sequence of points such that

$$x_n \in \mathcal{A}(t_1) \quad \text{for all } n \geq 1 \quad \text{and} \quad \lim_{n\to\infty} x_n = x \tag{7.30}$$

for some $x \in \bar{B}_R(u_0)$. In order to prove that $x \in \mathcal{A}(t_1)$, we must construct a solution $(I, u) \in \Sigma$ such that $u(t_1) = x$. As $x_n \in \mathcal{A}(t_1)$ for all $n \geq 1$, there exists a sequence of solutions $(I_n, u_n) \in \Sigma$, $n \geq 1$, such that

$$u_n(t_1) = x_n, \quad n \geq 1. \tag{7.31}$$

Observe that, by the definition of δ_0 (see the proof of Theorem 7.15, if necessary), $\tilde{J} \subset I_n$ and $u_n(t) \in \bar{B}_R(u_0)$ for all $t \in \tilde{J}$ and $n \geq 1$. Then, we consider the set of functions

$$\mathfrak{F} := \{u_n|_{\tilde{J}} : n \geq 1\} \subset \mathcal{C}(\tilde{J}; \mathbb{R}^N).$$

Since $u_n(t) \in \bar{B}_R(u_0)$ for all $t \in \tilde{J}$ and $n \geq 1$, $\mathfrak{F}$ is bounded. Indeed, for every $t \in \tilde{J}$,

$$\|u_n(t)\| = \left\|u_0 + \int_{t_0}^{t} f(s, u_n(s))\, ds\right\| \leq \|u_0\| + \|f\|_\infty \, |t - t_0|.$$

Thus,

$$\|u_n\|_{\mathcal{C}(\tilde{J};\mathbb{R}^N)} \leq \|u_0\| + \|f\|_\infty\, T, \quad n \geq 1.$$

Moreover, for every $t, s \in \tilde{J}$, we have that

$$\|u_n(t) - u_n(s)\| = \left\|\int_{s}^{t} f(\tau, u_n(\tau))\, d\tau\right\| \leq \|f\|_\infty\, |t - s|$$

for all $n \geq 1$ and, hence, $\mathfrak{F}$ is equicontinuous by Proposition 7.4. Therefore, by Theorem 7.5, we can extract a subsequence, $\{u_{n_m}\}_{m\geq 1}$,

such that

$$\lim_{m \to \infty} \|u_{n_m} - u\|_{\mathcal{C}(\tilde{J};\mathbb{R}^N)} = 0 \qquad (7.32)$$

for some function $u \in \mathcal{C}(\tilde{J}; \mathbb{R}^N)$. As, for every $t \in \tilde{J}$ and $m \geq 1$,

$$u_{n_m}(t) = u_0 + \int_{t_0}^{t} f(s, u_{n_m}(s))\, ds,$$

letting $m \to \infty$ in this identity, it follows from (7.32) that

$$u(t) = u_0 + \int_{t_0}^{t} f(s, u(s))\, ds.$$

Thus, u solves (7.1) in the interval $\tilde{J}$. By Theorem 7.10, this solution admits an extension to a maximal solution $(I, u) \in \Sigma$. Since t_1 lies in the interior of $\tilde{J}$, we can infer from (7.32), (7.31) and (7.30) that

$$u(t_1) = \lim_{m \to \infty} u_{n_m}(t_1) = \lim_{m \to \infty} x_{n_m} = x.$$

Therefore, $x \in \mathcal{A}(t_1)$, as claimed. This shows that $\mathcal{A}(t_1)$ is a compact subset of $\bar{B}_R(u_0)$.

To prove that $\mathcal{A}(t_1)$ is connected, we argue by contradiction. So, suppose that there are two disjoint non-empty closed subsets, A^0 and A^1, such that

$$\mathcal{A}(t_1) = A^0 \cup A^1. \qquad (7.33)$$

As these sets are closed in $\mathcal{A}(t_1) \subset \bar{B}_R(u_0)$, they are compact. Thus,

$$d := \mathrm{dist}(A^0, A^1) = \min_{x^0 \in A^0,\, x^1 \in A^1} \left\| x^0 - x^1 \right\| > 0.$$

Subsequently, we consider the auxiliary function $D \colon \mathbb{R}^N \to \mathbb{R}$, defined as

$$D(x) := \mathrm{dist}(x, A^0) - \mathrm{dist}(x, A^1) = \min_{x^0 \in A^0} \left\| x - x^0 \right\| - \min_{x^1 \in A^1} \left\| x - x^1 \right\|$$

for all $x \in \mathbb{R}^N$. As $D(x)$ is a difference between two continuous functions, it is continuous. Moreover, for every $x^0 \in A^0$,

$$D(x^0) = \mathrm{dist}(x^0, A^0) - \mathrm{dist}(x^0, A^1)$$

$$= -\,\mathrm{dist}(x^0, A^1) \leq -\,\mathrm{dist}(A^0, A^1) = -d < 0,$$

whereas, for every $x^1 \in A^1$,

$$D(x^1) = \mathrm{dist}(x^1, A^0) - \mathrm{dist}(x^1, A^1)$$
$$= \mathrm{dist}(x^1, A^0) \geq \mathrm{dist}(A^0, A^1) = d > 0.$$

Thus,

$$D\big|_{A^0} \leq -d < 0 \quad \text{and} \quad D\big|_{A^1} \geq d > 0. \tag{7.34}$$

To complete the proof, it suffices to construct a maximal solution $(I, u) \in \Sigma$ such that $t_1 \in I$ and $D(u(t_1)) = 0$. Indeed, in such a case, $u(t_1) \in \mathcal{A}(t_1)$ and, owing to (7.34), $u(t_1) \notin A^0 \cup A^1$, which contradicts (7.33) and completes the proof.

The construction of this solution is the most delicate part of the proof, as it is based on the continuous dependence of solutions with respect to the nonlinearity of the Cauchy problem. This point has been analyzed in Section 5.6, where $f(t, u)$ was required to be Lipschitz with respect to u, uniformly in t. As here $f(t, u)$ is merely assumed to be continuous, we cannot directly apply the results in Section 5.6. Previously, $f(t, u)$ has to be approximated through a family of Lipschitz functions, $g_\varepsilon(t, u)$. This approximation is possible thanks to the following result of technical nature, where we use the notation introduced in the statement of Theorem 7.17.

Lemma 7.18. *For every* $(I, u) \in \Sigma$ *and* $\varepsilon > 0$, *there exists a function*

$$g = g_\varepsilon \in \mathcal{C}(\tilde{J} \times \bar{B}_R(u_0); \mathbb{R}^N)$$

such that:

(a) $\|f - g\|_{\mathcal{C}(\tilde{J} \times \bar{B}_R(u_0); \mathbb{R}^N)} \leq \varepsilon$;

(b) *there exists* $L = L_\varepsilon > 0$ *such that*

$$\|g(t, x) - g(t, y)\| \leq L\|x - y\|, \quad x, y \in \bar{B}_R(u_0), \quad t \in \tilde{J},$$

i.e. $g(t, u)$ *is globally Lipschitz with respect to* $u \in \bar{B}_R(u_0)$ *uniformly in* $t \in \tilde{J}$;

(c) u *is the unique solution in* $\tilde{J}$ *of the Cauchy problem*

$$\begin{cases} u' = g(t, u), \\ u(t_0) = u_0, \end{cases}$$

Proof. As we will show in Section 7.8.1 (see Theorem 7.19), every continuous function can be approximated uniformly on compact subsets by $\mathcal{C}^\infty$ functions. This, together with Proposition 5.2, ensures that there exist a function $g^* \in \mathcal{C}(\tilde{J} \times \bar{B}_R(u_0); \mathbb{R}^N)$ and a constant $L > 0$ such that

$$\|f - g^*\|_{\mathcal{C}(\tilde{J} \times \bar{B}_R(u_0); \mathbb{R}^N)} \leq \frac{\varepsilon}{2} \tag{7.35}$$

and

$$\|g^*(t, x) - g^*(t, y)\| \leq L\|x - y\|, \quad x, y \in \bar{B}_R(u_0), \quad t \in \tilde{J}. \tag{7.36}$$

Then, the function

$$g(t, x) := g^*(t, x) + f(t, u(t)) - g^*(t, u(t)), \quad (t, x) \in \tilde{J} \times \bar{B}_R(u_0),$$

satisfies all the requirements of the lemma. Indeed, by (7.35), for all $(t, x) \in \tilde{J} \times \bar{B}_R(u_0)$,

$$\|f(t, x) - g(t, x)\| \leq \|f(t, x) - g^*(t, x)\| + \|f(t, u(t)) - g^*(t, u(t))\|$$

$$\leq \frac{\varepsilon}{2} + \frac{\varepsilon}{2} = \varepsilon,$$

and, hence, (a) holds. Similarly, by (7.36), for every $x, y \in \bar{B}_R(u_0)$ and $t \in \tilde{J}$

$$\|g(t, x) - g(t, y)\| = \|g^*(t, x) - g^*(t, y)\| \leq L\|x - y\|,$$

which shows (b). Finally, (c) is also satisfied because, by definition,

$$g(t, u(t)) = f(t, u(t)) \quad \text{for all } t \in \tilde{J}.$$

This completes the proof of the lemma. $\qquad\square$

It is important to observe that the globally Lipschitz function g constructed in Lemma 7.18, as well as its Lipschitz constant L, depends not only on f and ε but also on the considered solution (I, u) of (7.1).

Now, fix any two points, $x^0 \in A^0$ and $x^1 \in A^1$. By the definition of $\mathcal{A}(t_1)$, there exist two solutions, (I^0, u^0) and $(I^1, u^1) \in \Sigma$, such that $t_1 \in I^0 \cap I^1$ and

$$u^0(t_1) = x^0 \in A^0, \quad u^1(t_1) = x^1 \in A^1.$$

Subsequently, for every $\varepsilon > 0$, we denote by $g_\varepsilon^0(t, u)$ and $g_\varepsilon^1(t, u)$ the globally Lipschitz functions given by Lemma 7.18 corresponding to the solutions $u^0(t)$ and $u^1(t)$, respectively. Then, for every $\lambda \in [0, 1]$ and $\varepsilon > 0$, we define the function $G_{[\lambda,\varepsilon]} \in \mathcal{C}(\tilde{J} \times \bar{B}_R(u_0); \mathbb{R}^N)$, for every $(t, u) \in \tilde{J} \times \bar{B}_R(u_0)$, as

$$G_{[\lambda,\varepsilon]}(t, u) := \lambda g_\varepsilon^1(t, u) + (1 - \lambda)g_\varepsilon^0(t, u). \tag{7.37}$$

It is continuous in λ and establishes a deformation between g_ε^0 and g_ε^1, in the sense that, for every $\varepsilon > 0$,

$$G_{[0,\varepsilon]} = g_\varepsilon^0 \quad \text{and} \quad G_{[1,\varepsilon]} = g_\varepsilon^1.$$

Then, for every $\lambda \in [0, 1]$ and $\varepsilon > 0$, we consider the Cauchy problems

$$\begin{cases} u' = G_{[\lambda,\varepsilon]}(t, u), \\ u(t_0) = u_0. \end{cases} \tag{7.38}$$

Now, we indicate the steps that will allow us to conclude the proof. The idea of the construction is sketched in Figure 7.8. The technical details will be given later.

(1) For every $\lambda \in [0, 1]$ and $\varepsilon > 0$, $G_{[\lambda,\varepsilon]}$ is globally Lipschitz in $u \in \bar{B}_R(u_0)$ uniformly in $t \in \tilde{J}$. As a consequence, Theorem 5.19 ensures that (7.38) possesses a unique maximal solution, which we denote by $(I_{[\lambda,\varepsilon]}, u_{[\lambda,\varepsilon]})$.
(2) There exists $\varepsilon_0 = \varepsilon_0(t_1) > 0$ such that, for every $\lambda \in [0, 1]$ and $\varepsilon \in (0, \varepsilon_0)$, $[t_0, t_1] \subset I_{[\lambda,\varepsilon]}$. In particular, the solutions $u_{[\lambda,\varepsilon]}(t)$ are defined for $t = t_1$ when $\varepsilon \in (0, \varepsilon_0)$.

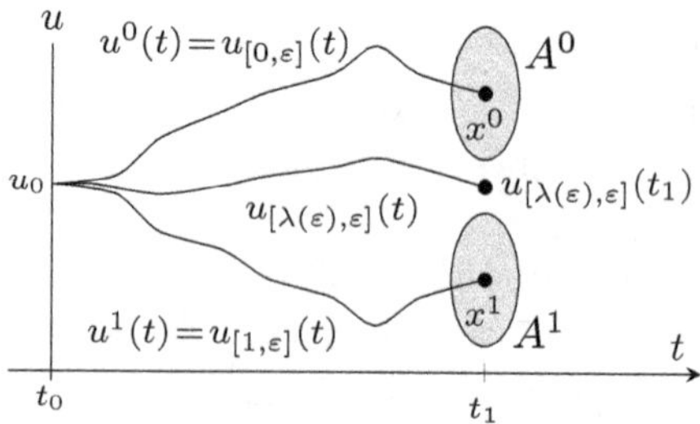

Fig. 7.8.   Scheme of the proof of the connectedness of $\mathcal{A}(t_1)$.

(3) For every $\varepsilon \in (0, \varepsilon_0)$, the function $\varphi \colon [0,1] \to \mathbb{R}$ defined as

$$\varphi(\lambda) := D(u_{[\lambda,\varepsilon]}(t_1))$$

is continuous. As a consequence, since, according to (7.34), we have

$$\varphi(0) = D(u_{[0,\varepsilon]}(t_1)) = D(u^0(t_1)) = D(x^0) < 0,$$
$$\varphi(1) = D(u_{[1,\varepsilon]}(t_1)) = D(u^1(t_1)) = D(x^1) > 0,$$

for all $\varepsilon \in (0, \varepsilon_0)$, by the intermediate value theorem, it becomes apparent that, for every $\varepsilon \in (0, \varepsilon_0)$, there exists $\lambda(\varepsilon) \in (0,1)$ such that

$$\varphi(\lambda(\varepsilon)) = D(u_{[\lambda(\varepsilon),\varepsilon]}(t_1)) = 0. \tag{7.39}$$

(4) Let $\{\varepsilon_n\}_{n\geq 1} \subset (0, \varepsilon_0)$ be a sequence converging to 0, and set

$$u_n := u_{[\lambda(\varepsilon_n),\varepsilon_n]}\big|_{[t_0,t_1]}.$$

Then, there exist a subsequence $\{\varepsilon_{n_m}\}_{m\geq 1}$ and a function $u \in \mathcal{C}([t_0,t_1]; \bar{B}_R(u_0))$ such that

$$\lim_{m\to\infty} \|u_{n_m} - u\|_{\mathcal{C}([t_0,t_1];\bar{B}_R(u_0))} = 0.$$

(5) Such a u is a solution of (7.1) in $[t_0, t_1]$, and evaluating (7.39) at $\varepsilon = \varepsilon_{n_m}$ and letting $m \to +\infty$ yield

$$D(u(t_1)) = 0,$$

which gives the desired contradiction since, by (7.33) and (7.34), $D(x) \neq 0$ for all $x \in \mathcal{A}(t_1)$.

We now proceed to prove each of these steps.

Proof of Step (1). To show that $G_{[\lambda,\varepsilon]}$ is globally Lipschitz in $u \in \bar{B}_R(u_0)$ uniformly in $t \in \tilde{J}$, recall the definition of $G_{[\lambda,\varepsilon]}$ given in (7.37). For every $\lambda \in [0,1]$, $t \in \tilde{J}$, and $x, y \in \bar{B}_R(u_0)$, we have that

$$\big\|G_{[\lambda,\varepsilon]}(t,x) - G_{[\lambda,\varepsilon]}(t,y)\big\|$$
$$= \big\|\lambda g_\varepsilon^1(t,x) + (1-\lambda)g_\varepsilon^0(t,x) - \lambda g_\varepsilon^1(t,y) - (1-\lambda)g_\varepsilon^0(t,y)\big\|$$

$$= \left\| \lambda \left(g_\varepsilon^1(t,x) - g_\varepsilon^1(t,y) \right) + (1-\lambda) \left(g_\varepsilon^0(t,x) - g_\varepsilon^0(t,y) \right) \right\|$$

$$\leq \lambda \left\| g_\varepsilon^1(t,x) - g_\varepsilon^1(t,y) \right\| + (1-\lambda) \| g_\varepsilon^0(t,x) - g_\varepsilon^0(t,y) \|$$

$$\leq \lambda L_1 \left\| x - y \right\| + (1-\lambda) L_0 \left\| x - y \right\|,$$

where L_0 and L_1 are the Lipschitz constants of g_ε^0 and g_ε^1, respectively. Thus,

$$\left\| G_{[\lambda,\varepsilon]}(t,x) - G_{[\lambda,\varepsilon]}(t,y) \right\|$$

$$\leq \lambda \max\{L_0, L_1\} \left\| x - y \right\| + (1-\lambda) \max\{L_0, L_1\} \left\| x - y \right\|$$

$$= \max\{L_0, L_1\} \left\| x - y \right\|,$$

as we claimed. Observe that the Lipschitz constant $\max\{L_0, L_1\}$ is independent of $\lambda \in [0,1]$, though it depends on $\varepsilon > 0$, as L_0 and L_1 do.

Proof of Step (2). Take any $\lambda \in [0,1]$ and $\varepsilon > 0$. Then, setting

$$\| G_{[\lambda,\varepsilon]} \|_\infty := \max_{(t,x) \in \tilde{J} \times \bar{B}_R(u_0)} \| G_{[\lambda,\varepsilon]}(\lambda,x) \|,$$

from Step (1) and Theorem 5.9, we have that

$$\left[t_0, t_0 + \min\left\{ T, \frac{R}{\|G_{[\lambda,\varepsilon]}\|_\infty} \right\} \right] \subset I_{[\lambda,\varepsilon]}.$$

Thus, to get $[t_0, t_1] \subset I_{[\lambda,\varepsilon]}$ for ε small enough, it suffices to show that

$$\| G_{[\lambda,\varepsilon]} \|_\infty \leq \varepsilon + \| f \|_\infty \quad \text{for all } \varepsilon > 0, \tag{7.40}$$

and then take a sufficiently small ε so that

$$t_1 < t_0 + \frac{R}{\varepsilon + \|f\|_\infty} \leq t_0 + \frac{R}{\|G_{[\lambda,\varepsilon]}\|_\infty},$$

which is possible thanks to (7.29).

From (7.37) and Lemma 7.18(a), we obtain that, for every $t \in \tilde{J}$ and $x \in \bar{B}_R(u_0)$,

$$\big\| G_{[\lambda,\varepsilon]}(t,x) - f(t,x) \big\|$$
$$= \big\| \lambda g_\varepsilon^1(t,x) + (1-\lambda) g_\varepsilon^0(t,x) - f(t,x) \big\|$$
$$= \big\| \lambda \left(g_\varepsilon^1(t,x) - f(t,x) \right) + (1-\lambda) \left(g_\varepsilon^0(t,x) - f(t,x) \right) \big\|$$
$$\le \lambda \big\| g_\varepsilon^1(t,x) - f(t,x) \big\| + (1-\lambda) \big\| g_\varepsilon^0(t,x) - f(t,x) \big\|$$
$$\le \lambda \varepsilon + (1-\lambda) \varepsilon = \varepsilon.$$

Thus, for every $\lambda \in [0,1]$ and $\varepsilon > 0$,

$$\big\| G_{[\lambda,\varepsilon]} - f \big\|_{\mathcal{C}(\tilde{J} \times \bar{B}_R(u_0); \mathbb{R}^N)} \le \varepsilon \tag{7.41}$$

and, hence, for every $\lambda \in [0,1]$ and $\varepsilon > 0$,

$$\big\| G_{[\lambda,\varepsilon]} \big\|_\infty = \big\| G_{[\lambda,\varepsilon]} - f + f \big\|_{\mathcal{C}(\tilde{J} \times \bar{B}_R(u_0); \mathbb{R}^N)}$$
$$\le \big\| G_{[\lambda,\varepsilon]} - f \big\|_{\mathcal{C}(\tilde{J} \times \bar{B}_R(u_0); \mathbb{R}^N)} + \big\| f \big\|_{\mathcal{C}(\tilde{J} \times \bar{B}_R(u_0); \mathbb{R}^N)}$$
$$\le \varepsilon + \| f \|_\infty,$$

which shows (7.40). Observe that, to obtain the last inequality, we have used that $\tilde{J} \subset J$ and, by definition,

$$\| f \|_\infty := \max_{(t,x) \in J \times \bar{B}_R(u_0)} \| f(t,x) \|.$$

Proof of Step (3). To prove this step, we use the continuous (actually Lipschitz) dependence of the solutions of (7.38) with respect to the nonlinearity $G_{[\lambda,\varepsilon]}$, which has been studied in Section 5.6.2. From (7.37) and (7.40), for every $\varepsilon > 0$ and $\lambda, \mu \in [0,1]$, we find that

$$\big\| G_{[\lambda,\varepsilon]} - G_{[\mu,\varepsilon]} \big\|_{\mathcal{C}([t_0,t_1] \times \bar{B}_R(u_0); \mathbb{R}^N)}$$
$$= \big\| \lambda g_\varepsilon^1 + (1-\lambda) g_\varepsilon^0 - \mu g_\varepsilon^1 - (1-\mu) g_\varepsilon^0 \big\|_{\mathcal{C}([t_0,t_1] \times \bar{B}_R(u_0); \mathbb{R}^N)}$$
$$= \big\| (\lambda - \mu) g_\varepsilon^1 - (\lambda - \mu) g_\varepsilon^0 \big\|_{\mathcal{C}([t_0,t_1] \times \bar{B}_R(u_0); \mathbb{R}^N)}$$
$$\le |\lambda - \mu| \left(\big\| g_\varepsilon^1 \big\|_{\mathcal{C}([t_0,t_1] \times \bar{B}_R(u_0); \mathbb{R}^N)} + \big\| g_\varepsilon^0 \big\|_{\mathcal{C}([t_0,t_1] \times \bar{B}_R(u_0); \mathbb{R}^N)} \right)$$
$$\le 2|\lambda - \mu| \left(\| f \|_\infty + \varepsilon \right).$$

Thus, for every $\varepsilon \le \varepsilon_0$, applying Theorem 5.24 in the interval $[t_0, t_1]$ yields

$$\left\| u_{[\lambda,\varepsilon]} - u_{[\mu,\varepsilon]} \right\|_{\mathcal{C}([t_0,t_1] \times \bar{B}_R(u_0); \mathbb{R}^N)}$$

$$\le 2\,|\lambda - \mu|\,(\|f\|_\infty + \varepsilon)\,\frac{e^{\max\{L_0, L_1\}(t_1 - t_0)} - 1}{\max\{L_0, L_1\}}$$

(recall that $\max\{L_0, L_1\}$ is a Lipschitz constant for $G_{[\lambda,\varepsilon]}$ and that it does not depend on λ). As a consequence, the map

$$[0,1] \longrightarrow \quad \mathbb{R}^N$$
$$\lambda \longmapsto u_{[\lambda,\varepsilon]}(t_1)$$

is continuous, and so is φ, as a composition of continuous functions.

Proof of Step (4). Setting

$$\mathfrak{F} := \{u_n \colon n \ge 1\} \subset \mathcal{C}([t_0, t_1]; \bar{B}_R(u_0)),$$

we claim that $\mathfrak{F}$ is bounded and equicontinuous. Then, the desired conclusion will follow from the Ascoli–Arzelà Theorem (see Theorem 7.5).

As u_n solves (7.38) in $t \in [t_0, t_1]$, with $\varepsilon = \varepsilon_n$ and $\lambda = \lambda(\varepsilon_n)$, we have

$$u_n(t) = u_0 + \int_{t_0}^{t} G_{[\lambda(\varepsilon_n), \varepsilon_n]}(\tau, u_n(\tau))\, d\tau \quad \text{for all } t \in [t_0, t_1].$$

Thus, it follows from (7.40) that

$$\|u_n(t)\| \le \|u_0\| + (\varepsilon_n + \|f\|_\infty)\,T \le \|u_0\| + (\varepsilon_0 + \|f\|_\infty)\,T$$

for all $n \ge 1$ and $t \in [t_0, t_1]$. Hence, $\mathfrak{F}$ is bounded.

Similarly, for every $n \ge 1$ and $s, t \in [t_0, t_1]$,

$$\|u_n(t) - u_n(s)\| = \left\| \int_{s}^{t} G_{[\lambda(\varepsilon_n), \varepsilon_n]}(\tau, u_n(\tau))\, d\tau \right\|$$

$$\le (\varepsilon_n + \|f\|_\infty)\,|t - s| \le (\varepsilon_0 + \|f\|_\infty)\,|t - s|.$$

So, $\mathfrak{F}$ is equicontinuous by Proposition 7.4. Therefore, Theorem 7.5 concludes the proof of this step.

Proof of Step (5). As, for every $m \geq 1$ and $t \in [t_0, t_1]$, we have that

$$u_{n_m}(t) = u_0 + \int_{t_0}^t G_{[\lambda(\varepsilon_{nm}),\varepsilon_{nm}]}(s, u_{n_m}(s))ds$$

$$= u_0 + \int_{t_0}^t f(s, u_{n_m}(s))ds \tag{7.42}$$

$$+ \int_{t_0}^t \left(G_{[\lambda(\varepsilon_{nm}),\varepsilon_{nm}]} - f\right)(s, u_{n_m}(s))ds$$

and, in addition, from (7.41), we obtain

$$\left\| \int_{t_0}^t \left(G_{[\lambda(\varepsilon_{nm}),\varepsilon_{nm}]} - f\right)(s, u_{n_m}(s))\, ds \right\| \leq \varepsilon_{nm}|t_1 - t_0|,$$

the last integral in (7.42) converges to zero as $m \to \infty$. Thus, letting $m \to \infty$ in (7.42) yields

$$u(t) = u_0 + \int_{t_0}^t f(s, u(s))\, ds$$

for all $t \in [t_0, t_1]$. Consequently, $u(t)$ is a solution of (7.1) in $[t_0, t_1]$. This completes the proof of the theorem. $\qquad\square$

7.8.1 *Regularization of continuous functions*

Consider $J = [t_0, t_0 + T]$, $R > 0$, $u_0 \in \mathbb{R}^N$, and $f \in \mathcal{C}(J \times \bar{B}_R(u_0); \mathbb{R}^N)$. Then, as we have seen in the proof of Theorem 7.7, the function

$$\tilde{f}(t, u) := \begin{cases} f(t, u) & \text{if } \|u - u_0\| \leq R, \\ f(t, \tilde{u}) & \text{if } \|u - u_0\| > R, \end{cases}$$

with

$$\tilde{u} = \tilde{u}(u) := u_0 + R\frac{u - u_0}{\|u - u_0\|}, \quad \|u - u_0\| > R,$$

is a continuous extension of $f(t, u)$ to $J \times \mathbb{R}^N$. Thus, the extension

$$\hat{f}(t, u) := \begin{cases} \tilde{f}(t_0, u) & \text{if } t < t_0, \\ \tilde{f}(t, u) & \text{if } t_0 \leq t \leq t_0 + T, \quad u \in \mathbb{R}^N, \\ \tilde{f}(t_0 + T, u) & \text{if } t > t_0 + T, \end{cases}$$

provides us with a continuous extension of $f(t, u)$ to $\mathbb{R} \times \mathbb{R}^N$. Moreover,

$$\|\hat{f}\|_{\mathcal{C}(\mathbb{R} \times \mathbb{R}^N; \mathbb{R}^N)} := \sup_{(t,u) \in \mathbb{R} \times \mathbb{R}^N} \|\hat{f}(t, u)\| = \|f\|_{\mathcal{C}(J \times \bar{B}_R(u_0); \mathbb{R}^N)} < +\infty.$$

The main goal of this section is to approximate, uniformly on compact subsets of $\mathbb{R} \times \mathbb{R}^N$, each of the components of $\hat{f}$, $\hat{f}_j$, $1 \leq j \leq N$, by means of $\mathcal{C}^\infty$ functions. In particular, this entails that $f(t, u)$ can be approximated, uniformly in $J \times \bar{B}_R(u_0)$, by $\mathcal{C}^\infty$ functions—a fact that has been used in the proof of Lemma 7.18.

To obtain the desired approximation, we consider the function $\psi : \mathbb{R} \to \mathbb{R}$ defined as

$$\psi(x) := \begin{cases} e^{\frac{1}{x^2 - 1}} & \text{if } |x| < 1, \\ 0 & \text{if } |x| \geq 1, \end{cases}$$

which is of class $\mathcal{C}^\infty(\mathbb{R}; \mathbb{R})$ and satisfies

$$\frac{d^k \psi}{dx^k}(\pm 1) = 0 \quad \text{for all } k \geq 1.$$

This function is the paradigm of a $\mathcal{C}^\infty$ function which is not real analytic since its Taylor series in a neighborhood of ± 1 is 0, and thus, it does not represent $\psi(x)$. Next, we consider the M-dimensional counterpart of $\psi(x)$, $\Psi(x) \colon \mathbb{R}^M \to \mathbb{R}$,

$$\Psi(x) := \begin{cases} e^{\frac{1}{\|x\|^2 - 1}} & \text{if } \|x\| < 1, \\ 0 & \text{if } \|x\| \geq 1. \end{cases}$$

As Ψ is continuous and has compact support, we have that

$$\int_{\mathbb{R}^M} \Psi(y)\, dy < +\infty;$$

thus, the function

$$\varphi(x) := \frac{\Psi(x)}{\int_{\mathbb{R}^M} \Psi(y)\, dy}, \quad x \in \mathbb{R}^M,$$

is well defined. Moreover, $\varphi(x)$ satisfies

$$\varphi \in C^{\infty}(\mathbb{R}^M; \mathbb{R}), \quad \varphi \geq 0, \quad \operatorname{supp} \varphi = \bar{B}_1(0), \quad \int_{\mathbb{R}^M} \varphi(x)\, dx = 1.$$

Therefore, for every $\varepsilon > 0$, the functions

$$\varphi_\varepsilon(x) := \varepsilon^{-M} \varphi\left(\frac{x}{\varepsilon}\right), \quad x \in \mathbb{R}^M,$$

satisfy

$$\varphi_\varepsilon \in C^{\infty}(\mathbb{R}^M; \mathbb{R}), \quad \varphi_\varepsilon \geq 0, \quad \operatorname{supp} \varphi_\varepsilon = \bar{B}_\varepsilon(0), \quad \int_{\mathbb{R}^M} \varphi_\varepsilon(x)\, dx = 1.$$

Indeed, performing the change of variable $x = \varepsilon y$ yields

$$dx = \varepsilon^M\, dy$$

and, hence,

$$\int_{\mathbb{R}^M} \varphi_\varepsilon(x)\, dx = \int_{\mathbb{R}^M} \varepsilon^{-M} \varphi(y) \varepsilon^M\, dy = \int_{\mathbb{R}^M} \varphi(y)\, dy = 1.$$

The remaining properties are straightforward. Thus, these functions concentrate their mass (which is equal to 1) in the closed ball $\bar{B}_\varepsilon(0)$. Figure 7.9 shows the plots of φ_ε for $M = 1$ and $\varepsilon \in \{1, 0.5, 0.2, 0.1\}$. Observe that, the smaller the value of $\varepsilon > 0$, the more concentrated φ_ε is close to the origin.

Now, we can proceed to introduce the approximating functions: given any continuous $F \colon \mathbb{R}^M \to \mathbb{R}$, in order to approximate it by C^{∞} functions, one can consider its convolution product with φ_ε:

$$F_\varepsilon(x) := F * \varphi_\varepsilon(x) = \int_{\mathbb{R}^M} F(x - y)\varphi_\varepsilon(y)\, dy. \tag{7.43}$$

Note that making the change of variable $x - y = z$ gives

$$F_\varepsilon(x) = \int_{\mathbb{R}^M} F(z)\varphi_\varepsilon(x - z)\, dz = \varphi_\varepsilon * F(x),$$

which shows that the convolution product is commutative. Since the function F is continuous and φ_ε has compact support, it is easily

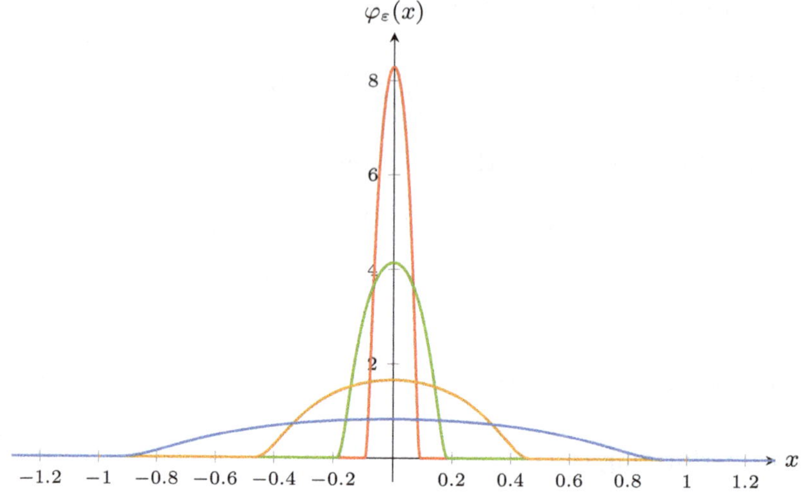

Fig. 7.9. Plots of $\varphi_\varepsilon(x)$ for $M = 1$ and $\varepsilon \in \{1, 0.5, 0.2, 0.1\}$.

seen that

$$D^\alpha F_\varepsilon(x) = \int_{\mathbb{R}^M} F(z) D^\alpha \varphi_\varepsilon(x - z)\, dz, \quad x \in \mathbb{R}^M,$$

for all multi-index

$$\alpha = (\alpha_1, \ldots, \alpha_M) \in \mathbb{N}^M,$$

where we denote

$$D^\alpha := \frac{\partial^{\alpha_1 + \cdots + \alpha_M}}{\partial x_1^{\alpha_1} \cdots \partial x_M^{\alpha_M}}.$$

Therefore, $F_\varepsilon \in \mathcal{C}^\infty(\mathbb{R}^M; \mathbb{R})$ for all $\varepsilon > 0$. The following result shows that F_ε approximates F as $\varepsilon \to 0$ uniformly on compact subsets of $\mathbb{R}^M$.

Theorem 7.19. *Let* $F \in \mathcal{C}(\mathbb{R}^M; \mathbb{R})$, *and for* $\varepsilon > 0$, *consider* $F_\varepsilon \in \mathcal{C}^\infty(\mathbb{R}^M; \mathbb{R})$, *defined in* (7.43). *Then, for any compact set* $K \subset \mathbb{R}^M$,

$$\lim_{\varepsilon \downarrow 0} \|F_\varepsilon - F\|_{\mathcal{C}(K; \mathbb{R})} = 0.$$

Proof. For every $x \in K$, we have that

$$|F_\varepsilon(x) - F(x)| = \left| \int_{\mathbb{R}^M} F(x-y)\varphi_\varepsilon(y)\, dy - F(x) \right|$$

$$= \left| \varepsilon^{-M} \int_{\mathbb{R}^M} F(x-y)\, \varphi\!\left(\frac{y}{\varepsilon}\right) dy - F(x) \right|$$

$$= \left| \varepsilon^{-M} \int_{\mathbb{R}^M} F(x-\varepsilon z)\varphi(z)\varepsilon^M\, dz - F(x) \right|$$

$$= \left| \int_{\mathbb{R}^M} F(x-\varepsilon z)\varphi(z)\, dz - F(x) \right|.$$

On the other hand,

$$F(x) = \int_{\mathbb{R}^M} F(x)\varphi(z)\, dz.$$

Thus,

$$|F_\varepsilon(x) - F(x)| = \left| \int_{\mathbb{R}^M} F(x-\varepsilon z)\varphi(z)\, dz - \int_{\mathbb{R}^M} F(x)\varphi(z)\, dz \right|$$

$$= \left| \int_{\mathbb{R}^M} \left(F(x-\varepsilon z) - F(x) \right) \varphi(z)\, dz \right|$$

$$\leq \int_{B_1(0)} |F(x-\varepsilon z) - F(x)|\varphi(z)\, dz.$$

Moreover, for every $x \in K$,

$$\|x - \varepsilon z - x\| = \varepsilon \, \|z\| \leq \varepsilon \quad \text{if } \|z\| \leq 1,$$

and since F is uniformly continuous in

$$K + \bar{B}_\varepsilon(0) = \left\{ x \in \mathbb{R}^M : \ \mathrm{dist}(x, K) \leq \varepsilon \right\},$$

it becomes apparent that, for any given $\eta > 0$, there exists $\varepsilon_0 > 0$ such that, for every $x \in K$, $z \in \bar{B}_1(0)$, and $\varepsilon \in (0, \varepsilon_0)$,

$$|F(x - \varepsilon z) - F(x)| \leq \eta.$$

Therefore, for every $x \in K$ and $\varepsilon \in (0, \varepsilon_0)$,

$$|F_\varepsilon(x) - F(x)| \leq \int_{B_1(0)} |F(x - \varepsilon z) - F(x)|\, \varphi(z)\, dz \leq \eta.$$

The proof is complete. $\square$

Summarizing, to regularize $f(t, u)$, one shall regularize each of the components of $\hat{f}(t, u)$, $\hat{f}_j(t, u)$, $1 \leq j \leq N$, through the previous scheme, i.e. by performing the convolution $\hat{f}_j * \varphi_\varepsilon(t, u)$ for $(t, u) \in \mathbb{R} \times \mathbb{R}^N$.

7.9   Exercises

1. With the notation in Section 7.1, prove that $\left(\mathcal{C}(K; \mathbb{R}^N), \|\cdot\|_\infty \right)$ is a real Banach space.

2. State and prove the backward counterparts of Theorems 7.10–7.14.

3. Give a precise statement of the fundamental theorems for the Cauchy problems (6.8), (6.45), and (6.47) when their respective nonlinearities are continuous.

4. Given $p > 0$ and $q > 0$ with $pq < 1$, consider the problem

$$\begin{cases} u' = |v|^p, \\ v' = |u|^q, \\ u(0) = 1, \quad v(0) = 0. \end{cases}$$

Analyze the existence, uniqueness, and global behavior of its solutions.

5. Given $p > 0$ and $q > 0$ with $pq < 1$, determine all the solutions of

$$\begin{cases} u' = |v|^p, \\ v' = |u|^q, \\ u(0) = v(0) = 0. \end{cases}$$

6. Let $f(\theta)$ denote the odd extension of the function

$$F(\theta) := \sqrt{\theta(\pi - \theta)}, \quad 0 \leq \theta \leq \pi,$$

to the interval $[-\pi, \pi]$, and let $\varphi \colon [0, \infty) \to \mathbb{R}$ be a continuous function such that $\varphi(1) = 1$. Denote

$$u(t) = \rho(t) \cos \theta(t), \quad v(t) = \rho(t) \sin \theta(t).$$

Prove that, for sufficiently large $t_1 > 0$, the set of points reached by the solutions of

$$\begin{cases} u' = -v\,\varphi(\rho)f(\theta), \\ v' = u\,\varphi(\rho)f(\theta), \\ (u(0), v(0)) = (1, 0), \end{cases}$$

after time t_1, $\mathcal{A}(t_1)$, is the unite circle. Therefore, this example shows that, in the setting of the Kneser theorem (see Theorem 7.17), $\mathcal{A}(t_1)$ is not necessarily convex when $N \geq 2$.

7. Suppose $J = [t_0, t_0 + T]$ and $f \in \mathcal{C}(J \times \mathbb{R}^N; \mathbb{R}^N)$ is bounded, i.e. (7.11) holds. Assume, in addition, that f is non-increasing in u in the sense that

$$\langle f(t, u) - f(t, v), u - v \rangle \leq 0$$

for all $t \in J$ and $u, v \in \mathbb{R}^N$, where $\langle \cdot, \cdot \rangle$ indicates the Euclidean inner product of $\mathbb{R}^N$. Prove that, for every $u_0 \in \mathbb{R}^N$, problem (7.1) has a unique solution which is defined in J.

8. Adapt Theorem 7.16 to get an optimal result establishing the continuous dependence of the solution of (7.1) with respect to f.

9. Prove Theorem 7.17 when $N = 1$ using the following argument. For every $x^0, x^1 \in \mathcal{A}(t_1)$, with $x^0 < x^1$ and $x \in (x^0, x^1)$, there exists a solution of (7.1), u, such that $u(t_1) = x$. [Hint: Going backward in time might be helpful.]

10. Suppose $J = [t_0, t_0 + T]$ and $f \in \mathcal{C}(J \times \mathbb{R}; \mathbb{R})$ is a bounded function. Prove that (7.1) possesses a minimal solution, $u_{\min}$, and a maximal solution, $u_{\max}$, defined in J, in the sense that any other solution of (7.1), u, satisfies

$$u_{\min} \leq u \leq u_{\max}.$$

Deduce that, for every $t \in J$,

$$\mathcal{A}(t) = [u_{\min}(t), u_{\max}(t)].$$

11. Under the same hypothesis of the global Peano theorem (see Theorem 7.6), consider the sequence of polygonal curves, P_n, constructed as follows: for every integer $n \geq 1$, take

$$t_i := t_0 + i\frac{T}{n}, \quad I_i := [t_{i-1}, t_i], \quad i \in \{1, \ldots, n\},$$

and define

$$P_n(t) := u_0 + f(t_0, u_0)(t - t_0), \quad t \in I_1.$$

Then, proceed recursively: once defined $P_n(t)$ for all $t \in I_i$, $i \in \{1, \ldots, n-1\}$, set

$$P_n(t) := P_n(t_i) + f(t_i, P_n(t_i))(t - t_i), \quad t \in (t_i, t_{i+1}].$$

Prove that there exists a subsequence of $\{P_n\}_{n \geq 1}$ that approximates a solution of (7.1).

12. Take $J = [t_0, t_0 + T]$, $u_0 \in \mathbb{R}^N$, $R > 0$ and $f \in \mathcal{C}(J \times \bar{B}_R(u_0); \mathbb{R}^N)$.

(a) Prove that there exists a sequence of functions $\{f_n\}_{n \geq 1}$ in the Banach space $\mathcal{C}(J \times \bar{B}_R(u_0); \mathbb{R}^N)$ equipped with the maximum norm that are globally Lipschitz with respect to $u \in \bar{B}_R(u_0)$ uniformly in $t \in J$ and satisfy

$$\lim_{n \to \infty} f_n = f \quad \text{uniformly in } J \times \bar{B}_R(u_0).$$

(b) Consider, for every $n \geq 1$, the unique maximal solution (I_n, u_n) of

$$\begin{cases} u' = f_n(t, u), \\ u(t_0) = u_0. \end{cases}$$

Prove that there exists a subsequence of $\{(I_n, u_n)\}_{n \geq 1}$ that approximates a solution of (7.1).

7.10　Final Comments

The necessity of equicontinuity in Theorem 7.5 goes back to Ascoli, who proved it in 1883/4. Some years later, in 1895, Arzelà established that equicontinuity was also sufficient. This is why the Ascoli–Arzelà theorem, which is the first result in modern Functional Analysis as we understand it today, bears the name of both these mathematicians. It represents a milestone for the development of new results in Analysis. Besides its important applications in Chapters 7 and 8 of this book, it is pivotal in the context of Partial Differential Equations, for, as in many cases, it guarantees the compactness of their associated integral equations (see, e.g. López-Gómez, 2013, Chapter 6). For example, it allows us to prove the celebrated Harnack–Montel theorem in classical potential theory, which establishes that relatively compact subsets of the set of harmonic functions on any open bounded domain are exactly the bounded subsets.

Theorem 7.6 was proven by Peano in 1890. Due to the huge importance of this theorem in the theory of Ordinary Differential Equations, the original proof given by Peano was considerably polished and refined by many other authors. The proof of the Peano theorem presented in Section 7.2 uses the elegant scheme (7.13) proposed by Tonelli in 1928. Peano is also renowned as one of the founders of Mathematical Logic and Set Theory. He played a central role in the axiomatization of Mathematics. Tonelli is the father of the modern Calculus of Variations since, for example, he introduced the concept of semicontinuity.

The example analyzed in Section 7.6 was proposed by Müller in 1927. It reveals that, in the absence of local Lipschitz continuity for $f(t, u)$ in u, the uniqueness of solutions for a Cauchy problem and the convergence of the iterates of the associated integral operator are independent phenomena. In Chapter 8, it will be shown that the iterates can converge even when $f(t, u)$ is only assumed to be continuous in u, provided it is non-decreasing in u. But this is a result of a completely different nature.

Theorem 7.16 establishes that the continuous dependence relies, simply, on the uniqueness of the solution. This result originated from a remark by Hartman (2002). Besides Theorem 7.16 and the results in Section 5.6, there is another celebrated theorem by Peano (1897) establishing that if $f(t, u)$ is of class C^1, then the dependence of $u(t; u_0, f)$ with respect to u_0 and f is also differentiable, but this result, of a more advanced nature, remains outside the general scope of this introductory textbook. Some previous one-dimensional versions of the differentiability theorem of Peano had already been given by Picard (see the Notes on p. 116 of Hartman, 2002) and Bendixson (1896). Essentially, the results on the dependence of the solution of

(7.1) on u_0 and f establish that, as soon as there is uniqueness, the sensitivity of the solution inherits the same regularity as $f(t, u)$.

Theorem 7.17 was proven by Kneser in 1923. Some time later, Fukuhara (1928) and Fukuhara and Nagumo (1930) established it independently. In the special one-dimensional case, M. Fukuhara could also establish the existence of a maximal and a minimal solution (see Exercise 10 of Chapter 7). In most of the one-dimensional multiplicity examples given in this book, the minimal solution was $u_{\min} = 0$, while the maximal solution was $u_{\max} = \varphi$. Thus, $\mathcal{A}(t_1) = [0, \varphi(t_1)]$. The proof of Theorem 7.17 given in this book is due to Müller (1928). However, in this book, the proof has been considerably polished and completed from the scheme of Hartman (2002).

The example of the Exercise 6 of Chapter 7 has been taken, almost exactly, from the Hints in Exercise II.4.2 on p. 557 of Hartman (2002). In the Notes of Hartman (2002, p. 23) it is recognized that it is due to C. C. Pugh (unpublished).

Exercises 11 and 12 of Chapter 7 provide two additional proofs for the global Peano theorem. The first one uses the classical Euler's scheme. The second one uses the regularization techniques introduced in Section 7.8.1 and the global Cauchy–Lipschitz theorem. Naturally, these two supplementary proofs are also based on the Ascoli–Arzelà theorem.

Finally, we sketch the construction of the non-uniqueness example given by Hartman (2002, p. 18), which the reader can refer for any further required details.

As a first step, one starts by considering the set of curves—or, with an abuse of notation, the set of points on those curves—defined as

$$S_0 := \{(t, 4j + \cos(\pi t)) : t \in \mathbb{R}, j \in \mathbb{Z}\}$$
$$\cup \{(t, 4j + 2 - \cos(\pi t)) : t \in \mathbb{R}, j \in \mathbb{Z}\}.$$

These arcs determine certain cells that "tessellate" periodically the plane, as shown in the first row in Figure 7.10. The left-hand plot shows one of such cells, delimited by red curves, while the right-hand one shows their global configuration in the plane. Observe, in particular, that these curves intersect, being tangent, at points of the form

$$(2k, 4j + 1) \cup (2k + 1, 4j + 3), \quad k \in \mathbb{Z}, \quad j \in \mathbb{Z}.$$

As a second step, one defines a set of curves S_1 consisting of S_0 plus, inside each cell determined by S_0, finitely many additional arcs. This construction is represented in the second row in Figure 7.10. In the left-hand plot, the additional arcs in one base cell are plotted in red, while those already present in the previous generation are plotted with black dashed curves. The same construction is repeated by periodicity and translations in all other cells, as shown in the corresponding right-hand plot. The key property in the construction of S_1 is that any two adjacent arcs possibly intersect at points whose abscissa is $\frac{k}{2}$ for some $k \in \mathbb{Z}$. Moreover, they are tangent whenever they intersect.

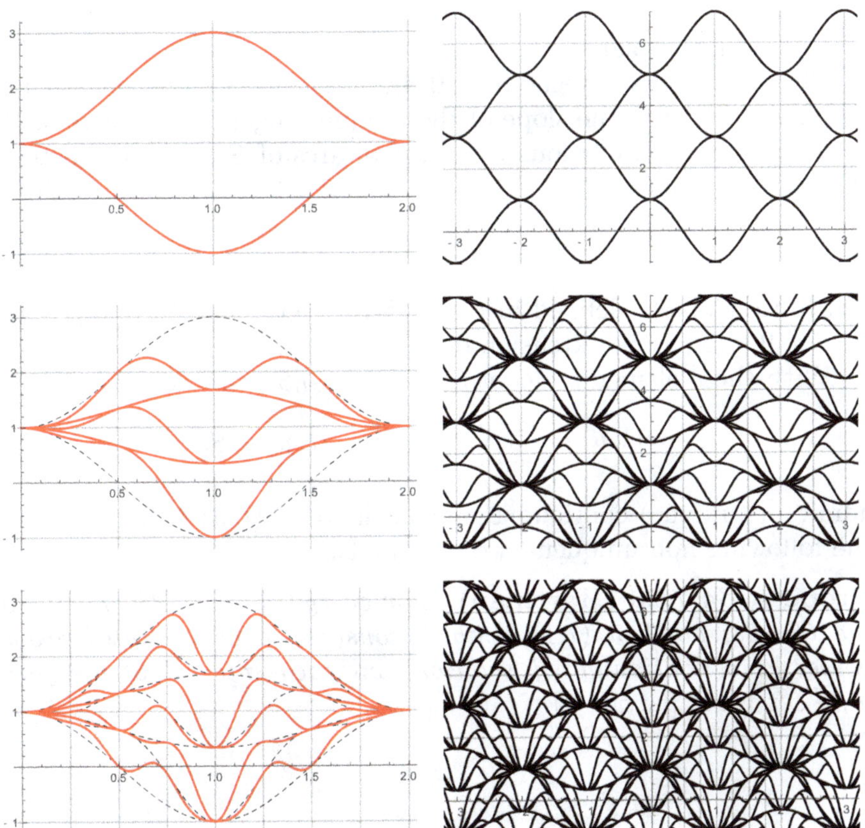

Fig. 7.10. Sketch of the construction of the curves in Hartman's multiplicity example.

Inductively, for all $n \geq 0$, one defines S_{n+1} by adding to the curves in S_n a finite number of arcs so that the same key property is satisfied, with the difference that now the tangency points have abscissas $\frac{k}{2^{n+1}}$, $k \in \mathbb{Z}$. The last row in Figure 7.10 illustrates the construction of S_2: in the left-hand plot, the newly introduced arcs inside the base cell are plotted in red, while the black dashed curves are those already in S_1; in the right-hand plot, the whole construction on the plane is depicted.

If one labels the arcs of S_n as $u_{j,k}$, $k,j \in \mathbb{Z}$, the described construction, based on this increasingly finer tessellation of the plane, can ensure that the points

$$\left(\tfrac{k}{2^n}, u_{j,k}\left(\tfrac{k}{2^n}\right) \right), \quad k, j \in \mathbb{Z}, \quad n \geq 0,$$

are dense on the plane.

Then, for every integer $n \geq 0$ and $(t, u) \in S_n$, one defines the function $f_n(t, u)$ as the slope of the unique tangent to the arcs of S_n at the point (t, u). Obviously, the set of arcs of S_n provides us with the set of solutions of

$$u' = f_n(t, u),$$

which are globally defined in $t \in \mathbb{R}$. Moreover, since $S_n \subset S_{n+1}$, for all $n \geq 0$, f_{n+1} is an extension of f_n.

With these elements, denoting $S := \cup_{n \geq 0} S_n$, one can show that

$$f(t, u) := \lim_{n \to \infty} f_n(t, u), \quad (t, u) \in S,$$

admits a (unique) continuous extension to all $(t, u) \in \mathbb{R}^2$, and that the following non-uniqueness result holds.

Proposition 7.20 (Hartman). *For every $(t_0, u_0) \in \mathbb{R}^2$ and $\varepsilon > 0$, the Cauchy problem (7.1), with f constructed as indicated above, possesses a continuum of solutions on each interval of the form $[t_0 - \varepsilon, t_0]$ and $[t_0, t_0 + \varepsilon]$.*

Chapter 8

Method of Sub- and Supersolutions

Müller's example presented in Section 7.6 shows that the $\mathcal{K}$-iterates of the Cauchy problem

$$\begin{cases} u' = f(t, u), \\ u(t_0) = u_0, \end{cases} \tag{8.1}$$

do not necessarily approximate any solution of this problem. In such an example, $f(t, u)$ was non-increasing in u. In this chapter, we instead assume that $f(t, u)$ is non-decreasing in u, and our main goal is to present some general sufficient conditions for the $\mathcal{K}$-iterates to approximate a solution. To do so, we begin by introducing in Section 8.1 some pivotal concepts in the theory of *positive operators* and the concept of sub- and supersolution for the Cauchy problem (8.1). In Section 8.2, the main theorem of the chapter is described; it is a result proved by Kamke in 1942, establishing that between any pair of ordered sub- and supersolutions of (8.1) in $J = [t_0, t_0 + T]$, there is a solution to (8.1) in J provided $f(t, u)$ is non-decreasing in u for all $t \in J$. Then, in Sections 8.3–8.5, we apply Kamke's theorem to estimate the existence and blow up times of the explosive solutions of some important families of cooperative systems. Finally, Section 8.6 applies Kamke's theorem to establish the extensibility criterion by Wintner, and Section 8.7 deduces from Wintner's criterion the uniqueness criterion by Osgood.

433

8.1 Positivity. Concept of Sub- and Supersolution

We begin by introducing some basic notations that will be used throughout this chapter. For every integer $N \geq 1$ and

$$x = (x_1, \ldots, x_N)^T, \quad y = (y_1, \ldots, y_N)^T \in \mathbb{R}^N,$$

we say that $x \leq y$ if $x_i \leq y_i$ for all $i \in \{1, \ldots, N\}$.

Given two integers $N, M \geq 1$, an arbitrary subset $\Omega \subset \mathbb{R}^N$, and a function $f \colon \Omega \to \mathbb{R}^M$, it is said that f is *non-decreasing* in Ω when, for all $x, y \in \Omega$,

$$x \leq y \quad \text{implies} \quad f(x) \leq f(y).$$

Given an interval $J \subset \mathbb{R}$, an arbitrary subset $\Omega \subset \mathbb{R}^N$ and a function $f \colon J \times \Omega \to \mathbb{R}^M$, it is said that $f(t, u)$ is non-decreasing in $u \in \Omega$ when, for every $t \in J$, the function $f(t, \cdot)$ is non-decreasing in Ω.

Given an interval J and two functions $h_1, h_2 \in \mathcal{C}(J; \mathbb{R}^M)$, we use the notation $h_1 \geq h_2$ whenever $h_1(t) \geq h_2(t)$ for all $t \in J$. When, in addition, $h_1 \neq h_2$, we use $h_1 \gneq h_2$.

The following result establishes an important property of the integral operator $\mathcal{K}$ associated with problem (8.1).

Proposition 8.1. *Let $J = [t_0, t_0 + T]$, Ω be an open subset of $\mathbb{R}^N$, $N \geq 1$, $u_0 \in \Omega$, and assume that $f \colon J \times \Omega \to \mathbb{R}^N$ is continuous and non-decreasing in $u \in \Omega$. Then, the integral operator associated with* (8.1),

$$\mathcal{K}h(t) := u_0 + \int_{t_0}^{t} f(s, h(s))\, ds, \qquad t \in J, \ h \in \mathcal{C}(J; \mathbb{R}^N),$$

is order-preserving, in the sense that

$$h_1 \geq h_2 \quad \text{implies} \quad \mathcal{K}h_1 \geq \mathcal{K}h_2.$$

Moreover, for every integer $n \geq 1$, $\mathcal{K}^n$ is also order-preserving.

Proof. Suppose $h_1 \geq h_2$, i.e. $h_1(t) \geq h_2(t)$ for all $t \in J$. Since $f(t, u)$ is non-decreasing in $u \in \Omega$,

$$f(t, h_1(t)) \geq f(t, h_2(t)) \quad \text{for all } t \in J.$$

Thus,

$$\mathcal{K}h_1(t) = u_0 + \int_{t_0}^{t} f(s, h_1(s))\, ds \geq u_0 + \int_{t_0}^{t} f(s, h_2(s))\, ds = \mathcal{K}h_2(t)$$

for all $t \in J$. Equivalently, $\mathcal{K}h_1 \geq \mathcal{K}h_2$, as we wanted. Naturally, this also implies that $\mathcal{K}^2 h_1 \geq \mathcal{K}^2 h_2$, and reasoning inductively, it is easily seen that, for every integer $n \geq 1$, $\mathcal{K}^n h_1 \geq \mathcal{K}^n h_2$. The proof is complete. $\qquad\square$

The following concept is pivotal not only in the context of the theory of ordinary differential equations but also in that of partial differential equations.

Definition 8.2 (Subsolution and supersolution). Let $J = [t_0, t_0 + T]$, Ω be an open subset of $\mathbb{R}^N$, $N \geq 1$, $u_0 \in \Omega$, $f \in C(J \times \Omega; \mathbb{R}^N)$, and $\mathcal{K}$ be the integral operator associated with (8.1). Then:

- a function $\underline{u} \in C(J; \Omega)$ is said to be a *subsolution* of (8.1) in J if

$$\underline{u} \leq \mathcal{K}\underline{u} \quad \text{in } J;$$

- a function $\overline{u} \in C(J; \Omega)$ is said to be a *supersolution* of (8.1) in J if

$$\overline{u} \geq \mathcal{K}\overline{u} \quad \text{in } J.$$

Thus, by definition, the $\mathcal{K}$-iterate of a subsolution increases with respect to the subsolution itself, while the $\mathcal{K}$-iterate of a supersolution decreases. The following result gives an extremely useful tool for constructing sub- and supersolutions.

Lemma 8.3. *Suppose $J = [t_0, t_0 + T]$ and $f \in C(J \times \Omega; \mathbb{R}^N)$.*

(a) *Let $\underline{u} \in C^1(J; \Omega)$ be a function satisfying*

$$\begin{cases} \underline{u}'(t) \leq f(t, \underline{u}(t)) & \text{for all } t \in J, \\ \underline{u}(t_0) \leq u_0. \end{cases} \tag{8.2}$$

Then, $\underline{u}$ is a subsolution of (8.1) in J.

(b) *Let $\overline{u} \in \mathcal{C}^1(J;\Omega)$ be a function satisfying*

$$\begin{cases} \overline{u}'(t) \geq f(t,\overline{u}(t)) & \text{for all } t \in J, \\ \overline{u}(t_0) \geq u_0. \end{cases}$$

Then, $\overline{u}$ is a supersolution of (8.1) in J.

Proof. To prove (a), suppose that $\underline{u} \in \mathcal{C}^1(J;\Omega)$ satisfies (8.2). Then,

$$\int_{t_0}^{t} \underline{u}'(s)\, ds \leq \int_{t_0}^{t} f(s,\underline{u}(s))\, ds$$

for all $t \in J$. Thus,

$$\underline{u}(t) = \underline{u}(t_0) + \int_{t_0}^{t} \underline{u}'(s)\, ds$$

$$\leq \underline{u}(t_0) + \int_{t_0}^{t} f(s,\underline{u}(s))\, ds$$

$$\leq u_0 + \int_{t_0}^{t} f(s,\underline{u}(s))\, ds = \mathcal{K}\underline{u}(t),$$

and hence $\underline{u} \leq \mathcal{K}\underline{u}$ in J, i.e. $\underline{u}$ is a subsolution of (8.1) in J. The proof of Part (b) follows similarly. $\qquad\square$

8.2 Kamke's Theorem

The following result is the main theorem of this chapter. It gives a sufficient condition for the $\mathcal{K}$-iterates of (8.1) to converge to a solution in the context of Peano's theory.

Theorem 8.4 (Kamke). *Let $J = [t_0, t_0+T]$ and $f \in \mathcal{C}(J \times \mathbb{R}^N; \mathbb{R}^N)$ be non-decreasing in $u \in \mathbb{R}^N$. Assume that the Cauchy problem (8.1) possesses a subsolution, $\underline{u}(t)$, and a supersolution, $\overline{u}(t)$, in J, ordered in the sense that*

$$\underline{u}(t) \leq \overline{u}(t) \quad \text{for all } t \in J.$$

Then, the sequences of $\mathcal{K}$-iterates from $\underline{u}$ and $\overline{u}$, defined as

$$\begin{cases} \underline{u}_n := \mathcal{K}\underline{u}_{n-1}, & n \geq 1, \quad \underline{u}_0 := \underline{u}, \\ \overline{u}_n := \mathcal{K}\overline{u}_{n-1}, & n \geq 1, \quad \overline{u}_0 := \overline{u}, \end{cases}$$

satisfy the following properties:

(a) *For every $n \geq 1$,*

$$\underline{u} \leq \underline{u}_n \leq \underline{u}_{n+1} \leq \overline{u}_{n+1} \leq \overline{u}_n \leq \overline{u}. \tag{8.3}$$

(b) *The point-wise limits*

$$u_* := \lim_{n \to \infty} \underline{u}_n \quad and \quad u^* := \lim_{n \to \infty} \overline{u}_n$$

are well-defined and solve problem (8.1) in J. Actually, these limits are uniform in the interval J and satisfy

$$\underline{u} \leq u_* \leq u^* \leq \overline{u} \quad in \ J.$$

(c) *u_* and u^* provide us with the minimal and maximal solutions, respectively, of (8.1) in the interval $[\underline{u}, \overline{u}]$, in the sense that any other solution, u, of (8.1) such that $\underline{u} \leq u \leq \overline{u}$ in J satisfies*

$$u_* \leq u \leq u^* \quad in \ J.$$

Figure 8.1 illustrates the result of Theorem 8.4. Observe that Theorem 8.4 is extremely useful because it helps us locate and estimate the solutions of (8.1). In particular, in the context of the Cauchy–Lipschitz theory, necessarily, $u_* = u^*$ by uniqueness.

Proof. The crucial point in this proof is the fact that, for every integer $n \geq 1$, $\mathcal{K}^n$ is order-preserving, as shown in Proposition 8.1. By definition, since $\underline{u}_0 = \underline{u}$ is a subsolution,

$$\underline{u}_0 \leq \mathcal{K}\underline{u}_0 = \underline{u}_1.$$

Similarly, as $\overline{u}_0 = \overline{u}$ is a supersolution,

$$\overline{u}_0 \geq \mathcal{K}\overline{u}_0 = \overline{u}_1.$$

Moreover, as we are assuming that $\underline{u} = \underline{u}_0 \leq \overline{u}_0 = \overline{u}$ in J and $\mathcal{K}$ is order-preserving, it is apparent that

$$\underline{u}_1 = \mathcal{K}\underline{u}_0 \leq \mathcal{K}\overline{u}_0 = \overline{u}_1.$$

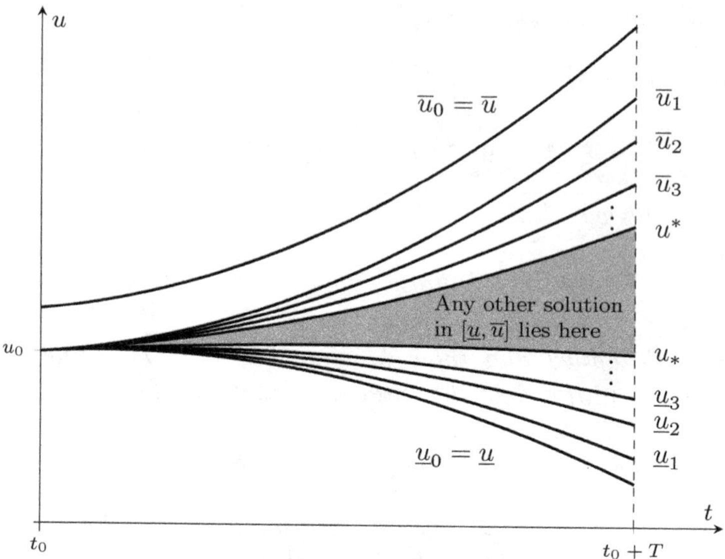

Fig. 8.1. Sequences of the $\mathcal{K}$-iterates in Kamke's theorem. They are constructed starting from the given ordered sub- and supersolution.

Thus, we have shown that

$$\underline{u} = \underline{u}_0 \leq \underline{u}_1 \leq \overline{u}_1 \leq \overline{u}_0 = \overline{u} \quad \text{in } J.$$

Since $\mathcal{K}^n$ is order-preserving, we find that, for every integer $n \geq 1$,

$$\underline{u}_n = \mathcal{K}^n \underline{u}_0 \leq \mathcal{K}^n \underline{u}_1 = \underline{u}_{n+1} \leq \overline{u}_{n+1} = \mathcal{K}^n \overline{u}_1 \leq \mathcal{K}^n \overline{u}_0 = \overline{u}_n \quad \text{in } J,$$

which concludes the proof of Part (a).

From (a), we know that the sequence of iterates $\{\underline{u}_n\}_{n \geq 1}$ is point-wise non-decreasing, and it is bounded from above, e.g. by $\overline{u}$. Thus, the point-wise limit

$$u_*(t) := \lim_{n \to \infty} \underline{u}_n(t), \qquad t \in J,$$

is well-defined and finite. Similarly, the sequence $\{\overline{u}_n\}_{n \geq 1}$ is non-increasing and bounded from below, e.g. by $\underline{u}$. Thus,

$$u^*(t) := \lim_{n \to \infty} \overline{u}_n(t), \qquad t \in J,$$

is also well-defined and finite. Moreover, by letting $n \to +\infty$ in (8.3), $u_* \leq u^*$. To show that these limits are uniform in the interval J, we are going to use the Ascoli–Arzelà theorem (see Theorem 7.5). From (a), we have that

$$\underline{u} \leq \underline{u}_n \leq \overline{u}_n \leq \overline{u}$$

in J for all $n \geq 1$. Thus, $\{\underline{u}_n\}_{n\geq 1}$ and $\{\overline{u}_n\}_{n\geq 1}$ are bounded in $\mathcal{C}(J; \mathbb{R}^N)$. So, to apply Theorem 7.5, it suffices to show that these sequences are also equicontinuous. Since the set

$$\mathcal{Q} := \{(t, u) \in J \times \mathbb{R}^N : \underline{u}(t) \leq u \leq \overline{u}(t)\} \tag{8.4}$$

is a compact subset of $J \times \mathbb{R}^N$, as it is closed and bounded, the constant

$$\|f\|_\infty := \max_{(t,u)\in\mathcal{Q}} \|f(t, u)\| < +\infty$$

is well-defined because f is continuous. Thus, for every $n \geq 1$ and $t, s \in J$, we have that

$$\begin{aligned}
\|\underline{u}_n(t) - \underline{u}_n(s)\| &= \|\mathcal{K}\underline{u}_{n-1}(t) - \mathcal{K}\underline{u}_{n-1}(s)\| \\
&= \left\| u_0 + \int_{t_0}^{t} f(\tau, \underline{u}_{n-1}(\tau))\, d\tau \right. \\
&\qquad \left. - u_0 - \int_{t_0}^{s} f(\tau, \underline{u}_{n-1}(\tau))\, d\tau \right\| \\
&= \left\| \int_{s}^{t} f(\tau, \underline{u}_{n-1}(\tau))\, d\tau \right\| \leq \|f\|_\infty\, |t - s|.
\end{aligned}$$

Consequently, by Proposition 7.4, the set

$$\mathfrak{F} := \{\underline{u}_n : n \geq 1\} \subset \mathcal{C}(J; \mathbb{R}^N)$$

is equicontinuous. Therefore, owing to Theorem 7.5, there exist a subsequence of $\{\underline{u}_n\}_{n\geq 1}$, $\{\underline{u}_{n_m}\}_{m\geq 1}$, and a function $\underline{u}_\infty \in \mathcal{C}(J; \mathbb{R}^N)$

such that

$$\lim_{m \to \infty} \|\underline{u}_{n_m} - \underline{u}_\infty\|_{\mathcal{C}(J;\mathbb{R}^N)} = 0.$$

By the uniqueness of the point-wise limit, $u_* = \underline{u}_\infty \in \mathcal{C}(J;\mathbb{R}^N)$. Moreover, by the monotonicity of $\{\underline{u}_n\}_{n \geq 1}$, it becomes apparent that

$$\lim_{n \to \infty} \|\underline{u}_n - u_*\|_{\mathcal{C}(J;\mathbb{R}^N)} = 0.$$

Naturally, this argument can also be easily adapted to show that

$$\lim_{n \to \infty} \|\overline{u}_n - u^*\|_{\mathcal{C}(J;\mathbb{R}^N)} = 0.$$

Since, for every $n \geq 1$,

$$\underline{u}_n = \mathcal{K}\underline{u}_{n-1}, \qquad \overline{u}_n = \mathcal{K}\overline{u}_{n-1},$$

we have that, for every $t \in J$ and $n \geq 1$,

$$\underline{u}_n(t) = u_0 + \int_{t_0}^t f(s, \underline{u}_{n-1}(s))\, ds, \quad \overline{u}_n(t) = u_0 + \int_{t_0}^t f(s, \overline{u}_{n-1}(s))\, ds.$$

Therefore, letting $n \to +\infty$ in these identities, we find that

$$u_*(t) = u_0 + \int_{t_0}^t f(s, u_*(s))\, ds, \qquad u^*(t) = u_0 + \int_{t_0}^t f(s, u^*(s))\, ds,$$

for all $t \in J$. So, u_* and u^* solve (8.1), which concludes the proof of Part (b).

Finally, let $u \in \mathcal{C}^1(J;\mathbb{R}^N)$ be a solution of (8.1) such that

$$\underline{u} \leq u \leq \overline{u}$$

in J. Then, since u is a fixed point of $\mathcal{K}^n$ and $\mathcal{K}^n$ is order-preserving, for every $n \geq 1$,

$$\underline{u}_n = \mathcal{K}^n \underline{u} \leq \mathcal{K}^n u = u \leq \overline{u}_n = \mathcal{K}^n \overline{u}$$

and, letting $n \to +\infty$, gives

$$u_* \leq u \leq u^*,$$

which completes the proof of Part (c). $\square$

Remark 8.5.

(a) As the values of $f(t, u)$ outside $\mathcal{Q}$ are not used in the proof of Theorem 8.4, to obtain the results of the theorem, it suffices to impose that $f(t, \cdot)$ is non-decreasing in the region $\mathcal{Q}$ defined in (8.4).

(b) When f is of class $\mathcal{C}^1$ in u and $\frac{\partial f_i}{\partial u_j} \in \mathcal{C}(\mathcal{Q}; \mathbb{R}^N)$, the simplest criterion ensuring that $f(t, u)$ is non-decreasing in u is

$$\frac{\partial f_i}{\partial u_j} \geq 0 \text{ in } \mathcal{Q} \qquad \text{for all } i, j \in \{1, \ldots, N\}.$$

(c) Theorem 8.4 admits far more general versions in the context of the theory of partial differential equations (see, e.g. Amann, 1971/72, 1976, 2005; López-Gómez and Molina-Meyer, 1994; Molina-Meyer, 1995, 1996).

In the remainder of this chapter, we illustrate how Theorem 8.4 can be used in a series of practical applications, some of which are of great interest on their own.

8.3 Example of Blow up in a Single Equation

We begin this section by studying the Cauchy problem

$$\begin{cases} u' = u + \gamma a(t) u^3, \\ u(0) = x > 0, \end{cases} \tag{8.5}$$

where $a \in \mathcal{C}([0, +\infty); (0, +\infty))$, i.e. it is continuous and positive, and $\gamma > 0$ is regarded as a parameter. Thanks to Theorem 5.14, (8.5) possesses a unique maximal solution, (I, u). Moreover, since $u' = \left(1 + \gamma a(t) u^2\right) u$, we get

$$u(t) = x e^{\int_0^t [1 + \gamma a(s) u^2(s)] \, ds} > 0 \quad \text{for all } t \in I.$$

Actually, as the equation in (8.5) is of Bernoulli type, (I, u) can be easily determined, as described in Section 4.1: the change of variable $u = v^{-1/2}$ transforms the differential equation in (8.5) into

$$v' = -2v - 2\gamma a(t), \tag{8.6}$$

together with the initial condition $v(0) = x^{-2} > 0$. The variation of constants formula then gives

$$v(t) = e^{-2t}\left(x^{-2} - 2\gamma \int_0^t a(s)e^{2s}\,ds\right),$$

and hence the unique solution of (8.5) is

$$u(t) = \frac{e^t}{\sqrt{x^{-2} - 2\gamma \int_0^t a(s)e^{2s}\,ds}}, \qquad t \in I. \tag{8.7}$$

Subsequently, we emphasize the dependence of the maximal solution of (8.5) on the parameter γ, denoting it by (I_γ, u_γ). Obviously, I_γ consists of the set of $t \geq 0$ such that

$$x^{-2} - 2\gamma \int_0^t a(s)e^{2s}\,ds > 0,$$

which might significantly depend on the weight function $a(t)$. Hereinafter, we assume that

$$\int_0^\infty a(s)e^{2s}\,ds = +\infty. \tag{8.8}$$

Then, $I_\gamma = [0, T_{\max}(\gamma))$, where $T_{\max}(\gamma) > 0$ is the unique value of T for which

$$\int_0^T a(s)e^{2s}\,ds = \frac{1}{2\gamma x^2}.$$

Since $a(t) > 0$ for all $t > 0$, we have that

$$\frac{d}{dt}\int_0^t a(s)e^{2s}\,ds = a(t)e^{2t} > 0,$$

and hence the blow up time $T_{\max}(\gamma)$ is decreasing with respect to γ, and

$$\lim_{\gamma \downarrow 0} T_{\max}(\gamma) = +\infty, \qquad \lim_{\gamma \uparrow +\infty} T_{\max}(\gamma) = 0,$$

as illustrated in the left plot of Figure 8.2. Moreover, from (8.7), we have that $u_\gamma(t)$ is increasing with respect to γ, i.e. whenever $\gamma_1 < \gamma_2$,

$$u_{\gamma_1} \lneq u_{\gamma_2} \quad \text{in } I_{\gamma_2} = [0, T_{\max}(\gamma_2)) \subsetneq I_{\gamma_1} = [0, T_{\max}(\gamma_1)),$$

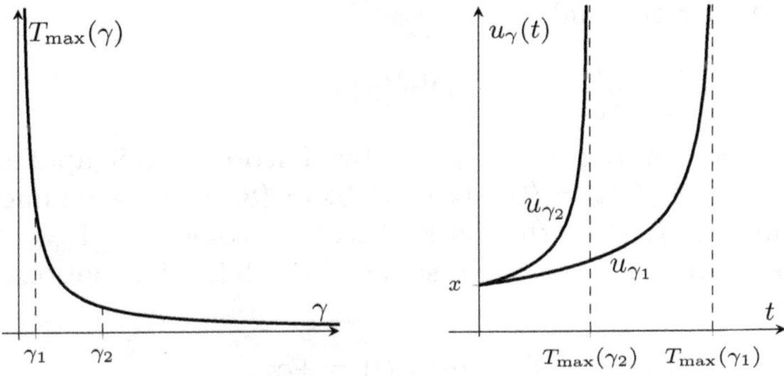

Fig. 8.2. Plots of $T_{\max}(\gamma)$ and u_γ for problem (8.5).

as illustrated in the right plot of Figure 8.2. Actually,

$$u_{\gamma_1}(t) < u_{\gamma_2}(t) \quad \text{for all } t \in (0, T_{\max}(\gamma_2)).$$

Now, we move on to the analysis of the more general Cauchy problem

$$\begin{cases} u' = u + a(t)p(u)u, \\ u(0) = x > 0, \end{cases} \tag{8.9}$$

where $a(t)$ satisfies the above assumptions, particularly (8.8), and the function $p \in \mathcal{C}^1([0, +\infty); \mathbb{R})$ satisfies $p'(u) > 0$ for all $u > 0$, and the global estimate

$$\alpha u^2 \le p(u) \le \beta u^2 \quad \text{for all } u \ge 0, \tag{8.10}$$

for some constants $0 < \alpha < \beta$. For instance, the function

$$p(u) := \left(1 + \frac{u}{1+u}\right) u^2, \qquad u \ge 0,$$

satisfies these requirements with $\alpha = 1$ and $\beta = 2$ because

$$p'(u) = \frac{u^2}{(1+u)^2} + \left(1 + \frac{u}{1+u}\right) 2u > 0 \quad \text{and} \quad 1 \le 1 + \frac{u}{1+u} \le 2,$$

for all $u > 0$. Subsequently, we denote

$$f(t, u) := u + a(t)p(u)u = [1 + a(t)p(u)]u, \qquad t \ge 0, \ u \ge 0.$$

Since $u(0) = x > 0$ and

$$\frac{\partial f}{\partial u} = 1 + a(t)[p'(u)u + p(u)] \geq 1$$

is continuous in $t, u \in [0, +\infty)$, by Theorem 5.18 applied with $J = [0, +\infty)$ and $\Omega = (0, +\infty)$, problem (8.9) has a unique maximal solution, (I, u), with $I = [0, T_{\max})$ for some $0 < T_{\max} \leq +\infty$. Moreover, when $T_{\max} < +\infty$, some of the following options occur. Either

$$\limsup_{t \uparrow T_{\max}} |u(t)| = +\infty,$$

or, since $\partial \Omega = \{0\}$, there exists a sequence of times $\{t_n\}_{n \geq 1} \subset I$ such that

$$\lim_{n \to \infty} t_n = T_{\max} \qquad \text{and} \qquad \lim_{n \to \infty} u(t_n) = 0.$$

As, for every $t \in I$,

$$u'(t) = [1 + a(t)p(u(t))]u(t),$$

it is apparent that

$$u(t) = x e^{\int_0^t [1 + a(s)p(u(s))]\,ds} > 0.$$

Thus, $u'(t) > 0$ for all $t \in I$, and hence either $T_{\max} = +\infty$ or $u(t)$ blows up at $T_{\max}$. Moreover, since $\frac{\partial f}{\partial u} \geq 1$, f is increasing in u for all $t \geq 0$, and Kamke's theorem 8.4 can be applied to problem (8.9).

Although, in general, the solution to this problem, as well as its maximal existence interval I, cannot be explicitly determined, prompted by (8.10), we estimate $T_{\max}$ using the following auxiliary problems:

$$\begin{cases} u' = u + \alpha a(t)u^3, \\ u(0) = x > 0, \end{cases} \tag{8.11}$$

and

$$\begin{cases} u' = u + \beta a(t)u^3, \\ u(0) = x > 0. \end{cases}$$

These problems are particular instances of (8.5), with $\gamma = \alpha$ and $\gamma = \beta$, respectively. As above, we denote their corresponding solutions by (I_α, u_α) and (I_β, u_β); nevertheless, to avoid confusion with the maximal time of existence $T_{\max}$ of problem (8.9), we slightly change the notation and set

$$I_\alpha = [0, T_\alpha), \qquad I_\beta = [0, T_\beta).$$

The comparison between (I_α, u_α), (I_β, u_β), and (I, u) is accomplished in two steps, as follows:

Step 1. In this step, we obtain an upper bound for u and a lower bound for $T_{\max}$ by means of $\overline{u} := u_\beta$. From (8.10), for every $t \in I_\beta$, we have

$$\overline{u}'(t) = \overline{u}(t) + \beta a(t)\overline{u}^3(t) \geq \overline{u}(t) + a(t)p(\overline{u}(t))\overline{u}(t).$$

Thus, since $\overline{u}(0) = u_\beta(0) = x$, $\overline{u}(t)$ is a supersolution of problem (8.9) in I_β. Obviously, the constant function $\underline{u} := 0$ provides us with a subsolution of (8.9) in $[0, +\infty)$ since, apart from satisfying the differential equation in (8.9), $\underline{u}(0) = 0 < x = u(0)$. Consequently, problem (8.9) admits a subsolution, $\underline{u}$, and a supersolution, $\overline{u}$, in the same interval I_β. Moreover, they are ordered, since $\underline{u} \leq \overline{u}$ in I_β. Then, by Theorem 8.4,

$$\underline{u} \leq u \leq \overline{u} = u_\beta \quad \text{in } I_\beta. \tag{8.12}$$

Therefore,

$$[0, T_\beta) = I_\beta \subset I = [0, T_{\max}).$$

Equivalently,

$$T_\beta \leq T_{\max}.$$

Step 2. In this step, we obtain a lower bound for u and an upper bound for $T_{\max}$. By (8.10), in the interval $I = [0, T_{\max})$, we have that

$$u'(t) = u(t) + a(t)p(u(t))u(t) \geq u(t) + \alpha a(t)u^3(t),$$

and hence $\overline{u} := u$ provides us with a supersolution of (8.11) in the interval I. As above, one can easily show that $\underline{u} = 0$ provides us with a subsolution in the same interval and that $\underline{u} \le \overline{u} = u$. Thus, according to Theorem 8.4,

$$u_\alpha \le u \quad \text{in } I = [0, T_{\max}). \tag{8.13}$$

Therefore, u_α must be defined in I, which implies

$$[0, T_{\max}) = I \subset I_\alpha = [0, T_\alpha).$$

Equivalently,

$$T_{\max} \le T_\alpha.$$

Summarizing, we have obtained the following bilateral estimate for the global existence time of (I, u):

$$T_\beta \le T_{\max} \le T_\alpha.$$

Furthermore, owing to (8.12) and (8.13), the graph of u, compared with the graphs of u_α and u_β, looks like that shown in Figure 8.3.

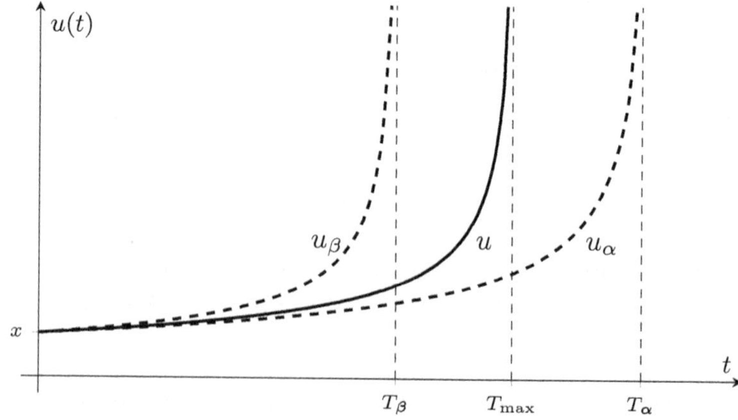

Fig. 8.3. Plots of the solution u of (8.9), compared with u_α and u_β.

8.4 Example of Blow up in a Cooperative System

The main goal of this section is to prove that the solution of the Cauchy problem

$$\begin{cases} u' = u + a(t)v^3, \\ v' = v + a(t)u^3, \\ u(0) = x > 0, \quad v(0) = y > 0, \end{cases} \tag{8.14}$$

blows up in a finite time, as well as to estimate its blow up time. Here, $a(t)$ satisfies the same assumptions as those in Section 8.3; in particular, it is positive, continuous and satisfies (8.8). Thus, if $v(t) > 0$, $u'(t)$ is larger than in the case of $v(t) = 0$, and the same occurs for $v'(t)$ in the presence of $u(t)$. For this reason, if we imagine that u and v describe the time evolution of two biological species, system (8.14) is referred to as a *cooperative system* because the presence of v increases the growth rate of u, and vice versa. Since

$$f(t, u, v) = \begin{pmatrix} u + a(t)v^3 \\ v + a(t)u^3 \end{pmatrix}$$

is of class C^1 in $(u, v) \in \mathbb{R}^2$, by Corollary 5.3, f is locally Lipschitz in $(u, v) \in \mathbb{R}^2$, uniformly in t on compact subsets of $J := [0, \infty)$. Thus, thanks to Theorem 5.17, the unique maximal solution, (I, u, v), of (8.14) satisfies $I = [0, T_{\max})$ for some $0 < T_{\max} \leq +\infty$. Moreover,

$$\limsup_{t \uparrow T_{\max}} (|u(t)| + |v(t)|) = +\infty \quad \text{if } T_{\max} < +\infty.$$

The main objective of this section is to show that $T_{\max} < +\infty$ and estimate it. Since the Jacobian

$$D_{(u,v)} f(t, u, v) = \begin{pmatrix} 1 & 3a(t)v^2 \\ 3a(t)u^2 & 1 \end{pmatrix}$$

has non-negative entries, it becomes apparent that $f(t, u, v)$ is increasing in $(u, v) \in \mathbb{R}^2$, according to the definition given in Section 8.1. Thus, Theorem 8.4 can be applied, and we use it to achieve our goals.

Observe that, in the special case of $x = y$, by the symmetry of the problem in u and v, thanks to uniqueness, the solution of (8.14) must be of the form (I_x, w_x, w_x), where (I_x, w_x) denotes the unique maximal solution of

$$\begin{cases} w' = w + a(t)w^3, \\ w(0) = x > 0. \end{cases}$$

By the analysis carried out in the previous section, we know that

$$w_x(t) = \frac{e^t}{\sqrt{x^{-2} - 2\int_0^t a(s)e^{2s}\, ds}}, \qquad t \in I_x = [0, T_x), \qquad (8.15)$$

where T_x is characterized by

$$\int_0^{T_x} a(s)e^{2s}\, ds = \frac{1}{2x^2}.$$

So, in this case, the problem has already been completely solved.

Next, we consider the more interesting case when the symmetry of the problem is lost. We can assume, without loss of generality, that $x > y$. The following result shows that u and v are increasing in I.

Lemma 8.6. *For every $t \in I$, we have that $u'(t) > 0$ and $v'(t) > 0$. Thus, $u(t) \geq x > 0$ and $v(t) \geq y > 0$ for all $t \in I$.*

Proof. Since

$$u'(0) = u(0) + a(0)v^3(0) = x + a(0)y^3 > x > 0,$$

$$v'(0) = v(0) + a(0)u^3(0) = y + a(0)x^3 > y > 0,$$

by continuity, there exists $\delta > 0$ such that

$$u'(t) > 0 \quad \text{and} \quad v'(t) > 0 \qquad \text{for all } t \in [0, \delta).$$

Suppose there exists $t_1 \in I$ such that $u'(t) > 0$ for all $t \in [0, t_1)$ and $u'(t_1) = 0$. Then,

$$0 = u'(t_1) = u(t_1) + a(t_1)v^3(t_1),$$

and, since $u(t_1) > 0$ and $a(t_1) > 0$, necessarily, $v(t_1) < 0$. Thus, there exists $t_2 \in (0, t_1)$ such that $v'(t) > 0$ for all $t \in (0, t_2)$ and $v'(t_2) = 0$. This implies

$$0 = v'(t_2) = v(t_2) + a(t_2)u^3(t_2),$$

and since $v(t_2) > 0$ and $a(t_2) > 0$, we find that $u(t_2) < 0$, which gives a contradiction. Therefore, $u'(t) > 0$ for all $t \in I$. Similarly, $v'(t) > 0$ for all $t \in I$. $\qquad\qquad \square$

Lemma 8.7. *Suppose $x > y$. Then, $u(t) > v(t)$ for all $t \in I = [0, T_{\max})$.*

Proof. Since

$$u(0) = x > y = v(0),$$

by continuity, there exists $\delta > 0$ such that

$$u(t) > v(t) \quad \text{for all } t \in [0, \delta).$$

Reasoning by contradiction, suppose there is a $t_1 \in I$ such that

$$u(t) > v(t) \ \text{ if } t \in [0, t_1), \quad \text{but } \ u(t_1) = v(t_1).$$

Then, setting

$$x_1 := u(t_1) = v(t_1),$$

it becomes apparent that (u, v) solves the problem

$$\begin{cases} u' = u + a(t)v^3, \\ v' = v + a(t)u^3, \\ u(t_1) = v(t_1) = x_1 > 0, \end{cases}$$

and by uniqueness, $u = v$ in I, which is impossible. This concludes the proof. $\qquad\square$

Lemma 8.8. *Suppose $x > y$ and $T_{\max} < +\infty$. Then,*

$$\lim_{t \uparrow T_{\max}} u(t) = +\infty = \lim_{t \uparrow T_{\max}} v(t).$$

Proof. Since $T_{\max} < +\infty$, owing to Theorem 5.17,

$$\limsup_{t \uparrow T_{\max}} (|u(t)| + |v(t)|) = +\infty.$$

Thus, since $x > y$, it follows from Lemmas 8.6 and 8.7 that

$$\lim_{t \uparrow T_{\max}} u(t) = +\infty. \tag{8.16}$$

Assume by contradiction that $v(t)$ does not blow up at $T_{\max}$. Then, since it is increasing, necessarily

$$v(t) \leq C, \qquad t \in [0, T_{\max}),$$

for some constant $C > 0$. Thus,

$$u'(t) = u(t) + a(t)v^3(t) \leq u(t) + M, \qquad t \in [0, T_{\max}),$$

where

$$M := C^3 \max_{t \in \bar{I}} a(t) < +\infty.$$

Setting

$$u(t) = e^t w(t), \qquad t \in I,$$

we find that

$$e^t w'(t) \leq M, \qquad t \in I,$$

and hence

$$w(t) \leq w(0) + M \int_0^t e^{-s}\, ds = x + M(1 - e^{-t}), \qquad t \in I.$$

Therefore,

$$u(t) \leq xe^t + M(e^t - 1), \qquad t \in I,$$

which contradicts (8.16) and concludes the proof. $\qquad\qquad\square$

Summarizing, the previous result shows that, if $T_{\max} < +\infty$, i.e. if the maximal solution blows up in a finite time, then both $u(t)$ and $v(t)$ blow up at $T_{\max}$, as sketched in Figure 8.4.

Subsequently, we use Kamke's theorem (see Theorem 8.4) to compare, when $x > y$, the solution of (8.14) with the solutions of the auxiliary problems

$$\begin{cases} u' = u + a(t)v^3, \\ v' = v + a(t)u^3, \\ u(0) = v(0) = x, \end{cases} \qquad (8.17)$$

and

$$\begin{cases} u' = u + a(t)v^3, \\ v' = v + a(t)u^3, \\ u(0) = v(0) = y. \end{cases} \qquad (8.18)$$

We already know that the solutions of (8.17) and (8.18) are, respectively, (I_x, w_x, w_x) and (I_y, w_y, w_y), where w_x is given in (8.15). As in

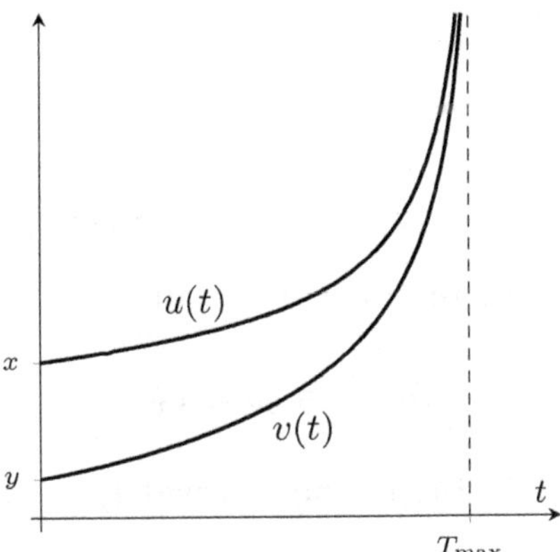

Fig. 8.4. Plot of the solution $(u(t), v(t))$ of (8.14) for $x > y$ in the case of blow up at $T_{\max} < +\infty$.

the previous section, the comparison of (I, u, v) with these solutions is accomplished in two steps:

Step 1. From Lemma 8.3, it is apparent that

$$(\overline{u}, \overline{v}) := (u, v)$$

is a supersolution of (8.18) in the interval $I = [0, T_{\max})$. Moreover,

$$(\underline{u}, \underline{v}) := (0, 0)$$

is a subsolution of (8.18) in the interval $[0, +\infty)$, and

$$(\underline{u}, \underline{v}) \leq (\overline{u}, \overline{v}) \quad \text{in } I = [0, T_{\max}).$$

Thus, by Theorem 8.4, we find that

$$(0, 0) = (\underline{u}, \underline{v}) \leq (w_y, w_y) \leq (\overline{u}, \overline{v}) = (u, v) \quad \text{in } I = [0, T_{\max}).$$

Therefore, w_y must be defined, at least, in the interval $[0, T_{\max})$, and hence

$$[0, T_{\max}) \subset [0, T_y).$$

Equivalently,

$$T_{\max} \leq T_y < +\infty.$$

Therefore, it follows from Lemma 8.8 that (u, v) blows up at $T_{\max}$.

Step 2. Similarly, by Lemma 8.3, the pair

$$(\overline{u}, \overline{v}) = (w_x, w_x)$$

is a supersolution of (8.14) in the interval $I_x = [0, T_x)$, while

$$(\underline{u}, \underline{v}) = (0, 0)$$

is a subsolution of (8.14) in $[0, +\infty)$. Moreover,

$$(\underline{u}, \underline{v}) \leq (\overline{u}, \overline{v}) \quad \text{in } I = [0, T_x).$$

Thus, owing to Theorem 8.4,

$$(0, 0) = (\underline{u}, \underline{v}) \leq (u, v) \leq (\overline{u}, \overline{v}) = (w_x, w_x) \quad \text{in } I = [0, T_x).$$

Consequently, (u, v) is defined, at least, in the interval $[0, T_x)$. Therefore,

$$[0, T_x) \subset [0, T_{\max}).$$

Equivalently,

$$T_x \leq T_{\max}.$$

Therefore,

$$T_x \leq T_{\max} \leq T_y, \tag{8.19}$$

and the graphs of $u(t)$ and $v(t)$ look like those shown in Figure 8.5.

Actually, we conjecture that both inequalities in (8.19) are strict. Since a proof of this fact is beyond the scope of this book, we conclude our analysis here.

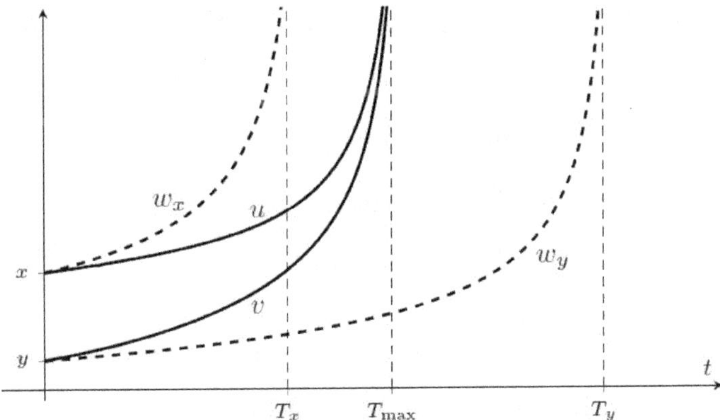

Fig. 8.5. Plot of the solution $(u(t), v(t))$ of (8.14) for $x > y$, compared with the solutions of (8.17) and (8.18).

8.5 The Synergy of Symbiosis

In this section, we analyze the Cauchy problem

$$\begin{cases} u' = u + buv, \\ v' = v + cuv, \\ u(0) = v(0) = x > 0, \end{cases} \tag{8.20}$$

where $b > 0$ and $c > 0$ are constants. In this example, the nonlinearity is

$$f(u, v) = \begin{pmatrix} u + buv \\ v + cuv \end{pmatrix},$$

which, if we think of u and v as two biological species, describes a symbiotic interaction between them, given by a term which is proportional to uv, together with an exponential growth for each of the species u and v, given by the linear factor u in the first component and v in the second one. In the literature, this system is known as *Lotka–Volterra symbiotic model*.

Since $f(u, v) \in \mathcal{C}^\infty(\mathbb{R}^2; \mathbb{R}^2)$, by Corollary 5.3, it is locally Lipschitz in $(u, v) \in \mathbb{R}^2$, uniformly in t on compact subsets of $\mathbb{R}$. Hence, according to Theorem 5.17, the maximal solution, (I, u, v), satisfies $I = [0, T_{\max})$ for some $T_{\max} \leq +\infty$. The main goal of the forthcoming discussion is to show that $T_{\max} < +\infty$ and estimate it. Since

$$u' = (1 + bv)u, \quad v' = (1 + cu)v,$$

and $u(0) = v(0) = x > 0$, it is apparent that

$$u(t) = xe^{\int_0^t [1+bv(s)]\,ds} > 0 \quad \text{and} \quad v(t) = xe^{\int_0^t [1+cu(s)]\,ds} > 0$$

for all $t \in I$. In particular, this implies that $(u(t), v(t)) \in \mathcal{Q}$ for all $t \in I$, where

$$\mathcal{Q} := \{(u, v) \in \mathbb{R}^2 : u \geq 0, \ v \geq 0\}$$

indicates the positive quadrant of the plane. As a consequence, we have that u' and v' are positive in I. Moreover, all the entries of

$$D_{(u,v)} f(u, v) = \begin{pmatrix} 1 + bv & bu \\ cv & 1 + cu \end{pmatrix}$$

are positive in $\mathcal{Q}$. Thus, by Remark 8.5(a), we can apply Kamke's theorem (see Theorem 8.4), provided we construct a suitable pair of sub- and supersolutions.

We begin by analyzing (8.20) for $b = c$. In this simpler case, by the symmetry of the problem, it becomes apparent that

$$(I, u, v) = (I_b, w_b, w_b),$$

where (I_b, w_b) denotes the unique solution of the problem

$$\begin{cases} w' = w + bw^2, \\ w(0) = x, \end{cases}$$

which is given by

$$w_b(t) = \frac{1}{\left(b + \frac{1}{x}\right) e^{-t} - b}, \qquad t \in I_b = [0, T_b),$$

where we have set

$$T_b := \ln\left(1 + \frac{1}{bx}\right).$$

Note that T_b decreases in b, and

$$\lim_{b \downarrow 0} T_b = +\infty, \qquad \lim_{b \uparrow \infty} T_b = 0.$$

Now, we study the general case $b \neq c$. Without loss of generality, in the remainder of this section, we assume that $b > c$. Then, the following result holds.

Lemma 8.9. *Suppose $b > c$. Then, $u(t) > v(t)$ for all $t \in (0, T_{\max})$.*

Proof. Since

$$u'(0) = u(0) + bu(0)v(0) = x(1 + bx),$$
$$v'(0) = v(0) + cu(0)v(0) = x(1 + cx),$$

it is apparent that $u'(0) > v'(0)$, and hence there exists $\delta > 0$ such that

$$u(t) > v(t) \quad \text{for all } t \in (0, \delta).$$

Assume by contradiction that there exists $t_1 \in (0, T_{\max})$ such that $u(t) > v(t)$ for all $t \in (0, t_1)$ and $u(t_1) = v(t_1)$. Then,

$$u'(t_1) \leq v'(t_1),$$

and hence

$$0 \geq u'(t_1) - v'(t_1) = u(t_1) + bu(t_1)v(t_1) - v(t_1) - cu(t_1)v(t_1)$$
$$= (b - c)u(t_1)v(t_1) > 0,$$

which is impossible. This ends the proof. $\qquad\square$

Lemma 8.10. *Suppose $b > c$ and $T_{\max} < +\infty$. Then,*

$$\lim_{t \uparrow T_{\max}} u(t) = +\infty = \lim_{t \uparrow T_{\max}} v(t).$$

Proof. Since $T_{\max} < +\infty$, thanks to Theorem 5.17,

$$\limsup_{t \uparrow T_{\max}} (|u(t)| + |v(t)|) = +\infty.$$

Thus, since $u(t)$ is increasing, we can infer from Lemma 8.9 that

$$\lim_{t \uparrow T_{\max}} u(t) = +\infty. \tag{8.21}$$

Moreover, since $v(t)$ is also increasing, if $v(t)$ does not blow up at $T_{\max}$, necessarily, there exists a constant $C > 0$ such that

$$v(t) \leq C, \qquad t \in [0, T_{\max}).$$

Thus,

$$u'(t) = u(t) + bu(t)v(t) \leq (1 + bC)\, u(t), \qquad t \in [0, T_{\max}),$$

and reasoning as in the proof of Lemma 8.8 leads to a contradiction with (8.21). This ends the proof. $\qquad\square$

Next, we compare the solution of problem (8.20) with the solutions of the problems

$$\begin{cases} u' = u + buv, \\ v' = v + buv, \\ u(0) = v(0) = x > 0, \end{cases} \tag{8.22}$$

and

$$\begin{cases} u' = u + cuv, \\ v' = v + cuv, \\ u(0) = v(0) = x > 0, \end{cases} \tag{8.23}$$

which are given by (I_b, w_b, w_b) and (I_c, w_c, w_c), respectively.

As $b > c$, $(\overline{u}, \overline{v}) = (w_b, w_b)$ is a supersolution of (8.20) in I_b and $(\underline{u}, \underline{v}) = (0, 0)$ is a subsolution of (8.20) in $[0, +\infty)$ such that $(0, 0) \leq (w_b, w_b)$ in I_b. Thus, we infer from Theorem 8.4 that

$$(u, v) \leq (w_b, w_b) \quad \text{in } I_b = [0, T_b),$$

which implies that (u, v) must be defined in I_b, i.e.

$$T_b \leq T_{\max}.$$

Similarly, (u, v) is a supersolution of (8.23) in I and $(0, 0)$ is a subsolution of (8.23) in $[0, +\infty)$. Thus, since $(0, 0) \leq (u, v)$, we infer from

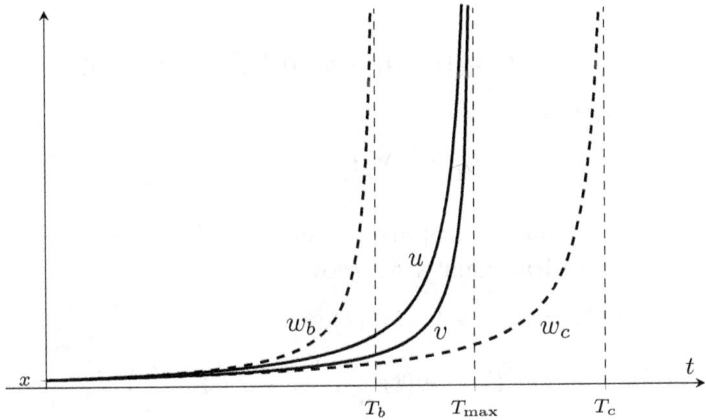

Fig. 8.6. Plot of the solution $(u(t), v(t))$ of (8.20) for $b > c$, compared with the solutions of (8.22) and (8.23).

Theorem 8.4 that

$$(w_c, w_c) \le (u, v)$$

in the interval $I = [0, T_{\max})$. Therefore, w_c is defined in I, and hence $T_{\max} \le T_c$. Consequently,

$$\ln\left(1 + \frac{1}{bx}\right) = T_b \le T_{\max} \le T_c = \ln\left(1 + \frac{1}{cx}\right). \qquad (8.24)$$

As a consequence, since $T_{\max} < +\infty$, Lemma 8.10 ensures that

$$\lim_{t \uparrow T_{\max}} u(t) = +\infty = \lim_{t \uparrow T_{\max}} v(t).$$

Figure 8.6 shows the plots of $u(t)$ and $v(t)$ when $b > c$.

In this example, we can compute $T_{\max}$ by proceeding as follows. Multiplying the first equation in (8.20) by c and the second one by b gives

$$cu' = cu + bcuv, \qquad bv' = bv + bcuv.$$

Thus, eliminating the term $bcuv$ yields

$$c\left(u' - u\right) = b\left(v' - v\right).$$

Equivalently,

$$(cu - bv)' = cu - bv,$$

and hence

$$cu(t) - bv(t) = (cu(0) - bv(0))\, e^t = x\,(c - b)\, e^t$$

for all $t \in I$. Therefore,

$$v(t) = \frac{c}{b} u(t) + x \left(1 - \frac{c}{b}\right) e^t, \quad t \in I = [0, T_{\max}). \tag{8.25}$$

From (8.25), it becomes apparent, once again, that $u(t)$ and $v(t)$ must blow up simultaneously. Moreover,

$$\begin{aligned}
u'(t) &= u(t) + bu(t)v(t) \\
&= u(t) + bu(t) \left[\frac{c}{b} u(t) + x \left(1 - \frac{c}{b}\right) e^t\right] \\
&= \left[1 + x\,(b - c)\, e^t\right] u(t) + cu^2(t),
\end{aligned}$$

which is an equation of Bernoulli type:

$$u' = a(t)u + cu^2, \quad \text{with } a(t) := 1 + x\,(b - c)\, e^t, \quad t \in I.$$

As usual, the change of variable $u = w^{-1}$ leads to the linear problem

$$\begin{cases} w' = -a(t)w - c, \\ w(0) = \frac{1}{x}. \end{cases} \tag{8.26}$$

Note that

$$\begin{aligned}
\int_0^t a(s)\, ds &= \int_0^t \left[1 + x\,(b - c)\, e^s\right] ds \\
&= \left[s + x\,(b - c)\, e^s\right]_{s=0}^{s=t} = t + x\,(b - c)\left(e^t - 1\right).
\end{aligned}$$

Moreover, the change of variable

$$w(t) = e^{-\int_0^t a(s)\, ds} \kappa(t)$$

transforms (8.26) into the problem

$$e^{-\int_0^t a(s)\, ds} \kappa'(t) = -c, \quad \kappa(0) = \frac{1}{x},$$

whose unique solution is

$$\kappa(t) = \frac{1}{x} - c \int_0^t e^{\int_0^s a(\sigma)\, d\sigma}\, ds.$$

As the first time where $\kappa(t) = 0$ coincides with the first time where $w(t) = 0$ and $u(t) = \frac{1}{w(t)}$, we have that $T_{\max}$ equals the first $T > 0$

such that

$$\int_0^T e^{\int_0^s a(\sigma)\,d\sigma}\,ds = \frac{1}{cx},$$

i.e.

$$\int_0^T e^{s+x(b-c)(e^s-1)}\,ds = \frac{1}{cx}. \tag{8.27}$$

When $b = c$, (8.27) reduces to

$$\frac{1}{cx} = \int_0^T e^s\,ds = e^T - 1,$$

which gives

$$T = \ln\left(1 + \frac{1}{cx}\right),$$

i.e. $T = T_c$ (see (8.24)). When $b \neq c$, (8.27) can be expressed as

$$\frac{1}{cx} = e^{-x(b-c)} \int_0^T e^s e^{x(b-c)e^s}\,ds,$$

and hence

$$\frac{e^{x(b-c)}}{cx} = \int_0^T e^s e^{x(b-c)e^s}\,ds = \frac{1}{x(b-c)} \int_0^T \frac{d}{ds} e^{x(b-c)e^s}\,ds$$

$$= \frac{1}{x(b-c)}\left[e^{x(b-c)e^T} - e^{x(b-c)}\right] = \frac{e^{x(b-c)}}{x(b-c)}\left[e^{x(b-c)(e^T-1)} - 1\right].$$

Therefore, solving for T in this identity gives the desired value:

$$T_{\max} = \ln\left(1 + \frac{1}{x}\frac{\ln b - \ln c}{b - c}\right).$$

Thanks to the mean value theorem, when $b > c$, there exists $\xi \in (c, b)$ such that

$$\ln b - \ln c = \frac{1}{\xi}(b - c).$$

Thus,

$$\frac{1}{b} < \frac{\ln b - \ln c}{b - c} = \frac{1}{\xi} < \frac{1}{c}$$

and

$$T_b = \ln\left(1 + \frac{1}{bx}\right) < T_{\max} < T_c = \ln\left(1 + \frac{1}{cx}\right),$$

i.e. we obtain again estimate (8.24), which was previously established using Theorem 8.4, with the improvement that now the inequalities are strict. This computation actually shows that (8.24) is a really sharp estimate of $T_{\max}$.

8.6 Wintner's Extensibility Criterion

The main goal of this section is to establish Wintner's extensibility criterion (see Theorem 8.14), which gives sufficient conditions for a solution of

$$\begin{cases} u' = f(t, u), \\ u(t_0) = u_0, \end{cases}$$

to be globally defined, where $f \in \mathcal{C}\left([t_0, +\infty) \times \mathbb{R}^N; \mathbb{R}^N\right)$ is non-decreasing with respect to $u \in \mathbb{R}^N$ and satisfies some additional conditions.

Before arriving at such a result, we need to prove some preliminary lemmas of technical nature, some of which are of their own interest. In the first one, as usual, given a vector $v \in \mathbb{R}^M$, we denote its components by $v = (v_1, \ldots, v_M)^T$.

Lemma 8.11. *Let $M \geq 1$ be a positive integer, $t_0 < t_1$, $J := [t_0, t_1]$, $g \in \mathcal{C}(J \times \mathbb{R}^M; \mathbb{R}^M)$ non-decreasing with respect to $v \in \mathbb{R}^M$, and $v(t)$, $t \in J$, a solution of the system*

$$v' = g(t, v) \tag{8.28}$$

in the interval J. Suppose that $\underline{v} \in \mathcal{C}(J; \mathbb{R}^N)$ satisfies

$$\underline{v}_j(t_0) < v_j(t_0), \qquad j \in \{1, \ldots, M\}, \tag{8.29}$$

and that, for every $t \in J$,

$$\underline{v}(t) \le \underline{v}(t_0) + \int_{t_0}^{t} g(s, \underline{v}(s)) \, ds. \qquad (8.30)$$

Then, for every $t \in J$ and $j \in \{1, \ldots, M\}$, we have that

$$\underline{v}_j(t) < v_j(t).$$

Proof. By continuity, it follows from (8.29) that there exists $\delta > 0$ such that, for every $j \in \{1, \ldots, M\}$,

$$\underline{v}_j(t) < v_j(t) \quad \text{for all } t \in [t_0, t_0 + \delta).$$

Thus, the set $\tilde{J}$, consisting of the times $\tilde{t} \in (t_0, t_1]$ for which

$$\underline{v}_j(t) < v_j(t) \quad \text{for all } j \in \{1, \ldots, M\} \text{ and } t \in [t_0, \tilde{t}), \qquad (8.31)$$

is non-empty. Consider

$$\tilde{t}_s := \sup \tilde{J} \in (t_0, t_1].$$

According to (8.30), for every $j \in \{1, \ldots, M\}$,

$$\underline{v}_j(\tilde{t}_s) \le \underline{v}_j(t_0) + \int_{t_0}^{\tilde{t}_s} g_j(s, \underline{v}(s)) \, ds.$$

Thus, by (8.29), (8.31), and the fact that $g(t, v)$ is non-decreasing in v,

$$\underline{v}_j(\tilde{t}_s) < v_j(t_0) + \int_{t_0}^{\tilde{t}_s} g_j(s, v(s)) \, ds = v_j(\tilde{t}_s)$$

because $v(t)$ solves (8.28) in J. This allows us to show that $\tilde{t}_s = t_1$. Indeed, if we assume by contradiction that $\tilde{t}_s < t_1$, again by continuity, we obtain the existence of $\delta_1 > 0$ such that, for every $j \in \{1, \ldots, M\}$,

$$\underline{v}_j(t) < v_j(t) \quad \text{for all } t \in [\tilde{t}_s, \tilde{t}_s + \delta_1),$$

which contradicts the definition of $\tilde{t}_s$ as the supremum of $\tilde{J}$. $\qquad \square$

 Ordinary Differential Equations

The following result provides an estimate for the norm of the solution of a system of ordinary differential equations in terms of a solution of a scalar differential equation whose nonlinearity somehow controls the nonlinearity of the system. Subsequently, we denote $\mathbb{R}_+ := [0, +\infty)$.

Lemma 8.12. *Let* $J = [t_0, t_0 + T]$ *for some* $T > 0$, $f \in C\left(J \times \mathbb{R}^N; \mathbb{R}^N\right)$ *and* $u(t)$ *be a solution of*

$$u' = f(t, u) \tag{8.32}$$

in the interval J. *Assume, in addition, that* $F \colon J \times \mathbb{R}_+ \to \mathbb{R}_+$ *is a continuous function, non-decreasing in* $u \in \mathbb{R}_+$, *such that*

$$\|f(t, u)\| \le F(t, \|u\|) \tag{8.33}$$

for all $(t, u) \in J \times \mathbb{R}^N$. *If* $x(t)$ *solves the scalar equation*

$$x' = F(t, x) \tag{8.34}$$

in J *and satisfies* $\|u(t_0)\| < x(t_0)$, *then*

$$\|u(t)\| < x(t) \quad \text{for all } t \in J.$$

Proof. Since $u(t)$ solves (8.32) in J, we have that

$$u(t) = u(t_0) + \int_{t_0}^{t} f(s, u(s))\, ds \quad \text{for all } t \in J.$$

Then, by taking norms, (8.33) implies that

$$\|u(t)\| \le \|u(t_0)\| + \int_{t_0}^{t} \|f(s, u(s))\|\, ds \le \|u(t_0)\| + \int_{t_0}^{t} F(s, \|u(s)\|)\, ds$$

for all $t \in J$. Thus, setting $\underline{x}(t) := \|u(t)\|$, $t \in J$, it becomes apparent that, for every $t \in J$,

$$\underline{x}(t) \le \underline{x}(t_0) + \int_{t_0}^{t} F(s, \underline{x}(s))\, ds.$$

Moreover, by assumption,

$$\underline{x}(t_0) = \|u(t_0)\| < x(t_0).$$

Therefore, thanks to Lemma 8.11, we find that, for every $t \in J$,

$$\|u(t)\| = \underline{x}(t) < x(t),$$

which concludes the proof. $\qquad\square$

The following result provides us with a global existence result for the scalar equation (8.34), provided $F(t, x)$ satisfies certain additional assumptions.

Lemma 8.13. *Suppose that*

$$F(t, x) = M(t)L(x),$$

where $M \in \mathcal{C}([t_0, +\infty); \mathbb{R}_+)$, and $L \in \mathcal{C}(\mathbb{R}_+; \mathbb{R}_+)$ satisfies $L(x) > 0$ if $x > 0$ and, for every $\varepsilon > 0$,

$$\int_\varepsilon^\infty \frac{1}{L(x)}\, dx = \infty. \tag{8.35}$$

Then, the solutions of the scalar differential equation

$$x' = M(t)L(x),$$

with $x(t_0) > 0$, are globally defined in $[t_0, +\infty)$.

Proof. Consider, for every $x_0 > 0$, the problem

$$\begin{cases} x' = M(t)L(x), \\ x(t_0) = x_0. \end{cases} \tag{8.36}$$

By adapting the proof of Theorem 4.2, it can be easily seen that (8.36) admits a unique solution, $x(t)$, as soon as $x(t) > 0$. Thus, since

$$x'(t) = M(t)L(x(t)) \geq 0,$$

we conclude that $x(t) \geq x_0 > 0$ for all times where the solution is defined and, in particular, that the solution is unique whenever it exists. In addition, by adapting the proof of Theorem 7.12, it becomes apparent that (8.36) has a unique maximal solution, (I, x), and that

$I = [t_0, T_{\max})$ for some $T_{\max} \in (t_0, +\infty]$. Moreover, since $x(t)$ is increasing,

$$\lim_{t \uparrow T_{\max}} x(t) = +\infty \quad \text{if } T_{\max} < +\infty.$$

Assume, by contradiction, that $T_{\max} < +\infty$. Then, (8.36) implies that, for every $t \in I = [t_0, T_{\max})$,

$$\frac{d}{dt} \int_{x_0}^{x(t)} \frac{1}{L(x)}\, dx = \frac{x'(t)}{L(x(t))} = M(t),$$

and hence, since I is bounded, there exists a constant $C > 0$ such that

$$\int_{x_0}^{x(t)} \frac{1}{L(x)}\, dx = \int_{t_0}^{t} M(s)\, ds \leq C \quad \text{for all } t \in I.$$

Therefore, letting $t \uparrow T_{\max}$ in this identity yields

$$\int_{x_0}^{\infty} \frac{1}{L(x)}\, dx \leq C < +\infty,$$

which contradicts (8.35). So, $T_{\max} = +\infty$. $\square$

We now have all the necessary ingredients to state and prove the main result of this section.

Theorem 8.14 (Wintner's extensibility criterion). *Assume that $J = [t_0, +\infty)$, $f \in \mathcal{C}(J \times \mathbb{R}^N; \mathbb{R}^N)$ is non-decreasing in $u \in \mathbb{R}^N$, and we have a function $M \in \mathcal{C}(J; \mathbb{R}_+)$ and a function $L \in \mathcal{C}(\mathbb{R}_+; \mathbb{R}_+)$ such that $L(x) > 0$ if $x > 0$, (8.35) holds for all $\varepsilon > 0$, and*

$$\|f(t, u)\| \leq M(t) L(\|u\|). \tag{8.37}$$

Then, for every $u_0 \in \mathbb{R}^N$, every maximal solution of the Cauchy problem

$$\begin{cases} u' = f(t, u) \\ u(t_0) = u_0 \end{cases} \tag{8.38}$$

is globally defined in $[t_0, +\infty)$.

Proof. Fix $x_0 > 0$ such that $\|u_0\| < x_0$, and consider the following scalar Cauchy problem:

$$\begin{cases} x' = M(t)L(x), \\ x(t_0) = x_0. \end{cases} \tag{8.39}$$

According to Lemma 8.13, (8.39) has a unique maximal solution, $x(t)$, which is globally defined in $[t_0, +\infty)$.

Let (I, u) be any maximal solution of (8.38). According to Theorem 7.12, $I = [t_0, T_{\max})$ for some $T_{\max} \in (t_0, +\infty]$, and

$$\limsup_{t \uparrow T_{\max}} \|u(t)\| = +\infty \quad \text{if } T_{\max} < +\infty. \tag{8.40}$$

Assume, by contradiction, that $T_{\max} < +\infty$. Then, since $\|u_0\| < x_0$, by Lemma 8.12,

$$\|u(t)\| \leq x(t) \quad \text{for all } t \in \tilde{J},$$

where $\tilde{J}$ is any compact interval contained in I. Since $x(t)$ is bounded in the compact interval $\bar{I}$, taking the limsup as $t \uparrow T_{\max}$ in the previous relation leads to a contradiction with (8.40). Therefore, $I = [t_0, +\infty)$, and the proof is complete. $\qquad\square$

8.7 Osgood Uniqueness Criterion

This section is devoted to the following uniqueness criterion, which is optimal in a sense to be discussed later.

Theorem 8.15 (Osgood uniqueness criterion). *Let $t_0 \in \mathbb{R}$, $T > 0$, $R > 0$, $u_0 \in \mathbb{R}^N$, $J := [t_0, t_0+T]$, and $f \in \mathcal{C}(J \times \bar{B}_R(u_0); \mathbb{R}^N)$. Assume that $F \colon [0, 2R] \to \mathbb{R}_+$ is a non-decreasing continuous function such that $F(0) = 0$, $F(x) > 0$ if $x > 0$,*

$$\lim_{\varepsilon \downarrow 0} \int_\varepsilon^{2R} \frac{1}{F(x)}\, dx = \infty, \tag{8.41}$$

and, for every $u, v \in \bar{B}_R(u_0)$ and $t \in J$,

$$\|f(t, u) - f(t, v)\| \leq F(\|u - v\|). \tag{8.42}$$

Then, the Cauchy problem

$$\begin{cases} u' = f(t, u), \\ u(t_0) = u_0, \end{cases} \tag{8.43}$$

has a unique solution in J.

Proof. Consider, for $\varepsilon \in (0, 2R)$, the problems

$$\begin{cases} x' = F(x), \\ x(t_0) = \varepsilon. \end{cases} \tag{8.44}$$

Since $x(t_0) = \varepsilon > 0$ and $F(x) > 0$ for all $x > 0$, it becomes apparent that $x'(t) = F(x(t)) > 0$, i.e. every solution of (8.44) is increasing, as long as it is defined. Therefore, by reasoning as in the proof of Lemma 8.13, it is possible to show that (8.44) has a unique maximal solution, which is denoted by $(I_\varepsilon, x_\varepsilon)$. Moreover, thanks to the fourth fundamental theorem (see Theorem 7.14), either $I_\varepsilon = J$, as illustrated in the left plot of Figure 8.7, or $I_\varepsilon = [t_0, T_{\max}(\varepsilon)]$ for some $T_{\max}(\varepsilon) < t_0 + T$ and, in such a case, $x_\varepsilon(T_{\max}(\varepsilon)) = 2R$, as illustrated in the right plot of Figure 8.7.

We claim that $T_{\max}(\varepsilon) = t_0 + T$ for sufficiently small $\varepsilon > 0$ and that

$$\lim_{\varepsilon \downarrow 0} x_\varepsilon(t_0 + T) = 0. \tag{8.45}$$

Assume, by contradiction, that there exists a sequence $\{\varepsilon_n\}_{n \geq 1} \subset (0, 2R)$ converging to 0 as $n \to \infty$ and such that $T_n := T_{\max}(\varepsilon_n) <$

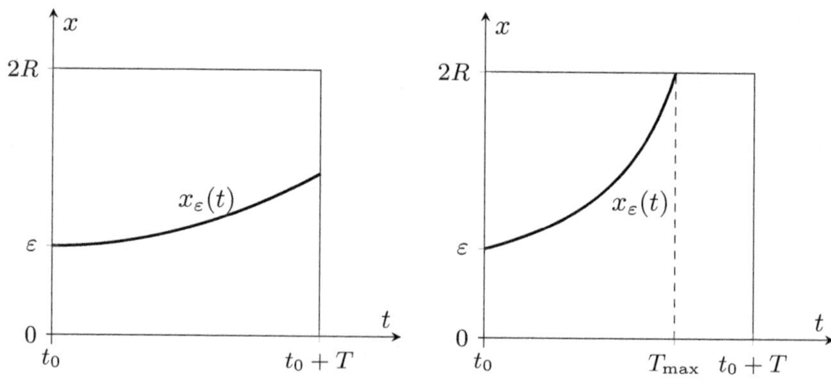

Fig. 8.7. Two possible behaviors for the maximal solution $(I_\varepsilon, x_\varepsilon)$ of (8.44).

$t_0 + T$. Then, $x_{\varepsilon_n}(T_n) = 2R$, and integrating and making the change of variable $s = x_{\varepsilon_n}(t)$ gives

$$T > T_n - t_0 = \int_{t_0}^{T_n} \frac{x'_{\varepsilon_n}(t)}{F(x_{\varepsilon_n}(t))}\, dt = \int_{\varepsilon_n}^{2R} \frac{1}{F(s)}\, ds.$$

Letting $n \to \infty$ in this relation gives a contradiction with (8.41). Therefore, $I_\varepsilon = J$ for sufficiently small $\varepsilon > 0$. Now, assume, by contradiction, that

$$\limsup_{\varepsilon \downarrow 0} x_\varepsilon(t_0 + T) > 0.$$

Then,

$$T = \int_\varepsilon^{x_\varepsilon(t_0+T)} \frac{1}{F(s)}\, ds = \int_\varepsilon^{2R} \frac{1}{F(s)}\, ds - \int_{x_\varepsilon(t_0+T)}^{2R} \frac{1}{F(s)}\, ds$$

and taking the limsup for $\varepsilon \downarrow 0$ gives a contradiction with (8.41). Thus, (8.45) holds. Since $x_\varepsilon(t)$ is increasing in J, we obtain that

$$\lim_{\varepsilon \downarrow 0} x_\varepsilon(t) = 0 \quad \text{uniformly in } t \in J. \tag{8.46}$$

Assume now that u_1 and u_2 are two solutions of (8.43) in J, and consider the scalar function

$$\underline{x}(t) := \|u_1(t) - u_2(t)\| \geq 0, \qquad t \in J.$$

Then, $u_1(t), u_2(t) \in \bar{B}_R(u_0)$ for all $t \in J$, and

$$u_1(t) = u_0 + \int_{t_0}^t f(s, u_1(s))\, ds, \qquad u_2(t) = u_0 + \int_{t_0}^t f(s, u_2(s))\, ds.$$

Thus, according to (8.42), we find that, for every $t \in J$,

$$\underline{x}(t) = \|u_1(t) - u_2(t)\| = \left\| \int_{t_0}^t [f(s, u_1(s)) - f(s, u_2(s))]\, ds \right\|$$

$$\leq \int_{t_0}^t \|f(s, u_1(s)) - f(s, u_2(s))\|\, ds$$

$$\leq \int_{t_0}^t F(\|u_1(s) - u_2(s)\|)\, ds = \int_{t_0}^t F(\underline{x}(s))\, ds.$$

Moreover, $\underline{x}(t_0) = 0 < \varepsilon$ for all $\varepsilon \in (0, 2R)$. Consequently, applying Lemma 8.11 to $x' = F(x)$ yields

$$\underline{x}(t) < x_\varepsilon(t) \quad \text{for all } t \in J.$$

Consequently, letting $\varepsilon \downarrow 0$, it follows from (8.46) that $\underline{x}(t) = 0$ for all $t \in J$. Therefore, $u_1 = u_2$ in J. This concludes the proof. $\quad\square$

Observe that, in this uniqueness theorem, condition (8.41) is optimal. Indeed, if we consider $F(x) = x^q$ with $q \in (0, 1)$, then, for all $R > 0$,

$$\int_0^R x^{-q}\, dx < +\infty,$$

and we know from Section 4.5 that $x' = x^q$ has a one-parameter family of non-negative solutions with $x(0) = 0$, all of which are globally defined in $[0, +\infty)$. Therefore, (8.41) is imperative even in the most simple examples.

8.8   Exercises

1. Analyze, as $x > 0$ varies, the behavior of the solution of (8.9), assuming, instead of (8.8), that

$$\int_0^\infty a(s)e^{2s}\, ds < +\infty.$$

2. Let $a \in \mathcal{C}\left([0, +\infty); (0, +\infty)\right)$ and $p \in \mathcal{C}^1([0, +\infty); \mathbb{R})$ such that $p'(u) > 0$ for all $u > 0$, and

$$\alpha u^n \le p(u) \le \beta u^n, \quad u \ge 0,$$

   for some integer $n \ge 3$ and positive constants $0 < \alpha < \beta$. Analyze, for every $x > 0$, the behavior of the solution of

$$\begin{cases} u' = u + a(t)p(u), \\ u(0) = x. \end{cases}$$

3. Consider the problem

$$\begin{cases} u' = u + bv^3, \\ v' = v + cu^3, \\ u(0) = v(0) = x, \end{cases}$$

   where x, b and c are positive constants.

(a) Show the existence of a unique maximal solution, (I, u, v).
(b) Show that $u(t)$ and $v(t)$ are positive and increasing in I.
(c) Show that $I = [0, T_{\max})$ for some $T_{\max} < +\infty$, and use Kamke's theorem to estimate $T_{\max}$ from above and from below.
(d) Show that

$$\lim_{t \uparrow T_{\max}} u(t) = +\infty = \lim_{t \uparrow T_{\max}} v(t).$$

4. Repeat the previous exercise considering the general initial conditions $u(0) = x > 0$ and $v(0) = y > 0$.

5. Consider the problem

$$\begin{cases} u' = u + (2 + \sin t)uv, \\ v' = v + (2 - \cos t)uv, \\ u(0) = v(0) = x, \end{cases}$$

where x is a positive constant.

(a) Show the existence of a unique maximal solution, (I, u, v).
(b) Show that $u(t)$ and $v(t)$ are positive and increasing in I.
(c) Show that $I = [0, T_{\max})$ for some $T_{\max} < +\infty$, and use Kamke's theorem to estimate $T_{\max}$ from above and from below.
(d) Show that

$$\lim_{t \uparrow T_{\max}} u(t) = +\infty = \lim_{t \uparrow T_{\max}} v(t).$$

6. Repeat the previous exercise considering the general initial conditions $u(0) = x > 0$ and $v(0) = y > 0$.

7. Let $g \in C(\mathbb{R}; \mathbb{R})$ be non-decreasing in $(-\eta, \eta)$ for some $\eta > 0$ and such that $g(0) > 0$. Consider, for some constant $q \in (0, 1)$, the Cauchy problem

$$\begin{cases} u' = u^q g(u), \\ u(0) = 0. \end{cases} \tag{8.47}$$

Note that $u = 0$ is a solution. Prove the existence of a positive solution for (8.47):

(a) first, when $g(u) = \alpha > 0$ is a positive constant in $(-\eta, \eta)$;

(b) then, in the general case, by using appropriate sub- and supersolutions.

8. Let $A = (a_{ij}(t))$, where, for every $i, j \in \{1, \ldots, N\}$, $a_{ij} \in \mathcal{C}(\mathbb{R}_+; [\alpha_{ij}, \beta_{ij}])$ for some constants $\beta_{ij} \geq \alpha_{ij} > 0$. Consider the Cauchy problem

$$\begin{cases} u_i' = \lambda_i u_i + \sum_{j=1}^{N} a_{ij}(t) u_j u_i, & i \in \{1, \ldots, N\}, \\ (u_1(0), \ldots, u_N(0)) = (x_1, \ldots, x_N), \end{cases}$$

where $\lambda_i \in \mathbb{R}$ and $x_i > 0$ for all $i \in \{1, \ldots, N\}$. Prove that the maximal solution of this problem is explosive, and estimate its blow up time from above and from below.

9. Let $t_0 \in \mathbb{R}$, $T > 0$, $J = (t_0, t_0 + T]$, $u_0 \in \mathbb{R}^N$, $R > 0$, and $f \in \mathcal{C}(\bar{J} \times \bar{B}_R(u_0); \mathbb{R}^N)$. Assume that $F(t, x)$ is a continuous scalar function defined on $J \times [0, 2R]$ such that $F(t, 0) = 0$, for which the constant function 0 is the only solution $x(t)$ of

$$x' = F(t, x)$$

that is defined in any interval $(t_0, t_0 + \varepsilon]$, with $\varepsilon \in (0, T)$, and satisfies

$$\lim_{t \downarrow t_0} x(t) = 0 \quad \text{and} \quad \lim_{t \downarrow t_0} \frac{x(t)}{t - t_0} = 0. \tag{8.48}$$

In addition, suppose that, for every $t \in J$ and $u, v \in \bar{B}_R(u_0)$,

$$\|f(t, u) - f(t, v)\| \leq F(t, \|u - v\|). \tag{8.49}$$

Prove that the initial value problem

$$\begin{cases} u' = f(t, u), \\ u(t_0) = u_0, \end{cases} \tag{8.50}$$

where the initial condition is given by

$$\lim_{t \downarrow t_0} u(t) = u_0,$$

has at most one solution on any interval $[t_0, t_0 + \varepsilon]$, with $\varepsilon \in (0, T)$.

Show that this result is false if (8.48) is replaced by

$$\lim_{t \downarrow t_0} x(t) = 0, \qquad \lim_{t \downarrow t_0} x'(t) = 0.$$

[Hint: See Hartman, 2002, p. 32, if necessary.]

10. With the notation of the previous exercise, take $t_0 = 0$ and $F(t,x) = x/t$. Show that (8.50) still has a unique solution if (8.49) is replaced by

$$\|f(t,u) - f(t,v)\| \leq F(t, \|u - v\|) = \frac{\|u - v\|}{t}$$

for all $t \in J$ and $u, v \in \bar{B}_R(u_0)$.

Prove that the previous statement is false if we replace $F(t,x) = x/t$ by $F(t,x) = Cx/t$ for any constant $C > 1$. [Hint: Observe that, for every $C > 1$, there exists a continuous function $f(t,u)$, $t \in [0,1]$, $|u| \leq 1$, such that

$$\|f(t,u) - f(t,v)\| \leq C\frac{\|u - v\|}{t}$$

for $t \in (0,1]$ but such that

$$\begin{cases} u' = f(t,u), \\ u(0) = 0, \end{cases}$$

has more than one solution.]

11. Consider $t_0 \in \mathbb{R}$, $T > 0$, $J = [t_0, t_0 + T]$, $u_0 \in \mathbb{R}^N$, $R > 0$, and let $f(t,u)$ be a continuous function defined on the cylinder $Q := J \times \bar{B}_R(u_0)$. Assume there exist $\rho \in (0, R)$ and a function $\eta(t; s, x)$ defined for

$$t, s \in J, \qquad \|x - u_0\| \leq \rho < R$$

such that:

(a) for (s, x) fixed, $t \mapsto \eta(t; s, x)$ is a solution of

$$\begin{cases} u' = f(t,u), \\ u(s) = x; \end{cases}$$

(b) $\eta(t; s, x)$ is globally Lipschitz with respect to x, uniformly in $t, s \in J$;

(c) if there exists $t_1 \in J$ such that $\eta(t_1; s, x) = \eta(t_1; \tilde{s}, \tilde{x})$, then, necessarily, $\eta(t; s, x) = \eta(t; \tilde{s}, \tilde{x})$ for all $t \in J$.

Prove that $t \mapsto \eta(t; t_0, u_0)$ is the unique solution of

$$\begin{cases} u' = f(t,u), \\ u(t_0) = u_0, \end{cases}$$

in J. [Hint: See Hartman, 2002, p. 35, if necessary.]

8.9   Final Comments

Theorem 8.4 refers to Kamke (1942). Kamke's work has been a cornerstone in the resurgence of German mathematics after World War II. Indeed, he played a fundamental role in the organization of the first scientific congress in Germany after the war, and in 1948, he re-established the German Mathematical Society. Actually, his personal life had been previously affected by the political history of Germany. In particular, he was forced to retire as a professor in 1937 because of his marriage to a Jewish woman and his personal opposition to National Socialism.

In this final section, we briefly sketch how Kamke's theorem, presented in Theorem 8.4, can be extended to nonlinear partial differential equations involving the Laplace operator in $\mathbb{R}^N$, which is given by

$$\Delta = \sum_{j=1}^{N} \frac{\partial^2}{\partial x_j^2}.$$

The most natural precursors of the concepts of subsolution and supersolution are the concepts of subharmonic and superharmonic functions in classical potential theory. Given an open bounded subset $\Omega \subset \mathbb{R}^N$, a function $u \in \mathcal{C}^2(\Omega; \mathbb{R})$ is said to be *superharmonic* in Ω if

$$-\Delta u \geq 0 \quad \text{in } \Omega,$$

while it is said to be *subharmonic* if $-\Delta u \leq 0$ in Ω. Note that u is superharmonic if and only if $-u$ is subharmonic.

Superharmonic functions satisfy a pivotal property in analysis, referred to as the *strong minimum principle*. According to it, whenever Ω is connected, for every superharmonic function $u \in \mathcal{C}^2(\Omega)$,

$$u(x) > m := \inf_{\Omega} u \quad \text{for all } x \in \Omega,$$

unless $u = m$ in Ω, i.e. u cannot attain m in Ω, unless $u = m$ in Ω (see, e.g. John, 1978, p. 82, and López-Gómez, 2013, Chapters 1 and 2, for some refinements). Based on this property, when the boundary of Ω, $\partial\Omega$, is of class $\mathcal{C}^2$, a theorem obtained by Perron in 1923 establishes the existence of a continuous function $G \in \mathcal{C}(\bar{\Omega} \times \bar{\Omega}; \mathbb{R})$, named the *Green function* of $-\Delta$ in Ω under Dirichlet boundary conditions after

G. Green, such that the (unique) solution of the boundary value problem

$$\begin{cases} -\Delta u = h & \text{in } \Omega, \\ u = 0 & \text{on } \partial\Omega, \end{cases}$$

is given by

$$u(x) = \int_{\Omega} G(x, y) h(y) \, dy$$

for all $x \in \bar{\Omega}$. The Green function satisfies $G(x, y) = G(y, x) > 0$, for all $x, y \in \Omega$, and $G(x, y) = 0$ if $x \in \partial\Omega$. Thus, $u(x) > 0$ for all $x \in \Omega$ if $h \gneq 0$, regardless of the size of the support of the function h. Moreover, according to Theorem 1.3 by López-Gómez (2013), which is a result derived by Hopf (1952) and Oleinik (1952), it turns out that the solution of the above boundary value problem satisfies $\frac{\partial u}{\partial n}(x) < 0$ for every $x \in \partial\Omega$, where n indicates the outward unit normal vector field to Ω along $\partial\Omega$. By introducing the linear integral operator

$$\mathcal{K}h := \int_{\Omega} G(\cdot, y) h(y) \, dy,$$

these properties show that $\mathcal{K}$ is a *strongly order-preserving* operator in $\mathcal{C}(\bar{\Omega}; \mathbb{R})$, in the sense that, if $h \gneq 0$, $u = \mathcal{K}h$ is strictly positive in Ω. In terms of $\mathcal{K}$, for any given $f \in \mathcal{C}(\bar{\Omega} \times \mathbb{R}; \mathbb{R})$, solving the boundary value problem

$$\begin{cases} -\Delta u = f(x, u) & \text{in } \Omega, \\ u = 0 & \text{on } \partial\Omega, \end{cases} \tag{8.51}$$

is equivalent to solve the integral equation

$$u = \mathcal{K}(f(\cdot, u(\cdot))) = \int_{\Omega} G(\cdot, y) f(y, u(y)) \, dy.$$

By analogy with Definition 8.2, a function $\underline{u} \in \mathcal{C}(\bar{\Omega}; \mathbb{R})$ is said to be a subsolution of (8.51) if

$$\underline{u} \leq \mathcal{K}(f(\cdot, \underline{u}(\cdot))) = \int_{\Omega} G(\cdot, y) f(y, \underline{u}(y)) \, dy,$$

while a function $\overline{u} \in \mathcal{C}(\bar{\Omega}; \mathbb{R})$ is said to be a supersolution of (8.51) if

$$\overline{u} \geq \mathcal{K}(f(\cdot, \overline{u}(\cdot))) = \int_{\Omega} G(\cdot, y) f(y, \overline{u}(y))\, dy.$$

When, in addition, $f(x, u)$ is non-decreasing in u and $\underline{u} \in \mathcal{C}^2(\Omega; \mathbb{R}) \cap \mathcal{C}(\bar{\Omega}; \mathbb{R})$, by adapting the proof of Lemma 8.3, it becomes apparent that $\underline{u}$ is a subsolution of (8.51) provided

$$\begin{cases} -\Delta \underline{u}(x) \leq f(x, \underline{u}(x)) & \text{for all } x \in \Omega, \\ \underline{u}(x) \leq 0 & \text{for all } x \in \partial\Omega. \end{cases}$$

Similarly, $\overline{u} \in \mathcal{C}^2(\Omega; \mathbb{R}) \cap \mathcal{C}(\bar{\Omega}; \mathbb{R})$ is a supersolution of (8.51) if

$$\begin{cases} -\Delta \overline{u}(x) \geq f(x, \overline{u}(x)) & \text{for all } x \in \Omega, \\ \overline{u}(x) \geq 0 & \text{for all } x \in \partial\Omega. \end{cases}$$

A theorem by Amann (1971/72) shows that if $\underline{u}$ is a subsolution and $\overline{u}$ is a supersolution of (8.51) such that $\underline{u} \leq \overline{u}$ in Ω, then, much like in Theorem 8.4, problem (8.51) has a minimal solution, $u_{\min}$, and a maximal solution, $u_{\max}$, such that $\underline{u} \leq u_{\min} \leq u_{\max} \leq \overline{u}$.

Moreover, any other solution u of (8.51) lying in the interval $[\underline{u}, \overline{u}]$ satisfies $u_{\min} \leq u \leq u_{\max}$.

Naturally, once (8.51) is rewritten as an equivalent fixed point equation involving an integral operator, the proof of this result follows the same general patterns as the one of Theorem 8.4. These results were later extended to cover some general classes of time-dependent problems, such as (6.57) by Sattinger (1973), and admit extensions to general systems of cooperative type, such as those covered by Theorem 8.4 (see López-Gómez and Molina-Meyer, 1994; Antón and López-Gómez, 2019, for the strongly order-preserving property of the associated integral operators and Amann, 2005, for the method of sub- and supersolutions). In all these settings, the method of sub- and supersolutions obeys the same general patterns of Theorem 8.4.

Theorem 8.14 was established by Wintner (1945, 1946), and Theorem 8.15 was established by Osgood (1898).

The result of Exercise 9 of Chapter 8 is often referred to as Kamke's general uniqueness theorem. It was proved by Kamke in 1930. According to Hartman (2002, p. 44), an earlier version of this theorem, where $F(t, x)$ was assumed to be continuous for $t = t_0$ as well, had been previously given by Perron in 1926. A result by Olech, published in 1956, shows that, somehow, Perron's result is not less general than Kamke's (see Exercise 6.5 in Hartman, 2002, p. 33).

The result of Exercise 10 of Chapter 8 is often referred to as Nagumo's criterion. It was established by Nagumo in 1926; a less sharp result had been previously proved by Rosenblatt in 1909, with $F(t, u) = Cu/t$ for some $C \in (0, 1)$.

The result of Exercise 11 of Chapter 8 is referred to as Van Kampen's uniqueness theorem. It was proved by Kampen in 1937.

Part 3

Dynamical Systems

Chapter 9

Some Paradigmatic Dynamical Systems

One of the main goals of the theory of Dynamical Systems is describing the asymptotic behavior of the solutions of the system $u' = f(t, u)$, where $f \colon \mathbb{R} \times \mathbb{R}^N \to \mathbb{R}^N$ is locally Lipschitz in $u \in \mathbb{R}^N$ uniformly on compact subintervals of $t \in \mathbb{R}$. Under this assumption, by Theorem 5.14, for every $t_0 \in \mathbb{R}$ and $x \in \mathbb{R}^N$, the initial value problem

$$\begin{cases} u' = f(t, u), \\ u(t_0) = x, \end{cases} \tag{9.1}$$

possesses a unique (bilateral) maximal solution, (I_x, u_x). By the asymptotic behavior of the solution, we mean the behavior of $u_x(t)$ as $t \uparrow \sup I_x$ and $t \downarrow \inf I_x$.

Indeed, from the point of view of applications, it is desirable to ascertain whether $I_x = \mathbb{R}$ and $u_x(t)$ approximates, as $t \uparrow +\infty$, an equilibrium, a periodic solution, or a quasi-periodic one. The main goal of Part 3 is to address these questions and find some answers.

According to Theorem 5.17,

$$I_x = (T_{\min}(x), T_{\max}(x)),$$

and $u_x(t)$ blows up at $T_{\max}(x)$ (respectively, $T_{\min}(x)$) if $T_{\max}(x) < +\infty$ (respectively, $T_{\min}(x) > -\infty$).

Thus, the real new problem in this part will involve determining, when $T_{\max}(x) = +\infty$, or $T_{\min}(x) = -\infty$, whether $\lim_{t\uparrow+\infty} u_x(t)$,

479

or $\lim_{t \downarrow -\infty} u_x(t)$, exist and, in such a case, establishing their value. Naturally, these limits might not exist since $u_x(t)$ might oscillate, which makes the analysis more difficult. As we have seen in Chapter 4, even in the simplest prototypes, the qualitative behavior of $u_x(t)$ might dramatically change according to the values of x and t_0. For this reason, we set

$$u_x(t) = u(t; t_0, x) \quad \text{for all } t \in I_x.$$

In this chapter and in Chapter 10, we focus on autonomous systems of differential equations

$$\begin{cases} u' = f(u), \\ u(t_0) = x. \end{cases} \tag{9.2}$$

This substantially simplifies the structure of the solution set of $u' = f(u)$ because, as already remarked in Section 4.1, whenever $t \mapsto u(t)$ solves $u' = f(u)$, then, for every $s \in \mathbb{R}$, $t \mapsto u(t + s)$ also solves $u' = f(u)$ in the appropriate domain. Thus, in autonomous systems, the time variable does not play a special role in the asymptotic behavior of solutions.

We now introduce some notations and concepts that are relevant in the theory of Dynamical Systems. The *orbit*, or *trajectory*, of the unique solution of (9.2) is defined as

$$\Gamma_x := \{u(t; t_0, x) \colon t \in I_x\},$$

while the positive (or forward) and negative (or backward) *semi-orbits* are, respectively,

$$\Gamma_x^+ := \{u(t; t_0, x) \colon t \in [t_0, T_{\max}(x))\},$$

$$\Gamma_x^- := \{u(t; t_0, x) \colon t \in (T_{\min}(x), t_0]\}.$$

Analyzing the dynamics of $u' = f(u)$ consists of constructing in the phase space $\mathbb{R}^N$ the corresponding *phase portrait*, or *phase diagram*, which is the set of all trajectories Γ_x, with x varying in $\mathbb{R}^N$. In this context, an *integral curve* $\mathcal{C}$ of $u' = f(u)$ is any union of (complete) trajectories of solutions of (9.2). Thus, by definition, any integral curve $\mathcal{C}$ satisfies $u(t; t_0, x) \in \mathcal{C}$ for all $t \in I_x$ if $x \in \mathcal{C}$, i.e. $\mathcal{C}$ is an *invariant subset* of $\mathbb{R}^N$. More generally, a set $S \subset \mathbb{R}^N$ is said to be invariant for the flow induced by (9.2) if, whenever $x \in S$, one has $u(t; t_0, x) \in S$ for all $t \in I_x$.

The main goal of this chapter is to analyze the dynamics of some special, very important, classes of planar autonomous systems. In all these cases, we construct the integral curves through any point $x \in \mathbb{R}^2$. Then, from the integral curves, we properly identify the trajectories. This is the best methodology to perform, in a systematic way, the analysis of the dynamics of (9.2).

9.1 Linear Autonomous Planar Systems

In this section, we analyze the dynamics of the following family of linear planar systems:

$$\begin{cases} u' = au + bv, \\ v' = cu + dv, \end{cases} \tag{9.3}$$

where $a, b, c, d \in \mathbb{R}$ are constants such that $ad \neq bc$, i.e. the coefficient matrix

$$A := \begin{pmatrix} a & b \\ c & d \end{pmatrix}$$

is invertible. This implies that $(0,0)$ is the unique equilibrium of (9.3). Since

$$f(u, v) := \begin{pmatrix} au + bv \\ cu + dv \end{pmatrix}$$

is globally Lipschitz in $\mathbb{R}^2$ (see Section 5.1), for every $(u_0, v_0) \in \mathbb{R}^2$, the solution of the associated Cauchy problem

$$\begin{cases} u' = au + bv, \\ v' = cu + dv, \\ u(0) = u_0, \quad v(0) = v_0, \end{cases} \tag{9.4}$$

is defined for all $t \in \mathbb{R}$. Subsequently, we distinguish several cases, according to the nature of the spectrum of the matrix A, $\sigma(A)$. Each case is treated in a separate section.

9.1.1   *Case of real distinct eigenvalues*

If $\sigma(A) = \{\lambda, \mu\} \subset \mathbb{R}$ with $\lambda < \mu$, there exist two linearly independent vectors $(\alpha, \beta), (\gamma, \delta) \in \mathbb{R}^2$ such that

$$N[A - \lambda I] = \mathrm{span}\left[\begin{pmatrix}\alpha \\ \beta\end{pmatrix}\right], \quad N[A - \mu I] = \mathrm{span}\left[\begin{pmatrix}\gamma \\ \delta\end{pmatrix}\right], \qquad (9.5)$$

and the change of variable

$$\begin{pmatrix}u \\ v\end{pmatrix} := P\begin{pmatrix}x \\ y\end{pmatrix}, \quad \text{with } P := \begin{pmatrix}\alpha & \gamma \\ \beta & \delta\end{pmatrix}, \qquad (9.6)$$

transforms (9.3) into

$$P\begin{pmatrix}x \\ y\end{pmatrix}' = AP\begin{pmatrix}x \\ y\end{pmatrix},$$

which can be equivalently expressed as

$$\begin{pmatrix}x \\ y\end{pmatrix}' = P^{-1}AP\begin{pmatrix}x \\ y\end{pmatrix}. \qquad (9.7)$$

On the other hand, by (9.5),

$$P^{-1}AP = P^{-1}\left(A\begin{pmatrix}\alpha \\ \beta\end{pmatrix}\,\middle|\,A\begin{pmatrix}\gamma \\ \delta\end{pmatrix}\right) = P^{-1}\left(\lambda\begin{pmatrix}\alpha \\ \beta\end{pmatrix}\,\middle|\,\mu\begin{pmatrix}\gamma \\ \delta\end{pmatrix}\right)$$

$$= P^{-1}P\begin{pmatrix}\lambda & 0 \\ 0 & \mu\end{pmatrix} = \begin{pmatrix}\lambda & 0 \\ 0 & \mu\end{pmatrix},$$

which is the Jordan canonical form of the matrix A. Thus, the change of variable (9.6) transforms the Cauchy problem (9.4) into

$$\begin{cases} x' = \lambda x, \\ y' = \mu y, \\ x(0) = x_0, \ y(0) = y_0, \end{cases} \qquad (9.8)$$

where

$$\begin{pmatrix}x_0 \\ y_0\end{pmatrix} = P^{-1}\begin{pmatrix}u_0 \\ v_0\end{pmatrix}. \qquad (9.9)$$

As the two differential equations of (9.8) are uncoupled, the trajectories of (9.8) can be easily determined. Indeed, for every $(x_0, y_0) \in \mathbb{R}^2$,

$$\Gamma_{(x_0, y_0)} = \left\{ \left(x_0 e^{\lambda t}, y_0 e^{\mu t} \right) : t \in \mathbb{R} \right\}. \tag{9.10}$$

In particular, $\Gamma_{(0,0)} = \{(0,0)\}$, and for every $x_0 \in \mathbb{R} \setminus \{0\}$,

$$\Gamma_{(x_0, 0)} = \left\{ \left(x_0 e^{\lambda t}, 0 \right) : t \in \mathbb{R} \right\};$$

thus, it is the positive x-half axis if $x_0 > 0$ and the negative one if $x_0 < 0$. Similarly,

$$\Gamma_{(0, y_0)} = \left\{ \left(0, y_0 e^{\mu t} \right) : t \in \mathbb{R} \right\}$$

is the positive y-half axis if $y_0 > 0$ and the negative one if $y_0 < 0$. Moreover, from (9.10), it is apparent that, whenever $x_0 \neq 0$ and $y_0 \neq 0$, $\Gamma_{(x_0, y_0)}$ stays in the same quadrant as (x_0, y_0).

In order to determine the global structure of the set of trajectories, as well as the direction of the flow along them as time increases, we must further differentiate three cases, according to the signs of λ and μ:

Case A. $\lambda < 0 < \mu$. Then,

$$\lim_{t \uparrow +\infty} \left(x_0 e^{\lambda t} \right) = 0, \quad \lim_{t \downarrow -\infty} \left(x_0 e^{\lambda t} \right) = \begin{cases} +\infty & \text{if } x_0 > 0, \\ -\infty & \text{if } x_0 < 0, \end{cases}$$

whereas

$$\lim_{t \downarrow -\infty} \left(y_0 e^{\mu t} \right) = 0, \quad \lim_{t \uparrow +\infty} \left(y_0 e^{\mu t} \right) = \begin{cases} +\infty & \text{if } y_0 > 0, \\ -\infty & \text{if } y_0 < 0. \end{cases}$$

Therefore, the coordinate axes are covered by five different trajectories of system (9.8). Namely:

  (i) the trajectory of the equilibrium $(0,0)$,
 (ii) the trajectory of any solution starting at $(x_0, 0)$ with $x_0 > 0$,
(iii) the trajectory of any solution starting at $(x_0, 0)$ with $x_0 < 0$,
 (iv) the trajectory of any solution starting at $(0, y_0)$ with $y_0 > 0$,
  (v) the trajectory of any solution starting at $(0, y_0)$ with $y_0 < 0$.

These five trajectories form the integral curve $xy = 0$.

When $x_0 \neq 0$ and $y_0 \neq 0$, instead, the solution is

$$x(t) = x_0 e^{\lambda t}, \quad y(t) = y_0 e^{\mu t}, \quad t \in \mathbb{R}.$$

Thus, for every $t \in \mathbb{R}$,

$$t = \frac{1}{\lambda} \ln \frac{x(t)}{x_0} = \ln \left(\frac{x(t)}{x_0} \right)^{\frac{1}{\lambda}}, \quad t = \frac{1}{\mu} \ln \frac{y(t)}{y_0} = \ln \left(\frac{y(t)}{y_0} \right)^{\frac{1}{\mu}},$$

and hence,

$$\left(\frac{x(t)}{x_0} \right)^{\mu} = \left(\frac{y(t)}{y_0} \right)^{\lambda} \quad \text{for all } t \in \mathbb{R}.$$

Consequently, the solution $(x(t), y(t))$ lies on the integral curve through the point (x_0, y_0), denoted by $\mathcal{C}_{(x_0, y_0)}$, and defined as

$$\left\{ (x, y) \in \mathbb{R}^2 : \left(\tfrac{x}{x_0} \right)^{\mu} = \left(\tfrac{y}{y_0} \right)^{\lambda}, \text{ with } xx_0 > 0 \text{ and } yy_0 > 0 \right\}.$$

Since $\mu > 0$ and $-\lambda > 0$, $\mathcal{C}_{(x_0, y_0)}$ is a branch of a (generalized) hyperbola passing through the point (x_0, y_0). Actually, $\mathcal{C}_{(x_0, y_0)} = \Gamma_{(x_0, y_0)}$ in this case.

The left-hand plot of Figure 9.1 shows the set of trajectories $\Gamma_{(x_0, y_0)}$ through any point $(x_0, y_0) \in \mathbb{R}^2$. The right-hand one shows the corresponding phase diagram in the (u, v) variables. Note that the direction $(1, 0)$ is transformed into the direction of the eigenvector (α, β), while the direction $(0, 1)$ is transformed into the direction of the eigenvector (γ, δ). Actually, the second plot in Figure 9.1 represents the set of all curves

$$\begin{pmatrix} u(t) \\ v(t) \end{pmatrix} = x_0 e^{\lambda t} \begin{pmatrix} \alpha \\ \beta \end{pmatrix} + y_0 e^{\mu t} \begin{pmatrix} \gamma \\ \delta \end{pmatrix}, \quad t \in \mathbb{R},$$

for $(x_0, y_0) \in \mathbb{R}^2$.

In this case, the equilibrium $(0, 0)$ is said to be a *saddle point*. It should be noted that the set of integral curves in this case is of saddle type, as in Section 4.8.2 (see Figure 4.12).

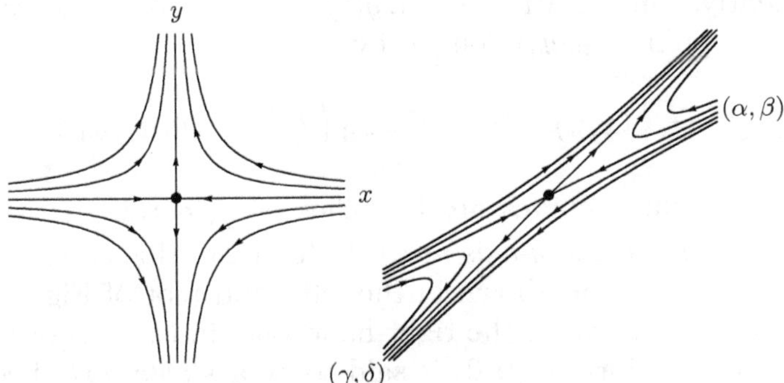

Fig. 9.1. Dynamics of (9.7) (left) and (9.3) (right), under the assumptions of Section 9.1.1, for $\lambda < 0 < \mu$.

The invariant axis $N[A - \lambda I]$, where the trajectories converge to the equilibrium $(0,0)$ as $t \uparrow +\infty$, is known as the *stable manifold* of $(0,0)$, and it is denoted by $W^s(0,0)$, while $N[A - \mu I]$, where the trajectories converge to the equilibrium $(0,0)$ as $t \downarrow -\infty$, is known as the *unstable manifold* of $(0,0)$ and is denoted by $W^u(0,0)$.

Case B. $\lambda < \mu < 0$. Then,

$$\lim_{t\uparrow+\infty} \left(x_0 e^{\lambda t}\right) = 0, \quad \lim_{t\downarrow-\infty} \left(x_0 e^{\lambda t}\right) = \begin{cases} +\infty & \text{if } x_0 > 0, \\ -\infty & \text{if } x_0 < 0, \end{cases}$$

and similarly,

$$\lim_{t\uparrow+\infty} \left(y_0 e^{\mu t}\right) = 0, \quad \lim_{t\downarrow-\infty} \left(y_0 e^{\mu t}\right) = \begin{cases} +\infty & \text{if } y_0 > 0, \\ -\infty & \text{if } y_0 < 0. \end{cases}$$

Moreover, since the components of the solutions are

$$x(t) = x_0 e^{\lambda t} \quad \text{and} \quad y(t) = y_0 e^{\mu t} \quad \text{for all } t \in \mathbb{R},$$

arguing as above, it becomes apparent that, whenever $x_0 \neq 0$ and $y_0 \neq 0$,

$$\frac{x(t)}{x_0} = \left(\frac{y(t)}{y_0}\right)^{\frac{\lambda}{\mu}} \quad \text{for all } t \in \mathbb{R}.$$

Consequently, the solution $(x(t), y(t))$ lies on the integral curve through the point (x_0, y_0) defined by

$$\mathcal{C}_{(x_0,y_0)} := \left\{ (x, y) \in \mathbb{R}^2 : \; x = x_0 \left(\frac{y}{y_0} \right)^{\frac{\lambda}{\mu}}, \; \text{with } yy_0 > 0 \right\}.$$

Since $\frac{\lambda}{\mu} > 1$, these curves are branches of (generalized) parabolas that are tangent to the y-axis at $(0,0)$. Thus, in this case, the phase portrait looks like that shown in the left-hand plot of Figure 9.2, in (x, y)-coordinates, and in the right-hand one, in (u, v)-coordinates.

Now, the equilibrium $(0,0)$ is said to be a *stable node* because it is a *global attractor* for all the solutions of the system, in the sense that

$$\lim_{t\uparrow+\infty} (u(t), v(t)) = (0, 0) \quad \text{for all } (u_0, v_0) \in \mathbb{R}^2.$$

As in Section 4.8.2, the set of integral curves in this case is of node type (see Figure 4.13).

Case C. $0 < \lambda < \mu$. Then,

$$\lim_{t\downarrow-\infty} \left(x_0 e^{\lambda t} \right) = 0, \quad \lim_{t\uparrow+\infty} \left(x_0 e^{\lambda t} \right) = \begin{cases} +\infty & \text{if } x_0 > 0, \\ -\infty & \text{if } x_0 < 0, \end{cases}$$

and similarly,

$$\lim_{t\downarrow-\infty} \left(y_0 e^{\mu t} \right) = 0, \quad \lim_{t\uparrow+\infty} \left(y_0 e^{\mu t} \right) = \begin{cases} +\infty & \text{if } y_0 > 0, \\ -\infty & \text{if } y_0 < 0, \end{cases}$$

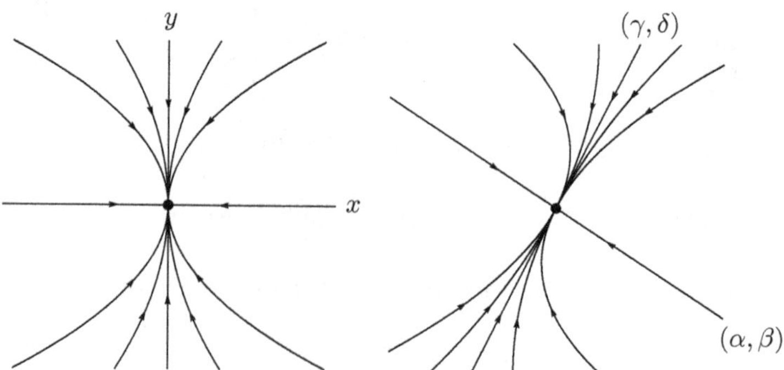

Fig. 9.2. Dynamics of (9.7) (left) and (9.3) (right), under the assumptions of Section 9.1.1, for $\lambda < \mu < 0$.

though the decay of $y(t)$ to zero as $t \downarrow -\infty$ is faster. Thus, all trajectories emanate from $(0,0)$ at $t = -\infty$ and separate away from $(0,0)$ as time increases, in strong contrast to the situation sketched in Figure 9.2.

Now, the solution $(x(t), y(t))$ through any point (x_0, y_0) with $x_0 \neq 0$ and $y_0 \neq 0$ satisfies

$$\frac{y(t)}{y_0} = \left(\frac{x(t)}{x_0}\right)^{\frac{\mu}{\lambda}} \qquad \text{for all } t \in \mathbb{R}.$$

Thus, it lies on the integral curve

$$\mathcal{C}_{(x_0, y_0)} := \left\{ (x, y) \in \mathbb{R}^2 : \ y = y_0 \left(\frac{x}{x_0}\right)^{\frac{\mu}{\lambda}}, \ \text{with } x x_0 > 0 \right\}.$$

Since $\frac{\mu}{\lambda} > 1$, these curves are branches of (generalized) parabolas that are tangent to the x-axis at $(0,0)$. Consequently, the phase portrait looks like that shown in the left-hand plot of Figure 9.3 in (x, y)-coordinates, and in the right-hand one in (u, v)-coordinates.

In this case, it is said that $(0,0)$ is an *unstable node* since it repels all solutions. Actually, all solutions are unbounded as t increases to $+\infty$. Note that, as in Section 4.8.2, the set of integral curves is of node type.

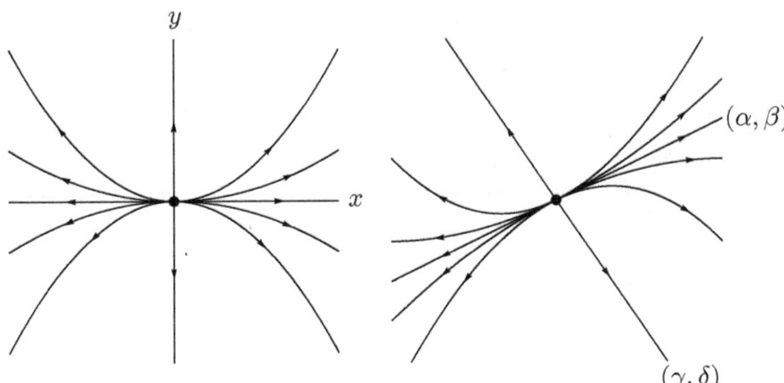

Fig. 9.3. Dynamics of (9.7) (left) and (9.3) (right), under the assumptions of Section 9.1.1, for $0 < \lambda < \mu$.

9.1.2 *Case of a real double eigenvalue*

If $\sigma(A) = \{\lambda\}$ with $\lambda \in \mathbb{R}$, $\lambda \neq 0$, two different cases have to be considered, depending on the geometric multiplicity of λ, or, equivalently in this case, on its algebraic ascent, $\nu(\lambda)$ (see Definition 2.4). Indeed, if

$$N[A - \lambda I] = \mathbb{R}^2,$$

i.e. $\nu(\lambda) = 1$, then $A = \lambda I$, and (9.4) becomes

$$\begin{cases} u' = \lambda u, \\ v' = \lambda v, \\ u(0) = u_0, \quad v(0) = v_0. \end{cases}$$

Thus, for every $t \in \mathbb{R}$,

$$(u(t), v(t)) = e^{\lambda t}(u_0, v_0).$$

Since $\lambda \neq 0$, $e^{\lambda t}$ takes all the values in $(0, +\infty)$ as $t \in \mathbb{R}$, and $\Gamma_{(u_0, v_0)}$ consists of the half-line passing through (u_0, v_0) and $(0, 0)$, with the latter point excluded. Therefore, the phase portrait in this case looks like that shown in Figure 9.4. The left-hand plot shows it for $\lambda < 0$, while the right-hand plot shows it for $\lambda > 0$.

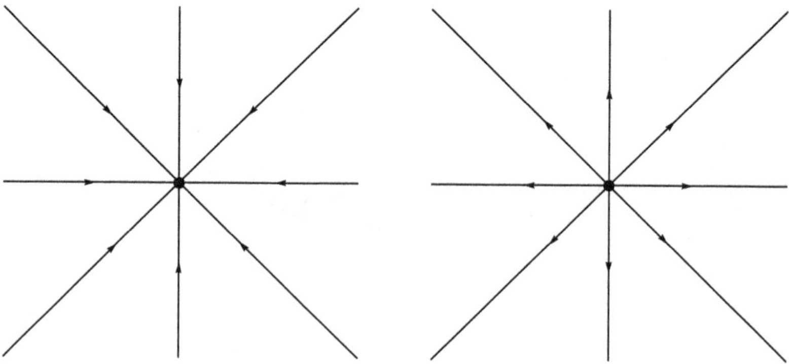

Fig. 9.4. Dynamics of (9.7) when λ is a double eigenvalue with $\nu(\lambda) = 1$: $\lambda < 0$ (left) and $\lambda > 0$ (right).

If, instead,

$$\dim N[A - \lambda I] = 1,$$

then, $N[(A - \lambda I)^2] = \mathbb{R}^2$, i.e. $\nu(\lambda) = 2$, and there exists $(\alpha, \beta) \in \mathbb{R}^2$ such that

$$\begin{pmatrix} \alpha \\ \beta \end{pmatrix} \in N[(A - \lambda I)^2] \setminus N[A - \lambda I].$$

Thus, setting

$$\begin{pmatrix} \gamma \\ \delta \end{pmatrix} := (A - \lambda I) \begin{pmatrix} \alpha \\ \beta \end{pmatrix} \in N[A - \lambda I],$$

the change of variable (9.6) leads to (9.7), where now

$$P^{-1}AP = P^{-1} \left(A \begin{pmatrix} \alpha \\ \beta \end{pmatrix} \middle| A \begin{pmatrix} \gamma \\ \delta \end{pmatrix} \right)$$

$$= P^{-1} \left(\lambda \begin{pmatrix} \alpha \\ \beta \end{pmatrix} + \begin{pmatrix} \gamma \\ \delta \end{pmatrix} \middle| \lambda \begin{pmatrix} \gamma \\ \delta \end{pmatrix} \right)$$

$$= P^{-1}P \begin{pmatrix} \lambda & 0 \\ 1 & \lambda \end{pmatrix} = \begin{pmatrix} \lambda & 0 \\ 1 & \lambda \end{pmatrix},$$

which provides us with the Jordan canonical form of A in this case. Thus, after this change of variables, (9.4) becomes

$$\begin{cases} x' = \lambda x, \\ y' = x + \lambda y, \\ x(0) = x_0, \ y(0) = y_0, \end{cases} \tag{9.11}$$

where (x_0, y_0) is given by (9.9). From the first equation in (9.11), we obtain

$$x(t) = x_0 e^{\lambda t} \quad \text{for all } t \in \mathbb{R}. \tag{9.12}$$

Thus, if $x_0 = 0$, $x(t) = 0$ for all $t \in \mathbb{R}$. In such a case, since the second equation reduces to $y' = \lambda y$, we have that

$$\Gamma_{(0,y_0)} = \left\{ \left(0, y_0 e^{\lambda t} \right) : t \in \mathbb{R} \right\}.$$

Therefore, the y-axis consists of three trajectories:

(i) the trajectory of the equilibrium $(0,0)$;
(ii) the positive y-axis, which is the trajectory of any solution starting on it;
(iii) the negative y-axis, which is the trajectory of any solution starting on it.

Suppose now that $x_0 \neq 0$. Then, $y(t)$ satisfies

$$\begin{cases} y' = \lambda y + x_0 e^{\lambda t}, \\ y(0) = y_0, \end{cases}$$

and by varying the coefficients, it is natural to look for $y(t)$ of the form $y(t) = e^{\lambda t}\kappa(t)$ for some function $\kappa(t)$. Necessarily, $\kappa(0) = y_0$ and

$$\lambda e^{\lambda t}\kappa(t) + e^{\lambda t}\kappa'(t) = \lambda e^{\lambda t}\kappa(t) + x_0 e^{\lambda t}.$$

Thus, $\kappa'(t) = x_0$, and $\kappa(t) = x_0 t + y_0$ for all $t \in \mathbb{R}$. Therefore,

$$y(t) = e^{\lambda t}(x_0 t + y_0) \quad \text{for all } t \in \mathbb{R}. \tag{9.13}$$

On the other hand, it follows from (9.12) that

$$e^{\lambda t} = \frac{x(t)}{x_0}, \quad t = \frac{1}{\lambda}\ln\frac{x(t)}{x_0},$$

for all $t \in \mathbb{R}$. Thus, by substituting in (9.13), we find that, for every $t \in \mathbb{R}$,

$$y(t) = \frac{x(t)}{x_0}\left(\frac{x_0}{\lambda}\ln\frac{x(t)}{x_0} + y_0\right) = \frac{x(t)}{\lambda}\ln\frac{x(t)}{x_0} + \frac{y_0}{x_0}x(t).$$

Consequently,

$$\mathcal{C}_{(x_0,y_0)} := \left\{(x,y) \in \mathbb{R}^2 : y = \frac{x}{\lambda}\ln\frac{x}{x_0} + \frac{y_0}{x_0}x, \text{ with } xx_0 > 0\right\}$$

provides us with an integral curve through (x_0, y_0) containing the trajectory $\Gamma_{(x_0,y_0)}$. To represent these integral curves, it is useful to introduce the function $\varphi(x)$ defined as

$$\varphi(x) := \frac{x}{\lambda} \ln \frac{x}{x_0} + \frac{y_0}{x_0}x, \qquad \operatorname{sign} x = \operatorname{sign} x_0,$$

and to distinguish two cases, according to the sign of λ:

Case A. $\lambda < 0$. In this case, according to (9.12) and (9.13),

$$\lim_{t\uparrow+\infty} (x(t), y(t)) = (0,0).$$

In particular, the three trajectories along the y-axis look like those shown in the first picture in Figure 9.5. Suppose $x_0 > 0$. Then, $\varphi(x)$ is defined for all $x > 0$. Moreover,

$$\lim_{x\downarrow 0} \varphi(x) = 0, \qquad \lim_{x\uparrow+\infty} \varphi(x) = -\infty$$

because $\lambda < 0$, and for every $x > 0$,

$$\varphi'(x) = \frac{1}{\lambda} \ln \frac{x}{x_0} + \frac{1}{\lambda} + \frac{y_0}{x_0}.$$

Thus,

$$\lim_{x\downarrow 0} \varphi'(x) = +\infty,$$

φ' vanishes at

$$x_M := x_0 e^{-1 - \frac{y_0}{x_0}\lambda} > 0,$$

$\varphi'(x) > 0$ if $x \in (0, x_M)$, and $\varphi'(x) < 0$ if $x > x_M$.
When $x_0 < 0$, $\varphi(x)$ is defined for $x < 0$, and

$$\lim_{x\uparrow 0} \varphi(x) = 0, \qquad \lim_{x\downarrow-\infty} \varphi(x) = +\infty$$

because $\lambda < 0$. Moreover,

$$\lim_{x\uparrow 0} \varphi'(x) = +\infty,$$

φ' vanishes at

$$x_m := x_0 e^{-1 - \frac{y_0}{x_0}\lambda} < 0,$$

$\varphi'(x) < 0$ if $x \in (-\infty, x_m)$, and $\varphi'(x) > 0$ if $x \in (x_m, 0)$. Therefore, the integral curve $y = \varphi(x)$ looks like that shown in the left-hand

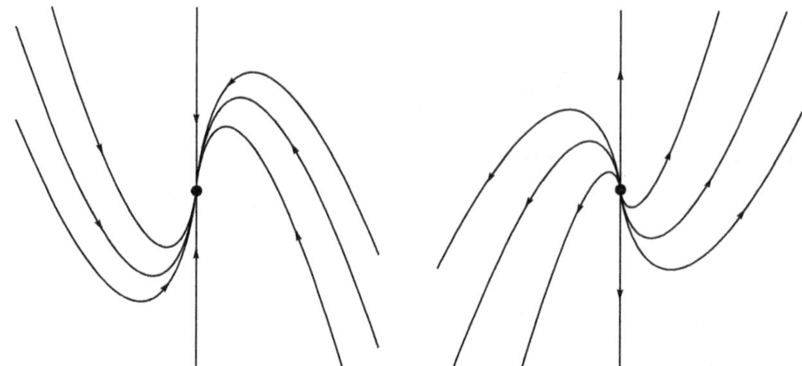

Fig. 9.5. Dynamics of (9.7) when λ is a double eigenvalue with $\nu(\lambda) = 2$: $\lambda < 0$ (left) and $\lambda > 0$ (right). The equilibrium is a stable (respectively, unstable) degenerate node.

plot of Figure 9.5. In this case, $(0,0)$ is said to be a *stable degenerate node* (compare with the case treated in Section 4.8.2 and represented in Figure 4.14).

Case B. $\lambda > 0$. In this case, according to (9.12) and (9.13), we have

$$\lim_{t\downarrow -\infty} (x(t), y(t)) = (0,0).$$

Moreover, by adapting the previous analysis of $\varphi(x)$, it is possible to show that the trajectories of the solutions through (x_0, y_0) with $x_0 \neq 0$ look like those shown in the right-hand plot of Figure 9.5. In this case, $(0,0)$ is said to be an *unstable degenerate node*.

Naturally, in the original variables (u, v), these phase portraits look like those shown in Figure 9.6.

9.1.3 *Case of complex eigenvalues*

If $\sigma(A) = \{\lambda \pm i\omega\}$ with $\omega > 0$, there exist two linearly independent vectors $(\alpha, \beta), (\gamma, \delta) \in \mathbb{R}^2$ such that

$$A\left(\begin{pmatrix}\alpha\\\beta\end{pmatrix} + i\begin{pmatrix}\gamma\\\delta\end{pmatrix}\right) = (\lambda + i\omega)\left(\begin{pmatrix}\alpha\\\beta\end{pmatrix} + i\begin{pmatrix}\gamma\\\delta\end{pmatrix}\right)$$

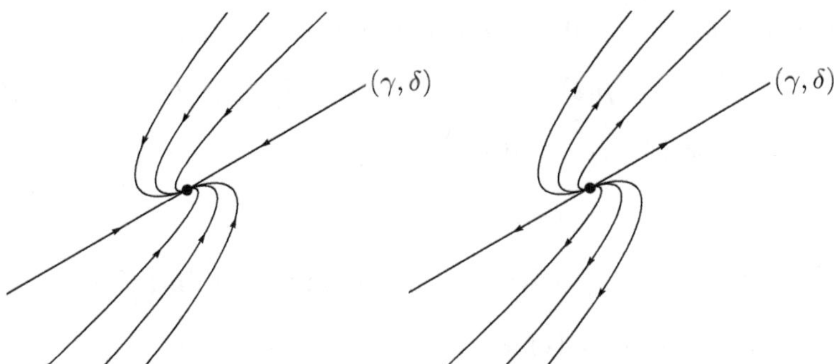

Fig. 9.6. Dynamics of (9.3) when λ is a double eigenvalue with $\nu(\lambda) = 2$: $\lambda < 0$ (left) and $\lambda > 0$ (right). The equilibrium is a stable (respectively, unstable) degenerate node.

and hence, by identifying the real and imaginary parts, it becomes apparent that

$$A \begin{pmatrix} \alpha \\ \beta \end{pmatrix} = \lambda \begin{pmatrix} \alpha \\ \beta \end{pmatrix} - \omega \begin{pmatrix} \gamma \\ \delta \end{pmatrix},$$

$$A \begin{pmatrix} \gamma \\ \delta \end{pmatrix} = \omega \begin{pmatrix} \alpha \\ \beta \end{pmatrix} + \lambda \begin{pmatrix} \gamma \\ \delta \end{pmatrix}. \tag{9.14}$$

Thus, performing the change of variable (9.6), we are led to (9.7), and (9.14) now implies that

$$
\begin{aligned}
P^{-1}AP &= P^{-1}\left(A\begin{pmatrix} \alpha \\ \beta \end{pmatrix} \middle| A\begin{pmatrix} \gamma \\ \delta \end{pmatrix} \right) \\
&= P^{-1}\left(\lambda \begin{pmatrix} \alpha \\ \beta \end{pmatrix} - \omega \begin{pmatrix} \gamma \\ \delta \end{pmatrix} \middle| \omega \begin{pmatrix} \alpha \\ \beta \end{pmatrix} + \lambda \begin{pmatrix} \gamma \\ \delta \end{pmatrix} \right) \\
&= P^{-1}P \begin{pmatrix} \lambda & \omega \\ -\omega & \lambda \end{pmatrix} = \begin{pmatrix} \lambda & \omega \\ -\omega & \lambda \end{pmatrix},
\end{aligned}
$$

which is the real Jordan canonical form of A. To construct the trajectories of the solutions of

$$\begin{cases} x' = \lambda x + \omega y, \\ y' = -\omega x + \lambda y, \\ x(0) = x_0, \quad y(0) = y_0, \end{cases} \tag{9.15}$$

for every $(x_0, y_0) \in \mathbb{R}^2$, it is appropriate to work in polar coordinates,

$$x(t) = \rho(t) \cos \theta(t), \quad y(t) = \rho(t) \sin \theta(t). \tag{9.16}$$

By adding these identities after taking squares, we get

$$\rho^2(t) = x^2(t) + y^2(t).$$

Thus, by differentiating with respect to time, we find that

$$\begin{aligned}
\rho(t)\rho'(t) &= x(t)x'(t) + y(t)y'(t) \\
&= x(t)\left[\lambda x(t) + \omega y(t)\right] + y(t)\left[-\omega x(t) + \lambda y(t)\right] \\
&= \lambda\left[x^2(t) + y^2(t)\right] = \lambda\rho^2(t).
\end{aligned}$$

Observe that the constant function $\rho(t) = 0$ for all $t \in \mathbb{R}$ solves this relation. Moreover, $\rho'(t) = \lambda\rho(t)$ as soon as $\rho(t) \neq 0$, which implies

$$\rho(t) = \rho(0)e^{\lambda t} = \sqrt{x_0^2 + y_0^2}\, e^{\lambda t}, \quad t \in \mathbb{R}. \tag{9.17}$$

This analysis shows that, either $\rho(0) = 0$ (equivalently, $(x_0, y_0) = (0,0)$) and, in such a case $\rho(t) = 0$ for all $t \in \mathbb{R}$, or $\rho(0) \neq 0$ (equivalently, $(x_0, y_0) \neq (0,0)$) and $\rho(t)$ satisfies (9.17). Subsequently, we further analyze the latter situation. Differentiating (9.16) with respect to t yields

$$\begin{aligned}
x'(t) &= \rho'(t) \cos \theta(t) - \rho(t)\theta'(t) \sin \theta(t), \\
y'(t) &= \rho'(t) \sin \theta(t) + \rho(t)\theta'(t) \cos \theta(t).
\end{aligned}$$

Thus,

$$\begin{aligned}
y(t)x'(t) &= \rho'(t)y(t) \cos \theta(t) - \rho(t)\theta'(t)y(t) \sin \theta(t), \\
x(t)y'(t) &= \rho'(t)x(t) \sin \theta(t) + \rho(t)\theta'(t)x(t) \cos \theta(t),
\end{aligned}$$

and using again (9.16) leads to

$$\begin{aligned}
y(t)x'(t) &= \rho(t)\rho'(t) \sin \theta(t) \cos \theta(t) - \rho^2(t)\theta'(t) \sin^2 \theta(t), \\
x(t)y'(t) &= \rho(t)\rho'(t) \cos \theta(t) \sin \theta(t) + \rho^2(t)\theta'(t) \cos^2 \theta(t).
\end{aligned}$$

Subtracting these identities gives

$$x(t)y'(t) - y(t)x'(t) = \rho^2(t)\theta'(t).$$

Therefore, since we are considering $(x_0, y_0) \neq (0,0)$, which implies, $\rho(t) \neq 0$,

$$\theta'(t) = \frac{x(t)y'(t) - y(t)x'(t)}{\rho^2(t)}$$

$$= \frac{x(t)\left[-\omega x(t) + \lambda y(t)\right] - y(t)\left[\lambda x(t) + \omega y(t)\right]}{\rho^2(t)} = -\omega.$$

Consequently,

$$\theta(t) = -\omega t + \theta_0 \quad \text{for all } t \in \mathbb{R}.$$

In particular, in polar coordinates,

$$\theta\left(\frac{2\pi}{\omega}\right) = -2\pi + \theta_0 = \theta_0 \pmod{2\pi}.$$

Therefore, for $(x_0, y_0) \neq (0,0)$, the trajectories are logarithmic spirals if $\lambda \neq 0$ and closed circles if $\lambda = 0$. From (9.17), it is apparent that the spirals approximate to zero as $t \uparrow +\infty$ if $\lambda < 0$, while they approximate to $(0,0)$ as $t \downarrow -\infty$ if $\lambda > 0$, as sketched in Figure 9.7.

The equilibrium $(0,0)$ is said to be an *unstable focus* if $\lambda > 0$, a *center* if $\lambda = 0$, and a *stable focus* if $\lambda < 0$. This terminology is consistent with the one used in Section 4.8.2 (cf. Figure 4.15). Observe that, in the case of a center, all non-trivial solutions are $\frac{2\pi}{\omega}$-periodic. Indeed,

$$\begin{pmatrix} x(t) \\ y(t) \end{pmatrix} = e^{t\begin{pmatrix} 0 & \omega \\ -\omega & 0 \end{pmatrix}} \begin{pmatrix} x_0 \\ y_0 \end{pmatrix} = \begin{pmatrix} \cos(\omega t) & \sin(\omega t) \\ -\sin(\omega t) & \cos(\omega t) \end{pmatrix} \begin{pmatrix} x_0 \\ y_0 \end{pmatrix}.$$

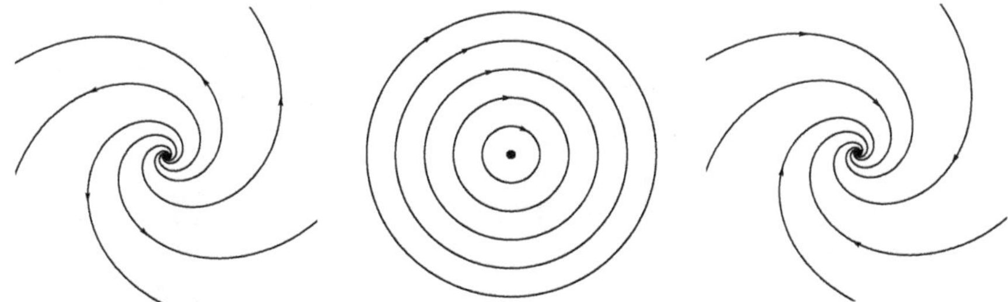

Fig. 9.7. Dynamics of (9.15) for $\lambda > 0$ (left), $\lambda = 0$ (center), and $\lambda < 0$ (right). The equilibrium is an unstable focus, a center, and a stable focus, respectively.

Finally, to get the phase portrait in the original variables (u, v), one performs the change of variable (9.6). Observe that, if $\det P < 0$, the orientation of the trajectories in the phase portrait changes with respect to the ones plotted in Figure 9.7.

To determine the direction of rotation directly in the original variables (u, v), one might also plot the vector of the vector field $(u, v) \mapsto A(u, v)^T$ at some chosen point $(u_0, v_0) \in \mathbb{R}^2$, which gives the tangent vector to the solution of (9.3) through (u_0, v_0) and provides us with the direction of rotation.

As soon as $x(t) \neq 0$, the angular coordinate can be expressed as

$$\theta(t) = \arctan \frac{y(t)}{x(t)}$$

and, hence,

$$\theta'(t) = \frac{\left(\frac{y(t)}{x(t)}\right)'}{1 + \left(\frac{y(t)}{x(t)}\right)^2} = \frac{y'(t)x(t) - x'(t)y(t)}{x^2(t) + y^2(t)}.$$

Actually, our previous (rigorous) calculation of $\theta'(t)$ shows the validity of this identity even when $x(t) = 0$ if $y(t) \neq 0$. Thus, in practice, in order to get $\theta'(t)$, it suffices to differentiate (formally) the arctan function.

Remark 9.1. By comparing the phase planes studied in this section for the autonomous linear planar system (9.3) with the solutions of the homogeneous equations (4.44) integrated in Section 4.8.2, one observes a rather direct correspondence. It turns out that the solutions of (9.3) are parametric curves $t \mapsto (u(t), v(t))$ satisfying the two differential equations of the system. Thus, by searching for the Cartesian expression of these curves, for example, as a graph $u = U(v)$, one should have that

$$u(t) = U(v(t)) \quad \text{for all } t \in \mathbb{R},$$

and the chain rule shows that

$$u'(t) = \frac{dU}{dv}(v(t))v'(t).$$

Hence, as soon as $v'(t) \neq 0$,

$$\frac{dU}{dv}(v(t)) = \frac{u'(t)}{v'(t)} = \frac{au(t) + bv(t)}{cu(t) + dv(t)} = \frac{aU(v(t)) + bv(t)}{cU(v(t)) + dv(t)}.$$

Consequently,

$$U'(v) = \frac{aU(v) + bv}{cU(v) + dv},\tag{9.18}$$

where now $'$ indicates differentiation with respect to v. Therefore, the function $U(v)$ satisfies a homogeneous equation of first order, v being the independent variable and $u = U(v)$ the dependent one. In other words, v in (9.18) plays the same role as t in (4.44).

When $v'(t) = 0$ and $u'(t) \neq 0$, one should search for integral curves of the form $v = V(u)$. In such a case, the function V satisfies

$$V'(u) = \frac{cu + dV(u)}{au + bV(u)},$$

with $' = \frac{d}{du}$. Obviously, if $u'(t) = v'(t) = 0$, we remain at the rest point $(0,0)$ for all time.

As the Cartesian expression of a curve does not give information about the flow of time, while the parametric expression does, the arrows in the phase diagrams in this section indicate the direction of the motion along the trajectories, while no arrows are present in the figures in Section 4.8.2.

9.2 General Properties of Trajectories and Integral Curves

A careful analysis of the discussion carried out in Section 9.1 reveals that, in all the analyzed cases related to system (9.3), the trajectories share some common important properties. The main goal of this section is to summarize and prove these properties for the general autonomous system

$$\begin{cases} u' = f(u), \\ u(0) = x. \end{cases}\tag{9.19}$$

Observe that, with respect to (9.2), here we take $t_0 = 0$ with no loss of generality due to the invariance under time translations of the solutions of (9.19). Moreover, we assume $f \in \mathcal{C}^1(\mathbb{R}^N; \mathbb{R}^N)$. Thus, f is locally Lipschitz with respect to $u \in \mathbb{R}^N$ uniformly on compact subsets of $t \in \mathbb{R}$, and Theorem 5.14 guarantees that, for every $x \in \mathbb{R}^N$,

(9.19) possesses a unique bilateral maximal solution, denoted by $(I_x, u(\cdot\,; x))$. Once again, here we are not emphasizing the initial time because the system is autonomous. Moreover, thanks to Theorem 5.17, $I_x = (T_{\min}(x), T_{\max}(x))$ for some $T_{\min}(x) \in [-\infty, 0)$ and $T_{\max}(x) \in (0, +\infty]$, and the solution blows up at $T_{\max}(x)$ (resp., $T_{\min}(x)$) if $T_{\max}(x) < +\infty$ (resp., $T_{\min}(x) > -\infty$). In this general framework, the following fundamental properties hold:

Property 1. *Two trajectories either coincide or are disjoint, i.e.*

$$\Gamma_x = \Gamma_y \quad \text{if } \Gamma_x \cap \Gamma_y \neq \emptyset.$$

Proof. Suppose $z \in \Gamma_x \cap \Gamma_y$. Then, there exist $t_x \in I_x$ and $t_y \in I_y$ such that

$$z = u(t_x; x) = u(t_y; y).$$

Now, consider the function defined by

$$v(t) := u(t + t_x - t_y; x), \quad t \in t_y - t_x + I_x.$$

As the system is autonomous,

$$v'(t) = f(v(t)) \quad \text{for all } t \in t_y - t_x + I_x.$$

Moreover,

$$v(t_y) = u(t_x; x) = z = u(t_y; y).$$

Thus, by the uniqueness of the maximal solution,

$$v(t) = u(t; y) \quad \text{for all } t \in I_y,$$

and I_y is the maximal interval of definition of v. Therefore,

$$u(t; y) = u(t + t_x - t_y; x) \quad \text{for all } t \in I_y = t_y - t_x + I_x.$$

In particular, $\Gamma_y = \Gamma_x$, which completes the proof of Property 1. $\quad\square$

Property 2. *Any compact arc of integral curve of* (9.19) *is traveled in a finite time, provided that it does not contain any equilibrium.*

Proof. Consider an integral curve associated with (9.19) and assume, on the contrary, that there exists a point y on it such that the compact arc of the integral curve from x to y, denoted by $\mathcal{A}$, is traveled in infinite time, i.e. $u(t; x) \in \mathcal{A}$ for all $t \in [0, +\infty)$. Then, as any compact arc of $\mathcal{C}^1$-curve is rectifiable, we have that

$$\int_0^{+\infty} \|f(u(t; x))\| \, dt = \int_0^{+\infty} \|u'(t; x)\| \, dt = \text{length}\, \mathcal{A} < +\infty.$$

As a consequence, there exists a sequence of times, $\{t_n\}_{n \geq 1}$, such that

$$\lim_{n \to \infty} t_n = +\infty, \quad \lim_{n \to \infty} f(u(t_n; x)) = 0.$$

Since the sequence $\{u(t_n; x)\}_{n \geq 1} \subset \mathcal{A}$ is bounded, by the Bolzano–Weierstrass theorem, we can extract a subsequence, labeled again with n, such that

$$\lim_{n \to \infty} u(t_n; x) = \omega \in \mathcal{A}.$$

Since f is continuous,

$$0 = \lim_{n \to \infty} f(u(t_n; x)) = f(\omega).$$

Therefore, ω is an equilibrium of f in $\mathcal{A}$, which contradicts the assumption that $\mathcal{A}$ does not contain any equilibrium. $\square$

Property 3. *Any integral curve of (9.19) that is a Jordan curve and has no equilibria on it is the orbit of a non-trivial periodic solution.*

Proof. Recall that a Jordan curve is a simple closed curve; in particular, it is compact. By Property 2, there exists a minimal $T > 0$ such that $u(T; x) = x$, as represented in Figure 9.8.

Since the differential equation of (9.19) is autonomous, the function

$$v(t) := u(t + T; x), \quad t \in -T + I_x,$$

solves the differential equation too. In addition,

$$v(0) = u(T; x) = x = u(0; x).$$

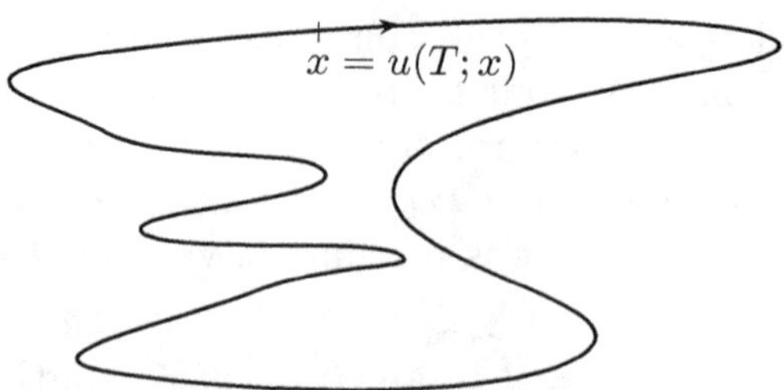

Fig. 9.8. A simple closed curve that is completely traveled after a minimal time T. It corresponds to a T-periodic solution of (9.19).

Thus, by uniqueness, it becomes apparent that

$$u(t + T; x) = v(t) = u(t; x) \quad \text{for all } t \in I_x.$$

Moreover, this argument shows that $-T + I_x = I_x$, which entails $I_x = \mathbb{R}$, and concludes the proof that u is periodic with the minimal period T. $\qquad\square$

Property 4. *No equilibrium point on an integral curve of (9.19) can be reached in a finite time—neither forward, nor backward.*

This property immediately follows from the uniqueness of solutions for problem (9.19) since an equilibrium corresponds to a constant solution. To illustrate this property, consider the situation sketched in Figure 9.9. Property 4 establishes that an infinite time

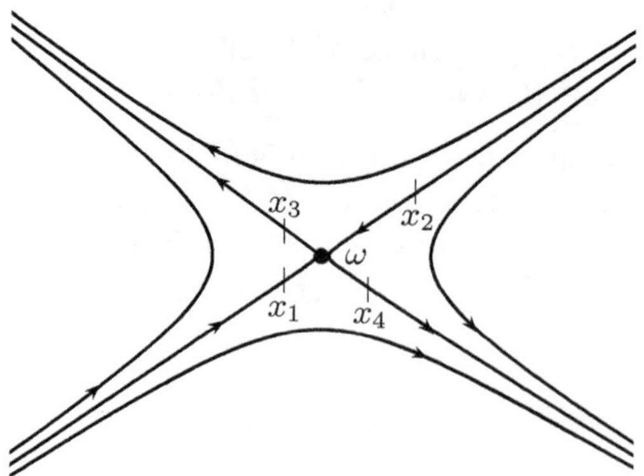

Fig. 9.9. According to Property 4, the equilibrium ω cannot be reached in a finite time.

is necessary to reach, or abandon, the equilibrium ω, i.e.

$$\lim_{t\uparrow+\infty} u(t;x_1) = \omega = \lim_{t\uparrow+\infty} u(t;x_2), \quad \lim_{t\downarrow-\infty} u(t;x_3) = \omega = \lim_{t\downarrow-\infty} u(t;x_4).$$

Property 5. *For $N = 2$, any orbit corresponding to a non-trivial periodic solution of* (9.19), *i.e. not an equilibrium, encloses, at least, one equilibrium.*

Proof. It is important to observe that this property only holds true in the plane because in $\mathbb{R}^N$, with $N > 2$, a Jordan curve cannot separate the whole space into two connected components. A completely rigorous proof of Property 5 makes use of the *topological degree*, which remains outside the scope of this introductory textbook. The topological degree is a generalized counter of the number of zeroes that a continuous function can have on an open bounded subset (see, e.g. López-Gómez and Mora-Corral, 2007, Section 12.2). In particular, the topological degree of a polynomial of degree $d \geq 1$ in a sufficiently large open bounded set of $\mathbb{C}$ equals d. In this introductory textbook, we give an heuristic rather direct proof of Property 5 based on the *winding number* of the vector field f along the orbit

$$\Gamma_x := \{u(t;x) : t \in [0,T]\},$$

where $T > 0$ is the minimal period of a non-trivial periodic solution $u(t;x)$ of (9.19). The winding number in the plane equals the topological degree, of course.

In rough terms, the winding number of the planar vector field f around the closed curve Γ_x, denoted by $W_{\Gamma_x}(f)$, is the number of rounds given by the free vector $u'(t;x) = f(u(t;x))$, as t ranges from 0 to T, around any point lying inside Γ_x. A well-known result in the theory of complex variables—the *argument principle*—shows that it equals the number of zeroes of f in the bounded component enclosed by Γ_x, Ω_x, provided f is analytic in a neighborhood of $\bar{\Omega}_x$. More generally, when f is C^1, the winding number provides us with a generalized counter of the number of zeroes of f in the component Ω_x (see, e.g. López-Gómez, 2001a, Chapter 11, where the topological degree in the plane is constructed from the winding number). As T

is the minimal period of the solution, it becomes apparent that

$$W_{\Gamma_x}(f) = \pm 1,$$

depending on the orientation induced by $u(t; x)$ on Γ_x. For example, in the case sketched in the left-hand plot of Figure 9.10, $W_{\Gamma_x}(f) = -1$ since the orbit is traveled clockwise as t varies from 0 to T.

The proof of Property 5 follows by contradiction. Suppose $f(z) \neq 0$ for all $z \in \Omega_x$. As the winding number is continuous with respect to the curve and is an integer, it must be constant if we continuously deform the curve Γ_x; in other words, the winding number is invariant under homotopy. Thus, for every $z \in \Omega_x$ and sufficiently small $\varepsilon > 0$,

$$W_{\Gamma_x}(f) = W_{C_\varepsilon(z)}(f) = \pm 1,$$

where $C_\varepsilon(z)$ denotes the circle centered at z with radius $\varepsilon > 0$ having the same orientation as Γ_x (ε should be sufficiently small so that $C_\varepsilon(z) \subset \Omega_x$), as is sketched in the left-hand plot of Figure 9.10.

However, $f(z) \neq 0$ because z is not an equilibrium. Thus, by continuity, for sufficiently small $\varepsilon > 0$, all vectors of the vector field f along $C_\varepsilon(z)$ should be close to $f(z)$, as illustrated in Figure 9.11.

This entails

$$W_{C_\varepsilon(z)}(f) = 0,$$

and this contradiction ends the proof. $\qquad\square$

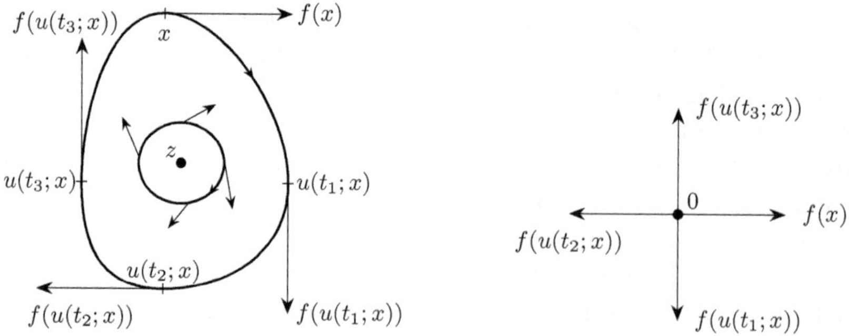

Fig. 9.10. By invariance by homotopy, the winding number $W_{\Gamma_x}(f)$ is the same on the original trajectory and on a circle centered at a point z lying inside Γ_x and with sufficiently small radius ε (left-hand plot). The right-hand plot illustrates how the free vectors $f(x)$, $f(u(t_1; x))$, $f(u(t_2; x))$ and $f(u(t_3; x))$ rotate clockwise around the origin in the plane, making one complete turn.

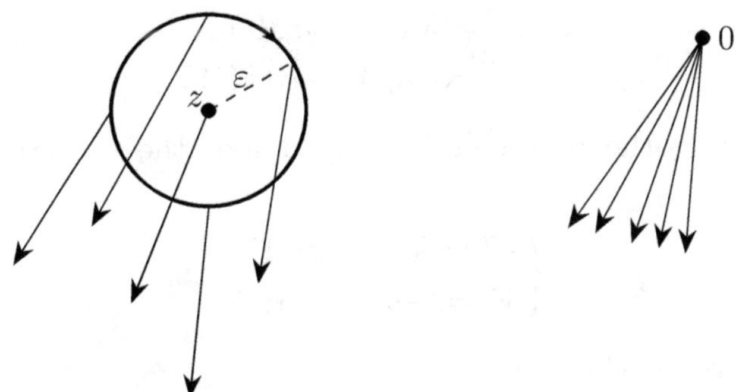

Fig. 9.11. The local winding number when $f(z) \neq 0$. In this case, the vectors of f along $C_\varepsilon(z)$ cannot make any turns if the circle is traveled once. Thus, $W_{C_\varepsilon(z)}(f) = 0$. Naturally, this contradicts the behavior sketched in Figure 9.10.

9.3   A Predator-Prey Model

In this section, we analyze the dynamics of the non-negative solutions, i.e. those with $u \geq 0$ ad $v \geq 0$, of the planar system

$$\begin{cases} u' = \lambda u - buv, \\ v' = -\mu v + cuv, \end{cases} \tag{9.20}$$

where λ, μ, b, and c are positive constants. In Ecology, $u(t)$ and $v(t)$ measure the number of individuals at time $t \geq 0$ of two interacting species, u and v, which might be considered preys and predators, respectively. Indeed, the interactions between the two species, measured by the factor proportional to uv, have a negative effect on u if $v > 0$, while they have a positive effect on v, provided that u is present. Moreover, if $v = 0$, then the first differential equation reduces to $u' = \lambda u$. Thus, the number of individuals of u grows exponentially if $v = 0$. Instead, if $u = 0$, then the second equation reduces to $v' = -\mu v$, and hence, the number of individuals of v decays exponentially in the absence of u.

Before beginning, we show that, without loss of generality, one can take $b = c = 1$. Indeed, the change of variable

$$u = \alpha U, \quad v = \beta V,$$

where α and β are positive constants, transforms (9.20) into

$$\begin{cases} \alpha U' = \lambda \alpha U - b\alpha\beta UV, \\ \beta V' = -\mu\beta V + c\alpha\beta UV. \end{cases}$$

Thus, dividing the first equation by α and the second one by β leads to

$$\begin{cases} U' = \lambda U - b\beta UV, \\ V' = -\mu V + c\alpha UV. \end{cases} \tag{9.21}$$

Consequently, by choosing

$$\alpha = \frac{1}{c}, \quad \beta = \frac{1}{b},$$

system (9.21) becomes

$$\begin{cases} U' = \lambda U - UV, \\ V' = -\mu V + UV, \end{cases}$$

and renaming $u = U$ and $v = V$ gives

$$\begin{cases} u' = \lambda u - uv, \\ v' = -\mu v + uv. \end{cases} \tag{9.22}$$

Thus, we finally focus on this simpler problem, as well as on the associated Cauchy problem

$$\begin{cases} u' = \lambda u - uv, \\ v' = -\mu v + uv, \\ u(0) = u_0, \quad v(0) = v_0, \end{cases} \tag{9.23}$$

with $u_0 \geq 0$ and $v_0 \geq 0$, since we want to consider component-wise non-negative solutions. As the function

$$f(u,v) := \begin{pmatrix} \lambda u - uv \\ -\mu v + uv \end{pmatrix}, \quad (u,v) \in \mathbb{R}^2,$$

is of class $\mathcal{C}^\infty(\mathbb{R}^2; \mathbb{R}^2)$, by Corollary 5.3, f is locally Lipschitz with respect to $(u,v) \in \mathbb{R}^2$ uniformly in $t \in \mathbb{R}$ since f is autonomous. Thus, owing to Theorem 5.14, (9.23) has a unique maximal bilateral solution, (I, u, v). Moreover, by Theorem 5.17, there exist $T_{\min} \in$

$[-\infty, 0)$ and $T_{\max} \in (0, +\infty]$ such that $I = (T_{\min}, T_{\max})$, and (u, v) blows up at $T_{\max}$ (respectively, $T_{\min}$) if $T_{\max} < +\infty$ (respectively, $T_{\min} > -\infty$). Furthermore, as (9.22) is of the Kolmogorov type (see Exercise 13 of Chapter 5, if necessary), we have that

$$\operatorname{sign} u(t) = \operatorname{sign} u_0 \quad \text{and} \quad \operatorname{sign} v(t) = \operatorname{sign} v_0 \quad \text{for all } t \in I.$$

Thus, if $(u_0, v_0) = (0, 0)$, then $I = \mathbb{R}$ and $(u(t), v(t)) = (0, 0)$ for all $t \in \mathbb{R}$. This shows that $(0, 0)$ is an equilibrium of (9.22). Similarly, if $u_0 = 0$ and $v_0 > 0$, then $u(t) = 0$ for all $t \in I$ and, hence, v satisfies $v' = -\mu v$ in I. So, $I = \mathbb{R}$ and

$$(u(t), v(t)) = (0, v_0 e^{-\mu t}) \quad \text{for all } t \in \mathbb{R},$$

which describes the evolution of the predator in the absence of preys. In particular, the predator becomes extinct in the long run, in the sense that

$$\lim_{t \uparrow +\infty} \left(v_0 e^{-\mu t} \right) = 0.$$

Analogously, if $u_0 > 0$ and $v_0 = 0$, then $v(t) = 0$ for all $t \in I$, and u satisfies $u' = \lambda u$ in I. Thus, $I = \mathbb{R}$, and

$$(u(t), v(t)) = (u_0 e^{\lambda t}, 0) \quad \text{for all } t \in \mathbb{R},$$

which provides us with the evolution of the preys in the absence of predators. In this case, the number of preys grows exponentially because

$$\lim_{t \uparrow +\infty} \left(u_0 e^{\lambda t} \right) = +\infty.$$

The problem of ascertaining the behavior of the solution when $u_0 > 0$ and $v_0 > 0$ is much more subtle. In such a case, since $u(t) > 0$ and $v(t) > 0$ for all $t \in I$, one has that, for every $t \in [0, T_{\max})$,

$$u(t) = u_0 e^{\int_0^t (\lambda - v(s)) \, ds} \leq u_0 e^{\lambda t} \tag{9.24}$$

and hence,

$$v(t) = v_0 e^{\int_0^t (-\mu + u(s)) \, ds} \leq v_0 e^{-\mu t + \frac{u_0}{\lambda}(e^{\lambda t} - 1)}. \tag{9.25}$$

Therefore, as we have these bounds for $(u(t), v(t))$ on any compact subinterval of $[0, +\infty)$, it becomes apparent that $T_{\max} = +\infty$.

Indeed, if $T_{\max} < +\infty$, then, thanks to Theorem 5.17, one should have that

$$\limsup_{t \uparrow T_{\max}} (u(t) + v(t)) = +\infty,$$

which is incompatible with (9.24) and (9.25). For the time being, we postpone the problem of ascertaining the value of $T_{\min}$.

By solving the system

$$\begin{cases} \lambda u - uv = (\lambda - v)u = 0, \\ -\mu v + uv = (-\mu + u)v = 0, \end{cases}$$

it is clear that $(0,0)$ and (μ, λ) are the unique equilibria of (9.22). Moreover, by (9.22),

$$u' = (\lambda - v)\, u \begin{cases} > 0 & \text{if } v < \lambda, \\ = 0 & \text{if } v = \lambda, \\ < 0 & \text{if } v > \lambda, \end{cases} \tag{9.26}$$

and

$$v' = (-\mu + u)\, v \begin{cases} > 0 & \text{if } u > \mu, \\ = 0 & \text{if } u = \mu, \\ < 0 & \text{if } u < \mu. \end{cases} \tag{9.27}$$

Consequently, the tangent vectors to the trajectories of non-negative solutions of (9.22) look like those shown in the *direction field* sketched in Figure 9.12; the left-hand plot provides a rough primary sketch of the direction field, while the right-hand plot shows the complete direction field.

More in detail, the dynamics of preys in the absence of predators is depicted on the u-axis of Figure 9.12, while the dynamics of predators in the absence of preys is sketched on the v-axis. Moreover, at each point (u, v) of the interior of the quadrant, the arrows represent the vector (u', v') according to the location of (u, v) and the differential equations (9.22). Naturally, $(u', v') = (0, 0)$ at the equilibria $(0, 0)$ and (μ, λ), $u' = 0$ on the line $v = \lambda$, and $v' = 0$ on the line $u = \mu$. In the remaining points, the signs of u' and v' can be determined

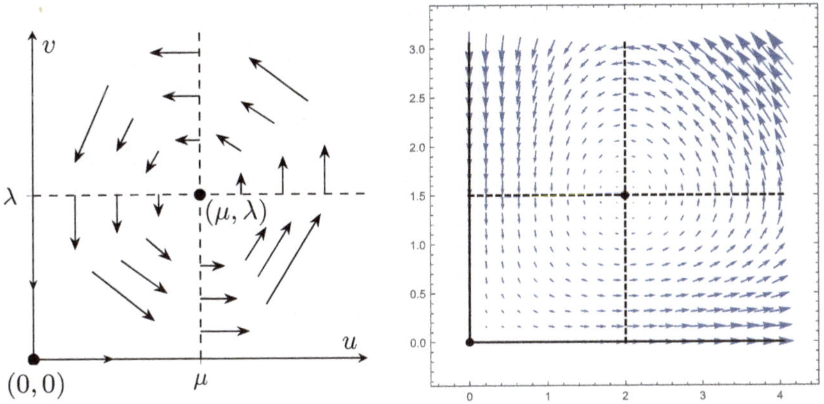

Fig. 9.12. Direction field associated with system (9.22): sketch (left) and complete plot (right).

from (9.26) and (9.27). Precisely, we have that, as soon as $u > 0$ and $v > 0$,

$$
\begin{aligned}
u' < 0 \quad &\text{and} \quad v' > 0 \quad \text{in } \{u > \mu \text{ and } v > \lambda\}, \\
u' < 0 \quad &\text{and} \quad v' < 0 \quad \text{in } \{u < \mu \text{ and } v > \lambda\}, \\
u' > 0 \quad &\text{and} \quad v' < 0 \quad \text{in } \{u < \mu \text{ and } v < \lambda\}, \\
u' > 0 \quad &\text{and} \quad v' > 0 \quad \text{in } \{u > \mu \text{ and } v < \lambda\}.
\end{aligned}
\tag{9.28}
$$

Naturally, the amplitudes of u' and v' depend on the specific chosen values of u and v. Accordingly, both components are very small in a neighborhood of the equilibria, though they can be very large (in absolute value) far away from them. For example, the v' component of the tangent vector (u', v') when calculated along $v = \lambda$ is given by $v' = (-\mu + u)\,\lambda$; thus, it grows from $v' = -\mu\lambda$ at $u = 0$ up to reaching $v' = 0$ at $u = \mu$, and it further grows from the value $v' = 0$ at $u = \mu$, reaching arbitrarily large values as $u \uparrow +\infty$. Similarly, the u' component along $u = \mu$ is given by $u' = (\lambda - v)\,\mu$; thus, it decreases from $u' = \mu\lambda$ at $v = 0$ up to reaching $u' = 0$ at $v = \lambda$, and it further decreases from $u' = 0$ at $v = \lambda$, reaching arbitrarily big negative values as $v \uparrow +\infty$. In the regions indicated in (9.28), the components of the tangent vectors (u', v') exhibit a composition of these two basic effects, as is apparent from Figure 9.12.

The construction of the direction field is imperative to get a first idea about the dynamics of (9.22). Nevertheless, it does not give any information about the precise nature of the equilibrium (μ, λ) because it does not allow us to determine whether it is a focus (stable or unstable) or a center, or has some other different behavior. This sharper analysis requires studying the trajectories through any point (u_0, v_0), with $u_0 > 0$ and $v_0 > 0$.

By taking cross products in (9.22), it follows that, for every $t \in I$,

$$u'(t)\,(-\mu + u(t))\,v(t) = v'(t)\,(\lambda - v(t))\,u(t).$$

Thus, dividing by $u(t)v(t)$ yields

$$-\mu\frac{u'(t)}{u(t)} + u'(t) = \lambda\frac{v'(t)}{v(t)} - v'(t) \quad \text{for all } t \in I,$$

which can be rewritten as

$$\frac{d}{dt}\left[\mu \ln u(t) - u(t) + \lambda \ln v(t) - v(t)\right] = 0 \quad \text{for all } t \in I.$$

Hence, for every $t \in I$,

$$\mu \ln u(t) - u(t) + \lambda \ln v(t) - v(t) = \mu \ln u_0 - u_0 + \lambda \ln v_0 - v_0.$$

Equivalently,

$$\mu \ln \frac{u(t)}{u_0} - (u(t) - u_0) = -\lambda \ln \frac{v(t)}{v_0} + v(t) - v_0.$$

Thus, taking exponentials gives

$$\left(\frac{u(t)}{u_0}\right)^{\mu} e^{-(u(t)-u_0)} = \left(\frac{v(t)}{v_0}\right)^{-\lambda} e^{v(t)-v_0} \quad \text{for all } t \in I.$$

Consequently, the trajectory of the solution of (9.23),

$$\Gamma_{(u_0,v_0)} := \{(u(t), v(t)) : t \in I\},$$

lies on the integral curve through the point (u_0, v_0), denoted by $\mathcal{C}_{(u_0,v_0)}$, which consists of the set of points $(u, v) \in (0, +\infty)^2$ such that

$$\left(\frac{u}{u_0}\right)^{\mu} e^{-(u-u_0)} = \left(\frac{v}{v_0}\right)^{-\lambda} e^{v-v_0}.$$

In other words, introducing the functions $\varphi(u)$ and $\psi(v)$, defined as

$$\varphi(u) := \left(\frac{u}{u_0}\right)^{\mu} e^{-(u-u_0)}, \quad \psi(v) := \left(\frac{v}{v_0}\right)^{-\lambda} e^{v-v_0}, \quad u, v \in (0, +\infty),$$

we have shown that

$$\Gamma_{(u_0,v_0)} \subset \mathcal{C}_{(u_0,v_0)} := \left\{ (u,v) \in (0,+\infty)^2 : \varphi(u) = \psi(v) \right\}. \qquad (9.29)$$

Although, in general, $\Gamma_{(u_0,v_0)}$ might be a proper subset of $\mathcal{C}_{(u_0,v_0)}$, in this particular model, they coincide due to the special structure of the orbit $\Gamma_{(u_0,v_0)}$, as will become apparent later.

The function $\varphi(u)$ satisfies $\varphi(0) = 0$, $\varphi(u) > 0$ if $u > 0$, and

$$\lim_{u\uparrow+\infty} \varphi(u) = 0.$$

Moreover, for every $u > 0$,

$$\varphi'(u) = \left(\frac{u}{u_0}\right)^{\mu-1} e^{-(u-u_0)} \frac{\mu - u}{u_0}$$

and, hence, $\varphi'(u) > 0$ if $u \in (0,\mu)$, $\varphi'(\mu) = 0$, and $\varphi'(u) < 0$ if $u > \mu$. Therefore, $\varphi(u)$ is increasing in $(0,\mu)$ and decreasing in $(\mu,+\infty)$, as illustrated in the left-hand plot of Figure 9.13. The function $\psi(v)$, instead, satisfies

$$\lim_{v\downarrow 0} \psi(v) = +\infty, \qquad \lim_{v\uparrow+\infty} \psi(v) = +\infty,$$

and it can be shown that it is decreasing in $(0,\lambda)$ and increasing in $(\lambda,+\infty)$. Thus, its graph looks like that shown in the right-hand plot of Figure 9.13.

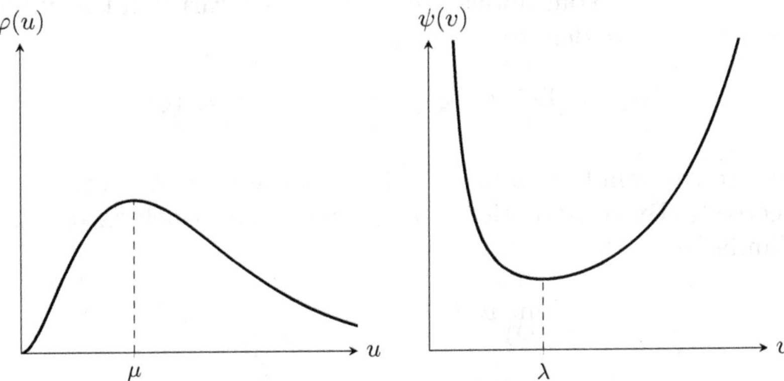

Fig. 9.13. The graphs of $\varphi(u)$ and $\psi(v)$.

Since $(u_0, v_0) \in \Gamma_{(u_0,v_0)}$, by (9.29) $\mathcal{C}_{(u_0,v_0)} \neq \emptyset$. Thus, $\varphi(\mu) \geq \psi(\lambda)$. Consequently, two different cases can arise. If $\varphi(\mu) = \psi(\lambda)$, then

$$\mathcal{C}_{(u_0,v_0)} = \{(\mu, \lambda)\}$$

and, hence,

$$\Gamma_{(u_0,v_0)} = \mathcal{C}_{(u_0,v_0)} = \{(\mu, \lambda)\},$$

which entails $(u_0, v_0) = (\mu, \lambda)$. In such a case, $\Gamma_{(u_0,v_0)} = \mathcal{C}_{(u_0,v_0)}$ reduces to the equilibrium point (μ, λ).

If $\varphi(\mu) > \psi(\lambda)$, we now show that $\mathcal{C}_{(u_0,v_0)}$ consists of a simple closed curve surrounding the equilibrium point (μ, λ). Indeed, as sketched in Figure 9.14, for every $\xi \in (\psi(\lambda), \varphi(\mu))$, there are two values of u and two values of v,

$$u_-(\xi) < \mu < u_+(\xi), \quad v_-(\xi) < \lambda < v_+(\xi),$$

such that

$$\xi = \varphi(u_\pm(\xi)) = \psi(v_\pm(\xi)).$$

Thus, each of the four points

$$(u_-(\xi), v_-(\xi)), \quad (u_-(\xi), v_+(\xi)), \quad (u_+(\xi), v_+(\xi)), \quad (u_+(\xi), v_-(\xi))$$

belongs to the integral curve $\mathcal{C}_{(u_0,v_0)}$. Since $u_\pm(\xi)$ and $v_\pm(\xi)$ vary continuously with respect to $\xi \in (\psi(\lambda), \varphi(\mu))$, each of these four points describes a continuous arc of curve as ξ varies in this interval. In Figure 9.14, we denote

$$v_L := \lim_{\xi \uparrow \varphi(\mu)} v_-(\xi), \quad v_M := \lim_{\xi \uparrow \varphi(\mu)} v_+(\xi),$$

which are the smallest value of $v_-(\xi)$ and the biggest value of $v_+(\xi)$, respectively. By construction, (μ, v_L) and (μ, v_M) belong to $\mathcal{C}_{(u_0,v_0)}$.

Similarly,

$$u_L := \lim_{\xi \downarrow \psi(\lambda)} u_-(\xi), \quad u_M := \lim_{\xi \downarrow \psi(\lambda)} u_+(\xi),$$

provide us with the smallest value of $u_-(\xi)$ and the biggest value of $u_+(\xi)$, respectively. By construction, we also have that (u_L, λ),

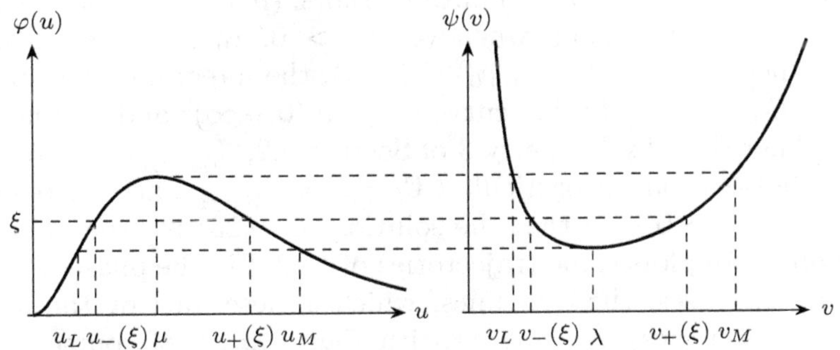

Fig. 9.14. Construction of $\mathcal{C}_{(u_0,v_0)}$ when $\varphi(\mu) > \psi(\lambda)$.

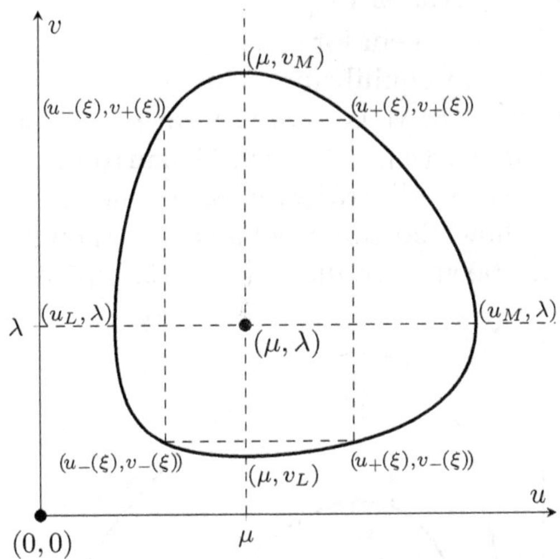

Fig. 9.15. The integral curve $\mathcal{C}_{(u_0,v_0)}$ when $\varphi(\mu) > \psi(\lambda)$.

$(u_M, \lambda) \in \mathcal{C}_{(u_0,v_0)}$. Observe that possible alternative definitions for these quantities are

$$u_L = u_-(\psi(\lambda)), \quad u_M = u_+(\psi(\lambda)), \quad v_L = v_-(\varphi(\mu)), \quad v_M = v_+(\varphi(\mu)).$$

Figure 9.15 represents the points $(u_\pm(\xi), v_\pm(\xi))$ as ξ varies in $[\psi(\lambda), \varphi(\mu)]$. They describe four compact arcs of continuous curves,

$$(u_-(\xi), v_-(\xi)), \quad (u_+(\xi), v_-(\xi)), \quad (u_+(\xi), v_+(\xi)), \quad (u_+(\xi), v_-(\xi))$$

that intersect at some of the limiting points: (μ, v_L), (u_M, λ), (μ, v_M), or (λ, u_L). Summarizing, whenever $u_0 > 0$, $v_0 > 0$, and $\varphi(\mu) > \psi(\lambda)$, which amounts to $(u_0, v_0) \neq (\mu, \lambda)$, the integral curve through (u_0, v_0), $\mathcal{C}_{(u_0, v_0)}$, is a Jordan curve lying in $(0, +\infty)^2$ and surrounding (μ, λ). Therefore, by Property 3 of Section 9.2, $\mathcal{C}_{(u_0, v_0)}$ is the orbit of a periodic solution. In particular, $\Gamma_{(u_0, v_0)} = \mathcal{C}_{(u_0, v_0)}$, and there exists $T = T(u_0, v_0) > 0$ such that the solution of (9.23) is T-periodic.

Figure 9.16 plots some trajectories of (9.22) in the phase portrait. As time increases, the dynamics, which is described by the arrows on the trajectories, must agree with Figure 9.12, of course. Moreover, if $(\tilde{u}_0, \tilde{v}_0)$ lies in the exterior component of the Jordan curve $\Gamma_{(u_0, v_0)}$ in $(0, +\infty)^2$, then, by Property 1 of Section 9.2, $\Gamma_{(\tilde{u}_0, \tilde{v}_0)}$ must entirely surround $\Gamma_{(u_0, v_0)}$. Finally, in agreement with Property 5 of Section 9.2, the trajectories $\Gamma_{(u_0, v_0)}$, which are Jordan curves without any equilibrium on them for every $(u_0, v_0) \in (0, +\infty)^2 \setminus \{(\mu, \lambda)\}$, enclose, necessarily, an equilibrium—in this case, (μ, λ). Summarizing, the equilibrium (μ, λ) is a *nonlinear center* since trajectories with $(u_0, v_0) \neq (\mu, \lambda)$, $(u_0, v_0) \sim (\mu, \lambda)$, surround the equilibrium. Actually, in this case, all trajectories in the first quadrant with $(u_0, v_0) \neq (\mu, \lambda)$ show the same behavior. Although the concept of center has already been introduced in Section 9.1.3 for linear planar

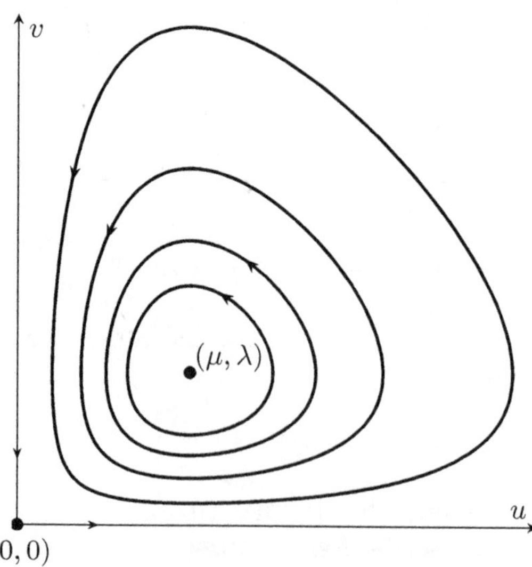

Fig. 9.16. The phase portrait of the predator-prey system (9.22).

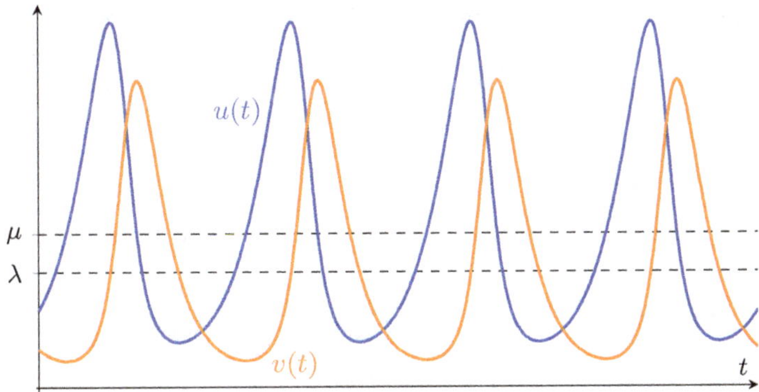

Fig. 9.17. Periodic fluctuations of the prey $u(t)$ and the predator $v(t)$, governed by system (9.22).

systems, this is the first example in this book where a center arises in a nonlinear context. Nonlinear centers are often structurally unstable, in the sense that certain small perturbations of the nonlinearity destroy them, leading to a completely different dynamics.

Figure 9.17 shows the plots of u and v for the special choices $\lambda = 1.5, \mu = 2$, and $u_0 = 1$, $v_0 = 0.5$. According to the analysis carried out in this section, the numbers of predators and preys fluctuate periodically in time, regardless of their initial sizes. In many circumstances, thought not always, this is a rather realistic phenomenon in Ecology and explains the huge success of this model in the scientific literature, where it is known as the *Lotka–Volterra predator-prey system* (see the discussion in Section 9.6).

9.4   A Competing Species Model

The main goal of this section is to ascertain the dynamics of the non-negative solutions of the system

$$\begin{cases} u' = \lambda u - uv, \\ v' = \mu v - uv, \end{cases} \tag{9.30}$$

where λ and μ are positive constants. This prototype can be regarded as a competition model of two species, u and v, which grow exponentially in the absence of their antagonist at rates λ and μ, respectively.

The Cauchy problem associated with (9.30) is

$$\begin{cases} u' = \lambda u - uv, \\ v' = \mu v - uv, \\ u(0) = u_0, \quad v(0) = v_0, \end{cases} \qquad (9.31)$$

where $u_0 \geq 0$ and $v_0 \geq 0$ indicate the initial numbers of individuals of each species.

As in Section 9.3, since

$$f(u, v) = \begin{pmatrix} \lambda u - uv \\ \mu v - uv \end{pmatrix}, \quad (u, v) \in \mathbb{R}^2,$$

is of class $\mathcal{C}^\infty(\mathbb{R}^2; \mathbb{R}^2)$, owing to Corollary 5.3, f is locally Lipschitz with respect to $(u, v) \in \mathbb{R}^2$, uniformly in $t \in \mathbb{R}$. Thus, for every $u_0 \geq 0$ and $v_0 \geq 0$, (9.31) has a unique maximal bilateral solution, (I, u, v). Moreover, by Theorem 5.17, there exist $T_{\min} \in [-\infty, 0)$ and $T_{\max} \in (0, +\infty]$ such that $I = (T_{\min}, T_{\max})$, and blow up occurs at $T_{\max}$ (respectively, $T_{\min}$), if $T_{\max} < +\infty$ (respectively, $T_{\min} > -\infty$). Furthermore, since (9.30) is of the Kolmogorov type,

$$\operatorname{sign} u(t) = \operatorname{sign} u_0 \quad \text{and} \quad \operatorname{sign} v(t) = \operatorname{sign} v_0 \quad \text{for all } t \in I.$$

Thus, if $(u_0, v_0) = (0, 0)$, then $(u(t), v(t)) = (0, 0)$ for all $t \in I = \mathbb{R}$, i.e. $(0, 0)$ is an equilibrium of (9.30). Moreover, if $u_0 = 0$ and $v_0 > 0$, then $u(t) = 0$ for all $t \in I$ and, hence, v satisfies $v' = \mu v$ in I. Thus,

$$(u(t), v(t)) = (0, v_0 e^{\mu t}) \quad \text{for all } t \in I = \mathbb{R},$$

which provides us with the evolution of v in the absence of u. Similarly, if $u_0 > 0$ and $v_0 = 0$, then $v(t) = 0$ for all $t \in I$ and, hence, u satisfies $u' = \lambda u$ in I. Thus,

$$(u(t), v(t)) = (u_0 e^{\lambda t}, 0) \quad \text{for all } t \in I = \mathbb{R},$$

which provides us with the evolution of u in the absence of v.

When $u_0 > 0$ and $v_0 > 0$, then, for every $t \in [0, T_{\max})$,

$$u(t) = u_0 e^{\int_0^t (\lambda - v(s))\, ds} \leq u_0 e^{\lambda t},$$

$$v(t) = v_0 e^{\int_0^t (\mu - u(s))\, ds} \leq v_0 e^{\mu t}.$$

Consequently, reasoning as in Section 9.3 shows that $T_{\max} = +\infty$.

Nevertheless, observe that the solution might not be globally defined for negative times. Indeed, if $\lambda = \mu$ and $u_0 = v_0 = x > 0$, then, by uniqueness and symmetry, the solution of (9.31) is given by $(u, v) = (w, w)$, where w is the unique solution of the logistic problem

$$\begin{cases} w' = \lambda w - w^2, \\ w(0) = x, \end{cases}$$

whose solution blows up at some $T_{\min} = T_{\min}(x) < 0$ whenever $x > \lambda$, as discussed in Section 4.1. Moreover, we already know that

$$\lim_{x \downarrow \lambda} T_{\min}(x) = -\infty, \qquad \lim_{x \uparrow +\infty} T_{\min}(x) = 0.$$

Thus, the maximal existence interval to the left of $t_0 = 0$ can have any size, depending on the parameters of the problem. Of course, if one instead takes $x \in (0, \lambda)$, the solution of the logistic problem is also globally defined for $t < 0$, i.e. $T_{\min} = -\infty$.

By solving the algebraic equations

$$\begin{cases} \lambda u - uv = (\lambda - v)\, u = 0, \\ \mu v - uv = (\mu - u)\, v = 0, \end{cases}$$

it is apparent that $(0, 0)$ and (μ, λ) are the unique equilibria of (9.30). Moreover,

$$u' = (\lambda - v)\, u \begin{cases} > 0 & \text{if } v < \lambda, \\ = 0 & \text{if } v = \lambda, \\ < 0 & \text{if } v > \lambda, \end{cases}$$

and

$$v' = (\mu - u)\, v \begin{cases} > 0 & \text{if } u < \mu, \\ = 0 & \text{if } u = \mu, \\ < 0 & \text{if } u > \mu. \end{cases}$$

Thus, the tangent vectors to the trajectories of the non-negative solutions of (9.30) look like those shown in the direction field sketched in Figure 9.18.

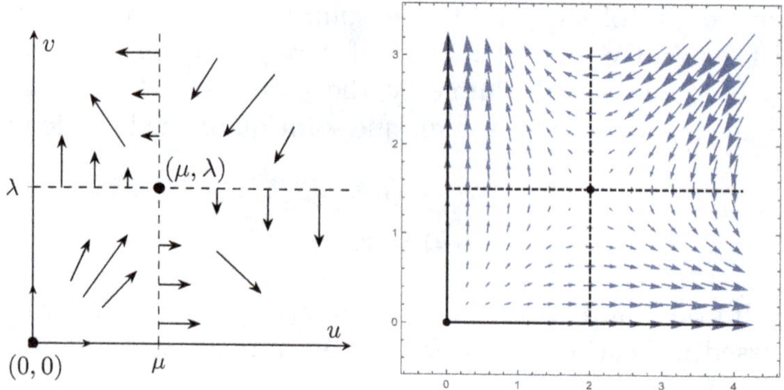

Fig. 9.18. Direction field of system (9.30): sketch (left) and complete plot (right).

To determine the trajectories of the system when $u_0 > 0$ and $v_0 > 0$, we take, as in Section 9.3, the cross products in (9.30). Then, for every $t \in I$,

$$u'(t)\left(\mu v(t) - u(t)v(t)\right) = v'(t)\left(\lambda u(t) - u(t)v(t)\right).$$

Thus, dividing by $u(t)v(t)$ yields

$$\mu \frac{u'(t)}{u(t)} - u'(t) = \lambda \frac{v'(t)}{v(t)} - v'(t) \quad \text{for all } t \in I,$$

which can be rewritten as

$$\frac{d}{dt}\left[\mu \ln u(t) - u(t) - \lambda \ln v(t) + v(t)\right] = 0 \quad \text{for all } t \in I.$$

Consequently, for every $t \in I$, we find that

$$\mu \ln u(t) - u(t) - \lambda \ln v(t) + v(t) = \mu \ln u_0 - u_0 - \lambda \ln v_0 + v_0.$$

Equivalently,

$$\mu \ln \frac{u(t)}{u_0} - (u(t) - u_0) = \lambda \ln \frac{v(t)}{v_0} - (v(t) - v_0).$$

Thus, by taking exponentials, we obtain that

$$\left(\frac{u(t)}{u_0}\right)^{\mu} e^{-(u(t)-u_0)} = \left(\frac{v(t)}{v_0}\right)^{\lambda} e^{-(v(t)-v_0)}$$

for all $t \in I$. Therefore, the trajectory of the solution of (9.31),

$$\Gamma_{(u_0,v_0)} := \{(u(t), v(t)) : t \in I\},$$

lies on the integral curve through (u_0, v_0), defined by

$$C_{(u_0,v_0)} := \left\{(u, v) \in (0, +\infty)^2 : \left(\frac{u}{u_0}\right)^{\mu} e^{-(u-u_0)} = \left(\frac{v}{v_0}\right)^{\lambda} e^{-(v-v_0)}\right\}.$$

In other words, introducing the functions $\varphi(u)$ and $\psi(v)$, defined as

$$\varphi(u) := \left(\frac{u}{u_0}\right)^{\mu} e^{-(u-u_0)}, \quad \psi(v) := \left(\frac{v}{v_0}\right)^{\lambda} e^{-(v-v_0)}, \quad u, v \in [0, +\infty),$$

we have shown that

$$\Gamma_{(u_0,v_0)} \subset C_{(u_0,v_0)} = \left\{(u, v) \in (0, +\infty)^2 : \varphi(u) = \psi(v)\right\}.$$

The graphs of the functions $\varphi(u)$ and $\psi(v)$ have already been analyzed in Section 9.3 (see the left-hand plot of Figure 9.13). To determine the precise structure of $C_{(u_0,v_0)}$, for a fixed (arbitrary) pair (u_0, v_0), we distinguish three different cases according to whether $\varphi(\mu) = \psi(\lambda)$, $\varphi(\mu) > \psi(\lambda)$, or $\varphi(\mu) < \psi(\lambda)$.

First, suppose that $\varphi(\mu) = \psi(\lambda)$, i.e. $(\mu, \lambda) \in C_{(u_0,v_0)}$. As sketched in Figure 9.19, in this case, for every $\xi \in (0, \varphi(\mu))$, there exist two values of u, $u_{\pm}(\xi)$, with $u_-(\xi) < \mu < u_+(\xi)$, and two values of v,

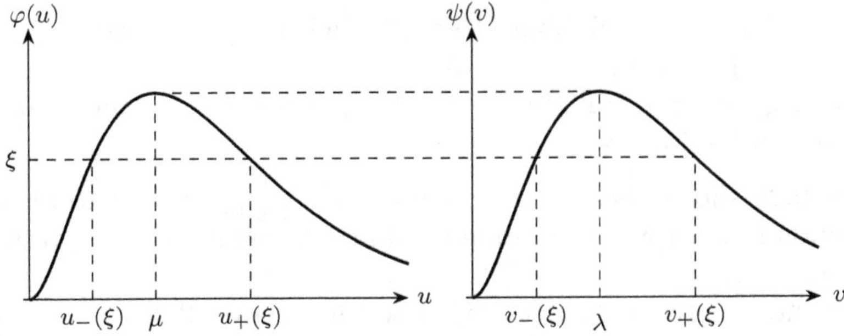

Fig. 9.19. The functions $\varphi(u)$ and $\psi(v)$ when $\varphi(\mu) = \psi(\lambda)$.

$v_\pm(\xi)$, with $v_-(\xi) < \lambda < v_+(\xi)$, such that

$$\varphi(u_\pm(\xi)) = \psi(v_\pm(\xi)) = \xi.$$

Thus, for every $\xi \in (0, \varphi(\mu))$, the four points

$$(u_-(\xi), v_-(\xi)), \quad (u_-(\xi), v_+(\xi)), \quad (u_+(\xi), v_-(\xi)), \quad (u_+(\xi), v_+(\xi)),$$

belong to $\mathcal{C}_{(u_0,v_0)}$. Since φ is continuous and increasing in $(0, \mu)$, $u_-(\xi)$ is continuous and increasing with respect to ξ, and

$$\lim_{\xi \downarrow 0} u_-(\xi) = 0, \quad \lim_{\xi \uparrow \varphi(\mu)} u_-(\xi) = \mu.$$

Since φ is continuous and decreasing in $(\mu, +\infty)$, $u_+(\xi)$ is continuous and decreasing with respect to ξ, and

$$\lim_{\xi \downarrow 0} u_+(\xi) = +\infty, \quad \lim_{\xi \uparrow \varphi(\mu)} u_+(\xi) = \mu.$$

Similarly, $v_-(\xi)$ is continuous and increasing with respect to ξ, $v_+(\xi)$ is continuous and decreasing with respect to ξ, and

$$\lim_{\xi \downarrow 0} v_-(\xi) = 0, \quad \lim_{\xi \downarrow 0} v_+(\xi) = +\infty, \quad \lim_{\xi \uparrow \psi(\lambda)} v_-(\xi) = \lim_{\xi \uparrow \psi(\lambda)} v_+(\xi) = \lambda.$$

Consequently, as ξ varies in the interval $(0, \varphi(\mu)]$, the integral curve $\mathcal{C}_{(u_0,v_0)}$ consists of four continuous arcs of curve:

(i) $(u_-(\xi), v_-(\xi))$, linking $(0,0)$ with (μ, λ) and lying in $(0, \mu] \times (0, \lambda]$,

(ii) $(u_-(\xi), v_+(\xi))$, linking $(0, +\infty)$ with (μ, λ) and lying in $(0, \mu] \times [\lambda, +\infty)$,

(iii) $(u_+(\xi), v_-(\xi))$, linking $(+\infty, 0)$ with (μ, λ) and lying in $[\mu, +\infty) \times (0, \lambda]$,

(iv) $(u_+(\xi), v_+(\xi))$, linking $(+\infty, +\infty)$ with (μ, λ) and lying in $[\mu, +\infty) \times [\lambda, +\infty)$.

Figure 9.20 shows these four arcs forming $\mathcal{C}_{(u_0,v_0)}$, together with the corresponding arrows determined from the direction field represented in Figure 9.18.

By the general properties discussed in Section 9.2, when $\varphi(\mu) = \psi(\lambda)$, the integral curve $\mathcal{C}_{(u_0,v_0)}$ consists of five different types of trajectories. Namely:

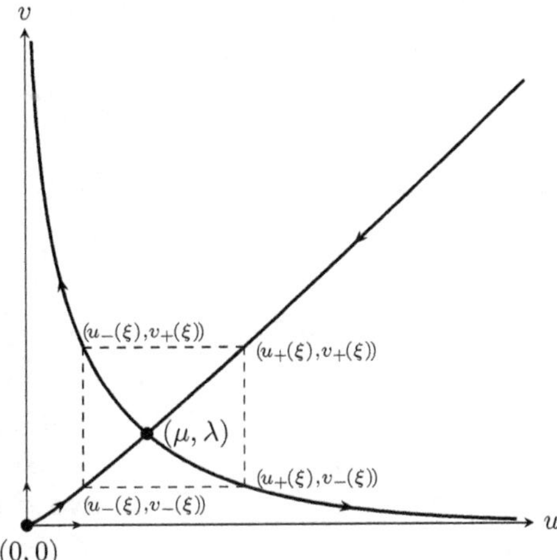

Fig. 9.20. The integral curve $\mathcal{C}_{(u_0,v_0)}$ of (9.31), in the case $\varphi(\mu) = \psi(\lambda)$.

(i) the trajectory of the equilibrium point (μ, λ);

(ii) the trajectory of any solution of (9.31), $(u(t), v(t))$, with $(u_0, v_0) = (u_+(\xi), v_+(\xi))$ for some $\xi \in (0, \varphi(\mu))$,

$$\Gamma_{(u_0,v_0)} = \{(u_+(\xi), v_+(\xi)) : \xi \in (0, \varphi(\mu))\},$$

which satisfies

$$\lim_{t \downarrow T_{\min}} (u(t), v(t)) = (+\infty, +\infty), \qquad \lim_{t \uparrow +\infty} (u(t), v(t)) = (\mu, \lambda);$$

(iii) the trajectory of any solution of (9.31), $(u(t), v(t))$, with $(u_0, v_0) = (u_-(\xi), v_-(\xi))$ for some $\xi \in (0, \varphi(\mu))$,

$$\Gamma_{(u_0,v_0)} = \{(u_-(\xi), v_-(\xi)) : \xi \in (0, \varphi(\mu))\},$$

which satisfies

$$\lim_{t \downarrow -\infty} (u(t), v(t)) = (0, 0), \qquad \lim_{t \uparrow +\infty} (u(t), v(t)) = (\mu, \lambda);$$

(iv) the trajectory of any solution of (9.31), $(u(t), v(t))$, with $(u_0, v_0) = (u_-(\xi), v_+(\xi))$ for some $\xi \in (0, \varphi(\mu))$,

$$\Gamma_{(u_0,v_0)} = \{(u_-(\xi), v_+(\xi)) : \xi \in (0, \varphi(\mu))\},$$

which satisfies

$$\lim_{t \downarrow -\infty} (u(t), v(t)) = (\mu, \lambda), \qquad \lim_{t \uparrow +\infty} (u(t), v(t)) = (0, +\infty);$$

(v) the trajectory of any solution of (9.31), $(u(t), v(t))$, with $(u_0, v_0) = (u_+(\xi), v_-(\xi))$ for some $\xi \in (0, \varphi(\mu))$,

$$\Gamma_{(u_0, v_0)} = \{(u_+(\xi), v_-(\xi)) : \xi \in (0, \varphi(\mu))\},$$

which satisfies

$$\lim_{t \downarrow -\infty} (u(t), v(t)) = (\mu, \lambda), \qquad \lim_{t \uparrow +\infty} (u(t), v(t)) = (+\infty, 0).$$

This is a general fact: the equilibria on any integral curve divide the whole integral curve into several trajectories because equilibria cannot be reached in a finite time, by Property 4 of Section 9.2.

In the context of the theory of Dynamical Systems, the third trajectory, which links (in infinite time) the two different equilibria $(0, 0)$ and (μ, λ), is referred to as a *heteroclinic connection* since the word "clinic", in Greek makes reference to "lying", or "coming to an end".

Now, suppose that $\varphi(\mu) > \psi(\lambda)$, as sketched in Figure 9.21. Then, as in the previous case, for every $\xi \in (0, \psi(\lambda))$, there are two values of u, $u_\pm(\xi)$, with $u_-(\xi) < \mu < u_+(\xi)$, and two values of v, $v_\pm(\xi)$, with $v_-(\xi) < \lambda < v_+(\xi)$, such that

$$\varphi(u_\pm(\xi)) = \psi(v_\pm(\xi)) = \xi.$$

Hence, for every $\xi \in (0, \psi(\lambda))$, we have that

$$(u_-(\xi), v_-(\xi)), \quad (u_-(\xi), v_+(\xi)), \quad (u_+(\xi), v_-(\xi)), \quad (u_+(\xi), v_+(\xi)),$$

belong to $\mathcal{C}_{(u_0, v_0)}$. In addition, $u_\pm(\xi)$ and $v_\pm(\xi)$ are continuous functions of ξ, $u_-(\xi)$ and $v_-(\xi)$ are increasing with respect to ξ, while

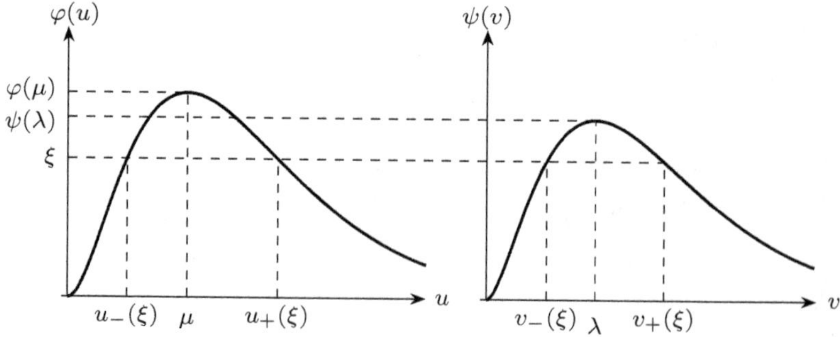

Fig. 9.21. The functions $\varphi(u)$ and $\psi(v)$ in the case $\varphi(\mu) > \psi(\lambda)$.

$u_+(\xi)$ and $v_+(\xi)$ are decreasing. Moreover, in this case,

$$\lim_{\xi\uparrow\psi(\lambda)} u_\pm(\xi) = u_\pm(\psi(\lambda)) \quad \text{and} \quad \lim_{\xi\uparrow\psi(\lambda)} v_\pm(\xi) = \lambda,$$

where $u_\pm(\psi(\lambda))$, with $u_-(\psi(\lambda)) < \mu < u_+(\psi(\lambda))$, are the unique values of u for which

$$\varphi(u_\pm(\psi(\lambda))) = \psi(\lambda).$$

Thus, $(u_\pm(\psi(\lambda)), \lambda) \in \mathcal{C}_{(u_0,v_0)}$. Also, it is apparent that

$$\lim_{\xi\downarrow 0} u_-(\xi) = \lim_{\xi\downarrow 0} v_-(\xi) = 0 \quad \text{and} \quad \lim_{\xi\downarrow 0} u_+(\xi) = \lim_{\xi\downarrow 0} v_+(\xi) = +\infty.$$

Therefore, the integral curve $\mathcal{C}_{(u_0,v_0)}$ consists of the four arcs of curve $(u_\pm(\xi), v_\pm(\xi))$, with $\xi \in (0, \psi(\lambda)]$, represented in Figure 9.22. Consequently, in this case, $\mathcal{C}_{(u_0,v_0)}$ consists of two trajectories:

(i) the trajectory of any solution of (9.31), $(u(t), v(t))$, with $(u_0, v_0) = (u_-(\xi), v_\pm(\xi))$ for some $\xi \in (0, \psi(\lambda)]$,

$$\Gamma_{(u_0,v_0)} = \{(u_-(\xi), v_\pm(\xi)) : \xi \in (0, \psi(\lambda)]\},$$

which satisfies

$$\lim_{t\downarrow-\infty} (u(t), v(t)) = (0,0), \quad \lim_{t\uparrow+\infty} (u(t), v(t)) = (0, +\infty),$$

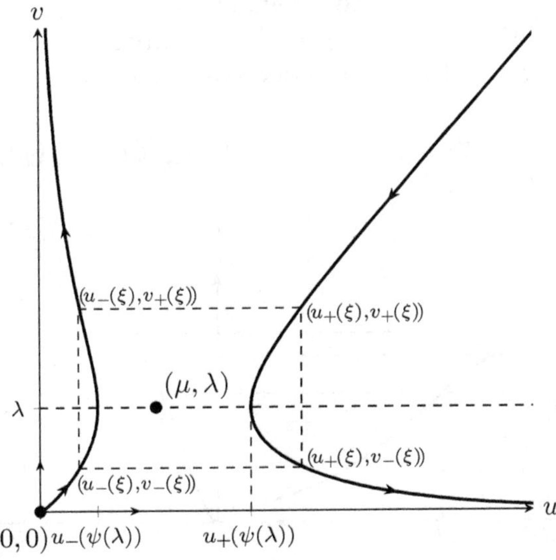

Fig. 9.22. The integral curve $\mathcal{C}_{(u_0,v_0)}$ when $\varphi(\mu) > \psi(\lambda)$.

(ii) the trajectory of any solution of (9.31), $(u(t), v(t))$, with $(u_0, v_0) = (u_+(\xi), v_\pm(\xi))$ for some $\xi \in (0, \psi(\lambda)]$,

$$\Gamma_{(u_0, v_0)} = \{(u_+(\xi), v_\pm(\xi)) : \xi \in (0, \psi(\lambda)]\},$$

which satisfies

$$\lim_{t \downarrow T_{\min}} (u(t), v(t)) = (+\infty, +\infty), \quad \lim_{t \uparrow +\infty} (u(t), v(t)) = (+\infty, 0).$$

Finally, suppose that $\varphi(\mu) < \psi(\lambda)$, as sketched in Figure 9.23. Arguing as before, it is apparent that, for every $\xi \in (0, \varphi(\mu))$, there are two values of u, $u_\pm(\xi)$, with $u_-(\xi) < \mu < u_+(\xi)$, and two values of v, $v_\pm(\xi)$, with $v_-(\xi) < \lambda < v_+(\xi)$, such that

$$\varphi(u_\pm(\xi)) = \psi(v_\pm(\xi)) = \xi.$$

Thus, for every $\xi \in (0, \varphi(\mu))$, we have that

$$(u_-(\xi), v_-(\xi)), \quad (u_-(\xi), v_+(\xi)), \quad (u_+(\xi), v_-(\xi)), \quad (u_+(\xi), v_+(\xi)),$$

belong to $\mathcal{C}_{(u_0, v_0)}$. As in the previous cases, $u_\pm(\xi)$ and $v_\pm(\xi)$ are continuous functions of ξ, $u_-(\xi)$ and $v_-(\xi)$ are increasing with respect to ξ, while $u_+(\xi)$ and $v_+(\xi)$ are decreasing. In addition,

$$\lim_{\xi \uparrow \varphi(\mu)} u_\pm(\xi) = \mu \quad \text{and} \quad \lim_{\xi \uparrow \varphi(\mu)} v_\pm(\xi) = v_\pm(\varphi(\mu)),$$

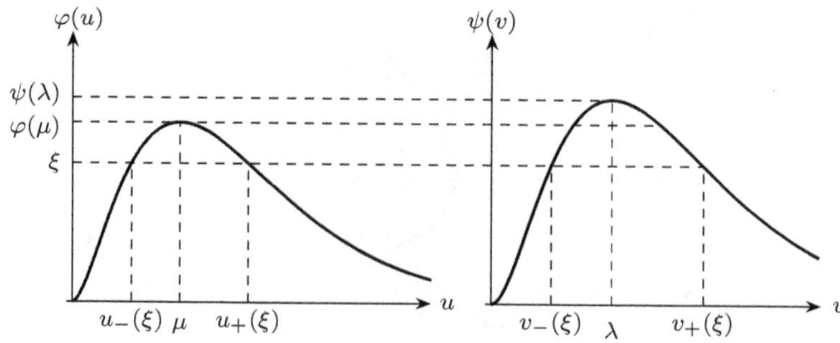

Fig. 9.23. The functions $\varphi(u)$ and $\psi(v)$ in the case $\varphi(\mu) < \psi(\lambda)$.

where $v_\pm(\varphi(\mu))$, with $v_-(\varphi(\mu)) < \lambda < v_+(\varphi(\mu))$, are the unique values of v for which

$$\psi(v_\pm(\varphi(\mu))) = \varphi(\mu).$$

Thus, $(\mu, v_\pm(\varphi(\mu))) \in \mathcal{C}_{(u_0, v_0)}$. Moreover, as in the previous cases,

$$\lim_{\xi \downarrow 0} u_-(\xi) = \lim_{\xi \downarrow 0} v_-(\xi) = 0 \quad \text{and} \quad \lim_{\xi \downarrow 0} u_+(\xi) = \lim_{\xi \downarrow 0} v_+(\xi) = +\infty.$$

Therefore, the integral curve $\mathcal{C}_{(u_0, v_0)}$ consists of the four arcs of curve $(u_\pm(\xi), v_\pm(\xi))$, $\xi \in (0, \varphi(\mu)]$, represented in Figure 9.24. Also in this case, $\mathcal{C}_{(u_0, v_0)}$ consists of two trajectories:

(i) the trajectory of any solution of (9.31), $(u(t), v(t))$, with $(u_0, v_0) = (u_\pm(\xi), v_-(\xi))$ for some $\xi \in (0, \varphi(\mu)]$,

$$\Gamma_{(u_0, v_0)} = \{(u_\pm(\xi), v_-(\xi)) : \xi \in (0, \varphi(\mu)]\},$$

which satisfies

$$\lim_{t \downarrow -\infty} (u(t), v(t)) = (0, 0), \quad \lim_{t \uparrow +\infty} (u(t), v(t)) = (+\infty, 0);$$

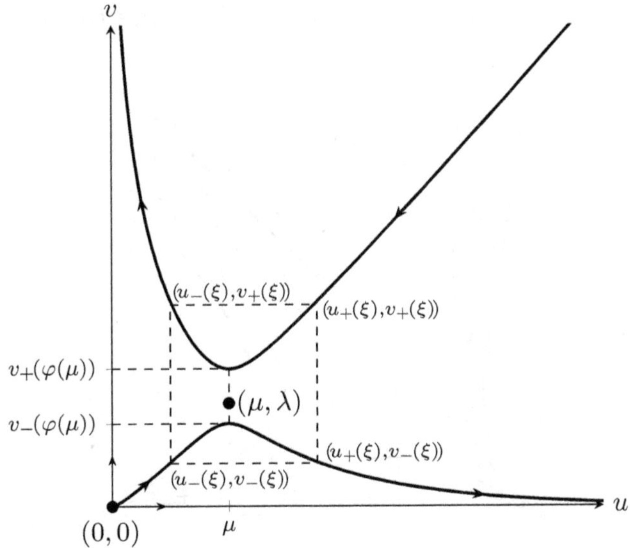

Fig. 9.24. The integral curve $\mathcal{C}_{(u_0, v_0)}$ when $\varphi(\mu) < \psi(\lambda)$.

(ii) the trajectory of any solution of (9.31), $(u(t), v(t))$, with $(u_0, v_0) = (u_\pm(\xi), v_+(\xi))$ for some $\xi \in (0, \varphi(\mu)]$,

$$\Gamma_{(u_0, v_0)} = \{(u_\pm(\xi), v_+(\xi)) : \xi \in (0, \varphi(\mu)]\},$$

which satisfies

$$\lim_{t \downarrow T_{\min}} (u(t), v(t)) = (+\infty, +\infty), \qquad \lim_{t \uparrow +\infty} (u(t), v(t)) = (0, +\infty).$$

Naturally, combining the three possible cases leads us to the phase portrait of the non-negative solutions of (9.30), which is sketched in Figure 9.25. According to it, we see that $(0, 0)$ is an *unstable node*, while (μ, λ) is a nonlinear *saddle point* (cf. Figure 9.1 for the case of a linear saddle point in a planar linear system).

The invariant curve through (μ, λ), whose solutions converge to (μ, λ) as $t \uparrow +\infty$, is refereed to as the *stable manifold* of the saddle point (μ, λ), and it is denoted by $W^s(\mu, \lambda)$. The *unstable manifold* of (μ, λ) can be defined as

$$W^u(\mu, \lambda) := \mathcal{C}_{(\mu, \lambda)} \setminus W^s(\mu, \lambda),$$

and it consists of two invariant curves whose solutions approximate (μ, λ) as $t \downarrow -\infty$, without being constant. On taking a careful look at Figure 9.25, it is easily seen how $W^s(\mu, \lambda)$ divides the first quadrant

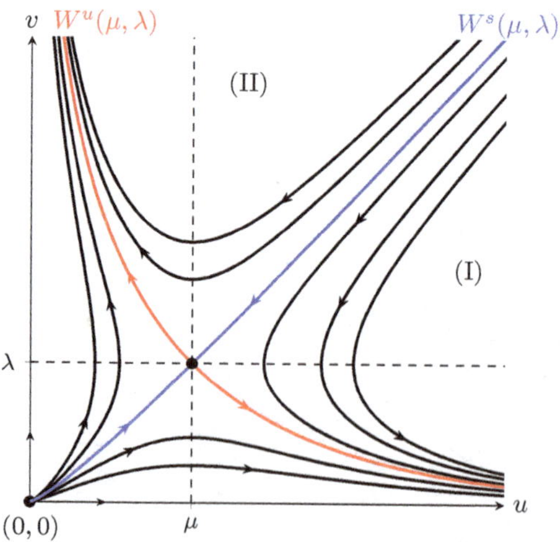

Fig. 9.25. The dynamics of non-negative solutions of (9.30).

of the phase plane into two regions: the lower one, denoted by (I), and the upper one, denoted by (II). If (u_0, v_0) lies in region (I), the solution of (9.31) satisfies

$$\lim_{t\uparrow+\infty} (u(t), v(t)) = (+\infty, 0).$$

Thus, v is driven to extinction by u. Similarly, for every (u_0, v_0) lying in region (II),

$$\lim_{t\uparrow+\infty} (u(t), v(t)) = (0, +\infty).$$

So, in this case, u is driven to extinction by v.

Therefore, according to this model, unless $(u_0, v_0) \in W^s(\mu, \lambda)$, one of the two species wipes out the antagonist, and the losing species is only determined by the initial sizes of the populations, u_0 and v_0.

9.5 Exercises

1. Construct the dynamics of the linear planar system $u' = Au$ for each of the following choices of A:

$$\begin{pmatrix} 1 & 3 \\ 3 & 1 \end{pmatrix}, \quad \begin{pmatrix} 3 & 1 \\ 1 & 3 \end{pmatrix}, \quad \begin{pmatrix} 7 & 2 \\ -5 & 5 \end{pmatrix},$$

$$\begin{pmatrix} 1 & 2 \\ -5 & -1 \end{pmatrix}, \quad \begin{pmatrix} 1 & -1 \\ 5 & -3 \end{pmatrix}, \quad \begin{pmatrix} 3 & 4 \\ -1 & 7 \end{pmatrix}.$$

2. Represent the phase diagram of the system $u' = Au$, with $A \in \mathcal{M}_3(\mathbb{R})$, when the eigenvalues of A, λ_1, λ_2 and λ_3, satisfy:
   (a) $\lambda_1 < \lambda_2 < \lambda_3 < 0$;
   (b) $\lambda_1 < \lambda_2 < 0 < \lambda_3$;
   (c) $\lambda_1 = \alpha + i\beta$, with $\beta > 0$, $\lambda_2 = \alpha - i\beta$, and $\alpha < \lambda_3 < 0$;
   (d) $\lambda_1 = \alpha + i\beta$, with $\beta > 0$, $\lambda_2 = \alpha - i\beta$, and $\lambda_3 < \alpha < 0$;
   (e) $\lambda_1 = \alpha + i\beta$, with $\beta > 0$, $\lambda_2 = \alpha - i\beta$, and $\alpha < 0 < \lambda_3$.

3. Compare the phase portraits of each of the systems

$$\begin{cases} u' = -v, \\ v' = u, \end{cases} \qquad \begin{cases} u' = -uv, \\ v' = u^2, \end{cases} \qquad \begin{cases} u' = -v^2, \\ v' = uv. \end{cases}$$

Observe that, in all these cases, the family of circles around the origin, $u^2 + v^2 = R^2$, consists of integral curves, though the three dynamics are substantially different.

4. Construct the phase portrait of the planar system

$$\begin{cases} u' = -u^3 v^4, \\ v' = -u^8 v^7. \end{cases}$$

5. For $\omega > 0$ fixed, use polar coordinates to construct the phase portrait of the following planar systems,

$$\begin{pmatrix} u' \\ v' \end{pmatrix} = \begin{pmatrix} \varepsilon & \omega \\ -\omega & \varepsilon \end{pmatrix} \begin{pmatrix} u \\ v \end{pmatrix} - \begin{pmatrix} u\left(u^2 + v^2\right) \\ v\left(u^2 + v^2\right) \end{pmatrix},$$

$$\begin{pmatrix} u' \\ v' \end{pmatrix} = \begin{pmatrix} \varepsilon^2 & \omega \\ -\omega & \varepsilon^2 \end{pmatrix} \begin{pmatrix} u \\ v \end{pmatrix} - \begin{pmatrix} u\left(u^2 + v^2\right) \\ v\left(u^2 + v^2\right) \end{pmatrix},$$

according to the values of the parameter $\varepsilon \in \mathbb{R}$. Analyze how the behavior changes when ε crosses zero in each of these examples.

6. For any given $\lambda > 0$ and $\mu > 0$, consider the following Cauchy problem related to the (normalized) symbiotic Lotka–Volterra model for the species u and v:

$$\begin{cases} u' = -\lambda u + uv, \\ v' = -\mu v + uv, \\ u(0) = u_0 \geq 0, \quad v(0) = v_0 \geq 0. \end{cases} \tag{9.32}$$

   (a) Construct the dynamics of each of the species when it is isolated, i.e. in the absence of the other.
   (b) Determine the equilibria of the system and sketch the direction field.
   (c) Construct the trajectory through (u_0, v_0) for every $u_0 > 0$ and $v_0 > 0$.
   (d) Characterize the set of initial data (u_0, v_0) for which the solution of (9.32) is globally defined in $t \in [0, +\infty)$ and characterize its asymptotic behavior as $t \uparrow +\infty$.
   (e) Study, for every $\gamma > 0$, the dynamics of the non-negative solutions of $w' = -\gamma w + w^2$. Show that its solutions blow up in a finite time if $w(0) > \gamma$.
   (f) Characterize the set of values (u_0, v_0) for which the maximal forward solution of (9.32) blows up in a finite time, $T_{\max}$, and

analyze the behavior of the solution as $t \uparrow T_{\max}$. [Hint: Allow the solution to travel the necessary time along its trajectory until it lies in the appropriate region of the phase plane in order to apply the method of sub- and supersolutions based on Kamke's theorem.]

(g) Interpret all this analysis from the point of view of Population Dynamics.

9.6 Final Comments

The main precursor of the modern theory of Dynamical Systems is Poincaré, who is also the father of Topology as we know it today. In 1885, as part of the celebrations for the sixtieth birthday of King Oscar II of Sweden and Norway, to be celebrated in January 1889, a mathematical competition was promoted in the journal *Acta Mathematica*, which had been founded in 1882 by G. Mittag-Leffler under the auspices of King Oscar II himself, and in the scientific journal *Nature*. The competition rules established four problems. The first one, known as the n-body problem, was proposed by K. Weirstrass, and it was closely related to the stability of the solar system. In May 1888, Poincaré presented an essay on the special case of $n = 3$, as he found the general problem to be almost unsolvable. His work was so successful that the jury gave him the prize.

Essentially, the main conclusion drawn by Poincaré was that the three-body system has an extremely unpredictable behavior, in the sense that small perturbations in the initial data, for example a small variation in the position of a planet, might bring the system to a completely different state. Thus, if measuring instruments cannot detect these variations, the final behavior of the system becomes unpredictable. Weierstrass recognized the huge importance of Poincaré's result in Celestial Mechanics. Incidentally, Poincaré had to withdraw the printed version of the journal with his paper, as he detected an error in a later stage of the publication process. The prize money was not enough to cover the expenses of withdrawing the printed wrong version.

Some years later, H. Poincaré published his celebrated synthesis *Les méthodes nouvelles de la mécanique céleste* (Poincaré, 1892), a pioneering book that opened up completely new horizons in the theory of Dynamical Systems, starting with his seminal work on the integral curves of differential equations (Poincaré, 1880).

The terminology of *Dynamical Systems* was coined 25 years later by Birkhoff in 1917. Among his results, Birkhoff obtained a proof for the *twist theorem*, which Poincaré had stated without a rigorous proof shortly before his death. For this reason, this result is nowadays known as *Poincaré's last theorem*, or the *Poincaré–Birkhoff theorem*.

From a different perspective, Lyapunov was the first scientist to address and solve the general problem of the stability of movement in a celebrated monograph titled *Problème général de la stabilité du mouvement* (Lyapunov, 1907). Some of the main results of this monograph will be presented in Chapter 11.

The model studied in Section 9.3 was introduced (among other models), independently, by Lotka (1932) and Volterra (1931). It was analyzed by Volterra in Chapter I, Section II, of his celebrated monograph *Théorie Mathématique de la lutte pour la vie* (Volterra, 1931). The phase portrait of the competing species model analyzed in Section 9.4 was actually given in the fifth picture of Figure 7 in Volterra (1931). The contributions of these two authors are seminal in Population Dynamics, Environmental Sciences, and Ecology and in fact pioneered modern Mathematical Biology.

Chapter 10

Newtonian Planar Conservative Systems

The main goal of this chapter is to study the solutions and global dynamics of scalar second-order differential equations of the type

$$-u'' = f(u), \tag{10.1}$$

where $f \in \mathcal{C}^1(\mathbb{R}; \mathbb{R})$. We refer to these equations as *of Newtonian type* since Newton's second law of Mechanics establishes that the acceleration of a moving body, u'', is proportional to the resulting force acting on the body, $-f(u)$. Thus, if the mass of the body is 1, (10.1) gives the equation of its motion.

Some paradigmatic examples of $f(u)$ that have already been discussed in Chapter 6 will be revisited here with the techniques introduced in Chapter 9. We already know that, in terms of the velocity $v := u'$, the Newtonian equation (10.1) can be equivalently expressed as a planar first-order system. Namely,

$$\begin{cases} u' = v, \\ v' = -f(u), \end{cases} \tag{10.2}$$

whose associated Cauchy problem is

$$\begin{cases} u' = v, \\ v' = -f(u), \\ u(0) = u_0, \quad v(0) = v_0, \end{cases} \tag{10.3}$$

for every $(u_0, v_0) \in \mathbb{R}^2$. The crucial fact that f does not depend on v, entailing the absence of *friction*, makes these systems *conservative*, as explained in the following section.

10.1   Conservation of Total Energy

According to Theorems 5.14 and 5.17, for every $(u_0, v_0) \in \mathbb{R}^2$, problem (10.3) has a unique maximal solution, (I, u, v), and there exist $T_{\min} \in [-\infty, 0)$ and $T_{\max} \in (0, +\infty]$ such that $I = (T_{\min}, T_{\max})$. Moreover,

$$\limsup_{t \uparrow T_{\max}} \left(|u(t)| + |u'(t)| \right) = +\infty \quad \text{if } T_{\max} < +\infty,$$

and similarly,

$$\limsup_{t \downarrow T_{\min}} \left(|u(t)| + |u'(t)| \right) = +\infty \quad \text{if } T_{\min} > -\infty.$$

Taking the cross products in (10.2) yields

$$v(t)v'(t) + u'(t)f(u(t)) = 0 \quad \text{for all } t \in I,$$

or, equivalently,

$$\frac{d}{dt}\left(\frac{v^2(t)}{2} + \int_0^{u(t)} f(s)\, ds \right) = 0.$$

Thus, the solution of (10.3) satisfies

$$\frac{v^2(t)}{2} + \int_0^{u(t)} f(s)\, ds = \frac{v_0^2}{2} + \int_0^{u_0} f(s)\, ds \quad \text{for all } t \in I. \tag{10.4}$$

In Newtonian Mechanics, the function

$$E(u, v) := \frac{v^2}{2} + \int_0^u f(s)\, ds, \quad (u, v) \in \mathbb{R}^2,$$

is known as the *total energy* of (10.1), while

$$E_k(v) := \frac{v^2}{2}, \quad \varphi(u) := E_p(u) = \int_0^u f(s)\, ds, \quad (u, v) \in \mathbb{R}^2,$$

are the *kinetic* and *potential* energies of (10.1), respectively. In terms of the total energy, (10.4) can be equivalently expressed as

$$E(u(t), v(t)) = E(u_0, v_0) =: E_0 \quad \text{for all } (u_0, v_0) \in \mathbb{R}^2,$$

i.e. the total energy of the movement is conserved along the trajectory of the solution of (10.3)

$$\Gamma_{(u_0, v_0)} := \{(u(t), v(t)) : t \in I\}.$$

For this reason, systems such as (10.3) are called *conservative*. The energy conservation can be equivalently expressed as

$$\frac{v^2(t)}{2} + \varphi(u(t)) = E_0 \quad \text{for all } t \in I,$$

and, hence,

$$v(t) = \pm\sqrt{2\left(E_0 - \varphi(u(t))\right)} \quad \text{for all } t \in I,$$

which provides us with the admissible values of the velocity $v(t)$ at the position $u(t)$. Therefore, the trajectory of the solution of (10.3), $\Gamma_{(u_0, v_0)}$, is a subset of the integral curve through the point (u_0, v_0), which is defined as the following geometrical set:

$$\mathcal{C}_{(u_0, v_0)} := \left\{(u, v) \in \mathbb{R}^2 : \ v = \pm\sqrt{2\left(E_0 - \varphi(u)\right)}\right\}, \tag{10.5}$$

where

$$E_0 = E(u_0, v_0) = \frac{v_0^2}{2} + \varphi(u_0).$$

As in Chapter 9, in some cases, $\Gamma_{(u_0, v_0)}$ is a proper subset of $\mathcal{C}_{(u_0, v_0)}$, which may indeed consist of several trajectories, while in other cases, these two sets coincide. Actually, $\mathcal{C}_{(u_0, v_0)}$ must consist of several trajectories in the following cases:

- if it includes some equilibrium but is not an equilibrium;
- if it is not a connected subset of $\mathbb{R}^2$.

According to (10.5), $\mathcal{C}_{(u_0, v_0)}$ is symmetric with respect to the u-axis in the phase plane. Moreover, the times taken to cover a compact arc of trajectory without equilibria along $v = +\sqrt{2\left(E_0 - \varphi(u)\right)}$ and the symmetric arc along $v = -\sqrt{2\left(E_0 - \varphi(u)\right)}$ coincide. Indeed, suppose that, for some $u_1, u_2 \in \mathbb{R}$, with $u_1 < u_2$, the arc of the integral curve

$$\Gamma_+ := \left\{(u, v) \in [u_1, u_2] \times \mathbb{R} \colon v = +\sqrt{2\left(E_0 - \varphi(u)\right)}\right\}$$

does not contain any equilibrium of (10.2). Then, the time τ_+ taken by the solution of

$$\begin{cases} u' = v, \\ v' = -f(u), \\ u(0) = u_1, \quad v(0) = +\sqrt{2\left(E_0 - \varphi(u_1)\right)}, \end{cases}$$

to reach the point $\left(u_2, +\sqrt{2\left(E_0 - \varphi(u_2)\right)}\right)$ along Γ_+ is given by

$$\tau_+ = \int_0^{\tau_+} dt = \int_0^{\tau_+} \frac{u'(t)}{v(t)}\, dt = \int_{u_1}^{u_2} \frac{d\xi}{\sqrt{2\left(E_0 - \varphi(\xi)\right)}}.$$

Similarly, the time τ_- taken by the solution of

$$\begin{cases} u' = v, \\ v' = -f(u), \\ u(0) = u_2, \quad v(0) = -\sqrt{2\left(E_0 - \varphi(u_2)\right)}, \end{cases}$$

to reach the point $\left(u_1, -\sqrt{2\left(E_0 - \varphi(u_1)\right)}\right)$ along

$$\Gamma_- := \left\{(u, v) \in [u_1, u_2] \times \mathbb{R} \colon v = -\sqrt{2\left(E_0 - \varphi(u)\right)}\right\}$$

is given by

$$\tau_- = \int_0^{\tau_-} dt = \int_0^{\tau_-} \frac{u'(t)}{v(t)}\, dt = \int_{u_2}^{u_1} \frac{d\xi}{-\sqrt{2\left(E_0 - \varphi(\xi)\right)}}.$$

Therefore, $\tau_+ = \tau_-$, as claimed.

10.2   Critical Points of the Potential Energy

Equilibria, i.e. constant solutions of (10.1) correspond to the zeros of $f(u)$. Indeed, ω is a constant solution of (10.1) if and only if $f(\omega) = 0$. Thus, since the potential energy has been defined as

$$\varphi(u) := \int_0^u f,$$

we find that $\varphi'(u) = f(u)$ for all $u \in \mathbb{R}$. Hence, the equilibria of (10.1), are the critical points of the potential energy. Naturally, the equilibria of the associated planar system (10.2) are of the form $(\omega, 0)$, where ω is a critical point of φ.

In this section, assuming that all critical points of φ are isolated, we show how the qualitative behavior of the solutions of (10.2) near an equilibrium, $(\omega, 0)$, is closely related to the local nature of the potential $\varphi(u)$ at its critical point ω. There are three possibilities, which are illustrated in Figure 10.1:

- ω is a local minimum;
- ω is a local maximum;
- ω is an inflection point with horizontal tangent, which entails

$$\varphi'(\omega) = \varphi''(\omega) = 0.$$

Subsequently, we construct the local phase diagrams in a neighborhood of $(\omega, 0)$ in each of these situations.

Case A. Strict local minimum. Suppose that ω is a strict local minimum of $\varphi(u)$, as illustrated in the left picture in Figure 10.1. Then, if the total energy E_0 satisfies $E_0 < \varphi(\omega)$, by continuity, there exists $\varepsilon > 0$ such that

$$E_0 < \varphi(u) \quad \text{for all } u \in J_\varepsilon := [\omega - \varepsilon, \omega + \varepsilon].$$

Thus, in this neighborhood, $\mathcal{C}_{(u_0, v_0)} = \emptyset$. Indeed, by (10.5), $\varphi(u) \leq E_0$ in J_ε is a necessary condition for $\mathcal{C}_{(u_0, v_0)}$ to be non-empty.

Next, suppose that $E_0 = \varphi(\omega)$. Then, as is clear from the left picture in Figure 10.1 and (10.5), $\mathcal{C}_{(u_0, v_0)} = \{(\omega, 0)\}$ in a neighborhood of $(\omega, 0)$, i.e. the integral curve consists of the equilibrium $(\omega, 0)$.

Finally, suppose that $E_0 > \varphi(\omega)$, as illustrated in the upper plot in Figure 10.2. Then, in a neighborhood of ω, the set of values of

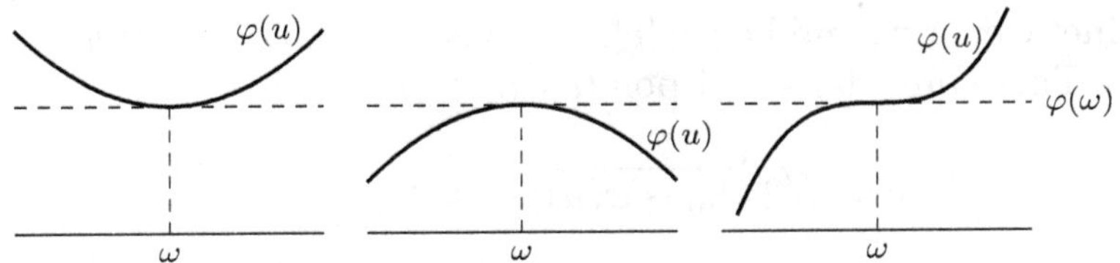

Fig. 10.1. The three possible types of isolated critical points of $\varphi(u)$.

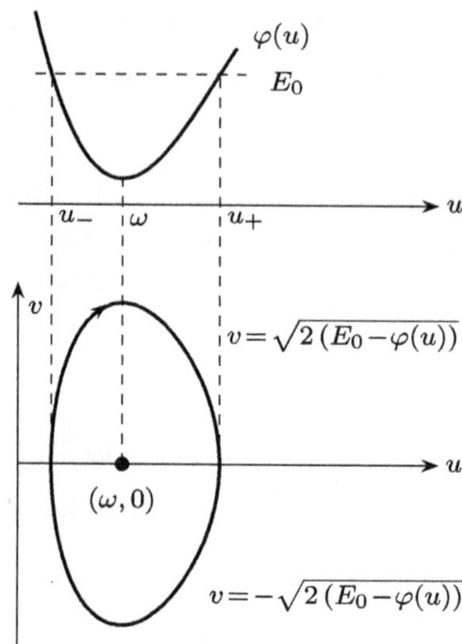

Fig. 10.2.   Integral curve corresponding to a strict local minimum of the potential energy, with $E_0 > \varphi(\omega)$.

u such that $\varphi(u) \le E_0$ is the closed interval $[u_-, u_+]$, where $u_\pm$ are characterized by the fact that

$$\varphi(u_\pm) = E_0 \quad \text{and} \quad u_- < \omega < u_+$$

(see Figure 10.2). At $u_\pm$, the potential energy equals the total energy. Thus, the kinetic energy must vanish, and we conclude that $v = 0$. Hence, $(u_-, 0)$ and $(u_+, 0)$ provide us with two points of the integral curve (10.5). As u increases from u_- to ω, $\varphi(u)$ decreases. Consequently, the kinetic energy, and hence $|v|$, grows until it reaches its maximum value at $u = \omega$; recall that the total energy is constant, and we are assuming that the potential energy is minimal at ω. As $u \in (\omega, u_+)$ increases, the potential energy increases. Thus, the kinetic energy, and hence $|v|$, decreases, until u reaches u_+, where $v = 0$. Therefore, the set of points (u, v) for which

$$v = \sqrt{2\,(E_0 - \varphi(u))}, \quad u \in [u_-, u_+],$$

is the upper arc of curve plotted for $v \ge 0$ in the lower picture in Figure 10.2. The lower arc corresponds to its reflection across the

u-axis:

$$v = -\sqrt{2\left(E_0 - \varphi(u)\right)}, \quad u \in [u_-, u_+].$$

Consequently, $\mathcal{C}_{(u_0,v_0)}$ is the closed curve shown in the lower picture in Figure 10.2. As no equilibrium lies on it, according to Property 3 in Section 9.2, it is the orbit of a non-trivial periodic solution. Since $u' = v$, u increases if $v > 0$, while it decreases if $v < 0$. Therefore, this closed orbit is traveled clockwise, as indicated by the arrow in Figure 10.2.

As this analysis can be repeated for any value of E_0 greater than and sufficiently close to $\varphi(\omega)$, it becomes apparent that the local phase portrait of (10.2), or (10.1), in a neighborhood of $(\omega, 0)$, is the one of a center, as shown in Figure 10.3. Therefore, *any strict local minimum of the potential energy $\varphi(u)$ provides us with a center in the phase plane.*

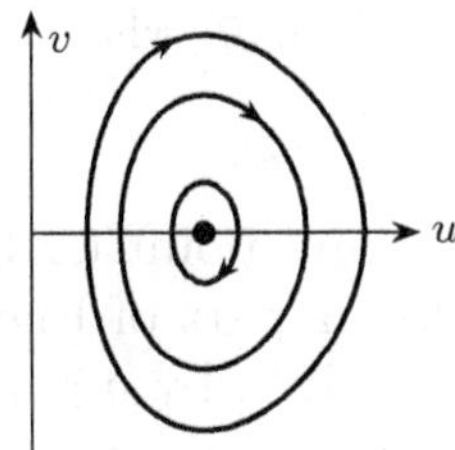

Fig. 10.3. Local phase portrait at $(\omega, 0)$, when ω is a strict local minimum of the potential energy.

Case B. Strict local maximum. Next, we assume that ω is a strict local maximum of $\varphi(u)$, as sketched in the middle picture in Figure 10.1. If $E_0 = \varphi(\omega)$, then, for every $u \in [\alpha, \beta]$, which is the interval where we have plotted $\varphi(u)$ in Figure 10.4, $(u, v) \in \mathcal{C}_{(u_0,v_0)}$ if and only if $v = \pm v(u)$, with

$$v(u) := \sqrt{2\left(E_0 - \varphi(u)\right)}, \quad u \in [\alpha, \beta].$$

In particular, $(\omega, 0) \in \mathcal{C}_{(u_0,v_0)}$ because $v(\omega) = 0$. Thus, in this case, the equilibrium $(\omega, 0)$ belongs to the integral curve $\mathcal{C}_{(u_0,v_0)}$. Besides $(\omega, 0)$, the integral curve consists of four additional trajectories. Indeed, as $\varphi(u)$ increases for $u \in [\alpha, \omega]$, for the same values of u, $v(u)$ decreases until it equals 0 at $u = \omega$. Then, as $\varphi(u)$ decreases for $u \in [\omega, \beta]$, $v(u)$ increases for the same values of u. Thus, the integral

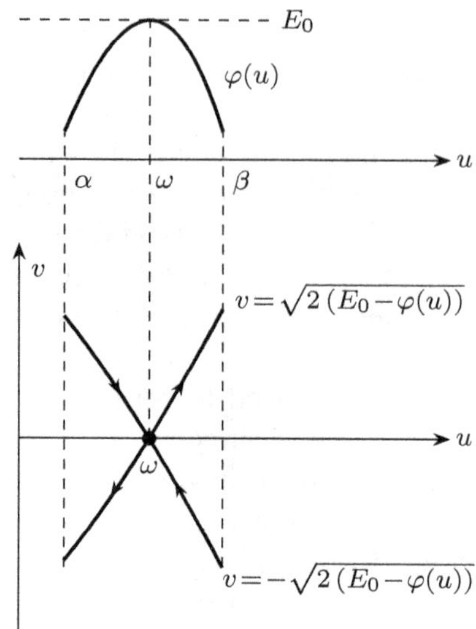

Fig. 10.4. Integral curve through $(\omega,0)$ when ω is a strict local maximum of the potential energy.

curve consists of the equilibrium point $(\omega,0)$ plus four arcs of curve like those sketched in the lower part of Figure 10.4. The upper arcs correspond to the values of v given by $v=v(u)$, while the lower ones are their reflections across the u-axis, i.e. those associated with the values $v=-v(u)$.

By the properties in Section 9.2, in this particular case, the integral curve $\mathcal{C}_{(u_0,v_0)}$ consists of five different types of trajectories. Two of them are the trajectories of the solutions approximating $(\omega,0)$ as $t\uparrow+\infty$, while another two are the trajectories of the solutions approximating $(\omega,0)$ as $t\downarrow-\infty$, and the fifth one is the equilibrium point. This is a paradigmatic example illustrating how an equilibrium on an integral curve divides it into several trajectories.

Figure 10.5 sketches the construction of the corresponding integral curves when $E_0>\varphi(\omega)$. As in the previous cases, we represent the set of points $(u,\pm v(u))$ for every $u\in[\alpha,\beta]$.

Since $\varphi(u)$ increases for $u\in[\alpha,\omega]$, $v(u)$ is decreasing in that interval. Similarly, since $\varphi(u)$ decreases for $u\in[\omega,\beta]$, $v(u)$ is increasing in such an interval. Thus, the graph of $v(u)$ looks like the upper curve represented in the lower picture in Figure 10.5. As it is a compact arc of integral curve without any equilibria on it, Property 2 in

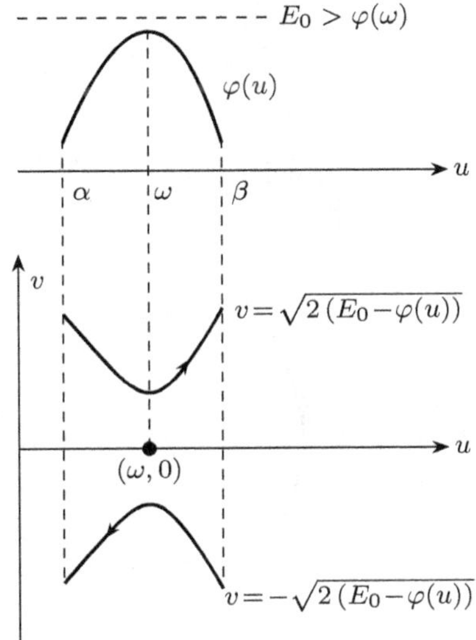

Fig. 10.5. Integral curve corresponding to a strict local maximum of the potential energy, when $E_0 > \varphi(\omega)$.

Section 9.2 ensures that it is the trajectory of any solution of (10.3) starting on it. Since $u' = v > 0$ in the upper half-plane, $u(t)$ increases with time, which explains the direction of the arrow in Figure 10.5, indicating the direction of the dynamics. Naturally, in this case, the second trajectory of the integral curve is the reflection of the previous one across the u-axis, which is the lower arc of trajectory plotted in Figure 10.5, and it is traveled from right to left since, in this case, $u' = v < 0$ because this trajectory lies in the lower half-plane.

Naturally, the closer E_0 is to $\varphi(\omega)$, the closer these arcs are to the integral curve sketched in Figure 10.4. Actually, they perturb from it as $E_0 > \varphi(\omega)$ separates away from $\varphi(\omega)$.

Finally, suppose that

$$\max\{\varphi(\alpha), \varphi(\beta)\} < E_0 < \varphi(\omega),$$

as illustrated in the upper plot in Figure 10.6. Then, $\varphi(u) \leq E_0$ if and only if

$$u \in J := [\alpha, u_-] \cup [u_+, \beta],$$

where $u_\pm$ are the unique values of $u \in [\alpha, \beta]$ such that

$$\varphi(u_\pm) = E_0 \quad \text{with} \quad u_- < \omega < u_+.$$

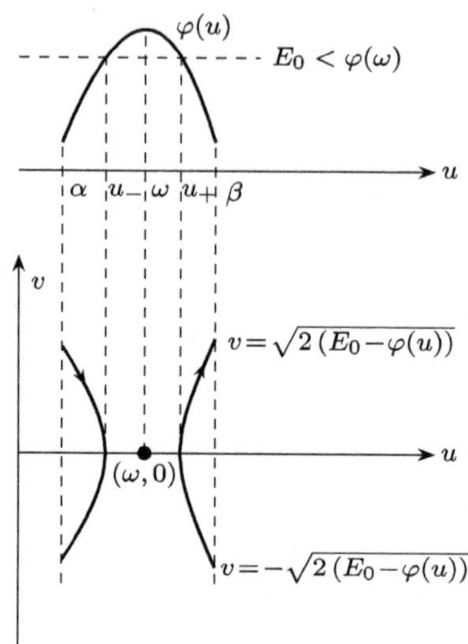

Fig. 10.6. Integral curve corresponding to a strict local maximum of the potential energy, when $E_0 < \varphi(\omega)$.

Thus, $(u_\pm, 0) \in \mathcal{C}_{(u_0, v_0)}$. Since $\varphi(u)$ increases for $u \in [\alpha, u_-]$, $v(u)$ decreases in this interval, from the value $v(\alpha)$ to $v(u_-) = 0$. Similarly, since $\varphi(u)$ decreases for $u \in [u_+, \beta]$, $v(u)$ increases in this interval, from $v(u_+) = 0$ up to reach the value $v(\beta) > 0$. Therefore, in this case, the integral curve $\mathcal{C}_{(u_0, v_0)}$ consists of the two arcs $(u, \pm v(u))$ and $u \in J$, which are plotted in the lower part of Figure 10.6. As any arc of the integral curve not containing any equilibrium must be traveled in a finite time, as established by Property 2 in Section 9.2, these two arcs are the trajectories of any solution of (10.3) starting on each of them. As usual, the direction of the arrows indicates the direction of the dynamics: since $u' = v$, u increases if $v > 0$, while it decreases if $v < 0$.

As above, these two curves perturb from the integral curve represented in Figure 10.4 as $E_0 < \varphi(\omega)$ separates away from $\varphi(\omega)$, in the sense that the closer E_0 is to $\varphi(\omega)$, the closer are the corresponding integral curves.

By bringing together the different types of trajectories that we have found in the previous analysis, it becomes apparent that *any*

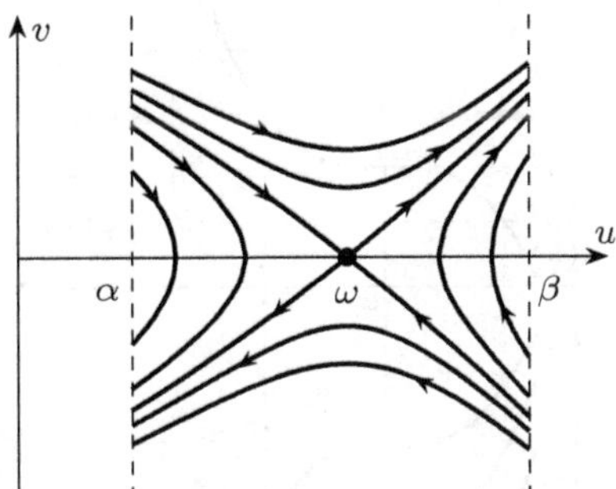

Fig. 10.7. Local phase portrait at $(\omega, 0)$, when ω is a strict local maximum of the potential energy.

strict local maximum of the potential energy provides us with a saddle point of the phase portrait of (10.2), or (10.1), as sketched in Figure 10.7.

Case C. Inflection point. Finally, suppose that ω is an inflection point of $\varphi(u)$ as the one sketched in the right picture in Figure 10.1, i.e.

$$\varphi(u) \begin{cases} < \varphi(\omega) & \text{if } u \in [\alpha, \omega), \\ > \varphi(\omega) & \text{if } u \in (\omega, \beta], \end{cases}$$

and

$$\varphi'(\omega) = \varphi''(\omega) = 0.$$

In this case, for every $E_0 \in (\varphi(\alpha), \varphi(\beta))$, there exists a unique $u_0 \in (\alpha, \beta)$ such that $\varphi(u_0) = E_0$, and $\varphi(u) \leq E_0$ if and only if $u \in [\alpha, u_0]$. Moreover,

$$u_0 \begin{cases} < \omega & \text{if } E_0 < \varphi(\omega), \\ = \omega & \text{if } E_0 = \varphi(\omega), \\ > \omega & \text{if } E_0 > \varphi(\omega). \end{cases}$$

The scheme sketched in Figure 10.8, where we have plotted three representative integral curves, together with the trajectories that conform them, provides us with the local phase portrait of (10.2) at these critical points, which are referred to as *degenerate saddle*

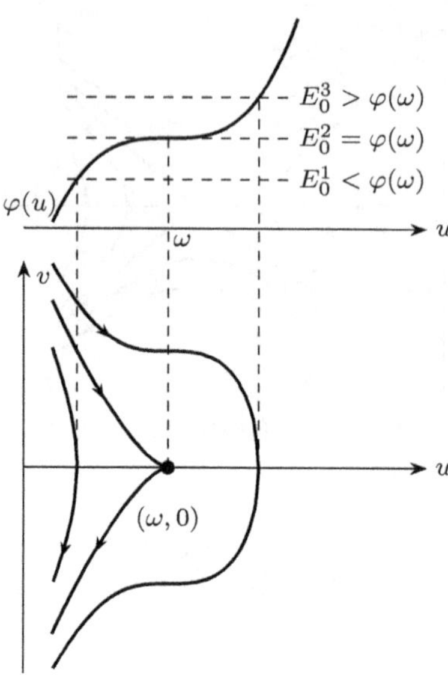

Fig. 10.8. Local phase portrait at an inflection point of the potential energy like the one represented in the right plot of Figure 10.1.

points in this book because two of the previous half-branches of the stable and unstable manifolds of a genuine saddle point are lost for these critical points. Another way of looking at these critical points is as being generated by the collision of a saddle and a center.

Note that, when $\varphi(\omega) = E_0$, we have that

$$\varphi(u) = E_0 + \varphi'(\omega)(u - \omega) + \frac{\varphi''(\omega)}{2}(u - \omega)^2 + o(|u - \omega|^2) \quad \text{as } u \to \omega.$$

Thus, since $\varphi'(\omega) = \varphi''(\omega) = 0$,

$$\varphi(u) = E_0 + o(|u - \omega|^2) \quad \text{as } u \to \omega$$

and, hence,

$$v(u) = \pm\sqrt{2\left(E_0 - \varphi(u)\right)} = \pm|u - \omega|\, o(1) \quad \text{as } u \to \omega.$$

Therefore, $v'(\omega) = 0$. So, the integral curve through $(\omega, 0)$ must be tangent to the u-axis in the phase plane, as shown in Figure 10.8.

10.3 The Simple Gravity Pendulum

As explained in Section 6.4, the simple gravity pendulum is governed by the Newtonian equation

$$u'' + \frac{G}{L} \sin u = 0, \tag{10.6}$$

whose solutions are globally defined in time $t \in \mathbb{R}$, regardless the values of $t_0 \in \mathbb{R}$ and of the initial conditions

$$u(t_0) = u_0, \quad v(t_0) := u'(t_0) = v_0.$$

In other words, for every $(u_0, v_0) \in \mathbb{R}^2$, the maximal solution of the associated Cauchy problem, (I, u), satisfies $I = (-\infty, +\infty)$. The equivalent first-order system is

$$\begin{cases} u' = v, \\ v' = -\frac{G}{L} \sin u. \end{cases}$$

Thus, taking the cross product, we find that, for every $t \in \mathbb{R}$,

$$v(t)v'(t) + \frac{G}{L} u'(t) \sin u(t) = 0.$$

Equivalently,

$$\frac{d}{dt} \left(\frac{v^2(t)}{2} - \frac{G}{L} \cos u(t) \right) = 0, \quad t \in \mathbb{R}.$$

Thus, the trajectory of the solution through $(u_0, v_0) \in \mathbb{R}^2$ satisfies

$$\frac{v^2(t)}{2} - \frac{G}{L} \cos u(t) = \frac{v_0^2}{2} - \frac{G}{L} \cos u_0 \quad \text{for all } t \in \mathbb{R}. \tag{10.7}$$

In particular,

$$\varphi(u) := -\frac{G}{L} \cos u, \quad u \in \mathbb{R},$$

can be seen as the potential energy. In terms of $\varphi(u)$, (10.7) reads

$$\frac{v^2(t)}{2} + \varphi(u(t)) = \frac{v_0^2}{2} + \varphi(u_0) \quad \text{for all } t \in \mathbb{R}.$$

Thus,

$$v(t) = \pm\sqrt{2\left(E_0 - \varphi(u(t))\right)} \quad \text{for all } t \in \mathbb{R},$$

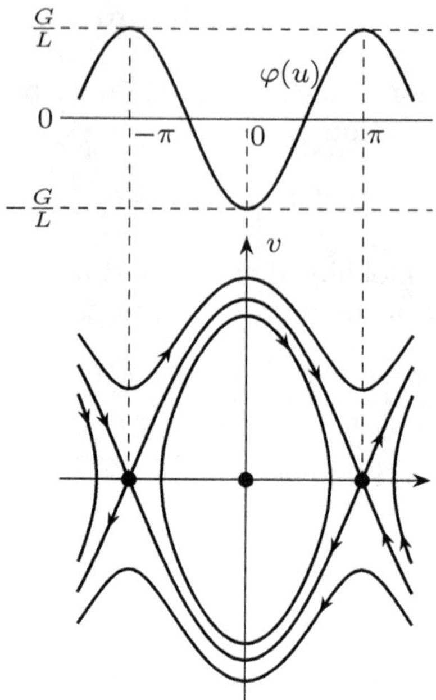

Fig. 10.9. Phase portrait of the free pendulum, obtained from the corresponding potential energy.

where

$$E_0 := \frac{v_0^2}{2} + \varphi(u_0) = \frac{v_0^2}{2} - \frac{G}{L}\cos u_0.$$

The upper picture in Figure 10.9 plots the potential energy $\varphi(u)$, while the lower one shows the corresponding phase portrait.

The set of equilibria of this system is given by

$$\mathcal{E} := \{(n\pi, 0)\colon n \in \mathbb{Z}\},$$

whose first components are the critical points of the potential energy. From a physical point of view, being an equilibrium means that the pendulum remains in the same rest position if the hanging mass is set with zero velocity and vertically, so that gravity is parallel to the rod.

When $n = 2m$ for some $m \in \mathbb{Z}$, then $(2m\pi, 0)$ is a strict local minimum of $\varphi(u)$ and, hence, by the analysis in Section 10.2, it gives rise, locally, to a center in the phase plane (u, v). When $n = 2m + 1$

for some $m \in \mathbb{Z}$, instead, $((2m+1)\pi, 0)$ is a strict local maximum of $\varphi(u)$. So, these points are, locally, saddle points in the phase plane (u, v), as sketched in the lower picture in Figure 10.9.

We now analyze the different trajectories in the phase portrait sketched in Figure 10.9. For $E_0 = -\frac{G}{L}$, the integral curve is given by the union of the equilibria $(2m\pi, 0)$, with $m \in \mathbb{Z}$; thus, the associated trajectories are the constant solutions corresponding to each of these equilibria.

For $E_0 = \frac{G}{L}$, the integral curve contains the equilibria of saddle type $((2m+1)\pi, 0)$, $m \in \mathbb{Z}$, as well as the stable and unstable manifolds emanating from each of them. Observe that one of the trajectories of the unstable manifold of each saddle point provides us with half of the stable manifold of the next saddle point, while the other trajectory of its unstable manifold provides us with half of the stable manifold of the previous saddle point. These *heteroclinic connections* between two consecutive saddle points cannot be maintained in *dissipative dynamical systems*, i.e. systems where the total (mechanical) energy decreases, for example because part of it is transformed into heat by the action of friction. This feature will be discussed in Section 11.8.

For $E_0 > \frac{G}{L}$, the integral curve consists of two trajectories: one with $v > 0$ and another one with $v < 0$. Along the former, since $u' = v > 0$, the solution u is increasing, while u is decreasing along the latter. Physically, these trajectories correspond to the case in which the pendulum always rotate in the same direction.

Finally, if the total energy, E_0, satisfies

$$E_0 = \frac{v_0^2}{2} - \frac{G}{L}\cos u_0 \in \left(-\frac{G}{L}, \frac{G}{L}\right),$$

then, the trajectories are simple closed curves around every equilibrium position $u = 2m\pi$, for all $m \in \mathbb{Z}$. In this case, the pendulum oscillates periodically around each of these equilibria, changing the direction of oscillation, which is reflected by the fact that v (periodically) changes sign along these trajectories. It should be noted that, since $2m\pi \equiv 0 \pmod{2\pi}$ for all $m \in \mathbb{Z}$, all these equilibria are equivalent from a physical perspective. Thus, we consider, without loss of generality, the case $m = 0$.

The amplitude of these periodic oscillations, defined as the maximal separation of the solution from the equilibrium position $u = 0$,

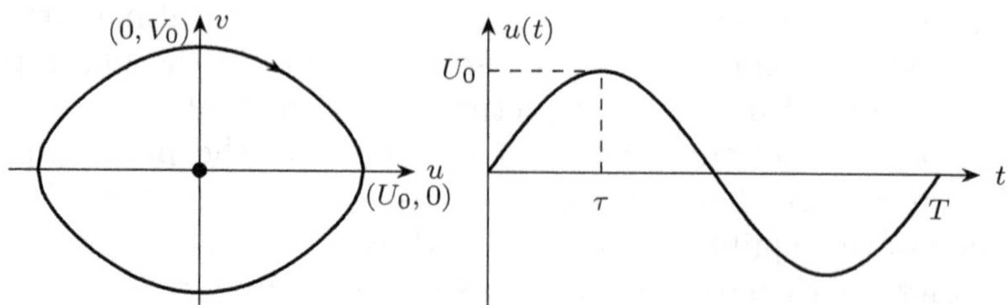

Fig. 10.10. A periodic orbit of the free pendulum around the equilibrium $(0,0)$ (left) and the graph of the corresponding solution $u(t)$ (right).

equals the u-component, U_0, of the point $(U_0, 0)$ on the closed orbit represented in the left picture in Figure 10.10. The right picture plots the graph of the unique solution of (10.6), $u(t)$, such that

$$u(0) = 0, \quad u'(0) = V_0 > 0.$$

This solution needs a time τ to leave the v-axis in the phase plane at $(0, V_0)$ and reach the u-axis, for the first time, at $(U_0, 0)$. Thus,

$$u(\tau) = U_0, \quad v(\tau) = u'(\tau) = 0.$$

As the total energy at time t,

$$E(u(t), v(t)) := \frac{v^2(t)}{2} - \frac{G}{L} \cos u(t),$$

is constant for all time $t \in \mathbb{R}$, it becomes apparent that

$$E_0 = E(U_0, 0) = E(0, V_0).$$

In other words,

$$E_0 = -\frac{G}{L} \cos U_0 = \frac{V_0^2}{2} - \frac{G}{L}.$$

Thus, the amplitude of the movement, U_0, can be expressed in terms of the total energy, E_0, or of the maximal velocity, V_0, as

$$U_0 = \arccos\left(-\tfrac{L}{G} E_0\right) = \arccos\left(1 - \tfrac{L}{G} \tfrac{V_0^2}{2}\right).$$

In particular, it increases with respect to the total energy, E_0.

The minimal period T of these solutions is the total time taken by $(u(t), v(t))$ to give a complete round along its orbit and return to the initial point of the phase portrait for the first time. As a direct consequence of the symmetries of (10.6), specifically the fact that the nonlinearity

$$f(u) := \frac{G}{L} \sin u$$

is odd, i.e. $f(-u) = -f(u)$ for all $u \in \mathbb{R}$, we show that

$$T = 4\tau. \tag{10.8}$$

The first step to prove (10.8) relies on the fact that the integral curves of Newtonian equations are symmetric across the u-axis. As established at the end of Section 10.1, this entails that the time τ taken by the solution

$$(u(t), v(t)) := (u(t; 0, V_0), v(t; 0, V_0))$$

to reach $(U_0, 0)$ starting at $(0, V_0)$ in the phase portrait equals the time taken to reach $(0, -V_0)$ starting from $(U_0, 0)$. The following result shows that, actually, in $[\tau, 2\tau]$, $u(t)$ is the even extension of $u(t)$ in $[0, \tau]$.

Lemma 10.1. *Let $u(t)$ be the unique solution of (10.6) such that $u(0) = 0$ and $u'(0) = V_0$, with*

$$E_0 = \frac{V_0^2}{2} - \frac{G}{L} \in \left(-\frac{G}{L}, \frac{G}{L}\right).$$

Let $\tau > 0$ be the minimal time for which $v(\tau) = u'(\tau) = 0$. Then,

$$u(t) = u(2\tau - t) \quad \text{for all } t \in [\tau, 2\tau]. \tag{10.9}$$

In practice, Lemma 10.1 establishes that the graph of $u(t)$ in $[\tau, 2\tau]$ is the reflection across $t = \tau$ of its graph in $[0, \tau]$. An immediate consequence of this fact is that the interior zero of $u(t)$ in the right picture in Figure 10.10 occurs at $t = 2\tau$, and

$$v(2\tau) = u'(2\tau) = -u'(0) = -V_0.$$

Proof of Lemma 10.1. Consider the auxiliary function $\tilde{u}(t)$, defined as

$$\tilde{u}(t) := \begin{cases} u(t) & \text{if } t \in [0, \tau], \\ u(2\tau - t) & \text{if } t \in (\tau, 2\tau], \end{cases}$$

which is obtained as the extension to the interval $[0, 2\tau]$ of $u|_{[0,\tau]}$ by reflection across $t = \tau$. By definition,

$$\tilde{u}(0) = u(0) = 0, \quad \tilde{u}'(0) = u'(0) = V_0.$$

Moreover,

$$\lim_{t \uparrow \tau} \tilde{u}(t) = \lim_{t \uparrow \tau} u(t) = u(\tau) = U_0,$$

$$\lim_{t \downarrow \tau} \tilde{u}(t) = \lim_{t \downarrow \tau} u(2\tau - t) = u(\tau) = U_0.$$

This shows that $\tilde{u}$ is continuous at τ and, hence, $\tilde{u} \in \mathcal{C}([0, 2\tau]; \mathbb{R})$. In addition, differentiating with respect to t gives

$$\tilde{u}'(t) = \begin{cases} u'(t) & \text{if } t \in [0, \tau], \\ -u'(2\tau - t) & \text{if } t \in (\tau, 2\tau]. \end{cases}$$

Consequently,

$$\lim_{t \uparrow \tau} \tilde{u}'(t) = \lim_{t \uparrow \tau} u'(t) = u'(\tau) = 0,$$

$$\lim_{t \downarrow \tau} \tilde{u}'(t) = -\lim_{t \downarrow \tau} u'(2\tau - t) = -u'(\tau) = 0.$$

So, $\tilde{u}'$ is also continuous at τ and, hence, $\tilde{u} \in \mathcal{C}^1([0, 2\tau]; \mathbb{R})$. Finally, since

$$\tilde{u}''(t) = \begin{cases} u''(t) & \text{if } t \in [0, \tau], \\ u''(2\tau - t) & \text{if } t \in (\tau, 2\tau], \end{cases}$$

and $-u'' = f(u)$, it becomes apparent that, due to the continuity of $f(u)$,

$$\lim_{t \uparrow \tau} \tilde{u}''(t) = \lim_{t \uparrow \tau} u''(t) = -\lim_{t \uparrow \tau} f(u(t)) = -f(U_0).$$

Similarly,

$$\lim_{t \downarrow \tau} \tilde{u}''(t) = \lim_{t \downarrow \tau} u''(2\tau - t) = -\lim_{t \downarrow \tau} f(u(2\tau - t)) = -f(U_0).$$

Therefore, $\tilde{u} \in \mathcal{C}^2([0, 2\tau]; \mathbb{R})$, and by definition,

$$-\tilde{u}''(t) = f(\tilde{u}(t)) \quad \text{for all } t \in [0, 2\tau].$$

Since

$$(\tilde{u}(0), \tilde{u}'(0)) = (0, V_0) = (u(0), u'(0)),$$

by uniqueness, we infer that $\tilde{u}(t) = u(t)$ for all $t \in [0, 2\tau]$. So, (10.9) holds. This completes the proof. $\qquad\square$

By analyzing the proof of Lemma 10.1, it is easily realized that the specific value of the initial conditions $(u(0), u'(0))$ and of the specific nonlinearity f are not relevant. Actually, the following generalized result holds.

Lemma 10.2. *Let $f\colon \mathbb{R} \to \mathbb{R}$ be a locally Lipschitz function, and assume that $u(t)$ is a solution of $-u'' = f(u)$ satisfying the following:*

- *there exists $\tau > 0$ such that $u(t)$ is defined in $[0, 2\tau]$;*
- *$u'(\tau) = 0$ and $u'(t) \neq 0$ for all $t \in [0, \tau)$.*

Then, $u(t) = u(2\tau - t)$ for all $t \in [\tau, 2\tau]$.

In particular, Lemma 10.2 holds in a neighborhood of any equilibrium $\omega \in f^{-1}(0)$ where the potential energy has a minimum.

The key point that will allow us to obtain $T = 4\tau$ in the case of the pendulum is the oddity of $f(u)$, which entails that $u(t)$ solves (10.6) if and only if $-u(t)$ solves it. This will force the solution in $[2\tau, 4\tau]$ to be the odd extension of the solution in $[0, 2\tau]$, as the following result shows.

Lemma 10.3. *Suppose that $f\colon \mathbb{R} \to \mathbb{R}$ is locally Lipschitz and that, for every $u \in \mathbb{R}$, $f(-u) = -f(u)$. Let $u(t)$ be a T-periodic solution of $-u'' = f(u)$ oscillating around the equilibrium 0, and let τ be the necessary time to reach $(U_0, 0)$ for the first time starting from $(0, V_0)$, with $U_0, V_0 > 0$, along the orbit of $u(t)$. Then,*

$$u(t) = -u(4\tau - t) \quad \text{for all } t \in (2\tau, 4\tau]. \tag{10.10}$$

As a consequence, $T = 4\tau$.

Proof. In this case, $u(t)$ is determined by the initial conditions

$$u(0) = 0, \quad u'(0) = V_0 > 0.$$

Subsequently, we define the odd extension of $u(t)$ from $[0, 2\tau]$ to $[0, 4\tau]$:

$$\hat{u}(t) := \begin{cases} u(t) & \text{if } t \in [0, 2\tau], \\ -u(4\tau - t) & \text{if } t \in (2\tau, 4\tau]. \end{cases}$$

By reasoning as in the proof of Lemma 10.1, it can be shown that $\hat{u} = u$ in $\mathbb{R}$. Moreover, it follows from the definition of τ and Lemma 10.1 that

$$u(\tau) = U_0, \quad u'(\tau) = 0, \quad u(2\tau) = 0, \quad u'(2\tau) = -V_0.$$

Thus, (10.10) entails that $T = 4\tau$ is the period of $u(t)$ since the necessary time to reach $(0, -V_0)$ from $(0, V_0)$ equals the time taken to reach $(0, V_0)$ from $(0, -V_0)$. $\square$

Naturally, the period of the solution, T, depends on the maximal (initial) velocity V_0, or, equivalently, on the amplitude of the solution:

$$U_0 = u(\tau) = \arccos\left(1 - \frac{L}{G}\frac{V_0^2}{2}\right).$$

Observe that, as V_0 increases from 0 to $2\sqrt{\frac{G}{L}}$, U_0 increases from 0 to π. This is also apparent from the phase portrait sketched in Figure 10.9.

In order to determine T as a function of U_0 (or V_0), we can proceed as follows. Since $u(t)$ is increasing in $[0, \tau]$, (10.7) implies that

$$\tau(U_0) = \tau = \int_0^\tau dt = \int_0^\tau \frac{u'(t)}{v(t)}\, dt = \int_0^\tau \frac{u'(t)}{\sqrt{2\left(\varphi(U_0) - \varphi(u(t))\right)}}\, dt$$

$$= \int_0^{U_0} \frac{d\xi}{\sqrt{2\left(\varphi(U_0) - \varphi(\xi)\right)}} = \sqrt{\frac{L}{2G}} \int_0^{U_0} \frac{d\xi}{\sqrt{\cos\xi - \cos U_0}}.$$

The change of variable $\xi = U_0\theta$ then gives

$$\tau(U_0) = \sqrt{\frac{L}{2G}} \int_0^1 \frac{U_0\, d\theta}{\sqrt{\cos(U_0\theta) - \cos U_0}}, \quad U_0 \in (0, \pi). \quad\quad (10.11)$$

Moreover, since

$$\cos(U_0\theta) - \cos U_0 = 1 - \frac{U_0^2}{2}\theta^2 + \frac{U_0^4}{4!}\theta^4 - \cdots - \left(1 - \frac{U_0^2}{2} + \frac{U_0^4}{4!} - \cdots\right)$$

$$= \frac{U_0^2}{2}(1 - \theta^2) - \frac{U_0^4}{4!}(1 - \theta^4) + O(U_0^6)$$

$$= \frac{U_0^2}{2}\left(1 - \theta^2 + O(U_0^2)\right) \quad \text{as} \quad U_0 \downarrow 0,$$

it readily follows from (10.11) that

$$\lim_{U_0 \downarrow 0} \tau(U_0) = \sqrt{\frac{L}{G}} \int_0^1 \frac{d\theta}{\sqrt{1 - \theta^2}} = \sqrt{\frac{L}{G}}\frac{\pi}{2}.$$

Therefore, the limiting period of the small oscillations of the simple gravity pendulum around its equilibrium position $u = 0$ is given by

$$\lim_{U_0 \downarrow 0} T(U_0) = 4 \lim_{U_0 \downarrow 0} \tau(U_0) = 2\pi\sqrt{\frac{L}{G}}, \qquad (10.12)$$

which coincides with the period of the solutions of the linear equation

$$u'' = -\frac{G}{L}u.$$

Observe that this equation is obtained from (10.6) by replacing $\sin u$ with u, an approximation that is good for small values of u since

$$\lim_{u \downarrow 0} \frac{\sin u}{u} = 1.$$

In particular, we have shown that the behavior of the nonlinear center $(0,0)$ associated with (10.6) mimics the behavior of the linear center associated with $u'' = -\frac{G}{L}u$ in a neighborhood of $(0,0)$.

Observe that both the limiting period (10.12) and the period of general, not necessarily small, oscillations given by (10.11) are increasing with respect to L, which reflects the physical evidence that the period of the simple gravity pendulum is increasing with respect to the length of the rod.

Moreover, by Property 4 in Section 9.2 and Theorem 5.26, it becomes apparent that

$$\lim_{U_0 \uparrow \pi} T(U_0) = +\infty$$

because the periodic solutions approximate the heteroclinic connections between the saddle points $(-\pi, 0)$ and $(\pi, 0)$.

The fact that $T(U_0)$ is a continuous function of $U_0 \in (0, \pi)$ follows, not only from Theorem 5.26 but also from the explicit expression of $T(U_0)$ given in (10.11).

10.4   A Generalized Diffusive Logistic Equation

This section analyzes the diffusive equation of logistic type

$$-u'' = \lambda u - mu^3,$$

where $\lambda > 0$ is regarded as a parameter and $m > 0$ is a constant. Since the change of variable $u := U/\sqrt{m}$ transforms this equation into

$$-U'' = \lambda U - U^3,$$

we consider the second-order differential equation

$$-u'' = \lambda u - u^3. \tag{10.13}$$

The main goal of this section is to construct all non-zero solutions of the associated boundary value problem

$$\begin{cases} -u'' = \lambda u - u^3 & \text{in } (0, L), \\ u(0) = u(L) = 0, \end{cases} \tag{10.14}$$

for every $L > 0$. Observe that these solutions provide us with the steady states of a generalized diffusive logistic equation. The term "generalized" is used since, unlike the classical diffusive logistic equation discussed in Section 6.7, the nonlinear term considered here is $-u^3$, instead of $-u^2$, in order to simplify the analysis. Indeed, in this way, the underlying nonlinearity $f(u) := \lambda u - u^3$, $u \in \mathbb{R}$, is odd, and this will be exploited similarly to the previous section.

Since f is locally Lipschitz with respect to u, uniformly in $t \in \mathbb{R}$, because $f \in \mathcal{C}^\infty(\mathbb{R}; \mathbb{R})$ is autonomous, Theorem 6.7 ensures that, for every $t_0 \in \mathbb{R}$ and $(u_0, v_0) \in \mathbb{R}^2$, the Cauchy problem

$$\begin{cases} -u'' = \lambda u - u^3, \\ u(t_0) = u_0, \quad u'(t_0) = v_0, \end{cases} \tag{10.15}$$

possesses a unique maximal solution, (I, u), with $I = (T_{\min}, T_{\max})$, for some $T_{\min} \in [-\infty, t_0)$ and $T_{\max} \in (t_0, +\infty]$. Moreover, by Theorem 6.9, the solution blows up at time $T_{\min}$ (respectively, $T_{\max}$) if $T_{\min} > -\infty$ (respectively, $T_{\max} < +\infty$). As usual, (10.15) is equivalent to the first-order system

$$\begin{cases} u' = v, \\ v' = -\lambda u + u^3, \\ u(t_0) = u_0, \quad v(t_0) = v_0, \end{cases} \tag{10.16}$$

whose maximal solution is (I, u, u').

10.4.1 *Construction of the phase portrait*

By taking the cross product in (10.16), we obtain that, for every $t \in I$,

$$v(t)v'(t) + \lambda u(t)u'(t) - u^3(t)u'(t) = 0.$$

Equivalently,

$$\frac{d}{dt}\left(\frac{v^2(t)}{2} + \lambda \frac{u^2(t)}{2} - \frac{u^4(t)}{4} \right) = 0, \quad t \in I.$$

Thus, the solution of (10.16) satisfies

$$\frac{v^2(t)}{2} + \lambda \frac{u^2(t)}{2} - \frac{u^4(t)}{4} = \frac{v_0^2}{2} + \lambda \frac{u_0^2}{2} - \frac{u_0^4}{4} \quad \text{for all } t \in I. \tag{10.17}$$

In particular,

$$\varphi(u) := \lambda \frac{u^2}{2} - \frac{u^4}{4}, \quad u \in \mathbb{R},$$

can be chosen as the potential energy. In terms of $\varphi(u)$, (10.17) reads

$$\frac{v^2(t)}{2} + \varphi(u(t)) = \frac{v_0^2}{2} + \varphi(u_0) \quad \text{for all } t \in I.$$

Thus, as in the example in Section 10.3,

$$v(t) = \pm\sqrt{2\left(E_0 - \varphi(u(t))\right)} \quad \text{for all } t \in I,$$

where

$$E_0 := \frac{v_0^2}{2} + \varphi(u_0) = \frac{v_0^2}{2} + \lambda\frac{u_0^2}{2} - \frac{u_0^4}{4}$$

is the total energy, which is conserved along the trajectory. Since

$$\varphi'(u) = \lambda u - u^3 = f(u),$$

the set of equilibria of the first-order system associated with (10.13) is

$$\mathcal{E} := \left\{ (-\sqrt{\lambda}, 0), (0, 0), (\sqrt{\lambda}, 0) \right\}.$$

Moreover, since $\varphi''(u) = \lambda - 3u^2$ and

$$\varphi''(0) = \lambda > 0, \quad \varphi''(\pm\sqrt{\lambda}) = \lambda - 3\lambda = -2\lambda < 0,$$

it becomes apparent that 0 is a strict local minimum of φ, while $\pm\sqrt{\lambda}$ are strict local maxima. Thus, since φ is an even function vanishing at 0 (a double root) and at $\pm\sqrt{2\lambda}$ (simple roots), the graph of $\varphi(u)$ looks like that shown in the upper plot of Figure 10.11. Hence, the phase portrait of (10.13) is "eye-shaped", as represented in the lower part of the figure.

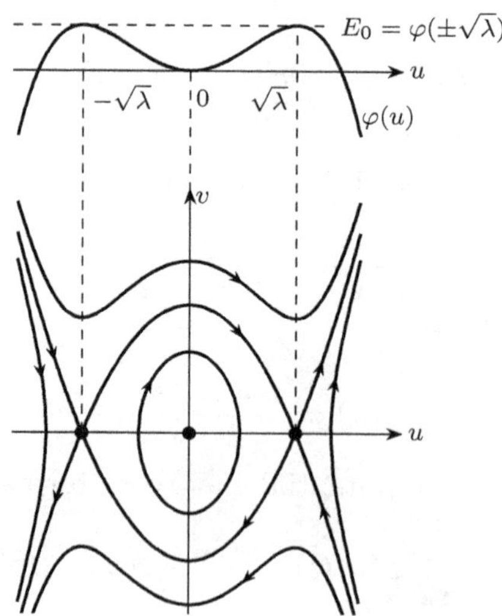

Fig. 10.11. Potential energy and phase portrait of (10.13).

This phase portrait is reminiscent of the one found for the free pendulum, except for the lack of periodicity. This can be explained by observing that, for the special choice of $\lambda = 1$ and $m = \frac{1}{6}$, the function $\lambda u - m u^3$ gives an approximation of $\sin u$ in a neighborhood of $u = 0$ up to terms of order five. Naturally, the phase portraits, as well as the nature of their solutions, are substantially different far away from the origin. For example, as will become apparent later, the lateral trajectories for $E_0 < \varphi(\pm\sqrt{\lambda})$ correspond to large solutions of (10.13) in the interval $(-L, L)$ for certain values of $L > 0$ (cf. Section 6.5.2), while they are pieces of periodic orbits in the simple gravity pendulum.

When $E_0 = \varphi(\pm\sqrt{\lambda})$, the integral curve through (u_0, v_0) consists of eight different types of trajectories: the two equilibria of saddle type $(\pm\sqrt{\lambda}, 0)$, two heteroclinic connections between them, and four additional trajectories consisting of the parts of the stable and unstable manifolds of these saddle points that do not lie on their heteroclinic connections. When $E_0 \in (0, \varphi(\pm\sqrt{\lambda}))$, the integral curve consists of three trajectories: the central one, which is the orbit of a periodic solution, and the lateral ones, which are the trajectories described by two large solutions, as will become apparent later. When $E_0 = 0$, the previous central orbit shrinks to the origin, providing us with the equilibrium $(0, 0)$. When $E_0 < 0$, the integral curves consist of the two lateral trajectories. Similarly, when $E_0 > \varphi(\pm\sqrt{\lambda})$, the integral curve consists of two trajectories: the upper one travels rightward, while the lower one travels leftward.

10.4.2 *The nature of the solutions*

A careful analysis of the phase portrait sketched in the lower plot of Figure 10.11 reveals the nature of the admissible solutions of the differential equation (10.13). Subsequently, we denote by $V_0^* > 0$ the unique value such that $(0, V_0^*)$ lies on the heteroclinic connection linking $(-\sqrt{\lambda}, 0)$ to $(\sqrt{\lambda}, 0)$. Observe that applying (10.17) with $(u_0, v_0) = (0, V_0^*)$ and $(u_0, v_0) = (\sqrt{\lambda}, 0)$ gives

$$V_0^* = \frac{\lambda}{\sqrt{2}}.$$

According to the phase portrait, the unique solution of the initial value problem

$$\begin{cases} -u'' = \lambda u - u^3, \\ u(0) = 0, \quad u'(0) = V_0^*, \end{cases}$$

denoted by $u(t; 0, V_0^*)$, satisfies $I = \mathbb{R}$, and

$$\lim_{t\downarrow-\infty} u(t; 0, V_0^*) = -\sqrt{\lambda}, \quad \lim_{t\uparrow+\infty} u(t; 0, V_0^*) = \sqrt{\lambda}.$$

In other words, V_0^* can be regarded as the *shooting speed* that, starting from the value $u(0) = 0$, allows to reach, in infinite time, the equilibria $(\pm\sqrt{\lambda}, 0)$. By the oddity of the nonlinearity, we have that

$$u(t; 0, -V_0^*) = -u(t; 0, V_0^*) \quad \text{for all } t \in \mathbb{R},$$

and, hence,

$$\lim_{t\downarrow-\infty} u(t; 0, -V_0^*) = \sqrt{\lambda}, \quad \lim_{t\uparrow+\infty} u(t; 0, -V_0^*) = -\sqrt{\lambda}.$$

Figure 10.12 shows the three constant solutions of the differential equation, together with these two heteroclinic connections between $\sqrt{\lambda}$ and $-\sqrt{\lambda}$. Observe that, since the differential equation is autonomous, for every $c \in \mathbb{R}$, the solution $u_c(t) := u(t - c; 0, V_0^*)$ is also globally defined and links $-\sqrt{\lambda}$ to $\sqrt{\lambda}$.

Note that, when v decreases along a trajectory, we have $u'' = v' < 0$ and hence u is concave, whereas u is convex if v increases because $u'' = v' > 0$. This explains the concavity/convexity of the heteroclinic connections between $-\sqrt{\lambda}$ and $\sqrt{\lambda}$ in Figure 10.12 and, more

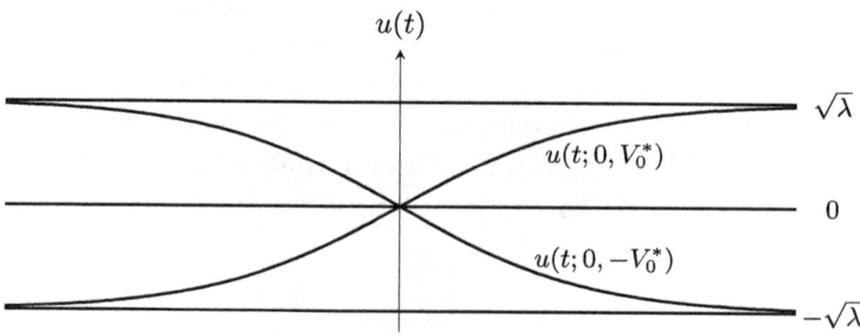

Fig. 10.12. The three equilibria and the two heteroclinic connections of (10.13).

generally, of any other solution of (10.13) analyzed in this section. For example, $u(t; 0, V_0^*)$ is concave for $t > 0$ and convex for $t < 0$.

By reasoning as in Section 10.3, it becomes apparent that, for every $V_0 \in (0, V_0^*)$, the maximal solution of

$$\begin{cases} -u'' = \lambda u - u^3, \\ u(0) = 0, \quad u'(0) = V_0, \end{cases} \tag{10.18}$$

satisfies $I = \mathbb{R}$ and is T-periodic, with $T = 4\tau$, where τ is the smallest time such that

$$u(\tau; 0, V_0) =: U_0 \in (0, \sqrt{\lambda}), \quad u'(\tau; 0, V_0) = 0.$$

Moreover, applying (10.17) with $(u_0, v_0) = (0, V_0)$ and $(u_0, v_0) = (U_0, 0)$ yields

$$\frac{V_0^2}{2} = \lambda \frac{U_0^2}{2} - \frac{U_0^4}{4}.$$

The value of the period in terms of U_0 can be determined with a similar computation as the one performed in Section 10.3. Indeed, since

$$\tau = \int_0^\tau dt = \int_0^\tau \frac{u'(t)}{v(t)}\, dt = \int_0^\tau \frac{u'(t)}{\sqrt{2\,(\varphi(U_0) - \varphi(u(t)))}}\, dt$$

$$= \int_0^{U_0} \frac{d\xi}{\sqrt{2\,(\varphi(U_0) - \varphi(\xi))}} = \int_0^{U_0} \frac{d\xi}{\sqrt{\lambda(U_0^2 - \xi^2) - \frac{1}{2}(U_0^4 - \xi^4)}},$$

the change of variable $\xi = U_0 \theta$ leads to

$$\tau = \int_0^1 \frac{d\theta}{\sqrt{\lambda(1 - \theta^2) - \frac{U_0^2}{2}(1 - \theta^4)}} \quad \text{for all } U_0 \in (0, \sqrt{\lambda}).$$

Therefore, the period of the periodic solution with amplitude $U_0 \in (0, \sqrt{\lambda})$ is given by

$$T(U_0) = 4\tau = 4 \int_0^1 \frac{d\theta}{\sqrt{\lambda(1 - \theta^2) - \frac{U_0^2}{2}(1 - \theta^4)}}. \tag{10.19}$$

According to (10.19), T is continuous and increasing with respect to $U_0 \in (0, \sqrt{\lambda})$. Actually, as discussed in the previous section, the continuity also follows, from a theoretical point of view, from Theorem 5.26. Moreover, by Theorem 5.26,

$$\lim_{U_0 \uparrow \sqrt{\lambda}} T(U_0) = +\infty$$

because, according to Property 4 in Section 9.2, the necessary time for $u(t; 0, V_0^*)$ to reach $\sqrt{\lambda}$ is infinite. Furthermore, letting $U_0 \downarrow 0$ in (10.19) gives

$$T(0) := \lim_{U_0 \downarrow 0} T(U_0) = 4 \int_0^1 \frac{d\theta}{\sqrt{\lambda(1 - \theta^2)}} = \frac{2\pi}{\sqrt{\lambda}}.$$

Therefore, the graph of the period map as a function of U_0 looks like that shown in the first picture in Figure 10.13.

An important feature is the continuity of the period with respect to the parameter λ. From (10.19), it is easily seen that, for every $U_0 \in (0, \sqrt{\lambda})$ fixed, $T(U_0)$ is decreasing (in its domain of definition) with respect to λ. The right-hand plot of Figure 10.13 shows two superimposed graphs of the period map, for $\lambda = \lambda_1$ and $\lambda = \lambda_2$, with $\lambda_1 < \lambda_2$.

Figure 10.14 shows three periodic solutions of (10.13), corresponding to different values of $V_0 \in (0, V_0^*)$. The first one, in the upper picture, is a periodic solution for a shooting speed V_0 very close to V_0^*, which is the (critical) shooting speed of the heteroclinic connection from $-\sqrt{\lambda}$ to $\sqrt{\lambda}$. These solutions have a very large period; observe that the period can be as large as we wish by taking V_0 sufficiently close to V_0^*, a continuous dependence phenomenon which is reflected

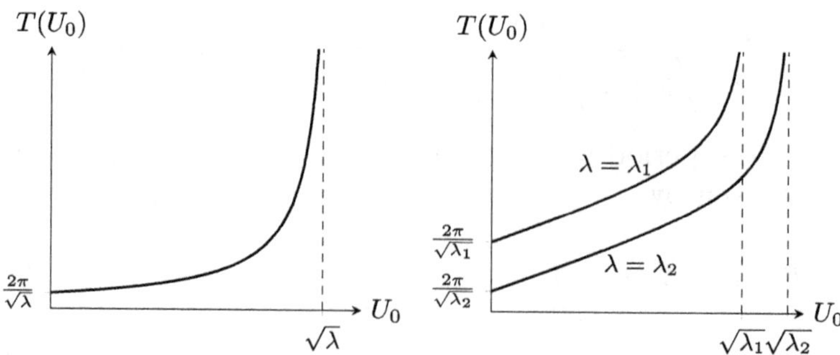

Fig. 10.13. The period map $T(U_0)$ for the generalized diffusive logistic equation.

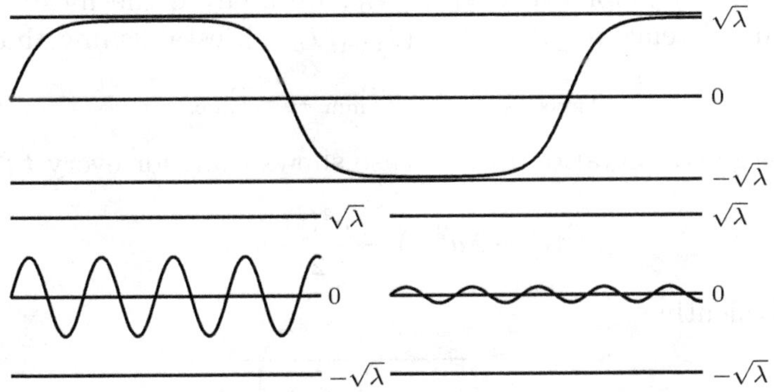

Fig. 10.14. Three periodic solutions of (10.13).

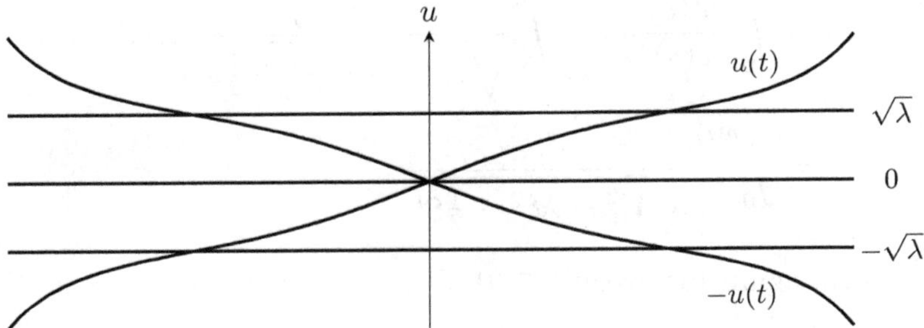

Fig. 10.15. Solution of (10.18) with $V_0 > V_0^*$ and its opposite.

in the fact that the solution can remain close to $\sqrt{\lambda}$, or $-\sqrt{\lambda}$, for an arbitrarily long time.

The lower part of Figure 10.14 shows two periodic solutions with smaller amplitudes (and smaller periods). In the first plot, $U_0 \sim \sqrt{\lambda}/2$, while in the second one, $U_0 \sim 0$. Thus, the period of the latter is close to $T(0) = 2\pi/\sqrt{\lambda}$.

Now, we analyze the solutions of (10.18) with $V_0 > V_0^*$. The associated trajectories in the phase plane are those lying above the upper heteroclinic connection between $-\sqrt{\lambda}$ and $\sqrt{\lambda}$ in Figure 10.11. Figure 10.15 represents one of these solutions, $u(t)$, as well as $-u(t)$. Observe that, if V_0 is taken above but sufficiently close to V_0^*, the solutions can remain close to $\pm\sqrt{\lambda}$ for an arbitrarily large time interval before eventually separating away from such values.

Subsequently, for every $V_0 > V_0^*$, we analyze the nature of the maximal existence interval, $I = (T_{\min}, T_{\max})$, establishing that

$$T_{\max} < +\infty, \quad T_{\min} = -T_{\max}. \qquad (10.20)$$

The energy conservation in this case shows that, for every $t \in I$,

$$v^2(t) + \lambda u^2(t) - \frac{u^4(t)}{2} = V_0^2,$$

or, equivalently,

$$v(t) = \sqrt{V_0^2 - \lambda u^2(t) + \frac{1}{2}u^4(t)}.$$

Thus, for every $t \in (0, T_{\max})$, we have that

$$t = \int_0^t \frac{u'(s)}{v(s)}\, ds = \int_0^t \frac{u'(s)}{\sqrt{V_0^2 - \lambda u^2(s) + \frac{1}{2}u^4(s)}}\, ds$$

$$= \int_0^{u(t)} \frac{d\xi}{\sqrt{V_0^2 - \lambda \xi^2 + \frac{1}{2}\xi^4}}$$

and, hence, again for every $t \in (0, T_{\max})$,

$$t = \int_0^{u(t)} \frac{d\xi}{\sqrt{V_0^2 - \lambda \xi^2 + \frac{1}{2}\xi^4}} < \int_0^{+\infty} \frac{d\xi}{\sqrt{V_0^2 - \lambda \xi^2 + \frac{1}{2}\xi^4}} =: \imath.$$

$$(10.21)$$

We now show that $\imath < +\infty$. This, together with (10.21), allows us to conclude that $T_{\max} < +\infty$. The convergence of the improper integral $\imath$ can be shown as follows. Since

$$\lim_{\xi \uparrow +\infty} \frac{\dfrac{1}{\sqrt{V_0^2 - \lambda \xi^2 + \frac{1}{2}\xi^4}}}{\sqrt{2}\,\xi^{-2}} = 1,$$

there exists $\bar{\xi} > 0$ such that

$$\frac{1}{\sqrt{V_0^2 - \lambda \xi^2 + \frac{1}{2}\xi^4}} < 2\xi^{-2} \quad \text{for all } \xi > \bar{\xi}.$$

Thus,

$$\imath = \int_0^{\bar{\xi}} \frac{d\xi}{\sqrt{V_0^2 - \lambda\xi^2 + \frac{1}{2}\xi^4}} + \int_{\bar{\xi}}^{+\infty} \frac{d\xi}{\sqrt{V_0^2 - \lambda\xi^2 + \frac{1}{2}\xi^4}}$$

$$< \int_0^{\bar{\xi}} \frac{d\xi}{\sqrt{V_0^2 - \lambda\xi^2 + \frac{1}{2}\xi^4}} + 2\bar{\xi}^{-1} < +\infty,$$

where we have used that $\int_{\bar{\xi}}^{+\infty} \xi^{-2}\, d\xi = \bar{\xi}^{-1}$.

Once $T_{\max} < +\infty$ is established, since $v(t) = u'(t) > 0$ for all $t \in I$, we have that u is increasing. Moreover, from the phase portrait in Figure 10.11 and Property 2 in Section 9.2, it becomes apparent that, for every $V_0 > V_0^*$, there exists $\bar{t} \in (0, T_{\max})$ for which the unique solution of (10.18) satisfies

$$u(\bar{t}) = \sqrt{\lambda}, \quad u'(\bar{t}) = v_0 \in (0, V_0).$$

Therefore, $u(t) > \sqrt{\lambda}$ for every $t \in (\bar{t}, T_{\max})$, and, hence,

$$v'(t) = u''(t) = -\lambda u(t) + u^3(t) = u(t)\left(-\lambda + u^2(t)\right) > 0,$$

i.e. $v(t)$ is also increasing for such times. As a consequence, we infer from Theorem 5.17 that

$$\lim_{t \uparrow T_{\max}} (u(t) + v(t)) = +\infty. \tag{10.22}$$

Suppose there exists a positive constant C such that $u(t) \in (0, C)$ for all $t \in (0, T_{\max})$. Then,

$$v'(t) = -\lambda u(t) + u^3(t) < C^3$$

for all $t \in (0, T_{\max})$ and, hence, $v(t) < C^3 T_{\max} + V_0$, leading to a contradiction with (10.22). Therefore,

$$\lim_{t \uparrow T_{\max}} u(t) = +\infty.$$

Consequently, letting $t \uparrow T_{\max}$ in (10.21), we find that

$$T_{\max} = \int_0^{+\infty} \frac{d\xi}{\sqrt{V_0^2 - \lambda\xi^2 + \frac{1}{2}\xi^4}}. \tag{10.23}$$

The fact that $T_{\min} = -T_{\max}$ easily follows by adapting the argument in the proof of Lemma 10.3. Indeed, since $f(u)$ is odd, the auxiliary function $\hat{u}(t)$ defined by

$$\hat{u}(t) := \begin{cases} u(t) & \text{if } t \in [0, T_{\max}), \\ -u(-t) & \text{if } t \in (-T_{\max}, 0), \end{cases}$$

is a solution of (10.18). By uniqueness, $u = \hat{u}$ in I; in particular, $T_{\min} = -T_{\max}$. This concludes the proof of (10.20). Figure 10.16 shows the complete graph of the solution $u(t)$ of (10.18) when $V_0 > V_0^*$, together with its opposite, $-u(t)$.

Naturally, by Theorem 5.26 and (10.23), it is apparent that

$$\lim_{V_0 \downarrow V_0^*} T_{\max} = +\infty, \qquad \lim_{V_0 \uparrow +\infty} T_{\max} = 0.$$

Now, we move on analyzing the lateral trajectories in the phase portrait in Figure 10.11 to show that the corresponding solutions also blow up in a finite time. To do so, we consider the Cauchy problem

$$\begin{cases} -u'' = \lambda u - u^3, \\ u(0) = U_0, \quad u'(0) = 0, \end{cases} \tag{10.24}$$

with $|U_0| > \sqrt{\lambda}$. Suppose $U_0 > \sqrt{\lambda}$. Then, from the phase portrait in Figure 10.11, it becomes apparent that $u(t) > U_0$ and

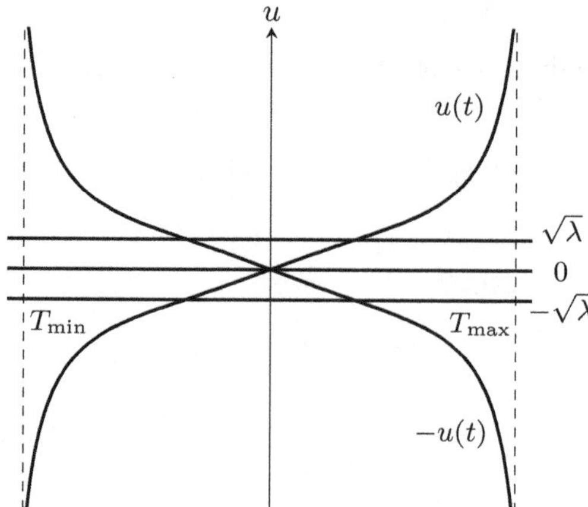

Fig. 10.16. Maximal solution $u(t)$ of (10.18) with $V_0 > V_0^*$, and its opposite $-u(t)$.

$v(t) = u'(t) > 0$ for all $t \in (0, T_{\max})$. Thus,

$$t = \int_0^t \frac{u'(s)}{v(s)}\, ds = \int_0^t \frac{u'(s)}{\sqrt{\lambda\left(U_0^2 - u^2(s)\right) - \frac{1}{2}\left(U_0^4 - u^4(s)\right)}}\, ds$$

$$= \int_{U_0}^{u(t)} \frac{d\xi}{\sqrt{\lambda\left(U_0^2 - \xi^2\right) - \frac{1}{2}\left(U_0^4 - \xi^4\right)}}$$

$$= \int_1^{\frac{u(t)}{U_0}} \frac{d\theta}{\sqrt{\lambda(1 - \theta^2) - \frac{U_0^2}{2}(1 - \theta^4)}}$$

and, hence,

$$t = \int_1^{\frac{u(t)}{U_0}} \frac{d\theta}{\sqrt{\frac{U_0^2}{2}\left(\theta^4 - 1\right) - \lambda\left(\theta^2 - 1\right)}}$$

$$< \int_1^{+\infty} \frac{d\theta}{\sqrt{\frac{U_0^2}{2}\left(\theta^4 - 1\right) - \lambda\left(\theta^2 - 1\right)}} =: \jmath. \tag{10.25}$$

We now show that $\jmath < +\infty$. Thus, (10.25) entails that $T_{\max} < +\infty$. The convergence of this integral can be established as follows. The function

$$g(\theta) := \frac{U_0^2}{2}(\theta^4 - 1) - \lambda(\theta^2 - 1) = (\theta^2 - 1)\left(\frac{U_0^2}{2}(\theta^2 + 1) - \lambda\right)$$

satisfies

$$\lim_{\theta \uparrow +\infty} \frac{\frac{U_0^2}{2}\theta^4}{g(\theta)} = 1.$$

Hence, we can fix $\bar{\theta} > 1$ such that

$$\frac{1}{\sqrt{g(\theta)}} < \frac{2}{U_0}\theta^{-2} \quad \text{for all } \theta > \bar{\theta}.$$

Then,

$$\int_{\bar{\theta}}^{+\infty} \frac{d\theta}{\sqrt{g(\theta)}} < \frac{2}{U_0}\int_{\bar{\theta}}^{+\infty} \theta^{-2}\, d\theta = \frac{2}{U_0}\bar{\theta}^{-1} < +\infty.$$

Moreover, for every $\theta > 1$,

$$g(\theta) > \left(\theta^2 - 1\right)\left(U_0^2 - \lambda\right) > 2\left(U_0^2 - \lambda\right)(\theta - 1),$$

which implies

$$\int_1^{\bar{\theta}} \frac{d\theta}{\sqrt{g(\theta)}} < \int_1^{\bar{\theta}} \frac{d\theta}{\sqrt{2\left(U_0^2 - \lambda\right)(\theta - 1)}} = \sqrt{\frac{2\left(\bar{\theta} - 1\right)}{U_0^2 - \lambda}} < +\infty.$$

Therefore,

$$\jmath = \int_1^{+\infty} \frac{d\theta}{\sqrt{g(\theta)}} = \int_1^{\bar{\theta}} \frac{d\theta}{\sqrt{g(\theta)}} + \int_{\bar{\theta}}^{+\infty} \frac{d\theta}{\sqrt{g(\theta)}} < +\infty.$$

Once we know that $T_{\max} < +\infty$, since $v(t) = u'(t) > 0$, we find that u is increasing, and

$$v'(t) = u''(t) = -\lambda u(t) + u^3(t)$$
$$= u(t)\left(-\lambda + u^2(t)\right) > u(t)\left(-\lambda + U_0^2\right) > 0$$

for all $t \in (0, T_{\max})$. Then, Theorem 5.17 allows us to obtain (10.22) in this case as well. Moreover, by reasoning as above, it readily follows that $\lim_{t \uparrow T_{\max}} u(t) = +\infty$. Thus, letting $t \uparrow T_{\max}$ in (10.25) yields

$$T_{\max} = \int_1^{+\infty} \frac{d\theta}{\sqrt{\frac{U_0^2}{2}(\theta^4 - 1) - \lambda(\theta^2 - 1)}} < +\infty. \tag{10.26}$$

In this case, the fact that $T_{\min} = -T_{\max}$ relies on the symmetry of the trajectories with respect to the u-axis, as in the proof of Lemma 10.1. Figure 10.17 shows the graphs of $u(t)$ and $-u(t)$ for $U_0 > \sqrt{\lambda}$.

Note that, thanks to Theorem 5.26 and (10.26),

$$\lim_{U_0 \downarrow \sqrt{\lambda}} T_{\max} = +\infty, \qquad \lim_{U_0 \uparrow +\infty} T_{\max} = 0,$$

and $T_{\max}$ is continuous and decreasing with respect to $U_0 > \sqrt{\lambda}$. Therefore, for every $L > 0$, there exists a unique $U_0 = U_0(L) > \sqrt{\lambda}$

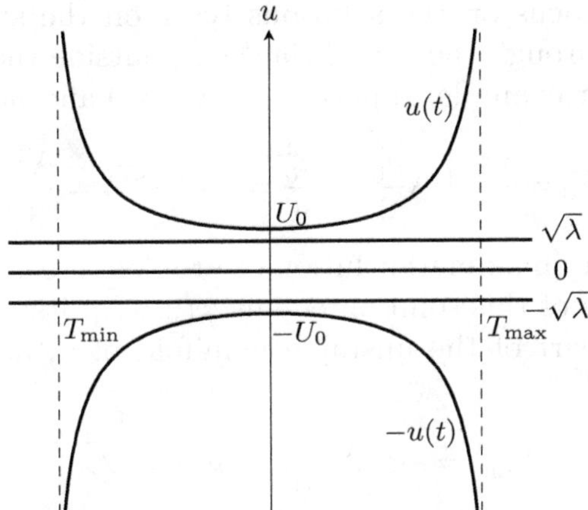

Fig. 10.17. Solutions of (10.24) with $|U_0| > \sqrt{\lambda}$.

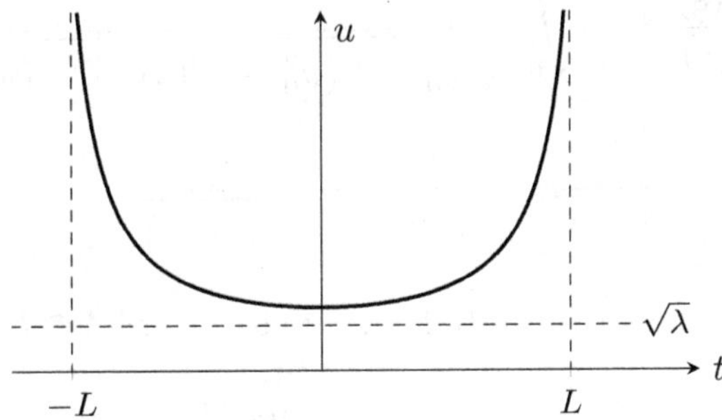

Fig. 10.18. Positive large solution of (10.13).

such that $T_{\max}(U_0) = L$. As a byproduct, for every $L > 0$, the singular problem

$$\begin{cases} -u'' = \lambda u - u^3 & \text{in } (-L, L), \\ u(-L) = u(L) = +\infty, \end{cases}$$

has a unique positive solution, u_L, which is referred to as a *large positive solution* and is represented in Figure 10.18.

Once again, by the oddity of $f(u)$, we have that $-u_L$ is the unique negative solution of the singular boundary value problem

$$\begin{cases} -u'' = \lambda u - u^3 & \text{in } (-L, L), \\ u(-L) = u(L) = -\infty. \end{cases}$$

Finally, we focus on the solutions lying on the stable or unstable manifolds through the equilibria $\pm\sqrt{\lambda}$, outside their heteroclinic connections. For example, suppose that $v_0 > 0$ and $u_0 > \sqrt{\lambda}$ satisfy

$$E_0 = \frac{v_0^2}{2} + \lambda\frac{u_0^2}{2} - \frac{u_0^4}{4} = \varphi(\pm\sqrt{\lambda}) = \frac{\lambda^2}{4},$$

and consider the (maximal) solution of problem (10.15), (I, u). Since the total energy of this solution equals $\varphi(\pm\sqrt{\lambda})$, its trajectory must be the lateral part of the unstable manifold of $(\sqrt{\lambda}, 0)$ (see Figure 10.11). Thus,

$$T_{\min} = -\infty, \qquad \lim_{t\downarrow-\infty} u(t) = \sqrt{\lambda},$$

and $v(t) = u'(t) > 0$ for all $t \in I$. Hence, for every $t \in I$,

$$t - t_0 = \int_{t_0}^t \frac{u'(s)}{v(s)}\, ds = \int_{t_0}^t \frac{u'(s)}{\sqrt{v_0^2 + \lambda\left(u_0^2 - u^2(s)\right) - \frac{1}{2}\left(u_0^4 - u^4(s)\right)}}\, ds$$

$$= \int_{u_0}^{u(t)} \frac{d\xi}{\sqrt{v_0^2 + \lambda\left(u_0^2 - \xi^2\right) - \frac{1}{2}\left(u_0^4 - \xi^4\right)}}.$$

Therefore, it follows from (10.25) that, for every $t \in I$,

$$t - t_0 = \int_1^{\frac{u(t)}{u_0}} \frac{u_0\, d\theta}{\sqrt{v_0^2 + \frac{u_0^4}{2}\left(\theta^4 - 1\right) - \lambda u_0^2\left(\theta^2 - 1\right)}}$$

$$< \int_1^{+\infty} \frac{d\theta}{\sqrt{\frac{u_0^2}{2}\left(\theta^4 - 1\right) - \lambda\left(\theta^2 - 1\right)}} < +\infty.$$

Consequently, $T_{\max} < +\infty$, and letting $t \uparrow T_{\max}$ in the previous expression, it becomes apparent that

$$T_{\max} = t_0 + \int_1^{+\infty} \frac{u_0\, d\theta}{\sqrt{v_0^2 + \frac{u_0^4}{2}\left(\theta^4 - 1\right) - \lambda u_0^2\left(\theta^2 - 1\right)}}.$$

Figure 10.19 shows the graph of this solution, together with the superimposed graphs of the solutions starting at $(u_0, -v_0)$ and $(-u_0, \pm v_0)$.

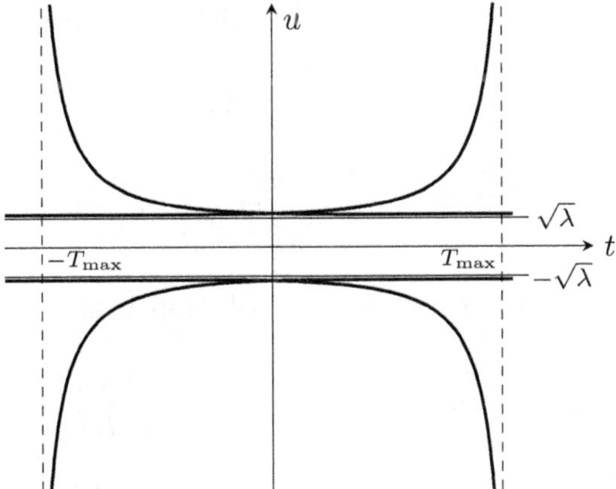

Fig. 10.19. Solutions along the stable and the unstable manifolds of $(\pm\sqrt{\lambda}, 0)$ outside the heteroclinic connections.

10.4.3 *Positive solutions of* (10.14)

In this section, for any fixed $L > 0$, we construct the set of positive solutions of the boundary value problem (10.14). This will be accomplished through a *shooting method*: it consists of considering problem (10.18) and shooting from the value $u = 0$ at $t = 0$ with the appropriate initial velocity, V_0, so that its solution is positive for all $t \in (0, L)$ and vanishes at $t = L$.

As $u = 0$ solves (10.14) for all $\lambda \in \mathbb{R}$, such a solution will be referred to as the *trivial solution*. A direct analysis of the phase portrait sketched in Figure 10.11 reveals that a solution of (10.13), u, is a positive solution of (10.14) if and only if $u(0) = 0$ and $u'(0) = V_0 \in (0, V_0^*)$ with $L = \frac{T}{2}$, where T is the period of this solution. According to (10.19), this can be equivalently expressed as

$$\tau(U_0) := \int_0^1 \frac{d\theta}{\sqrt{\lambda(1 - \theta^2) - \frac{U_0^2}{2}(1 - \theta^4)}} = \frac{L}{2}.$$

Subsequently, the dependence of τ on the parameter $\lambda > 0$ will be emphasized by setting

$$\tau(\lambda, U_0) := \int_0^1 \frac{d\theta}{\sqrt{\lambda(1 - \theta^2) - \frac{U_0^2}{2}(1 - \theta^4)}}.$$

Since $\tau = \frac{T}{4}$, τ inherits all the monotonicity properties of the period map T, which have been analyzed in the previous section, particularly those sketched in Figure 10.13. Thus, since

$$\inf_{U_0 > 0} \tau(\lambda, U_0) = \tau(\lambda, 0) = \int_0^1 \frac{d\theta}{\sqrt{\lambda(1 - \theta^2)}} = \frac{\pi}{2\sqrt{\lambda}} \quad \text{for all } \lambda > 0,$$

problem (10.14) possesses a positive solution if and only if

$$\frac{L}{2} > \frac{\pi}{2\sqrt{\lambda}}, \quad \text{i.e. } \lambda > \sigma_1 := \left(\frac{\pi}{L}\right)^2.$$

Indeed, as illustrated by the left-hand plot of Figure 10.20, if this condition holds, there exists a unique $U_0 = U_0(\lambda) \in (0, \sqrt{\lambda})$ such that

$$\tau(\lambda, U_0(\lambda)) = \frac{L}{2},$$

whereas $\tau(\lambda, U_0) > \frac{L}{2}$ for all $U_0 \in (0, \sqrt{\lambda})$ if $\lambda \leq \sigma_1$.

By the continuity properties of $\tau(\lambda, U_0)$ with respect to λ and U_0, $U_0(\lambda)$ varies continuously with respect to λ. Moreover,

$$\lim_{\lambda \downarrow \sigma_1} U_0(\lambda) = 0, \tag{10.27}$$

and the map $\lambda \mapsto U_0(\lambda)$ is increasing, as sketched in the right-hand plot of Figure 10.20. With some additional effort, one might even

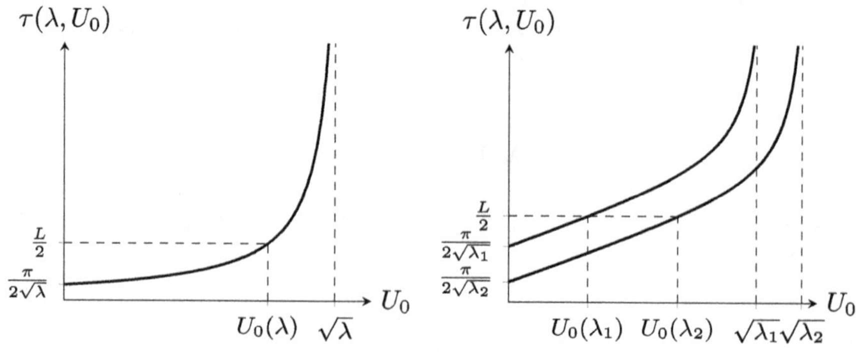

Fig. 10.20. The amplitude $U_0(\lambda)$ of the positive solutions of (10.14).

show that

$$\lim_{\lambda \uparrow +\infty} \frac{U_0(\lambda)}{\sqrt{\lambda}} = 1$$

(see, e.g. Furter and López-Gómez, 1997, or Fraile *et al.*, 1995, for further details on this particular point), which reveals that $U_0(\lambda) \sim \sqrt{\lambda}$ as $\lambda \uparrow +\infty$. Therefore, the graph of the curve $(\lambda, U_0(\lambda))$, $\lambda > \sigma_1$, looks like that shown in the upper half-curve in Figure 10.21.

Thanks to the oddity of $f(u)$, the lower half-curve, $(\lambda, -U_0(\lambda))$, $\lambda > \sigma_1$, provides us with the set of negative solutions of (10.14). Each point on these curves represents a solution of (10.14), either $(\lambda, U_0(\lambda))$ or $(\lambda, -U_0(\lambda))$, for every $\lambda > \sigma_1 = \left(\frac{\pi}{L}\right)^2$. Subsequently, for each $\lambda > \sigma_1$, the unique positive solution of (10.14) is denoted by $u_{0,\lambda}$. Naturally, $-u_{0,\lambda}$ provides us with its unique negative solution. Then,

$$U_0(\lambda) = \|u_{0,\lambda}\|_\infty,$$

and owing to (10.27), we have that

$$\lim_{\lambda \downarrow \sigma_1} \|u_{0,\lambda}\|_\infty = 0. \tag{10.28}$$

Thus, the positive solution bifurcates, or emanates, from the curve of the trivial solutions $\{(\lambda, u) \colon u = 0, \ \lambda \in \mathbb{R}\}$ at the critical value of the parameter $\lambda = \sigma_1$, which explains why these kinds of solution diagrams are usually referred to as *bifurcation diagrams*. Naturally,

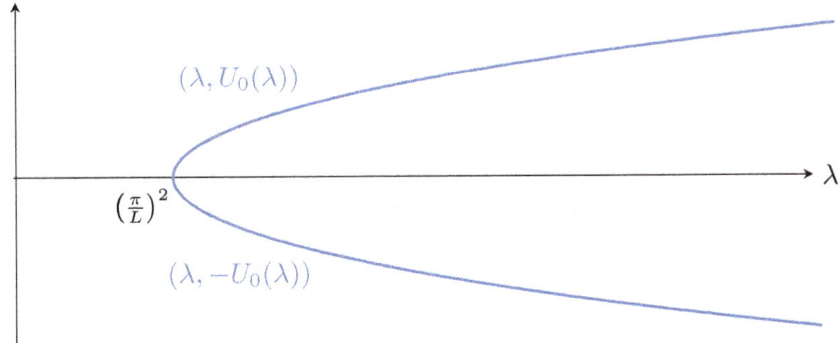

Fig. 10.21. Bifurcation diagram of positive and negative solutions of the boundary value problem (10.14).

in a bi-dimensional diagram, like the one given in Figure 10.21, one cannot represent a function, $u_{0,\lambda}$, on the vertical axis. So, one shall choose some representative scalar quantity that illustrates the properties of the solutions, depending on the parameter λ on the abscissas. The chosen quantity for the analysis in this section is $U_0(\lambda)$, the maximum of the positive solution $u_{0,\lambda}(t)$, or $-U_0(\lambda)$, the minimum of the negative solution $-u_{0,\lambda}(t)$, which are attained at $t = \frac{L}{2}$.

These results can be summarized as follows.

Theorem 10.4. *For every $L > 0$, the following assertions hold true.*

(a) *The boundary value problem*

$$\begin{cases} -u'' = \lambda u - u^3 & \text{in } (0, L), \\ u(0) = u(L) = 0, \end{cases} \tag{10.29}$$

has a positive solution if and only if $\lambda > \sigma_1 = \left(\frac{\pi}{L}\right)^2$, and it is unique if it exists. Moreover, if we denote it by $u_{0,\lambda}$, then

$$\|u_{0,\lambda}\|_\infty = u_{0,\lambda}\left(\tfrac{L}{2}\right) = U_0(\lambda) \quad \text{for all } \lambda > \sigma_1.$$

(b) *Identifying the solution $u_{0,\lambda}(t)$ with its maximum $U_0(\lambda)$, the set of positive solutions of* (10.29),

$$\mathcal{C}_0^+ := \{(\lambda, U_0(\lambda)): \lambda > \sigma_1\},$$

is a continuous increasing curve, parameterized by $\lambda > \sigma_1$, which bifurcates from the trivial solution $u = 0$ at $\lambda = \sigma_1$, in the sense that (10.28) *holds.*

(c) *Since $f(u)$ is odd, the set of negative solutions of* (10.29) *is given by*

$$\mathcal{C}_0^- := \{(\lambda, -U_0(\lambda)): \lambda > \sigma_1\}.$$

The continuity of the curve $\lambda \mapsto U_0(\lambda)$ is a direct consequence of the fact that the function $(\lambda, U_0) \mapsto \tau(\lambda, U_0)$ is continuous with respect to both variables in its domain of definition and that, for every $\lambda > \sigma_1$, there is a unique value of U_0, $U_0(\lambda)$, such that

$$\tau(\lambda, U_0(\lambda)) = \frac{L}{2}.$$

With some additional (substantial) work, which remains outside the scope of this introductory textbook, it can be shown that, actually, the map $\lambda \mapsto u_{0,\lambda}$ is point-wise increasing in $(0, L)$ and real analytic (see, e.g. López-Gómez, 2015).

10.4.4 *Nodal solutions of* (10.29)

A nodal solution of (10.29) is any solution (λ, u), with $u \neq 0$, such that u changes sign in $(0, L)$. By analyzing the phase portrait of (10.13) sketched in Figure 10.11, it becomes apparent that the trajectory of any nodal solution of (10.29) must lie on some bounded orbit that corresponds to a periodic solution of (10.13). Thus, in particular, any nodal solution, u, has finitely many *interior zeroes*; i.e. there exist finitely many values of $z \in (0, L)$ such that $u(z) = 0$, which will be referred to as *interior nodes*. As the solutions of (10.13) arise in pairs, u and $-u$, any nodal solution satisfies either $u'(0) > 0$ or $u'(0) < 0$. Subsequently, we only consider solutions of (10.29) such that $u(0) = 0$ and $u'(0) = V_0 > 0$.

By the symmetries of the solutions of (10.13) discussed in the previous sections, for every integer $n \geq 2$, problem (10.29) possesses a solution with $n - 1$ interior nodes if and only if the problem

$$\begin{cases} -u'' = \lambda u - u^3 & \text{in } \left(0, \frac{L}{n}\right), \\ u(0) = u(\frac{L}{n}) = 0, \end{cases} \tag{10.30}$$

admits a positive solution. In such a case, the $n - 1$ interior nodes of the solution are

$$z_j := \frac{j}{n}L, \quad j \in \{1, \ldots, n - 1\}.$$

Moreover, denoting by $u_{n-1,\lambda}$ the unique positive solution of (10.30), which, owing to Theorem 10.4(a), exists if and only if

$$\lambda > \left(\frac{\pi}{L/n}\right)^2 = \left(\frac{n\pi}{L}\right)^2 =: \sigma_n,$$

by the symmetries of the problem and the oddity of $f(u)$, it turns out that the nodal solution equals $u_{n-1,\lambda}$ in $[0, \frac{L}{n}]$; it equals $-u_{n-1,\lambda}(t - \frac{L}{n})$ if $t \in [\frac{L}{n}, \frac{2L}{n}]$; and, in general, for every $j \in \{1, \ldots, n - 1\}$, it equals

$$(-1)^j \, u_{n-1,\lambda}\left(t - \frac{jL}{n}\right) \quad \text{for all } t \in \left[\frac{jL}{n}, \frac{(j+1)L}{n}\right].$$

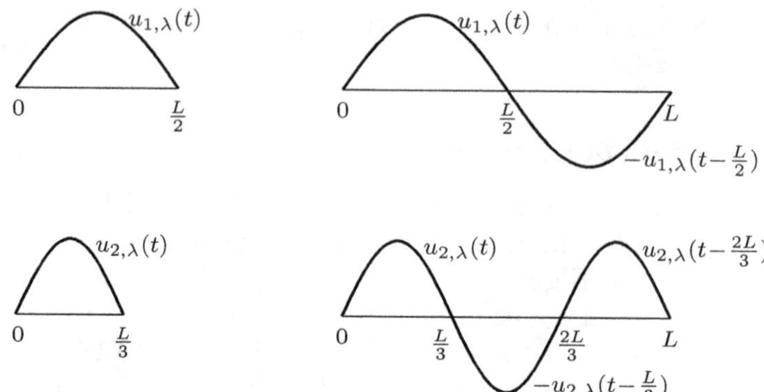

Fig. 10.22. Construction of the nodal solutions of (10.29) with one and two interior nodes.

Therefore, the value of the nodal solution in the first nodal interval $[0, \frac{L}{n}]$, $u_{n-1,\lambda}$, entirely determines its value in the whole interval $[0, L]$. Figure 10.22 illustrates the construction of these solutions for the special cases of $n = 2$ and $n = 3$, starting from the values in their first nodal intervals.

With such an extension process in mind, we denote by $u_{n-1,\lambda}$ the unique solution of (10.29) with $n - 1$ interior zeroes such that $u'(0) > 0$. We stress once again that the uniqueness is a direct consequence of the symmetries of the problem and the fact that the period map $T(U_0)$ is increasing with respect to U_0.

Subsequently, we denote

$$U_{n-1}(\lambda) := \max_{[0,\frac{L}{n}]} u_{n-1,\lambda} = \|u_{n-1,\lambda}\|_\infty$$

for all integers $n \geq 1$ and $\lambda > \sigma_n$, which is coherent with the notations introduced in the previous section concerning the positive solutions of (10.29). By simply looking at Figure 10.20, it can be easily realized that $U_0(\lambda)$ increases with respect to L. Thus,

$$U_{n-1}(\lambda) > U_n(\lambda) \quad \text{for all integer } n \geq 1 \text{ and } \lambda > \sigma_n.$$

Consequently, the global bifurcation diagram of the set of solutions of (10.29) can be represented as in Figure 10.23, where the set of solutions of (10.29), $(\lambda, \pm u_{n-1,\lambda})$, with $n - 1$ interior nodes, such

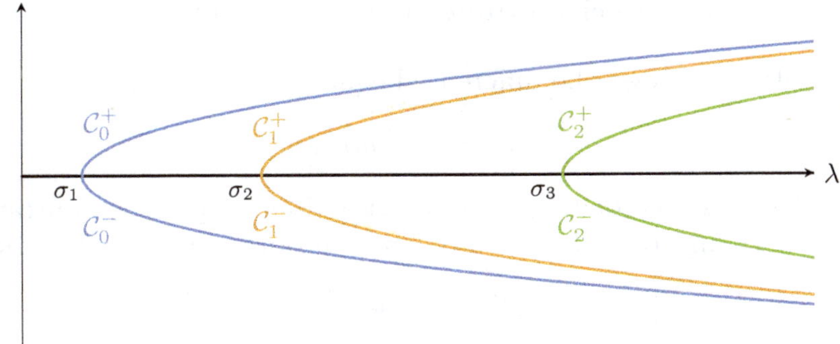

Fig. 10.23. Global bifurcation diagram of the solutions of the boundary value problem (10.29).

that $\pm u'_{n-1,\lambda}(0) > 0$, is identified with the curve

$$\mathcal{C}^{\pm}_{n-1} := \{(\lambda, \pm U_{n-1}(\lambda)) : \lambda > \sigma_n\}.$$

These facts can be summarized in the following result, which generalizes Theorem 10.4.

Theorem 10.5. *For every $L > 0$ and any integer $n \geq 1$, the following assertions hold true:*

(a) *The boundary value problem (10.29) admits a solution (λ, u) with $n - 1$ interior nodes if and only if $\lambda > \sigma_n$. Moreover, in such a case, it admits exactly two solutions with $n - 1$ interior nodes: $(\lambda, \pm u_{n-1,\lambda})$.*

(b) *The set of solutions of (10.29) with $n - 1$ interior nodes and $u'(0) > 0$, $\mathcal{C}^{+}_{n-1}$, is a continuous increasing curve, parameterized by $\lambda > \sigma_n$, that bifurcates from the trivial solution $u = 0$ at $\lambda = \sigma_n$, in the sense that*

$$\lim_{\lambda \downarrow \sigma_n} \|u_{n-1,\lambda}\|_\infty = 0.$$

(c) *Since $f(u)$ is odd, the set of solutions of (10.29) with $n - 1$ interior nodes and $u'(0) < 0$ is*

$$\mathcal{C}^{-}_{n-1} = \{(\lambda, -U_{n-1}(\lambda)) : \lambda > \sigma_n\}.$$

10.5   A Parameterized Superlinear Problem

This section analyzes the differential equation

$$-u'' = \lambda u + mu^2,$$

where $\lambda \in \mathbb{R}$ is regarded as a parameter and m is a positive constant. Since the change of variable $u := U/m$ transforms this equation into

$$-U'' = \lambda U + U^2,$$

it is not restrictive to consider the second-order equation

$$-u'' = \lambda u + u^2. \tag{10.31}$$

This equation is said to be of *superlinear type* because $\lambda u + u^2 \geq \lambda u$, while the differential equation (10.13) analyzed in the previous section is *sublinear* since $\lambda u - u^3 \leq \lambda u$. Another remarkable difference with respect to the previous section is that the nonlinearity of (10.31), $f(u) := \lambda u + u^2$, $u \in \mathbb{R}$, is not odd. Actually, it is even if $\lambda = 0$.

Arguing as in Section 10.4, for every $t_0 \in \mathbb{R}$ and $(u_0, v_0) \in \mathbb{R}^2$, the Cauchy problem

$$\begin{cases} -u'' = \lambda u + u^2, \\ u(t_0) = u_0, \quad u'(t_0) = v_0, \end{cases}$$

possesses a unique maximal solution, (I, u), with $I = (T_{\min}, T_{\max})$ for some $T_{\min} \in [-\infty, t_0)$ and $T_{\max} \in (t_0, +\infty]$. Moreover, the solution blows up at time $T_{\min}$ (respectively, $T_{\max}$) if $T_{\min} > -\infty$ (respectively, $T_{\max} < +\infty$). As usual, this solution is given by the unique maximal solution of the associated problem

$$\begin{cases} u' = v, \\ v' = -\lambda u - u^2, \\ u(t_0) = u_0, \quad v(t_0) = v_0, \end{cases} \tag{10.32}$$

which is given by $(u(t), u'(t))$, $t \in I$.

By taking cross products in (10.32), we find that, for every $t \in I$,

$$v(t)v'(t) = -\lambda u(t)u'(t) - u^2(t)u'(t),$$

or, equivalently,

$$\frac{d}{dt}\left(\frac{v^2(t)}{2} + \lambda\frac{u^2(t)}{2} + \frac{u^3(t)}{3}\right) = 0.$$

Thus,

$$\frac{v^2(t)}{2} + \lambda\frac{u^2(t)}{2} + \frac{u^3(t)}{3} = \frac{v_0^2}{2} + \lambda\frac{u_0^2}{2} + \frac{u_0^3}{3} \quad \text{for all } t \in I. \quad (10.33)$$

In particular,

$$\varphi(u) := \lambda\frac{u^2}{2} + \frac{u^3}{3}, \quad u \in \mathbb{R},$$

can be seen as the associated potential energy. In terms of $\varphi(u)$, the identity (10.33) reads

$$\frac{v^2(t)}{2} + \varphi(u(t)) = \frac{v_0^2}{2} + \varphi(u_0) \quad \text{for all } t \in I.$$

Thus, as in all the cases treated in this chapter,

$$v(t) = \pm\sqrt{2\left(E_0 - \varphi(u(t))\right)} \quad \text{for all } t \in I,$$

where

$$E_0 := \frac{v_0^2}{2} + \varphi(u_0) = \frac{v_0^2}{2} + \lambda\frac{u_0^2}{2} + \frac{u_0^3}{3}$$

is the total energy. Since

$$\varphi'(u) = \lambda u + u^2 = f(u),$$

it is apparent that the set of equilibria of (10.31) is

$$\mathcal{E} := \{(-\lambda, 0), (0, 0)\}$$

for every $\lambda \in \mathbb{R}$. Thus, (10.31) has two equilibria if $\lambda \neq 0$ and one if $\lambda = 0$. Moreover, as $\varphi''(u) = \lambda + 2u$,

$$\varphi''(0) = \lambda \quad \text{and} \quad \varphi''(-\lambda) = -\lambda. \quad (10.34)$$

Thus, the nature of the critical points 0 and λ and, as a consequence, that of the equilibria will depend on the sign of λ. For this reason, we distinguish three different cases.

10.5.1 *Phase portrait and solution types for* $\lambda > 0$

Throughout this section we assume that $\lambda > 0$. Then, by (10.34), $\varphi''(0) > 0$ and $\varphi''(-\lambda) < 0$. Thus, 0 is a local minimum of $\varphi(u)$, while $-\lambda$ is a local maximum, and the graph of the potential energy looks like that shown in the upper plot of Figure 10.24. As a consequence, the phase portrait is the one sketched in the lower plot.

In this case, the integral curve corresponding to $E_0 = \varphi(-\lambda)$ consists of four trajectories: the equilibrium $(-\lambda, 0)$, the *homoclinic connection* of $(-\lambda, 0)$, plus the two remaining pieces of the unstable and stable manifolds of the saddle point $(-\lambda, 0)$. The name *homoclinic connection* refers to the fact that the trajectory connects the equilibrium to itself. The homoclinic connection encloses a family of periodic solutions around $(0, 0)$, associated with the potential well at $u = 0$. The remaining trajectories are orbits of explosive solutions, as it will become apparent later.

Let $U_0^* > 0$ denote the amplitude of the homoclinic orbit. Observe that, since $\varphi(U_0^*) = \varphi(-\lambda)$, one can easily obtain that $U_0^* = \frac{\lambda}{2}$.

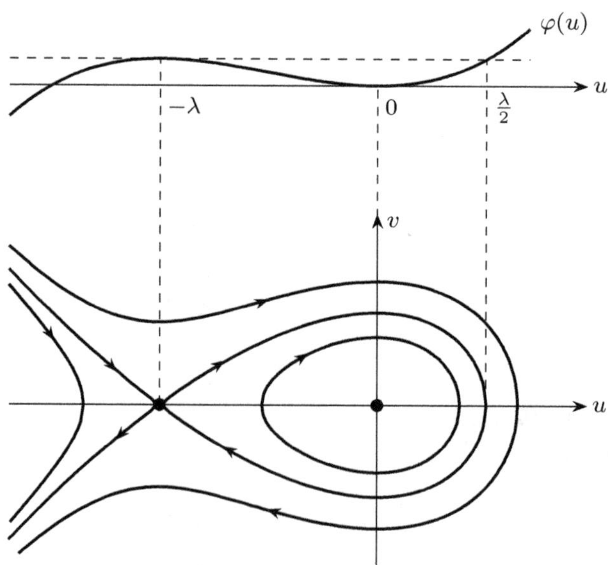

Fig. 10.24. Potential energy and phase portrait of (10.31) for $\lambda > 0$.

According to the phase portrait, the (maximal) solution of

$$\begin{cases} -u'' = \lambda u + u^2, \\ u(0) = U_0^*, \quad u'(0) = 0, \end{cases}$$

denoted by $u(t; U_0^*, 0)$, satisfies $I = \mathbb{R}$ and

$$\lim_{t\downarrow-\infty} u(t; U_0^*, 0) = \lim_{t\uparrow+\infty} u(t; U_0^*, 0) = -\lambda.$$

Since $\varphi(0) = 0$, the ordinate of the intersection of the homoclinic connection with the positive v-axis is

$$V_0^* = \sqrt{2\varphi(U_0^*)}.$$

Figure 10.25 shows the graphs of the solutions of

$$\begin{cases} -u'' = \lambda u + u^2, \\ u(0) = U_0, \quad u'(0) = 0, \end{cases} \tag{10.35}$$

for three different values of U_0: on the left for $U_0 \in (0, U_0^*)$, where $u(t; U_0, 0)$ is periodic; in the middle for $U = U_0^*$, when $u(t; U_0^*, 0)$ lies on the homoclinic connection of $(-\lambda, 0)$, and, on the right, for some $U_0 > U_0^*$. In the latter case, we will show that $u(t; U_0, 0)$ blows up, bilaterally, in a finite time. Actually, $I = (T_{\min}, T_{\max})$, with $T_{\min} \in (-\infty, 0)$ and $T_{\max} = -T_{\min}$. According to Theorem 5.26, by taking U_0 sufficiently close to U_0^*, these explosive solutions can be taken as close

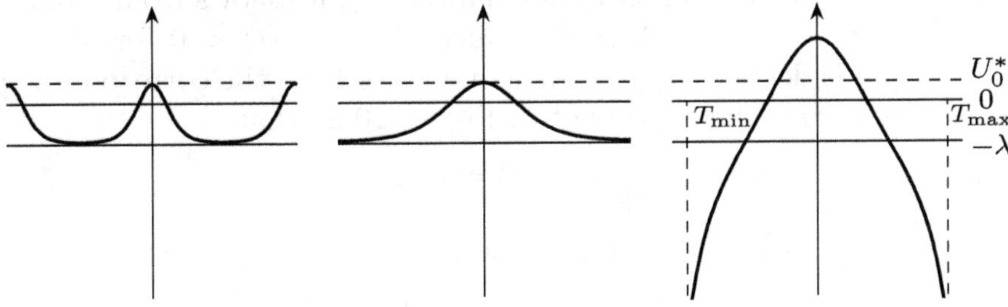

Fig. 10.25. Three different types of solutions of (10.35) with $U_0 > 0$ and $\lambda > 0$.

as we wish, in an arbitrarily large time interval, to the homoclinic connection. Thus,

$$\lim_{U_0 \downarrow U_0^*} T_{\min} = -\infty, \qquad \lim_{U_0 \downarrow U_0^*} T_{\max} = +\infty.$$

Arguing as in the proof of Lemma 10.1, it becomes apparent that, for every $U_0 > 0$, the maximal solution of (10.35) satisfies $u(-t) = u(t)$ for all $t \in (0, T_{\max})$. Thus, $T_{\min} = -T_{\max}$.

Next, we will show that $u(t; U_0, 0)$ blows up in a finite time if $U_0 > U_0^*$. Indeed, according to the phase portrait of Figure 10.24, we have that $v(t) = u'(t) < 0$ for all $t \in (0, T_{\max})$. Thus,

$$t = \int_0^t \frac{u'(s)}{v(s)} \, ds = -\int_0^t \frac{u'(s)}{\sqrt{\lambda \left(U_0^2 - u^2(s)\right) + \frac{2}{3}\left(U_0^3 - u^3(s)\right)}} \, ds$$

$$= -\int_{U_0}^{u(t)} \frac{d\xi}{\sqrt{\lambda \left(U_0^2 - \xi^2\right) + \frac{2}{3}\left(U_0^3 - \xi^3\right)}}$$

$$= \int_{u(t)}^{U_0} \frac{d\xi}{\sqrt{\lambda \left(U_0^2 - \xi^2\right) + \frac{2}{3}\left(U_0^3 - \xi^3\right)}}.$$

Consequently, for every $t \in (0, T_{\max})$,

$$\begin{aligned}
t &= \int_{\frac{u(t)}{U_0}}^1 \frac{d\theta}{\sqrt{\lambda (1 - \theta^2) + \frac{2U_0}{3}(1 - \theta^3)}} \\
&< \int_{-\infty}^1 \frac{d\theta}{\sqrt{\lambda (1 - \theta^2) + \frac{2U_0}{3}(1 - \theta^3)}} =: \kappa.
\end{aligned} \tag{10.36}$$

Since $\kappa < +\infty$ (see Exercise 2 of Chapter 10), it follows from (10.36) that $T_{\max} \le \kappa < +\infty$. Actually, since $u'(t) = v(t) < 0$ for all $t \in (0, T_{\max})$, u is decreasing, and we can deduce from the corresponding trajectory in the phase portrait of Figure 10.24 that

$$\lim_{t \uparrow T_{\max}} u(t) = -\infty.$$

Thus, letting $t \uparrow T_{\max}$ in (10.36) gives

$$T_{\max} = \int_{-\infty}^1 \frac{d\theta}{\sqrt{\lambda(1 - \theta^2) + \frac{2U_0}{3}(1 - \theta^3)}}.$$

Finally, Figure 10.26 shows the graphs of the solution of

$$\begin{cases} -u'' = \lambda u + u^2, \\ u(0) = u_0, \quad u'(0) = v_0, \end{cases} \tag{10.37}$$

for $u_0 < -\lambda$ and three different values of v_0. As usual, we denote by $u(t) := u(t; u_0, v_0)$ the unique (maximal) solution of (10.37), and by $I = (T_{\min}, T_{\max})$ its existence interval. The left plot of Figure 10.26 shows the graph for the choice $v_0 = 0$, where $-T_{\min} = T_{\max} < +\infty$, the middle plot shows the graph of the solution satisfying $(u_0, v_0) \in W^s(-\lambda, 0)$, where $W^s(-\lambda, 0)$ is the stable manifold of the equilibrium $(-\lambda, 0)$, and the right plot corresponds to the case $(u_0, v_0) \in W^u(-\lambda, 0)$, where $W^u(-\lambda, 0)$ indicates the unstable manifold of the equilibrium $(-\lambda, 0)$.

Next, we will show that these solutions, with $u_0 < -\lambda < 0$, are explosive. Indeed, take $v_0 = 0$. Then, $u'(t) = v(t) < 0$ for all $t \in (0, T_{\max})$. Thus,

$$t = \int_0^t \frac{u'(s)}{v(s)}\, ds = -\int_0^t \frac{u'(s)}{\sqrt{\lambda\left(u_0^2 - u^2(s)\right) + \frac{2}{3}\left(u_0^3 - u^3(s)\right)}}\, ds$$

$$= -\int_{u_0}^{u(t)} \frac{d\xi}{\sqrt{\lambda\left(u_0^2 - \xi^2\right) + \frac{2}{3}\left(u_0^3 - \xi^3\right)}}$$

$$= \int_{u(t)}^{u_0} \frac{d\xi}{\sqrt{\lambda\left(u_0^2 - \xi^2\right) + \frac{2}{3}\left(u_0^3 - \xi^3\right)}}.$$

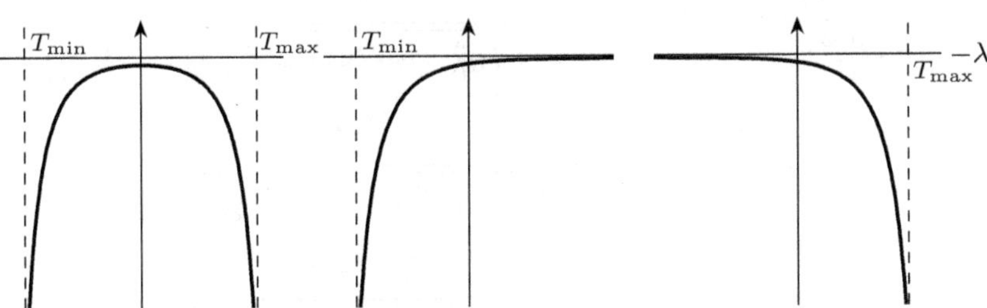

Fig. 10.26. Different types of solutions of (10.37), with $\lambda > 0$ and $u_0 < -\lambda$.

Consequently, since $u_0 < 0$, we find that, for every $t \in (0, T_{\max})$,

$$t = \int_1^{\frac{u(t)}{u_0}} \frac{d\theta}{\sqrt{-\frac{2u_0}{3}(\theta^3 - 1) - \lambda(\theta^2 - 1)}}$$

$$< \int_1^{+\infty} \frac{d\theta}{\sqrt{-\frac{2u_0}{3}(\theta^3 - 1) - \lambda(\theta^2 - 1)}} =: \tilde{\kappa}. \tag{10.38}$$

Note that $\frac{u(t)}{u_0} > 1$ because $u(t) < u_0 < 0$. Since $\tilde{\kappa} < +\infty$ (see Exercise 2 of Chapter 10), from (10.38) it follows that $T_{\max} < +\infty$. Moreover, from the phase portrait of Figure 10.24 we get

$$\lim_{t \uparrow T_{\max}} u(t) = -\infty.$$

Thus, letting $t \uparrow T_{\max}$ in (10.38) yields to

$$T_{\max} = \int_1^{+\infty} \frac{d\theta}{\sqrt{-\frac{2u_0}{3}(\theta^3 - 1) - \lambda(\theta^2 - 1)}}.$$

Now, suppose that $u_0 < -\lambda$ and $v_0 > 0$ are chosen so that $(u_0, v_0) \in W^s(-\lambda, 0)$, Then, by definition of stable manifold,

$$T_{\max} = +\infty \quad \text{and} \quad \lim_{t \uparrow +\infty} u(t; u_0, v_0) = -\lambda.$$

On the other hand, for every $t \in (T_{\min}, 0)$, since $u'(t) = v(t) > 0$, we have that

$$t = \int_0^t \frac{u'(s)}{v(s)} ds = \int_0^t \frac{u'(s)}{\sqrt{v_0^2 + \lambda\left(u_0^2 - u^2(s)\right) + \frac{2}{3}\left(u_0^3 - u^3(s)\right)}} ds$$

$$= \int_{u_0}^{u(t)} \frac{d\xi}{\sqrt{v_0^2 + \lambda\left(u_0^2 - \xi^2\right) + \frac{2}{3}\left(u_0^3 - \xi^3\right)}}$$

$$= -\int_{u(t)}^{u_0} \frac{d\xi}{\sqrt{v_0^2 + \lambda\left(u_0^2 - \xi^2\right) + \frac{2}{3}\left(u_0^3 - \xi^3\right)}}.$$

Thus,

$$
\begin{aligned}
-t &= \int_{u(t)}^{u_0} \frac{d\xi}{\sqrt{v_0^2 + \lambda\left(u_0^2 - \xi^2\right) + \frac{2}{3}\left(u_0^3 - \xi^3\right)}} \\
&< \int_{-\infty}^{u_0} \frac{d\xi}{\sqrt{v_0^2 + \lambda\left(u_0^2 - \xi^2\right) + \frac{2}{3}\left(u_0^3 - \xi^3\right)}} \\
&< \int_{-\infty}^{u_0} \frac{d\xi}{\sqrt{\lambda\left(u_0^2 - \xi^2\right) + \frac{2}{3}\left(u_0^3 - \xi^3\right)}} \\
&= \int_{1}^{+\infty} \frac{d\theta}{\sqrt{-\frac{2u_0}{3}\left(\theta^3 - 1\right) - \lambda\left(\theta^2 - 1\right)}} < +\infty,
\end{aligned}
$$

by Exercise 2 of Chapter 10. Consequently, $T_{\min} > -\infty$ and the graph of this solution looks like shown in the middle plot of Figure 10.26. This analysis can be easily adapted to show that, if $u_0 < -\lambda$ and $v_0 < 0$ satisfy $(u_0, v_0) \in W^u(-\lambda, 0)$, then $T_{\max} < +\infty$ and the graph of $u(t; u_0, v_0)$ looks like shown in the right plot of Figure 10.26.

10.5.2 *Phase portrait and solution types for $\lambda < 0$*

Throughout this section, we take $\lambda < 0$. Then, by (10.34), $\varphi''(0) < 0$ and $\varphi''(-\lambda) > 0$. Thus, 0 is a local maximum and $-\lambda > 0$ is a local minimum of $\varphi(u)$, and the graph of the potential energy looks like that shown in the upper plot of Figure 10.27. The corresponding phase portrait is represented in the lower plot. In this case, the integral curve corresponding to $E_0 = \varphi(0)$ consists of four trajectories: the equilibrium $(0,0)$, the *homoclinic connection* of $(0,0)$, plus the two remaining pieces of the unstable and stable manifolds of the saddle point $(0,0)$. The region enclosed by the homoclinic connection corresponds to a family of periodic solutions around $(-\lambda, 0)$, associated with the potential well at $u = -\lambda > 0$. By adapting the analysis of Section 10.5.1 (see Exercise 2 of Chapter 10), it is possible to show that the remaining trajectories are orbits of explosive solutions.

Figure 10.28 shows the graphs of the different types of admissible solutions of (10.31) when $\lambda < 0$. The first row plots a periodic solution, a solution on the homoclinic connection of $(0,0)$, and a solution

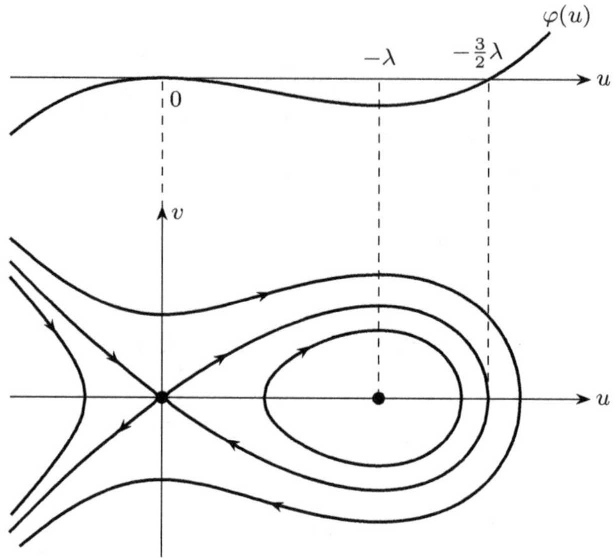

Fig. 10.27. Potential energy and phase portrait of (10.31) for $\lambda < 0$.

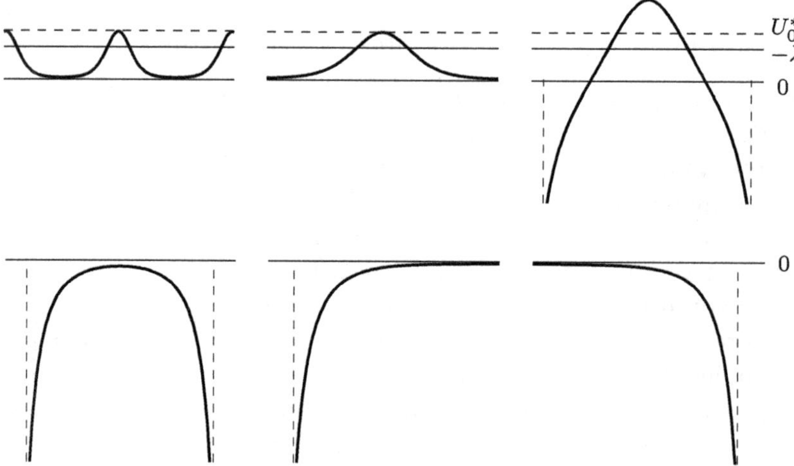

Fig. 10.28. Solutions of (10.31) for $\lambda < 0$.

on an exterior orbit. In these plots,

$$U_0^* = -\frac{3}{2}\lambda > 0$$

provides us with the abscissa of the unique pair of the form $(U_0^*, 0)$, with $U_0^* > 0$, lying on the homoclinic connection through $(0, 0)$ because

$$0 = \varphi(0) = \frac{\lambda}{2}(U_0^*)^2 + \frac{(U_0^*)^3}{3}.$$

The second row plots three solutions of (10.37) with $u_0 < 0$: the first one for $v_0 = 0$, the second one for $(u_0, v_0) \in W^s(0, 0)$, and the third one for $(u_0, v_0) \in W^u(0, 0)$.

In the light of this analysis, it becomes apparent how the case of $\lambda < 0$ is rather similar to the case of $\lambda > 0$, except for the roles of the equilibria $(0, 0)$ and $(-\lambda, 0)$, which have been interchanged.

10.5.3 *Phase portrait and solution types for $\lambda = 0$*

When $\lambda = 0$, the potential energy associated with (10.31) is $\varphi(u) = \frac{u^3}{3}$, $u \in \mathbb{R}$. Thus, the graph of $\varphi(u)$ and the corresponding phase portrait look like those shown in Figure 10.29. Now, 0 is the unique constant solution of (10.31), which becomes

$$-u'' = u^2.$$

The integral curve associated with the energy level $E_0 = \varphi(0) = 0$ consists of three trajectories: the equilibrium point $(0, 0)$ plus its stable and unstable manifolds.

Figure 10.30 shows the graphs of the solutions of

$$\begin{cases} -u'' = u^2, \\ u(0) = u_0, \quad u'(0) = v_0, \end{cases} \tag{10.39}$$

for different choices of $(u_0, v_0) \in \mathbb{R}^2$.

Let us denote by $(I, u(\cdot; u_0, v_0))$ the unique maximal solution of (10.39), with $I = (T_{\min}, T_{\max})$. We claim that, for $u_0 > 0$ and $v_0 = 0$, the solution $u(t; u_0, 0)$ satisfies

$$T_{\max} < +\infty, \quad T_{\min} = -T_{\max}. \tag{10.40}$$

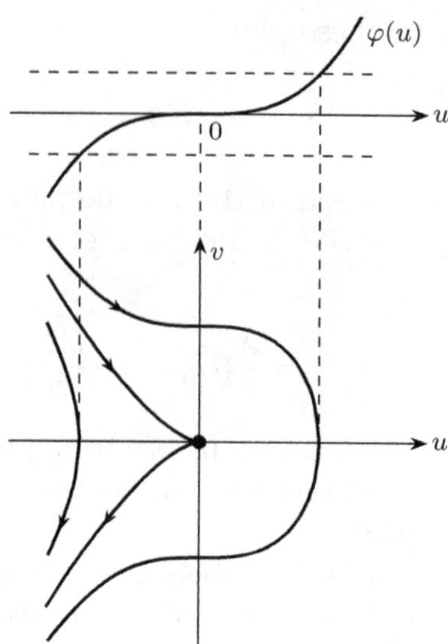

Fig. 10.29. Potential energy and phase portrait of (10.31) for $\lambda = 0$.

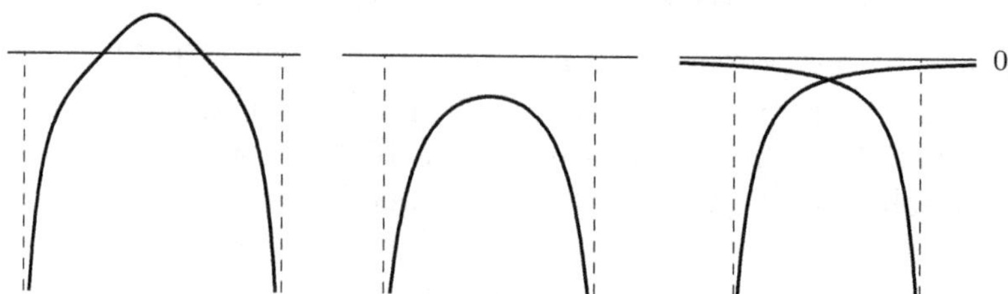

Fig. 10.30. Solutions of $-u'' = u^2$.

Indeed, since $u'(t) = v(t) < 0$ for all $t \in (0, T_{\max})$, we have that

$$t = \int_0^t \frac{u'(s)}{v(s)}\, ds = -\int_0^t \frac{u'(s)}{\sqrt{\frac{2}{3}\left(u_0^3 - u^3(s)\right)}}\, ds = \int_{u(t)}^{u_0} \frac{d\xi}{\sqrt{\frac{2}{3}\left(u_0^3 - \xi^3\right)}}$$

$$= \int_{\frac{u(t)}{u_0}}^{1} \frac{u_0\, d\theta}{\sqrt{\frac{2}{3}u_0^3\left(1 - \theta^3\right)}} < \sqrt{\frac{3}{2u_0}} \int_{-\infty}^{1} \frac{d\theta}{\sqrt{1 - \theta^3}} < +\infty.$$

Thus, $T_{\max} < +\infty$. Moreover, from the phase portrait, it is apparent that

$$\lim_{t \uparrow T_{\max}} u(t; u_0, 0) = -\infty.$$

Thus, letting $t \uparrow T_{\max}$ in the previous identity, we find that

$$T_{\max} = \sqrt{\frac{3}{2u_0}} \int_{-\infty}^{1} \frac{d\theta}{\sqrt{1 - \theta^3}}.$$

That $T_{\min} = -T_{\max}$ is a direct consequence of the fact that $u(-t; u_0, 0) = u(t; u_0, 0)$ for all $t \in I$. Note that

$$\lim_{u_0 \downarrow 0} T_{\max} = +\infty, \qquad \lim_{u_0 \uparrow +\infty} T_{\max} = 0.$$

These kinds of solutions have been represented in the left-hand plot of Figure 10.30.

The second plot shows the graph of $u(t; u_0, 0)$, with $u_0 < 0$. As in the previous case, (10.40) holds true. Indeed, arguing as above and taking into account that $u_0 < 0$, we find that, for all $t \in (0, T_{\max})$,

$$t = -\int_{1}^{\frac{u(t)}{u_0}} \frac{u_0 \, d\theta}{\sqrt{\frac{2}{3} u_0^3 (1 - \theta^3)}} = \sqrt{\frac{3}{2|u_0|}} \int_{1}^{\frac{u(t)}{u_0}} \frac{d\theta}{\sqrt{\theta^3 - 1}}.$$

Thus, reasoning as above leads to

$$T_{\max} = \sqrt{\frac{3}{2|u_0|}} \int_{1}^{+\infty} \frac{d\theta}{\sqrt{\theta^3 - 1}} < +\infty.$$

The right-hand plot of Figure 10.30 shows the equilibrium as well as two solutions on the stable and unstable manifolds of $(0,0)$. The previous arguments can be adapted to show that also these solutions blow up (backward and forward, respectively) in a finite time.

10.5.4 *Positive solutions of the associated BVP*

In this section, for a fixed $L > 0$, we determine the set of positive solutions of the boundary value problem

$$\begin{cases} -u'' = \lambda u + u^2 & \text{in } (0, L), \\ u(0) = u(L) = 0. \end{cases} \tag{10.41}$$

First, we consider the case of $\lambda > 0$. From the phase portrait in Figure 10.24, it becomes apparent that, for every $V_0 > 0$, there exists a minimal time, τ, for which the solution of

$$\begin{cases} -u'' = \lambda u + u^2 \\ u(0) = 0, \ u'(0) = V_0, \end{cases} \tag{10.42}$$

satisfies $u(\tau) = U_0$ and $u'(\tau) = 0$. In other words, τ is the time taken to travel from $(0, V_0)$ to $(U_0, 0)$ along the corresponding trajectory in the phase plane. We emphasize the dependence of τ on U_0 and λ by setting $\tau = \tau(\lambda, U_0)$. Note that, owing to (10.33),

$$\frac{v^2(t)}{2} + \lambda \frac{u^2(t)}{2} + \frac{u^3(t)}{3} = \frac{V_0^2}{2} = \lambda \frac{U_0^2}{2} + \frac{U_0^3}{3} \quad \text{for all } t \in I,$$

where I is the existence interval of the (maximal) solution of (10.42). To determine τ, we proceed as follows. Since $v(t) = u'(t) > 0$ for all $t \in (0, \tau)$,

$$\tau = \int_0^\tau \frac{u'(t)}{v(t)}\, dt = \int_0^\tau \frac{u'(t)}{\sqrt{\lambda\left(U_0^2 - u^2(t)\right) + \frac{2}{3}\left(U_0^3 - u^3(t)\right)}}\, dt$$

$$= \int_0^{U_0} \frac{d\xi}{\sqrt{\lambda\left(U_0^2 - \xi^2\right) + \frac{2}{3}\left(U_0^3 - \xi^3\right)}}$$

and, hence,

$$\tau(\lambda, U_0) = \int_0^1 \frac{d\theta}{\sqrt{\lambda\left(1 - \theta^2\right) + \frac{2U_0}{3}\left(1 - \theta^3\right)}}. \tag{10.43}$$

Thus, τ is continuous and decreasing with respect to $U_0 > 0$ and $\lambda > 0$. Moreover, for every $\lambda > 0$, we have that

$$\tau(\lambda, 0) := \lim_{U_0 \downarrow 0} \tau(\lambda, U_0) = \int_0^1 \frac{d\theta}{\sqrt{\lambda\left(1 - \theta^2\right)}} = \frac{\pi}{2\sqrt{\lambda}}$$

and

$$\lim_{U_0 \uparrow +\infty} \tau(\lambda, U_0) = 0. \tag{10.44}$$

Therefore, the graph of $\tau(\lambda, U_0)$ looks like that shown in the left-hand plot of Figure 10.31. The fact that τ decreases with respect to

λ is illustrated in the right-hand plot of the same figure, where we represent $\tau(\lambda_1, \cdot)$ and $\tau(\lambda_2, \cdot)$, with $\lambda_1 < \lambda_2$.

By the symmetries of the problem, the solution of (10.42) provides us with a positive solution of (10.41) if and only if

$$\tau(\lambda, U_0) = \frac{L}{2}.$$

Thus, by the first plot of Figure 10.31, problem (10.41) has a positive solution if and only if $\frac{L}{2} < \frac{\pi}{2\sqrt{\lambda}}$, i.e. $\lambda < \left(\frac{\pi}{L}\right)^2 =: \sigma_1$. Furthermore, by the monotonicity of τ with respect to U_0, the positive solution is unique if it exists. Suppose $\lambda < \sigma_1$ and denote by $u_{0,\lambda}$ the unique positive solution of (10.41). Then,

$$U_0(\lambda) = u_{0,\lambda}\left(\tfrac{L}{2}\right) = \|u_{0,\lambda}\|_\infty.$$

Moreover, as illustrated in the right-hand plot of Figure 10.31, since $\tau(\lambda, U_0)$ is decreasing in λ, $U_0(\lambda)$ also decreases in λ, and

$$\lim_{\lambda \uparrow \sigma_1} U_0(\lambda) = 0.$$

Therefore, the curve of positive solutions $(\lambda, u_{0,\lambda})$ bifurcates toward the left, i.e. sub-critically, from the curve of trivial solutions $(\lambda, 0)$ at $\lambda = \sigma_1$. Moreover, if $U_0(\lambda)$ is chosen to represent $u_{0,\lambda}$ schematically, it turns out that $U_0(\lambda)$ increases as λ decreases from σ_1 until it reaches $\lambda = 0$.

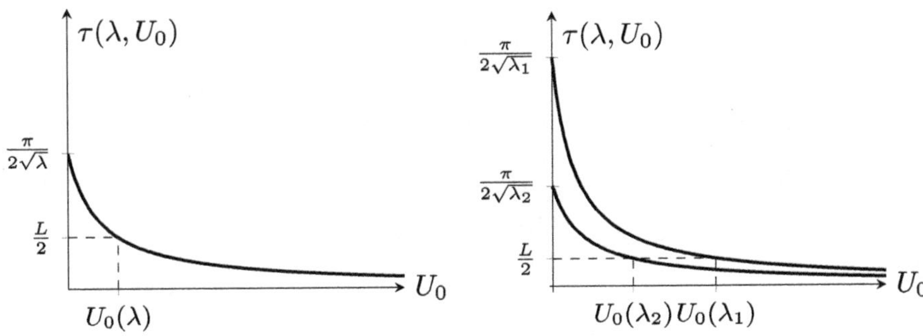

Fig. 10.31. The time map $\tau(\lambda, \cdot)$ for $\lambda > 0$.

Actually, problem (10.41) has a unique positive solution also for $\lambda = 0$. Indeed, since

$$\tau(0, U_0) = \int_0^1 \frac{d\theta}{\sqrt{\frac{2U_0}{3}(1 - \theta^3)}} = \sqrt{\frac{3}{2U_0}} \int_0^1 \frac{d\theta}{\sqrt{1 - \theta^3}},$$

the positive solution at $\lambda = 0$ is determined by

$$\frac{L}{2} = \sqrt{\frac{3}{2U_0}} \int_0^1 \frac{d\theta}{\sqrt{1 - \theta^3}},$$

which gives

$$U_0(0) := \frac{6}{L^2} \left(\int_0^1 \frac{d\theta}{\sqrt{1 - \theta^3}} \right)^2.$$

Observe that (10.43) provides us with the value of $\tau(\lambda, U_0)$ also for $\lambda < 0$, though, in this case, U_0 must be greater than $-\frac{3}{2}\lambda$ (see Figure 10.27). As, according to Theorem 5.26,

$$\lim_{U_0 \downarrow -\frac{3}{2}\lambda} \tau(\lambda, U_0) = +\infty,$$

the graph of τ in this case looks like that shown in the left-hand picture in Figure 10.32. Consequently, for every $\lambda < 0$, there exists a unique $U_0(\lambda) \in \left(-\frac{3}{2}\lambda, +\infty\right)$ such that $\tau(\lambda, U_0(\lambda)) = \frac{L}{2}$. Therefore, for all $\lambda < 0$, (10.41) possesses a unique positive solution, which

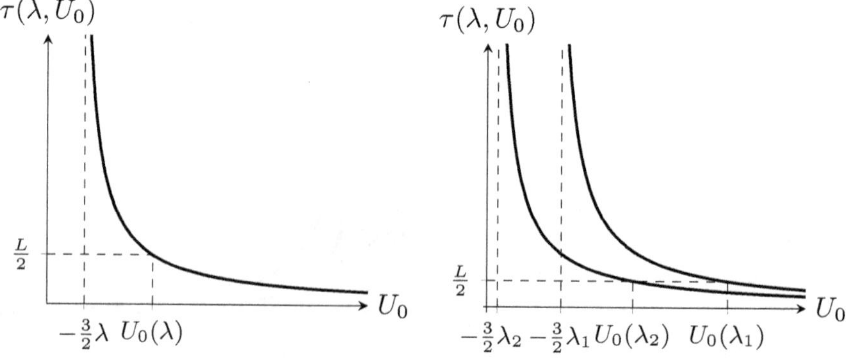

Fig. 10.32. The time map $\tau(\lambda, \cdot)$ for $\lambda < 0$.

we denote by $u_{0,\lambda}$. Moreover, as sketched in the right-hand picture in Figure 10.32, $U_0(\lambda_2) < U_0(\lambda_1)$ if $\lambda_1 < \lambda_2 < 0$. Thus, $U_0(\lambda)$ is decreasing with respect to $\lambda < 0$. In addition, since

$$U_0(\lambda) > -\frac{3}{2}\lambda \quad \text{for all } \lambda < 0,$$

we deduce that

$$\lim_{\lambda \downarrow -\infty} \|u_{0,\lambda}\|_\infty = \lim_{\lambda \downarrow -\infty} U_0(\lambda) = +\infty.$$

Actually, as $U_0(\lambda)$ is continuous up to $\lambda = 0$, by combining the analysis of the different cases, we obtain that $U_0(\lambda)$ is continuous in $(-\infty, \sigma_1)$.

Figure 10.33 shows the global bifurcation diagram of the positive solutions of (10.41). Contrarily to what happened in the sublinear problem treated in Section 10.4, the positive solutions arise for $\lambda < \sigma_1$, rather than for $\lambda > \sigma_1$. Basically, this difference is due to the fact that the time map $\tau = \tau(\lambda, U_0)$ is decreasing with respect to λ in the superlinear problem, while it was increasing for the sublinear problem in Section 10.4.

Another important feature of the superlinear problem analyzed in this section is that the larger the amplitude of the solution, measured by U_0, the shorter the time taken to connect $(0, V_0)$ with $(U_0, 0)$. Actually, by (10.44), such a time converges to zero as $U_0 \uparrow +\infty$. Therefore, the larger the orbit, the faster the arc of trajectory linking $(0, V_0)$ to $(U_0, 0)$ is traveled.

Summarizing, the following result holds.

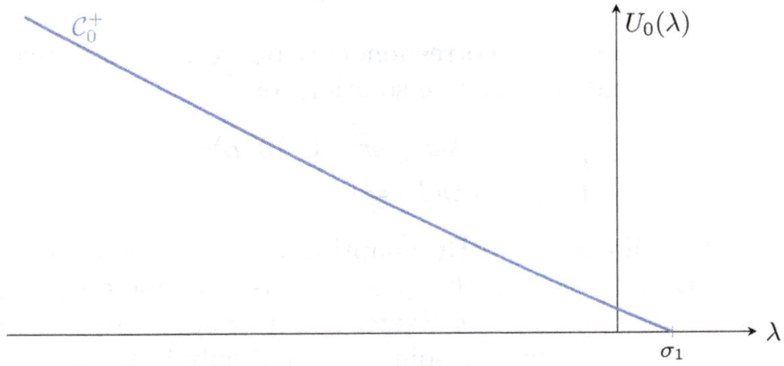

Fig. 10.33.   Global bifurcation diagram of positive solutions of (10.41).

Theorem 10.6. *For every $L > 0$, the following properties are satisfied:*

(a) *The boundary value problem (10.41) has a positive solution if and only if $\lambda < \sigma_1 =: \left(\frac{\pi}{L}\right)^2$. Moreover, it is unique if it exists. If we denote it by $u_{0,\lambda}$, then*

$$\|u_{0,\lambda}\|_\infty = u_{0,\lambda}\left(\tfrac{L}{2}\right) = U_0(\lambda) \quad \textit{for all } \lambda < \sigma_1.$$

(b) *Identifying $u_{0,\lambda}$ with $U_0(\lambda)$, the set of positive solutions of (10.41),*

$$\mathcal{C}_0^+ := \{(\lambda, U_0(\lambda)) \colon \lambda < \sigma_1\},$$

is a continuous curve, parameterized by $\lambda < \sigma_1$, which bifurcates from the trivial solution $u = 0$ at $\lambda = \sigma_1$, in the sense that

$$\lim_{\lambda \uparrow \sigma_1} \|u_{0,\lambda}\|_\infty = 0.$$

Moreover, $\lambda \mapsto U_0(\lambda) = \|u_{0,\lambda}\|_\infty$ is decreasing.

In order to characterize the existence of negative solutions for the boundary value problem (10.41) one cannot proceed as in Section 10.4 because

$$f(u) = \lambda u + u^2, \quad u \in \mathbb{R},$$

is not odd. In this case, we can perform the change of variable $u = -w$, which transforms (10.31) into

$$-w'' = \lambda w - w^2.$$

Thus, there is a one-to-one correspondence between the negative solutions of (10.41) and the positive solutions of

$$\begin{cases} -w'' = \lambda w - w^2 & \text{in } (0, L), \\ w(0) = w(L) = 0, \end{cases}$$

which is of sublinear type. By adapting the arguments in Section 10.4.3, it is easily seen that this problem has a unique positive solution if and only if $\lambda > \sigma_1$ (see Exercise 3 of Chapter 10). Therefore, (10.41) has a unique negative solution if and only if $\lambda > \sigma_1$.

The analysis of the existence of nodal solutions for the boundary value problem (10.41) is slightly more intricate than that in Section 10.4.4 because $f(u)$ is not odd. By analyzing the phase portraits in Figures 10.27 and 10.29, it readily follows that they cannot exist if $\lambda \leq 0$. The detailed analysis of this case is proposed in Exercise 10 of Chapter 10.

10.6 Exercises

1. Complete the technical details of the proof of Lemma 10.3.
2. Complete the technical details of the proof that the solutions of (10.31), with $\lambda > 0$, whose trajectories lie outside the integral curve through $(-\lambda, 0)$ blow up in a finite time. Precisely, show that κ defined in (10.36) and $\tilde{\kappa}$ defined in (10.38) are finite. Then, show that, for $\lambda < 0$, the solutions of (10.31) whose trajectories lie outside the homoclinic connection through $(0, 0)$ also blow up in a finite time.
3. Consider, for every $\lambda > 0$, the diffusive logistic equation

$$-u'' = \lambda u - u^2.$$

   (a) Construct the associated phase portrait.
   (b) Analyze the nature of its solutions.
   (c) Characterize the existence of positive solutions of the boundary value problem

$$\begin{cases} -u'' = \lambda u - u^2 & \text{in } (0, L), \\ u(0) = u(L) = 0, \end{cases}$$

   and analyze their global structure as λ varies.

4. Consider, for every $\lambda \in \mathbb{R}$, the differential equation

$$-u'' = \lambda u + u^3.$$

   (a) Construct the phase portrait for $\lambda > 0$, $\lambda < 0$, and $\lambda = 0$.
   (b) Analyze the different nature of its solutions according to the values of $\lambda \in \mathbb{R}$.

(c) Characterize the existence of positive solutions of the boundary value problem

$$\begin{cases} -u'' = \lambda u + u^3 & \text{in } (0, L), \\ u(0) = u(L) = 0, \end{cases} \tag{10.45}$$

and analyze their global structure as λ varies.

(d) Characterize the existence of nodal solutions for problem (10.45).

5. Determine the global bifurcation diagram of positive and nodal solutions for the following linear problem:

$$\begin{cases} -u'' = \lambda u & \text{in } (0, L), \\ u(0) = u(L) = 0. \end{cases}$$

Compare such a bifurcation diagram with the corresponding ones for (10.14) and (10.45).

6. Suppose that the graph of the potential energy, $\varphi(u)$, of the differential equation $-u'' = f(u)$ looks like that shown in Figure 10.34, with

$$\lim_{u \uparrow +\infty} \varphi(u) = 0.$$

Construct the phase portrait of this differential equation, and discuss the nature of all its solutions.

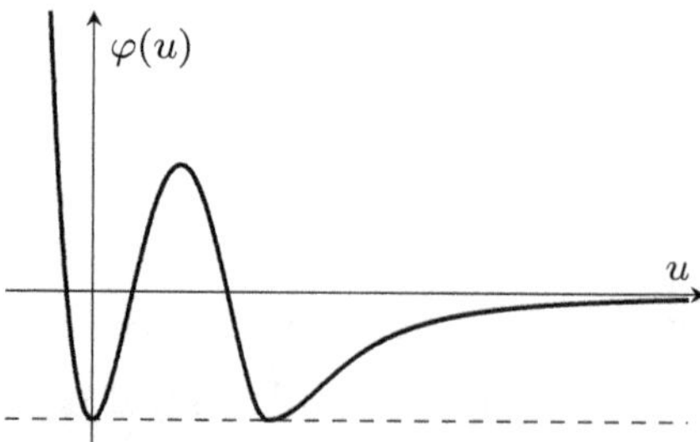

Fig. 10.34.   Potential energy of Exercise 6 of Chapter 10.

7. Consider the differential equation

$$u'' = e^u + e^{-u}.$$

(a) Construct the corresponding phase portrait.

(b) Discuss the nature of its solutions.

8. Fix $p > 1$, and for every $\alpha > 0$, consider the Cauchy problem

$$\begin{cases} -u'' = |u|^{p-1}u, \\ u(0) = \alpha, \ u'(0) = 0. \end{cases}$$

Let (I, u) denote its (maximal) solution.

(a) Prove that there exists $R_\alpha > 0$ such that $R_\alpha \in I$, $u(R_\alpha) = 0$, and $u(t) > 0$ for all $t \in [0, R_\alpha)$.
(b) Determine R_α as a function of α.
(c) Show that $\lim_{\alpha \uparrow +\infty} R_\alpha = 0$ and $\lim_{\alpha \downarrow 0} R_\alpha = +\infty$.

9. Consider the Cauchy problem

$$\begin{cases} u'' = -8B^2 u \left[1 + (u')^2\right]^{\frac{3}{2}}, \\ u(\tfrac{1}{2}) = \tfrac{1}{2B}, \quad u'(\tfrac{1}{2}) = 0, \end{cases}$$

where

$$B = \int_0^1 \frac{d\theta}{\sqrt{\theta^{-4} - 1}}.$$

Prove the existence of a unique maximal solution to the right, (I, u), and show that $I = [\tfrac{1}{2}, 1)$ and

$$\lim_{t \uparrow 1} u(t) = 0, \quad \lim_{t \uparrow 1} u'(t) = -\infty.$$

10. For $\lambda > 0$ fixed, analyze the existence of nodal solutions for the boundary value problem

$$\begin{cases} -u'' = \lambda u + u^2 \quad \text{in } (0, L), \\ u(0) = u(L) = 0. \end{cases}$$

[Hint: Determine the limiting period of the periodic solutions of the differential equation as their amplitudes converge to zero.]

11. Construct the set of positive and nodal solutions of the problem

$$\begin{cases} -u'' = \lambda u - |u|^{p-1}u \quad \text{in } (0, L), \\ u(0) = u(L) = 0, \end{cases}$$

for every $p > 1$.

12. Construct the set of positive and nodal solutions of the problem

$$\begin{cases} -u'' = \lambda u + |u|^{p-1}u & \text{in } (0, L), \\ u(0) = u(L) = 0, \end{cases}$$

for every $p > 1$.

13. Construct the set of positive and nodal solutions of the problem

$$\begin{cases} -u'' = \lambda u + |u|^{p-1}u & \text{in } (0, L), \\ u(0) = u(L) = 0, \end{cases}$$

for every $p \in (0, 1)$. [Hint: cf. Figure 3 of López-Gómez *et al.*, 2023, if necessary.]

14. Construct the set of positive and nodal solutions of the problem

$$\begin{cases} -u'' = \lambda u - |u|^{p-1}u & \text{in } (0, L), \\ u(0) = u(L) = 0, \end{cases}$$

for every $p \in (0, 1)$.

10.7   Final Comments

Although the contents of this chapter are rather classical, specially those coming from Classical Mechanics, the use of techniques from the theory of planar Dynamical Systems in other fields, such as Ecology or Population Dynamics, with the aim of constructing the global bifurcation diagrams of the solutions of

$$\begin{cases} -u'' = \lambda u + f(u) & \text{in } (0, L), \\ u(0) = u(L) = 0, \end{cases} \tag{10.46}$$

where f is a regular function such that

$$f(0) = f'(0) = 0,$$

is far from standard for an introductory textbook.

Analyzing (10.46) is mandatory as a first step in trying to understand the structure of the set of solutions of more general multidimensional parametric problems of the form

$$\begin{cases} -\Delta u = \lambda u + f(u) & \text{in } \Omega, \\ u = 0 & \text{on } \partial\Omega, \end{cases} \tag{10.47}$$

where Ω is a connected open bounded subset of $\mathbb{R}^N$, $N \geq 1$, with sufficiently smooth boundary, $\partial\Omega$, and Δ is the Laplace operator in $\mathbb{R}^N$.

In far more circumstances than desirable, there is currently no systematic way to get information about positive or nodal solutions for (10.47) unless we restrict ourselves to the simpler one-dimensional prototype (10.46). Actually, the structure of the solution set of the multidimensional problem (10.46) for the special choice of

$$f(u) = |u|^{p-1}u, \quad u \in \mathbb{R},$$

with $p > 1$, might depend on the size and geometry of Ω, as well as on the size of p, because, for $p \geq \frac{N+2}{N-2}$ and $N \geq 3$, *a priori* bounds for positive solutions on compact subintervals of the parameter λ are not available (see Amann and López-Gómez, 1998, and the references therein).

The phase portrait techniques introduced in this chapter for autonomous second-order differential equations can be adapted, not without technical difficulties, to study positive and nodal solutions of piece-wise autonomous equations of the type

$$-u'' = \lambda u + a(x)|u|^{p-1}u,$$

where $p > 1$ and $a(x)$ is a piecewise constant weight function. According to the terminology introduced in Section 10.5, the sign of $a(x)$ determines the nature of the equation. Some recent references for these non-autonomous differential equations are Cubillos *et al.* (2022) and López-Gómez and Rabinowitz (2020) for the sublinear and sublinear degenerate case ("degenerate" in the sense that $a(x)$ might vanish on some subintervals), López-Gómez *et al.* (2023) for the superlinear case, López-Gómez *et al.* (2014) and López-Gómez and Tellini (2014) for the superlinear indefinite case where $a(x)$ changes sign, thus simultaneously combining sublinear and superlinear features.

Exercise 9 of Chapter 10 goes back to Cano-Casanova *et al.* (2012), which the interested reader can refer for any further technical details.

Finally, we point out that ascertaining the monotonicity properties of the period of periodic solutions as a function of their amplitude can be, in general, a real challenge. In Section 10.4, thanks to the symmetries of the problem, we have been able to obtain (10.19), which shows that the period of the periodic solutions of (10.13) around $(0,0)$ is increasing with respect to the amplitude $U_0 \in (0, \sqrt{\lambda})$. However, the monotonicity of the period of the periodic solutions around (μ, λ) for the Lotka–Volterra predator-prey system considered in Section 9.3 is a recent result established by Rothe in 1985 and, independently, by Waldvogel in 1986.

Chapter 11

Non-Conservative Systems

In the previous two chapters, we have constructed the global dynamics of the considered systems by determining the integral curves and the trajectories through any point of the phase space. Nevertheless, in many cases, this approach is not possible, and more sophisticated techniques in the theory of Dynamical Systems are required to ascertain the global dynamics of the system.

As a paradigmatic example, we consider the free nonlinear pendulum with an additional friction term, assuming that the friction force is proportional to the (angular) velocity of the pendulum and denoting the proportionality constant by $\kappa > 0$. The corresponding equation reads

$$u'' = -\frac{\kappa}{M}u' - \frac{G}{L}\sin u, \tag{11.1}$$

where M is the mass of the pendulum, L is its length, and G is the gravity constant. Naturally, the constant κ depends on the viscosity of the medium in which the pendulum oscillates. For example, this coefficient in air is much smaller than in water and, in turn, that of water is much smaller than in gelatin. Observe that the frictionless case introduced in Chapter 6 and analyzed in Chapter 10 corresponds to the case $\kappa = 0$.

Let (I, u) be a (maximal) solution of (11.1), and consider the total energy introduced in Section 10.3:

$$E(t) = \frac{v^2(t)}{2} - \frac{G}{L}\cos u(t), \quad t \in I.$$

595

When $\kappa = 0$, we have seen in Section 10.3 that $E'(t) = 0$ for all $t \in I$. Thus, $E(t)$ retains the initial value $E(0)$ for all $t \in I$. When $\kappa > 0$ instead, a direct calculation shows that, for every $t \in I$,

$$E'(t) = u'(t)u''(t) + \frac{G}{L}u'(t)\sin u(t)$$

$$= u'(t)\left(-\frac{\kappa}{M}u'(t) - \frac{G}{L}\sin u(t)\right) + \frac{G}{L}u'(t)\sin u(t)$$

$$= -\frac{\kappa}{M}(u'(t))^2 \le 0.$$

Thus, the sum of the kinetic and potential energies decreases in time in the presence of friction. For this reason, this kind of system is referred to as *dissipative*. Physically, this corresponds to the fact that friction transforms part of the initial energy of the pendulum into heat, with the effect that the amplitude of its oscillations decreases, and the pendulum stabilizes toward an equilibrium position as time passes by. Moreover, the physical intuition leads us to conjecture that, generically, the pendulum will eventually stabilize to the lower vertical position. In other words, the equilibria of center type of the free pendulum become stable foci, or even stable nodes, in the presence of frictional forces. Observe that this occurs also in the case of the linear pendulum, considered in Section 1.11.

As the methods used in Chapter 10 are not suitable for non-conservative systems, in this chapter we present some new techniques for analyzing general non-conservative problems; among them are those of dissipative type such as (11.1).

The chapter is organized as follows. In Section 11.1, we present the concepts of stability, instability, and asymptotic stability, as introduced by Lyapunov (1907) in his seminal memoir. Then, we characterize, in the sense of Lyapunov, the character of linear systems with constant coefficients, $u' = Au$, by means of the sign and algebraic ascents of the eigenvalues of A.

In Section 11.2, we study the so-called *Lyapunov's first method*, which consists of getting information about the stability properties of an equilibrium from its linearized, or variational, equation. Once this method has been discussed, Lyapunov's stability and instability theorems are stated and proven. Finally, the theorem given by Grobman and Hartman is stated and discussed in depth, without proof.

In Section 11.3, we introduce the concepts of α- and ω-limit sets, going back to Birkhoff (1927), who also coined the modern concept of *Dynamical System*. Once these concepts are introduced, we analyze their most fundamental properties, including their compactness, connectedness, and invariance by the flow of $u' = f(u)$.

In Section 11.4, we present and prove the Poincaré–Bendixson theorem. This result was stated and proved by Poincaré in 1892, although some technical details in Poincaré's original proof were later completed by Bendixson in 1901, which is why the theorem bears both names. This result establishes an important property of the ω-limit sets of a planar system. Precisely, they are the orbit of a nontrivial periodic solution if they do not contain any equilibria.

In Section 11.5, we apply the previous theory to study the dynamics of positive solutions of the Lotka–Volterra competition model introduced in (5.60). This model goes back to Volterra (1931) and Lotka (1932).

The concept of competition, as used by most ecologists, describes the active demand for a common resource or need, which is actually or potentially limiting, by a number of individuals of the same species (intraspecific competition) or members of several species at the same trophic level (interspecific competition). We refer the interested reader to, e.g. Wilson (1980) and Begon *et al.* (1996) for a more complete presentation of this pivotal concept in Ecology and Population Dynamics. Despite its simplicity, the Lotka–Volterra competition model (5.60) is the basis of the mathematical modeling of competition due to its effectiveness.

In Section 11.6, we study a predator-prey model that takes into account saturation effects of the predator in the abundance of preys. Such a model, which is referred to as the *Holling–Tanner predator-prey model*, after the work of the mathematical ecologists Holling (1959) and Tanner (1975), represents a more realistic version of the classical Lotka–Volterra predator-prey model introduced in Exercise 8 of Chapter 5. The diffusive counterpart of this model has been analyzed by Casal *et al.* (1994).

In Section 11.7, we introduce, without proofs, *Lyapunov's second method*. It will allow us to analyze the local and global character of isolated equilibria in autonomous systems by means of the so-called *Lyapunov functions*. Finally, in Section 11.8, we derive and analyze the phase portrait of the damped pendulum.

11.1   Lyapunov Stability

Throughout this section, we take $t_0 \in \mathbb{R}$, set $J := [t_0, +\infty)$, and assume that $f : [t_0, +\infty) \times \mathbb{R}^N \to \mathbb{R}^N$ is locally Lipschitz in $u \in \mathbb{R}^N$ uniformly on compact subsets of J. Then, Theorem 5.17 guarantees that, for every $u_0 \in \mathbb{R}^N$, the Cauchy problem

$$\begin{cases} u' = f(t, u), \\ u(t_0) = u_0, \end{cases}$$

has a maximal solution, $u(t; t_0, u_0)$, defined in $I = [t_0, T_{\max})$ for some $T_{\max} \leq +\infty$. Suppose that $T_{\max} = +\infty$. Then, the solution $u(t; t_0, u_0)$ is said to be stable (to the right) in the sense of Lyapunov, if, by choosing x sufficiently close to u_0, $u(t; t_0, x)$ is defined in $[t_0, +\infty)$ and remains arbitrarily close to $u(t; t_0, u_0)$, i.e. if $u(t; t_0, \cdot)$ varies continuously with respect to $x \sim u_0$ for all $t \in [t_0, +\infty)$. The precise definition is the following.

Definition 11.1 (Lyapunov stability). Suppose that $u(t; t_0, u_0)$ is defined in $[t_0, +\infty)$. Then, $u(t; t_0, u_0)$ is said to be *stable (to the right) in the sense of Lyapunov*, or *(right) Lyapunov stable*, if:

(i)   there exists $\eta > 0$ such that, for every $x \in \mathbb{R}^N$ with $\|x - u_0\| \leq \eta$, also the solution $u(t; t_0, x)$ is defined for all $t \geq t_0$;

(ii)   for every $\varepsilon > 0$, there exists $\delta \in (0, \eta)$ such that, as soon as $\|x - u_0\| \leq \delta$, we have

$$\|u(t; t_0, u_0) - u(t; t_0, x)\| \leq \varepsilon \quad \text{for all } t \geq t_0.$$

Moreover, when $\delta > 0$ can be chosen in such a way that, for $\|x - u_0\| \leq \delta$,

$$\lim_{t \uparrow +\infty} \|u(t; t_0, u_0) - u(t; t_0, x)\| = 0,$$

then $u(t; t_0, u_0)$ is said to be *asymptotically stable*.

The solution $u(t; t_0, u_0)$ is said to be *exponentially asymptotically stable* if, in addition, there exist two constants $M > 0$ and $\omega > 0$ such that

$$\|u(t; t_0, u_0) - u(t; t_0, x)\| \leq M e^{-\omega(t - t_0)} \quad \text{for all } t \geq t_0.$$

Finally, the solution $u(t; t_0, u_0)$ is said to be *unstable* if it is not stable.

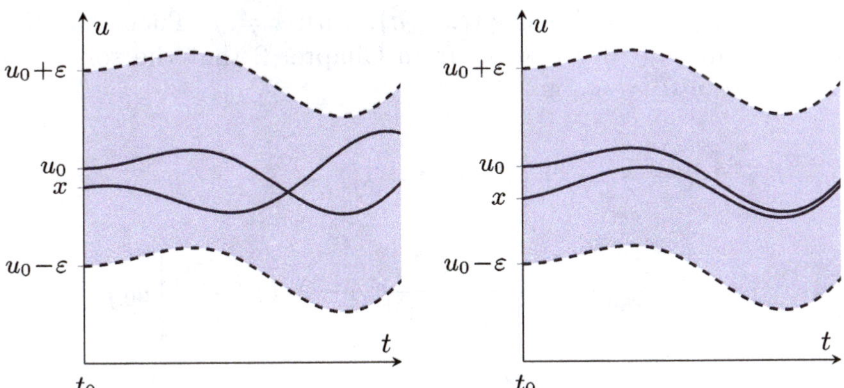

Fig. 11.1. Lyapunov stability (left) and asymptotic Lyapunov stability (right).

The left plot in Figure 11.1 shows an example of the right Lyapunov stability, while the second one shows an example of the asymptotic Lyapunov stability.

When $J = (-\infty, t_0]$, the concept of *stability to the left, in the sense of Lyapunov*, or *(left) Lyapunov stable*, can be defined in a similar way. In this chapter, we mainly focus on (asymptotic) stability to the right, although the results and analysis can be adapted to the case of left stability.

Simple examples, like the logistic equation analyzed in Section 4.1, reveal the stability/instability behavior to the left is independent from the behavior to the right. Indeed, the constant solution $u = 1$ of $u' = u - u^2$ is asymptotically stable to the right, but, for every $t_0 \in \mathbb{R}$ and $\varepsilon > 0$, the solution $u(t; t_0, 1 + \varepsilon)$ blows up at some $T_{\min} \in (-\infty, t_0)$, showing that $u = 1$ is Lyapunov unstable to the left.

Although, in principle, the concepts of Lyapunov stability/instability or asymptotic stability might depend on the specific value of t_0, there are many cases in which they do not. For example, consider the homogeneous linear system

$$u' = Au \tag{11.2}$$

where $A = (a_{ij})_{1 \le i,j \le N}$ is a matrix of order N with real coefficients, $a_{ij} \in \mathbb{R}$, and suppose that

$$\sigma(A) = \{\lambda_1, \ldots, \lambda_q\},$$

with $\lambda_i \neq \lambda_j$ for all $i, j \in \{1, \ldots, q\}$, with $i \neq j$. Then, for every $u_0 \in \mathbb{R}^N$ and $\tau \geq 0$, we know from Chapter 2 that the solution of (11.2), with $u(0) = u_0$, is

$$e^{\tau A} u_0 = \sum_{j=1}^{q} e^{\tau \lambda_j} \left[I + \tau(A - \lambda_j I) \right.$$
$$\left. + \cdots + \frac{\tau^{\nu(\lambda_j)-1}}{(\nu(\lambda_j) - 1)!}(A - \lambda_j I)^{\nu(\lambda_j)-1} \right] u_{0,j},$$

where

$$u_0 = \sum_{j=1}^{q} u_{0,j}, \quad \text{with } u_{0,j} \in N\left[(A - \lambda_j I)^{\nu(\lambda_j)}\right], \quad j \in \{1, \ldots, q\}.$$

Thus, for every $\tau \geq 0$, we have that

$$\|e^{\tau A} u_0\| \leq \sum_{j=1}^{q} e^{\tau \operatorname{Re} \lambda_j} P_j(\tau) \|u_{0,j}\|, \tag{11.3}$$

where

$$P_j(\tau) = \sum_{k=0}^{\nu(\lambda_j)-1} \frac{\|A - \lambda_j I\|^k}{k!} \tau^k, \quad j \in \{1, \ldots, q\}.$$

Subsequently, we distinguish three cases, according to the sign of the real parts of the eigenvalues of A:

Case 1. $\sigma(A) \subset \{z \in \mathbb{C} : \operatorname{Re} z < 0\}$. In other words,

$$\omega := \max_{j \in \{1, \ldots, q\}} \operatorname{Re} \lambda_j < 0.$$

Then, for every $\eta \in (0, -\omega)$, there exists a constant $C = C(\eta) > 0$ such that, for every $j \in \{1, \ldots, q\}$ and $\tau \geq 0$,

$$e^{\tau \operatorname{Re} \lambda_j} P_j(\tau) \leq e^{\tau \omega} P_j(\tau) \leq e^{\tau(\omega+\eta)} e^{-\tau \eta} P_j(\tau) \leq C e^{\tau(\omega+\eta)},$$

where we have used that, since $P_j(\tau)$ has polynomial growth in τ, the continuous function $e^{-\tau \eta} P_j(\tau)$, $\tau \geq 0$, is bounded in $[0, +\infty)$ since it converges to 0 as $\tau \to +\infty$. Thus, thanks to (11.3), we find that

$$\|e^{\tau A} u_0\| \leq C e^{\tau(\omega + \eta)} \sum_{j=1}^{q} \|u_{0,j}\| \quad \text{for all } \tau \geq 0 \text{ and } u_0 \in \mathbb{R}^N. \quad (11.4)$$

Since the map $\mathcal{N} \colon \mathbb{R}^N \to [0, +\infty)$ defined by

$$\mathcal{N}(u_0) := \sum_{j=1}^{q} \|u_{0,j}\|, \quad u_0 \in \mathbb{R}^N,$$

is a norm in $\mathbb{R}^N$, and all norms in $\mathbb{R}^N$ are equivalent, the constant C can be increased, if necessary, so that (11.4) implies

$$\|e^{\tau A} u_0\| \leq C e^{\tau(\omega + \eta)} \|u_0\| \quad \text{for all } \tau \geq 0 \text{ and } u_0 \in \mathbb{R}^N.$$

Equivalently,

$$\|e^{\tau A}\| \leq C e^{\tau(\omega + \eta)} \quad \text{for all } \tau \geq 0.$$

Therefore, for every $u_0 \in \mathbb{R}^N$, $t_0 \in \mathbb{R}$, and $t \geq t_0$, the unique solution of the Cauchy problem

$$\begin{cases} u' = Au, \\ u(t_0) = u_0, \end{cases} \quad (11.5)$$

denoted by $u(t; t_0, u_0)$, satisfies, for every $t \geq t_0$,

$$\|u(t; t_0, u_0)\| = \|e^{(t-t_0)A} u_0\| \leq C e^{(\omega + \eta)(t - t_0)} \|u_0\|. \quad (11.6)$$

As (11.5) is a linear problem, for every $x \in \mathbb{R}^N$ and $t, t_0 \in \mathbb{R}$, we have that

$$u(t; t_0, u_0) - u(t; t_0, x) = u(t; t_0, u_0 - x).$$

Thus, since $\omega + \eta < 0$, from (11.6), we find that, for every $t_0 \in \mathbb{R}$ and $t \geq t_0$,

$$\begin{aligned}
\|u(t; t_0, u_0) - u(t; t_0, x)\| &= \|u(t; t_0, u_0 - x)\| \\
&\leq C e^{(\omega + \eta)(t - t_0)} \|u_0 - x\| \quad (11.7) \\
&\leq C \|u_0 - x\|.
\end{aligned}$$

In particular, for every $\varepsilon > 0$, if we choose $\delta = \frac{\varepsilon}{C}$ and take $x \in \mathbb{R}^N$ such that $\|u_0 - x\| \leq \delta$, then

$$\|u(t; t_0, u_0) - u(t; t_0, x)\| \leq C\|u_0 - x\| \leq \varepsilon.$$

This shows that $u(t; t_0, u_0)$ is Lyapunov stable (to the right). Actually, by (11.7), we have that, for every $t \geq t_0$,

$$\|u(t; t_0, u_0) - u(t; t_0, x)\| \leq Ce^{(\omega+\eta)(t-t_0)}\|u_0 - x\|$$

and, therefore, since $\omega + \eta < 0$,

$$\lim_{t\uparrow+\infty} \|u(t; t_0, u_0) - u(t; t_0, x)\| = 0 \quad \text{exponentially.}$$

Consequently, in this case, $u(t; t_0, u_0)$ is exponentially asymptotically stable, for any value of $t_0 \in \mathbb{R}$ and $u_0 \in \mathbb{R}^N$.

Case 2. $\mathrm{Re}\,\lambda \leq 0$ for all $\lambda \in \sigma(A)$, and $\nu(\lambda) = 1$ if $\mathrm{Re}\,\lambda = 0$. In this case, we know from Chapter 2 that, for every $\tau \in \mathbb{R}$ and $u_0 \in \mathbb{R}^N$,

$$e^{\tau A} u_0 = \sum_{\substack{\lambda \in \sigma(A) \\ \mathrm{Re}\,\lambda < 0}} e^{\tau \lambda} \sum_{\kappa=0}^{\nu(\lambda)-1} \frac{\tau^k}{k!} (A - \lambda I)^\kappa u_{0,\lambda} + \sum_{\substack{\lambda \in \sigma(A) \\ \mathrm{Re}\,\lambda = 0}} e^{\tau \lambda} u_{0,\lambda},$$

where, for every $\lambda \in \sigma(A)$, $u_{0,\lambda}$ indicates the projection of u_0 onto the generalized eigenspace $N[(A - \lambda I)^{\nu(\lambda)}]$, corresponding to the ascent $\nu(\lambda)$. Thus, for every $\tau \geq 0$, we have that

$$\|e^{\tau A} u_0\| \leq \sum_{\substack{\lambda \in \sigma(A) \\ \mathrm{Re}\,\lambda < 0}} e^{\tau \mathrm{Re}\,\lambda} P_\lambda(\tau) \|u_{0,\lambda}\| + \sum_{\substack{\lambda \in \sigma(A) \\ \mathrm{Re}\,\lambda = 0}} e^{\tau \mathrm{Re}\,\lambda} \|u_{0,\lambda}\|$$

$$= \sum_{\substack{\lambda \in \sigma(A) \\ \mathrm{Re}\,\lambda < 0}} e^{\tau \mathrm{Re}\,\lambda} P_\lambda(\tau) \|u_{0,\lambda}\| + \sum_{\substack{\lambda \in \sigma(A) \\ \mathrm{Re}\,\lambda = 0}} \|u_{0,\lambda}\|,$$

where

$$P_\lambda(\tau) = \sum_{k=0}^{\nu(\lambda)-1} \frac{\|A - \lambda I\|^k}{k!} \tau^k$$

for all $\lambda \in \sigma(A)$ with $\mathrm{Re}\,\lambda < 0$. Consequently, arguing as in Case 1, it becomes apparent that there exists a constant $C > 0$ such that

$$\|e^{\tau A} u_0\| \le C \|u_0\| \quad \text{for all } u_0 \in \mathbb{R}^N \text{ and } \tau \ge 0.$$

Therefore, since (11.5) is a linear problem, we find that, for every $x \in \mathbb{R}^N$, $t_0 \in \mathbb{R}$ and $t \ge t_0$,

$$\|u(t; t_0, u_0) - u(t; t_0, x)\| = \|e^{(t-t_0)A}(u_0 - x)\| \le C \|u_0 - x\|.$$

As a byproduct, $u(t; t_0, u_0)$ is Lyapunov stable for all $u_0 \in \mathbb{R}^N$, for any $t_0 \in \mathbb{R}$. Nevertheless, in this case, the solutions cannot be asymptotically stable, unless $\mathrm{Re}\,\lambda < 0$ for all $\lambda \in \sigma(A)$. Indeed, assume, for example, that $\mathrm{Re}\,\lambda_1 = 0$, and let $x \in \mathbb{R}^N$ be such that

$$u_0 - x \in N[A - \lambda_1 I] \setminus \{0\}.$$

Then,

$$u(t; t_0, u_0) - u(t; t_0, x) = e^{(t-t_0)A}(u_0 - x) = e^{(t-t_0)\lambda_1}(u_0 - x)$$

and, hence,

$$\|u(t; t_0, u_0) - u(t; t_0, x)\| = e^{(t-t_0)\mathrm{Re}\,\lambda_1}\|u_0 - x\| = \|u_0 - x\|$$

for all $t \ge t_0$. Therefore, $u(t; t_0, u_0)$ cannot be asymptotically stable. Consequently, when we are in Case 2 but not in Case 1, all solutions of $u' = Au$ are stable but not asymptotically stable.

Case 3. There exists $\lambda \in \sigma(A)$ such that either $\mathrm{Re}\,\lambda = 0$ and $\nu(\lambda) > 1$ or $\mathrm{Re}\,\lambda > 0$. First, suppose that $\mathrm{Re}\,\lambda > 0$ and take $v \in N[A - \lambda I] \setminus \{0\}$. Then, for every $u_0 \in \mathbb{R}^N$ and $\gamma \in \mathbb{R} \setminus \{0\}$, the vector

$$x := u_0 - \gamma v \tag{11.8}$$

satisfies

$$u(t; t_0, u_0) - u(t; t_0, x) = e^{(t-t_0)A}(u_0 - x) = \gamma e^{(t-t_0)\lambda} v$$

and, hence,

$$\|u(t; t_0, u_0) - u(t; t_0, x)\| = |\gamma| e^{(t-t_0)\mathrm{Re}\,\lambda}\|v\|$$

for all $t \ge t_0$. Therefore, regardless of the size of $|\gamma|$, which might be arbitrarily small but is non-zero, we have that

$$\lim_{t \uparrow +\infty} \|u(t; t_0, u_0) - u(t; t_0, x)\| = +\infty,$$

and, hence, $u(t; t_0, u_0)$ is unstable. Similarly, if $\lambda \in \sigma(A)$ satisfies $\operatorname{Re}\lambda = 0$ with $\nu(\lambda) \geq 2$, then there is $v \in N[(A - \lambda I)^2]$ such that $v \notin N[A - \lambda I]$. Thus, taking x as in (11.8), we have that

$$u(t; t_0, u_0) - u(t; t_0, x) = e^{(t-t_0)A}(\gamma v)$$
$$= \gamma e^{(t-t_0)\lambda}\left[v + (t - t_0)(A - \lambda I)v\right].$$

Thus,

$$\|u(t; t_0, u_0) - u(t; t_0, x)\| = |\gamma|\,\|v + (t - t_0)(A - \lambda I)v\|$$
$$\geq |\gamma|\,|t - t_0|\,\|(A - \lambda I)v\| - |\gamma|\,\|v\|$$

for all $t \geq t_0$. Consequently, regardless of the size of γ,

$$\lim_{t\uparrow+\infty} \|u(t; t_0, u_0) - u(t; t_0, x)\| = +\infty.$$

Therefore, $u(t; t_0, u_0)$ is unstable for all $u_0 \in \mathbb{R}^N$ and any $t_0 \in \mathbb{R}$.

As a consequence of the previous analysis, it is apparent that the character of all the solutions of $u' = Au$ is the same and that it is independent of $t_0 \in \mathbb{R}$. In other words, all the solutions satisfy one of the following: all are Lyapunov unstable, all are stable, or all are asymptotically stable and, in such a case, they actually are exponentially asymptotically stable. Thus, instability, stability, or asymptotic stability is an inherent property of the system. Here and in the rest of the book, to simplify the nomenclature, we simply say that a solution is stable, unstable, or asymptotically stable, always referring to these concepts in the Lyapunov sense introduced above.

The analysis that we have just performed can be summarized as follows.

Theorem 11.2. *The following properties hold:*

(a) *all solutions of $u' = Au$ are unstable if there exists $\lambda \in \sigma(A)$ such that $\operatorname{Re}\lambda > 0$, or $\operatorname{Re}\lambda = 0$ with $\nu(\lambda) \geq 2$;*

(b) *all solutions of $u' = Au$ are stable if $\operatorname{Re}\lambda \leq 0$ for all $\lambda \in \sigma(A)$, and $\nu(\lambda) = 1$ whenever $\operatorname{Re}\lambda = 0$;*

(c) *all solutions of $u' = Au$ are exponentially asymptotically stable if and only if $\operatorname{Re}\lambda < 0$ for all $\lambda \in \sigma(A)$.*

More generally, one can show that the character, in the Lyapunov sense, of all the solutions of any linear system

$$u' = A(t)u + B(t),$$

where

$$A(t) = (a_{ij}(t))_{1 \leq i,j \leq N}, \quad B(t) = (b_j(t))_{1 \leq j \leq N},$$

with a_{ij}, $b_j \in \mathcal{C}(\mathbb{R})$ for all $i, j \in \{1, \ldots, N\}$, is the same and that it is independent of t_0 (see Exercise 1 of Chapter 11).

An homogeneous linear system with constant coefficients, $u' = Au$, is said to be *hyperbolic* when

$$\operatorname{Re} \lambda \neq 0 \quad \text{for all } \lambda \in \sigma(A).$$

In such a case, its *stable and unstable manifolds* are, respectively, defined as

$$W^s = W^s(A) := \bigoplus_{\substack{\lambda \in \sigma(A) \\ \operatorname{Re} \lambda < 0}} N[(A - \lambda I)^{\nu(\lambda)}],$$

$$W^u = W^u(A) := \bigoplus_{\substack{\lambda \in \sigma(A) \\ \operatorname{Re} \lambda > 0}} N[(A - \lambda I)^{\nu(\lambda)}].$$

Observe that these manifolds are invariant under the flow defined by $u' = Au$. Indeed, whenever $u_0 \in W^s$ (respectively, $u_0 \in W^u$), then $e^{\tau A} u_0 \in W^s$ (respectively, $e^{\tau A} u_0 \in W^u$) for all $\tau \in \mathbb{R}$, which is a direct consequence of the fact that, for each $\lambda \in \sigma(A)$, $N[(A - \lambda I)^{\nu(\lambda)}]$ is an invariant subspace by the action of A. Thanks to the analysis done to prove Theorem 11.2, it becomes apparent that, for every $t_0 \in \mathbb{R}$,

$$\lim_{t \uparrow +\infty} \|u(t; t_0, u_0)\| = 0 \quad \text{exponentially} \quad \text{if } u_0 \in W^s,$$

whereas

$$\lim_{t \downarrow -\infty} \|u(t; t_0, u_0)\| = 0 \quad \text{exponentially} \quad \text{if } u_0 \in W^u.$$

Therefore, if the system is hyperbolic, we have that

$$\mathbb{R}^N = W^s \oplus W^u,$$

i.e. every $u_0 \in \mathbb{R}^N$ admits a unique decomposition as

$$u_0 = u_{0,s} + u_{0,u}, \quad \text{with } u_{0,s} \in W^s \text{ and } u_{0,u} \in W^u.$$

Consequently, when, in addition, $W^s(A) \neq \{0\}$ and $W^u(A) \neq \{0\}$, the dynamics of $u' = Au$ can be schematically sketched as in Figure 11.2.

Naturally, we can also have $W^u = \{0\}$ or $W^s = \{0\}$. Indeed, if $\mathrm{Re}\,\lambda < 0$ for all $\lambda \in \sigma(A)$, then $W^s = \mathbb{R}^N$ and 0 is a global attractor, in the sense that all solutions converge to it as $t \uparrow +\infty$. Similarly, if $\mathrm{Re}\,\lambda > 0$ for all $\lambda \in \sigma(A)$, then $W^u = \mathbb{R}^N$ and

$$\lim_{t\uparrow+\infty} \|u(t;t_0,u_0)\| = +\infty \quad \text{for all } u_0 \in \mathbb{R}^N \setminus \{0\}.$$

Stable and unstable nodes and foci in the plane (cf. Figures 9.2–9.6 and the left and right plots in Figure 9.7) are the simplest cases of these hyperbolic limiting situations. Note that, in all the chapters of this book, where we previously used the terms stable/unstable, these concepts agree with those of Lyapunov stability/instability introduced in this section. As a particular case, when linear systems with constant coefficient were considered, their behavior was the same as described in Theorem 11.2.

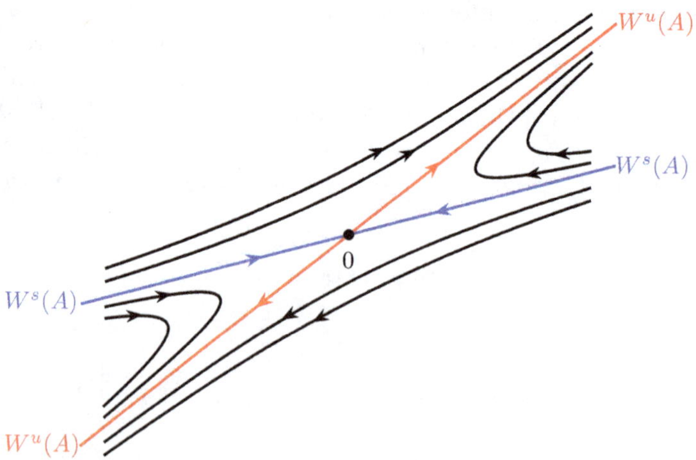

Fig. 11.2. Schematic representation of the dynamics of a linear hyperbolic system with $W^s(A) \neq \{0\}$ and $W^u(A) \neq \{0\}$.

In the theory of Dynamical Systems, it is important to also consider the following concept. A given solution $u(t; t_0, u_0)$ of $u' = f(t, u)$ is said to be *attractive* if there exists $\delta > 0$ such that,

$$\lim_{t\uparrow+\infty} \|u(t; t_0, u_0) - u(t; t_0, x)\| = 0 \quad \text{if } \|u_0 - x\| \le \delta.$$

Observe that the definition of asymptotic stability given in Definition 11.1 includes, besides Lyapunov stability, the attractivity of the solution.

Nevertheless, the concepts of attractivity and stability are independent, as shown by the example proposed by Vinograd in 1957 (see also Hahn, 1967, p. 191), which is given by $(u, v)' = (f_1(u, v), f_2(u, v))$, with

$$f_1(u, v) = \frac{u^2(v - u) + v^5}{(u^2 + v^2)(1 + (u^2 + v^2)^2)} \quad \text{if } (u, v) \ne (0, 0),$$

$$f_2(u, v) = \frac{v^2(v - 2u)}{(u^2 + v^2)(1 + (u^2 + v^2)^2)} \quad \text{if } (u, v) \ne (0, 0), \qquad (11.9)$$

$$f_1(0, 0) = f_2(0, 0) = 0,$$

whose phase portrait is sketched in Figure 11.3. Another example is the one given in Exercise 17 of Chapter 11. In these systems, the origin is an attractor, though it is unstable to the right in the sense of Lyapunov.

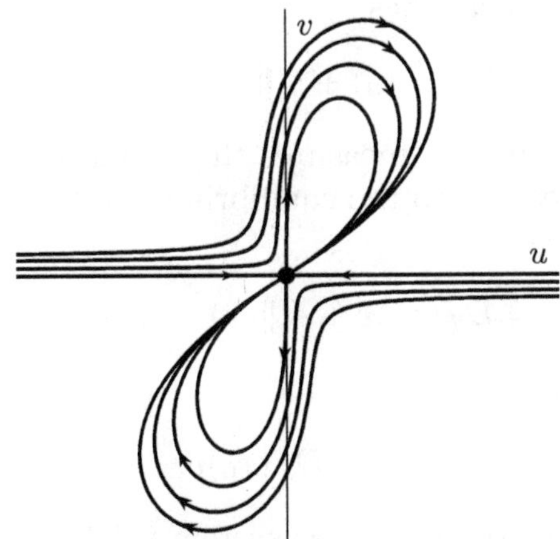

Fig. 11.3. Phase portrait of system (11.9), which gives an example of an attractive but unstable equilibrium.

11.2   Stability of Equilibria of Autonomous Systems

In this section, we present Lyapunov's first method, which consists of establishing the stability of the equilibria of an autonomous system by means of the associated linear variational system (also called linearization).

Consider an autonomous system $u' = f(u)$, with $f \in \mathcal{C}^1(\mathbb{R}^N; \mathbb{R}^N)$. Let $x_0 \in \mathbb{R}^N$ be such that $f(x_0) = 0$. Then, the constant function $u(t) = x_0$, $t \in \mathbb{R}$, is a solution (an equilibrium) of the system. Moreover, if $Df(x_0)$ denotes the Jacobian matrix of f at x_0, the differentiability of f at x_0 implies that the remainder function $R(u)$, defined as

$$R(u) := f(u) - f(x_0) - Df(x_0)(u - x_0)$$

$$= f(u) - Df(x_0)(u - x_0), \quad u \in \mathbb{R}^N,$$

satisfies $R(u) = o(\|u - x_0\|)$ as $u \to x_0$, i.e.

$$\lim_{u \to x_0} \frac{R(u)}{\|u - x_0\|} = 0. \tag{11.10}$$

Thus, as the system $u' = f(u)$ can be equivalently written as

$$(u - x_0)' = u' = f(u) = Df(x_0)(u - x_0) + R(u),$$

the difference $u - x_0$ satisfies

$$(u - x_0)' = Df(x_0)(u - x_0) + o(\|u - x_0\|) \quad \text{as } u \to x_0. \tag{11.11}$$

The function $v := u - x_0$ measures the variation of a solution u of the system with respect to the equilibrium x_0. In terms of v, (11.11) can be rewritten as

$$v' = Df(x_0)v + o(\|v\|) \quad \text{as } v \to 0. \tag{11.12}$$

The linear system

$$v' = Df(x_0)v, \tag{11.13}$$

which approximates the nonlinear system (11.12) in a neighborhood of $v = 0$, is usually refereed to as the *variational system*, or *linearization*, of $u' = f(u)$ at the equilibrium x_0.

The main results in this section relate the stability of x_0 as an equlibrium of $u' = f(u)$ with the stability of $v = 0$ as an equilibrium of the variational system at x_0, (11.13). Precisely, Lyapunov's stability theorem (see Theorem 11.3) establishes that x_0 is asymptotically stable as a solution of $u' = f(u)$ if (11.13) is asymptotically stable; thus, according to Theorem 11.2, this occurs if $\operatorname{Re} \lambda < 0$ for all $\lambda \in \sigma(Df(x_0))$. Instead, according to Lyapunov's instability theorem (see Theorem 11.4), x_0 is unstable as a solution of $u' = f(u)$ if $\operatorname{Re} \lambda > 0$ for some $\lambda \in \sigma(Df(x_0))$. These theorems will be discussed and proved in the following two sections.

11.2.1 *Lyapunov stability theorem*

The main result in this section is the following stability theorem.

Theorem 11.3 (Lyapunov stability theorem). *Suppose that $v = 0$ is asymptotically stable as a solution of $v' = Df(x_0)v$, i.e. according to Theorem 11.2, $\operatorname{Re} \lambda < 0$ for all $\lambda \in \sigma(Df(x_0))$. Then, x_0 is asymptotically stable as an equilibrium of $u' = f(u)$.*

Proof. Set

$$A := Df(x_0), \quad \omega := \max_{\lambda \in \sigma(A)} \operatorname{Re} \lambda < 0,$$

and fix $\varepsilon \in (0, -\omega/2)$. Then, as in the proof of Theorem 11.2 (see Case 1 on p. 600), there exists $C(\varepsilon) > 1$ such that

$$\|e^{\tau A}\| \leq C(\varepsilon)e^{(\omega+\varepsilon)\tau} \quad \text{for all } \tau \geq 0. \tag{11.14}$$

Moreover, thanks to (11.10), there exists $\delta = \delta(\varepsilon) > 0$ such that

$$\|R(u)\| \leq \frac{\varepsilon}{C(\varepsilon)}\|u - x_0\| \quad \text{if } \|u - x_0\| \leq \delta. \tag{11.15}$$

Fix $t_0 \in \mathbb{R}$ and $u_0 \in \mathbb{R}^N$ so that

$$\|u_0 - x_0\| < \frac{\delta}{C(\varepsilon)} < \delta. \tag{11.16}$$

By Theorem 5.17, the problem

$$\begin{cases} u' = f(u), \\ u(t_0) = u_0, \end{cases}$$

possesses a unique maximal solution, $(I, u(t; u_0))$, with $I = [t_0, T_{\max})$, for some $T_{\max} \leq +\infty$, and $u(t; u_0)$ blows up at $T_{\max}$ if $T_{\max} < +\infty$.

Furthermore, since we have that

$$\frac{d}{dt}(u(t; u_0) - x_0) = A(u(t; u_0) - x_0) + R(u(t; u_0)) \quad \text{for all } t \in I,$$

it follows from the variation of constants formula (see Corollary 2.10) that

$$u(t; u_0) - x_0 = e^{(t-t_0)A}(u_0 - x_0) + \int_{t_0}^{t} e^{(t-s)A} R(u(s; u_0))\, ds$$

for all $t \in I$. Thus, thanks to (11.14), we find that, for every $t \in I$,

$$\|u(t; u_0) - x_0\| \leq C(\varepsilon) e^{(\omega+\varepsilon)(t-t_0)} \|u_0 - x_0\|$$

$$+ C(\varepsilon) \int_{t_0}^{t} e^{(\omega+\varepsilon)(t-s)} \|R(u(s; u_0))\|\, ds.$$

Equivalently,

$$e^{-(\omega+\varepsilon)(t-t_0)} \|u(t; u_0) - x_0\| \leq C(\varepsilon) \|u_0 - x_0\|$$

$$+ C(\varepsilon) \int_{t_0}^{t} e^{-(\omega+\varepsilon)(s-t_0)} \|R(u(s; u_0))\|\, ds \tag{11.17}$$

for all $t \in I$. Since $\|u_0 - x_0\| < \delta$, by continuity, there exists $\theta \in (t_0, T_{\max})$ such that

$$\|u(t; u_0) - x_0\| < \delta \quad \text{for all } t \in [t_0, \theta].$$

Consequently, the following quantity is well defined:

$$T := \sup \left\{ \theta \in (t_0, T_{\max}) : \|u(t; u_0) - x_0\| < \delta \text{ for all } t \in [t_0, \theta] \right\}.$$

Moreover, either $T = +\infty$, in which case

$$\|u(t; u_0) - x_0\| < \delta \quad \text{for all } t \in [t_0, +\infty), \tag{11.18}$$

which entails $T_{\max} = +\infty$, or $T < +\infty$ and, in such a case, for every $t \in [t_0, T)$,

$$\|u(t; u_0) - x_0\| < \delta \quad \text{and} \quad \|u(T; u_0) - x_0\| = \delta. \tag{11.19}$$

In both cases, we have that

$$\|u(t; u_0) - x_0\| < \delta \quad \text{for all } t \in [t_0, T),$$

and, hence, (11.15) gives

$$\|R(u(t;u_0))\| \le \frac{\varepsilon}{C(\varepsilon)}\|u(t;u_0) - x_0\| \quad \text{for all } t \in [t_0, T). \qquad (11.20)$$

Consequently, using (11.20) in (11.17) yields

$$e^{-(\omega+\varepsilon)(t-t_0)}\|u(t;u_0) - x_0\| \le C(\varepsilon)\|u_0 - x_0\|$$

$$+ \varepsilon \int_{t_0}^{t} e^{-(\omega+\varepsilon)(s-t_0)}\|u(s;u_0) - x_0\|\, ds$$

for all $t \in [t_0, T)$. By setting

$$x(t) := e^{-(\omega+\varepsilon)(t-t_0)}\|u(t;u_0) - x_0\|, \quad t \in [t_0, T),$$

the previous relation can be rewritten as

$$x(t) \le C(\varepsilon)\|u_0 - x_0\| + \varepsilon \int_{t_0}^{t} x(s)\, ds \quad \text{for all } t \in [t_0, T).$$

Consequently, by Lemma 5.23, we find that, for every $t \in [t_0, T)$,

$$x(t) \le C(\varepsilon)\|u_0 - x_0\| \left(1 + \varepsilon \int_{t_0}^{t} e^{(t-s)\varepsilon}\, ds\right) = C(\varepsilon)\|u_0 - x_0\|e^{(t-t_0)\varepsilon}.$$

So, for every $t \in [t_0, T)$,

$$\|u(t;u_0) - x_0\| = e^{(\omega+\varepsilon)(t-t_0)}x(t) \le C(\varepsilon)\|u_0 - x_0\|e^{(\omega+2\varepsilon)(t-t_0)}.$$

Thus, it follows from (11.16) that

$$\|u(t;u_0) - x_0\| \le \delta e^{(\omega+2\varepsilon)(t-t_0)} \quad \text{for all } t \in [t_0, T), \qquad (11.21)$$

which allows us to exclude that (11.19) holds. Indeed, since $\omega + 2\varepsilon < 0$ by the choice of ε, should T be finite, then, letting $t \uparrow T$ in the previous relation would give

$$\|u(T;u_0) - x_0\| \le \delta e^{(\omega+2\varepsilon)(T-t_0)} < \delta,$$

which contradicts (11.19). Therefore, $T = +\infty$, (11.18) holds true, and (11.21) is valid for all $t \ge t_0$. Since $\omega + 2\varepsilon < 0$, this shows the exponential asymptotic stability of the equilibrium x_0, which completes the proof. $\qquad \square$

11.2.2　*Lyapunov instability theorem*

The main result in this section is the following instability theorem.

Theorem 11.4 (Lyapunov instability theorem). *Suppose that* $\operatorname{Re} \lambda > 0$ *for some* $\lambda \in \sigma(Df(x_0))$*. Then,* x_0 *is unstable as a solution of* $u' = f(u)$*.*

To prove this theorem, we need two technical preliminary lemmas. The first one is a direct consequence of the construction of the Jordan basis for the Jordan canonical form.

Lemma 11.5. *Let* $A \in \mathcal{M}_N(\mathbb{C})$*, and suppose that* $\sigma(A) = \{\mu_1, \ldots, \mu_N\}$*, where the eigenvalues might be repeated according to their algebraic multiplicities. Then, for every* $\varepsilon > 0$*, there exists a basis* $\mathcal{B} := \{p_1, \ldots, p_N\}$ *of* $\mathbb{C}^N$ *such that, with respect to* $\mathcal{B}$*,* A *can be expressed as*

$$A = \mathcal{D} + \mathcal{N}_\varepsilon,$$

where $\mathcal{D}$ *is a diagonal matrix having the eigenvalues on its principal diagonal, and*

$$\mathcal{N}_\varepsilon = (n_{ij})_{i,j \in \{1,\ldots,N\}} \in \mathcal{M}_N(\mathbb{R})$$

is a nilpotent matrix such that $n_{ij} = 0$ *if* $i \leq j$ *or* $i \geq j + 2$*, and* $n_{ij} = \varepsilon$ *or* $n_{ij} = 0$ *if* $i = j + 1$*.*

Proof. In the construction of the Jordan basis of A in Section 2.2, it suffices to take $\varepsilon^{-1}(A - \lambda I)v_{\lambda,\ell}$, instead of $(A - \lambda I)v_{\lambda,\ell}$, in (2.13), for all $\lambda \in \sigma(A)$. More precisely, by choosing the Jordan basis $\mathcal{B}_\lambda$ in this way, the restriction of A to $N[(A - \lambda I)^{\nu(\lambda)}]$,

$$A_\lambda := A|_{N[(A-\lambda I)^{\nu(\lambda)}]} \colon N[(A - \lambda I)^{\nu(\lambda)}] \to N[(A - \lambda I)^{\nu(\lambda)}],$$

can be expressed as

$$A_{\lambda,\mathcal{B}_\lambda} = \operatorname{diag}\{J_{\lambda,\varepsilon,i_1}, J_{\lambda,\varepsilon,i_2}, \ldots, J_{\lambda,\varepsilon,i_{n_\lambda}}\}$$

for some integer number $n_\lambda \geq 1$, where, for all $h \in \{1, \ldots, n_\lambda\}$, $i_h \in \{1, \ldots, \nu(\lambda)\}$, and, for each natural number $n \geq 1$, $J_{\lambda,\varepsilon,n} \in \mathcal{M}_n(\mathbb{C})$

denotes the matrix

$$J_{\lambda,\varepsilon,n} := \begin{pmatrix} \lambda & 0 & 0 & \cdots & 0 & 0 \\ \varepsilon & \lambda & 0 & \cdots & 0 & 0 \\ 0 & \varepsilon & \lambda & \cdots & 0 & 0 \\ \vdots & \vdots & \vdots & \ddots & \vdots & \vdots \\ 0 & 0 & 0 & \cdots & \lambda & 0 \\ 0 & 0 & 0 & \cdots & \varepsilon & \lambda \end{pmatrix}.$$

In this way, the basis of $\mathbb{C}^N$ defined as

$$\mathcal{B} := \cup_{\lambda \in \sigma(A)} \mathcal{B}_\lambda = \{p_1, \ldots, p_N\},$$

satisfies the requirements of the statement. In other words, if P is the matrix whose columns are p_j, $j \in \{1, \ldots, N\}$, changing basis gives

$$P^{-1}AP = \mathcal{D} + \mathcal{N}_\varepsilon.$$

This concludes the proof. $\qquad\square$

Once fixed $\varepsilon > 0$, if we construct the corresponding basis $\mathcal{B} = \{p_1, \ldots, p_N\}$ of $\mathbb{C}^N$, as described in Lemma 11.5, every $x,\, y \in \mathbb{C}^N$ can be uniquely expressed as

$$x = x_1 p_1 + x_2 p_2 + \cdots + x_N p_N,$$
$$y = y_1 p_1 + y_2 p_2 + \cdots + y_N p_N,$$

for some

$$(x_1, \ldots, x_N),\ (y_1, \ldots, y_N) \in \mathbb{C}^N,$$

and we can define the associated sesquilinear inner product as

$$\langle x, y \rangle := \sum_{j=1}^{N} x_j \bar{y}_j = x_1 \bar{y}_1 + x_2 \bar{y}_2 + \cdots + x_N \bar{y}_N. \qquad (11.22)$$

The induced Euclidean norm is

$$\|x\| := \sqrt{\langle x, x \rangle} = \sqrt{|x_1|^2 + \cdots + |x_N|^2},$$

where $|z|$ denotes the modulus of $z \in \mathbb{C}$ and, for every $i, j \in \{1, \ldots, N\}$, it satisfies

$$\|p_j\| = 1 \quad \text{and} \quad \langle p_i, p_j \rangle = 0 \quad \text{if } i \neq j.$$

Thus, $\mathcal{B}$ is an orthonormal basis with respect to the inner product (11.22). Moreover, for every $x, y \in \mathbb{C}^N$, the following Cauchy–Schwarz inequality holds:

$$|\langle x, y \rangle| \leq \|x\|\|y\|. \tag{11.23}$$

Indeed, for every $x, y \in \mathbb{C}^N$ and $t \in \mathbb{R}$, we have that

$$0 \leq \|tx + y\|^2 = \langle tx + y, tx + y \rangle = \|x\|^2 t^2 + 2\mathrm{Re}\,\langle x, y \rangle t + \|y\|^2.$$

Thus, the polynomial

$$Q(t) := \|x\|^2 t^2 + 2\mathrm{Re}\,\langle x, y \rangle t + \|y\|^2, \quad t \in \mathbb{R},$$

possesses at most a unique real root and, hence,

$$|\mathrm{Re}\,\langle x, y \rangle| \leq \|x\|\|y\|.$$

Let $\theta \in \mathbb{R}$ be such that

$$\langle x, y \rangle = |\langle x, y \rangle| e^{i\theta}.$$

Then,

$$|\langle x, y \rangle| = e^{-i\theta}\langle x, y \rangle = \langle e^{-i\theta}x, y \rangle$$
$$= \mathrm{Re}\,\langle e^{-i\theta}x, y \rangle \leq \|e^{-i\theta}x\|\|y\| = \|x\|\|y\|,$$

as we wanted. This inequality is known as the Cauchy–Schwarz inequality, named after A. L. Cauchy and H. A. Schwarz. As a consequence of such an inequality, for every $x \in \mathbb{C}^N$,

$$|\langle \mathcal{N}_\varepsilon x, x \rangle| \leq \|\mathcal{N}_\varepsilon x\|\|x\| \leq \varepsilon\|x\|\|x\| = \varepsilon\|x\|^2, \tag{11.24}$$

where $\mathcal{N}_\varepsilon$ is the matrix obtained in Lemma 11.5, and we have used that its operator norm equals, at most, ε. With these elements, we can prove the second preliminary lemma.

Lemma 11.6. *Let $A \in \mathcal{M}_N(\mathbb{C})$ be a matrix such that, for some $\alpha, \beta \in \mathbb{R}$ with $\alpha < \beta$,*

$$\operatorname{Re} \sigma(A) \subset (\alpha, \beta). \tag{11.25}$$

Then, for sufficiently small $\varepsilon > 0$,

$$\alpha \|x\|^2 \leq \operatorname{Re} \langle Ax, x \rangle \leq \beta \|x\|^2 \quad \text{for all } x \in \mathbb{C}^N,$$

where we have used the inner product (11.22) and the induced Euclidean norm.

Proof. If we denote $\sigma(A) = \{\mu_1, \ldots, \mu_N\}$, according to (11.25) we can fix $\varepsilon > 0$ so that

$$\alpha + \varepsilon \leq \min_{j \in \{1,\ldots,N\}} \operatorname{Re} \mu_j \leq \max_{j \in \{1,\ldots,N\}} \operatorname{Re} \mu_j \leq \beta - \varepsilon, \tag{11.26}$$

and consider the associated inner product (11.22). Then, for every $x \in \mathbb{C}^N$,

$$\operatorname{Re} \langle Ax, x \rangle = \operatorname{Re} \langle \mathcal{D}x, x \rangle + \operatorname{Re} \langle \mathcal{N}_\varepsilon x, x \rangle.$$

Due to (11.26),

$$(\alpha + \varepsilon)\|x\|^2 \leq \operatorname{Re} \langle \mathcal{D}x, x \rangle = \sum_{j=1}^{N} \operatorname{Re} \mu_j |x_j|^2 \leq (\beta - \varepsilon)\|x\|^2. \tag{11.27}$$

Moreover, thanks to (11.24),

$$|\operatorname{Re} \langle \mathcal{N}_\varepsilon x, x \rangle| \leq |\langle \mathcal{N}_\varepsilon x, x \rangle| \leq \varepsilon \|x\|^2$$

and, hence,

$$-\varepsilon \|x\|^2 \leq \operatorname{Re} \langle \mathcal{N}_\varepsilon x, x \rangle \leq \varepsilon \|x\|^2. \tag{11.28}$$

Therefore, adding (11.27) and (11.28) leads to

$$\alpha \|x\|^2 \leq \operatorname{Re} \langle Ax, x \rangle \leq \beta \|x\|^2,$$

which completes the proof. $\qquad \square$

We already have all the necessary ingredients to prove Theorem 11.4.

Proof of Theorem 11.4. Throughout this proof, we set $A := Df(x_0)$ and consider the linear subspaces of $\mathbb{C}^N$

$$V_- := \bigoplus_{\substack{\lambda \in \sigma(A) \\ \operatorname{Re}\lambda \leq 0}} N[(A - \lambda I)^{\nu(\lambda)}], \quad V_+ := \bigoplus_{\substack{\lambda \in \sigma(A) \\ \operatorname{Re}\lambda > 0}} N[(A - \lambda I)^{\nu(\lambda)}].$$

By assumption, $V_+ \neq \{0\}$. Moreover, according to Theorem 2.5,

$$\mathbb{C}^N = V_+ \oplus V_-, \quad A(V_+) \subset V_+, \quad A(V_-) \subset V_-.$$

Therefore, considering the restricted operators

$$A_+ := A|_{V_+}, \quad A_- := A|_{V_-},$$

we have that

$$A = A_+ + A_-,$$

and

$$\sigma(A_+) = \{\lambda \in \sigma(A) \colon \operatorname{Re}\lambda > 0\},$$
$$\sigma(A_-) = \{\lambda \in \sigma(A) \colon \operatorname{Re}\lambda \leq 0\}.$$

Moreover, for every $v \in \mathbb{C}^N$ there exist unique $v_\pm \in V_\pm$ such that $v = v_+ + v_-$. Thus, since V_+ and V_- are invariant under the action of A, it is apparent that

$$(Av)_+ + (Av)_- = Av = A(v_+ + v_-) = Av_+ + Av_- = A_+v_+ + A_-v_-,$$

for all $v \in \mathbb{C}^N$. Thus,

$$(Av)_+ = A_+v_+ \quad \text{and} \quad (Av)_- = A_-v_- \quad \text{for all } v \in \mathbb{C}^N.$$

In other words, if we denote by $P_\pm$ the projections onto $V_\pm$ defined, for every $v \in \mathbb{C}^N$, as $P_\pm v = v_\pm$, then

$$P_+ A = AP_+ = A_+P_+, \quad P_- A = AP_- = A_-P_-. \tag{11.29}$$

Subsequently, we fix $0 < \eta < \omega$ such that

$$\operatorname{Re}\lambda_- \leq 0 < \eta < \omega < \operatorname{Re}\lambda_+$$

for all $\lambda_- \in \sigma(A_-)$ and $\lambda_+ \in \sigma(A_+)$.

As a consequence, we can fix a sufficiently small $\varepsilon > 0$ so that, according to Lemmas 11.5 and 11.6, there are bases $\mathcal{B}_\pm$ of $V_\pm$ in such a way that the inner product $\langle \cdot, \cdot \rangle$ defined by the basis

$$\mathcal{B} = \mathcal{B}_+ \cup \mathcal{B}_-$$

of $\mathbb{C}^N$ satisfies

$$\langle v, w \rangle = \langle v_+, w_+ \rangle + \langle v_-, w_- \rangle \quad \text{for all } v, w \in \mathbb{C}^N \tag{11.30}$$

(recall that the vectors of $\mathcal{B}$ are orthogonal with respect to $\langle \cdot, \cdot \rangle$) and, for every $v \in \mathbb{C}^N$,

$$\begin{aligned}
\operatorname{Re} \langle Av_+, v_+ \rangle = \operatorname{Re} \langle A_+ v_+, v_+ \rangle \geq \omega \, \|v_+\|^2 , \\
\operatorname{Re} \langle Av_-, v_- \rangle = \operatorname{Re} \langle A_- v_-, v_- \rangle \leq \eta \, \|v_-\|^2 .
\end{aligned} \tag{11.31}$$

Note that, thanks to (11.30), we have that

$$\|v_+\| \leq \|v\| \quad \text{and} \quad \|v_-\| \leq \|v\| \quad \text{for all } v \in \mathbb{C}^N.$$

Thus,

$$\|P_+ v\| = \|v_+\| \leq \|v\|, \quad \|P_- v\| = \|v_-\| \leq \|v\|, \tag{11.32}$$

which entails $\|P_\pm\| = 1$ because $P_\pm$ leaves $V_\pm$ invariant.

Subsequently, we maintain the notations introduced at the beginning of Section 11.2 and consider the function $\Phi \colon \mathbb{R}^N \to \mathbb{R}$ defined as

$$\Phi(v) := \frac{\|v_+\|^2 - \|v_-\|^2}{2} = \frac{\|P_+ v\|^2 - \|P_- v\|^2}{2}, \quad v \in \mathbb{R}^N.$$

For any fixed $\gamma > 0$, which will be specified later, (11.10) ensures that there exists $\varepsilon = \varepsilon(\gamma)$ such that

$$\|R(u)\| \leq \gamma \|u - x_0\| \quad \text{if } \|u - x_0\| \leq \varepsilon. \tag{11.33}$$

We reason by contradiction, assuming that x_0 is stable. Then, there exists $\delta = \delta(\varepsilon(\gamma))$ such that, if $\|u_0 - x_0\| < \delta$, the maximal solution (I, u) of $u' = f(u)$ with $u(0) = u_0$ is defined for all $t > 0$, and

$$\|u(t) - x_0\| < \varepsilon \quad \text{for all } t > 0. \tag{11.34}$$

Now, we fix $u_0 \in \mathbb{R}^N$ satisfying

$$\|u_0 - x_0\| < \delta, \quad \Phi(u_0 - x_0) > 0. \tag{11.35}$$

Observe that the first requirement guarantees that the associated maximal solution satisfies (11.34). Moreover, the latter requirement can be achieved because $V_+ \neq \{0\}$. Set, for every $t > 0$,

$$v(t) := u(t) - x_0, \quad \varphi(t) := \Phi(v(t)) = \frac{\|P_+ v(t)\|^2 - \|P_- v(t)\|^2}{2}.$$

From (11.34), it follows that $\varphi(t)$ is bounded for $t > 0$. Moreover, differentiating with respect to t gives

$$\varphi'(t) = \frac{1}{2}\Big[\langle P_+ v'(t), P_+ v(t) \rangle + \langle P_+ v(t), P_+ v'(t) \rangle$$

$$- \langle P_- v'(t), P_- v(t) \rangle - \langle P_- v(t), P_- v'(t) \rangle \Big]$$

$$= \mathrm{Re}\,\langle P_+ v'(t), P_+ v(t) \rangle - \mathrm{Re}\,\langle P_- v'(t), P_- v(t) \rangle$$

$$= \mathrm{Re}\,\langle P_+ [Av(t) + R(u(t))], P_+ v(t) \rangle$$

$$- \mathrm{Re}\,\langle P_- [Av(t) + R(u(t))], P_- v(t) \rangle$$

$$= \mathrm{Re}\,\langle P_+ Av(t), P_+ v(t) \rangle - \mathrm{Re}\,\langle P_- Av(t), P_- v(t) \rangle$$

$$+ \mathrm{Re}\,\langle P_+ R(u(t)), P_+ v(t) \rangle - \mathrm{Re}\,\langle P_- R(u(t)), P_- v(t) \rangle$$

for all $t > 0$. From (11.29) and (11.31), we obtain, for every $t > 0$,

$$\mathrm{Re}\,\langle P_+ Av(t), P_+ v(t) \rangle - \mathrm{Re}\,\langle P_- Av(t), P_- v(t) \rangle$$

$$= \mathrm{Re}\,\langle AP_+ v(t), P_+ v(t) \rangle - \mathrm{Re}\,\langle AP_- v(t), P_- v(t) \rangle$$

$$\geq \omega \|P_+ v(t)\|^2 - \eta \|P_- v(t)\|^2.$$

Moreover, by (11.33), (11.32) and (11.23), we find that, for every $t > 0$,

$$|\mathrm{Re}\,\langle P_+ R(u(t)), P_+ v(t) \rangle| \leq |\langle P_+ R(u(t)), P_+ v(t) \rangle|$$

$$\leq \|R(u(t))\|\,\|P_+ v(t)\| \leq \gamma \|v(t)\|\,\|P_+ v(t)\|.$$

Similarly,

$$|\mathrm{Re}\,\langle P_- R(u(t)), P_- v(t) \rangle| \leq \gamma \|v(t)\|\,\|P_- v(t)\| \quad \text{for all } t > 0.$$

Thus, we obtain that, for every $t > 0$,

$$\varphi'(t) \geq \omega \|P_+ v(t)\|^2 - \eta \|P_- v(t)\|^2 - \gamma \|v(t)\|\,(\|P_+ v(t)\| + \|P_- v(t)\|).$$

To continue with the estimates, we define

$$t^* := \sup \left\{ T > 0 \colon \varphi(t) = \Phi(v(t)) > 0 \text{ for all } t \in [0, T] \right\}, \qquad (11.36)$$

which is well defined because, by (11.35) and by continuity, $\varphi(t) > 0$ for sufficiently small $t > 0$. As, for any $t \in [0, t^*)$, we have that $\Phi(v(t)) > 0$, it is apparent that

$$\|P_- v(t)\| < \|P_+ v(t)\| \quad \text{for all } t \in [0, t^*).$$

So,

$$\|v(t)\|^2 = \|P_+ v(t)\|^2 + \|P_- v(t)\|^2 \leq 2 \|P_+ v(t)\|^2$$

and, hence,

$$\|v(t)\| \leq 2 \|P_+ v(t)\| \quad \text{for all } t \in [0, t^*).$$

Thus, for every $t \in [0, t^*)$, we find that

$$\varphi'(t) \geq \omega \|P_+ v(t)\|^2 - \eta \|P_- v(t)\|^2 - 2\gamma \|v(t)\| \, \|P_+ v(t)\|$$

$$\geq \omega \|P_+ v(t)\|^2 - \eta \|P_- v(t)\|^2 - 4\gamma \|P_+ v(t)\|^2$$

$$= (\omega - 4\gamma) \|P_+ v(t)\|^2 - \eta \|P_- v(t)\|^2 .$$

Consequently, by choosing $\gamma = \frac{\omega - \eta}{4}$, we get

$$\varphi'(t) \geq 2\eta \varphi(t) \quad \text{for all } t \in [0, t^*),$$

which implies, for all $t \in [0, t^*)$,

$$\varphi(t) \geq e^{2\eta t} \varphi(0). \qquad (11.37)$$

As a consequence of this relation and (11.36), we conclude that $t^* = +\infty$, and (11.37) holds for all $t > 0$. Thus, (11.37) entails

$$\lim_{t \uparrow +\infty} \varphi(t) = +\infty$$

because, thanks to (11.35),

$$\varphi(0) = \varphi(v(0)) = \Phi(u_0 - x_0) > 0.$$

This contradicts the fact that $\varphi(t)$ is bounded if (11.34) holds. $\qquad \square$

11.2.3   *Grobman–Hartman theorem*

In the context of Theorems 11.3 and 11.4, when, in addition, $Df(x_0)$ is an hyperbolic matrix, in the sense that $\operatorname{Re}\lambda \neq 0$ for all $\lambda \in \sigma(Df(x_0))$, the following result holds. The statement that we give here was proved independently by Grobman in 1959 and 1962 and by Hartman in 1960 and 1963b, using different proofs. The interested reader is referred to Hartman (2002, pp. 244–251) for a self-contained proof. We do not provide it here since it is beyond the scope of this book.

Theorem 11.7 (Grobman and Hartman). *Suppose that f is of class $\mathcal{C}^1$ in a neighborhood of $x_0 \in \mathbb{R}^N$, that $f(x_0) = 0$, i.e. x_0 is an equilibrium of $u' = f(u)$, and that $Df(x_0)$ is an hyperbolic matrix. Let $T(t) := u(t; u_0)$ and $L(t) := e^{tDf(x_0)}v_0$ be the general solutions of $u' = f(u)$ and $v' = Df(x_0)v$, respectively, with u_0 in a neighborhood of x_0, and $v_0 \in \mathbb{R}^N$. Then, there exists a (local) homeomorphism Ψ, i.e. a continuous one-to-one map with continuous inverse, from a neighborhood of $x_0 \in \mathbb{R}^N$ to a neighborhood of $0 \in \mathbb{R}^N$, such that*

$$L(t) = \Psi T(t)\Psi^{-1} \text{ or, equivalently, } T(t) = \Psi^{-1}L(t)\Psi.$$

In particular, Ψ maps solutions of $u' = f(u)$ near x_0 onto solutions of $v' = Df(x_0)v$, preserving parametrizations.

As a consequence, this result ensures that the structure of the set of solutions of $u' = f(u)$ in a neighborhood of $u = x_0$ is identical (in a topological sense, i.e. up to homeomorphisms) to the one of the solutions of the linearized system at x_0 near $v = 0$. In other words, the qualitative behavior of $u' = f(u)$ near an hyperbolic equilibrium point x_0 is the same as that of the linearized system near x_0. Therefore, at any hyperbolic equilibrium x_0, the phase portrait of the linearization provides us with the (local) phase portrait of $u' = f(u)$ in a neighborhood of x_0.

In particular, Theorem 11.7 implies the existence of two (local) *nonlinear* invariant manifolds through x_0, $W_f^u(x_0)$ and $W_f^s(x_0)$, defined as

$$W_f^u(x_0) := \Psi^{-1}(W^u(Df(x_0))), \quad W_f^s(x_0) := \Psi^{-1}(W^s(Df(x_0))),$$

where $W^s(Df(x_0))$ and $W^u(Df(x_0))$ are the linear manifolds

$$W^s(Df(x_0)) = \bigoplus_{\substack{\lambda \in \sigma(Df(x_0)) \\ \operatorname{Re}\lambda < 0}} N[(Df(x_0) - \lambda I)^{\nu(\lambda)}],$$

$$W^u(Df(x_0)) = \bigoplus_{\substack{\lambda \in \sigma(Df(x_0)) \\ \operatorname{Re}\lambda > 0}} N[(Df(x_0) - \lambda I)^{\nu(\lambda)}],$$

which have been defined in Section 11.1 (in this case, $A = Df(x_0)$). $W_f^u(x_0)$ is referred to as the (nonlinear) unstable manifold of $u' = f(u)$ through x_0, while $W_f^s(x_0)$ is known as the (nonlinear) stable manifold through x_0. Figure 11.4 illustrates the general case in the plane, i.e. for $N = 2$, when both the stable and unstable manifolds are non-trivial. When $W^u(Df(x_0)) = \{0\}$, then $W^s(Df(x_0)) = \mathbb{R}^N$ and, hence, x_0 is a local attractor. In particular, for $N = 2$, the equilibrium can either be a stable (non-degenerate or degenerate) node, or a stable focus, as shown in Figure 11.5.

Similarly, when $W^s(Df(x_0)) = \{0\}$, then $W^u(Df(x_0)) = \mathbb{R}^N$, x_0 is a local repeller and, hence, in the plane ($N = 2$), it can either be an unstable (non-degenerate or degenerate) node, or an unstable focus, as illustrated in Figure 11.6.

The example in Exercise 3 of Chapter 11 shows that, in general, the homeomorphism Ψ in Theorem 11.7 cannot be required to be a diffeomorphism with non-vanishing Jacobian. As a consequence, one

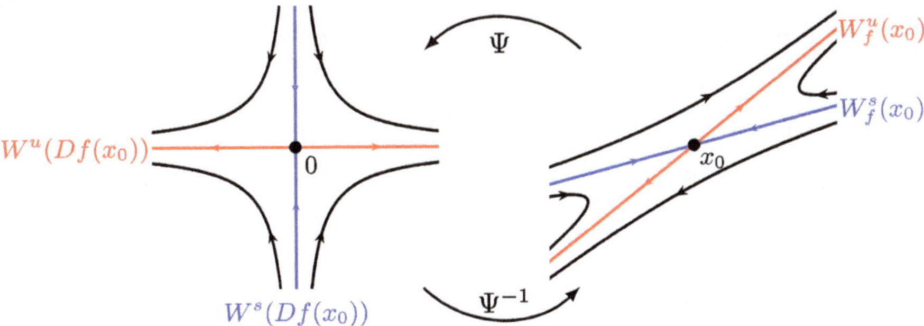

Fig. 11.4. Topological equivalence near an hyperbolic point between the solutions $L(t)$ of the linearized system (left) and the solutions $T(t)$ of the nonlinear system (right).

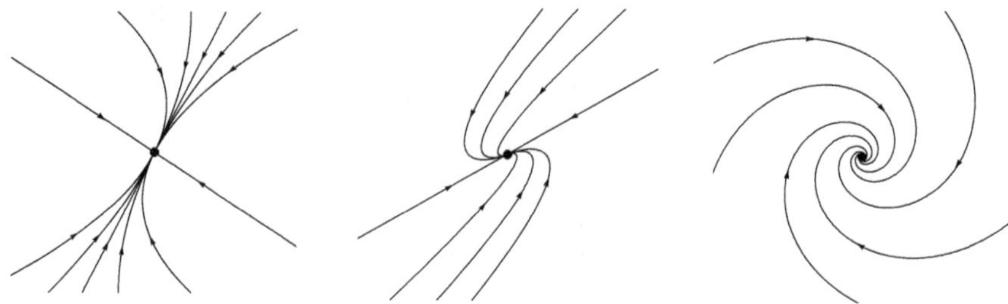

Fig. 11.5. Possible planar phase portraits with $W^u(Df(x_0)) = \{0\}$.

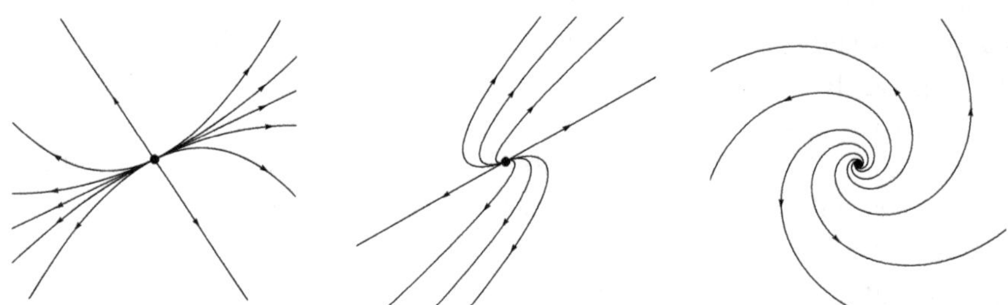

Fig. 11.6. Possible planar phase portraits with $W^s(Df(x_0)) = \{0\}$.

cannot guarantee, in general, that the nonlinear manifolds $W_f^s(x_0)$ and $W_f^u(x_0)$ are tangent at x_0 to their "linearizations" $W^s(Df(x_0))$ and $W^u(Df(x_0))$, respectively.

However, the following result, proved by S. Sternberg in 1958, gives a sufficient condition for Ψ to be of class $\mathcal{C}^\infty$ if f is of class $\mathcal{C}^\infty$.

Theorem 11.8 (Sternberg). *Suppose that f is of class $\mathcal{C}^\infty$ in a neighborhood of $x_0 \in \mathbb{R}^N$, that $f(x_0) = 0$, i.e. x_0 is an equilibrium of $u' = f(u)$, and that $Df(x_0)$ is an hyperbolic matrix. Suppose, in addition, that*

$$\sigma(Df(x_0)) = \{\lambda_1, \ldots, \lambda_q\},$$

with $\lambda_i \neq \lambda_j$ if $1 \leq i < j \leq q$, and that, for $j = 1, \ldots, q$,

$$\lambda_j \neq m_1\lambda_1 + \cdots + m_q\lambda_q \tag{11.38}$$

whenever all the m_i's are non-negative integers satisfying

$$2 \leq m_1 + \cdots + m_q.$$

Then, with the notations introduced in the statement of Theorem 11.7, there exists a (local) diffeomorphism of class $\mathcal{C}^\infty$, Ψ, between a

neighborhood of x_0 in $\mathbb{R}^N$ and a neighborhood of 0 in $\mathbb{R}^N$ such that
$T(t) = \Psi^{-1} L(t) \Psi$.

The eigenvalue condition (11.38) goes back to Poincaré (see Poincaré, 1879, pp. xcix–cx), who introduced it in 1879 to give the analytic counterpart of Theorem 11.8 (see Theorem 12.1 and the final notes on p. 271L of Hartman, 2002). Some further developments have been given by Nelson in 1969. The following $\mathcal{C}^1$ version of Theorem 11.7 for the plane goes back to Hartman (1962).

Theorem 11.9 (Hartman). *Suppose that $N = 2$ and that f is of class $\mathcal{C}^2$ in a neighborhood of $x_0 \in \mathbb{R}^N$, that $f(x_0) = 0$, i.e. x_0 is an equilibrium of $u' = f(u)$, and that $Df(x_0)$ is hyperbolic. Then, using the notations introduced in the statement of Theorem 11.7, there exists a (local) diffeomorphism of class $\mathcal{C}^1$, Ψ, between a neighborhood of x_0 in $\mathbb{R}^2$ and a neighborhood of 0 in $\mathbb{R}^2$ such that $T(t) = \Psi^{-1} L(t) \Psi$.*

The example of Exercise 4 of Chapter 11 shows that, in general, the diffeomorphism Ψ in the statement of Theorem 11.9 cannot be chosen of class $\mathcal{C}^2$ even if f is analytic. In the setting of Theorem 11.9, $W_f^u(x_0)$ and $W_f^s(x_0)$ are $\mathcal{C}^1$ curves through x_0 tangent to $W^s(Df(x_0))$ and $W^u(Df(x_0))$, respectively, at x_0.

The following examples show that, in general, Theorem 11.7 fails to be true when x_0 is not an hyperbolic equilibrium, though in some special cases the local phase portraits might be preserved.

Example 1. Consider the planar system

$$\begin{pmatrix} u' \\ v' \end{pmatrix} = \begin{pmatrix} 0 & 1 \\ -1 & 0 \end{pmatrix} \begin{pmatrix} u \\ v \end{pmatrix} + \begin{pmatrix} \varepsilon u(u^2 + v^2) \\ \varepsilon v(u^2 + v^2) \end{pmatrix}, \qquad (11.39)$$

where $\varepsilon \in \mathbb{R}$. Obviously, $(0,0)$ is an equilibrium for all $\varepsilon \in \mathbb{R}$. Moreover, the linear system

$$\begin{pmatrix} u' \\ v' \end{pmatrix} = \begin{pmatrix} 0 & 1 \\ -1 & 0 \end{pmatrix} \begin{pmatrix} u \\ v \end{pmatrix}$$

provides us with the linearization of (11.39) at $(0,0)$, whose eigenvalues, $\pm i$, have vanishing real parts, and for which $(0,0)$ is a global center. In this case, none of Theorems 11.3, 11.4, 11.7 and 11.9 can be used to determine the local character of $(0,0)$ as an equilibrium of (11.39). Instead, we proceed directly by passing to polar coordinates:

$$u(t) = \rho(t)\cos\theta(t), \quad v(t) = \rho(t)\sin\theta(t).$$

By differentiating the relation $u^2(t) + v^2(t) = \rho^2(t)$, we obtain

$$\rho(t)\rho'(t) = u(t)u'(t) + v(t)v'(t)$$
$$= u(t)\left(v(t) + \varepsilon u(t)\rho^2(t)\right) + v(t)\left(-u(t) + \varepsilon v(t)\rho^2(t)\right)$$
$$= \varepsilon\rho^4(t)$$

and, hence, $\rho' = \varepsilon\rho^3$. Similarly,

$$\theta'(t) = \frac{u(t)v'(t) - v(t)u'(t)}{\rho^2(t)}$$
$$= \frac{u(t)\left(-u(t) + \varepsilon v(t)\rho^2(t)\right) - v(t)\left(v(t) + \varepsilon u(t)\rho^2(t)\right)}{\rho^2(t)} = -1.$$

Thus, in polar coordinates, (11.39) can be expressed as

$$\theta' = -1, \quad \rho' = \varepsilon\rho^3.$$

Figure 11.7 shows the associated global phase portraits of (11.39) according to the sign of the parameter ε.

When $\varepsilon < 0$, the origin is a stable (nonlinear) focus. Thus, it is asymptotically stable. When $\varepsilon > 0$, instead, it becomes an unstable focus. This shows that the linearized system does not provide us with the local character of $(0,0)$ when $\varepsilon \neq 0$, showing also the optimality of Theorems 11.7 and 11.9.

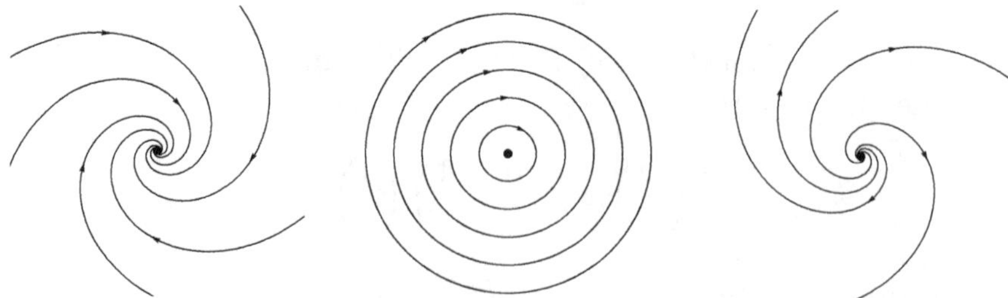

Fig. 11.7. Phase portrait of (11.39) according to the sign of ε: $\varepsilon < 0$ (left), $\varepsilon = 0$ (center), and $\varepsilon > 0$ (right).

Example 2. Now, we consider the predator-prey model (9.22), with $\lambda > 0$ and $\mu > 0$. As shown in Section 9.3, $x_0 = (\mu, \lambda)$ is its unique coexistence state. In this case, since

$$f(u, v) = \begin{pmatrix} \lambda u - uv \\ -\mu v + uv \end{pmatrix}, \quad (u, v) \in \mathbb{R}^2,$$

it is apparent that

$$Df(u, v) = \begin{pmatrix} \lambda - v & -u \\ v & -\mu + u \end{pmatrix}, \quad (u, v) \in \mathbb{R}^2,$$

and the variational system of (9.22) at (μ, λ) is the linear system

$$\begin{pmatrix} u' \\ v' \end{pmatrix} = Df(\mu, \lambda) \begin{pmatrix} u \\ v \end{pmatrix} = \begin{pmatrix} 0 & -\mu \\ \lambda & 0 \end{pmatrix} \begin{pmatrix} u \\ v \end{pmatrix}. \tag{11.40}$$

Since

$$\sigma(Df(\mu, \lambda)) = \left\{ \pm i \sqrt{\lambda \mu} \right\},$$

$(0, 0)$ is a center for (11.40). Therefore, in this case, the linearization maintains the local phase portrait of (9.22) at the equilibrium (μ, λ), even if (μ, λ) is not an hyperbolic equilibrium of (9.22). Nevertheless, one has to keep in mind that such a situation is truly exceptional!

Example 3. We conclude this section by illustrating Theorems 11.7 and 11.9 with a very simple, but extremely interesting, example of a planar system for which the unstable and stable manifolds, $W_f^u(x_0)$ and $W_f^s(x_0)$, can be easily constructed. Consider,

$$\begin{cases} u' = u, \\ v' = -v + g(u), \end{cases} \tag{11.41}$$

where $g \in C^2(\mathbb{R})$ satisfies

$$g(0) = g'(0) = 0.$$

Hence, $g(u) = O(u^2)$ as $u \to 0$. Actually,

$$\lim_{u \to 0} \frac{g(u)}{u^2} = \frac{g''(0)}{2}.$$

Since $g(0) = 0$, $(0,0)$ is the unique equilibrium of (11.41). The linearization at $(0,0)$ is the uncoupled linear system

$$\begin{cases} u' = u, \\ v' = -v. \end{cases} \tag{11.42}$$

Figure 11.8 shows the phase portrait of (11.42): as discussed in Section 9.1, it is a saddle point whose unstable and stable manifolds are, respectively, the u-axis and v-axis.

Fix now $(u_0, v_0) \in \mathbb{R}^2$. Since the nonlinearity of (11.41),

$$f(u,v) := \begin{pmatrix} u \\ -v + g(u) \end{pmatrix},$$

is autonomous and of class $\mathcal{C}^2$, by Corollary 5.3, it is locally Lipschitz in $(u,v) \in \mathbb{R}^2$, uniformly in $t \in \mathbb{R}$. Thus, according to Theorem 5.17 and its backward counterpart, system (11.41) has a unique maximal solution satisfying $(u(0), v(0)) = (u_0, v_0)$, denoted by $(I, (u,v))$. Moreover, there exist $T_{\min} \in [-\infty, 0)$ and $T_{\max} \in (0, +\infty]$ such that

$$I = (T_{\min}, T_{\max}),$$

and the solution (u,v) blows up at $T_{\min}$ (respectively, $T_{\max}$) if $-\infty < T_{\min}$ (respectively, $T_{\max} < +\infty$).

To determine the integral curve through the point (u_0, v_0), one can proceed as follows. Suppose first that $u_0 = 0$. Then, the first

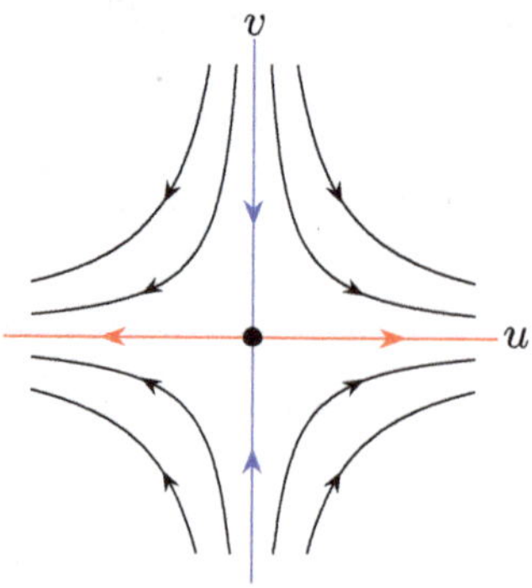

Fig. 11.8. Phase portrait of (11.42).

equation of the system, $u' = u$, implies $u(t) = 0$ for all $t \in I$, and, hence, $v' = -v$. Therefore, $v(t) = e^{-t}v_0$ and, by uniqueness,

$$(u(t), v(t)) = (0, e^{-t}v_0), \quad t \in \mathbb{R},$$

is the maximal solution. The corresponding orbit covers the positive v-axis if $v_0 > 0$, the negative v-axis if $v_0 < 0$, and it reduces to the equilibrium $(0, 0)$ if $v_0 = 0$.

Subsequently, we consider $u_0 \neq 0$. Then, by taking cross products in (11.41), we obtain that

$$u'(t)\,(-v(t) + g(u(t))) = v'(t)u(t) \quad \text{for all } t \in I,$$

or, equivalently,

$$\frac{d}{dt}\left(u(t)v(t) - \int_0^{u(t)} g(s)\,ds \right) = 0 \quad \text{for all } t \in I.$$

Consequently,

$$u(t)v(t) - \int_0^{u(t)} g(s)\,ds = u_0 v_0 - \int_0^{u_0} g(s)\,ds =: E \quad \text{for all } t \in I.$$

Moreover, since $u(t) = e^t u_0$ for all $t \in I$, it becomes apparent that $u(t) \neq 0$ for all $t \in I$ and, hence,

$$v(t) = \frac{E}{u(t)} + \frac{1}{u(t)} \int_0^{u(t)} g(s)\,ds \quad \text{for all } t \in I.$$

Therefore, the integral curve through (u_0, v_0) is given by

$$\mathcal{C}_{(u_0,v_0)} := \left\{ (u, v) \in \mathbb{R}^2 : u \neq 0 \text{ and } v = \frac{E}{u} + \frac{1}{u} \int_0^u g(s)\,ds \right\},$$

where

$$E := u_0 v_0 - \int_0^{u_0} g(s)\,ds.$$

Note that, if

$$v_0 = \frac{1}{u_0} \int_0^{u_0} g(s)\,ds, \tag{11.43}$$

one has that $E = 0$, and $\mathcal{C}_{(u_0,v_0)}$ becomes

$$\mathcal{C}_{(u_0,v_0)} := \left\{ (u,v) \in \mathbb{R}^2 : u \neq 0 \text{ and } v = \frac{1}{u} \int_0^u g(s)\, ds \right\}.$$

Subsequently, we set

$$h(u) := \int_0^u g(s)\, ds, \quad \varphi(u) := \frac{h(u)}{u}, \quad u \in \mathbb{R} \setminus \{0\}.$$

Since g is of class $\mathcal{C}^2$, by the fundamental theorem of calculus, h is of class $\mathcal{C}^3$, $h'(u) = g(u)$, and $h''(u) = g'(u)$. Thus,

$$h(0) = h'(0) = h''(0) = 0,$$

and, hence, $h(u) = O(u^3)$ as $u \to 0$. Thus, $\varphi(u)$ is of class $\mathcal{C}^2$ for $u \neq 0$, and $\varphi(u) = O(u^2)$, as $u \to 0$. As a consequence, $\varphi(u)$ can be extended to a $\mathcal{C}^2$-function up to $u = 0$ by setting

$$\varphi(0) = \varphi'(0) = 0, \quad \varphi''(0) = \lim_{u \to 0} \varphi''(u) = \frac{g''(0)}{3}.$$

Observe that the function $\varphi(u)$ can actually be quite arbitrary. Indeed, for any given $\varphi \in \mathcal{C}^3(\mathbb{R})$ such that $\varphi(0) = \varphi'(0) = 0$, if we define $h(u) := u\varphi(u)$, $u \in \mathbb{R}$, the function

$$g(u) := h'(u) \tag{11.44}$$

is of class $\mathcal{C}^2(\mathbb{R})$, and it is easily seen that $g(0) = g'(0) = 0$, because

$$h'(u) = \varphi(u) + u\varphi'(u), \quad h''(u) = 2\varphi'(u) + u\varphi''(u),$$

for all $u \in \mathbb{R}$, and, in particular,

$$h'(0) = h''(0) = 0.$$

Moreover, since $h(0) = 0$, (11.44) implies that

$$h(u) = \int_0^u g(s)\, ds \quad \text{for all } u \in \mathbb{R},$$

i.e. we are exactly in the setting described above, and we can conclude that the curve $v = \varphi(u)$ provides us with an integral curve through $(0,0)$ for system (11.41), with $g(u)$ given by (11.44). Precisely, it is the integral curve through any point (u_0, v_0), with $u_0 \neq 0$, satisfying (11.43). Furthermore, since $u(t) = e^t u_0$ for all $t \in \mathbb{R}$, $v = \varphi(u)$ must be the unstable manifold of $(0,0)$ because $u_0 \neq 0$, and $v = \varphi(u)$ is an invariant curve along which $u(t)$ separates away from zero as t increases.

Note that, if $E > 0$, we have that

$$v = \frac{E}{u} + \varphi(u) \begin{cases} > \varphi(u) & \text{if } u > 0, \\ < \varphi(u) & \text{if } u < 0, \end{cases}$$

whereas, if $E < 0$, we have that

$$v = \frac{E}{u} + \varphi(u) \begin{cases} < \varphi(u) & \text{if } u > 0, \\ > \varphi(u) & \text{if } u < 0. \end{cases}$$

All these integral curves approximate asymptotically to $v = \varphi(u)$ as $u \to \pm\infty$, because

$$\lim_{u \to \pm\infty} \frac{E}{u} = 0,$$

while they behave like $v = \frac{E}{u}$ as $u \to 0$ because $\varphi(0) = 0$. Therefore, according to the general properties discussed in Section 9.2, the phase portrait of (11.41) looks like the one sketched in Figure 11.9.

Although the curve $v = \varphi(u)$ is quite arbitrary, as commented above, observe that the phase portraits of (11.41) and (11.42) are topologically equivalent in a neighborhood of $(0,0)$. Thus, system (11.41) confirms the validity of Theorems 11.7 and 11.9, for any

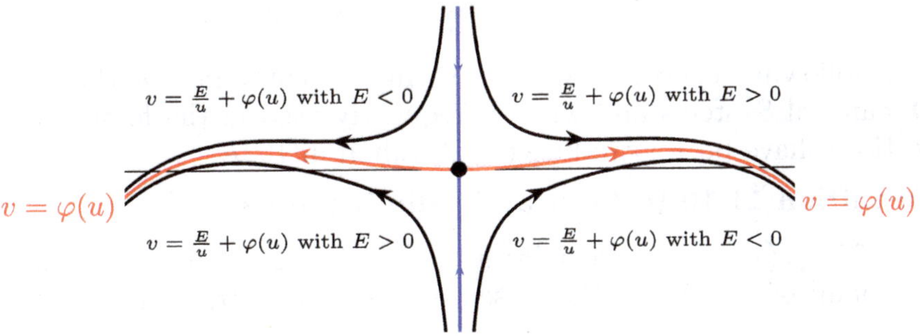

Fig. 11.9. Phase portrait of (11.41).

function $g(u)$, or equivalently $\varphi(u)$, satisfying the above-indicated assumptions. In addition, observe how, in this example, not only the local structure of the phase portrait around $(0,0)$ is preserved (up to diffeomorphisms) but also the global one. Finally, observe that (11.41) and the corresponding phase portrait reduce exactly to (11.42) and its associated diagram when $g(u) = \varphi(u) = 0$.

11.3 ω-Limit and α-Limit Sets

Throughout this section, $f \in \mathcal{C}^1(\mathbb{R}^N; \mathbb{R}^N)$ and, for every $x \in \mathbb{R}^N$, we denote by $(I_x, u(t;x))$ the unique maximal solution of the Cauchy problem

$$\begin{cases} u' = f(u), \\ u(0) = x. \end{cases} \tag{11.45}$$

According to Theorem 5.17, there exist

$$-\infty \leq T_{\min}(x) < 0 < T_{\max}(x) \leq +\infty,$$

such that $I_x = (T_{\min}(x), T_{\max}(x))$. When $T_{\max}(x) = +\infty$, we recall that the *positive semi-orbit through* x is defined as

$$\Gamma_x^+ := \{u(t;x)\colon t \geq 0\}.$$

Similarly, the *negative semi-orbit through* x is defined as

$$\Gamma_x^- := \{u(t;x)\colon t \leq 0\}$$

if $T_{\min}(x) = -\infty$. When $u(t;x)$ is defined for all $t \in \mathbb{R}$, the orbit through x is defined as

$$\Gamma_x := \Gamma_x^+ \cup \Gamma_x^-.$$

The following concepts, which are fundamental in the theory of Dynamical Systems and will be frequently used in the forthcoming sections, have been introduced by Birkhoff in 1927.

Definition 11.10 (α-limit and ω-limit points).

(a) If $T_{\max}(x) = +\infty$, a point $x_{+\infty} \in \mathbb{R}^N$ is said to be an ω-limit point of x if there exists a sequence of times, $\{t_n\}_{n\geq 1}$, such that

$$\lim_{n\to+\infty} t_n = +\infty \quad \text{and} \quad \lim_{n\to+\infty} u(t_n;x) = x_{+\infty}.$$

(b) If $T_{\min}(x) = -\infty$, a point $x_{-\infty} \in \mathbb{R}^N$ is said to be an α-limit point of x if there exists a sequence of times, $\{t_n\}_{n \geq 1}$, such that

$$\lim_{n \to +\infty} t_n = -\infty \quad \text{and} \quad \lim_{n \to +\infty} u(t_n; x) = x_{-\infty}.$$

Observe that, if $y \in \Gamma_x^+$ and $x_{+\infty}$ is an ω-limit point of x according to this definition, then $x_{+\infty}$ is an ω-limit point also of y. Indeed, since $y \in \Gamma_x^+$, there exists $\bar{t} \geq 0$ such that $y = u(\bar{t}; x)$. Moreover, since $x_{+\infty}$ is an ω-limit point of x, there exists a sequence $\{t_n\}_{n \geq 1}$ such that

$$\lim_{n \to +\infty} t_n = +\infty \quad \text{and} \quad \lim_{n \to +\infty} u(t_n; x) = x_{+\infty}.$$

Now, it suffices to observe that, since the problem (11.45) is autonomous and admits a unique solution for all $x \in \mathbb{R}$, necessarily,

$$u(t; y) = u(t + \bar{t}; x) \quad \text{for all } t \in [-\bar{t}, +\infty).$$

Thus, setting $s_n := t_n - \bar{t}$ for all $n \geq 1$, we obtain that

$$\lim_{n \to +\infty} s_n = +\infty \quad \text{and} \quad \lim_{n \to +\infty} u(s_n; y) = \lim_{n \to +\infty} u(t_n; x) = x_{+\infty},$$

i.e. $x_{+\infty}$ is also an ω-limit point of y. As a consequence, in Definition 11.10, it makes sense to define an ω-limit point of Γ_x^+, and not only of x. Of course, a similar discussion can be done for α-limit points and negative semi-orbits.

Subsequently, the set of ω-limit points of x (or, equivalently, of Γ_x^+) will be denoted by

$$\omega(x) = \omega(\Gamma_x^+)$$

if $T_{\max} = +\infty$, and it will be referred to as the ω-*limit set of* x. Similarly, the set of α-limit points of x (or, equivalently, of Γ_x^-) will be denoted by

$$\alpha(x) = \alpha(\Gamma_x^+)$$

if $T_{\min} = -\infty$, and it will be referred to as the α-*limit set of* x.

By definition, every ω-limit (respectively, α-limit) point of x belongs to the closure of Γ_x^+ (respectively, of Γ_x^-), i.e.

$$\omega(x) \subset \bar{\Gamma}_x^+, \quad \alpha(x) \subset \bar{\Gamma}_x^-. \tag{11.46}$$

Subsequently, we assume that $T_{\max} = +\infty$, and we study the most fundamental properties of $\omega(x)$ in the general case when $N \geq 1$. The following result establishes that $\omega(x)$ is a closed subset of $\mathbb{R}^N$ and that it is non-empty, compact, and connected if, in addition, Γ_x^+ is bounded.

Theorem 11.11. *Assume that $x \in \mathbb{R}^N$, $N \geq 1$, is such that the maximal solution $u(t; x)$ of (11.45) is defined in $[0, +\infty)$. Then, $\omega(x)$ is a closed subset of $\mathbb{R}^N$. If, in addition, Γ_x^+ is bounded, then $\omega(x)$ is nonempty, connected, and compact; i.e. it is a topological continuum.*

Proof. First, we show that $\omega(x)$ is closed. Thus, we take a sequence $\{x_{n,+\infty}\}_{n\geq 1} \subset \omega(x)$ of ω-limit points of x such that

$$\lim_{n\to+\infty} x_{n,+\infty} = x_{+\infty} \in \mathbb{R}^N. \tag{11.47}$$

We shall use a diagonal argument to show that also $x_{+\infty} \in \omega(x)$. Since $x_{n,+\infty} \in \omega(x)$, for every $n \geq 1$, there is a sequence $\{t_{n,m}\}_{m\geq 1}$ such that

$$\lim_{m\to+\infty} t_{n,m} = +\infty, \qquad \lim_{m\to+\infty} u(t_{n,m}; x) = x_{n,+\infty}. \tag{11.48}$$

For each $k \geq 1$, thanks to (11.47), there exists $n_0 = n_0(k) \in \mathbb{N}$ such that

$$|x_{n,+\infty} - x_{+\infty}| \leq \frac{1}{2k} \quad \text{for all } n \geq n_0. \tag{11.49}$$

Moreover, by (11.48), there exists $m_0 = m_0(k) \in \mathbb{N}$ such that, for every $m \geq m_0$,

$$t_{n_0,m} \geq k \quad \text{and} \quad |u(t_{n_0,m}; x) - x_{n_0,+\infty}| \leq \frac{1}{2k}. \tag{11.50}$$

Thus, combining (11.49) with (11.50), it becomes apparent that, for every $k \geq 1$, if we set $\tau_k := t_{n_0(k),m_0(k)}$, we have $\tau_k \geq k$ and

$$|u(\tau_k; x) - x_{+\infty}| \leq |u(\tau_k; x) - x_{n_0(k),+\infty}| + |x_{n_0(k),+\infty} - x_{+\infty}| \leq \frac{1}{k}.$$

Thus,

$$\lim_{k\to+\infty} \tau_k = +\infty \quad \text{and} \quad \lim_{k\to+\infty} u(\tau_k; x) = x_{+\infty},$$

which shows that $x_{+\infty} \in \omega(x)$. So, $\omega(x)$ is closed.

From now on we assume, in the rest of the proof, that Γ_x^+ is bounded. In this case, since $\omega(x)$ is a closed subset contained in $\bar{\Gamma}_x^+$ which is compact, $\omega(x)$ is compact too.

To show that $\omega(x)$ is nonempty, consider any sequence $\{t_n\}_{n\geq 1}$ such that

$$\lim_{n\to+\infty} t_n = +\infty.$$

Since $\{u(t_n; x)\}_{n\geq 1}$ is a bounded sequence, by the Bolzano–Weierstrass theorem, there exists $x_{+\infty} \in \mathbb{R}^N$ and a subsequence, $\{u(t_{n_m}; x)\}_{m\geq 1}$, such that

$$\lim_{m\to+\infty} u(t_{n_m}; x) = x_{+\infty}.$$

Moreover,

$$\lim_{m\to+\infty} t_{n_m} = +\infty.$$

Therefore, $x_{+\infty}$ belongs to $\omega(x)$, as we wanted to prove.

To show that $\omega(x)$ is connected, we argue by contradiction, assuming that there exist two disjoint non-empty compact sets C_1 and C_2 such that $\omega(x) = C_1 \cup C_2$. Then, setting

$$d := \mathrm{dist}\,(C_1, C_2) = \min_{x_1 \in C_1,\, x_2 \in C_2} \mathrm{dist}\,(x_1, x_2) > 0,$$

the open sets

$$C_{j,d} := \left\{ x \in \mathbb{R}^N : \mathrm{dist}\,(x, C_j) < \frac{d}{7} \right\}, \quad j \in \{1, 2\},$$

are disjoint. Since $C_1 \subset C_{1,d}$ contains some point of $\omega(x)$, there exists $t_1 > 1$ such that $u(t_1; x) \in C_{1,d}$. Similarly, since $C_2 \subset C_{2,d}$ contains some point of $\omega(x)$, there exists $t_2 > \max\{t_1, 2\}$ such that $u(t_2; x) \in C_{2,d}$. Arguing recursively, one can construct a sequence of times, $\{t_n\}_{n\geq 1}$, such that, for every $n \geq 1$, $t_{n+1} > \max\{t_n, n+1\}$ and

$$u(t_{2n-1}; x) \in C_{1,d}, \quad u(t_{2n}; x) \in C_{2,d}, \tag{11.51}$$

as illustrated in Figure 11.10.

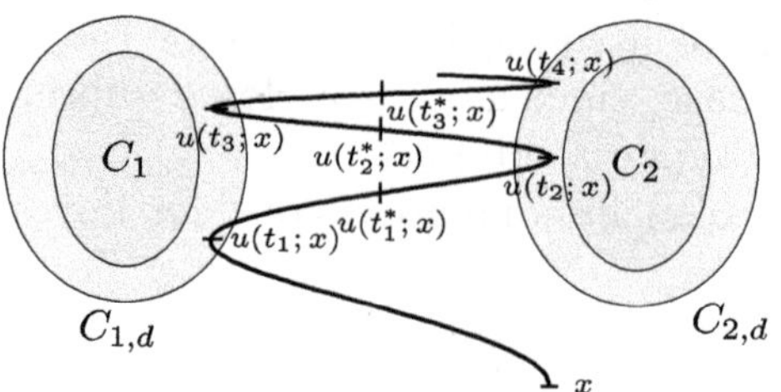

Fig. 11.10.   Proof of the connectedness of $\omega(x)$.

Consequently, thanks to the continuity of the functions $t \mapsto u(t;x)$ and $y \mapsto \mathrm{dist}\,(y, C_j)$, $j \in \{1,2\}$, for every $n \geq 1$, there exists $t_n^* \in (t_n, t_{n+1})$ such that

$$\mathrm{dist}\,(u(t_n^*;x), C_1) = \mathrm{dist}\,(u(t_n^*;x), C_2)$$

(see Figure 11.10). In particular,

$$u(t_n^*;x) \in K := \bar{\Gamma}_x^+ \setminus (C_{1,d} \cup C_{2,d}) \quad \text{for all } n \geq 1. \tag{11.52}$$

By (11.51), since $t_n^* \in (t_n, t_{n+1})$ for all $n \geq 1$, it is apparent that

$$\lim_{n \to +\infty} t_n^* = +\infty.$$

Moreover, since K is a compact subset of $\mathbb{R}^N$, by the Bolzano–Weierstrass theorem, there exists a subsequence of $\{u(t_n^*;x)\}_{n \geq 1}$, say $\{u(t_{n_m}^*;x)\}_{m \geq 1}$, such that

$$\lim_{m \to +\infty} u(t_{n_m}^*;x) = x_{+\infty}$$

for some $x_{+\infty} \in K$. By definition, $x_{+\infty} \in \omega(x)$, but (11.52) implies that

$$x_{+\infty} \notin C_1 \cup C_2 = \omega(x).$$

This contradiction shows that $\omega(x)$ must be connected if Γ_x^+ is bounded, and the proof is concluded. $\qquad\square$

Observe that the example in Figure 5.4 (see also Exercise 20 of Chapter 5) shows that, if Γ_x^+ is unbounded, $\omega(x)$ can be non-compact and can consist of several connected components, precisely the upper and lower dashed lines in the figure.

Moreover, if we set $s = -t$ and $v(s) = u(t)$, problem (11.45) is transformed into

$$\begin{cases} \dot{v} = -f(v), \\ v(0) = x, \end{cases} \tag{11.53}$$

where $\dot{v} = \frac{dv}{ds}$. Thus, the negative semi-orbit of $u(t;x)$ as a solution of (11.45), say, $\Gamma^-_{x,f}$, corresponds to the positive semi-orbit of $v(s;x)$ as a solution of (11.53), say $\Gamma^+_{x,-f}$. Therefore, we immediately obtain the following backward counterpart of Theorem 11.11.

Corollary 11.12. *Assume that $x \in \mathbb{R}^N$, $N \geq 1$, is such that the maximal solution $u(t;x)$ of (11.45) is defined in $(-\infty, 0]$. Then, $\alpha(x)$ is a closed subset of $\mathbb{R}^N$. If, in addition, Γ^-_x is bounded, then $\alpha(x)$ is nonempty, connected, and compact.*

The following result establishes that, whenever Γ^+_x is bounded, $\omega(x)$ consists of trajectories of solutions that are globally defined in time. Moreover, it shows that $\omega(x)$ is invariant for the flow induced by $u' = f(u)$.

Theorem 11.13. *Let $x \in \mathbb{R}^N$ be such that $u(t;x)$ is defined in $[0, +\infty)$ and Γ^+_x is bounded. Then, for every $x_{+\infty} \in \omega(x)$, the (maximal) solution of*

$$\begin{cases} u' = f(u), \\ u(0) = x_{+\infty}, \end{cases} \tag{11.54}$$

denoted by $(I_{x_{+\infty}}, u(t; x_{+\infty}))$, is globally defined in time, i.e. $I_{x_{+\infty}} = \mathbb{R}$. Moreover, the following invariance property holds:

$$u(t; x_{+\infty}) \in \omega(x) \quad \text{for all } t \in I_{x_{+\infty}} = \mathbb{R}.$$

Similarly, if $u(t;x)$ is defined in $(-\infty, 0]$ and Γ^-_x is bounded, then, for every $x_{-\infty} \in \alpha(x)$, the (maximal) solution of

$$\begin{cases} u' = f(u), \\ u(0) = x_{-\infty}, \end{cases}$$

denoted by $(I_{x_{-\infty}}, u(t; x_{-\infty}))$, is globally defined in time, i.e. $I_{x_{-\infty}} = \mathbb{R}$, and

$$u(t; x_{-\infty}) \in \alpha(x) \quad \text{for all } t \in I_{x_{-\infty}} = \mathbb{R}.$$

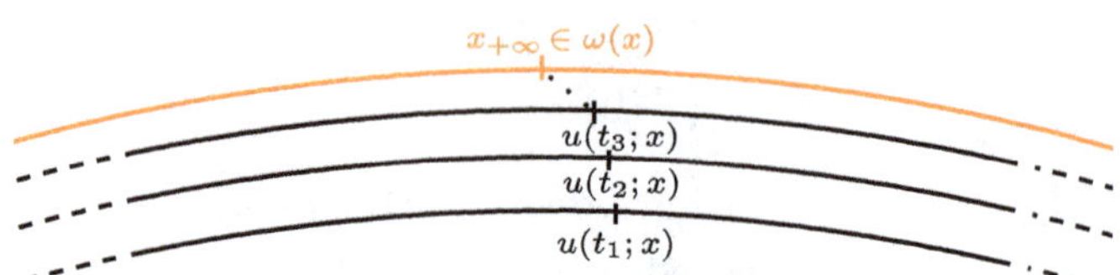

Fig. 11.11. Invariance property of the ω-limit set.

According to Theorem 11.13, when Γ_x is bounded, the sets $\omega(x)$ and $\alpha(x)$ consist of the trajectories of the (global) solutions of $u' = f(u)$ starting at the points of $\omega(x)$ and $\alpha(x)$, respectively. Figure 11.11 illustrates this feature: the set $\omega(x)$ consists of

$$x_{+\infty} = \lim_{n \to +\infty} u(t_n; x) \in \mathbb{R}^N,$$

together with $u(t; x_{+\infty})$, for all $t \in \mathbb{R}$. In such an example, $\lim_{n \to +\infty} t_n = +\infty$.

The properties established in Theorem 11.13 are usually referred to as the *invariance properties* of the ω-limit and the α-limit sets.

Proof of Theorem 11.13. Thanks to Theorem 11.11, $\omega(x)$ is a nonempty connected compact set. Fix any $x_{+\infty} \in \omega(x)$. By definition, there exists a sequence of times, $\{t_n\}_{n \geq 1}$, such that $t_n \geq 0$ for all $n \geq 1$,

$$\lim_{n \to +\infty} t_n = +\infty, \quad \text{and} \quad \lim_{n \to +\infty} u(t_n; x) = x_{+\infty}. \tag{11.55}$$

Subsequently, for every integer $n \geq 1$, we consider the shifted function

$$u_n(t) := u(t + t_n; x), \quad t \geq -t_n.$$

Since $u' = f(u)$ is autonomous, u_n also solves the system $u' = f(u)$. Actually, it solves the Cauchy problem

$$\begin{cases} u' = f(u), \\ u(0) = u(t_n; x). \end{cases}$$

By definition, u_n is well defined, at least, in $[-t_n, +\infty)$. Moreover, since Γ_x^+ is bounded and

$$\Gamma_x^+ = \{u(t; x)\colon t \geq 0\} = \{u_n(t)\colon t \geq -t_n\} \quad \text{for all } n \geq 1,$$

it becomes apparent that, for every integer $j \geq 1$, there exist $n_0 = n_0(j) \in \mathbb{N}$ and a positive constant $C = C(j)$ such that $t_n \geq j$ for all $n \geq n_0$ and

$$\|u_n(t)\| \leq C \quad \text{for all } t \in [-j, j] \text{ and } n \geq n_0. \tag{11.56}$$

In other words, the sequence $\{u_n\}_{n \geq n_0}$ is bounded in the space of continuous functions $\mathcal{C}([-j, j]; \mathbb{R}^N)$. Consequently, since f is bounded in the closed ball $\|u\| \leq C$ of $\mathbb{R}^N$, it follows from (11.56) that, for every $t, s \in [-j, j]$ and $n \geq n_0$,

$$\|u_n(t) - u_n(s)\| = \left\| \int_s^t u_n'(\tau)\, d\tau \right\| = \left\| \int_s^t f(u_n(\tau))\, d\tau \right\|$$

$$\leq \int_{\min\{t,s\}}^{\max\{t,s\}} \|f(u_n(\tau))\|\, d\tau \leq \max_{\|u\| \leq C} \|f(u)\|\,|t - s|.$$

Proposition 7.4 then ensures that the sequence $\{u_n\}_{n \geq n_0}$ is bounded and equicontinuous in $\mathcal{C}([-j, j]; \mathbb{R}^N)$. Thus, thanks to Theorem 7.5, we can extract a subsequence of $\{u_n\}_{n \geq n_0}$, say $\{u_{n_m}\}_{m \geq 1}$, such that, for some $u_\infty \in \mathcal{C}([-j, j]; \mathbb{R}^N)$,

$$\lim_{m \to +\infty} \|u_{n_m} - u_\infty\|_{\mathcal{C}([-j,j];\mathbb{R}^N)} = 0. \tag{11.57}$$

In particular, it follows from (11.55) and (11.57) that

$$x_{+\infty} = \lim_{m \to +\infty} u(t_{n_m}; x) = \lim_{m \to +\infty} u_{n_m}(0) = u_\infty(0). \tag{11.58}$$

Adapting the proof of Theorem 7.6, and based on (11.57) and (11.58), it is easily seen that u_∞ solves (11.54) in $[-j, j]$ for all integer $j \geq 1$. So, the maximal solution of (11.54), $u(t; x_{+\infty})$, is defined for all $t \in \mathbb{R}$. Moreover, by uniqueness, for every $t \in \mathbb{R}$,

$$u(t; x_{+\infty}) = u_\infty(t) = \lim_{m \to +\infty} u_{n_m}(t) = \lim_{m \to +\infty} u(t + t_{n_m}; x),$$

which shows that $u(t; x_{+\infty}) \in \omega(x)$ for all $t \in \mathbb{R}$ and concludes the proof of the invariance of $\omega(x)$.

Finally, the change of variables $s = -t$ and $v(s) = u(t)$ allows us to readily obtain the corresponding results for $\alpha(x)$. $\qquad\square$

The corollary we present now to conclude this section provides us with some immediate consequences of Theorem 11.13.

Corollary 11.14. *Let $x \in \mathbb{R}^N$ be such that $u(t; x)$ is defined in $[0, +\infty)$, Γ_x^+ is bounded, and assume that $\omega(x) = \{x_{+\infty}\}$. Then, $x_{+\infty}$ is an equilibrium, i.e. $f(x_{+\infty}) = 0$, and*

$$\lim_{t \to +\infty} u(t; x) = x_{+\infty}.$$

Similarly, if $u(t; x)$ is defined in $(-\infty, 0]$, Γ_x^- is bounded, and $\alpha(x) = \{x_{-\infty}\}$, then, $x_{-\infty}$ is an equilibrium, i.e. $f(x_{-\infty}) = 0$, and

$$\lim_{t \to -\infty} u(t; x) = x_{-\infty}.$$

Proof. By Theorem 11.13, $u(t; x_{+\infty}) = x_{+\infty}$ for all $t \in \mathbb{R}$. Thus,

$$0 = u'(t; x_{+\infty}) = f(u(t; x_{+\infty})) = f(x_{+\infty}) \quad \text{for all } t \in \mathbb{R}.$$

Hence, $x_{+\infty}$ is an equilibrium of $u' = f(u)$.

Let $\{t_n\}_{n \geq 1}$, with $t_n \geq 0$, satisfy $\lim_{n \to +\infty} t_n = +\infty$. Then, the sequence $\{u(t_n; x)\}_{n \geq 1}$ is bounded because it is contained in Γ_x^+, which is bounded. Thanks to the Bolzano–Weierstrass theorem, there exist $x_* \in \mathbb{R}^N$ and a subsequence, $\{u(t_{n_m}; x)\}_{m \geq 1}$, such that

$$\lim_{m \to +\infty} u(t_{n_m}; x) = x_*.$$

By definition, x_* belongs to $\omega(x) = \{x_{+\infty}\}$. Thus, $x_* = x_{+\infty}$. This completes the proof of the first part of the statement. The second one follows, as usual, by making the change of variables $s = -t$ and $v(s) = u(t)$. $\qquad\square$

11.4 The Poincaré–Bendixson Theorem

This section studies a very sharp property of the ω-limit set for systems in the plane ($N = 2$). Consider $f \in \mathcal{C}^1(\mathbb{R}^2; \mathbb{R}^2)$, and for every

$x \in \mathbb{R}^2$, denote by $(I_x, u(t; x))$ the unique (maximal) solution of the Cauchy problem

$$\begin{cases} u' = f(u), \\ u(0) = x. \end{cases} \tag{11.59}$$

The main result in this section is the following one.

Theorem 11.15 (Poincaré–Bendixson). *Assume that, for some $x \in \mathbb{R}^2$, the unique maximal solution of (11.59) satisfies the following:*

(i) *$u(t; x)$ is defined in $[0, +\infty)$, and it is not a periodic solution;*
(ii) *Γ_x^+ is bounded;*
(iii) *$\omega(x)$ does not contain any equilibria, i.e. $f(x_\infty) \neq 0$ for all $x_\infty \in \omega(x)$.*

Then, $\omega(x)$ is the orbit of a non-trivial periodic solution.

By "non-trivial" we mean that it is not an equilibrium point. Observe that the conclusion of the theorem trivially holds true also when $u(t; x)$ is periodic because, in such a case,

$$\omega(x) = \Gamma_x = \{u(t; x) \colon t \in \mathbb{R}\}.$$

This is why we require u to be not periodic in (i).

We start by analyzing an example where this theorem can be applied. The proof is postponed to the following section.

11.4.1 *An illustrative example*

In this section, we analyze the dynamics of the planar system

$$\begin{cases} u' = u + v - u^3, \\ v' = -u + v - v^3, \end{cases} \tag{11.60}$$

whose equilibria are the solutions of the algebraic system

$$\begin{cases} v = u^3 - u = u\left(u^2 - 1\right), \\ u = v - v^3 = v\left(1 - v^2\right). \end{cases} \tag{11.61}$$

Figure 11.12 shows the graphs of these curves.

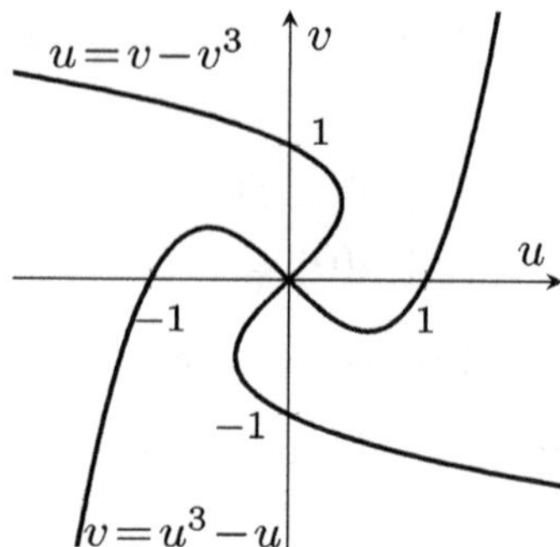

Fig. 11.12. Graphs of the curves (11.61).

Since the polynomial

$$v(u) = u^3 - u = u\left(u^2 - 1\right)$$

has three real roots, -1, 0, and 1, and

$$\frac{dv}{du} = 3u^2 - 1,$$

its critical points are located at $u = \pm\frac{\sqrt{3}}{3}$. Moreover,

$$v\left(\pm\tfrac{\sqrt{3}}{3}\right) = \pm\tfrac{\sqrt{3}}{3}\left(\tfrac{1}{3} - 1\right) = \mp\tfrac{2\sqrt{3}}{9},$$

which gives

$$\max_{u\in[-1,1]} |v(u)| = \frac{2\sqrt{3}}{9} < 1.$$

This analysis shows that the graphs of the curves in (11.61) only cross at the origin. Thus, $(u, v) = (0, 0)$ is the unique equilibrium of (11.60).

The linearization of (11.60) at $(0, 0)$ is the linear system

$$\begin{cases} u' = u + v, \\ v' = -u + v, \end{cases}$$

whose associated matrix is

$$A = \begin{pmatrix} 1 & 1 \\ -1 & 1 \end{pmatrix}.$$

Since its spectrum is $\sigma(A) = \{1 \pm i\}$, it turns out that $(0, 0)$ is hyperbolic, and Theorem 11.9 guarantees that $(0, 0)$ is an unstable

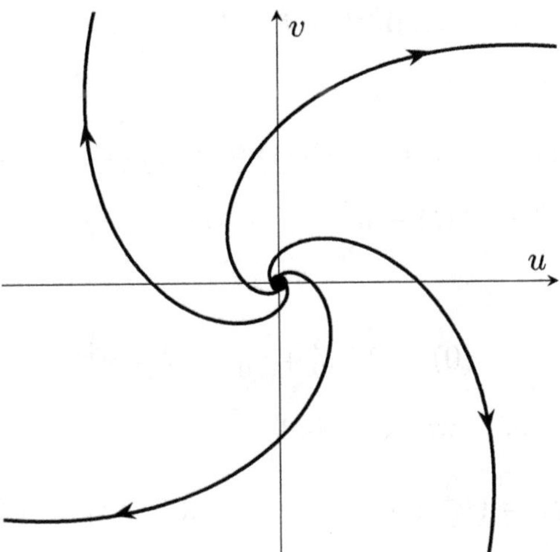

Fig. 11.13. Local phase portrait of (11.60) at $(0,0)$.

focus of (11.60), as illustrated in Figure 11.13. The clockwise rotation of the trajectories can be inferred since, for sufficiently small $\varepsilon > 0$, one has that, at $(u, v) = (0, \varepsilon)$,

$$u' = \varepsilon > 0, \quad v' = \varepsilon - \varepsilon^3 > 0.$$

In order to analyze the qualitative behavior of (11.60) far away from the equilibrium $(0,0)$, we study the temporal evolution of the function

$$\varphi(t) := u^2(t) + v^2(t), \quad t \in I,$$

where $(I, (u, v))$ is the unique maximal solution of the Cauchy problem

$$\begin{cases} u' = u + v - u^3, \\ v' = -u + v - v^3, \\ u(0) = u_0, \quad v(0) = v_0, \end{cases} \tag{11.62}$$

with

$$u_0^2 + v_0^2 = R^2, \quad R > \sqrt{2}. \tag{11.63}$$

By differentiating with respect to t, (11.62) gives that, for every $t \in I$,

$$\varphi'(t) = 2u(t)u'(t) + 2v(t)v'(t)$$
$$= 2u(t)\left(u(t) + v(t) - u^3(t)\right) + 2v(t)\left(-u(t) + v(t) - v^3(t)\right)$$
$$= 2\left(u^2(t) + v^2(t) - u^4(t) - v^4(t)\right).$$

In particular,

$$\varphi'(0) = 2\left(u_0^2 + v_0^2 - u_0^4 - v_0^4\right). \tag{11.64}$$

Moreover, by (11.63), we find that

$$R^2\left(u_0^2 + v_0^2\right) = \left(u_0^2 + v_0^2\right)^2 = u_0^4 + v_0^4 + 2u_0^2 v_0^2 \leq 2\left(u_0^4 + v_0^4\right)$$

and, hence,

$$u_0^2 + v_0^2 \leq \frac{2}{R^2}\left(u_0^4 + v_0^4\right).$$

Thus, (11.64) implies that

$$\varphi'(0) \leq 2\left(\frac{2}{R^2} - 1\right)\left(u_0^4 + v_0^4\right) < 0$$

because $R > \sqrt{2}$. Therefore, there exists $\varepsilon > 0$ such that

$$\varphi(t) = u^2(t) + v^2(t) < R^2 \quad \text{for all } t \in (0, \varepsilon].$$

Consequently, the solutions of (11.60) starting on the circle of radius R centered at $(0, 0)$ evolve toward the interior of the disk D_R enclosed by such a circle, as illustrated in Figure 11.14.

Since these solutions are confined to the disk D_R for all (forward) time where they are defined, they are bounded and, hence, cannot blow up in a finite time. Therefore, thanks to Theorem 5.17, all these solutions are globally defined in $[0, +\infty)$. As we have seen, for every $(u_0, v_0) \in \partial D_R$, we have that

$$\Gamma^+_{(u_0, v_0)} \subset \bar{D}_R,$$

thus $\Gamma^+_{(u_0, v_0)}$ is bounded, and Theorem 11.11 guarantees that $\omega(u_0, v_0) \neq \emptyset$. Moreover, $(0, 0) \notin \omega(u_0, v_0)$ because we have shown

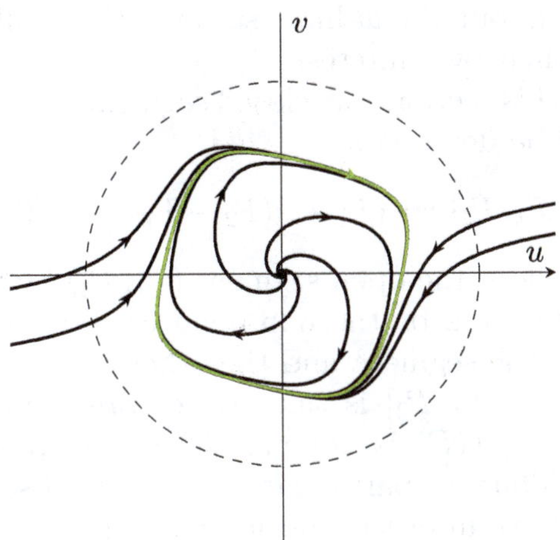

Fig. 11.14. Phase portrait of (11.60).

that $(0, 0)$ is unstable. As a consequence, no equilibrium of the planar system (11.60) lies in $\omega(u_0, v_0)$. In such a setting, the Poincaré–Bendixson theorem establishes that all the solutions starting on ∂D_R must spiral toward the orbit of some non-trivial periodic solution, as illustrated in Figure 11.14.

Similarly, the solutions starting in a punctured neighborhood of the origin must spiral toward a periodic solution. Although these periodic solutions might, in principle, be different, some numerical experiments suggest that (11.60) has a unique non-trivial periodic solution.

11.4.2 *Proof of the Poincaré–Bendixson theorem*

The proof of the Poincaré–Bendixson theorem, Theorem 11.15, relies on the Jordan curve theorem, which is valid only in the plane. This is why condition $N = 2$ is necessary for its validity. We recall, that we consider the following problem,

$$\begin{cases} u' = f(u), \\ u(0) = x, \end{cases} \tag{11.65}$$

with the assumptions and notations detailed at the beginning of Section 11.4. On our way to the proof, we present some preliminary technical results about the ω-limit set of x that will be used in the proof and have their own interest.

In the rest of this section, the closed segment defined by the points $P_1, P_2 \in \mathbb{R}^2$, will be denoted as

$$S = [P_1, P_2] := \{P_1 + s(P_2 - P_1) \colon s \in [0,1]\}.$$

Similarly, we consider the open segment (P_1, P_2) by taking $s \in (0,1)$ in the above definition, or the half-open case, in which one of the endpoints belong the segment and the other one not.

A segment $S = [P_1, P_2]$ is said to be *transversal* to the vector field f of (11.65) if $f(P)$ is not parallel to the vector $P_2 - P_1$ for all $P \in [P_1, P_2]$. Thus, in particular, a transversal segment does not possess any equilibrium of the considered system.

Suppose that $S = [P_1, P_2]$ is transversal to f. Then, according to Theorem 5.26 (see also Remark 5.27), for every $P \in S$, there exist $\tau(P) > 0$ and $\varepsilon(P) > 0$ such that, for each $y \in B_{\varepsilon(P)}(P)$, $[-\tau(P), \tau(P)] \subset I_y$, i.e. $u(t; y)$ is well defined in $[-\tau(P), \tau(P)]$. Since

$$S = [P_1, P_2] \subset \bigcup_{P \in S} B_{\varepsilon(P)}(P),$$

by the compactness of S, there exist an integer $k \geq 1$ and k points, $x_j \in S$, $j \in \{1, \dots, k\}$, such that

$$S = [P_1, P_2] \subset \bigcup_{j=1}^{k} B_{\varepsilon(x_j)}(x_j).$$

Consequently, for every $P \in S$, there exists $j = j(P) \in \{1, \dots, k\}$ such that $P \in B_{\varepsilon(x_j)}(x_j)$ and, hence,

$$[-\tau(x_j), \tau(x_j)] \subset I_P.$$

In other words, $u(t; P)$ is well defined in $[-\tau(x_j), \tau(x_j)]$. Therefore, setting

$$\tau := \min_{j \in \{1, \dots, q\}} \tau(x_j) > 0,$$

it becomes apparent that $u(t; P)$ is well defined in $[-\tau, \tau]$ for all $P \in S = [P_1, P_2]$. Actually, by Theorem 5.24, the solution operator of $u' = f(u)$ establishes a Lipschitz correspondence between

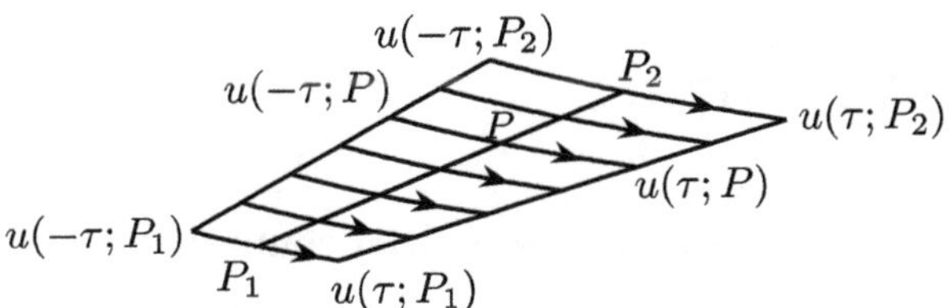

Fig. 11.15. Flow box through the transversal segment $[P_1, P_2]$ to the flow of (11.65).

$\{u(-\tau; P)\colon P \in S\}$ and $\{u(t; P)\colon P \in S\}$, for all $t \in [-\tau, \tau]$. Thus, up to shortening $\tau > 0$, if necessary, we can assume that $\{u(t; P)\colon P \in S\}$ does not intersect the transversal segment S for all $t \in [-\tau, \tau] \setminus \{0\}$. In such a case, the set

$$\Phi_S := \{u(t; P)\colon (t, P) \in [-\tau, \tau] \times S\}$$

is said to be a *flow box* through the transversal segment S (see Figure 11.15). Observe that, since S is transversal to f, the direction field f points always towards the same side of S. Moreover, by construction, for every $u_0 \in \Phi_S$, there exists a unique $t \in [-\tau, \tau]$ such that $u(t; u_0) \in [P_1, P_2]$.

Next, we study three fundamental properties of flow boxes through transversal segments that will be pivotal in the proof of Theorem 11.15. The first one can be stated as follows.

Lemma 11.16. *Let $S = [P_1, P_2]$ be a transversal segment to f, and $u(t)$ a solution of $u' = f(u)$ defined in $[0, T]$ for some $T > 0$. Then, the set*

$$[P_1, P_2] \cap \{u(t)\colon t \in [0, T]\}$$

is, at most, finite. Thus, for every $x \in \mathbb{R}^2$ such that $[0, +\infty) \subset I_x$, the set $[P_1, P_2] \cap \Gamma_x^+$ is, at most, countable.

Proof. Let $\Phi_S := \{u(t; P)\colon (t, P) \in [-\tau, \tau] \times S\}$ be a flow box through S. By construction of the flow box (see the discussion above), if $u(t_0) \in S$ for a certain $t_0 \in [0, T]$, then $u(t_0 + t) \notin S$ for all $t \in (0, 2\tau)$. Thus, the number of intersection points between S and the compact arc of trajectory $\{u(t)\colon t \in [0, T]\}$ is, at most, finite.

Therefore, since

$$[P_1, P_2] \cap \Gamma_x^+ = \bigcup_{n \in \mathbb{N}} ([P_1, P_2] \cap \{u(t) \colon t \in [0, n]\})$$

is a countable union of finite sets, it is countable. $\square$

The order on $[0, 1]$ induces an order on the segment

$$[P_1, P_2] := \{P_1 + s(P_2 - P_1) \colon s \in [0, 1]\}.$$

Indeed, it is said that $x_1 < x_2$ if

$$x_1 = P_1 + s_1(P_2 - P_1) \quad \text{and} \quad x_2 = P_1 + s_2(P_2 - P_1),$$

with $0 \le s_1 < s_2 \le 1$. Naturally, it is said that $x_1 > x_2$ if $x_2 < x_1$ according to the previous definition, and a sequence $\{x_n\}_{n \ge 1} \subset [P_1, P_2]$ is said to be monotone if $x_n < x_{n+1}$ or $x_n > x_{n+1}$ for all $n \ge 1$.

The second property of transversal segments essentially establishes that consecutive intersections of a solution with a transversal segment, if they do not coincide, must be ordered.

Lemma 11.17. *Let $S = [P_1, P_2]$ be a transversal segment to f, and $u(t)$ a solution of $u' = f(u)$ defined in $[0, +\infty)$ such that:*

- *$u(t_j) \in [P_1, P_2]$, $j \in \{1, 2, 3\}$, for some $0 \le t_1 < t_2 < t_3$;*
- *$u(t_1) \ne u(t_2)$;*
- *$u(t) \in \mathbb{R}^2 \setminus [P_1, P_2]$ for all $t \in (t_1, t_3) \setminus \{t_2\}$.*

Then, either $u(t_1) < u(t_2) < u(t_3)$, or $u(t_1) > u(t_2) > u(t_3)$.

For the validity of this result, the condition $N = 2$ is imperative. Actually, the higher complexity of the dynamics for $N \ge 3$ is provoked by the possible lack of monotonicity of consecutive intersections of an orbit with a transversal section.

Proof of Lemma 11.17. Let Φ_S be a flow box through the transversal segment S, and suppose, without loss of generality, that the flow is oriented as in Figure 11.15. Then, either $u(t_1) < u(t_2)$, or $u(t_2) < u(t_1)$. To fix ideas, suppose that $u(t_1) < u(t_2)$. Then, the trajectory of $u(t)$ in $[t_1, t_2]$ must look like one of the two types represented in Figure 11.16.

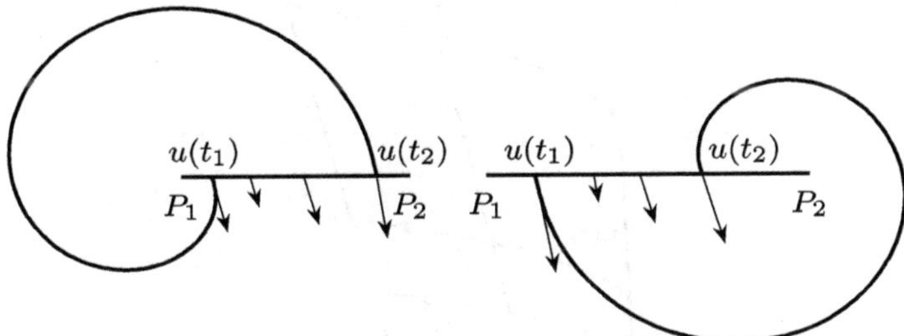

Fig. 11.16. The two possible behaviors of the orbit intersecting the transversal segment $[P_1, P_2]$ when $u(t_1) < u(t_2)$.

By Property 1 in Section 9.2, no orbit can self-intersect. Thus,

$$\gamma := \{u(t): t \in [t_1, t_2]\} \cup [u(t_1), u(t_2)]$$

is a simple closed curve which, according to the Jordan curve theorem, divides $\mathbb{R}^2 \setminus \gamma$ into two open connected components: a bounded one, D, and $\mathbb{R}^2 \setminus \bar{D}$.

Thanks to the direction of the flow on $[P_1, P_2]$ and, once again, to the impossibility of self-intersections on the orbit, it becomes apparent that $u(t_3) \in \mathbb{R}^2 \setminus \bar{D}$ in the first case described in Figure 11.16, while $u(t_3) \in D$ in the second one. Therefore, in both cases, $u(t_3) \in (u(t_2), P_2]$ and, hence,

$$u(t_1) < u(t_2) < u(t_3).$$

Similarly, one proves that $u(t_3) < u(t_2) < u(t_1)$ if $u(t_2) < u(t_1)$. $\square$

The last preliminary lemma we need shows that an ω-limit point of a non-periodic orbit is approached, on any transversal segment passing through it, by a sequence of points of the orbit intersecting the segment in a monotone way.

Lemma 11.18. *Suppose, for some $x \in \mathbb{R}^2$, that $u(t; x)$ is defined in $[0, +\infty)$ and is non-periodic (in particular, it is not an equilibrium). Assume there exists $x_\infty \in \omega(x)$ such that $f(x_\infty) \neq 0$, and consider any transversal segment, $[P_1, P_2]$, with $x_\infty \in (P_1, P_2)$. Then, there exists a monotone sequence $\{u(s_n; x)\}_{n \geq 1} \subset \Gamma_x^+ \cap [P_1, P_2]$ with $s_n < s_{n+1}$ and such that*

$$\lim_{n \to +\infty} s_n = +\infty \quad \text{and} \quad \lim_{n \to +\infty} u(s_n; x) = x_\infty.$$

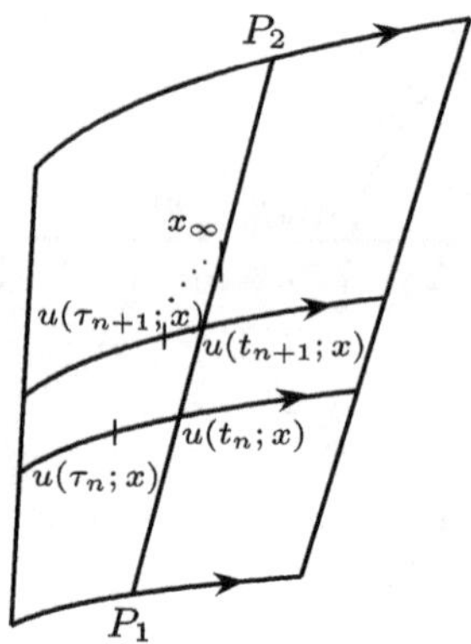

Fig. 11.17. Construction in the proof of Lemma 11.18.

Proof. Observe that the existence of a segment $S = [P_1, P_2]$ transversal to f with $x_\infty \in (P_1, P_2)$ is guaranteed because $f(x_\infty) \neq 0$. Let Φ_S be a flow box through S (see Figure 11.17). Since $x_\infty \in \omega(x) \cap (P_1, P_2)$, there exists a sequence of times, $\{\tau_n\}_{n \geq 1}$, which can be assumed to be increasing without loss of generality, such that

$$\lim_{n \to +\infty} \tau_n = +\infty, \qquad \lim_{n \to +\infty} u(\tau_n; x) = x_\infty. \qquad (11.66)$$

Then, for sufficiently large n, say $n \geq n_0$, $u(\tau_n; x) \in \Phi_S$. As Φ_S is a flow box for $u' = f(u)$, for every $n \geq n_0$, there exists t_n, with $\tau_n \in [t_n - \tau, t_n + \tau]$, such that $u(t_n; x) \in (P_1, P_2)$, as illustrated in Figure 11.17.

Thanks to (11.66), we also have that

$$\lim_{n \to +\infty} t_n = +\infty,$$

and, without loss of generality, we can assume that $\{t_n\}_{n \geq n_0}$ is increasing. Moreover, since $u(t_n; x) \in [P_1, P_2]$ and, by (11.66),

$$\lim_{n \to +\infty} \operatorname{dist}(u(\tau_n; x), [P_1, P_2]) = 0,$$

the continuity of the flow and the construction of the flow box ensure

$$\lim_{n \to +\infty} \operatorname{dist}(u(\tau_n; x), u(t_n; x)) = 0.$$

Thus, since

$$\operatorname{dist}(u(t_n; x), x_\infty) \leq \operatorname{dist}(u(t_n; x), u(\tau_n; x)) + \operatorname{dist}(u(\tau_n; x), x_\infty),$$

it is apparent that

$$\lim_{n \to +\infty} u(t_n; x) = x_\infty.$$

Lastly, observe that, for all $n \geq n_0$, $u(t_n; x) \neq u(t_{n+1}; x)$, because $u(t; x)$ is not periodic. Thus, by Lemma 11.17, we can take $s_n = t_{n+n_0}$ for $n \geq 1$, and $\{u(s_n; x)\}_{n \geq 1}$ gives a monotone sequence on $[P_1, P_2]$, with $s_n < s_{n+1}$ for all $n \geq 1$, converging to x_∞. $\square$

Another direct consequence of Lemma 11.17 is the fact that a transversal segment can admit, at most, a unique point of the ω-limit set of Γ_x^+, as established in the following result.

Corollary 11.19. *Let* $x \in \mathbb{R}^2$ *be such that* $u(t; x)$ *is defined in* $[0, +\infty)$ *and is non-periodic. Let* $x_\infty \in \omega(x)$ *be such that* $f(x_\infty) \neq 0$ *and consider a segment* $[P_1, P_2]$ *transversal to* f *with* $x_\infty \in (P_1, P_2)$. *Then,*

$$\omega(x) \cap (P_1, P_2) = \{x_\infty\}.$$

Proof. Assume by contradiction that there exists $\tilde{x}_\infty \in \omega(x) \cap (P_1, P_2)$ with $\tilde{x}_\infty \neq x_\infty$, for example $x_\infty < \tilde{x}_\infty$. Then, there exist two disjoint balls B and $\tilde{B}$, centered at x_∞ and $\tilde{x}_\infty$, respectively. In particular,

$$P < \tilde{P} \quad \text{for all } P \in B \cap (P_1, P_2) \text{ and } \tilde{P} \in \tilde{B} \cap (P_1, P_2). \tag{11.67}$$

Since $x_\infty \in \omega(x)$, there exists τ_1 such that $u(\tau_1; x) \in B$ and, by reasoning as in the proof of Lemma 11.18, there exists t_1 such that $u(t_1; x) \in B \cap (P_1, P_2)$. Now, since also $\tilde{x}_\infty \in \omega(x)$, there exists $\tau_2 > t_1$ such that $u(\tau_2; x) \in \tilde{B}$ and, again, there exists $t_2 > t_1$ with $u(t_2; x) \in \tilde{B} \cap (P_1, P_2)$. By repeating this argument, we obtain the existence of $t_3 > t_2$ with $u(t_3; x) \in B \cap (P_1, P_2)$. From (11.67), we obtain that

$$u(t_1; x) < u(t_2; x) > u(t_3; x),$$

against the monotonicity established in Lemma 11.17. $\square$

We are now ready to prove the Poincaré–Bendixson theorem.

Proof of Theorem 11.15. Since Γ_x^+ is bounded, by Theorem 11.11 $\omega(x)$ is nonempty; thus, there exists $x_\infty \in \omega(x)$. Since $\omega(x)$ does not contain any equilibrium, $f(x_\infty) \neq 0$. Hence, there is a transversal segment, $[P_1, P_2]$, with $x_\infty \in (P_1, P_2)$. As $u(t; x)$ is not periodic, by Corollary 11.19, $\omega(x) \cap (P_1, P_2) = \{x_\infty\}$.

According to Theorem 11.13, $u(t; x_\infty)$ is globally defined in time, i.e. $I_{x_\infty} = \mathbb{R}$, and $u(t; x_\infty) \in \omega(x)$ for all $t \in \mathbb{R}$. Hence, since $\omega(x)$ is compact by Theorem 11.11, the orbit

$$\Gamma_{x_\infty} := \{u(t; x_\infty) \colon t \in \mathbb{R}\} \subset \omega(x) \tag{11.68}$$

is bounded, and, thanks again to Theorem 11.11, $\omega(x_\infty)$ is nonempty. Consider a point $x_{\infty,\infty} \in \omega(x_\infty)$. By (11.46) and (11.68),

$$x_{\infty,\infty} \in \bar{\Gamma}_{x_\infty}^+ \subset \overline{\omega(x)} = \omega(x)$$

because $\omega(x)$ is compact. Since no equilibrium lies in $\omega(x)$, $f(x_{\infty,\infty}) \neq 0$. So, there exists a transversal segment, $[Q_1, Q_2]$, with $x_{\infty,\infty} \in (Q_1, Q_2)$.

We now show that $u(t; x_\infty)$ is periodic. If this was not the case, then, by Lemma 11.18, there would be a monotone sequence of points in $\Gamma_{x_\infty}^+ \cap (Q_1, Q_2)$ converging to $x_{\infty,\infty}$. Consequently, by (11.68), there would be infinitely many points of $\omega(x)$ on (Q_1, Q_2), which contradicts Corollary 11.19. This contradiction shows that $u(t; x_\infty)$ is periodic. Moreover, since $\omega(x)$ does not contain any equilibrium, $u(t; x_\infty)$ is a non-trivial periodic solution.

It remains to prove that $\omega(x) = \Gamma_{x_\infty}$. Recalling (11.68), assume by contradiction that Γ_{x_∞} is a proper subset of $\omega(x)$. Then, since $\omega(x)$ is connected, in order not to be disconnected, there should exist a sequence of points $\{x_n\}_{n \geq 1}$ in $\omega(x) \setminus \Gamma_{x_\infty}$ such that

$$x_n \neq x_m \quad \text{if} \quad n \neq m \quad \text{and} \quad \lim_{n \to +\infty} x_n = x_\omega \tag{11.69}$$

for some $x_\omega \in \Gamma_{x_\infty} \subsetneq \omega(x)$. As $\omega(x)$ does not contain any equilibrium, $f(x_\omega) \neq 0$. Thus, there is a transversal segment, $S := [R_1, R_2]$, such that $x_\omega \in (R_1, R_2)$. Let Φ_S be a flow box through S.

According to (11.69), there exists an integer $n_0 \geq 1$ such that $x_n \in \Phi_S$ for all $n \geq n_0$. Moreover, since $x_n \in \omega(x)$, it follows from Theorem 11.13 that, for every $n \geq 1$, $u(t; x_n)$ is well defined and $u(t; x_n) \in \omega(x)$ for all $t \in \mathbb{R}$. By reasoning as in the proof of Lemma 11.18, since $x_n \in \Phi_S$ for every $n \geq n_0$, there exist $\tau > 0$ and $t_n \in [-\tau, \tau]$ such that $\{u(t_n; x_n)\}_{n \geq n_0}$ is a monotone sequence in (R_1, R_2). But $u(t_n; x_n) \in \omega(x)$ for all $n \geq n_0$, contradicting Corollary 11.19. This contradiction concludes the proof. $\qquad\square$

The same ingredients used in the proof of Theorem 11.15 provide us with the following result, whose proof is proposed in Exercise 8 of Chapter 11.

Theorem 11.20. *Assume that the equilibria of $u' = f(u)$ are isolated, and that, for some $x \in \mathbb{R}^2$, the unique maximal solution of (11.65) satisfies the following:*

(i) $u(t; x)$ is defined in $[0, +\infty)$, and it is not a periodic solution;
(ii) Γ_x^+ is bounded;
(iii) $f(x_\infty) = 0$ for some $x_\infty \in \omega(x)$.

Then, $\omega(x)$ consists of finitely many equilibria, plus a certain number of heteroclinic, or homoclinic, connections linking such equilibria.

11.5 The Lotka–Volterra Competition Model

The main goal of this section is to use the theory previously presented in this chapter to study the dynamics of the non-negative solutions of the Lotka–Volterra competition model

$$\begin{cases} u' = \lambda u - \alpha u^2 - \beta uv, \\ v' = \mu v - \gamma uv - \delta v^2, \end{cases} \tag{11.70}$$

where λ, μ, α, β, γ, and δ are positive constants. This model describes the temporal evolution of two species, u and v, with birth rates λ and μ, intraspecific competition rates α and δ, and interaction coefficients β and γ, respectively. Such a system is obtained by coupling the two logistic equations

$$u' = \lambda u - \alpha u^2, \quad v' = \mu v - \delta v^2,$$

through the interaction terms $-\beta uv$ and $-\gamma uv$ that describe a competition between the two species. As the change of variables

$$u := \frac{U}{\alpha}, \quad v := \frac{V}{\delta},$$

transforms (11.70) into

$$\begin{cases} U' = \lambda U - U^2 - \frac{\beta}{\delta} UV, \\ V' = \mu V - \frac{\gamma}{\alpha} UV - V^2, \end{cases}$$

setting

$$b := \frac{\beta}{\delta}, \quad c := \frac{\gamma}{\alpha}, \quad (u, v) := (U, V),$$

it becomes apparent that analyzing (11.70) is equivalent to analyzing

$$\begin{cases} u' = \lambda u - u^2 - buv, \\ v' = \mu v - cuv - v^2. \end{cases} \tag{11.71}$$

In other words, one can take $\alpha = \delta = 1$ in (11.70) without loss of generality. Thus, in the rest of this section, we study the model (11.71).

As we want to ascertain the dynamics of non-negative solutions, we consider the following Cauchy problem:

$$\begin{cases} u' = \lambda u - u^2 - buv, \\ v' = \mu v - cuv - v^2, \\ u(0) = u_0 \geq 0, \quad v(0) = v_0 \geq 0. \end{cases} \tag{11.72}$$

According to the results in Section 5.5.1, (11.72) has a unique (maximal) solution $(u(t; u_0, v_0), v(t; u_0, v_0))$ defined in $t \in (T_{\min}, +\infty)$ for some $-\infty \leq T_{\min} < 0$. Moreover, for every $t \in (T_{\min}, +\infty)$,

$$\operatorname{sign} u(t; u_0, v_0) = \operatorname{sign} u_0, \quad \operatorname{sign} v(t; u_0, v_0) = \operatorname{sign} v_0.$$

In particular, we are interested in the component-wise non-negative steady-state solutions of (11.71), which are given by the non-negative solutions of the following algebraic system:

$$\begin{cases} \lambda u - u^2 - buv = 0, \\ \mu v - cuv - v^2 = 0. \end{cases} \tag{11.73}$$

Such steady states can be of three different types according to the nature of their components. Namely, the *coexistence states*, which are the equilibria (u, v) with $u > 0$ and $v > 0$, the *semitrivial states*, which are those with one vanishing component, i.e. $(\lambda, 0)$ and $(0, \mu)$, and the *trivial solution* $(0,0)$. The coexistence states, if they exist, must solve the linear system

$$\begin{pmatrix} 1 & b \\ c & 1 \end{pmatrix} \begin{pmatrix} u \\ v \end{pmatrix} = \begin{pmatrix} \lambda \\ \mu \end{pmatrix}, \tag{11.74}$$

which has a unique solution if $bc \neq 1$. Throughout this section, we assume that either $bc < 1$ or $bc > 1$. Then, the unique solution of (11.74) is

$$u_{co} = \frac{\lambda - b\mu}{1 - bc}, \qquad v_{co} = \frac{\mu - c\lambda}{1 - bc}.$$

Thus, (u_{co}, v_{co}) is a coexistence state if and only if both its components are positive, which is equivalent to

$$\lambda > b\mu \quad \text{and} \quad \mu > c\lambda, \quad \text{if } bc < 1, \tag{11.75}$$

$$\lambda < b\mu \quad \text{and} \quad \mu < c\lambda, \quad \text{if } bc > 1. \tag{11.76}$$

Since b and c measure the strength of the competition between the two species, in the specialized literature, $bc < 1$ is referred to as the *low-intensity competition* case, while $bc > 1$ is the case of *high-intensity competition*.

In the following section, we relate the existence of the coexistence state to the linearized stability of the semitrivial solutions $(\lambda, 0)$ and $(0, \mu)$.

11.5.1 *Local character of* $(0,0)$, $(\lambda, 0)$ *and* $(0, \mu)$

Setting

$$f(u, v) := \begin{pmatrix} \lambda u - u^2 - buv \\ \mu v - v^2 - cuv \end{pmatrix}, \qquad (u, v) \in \mathbb{R}^2,$$

we have that

$$Df(u, v) = \begin{pmatrix} \lambda - 2u - bv & -bu \\ -cv & \mu - 2v - cu \end{pmatrix}, \qquad (u, v) \in \mathbb{R}^2. \tag{11.77}$$

In particular,

$$Df(0,0) = \begin{pmatrix} \lambda & 0 \\ 0 & \mu \end{pmatrix}, \quad Df(\lambda,0) = \begin{pmatrix} -\lambda & -b\lambda \\ 0 & \mu-c\lambda \end{pmatrix},$$

$$Df(0,\mu) = \begin{pmatrix} \lambda-b\mu & 0 \\ -c\mu & -\mu \end{pmatrix}.$$

Thus, the corresponding eigenvalues are

$$\sigma\left(Df(0,0)\right) = \{\lambda,\mu\},$$

$$\sigma\left(Df(\lambda,0)\right) = \{-\lambda,\mu-c\lambda\}, \tag{11.78}$$

$$\sigma\left(Df(0,\mu)\right) = \{\lambda-b\mu,-\mu\},$$

and thanks to the analysis of Section 9.1 and to Theorem 11.9, the following result holds.

Proposition 11.21. *The following assertions hold true:*

- $(0,0)$ *is an unstable node;*
- $(\lambda,0)$ *is a stable node if* $\mu - c\lambda < 0$*;*
- $(\lambda,0)$ *is a saddle point if* $\mu - c\lambda > 0$*;*
- $(0,\mu)$ *is a stable node if* $\lambda - b\mu < 0$*;*
- $(0,\mu)$ *is a saddle point if* $\lambda - b\mu > 0$*.*

Observe that, thanks to this result, in the parameter plane (λ,μ), the stability of the semitrivial non-negative steady-states $(\lambda,0)$ and $(0,\mu)$ changes across the lines $\mu = c\lambda$ and $\lambda = b\mu$, respectively.

Moreover, by combining the information given by Proposition 11.21 with (11.75) and (11.76), the existence of a coexistence state can be characterized by means of the linearized character of the semitrivial non-negative solutions, as follows.

Proposition 11.22. *The following assertions are true:*

(a) *Suppose that* $bc < 1$*. Then, (11.72) has a (unique) coexistence state*

$$(u_{co}, v_{co}) = \left(\frac{\lambda - b\mu}{1 - bc}, \frac{\mu - c\lambda}{1 - bc} \right), \tag{11.79}$$

if both semitrivial non-negative steady-states $(\lambda,0)$ *and* $(0,\mu)$ *are saddle points.*

(b) *Suppose that bc > 1. Then, (11.72) has a (unique) coexistence state, given by (11.79), if both semitrivial non-negative steady-states $(\lambda, 0)$ and $(0, \mu)$ are stable nodes.*

11.5.2 Local character of (u_{co}, v_{co})

In this section, we assume that $bc \neq 1$ and that (11.72) has a (unique) coexistence state, (u_{co}, v_{co}), given by (11.79). To analyze its local character, we consider the corresponding linearization given by (11.77):

$$Df(u_{co}, v_{co}) = \begin{pmatrix} \lambda - 2u_{co} - bv_{co} & -bu_{co} \\ -cv_{co} & \mu - 2v_{co} - cu_{co} \end{pmatrix}. \qquad (11.80)$$

Moreover, since (u_{co}, v_{co}) solves (11.73), we have that

$$\lambda - 2u_{co} - bv_{co} = (\lambda - u_{co} - bv_{co}) - u_{co} = -u_{co},$$

$$\mu - 2v_{co} - cu_{co} = (\mu - v_{co} - cu_{co}) - v_{co} = -v_{co}.$$

Thus, (11.80) can be rewritten as

$$Df(u_{co}, v_{co}) = \begin{pmatrix} -u_{co} & -bu_{co} \\ -cv_{co} & -v_{co} \end{pmatrix}. \qquad (11.81)$$

Hence, the characteristic equation associated with $Df(u_{co}, v_{co})$ is

$$z^2 + (u_{co} + v_{co})z + (1 - bc)u_{co}v_{co} = 0,$$

and its eigenvalues are

$$z_{\pm} := \frac{1}{2}\left(-(u_{co} + v_{co}) \pm \sqrt{(u_{co} + v_{co})^2 - 4(1 - bc)u_{co}v_{co}}\right)$$

$$= \frac{1}{2}\left(-(u_{co} + v_{co}) \pm \sqrt{(u_{co} - v_{co})^2 + 4bcu_{co}v_{co}}\right).$$

Since the components of the coexistence state are positive, the previous expression shows that the eigenvalues are real and distinct. Moreover,

$$z_+ + z_- = -(u_{co} + v_{co}) < 0, \quad z_+ z_- = (1 - bc)u_{co}v_{co} \begin{cases} > 0 & \text{if } bc < 1, \\ < 0 & \text{if } bc > 1. \end{cases}$$

Thus, $Df(u_{co}, v_{co})$ has two negative eigenvalues if $bc < 1$, while one eigenvalue is positive and one negative if $bc > 1$. Therefore, thanks to

Theorem 11.3, (u_{co}, v_{co}) is asymptotically stable if $bc < 1$, while, due to Theorem 11.4, it is unstable if $bc > 1$. In addition, Theorem 11.9 entails the following result.

Theorem 11.23. *Assume that* (11.72) *has a unique coexistence state* (u_{co}, v_{co}). *Then:*

(a) (u_{co}, v_{co}) *is an asymptotically stable node if* $bc < 1$;
(b) (u_{co}, v_{co}) *is a saddle point if* $bc > 1$.

11.5.3　*Dynamics of system* (11.72) *when* $bc < 1$

Throughout this section we assume that $bc < 1$. As established in Section 11.5.1, the lines across which the stability of the semitrivial non-negative steady-states $(\lambda, 0)$ and $(0, \mu)$ change are $\mu = c\lambda$ and $\lambda = b\mu$. Thus, since $c < \frac{1}{b}$, these lines, in the first quadrant of the parameter plane $\lambda > 0$ and $\mu > 0$, look like those shown in Figure 11.18 and divide such quadrant into three regions:

$$P_1 := \{\lambda > 0, \mu > 0 \colon \lambda < b\mu \text{ and } \mu > c\lambda\},$$

$$P_2 := \{\lambda > 0, \mu > 0 \colon \lambda > b\mu \text{ and } \mu > c\lambda\},$$

$$P_3 := \{\lambda > 0, \mu > 0 \colon \lambda > b\mu \text{ and } \mu < c\lambda\}.$$

Thanks to Propositions 11.21, 11.22 and Theorem 11.23, the following result about the existence and stability of non-negative equilibria of (11.71) holds.

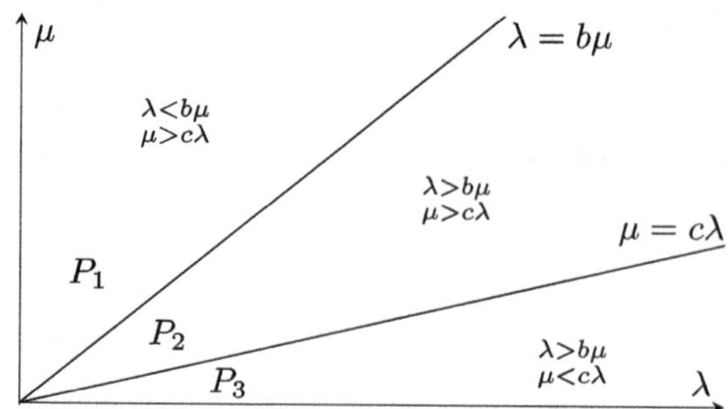

Fig. 11.18. Three parameter regions with different stability of the semitrivial steady-states when $bc < 1$.

Theorem 11.24. *Consider* (11.72) *with* $bc < 1$. *Then,* $(0,0)$ *is an unstable node, and:*

(i) *in the region* P_1, $(0, \mu)$ *is a stable node,* $(\lambda, 0)$ *is a saddle point, and there is no coexistence state;*

(ii) *in the region* P_2, $(\lambda, 0)$ *and* $(0, \mu)$ *are saddle points and there exists a unique coexistence state,* (u_{co}, v_{co}), *which is an asymptotically stable node;*

(iii) *in the region* P_3, $(\lambda, 0)$ *is a stable node,* $(0, \mu)$ *is a saddle point, and there is no coexistence state.*

The main result of this section establishes that, in the region P_1 of the parameters, $(0, \mu)$ is a global attractor for the solutions of (11.72) with $v_0 > 0$, while in P_3 $(\lambda, 0)$ is a global attractor for the solutions of (11.72) with $u_0 > 0$. Finally, in P_2 the coexistence state (u_{co}, v_{co}) is a global attractor for the solutions of (11.72) with $u_0 > 0$ and $v_0 > 0$. Precisely, it can be stated as follows.

Theorem 11.25. *Assume* $bc < 1$. *Then, the following holds.*

(a) *Suppose* $\lambda < b\mu$ *and* $\mu > c\lambda$. *Then, for each* $u_0 \geq 0$, $v_0 > 0$, *the solution of* (11.72), $(u(t; u_0, v_0), v(t; u_0, v_0))$, *satisfies*

$$\lim_{t \to +\infty} u(t; u_0, v_0) = 0, \qquad \lim_{t \to +\infty} v(t; u_0, v_0) = \mu.$$

In particular, $(0, \mu)$ *is a global attractor for the interior of the first quadrant in the phase plane* (u, v).

(b) *Suppose* $\lambda > b\mu$ *and* $\mu < c\lambda$. *Then, for each* $u_0 > 0$ *and* $v_0 \geq 0$,

$$\lim_{t \to +\infty} u(t; u_0, v_0) = \lambda, \qquad \lim_{t \to +\infty} v(t; u_0, v_0) = 0.$$

In particular, $(\lambda, 0)$ *is a global attractor for the interior of the first quadrant in the phase plane* (u, v).

(c) *Suppose* $\lambda > b\mu$ *and* $\mu > c\lambda$. *Then, for each* $u_0 > 0$ *and* $v_0 > 0$,

$$\lim_{t \to +\infty} u(t; u_0, v_0) = u_{co} = \frac{\lambda - b\mu}{1 - bc},$$

$$\lim_{t \to +\infty} v(t; u_0, v_0) = v_{co} = \frac{\mu - c\lambda}{1 - bc}.$$

Thus, (u_{co}, v_{co}) is a global attractor for the interior of the first quadrant in the phase plane (u, v).

Proof. (a) Suppose that $\lambda < b\mu$ and $\mu > c\lambda$, i.e. that the parameters λ and μ lie in the region P_1. Then, due to Theorem 11.24(i), $(0, \mu)$ is a stable node, $(\lambda, 0)$ a saddle point, and there is no coexistence state. Since the analysis is quite long, we divide the proof of this part into several steps.

Step 1. Analysis of the dynamics on the axes. Since

$$\begin{cases} u' = \lambda u - u^2 - buv = (\lambda - u - bv)\, u, \\ v' = \mu v - cuv - v^2 = (\mu - v - cu)\, v, \end{cases} \qquad (11.82)$$

it is apparent that:

- if $u = 0$, $(u', v') = (0, (\mu - v)v)$. Thus, in such a case, $v' > 0$ if $v < \mu$, while $v' < 0$ if $v > \mu$. Therefore, $(0, \mu)$ is a global attractor on the v-axis and the conclusion of Part (a) is achieved in this case;
- if $v = 0$, $(u', v') = ((\lambda - u)u, 0)$. Thus, in such a case, $u' > 0$ if $u < \lambda$, while $u' < 0$ if $u > \lambda$. Therefore, $(\lambda, 0)$ is a global attractor on the u-axis.

Step 2. Sketch of the direction field in the interior of the quadrant. From (11.82) it is apparent that, if $u > 0$ and $v > 0$, then

$$\operatorname{sign} u' = \operatorname{sign}(\lambda - u - bv), \quad \operatorname{sign} v' = \operatorname{sign}(\mu - v - cu).$$

Hence,

$$u' \begin{cases} > 0 & \text{if } \lambda - u - bv > 0, \\ = 0 & \text{if } \lambda - u - bv = 0, \\ < 0 & \text{if } \lambda - u - bv < 0, \end{cases} \qquad v' \begin{cases} > 0 & \text{if } \mu - v - cu > 0, \\ = 0 & \text{if } \mu - v - cu = 0, \\ < 0 & \text{if } \mu - v - cu < 0. \end{cases}$$

Figure 11.19 shows the direction field of (11.82). In particular we have represented the lines $u + bv = \lambda$ and $v + cu = \mu$, where $u' = 0$ and $v' = 0$, respectively, which are referred to as *nullclines*. In addition, the vector (u', v') is depicted at some representative base points (u, v) in the first quadrant.

The line $u + bv = \lambda$ intersects the u-axis at $(\lambda, 0)$ and the v-axis at $(0, \frac{\lambda}{b})$. Similarly, the line $v + cu = \mu$ crosses the u-axis at $(\frac{\mu}{c}, 0)$ and the v-axis at $(0, \mu)$. Note that

$$\frac{\lambda}{b} < \mu, \quad \lambda < \frac{\mu}{c},$$

because we are assuming that $\lambda < b\mu$ and $\mu > c\lambda$. Consequently, these lines divide the first quadrant of the (u, v)-plane into three regions:

$$R_1 := \{(u, v) \in (0, +\infty)^2 : u + bv < \lambda, \; v + cu < \mu\},$$

$$R_2 := \{(u, v) \in (0, +\infty)^2 : u + bv > \lambda, \; v + cu < \mu\},$$

$$R_3 := \{(u, v) \in (0, +\infty)^2 : u + bv > \lambda, \; v + cu > \mu\}.$$

From what we have seen above, in R_1 $u' > 0$ and $v' > 0$, and the direction field points right-upwards. In R_2, instead, $u' < 0$ and $v' > 0$, and the direction field points left-upwards. Finally, in R_3, $u' < 0$ and $v' < 0$; so, the direction field points left-downwards. This behavior is schematically shown in Figure 11.19.

Step 3. Local analysis of the equilibrium $(\lambda, 0)$. Although this step is not strictly necessary to complete the proof of Part (a), it is instructive to perform a complete local analysis of the dynamics in a neighborhood of the equilibria, since in other situations it could give important information. To illustrate this procedure, we perform

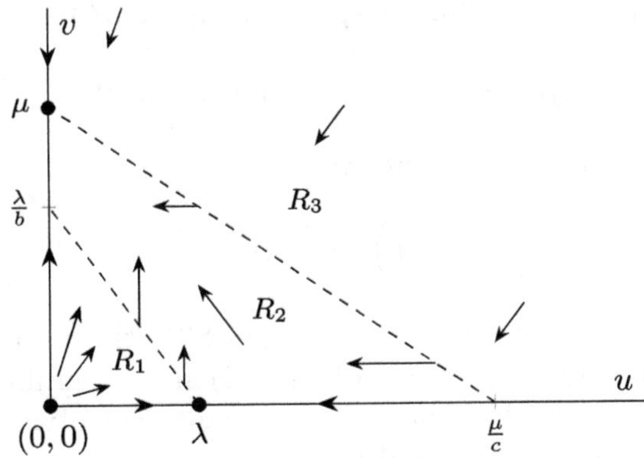

Fig. 11.19. Direction field of (11.72) when $bc < 1$ and (λ, μ) belongs to P_1.

the analysis in a neighborhood of $(\lambda, 0)$, which, according to Theorem 11.24(i) is a saddle point.

Necessarily, the stable manifold of $(\lambda, 0)$ is the u-axis, because it is invariant. Indeed, as observed in Step 1, $v_0 = 0$ implies that $v = 0$ whenever the solution is defined, and, hence, the first equation reduces to

$$u' = \lambda u - u^2,$$

for which $(\lambda, 0)$ is a global attractor if $u_0 > 0$. To determine the direction of its unstable manifold $W^u(\lambda, 0)$, according to Theorem 11.9, it is enough to find the unstable manifold of the associated linearized system. Thus, we have to compute the line of eigenvectors associated with $\mu - c\lambda$, which, according to (11.78), is the positive eigenvalue of $Df(\lambda, 0)$. To do this, we have to solve the linear system

$$\begin{pmatrix} -\lambda & -b\lambda \\ 0 & \mu - c\lambda \end{pmatrix} \begin{pmatrix} u \\ v \end{pmatrix} = (\mu - c\lambda) \begin{pmatrix} u \\ v \end{pmatrix},$$

which is equivalent to

$$-\lambda u - b\lambda v = (\mu - c\lambda)\, u.$$

Thus, the direction of the tangent to the unstable manifold of $(\lambda, 0)$ is given by the straight line through the origin

$$(\mu - c\lambda + \lambda)\, u + b\lambda v = 0,$$

and, hence, the tangent line to $W^u(\lambda, 0)$ at $(\lambda, 0)$ is

$$(\mu - c\lambda + \lambda)\, (u - \lambda) + b\lambda v = 0,$$

which can be equivalently written as

$$\left(\frac{\mu}{\lambda} - c + 1 \right) (u - \lambda) + bv = 0. \tag{11.83}$$

Figure 11.20 shows the straight line (11.83), the nullcline $u - \lambda + bv = 0$, and the phase portrait of (11.72) in a neighborhood of the saddle point $(\lambda, 0)$ within the first quadrant. Since we are assuming $(\lambda, \mu) \in P_1$, $\frac{\mu}{\lambda} - c + 1 > 1$, and the relative position of these lines is indeed the one represented in Figure 11.20.

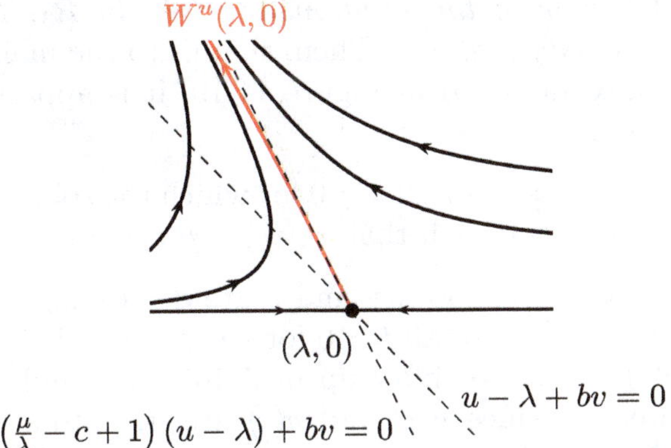

Fig. 11.20. Local phase portrait of (11.72) at $(\lambda,0)$ when $bc<1$ and (λ,μ) belongs to P_1.

Step 4. Behavior of the solutions starting in R_1. Assume that $(u_0,v_0)\in R_1$. Then, since $(0,0)$ is an unstable node, $(0,0)\notin \omega(u_0,v_0)$. Actually, since $u'>0$ and $v'>0$ in this region, it is apparent that the maximal solution of (11.72) is defined for all $t\in(-\infty,0]$, that $\alpha(u_0,v_0)=\{(0,0)\}$, and that

$$\lim_{t\downarrow-\infty}(u(t;u_0,v_0),v(t;u_0,v_0))=(0,0),$$

because the interior of this region does not contain any equilibrium.

Moreover, the previous analysis (see, e.g. Figure 11.20) guarantees that $(\lambda,0)\notin\omega(u_0,v_0)$, essentially because v is increasing for small values of v. We now show that the solution $(u(t),v(t)):=(u(t;u_0,v_0),v(t;u_0,v_0))$ leaves R_1 in finite time and enters R_2 through the nullcline $u+bv=\lambda$. Indeed, assume by contradiction that it remains in R_1 for all $t>0$. Then, $u(t)$ and $v(t)$ are increasing, and, since R_1 is bounded, there exists $(u_\infty,v_\infty)\in\bar{R}_1$ such that

$$\lim_{t\uparrow\infty}(u(t),v(t))=(u_\infty,v_\infty),$$

i.e. $(u_\infty,v_\infty)\in\omega(u_0,v_0)$. Thanks to Property 2 of Section 9.2, (u_∞,v_∞) is an equilibrium of the system, because it is not reached in finite time. But we have shown above that no equilibria lying in $\bar{R}_1$ may belong to $\omega(u_0,v_0)$, which gives the desired contradiction.

Step 5. Behavior of the solutions starting in R_3. Assume that $(u_0, v_0) \in R_3$, i.e. $cu_0 + v_0 > \mu$. Then, thanks to the main features of the direction field depicted in Figure 11.19, it is apparent that two options can occur:

- either $cu(t) + v(t) > \mu$ for all $t \geq 0$ for which the solution is defined,
- or there exists $t_1 > 0$ such that $cu(t_1) + v(t_1) = \mu$.

In the former case, since both u and v are decreasing, we have that $u(t) < u_0$ and $v(t) < v_0$ for all $t > 0$ for which the solution is defined. Thus, the solution cannot blow up in finite time, and it is globally defined.[1] Moreover, since $u(t)$ and $v(t)$ are monotone, there exists $(u_\infty, v_\infty) \in \bar{R}_3$ such that

$$\lim_{t \uparrow \infty} (u(t), v(t)) = (u_\infty, v_\infty).$$

As in Step 4, (u_∞, v_∞) is an equilibrium of the system. Since $(0, \mu)$ is the only equilibrium lying in the closure of R_3, necessarily $(u_\infty, v_\infty) = (0, \mu)$, i.e. we reach the conclusion of Part (a) in this case.

In the latter case, the solution enters the region R_2, and its future behavior will be analyzed in the next step.

Step 6. Behavior of the solutions starting in R_2. Assume that $(u_0, v_0) \in R_2$. Thanks to the properties of the direction field, it is apparent that the solution $(u(t), v(t)) := (u(t; u_0, v_0), v(t; u_0, v_0))$ cannot leave R_2. Thus, it is globally defined. We now show that

$$\omega(u_0, v_0) = \{(0, \mu)\}.$$

Although this can be proved by using the monotonicities of $u(t)$ and $v(t)$ as in Step 5, we give here a different proof, based on the Poincaré–Bendixson theorem, since these topological arguments are valid in more general contexts.

If, arguing by contradiction, we assume that $(0, \mu) \notin \omega(u_0, v_0)$, then, by Theorem 11.15, $\omega(u_0, v_0)$ must be the orbit of a nontrivial periodic solution. Consequently, Property 5 of Section 9.2 would give the existence of an equilibrium for (11.72) inside the orbit of such a periodic solution. As (11.72) does not admit any coexistence state

---

[1] Another proof of the fact that the positive semi-orbit $\Gamma^+_{(u_0,v_0)}$ is well defined and bounded will be given in Remark 11.26.

in this case, we conclude that $(0, \mu) \in \omega(u_0, v_0)$. Consequently, since $(0, \mu)$ is a stable node, we obtain that $\{(0, \mu)\} = \omega(u_0, v_0)$.

Observe that this argument shows in particular that the portion of $W^u(\lambda, 0)$ lying in the first quadrant of the phase plane completely lies in R_2 and actually establishes an heteroclinic connection from $(\lambda, 0)$ to $(0, \mu)$. As a consequence, every solution below $W^u(\lambda, 0)$ provides us with an heteroclinic connection linking $(0, 0)$ to $(0, \mu)$.

Moreover, since the system is autonomous, this analysis also applies to the solutions that enter R_2 after leaving R_1 or R_3 after a certain time, and concludes the proof of Part (a).

Figure 11.21 sketches the dynamics of (11.72) corresponding to the case $bc < 1$ and $(\lambda, \mu) \in P_1$ through the associated phase portrait.

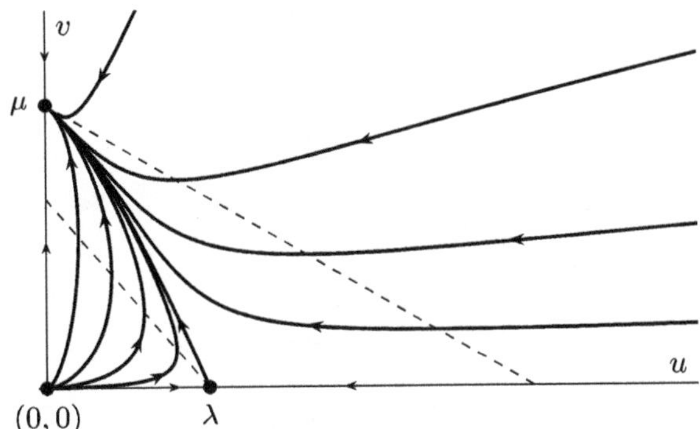

Fig. 11.21. Phase portrait of (11.72) when $bc < 1$ and (λ, μ) belongs to P_1: $(0, \mu)$ is a global attractor for the interior of the first quadrant.

The proof of Part (b), where the parameters λ and μ lie in the region P_3, can be easily adapted from the proof of Part (a). Actually, Part (b) can be directly derived from Part (a) by exchanging the roles of (u, λ, b) and (v, μ, c). Figure 11.22 shows the corresponding direction field on the left and the phase portrait on the right. By Theorem 11.24(iii), in this case, $(\lambda, 0)$ is a stable node, $(0, \mu)$ is a saddle point, and the system does not admit any coexistence state. Moreover, $(\lambda, 0)$ is a global attractor for the interior of the first quadrant. In addition, the unstable manifold of the saddle point $(0, \mu)$, $W^u(0, \mu)$, is an heteroclinic connection from the equilibrium $(0, \mu)$ to $(\lambda, 0)$, while the solutions below $W^u(0, \mu)$ are heteroclinic connections from $(0, 0)$ to $(\lambda, 0)$. Finally, the solutions above $W^u(0, \mu)$ approach $(\lambda, 0)$ as $t \uparrow +\infty$.

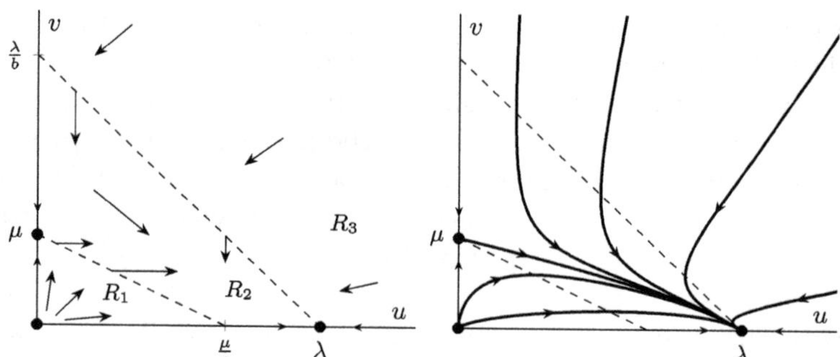

Fig. 11.22. Direction field (left) and phase portrait (right) of (11.72) when $bc < 1$ and (λ, μ) belongs to P_3: $(\lambda, 0)$ is a global attractor for the interior of the first quadrant.

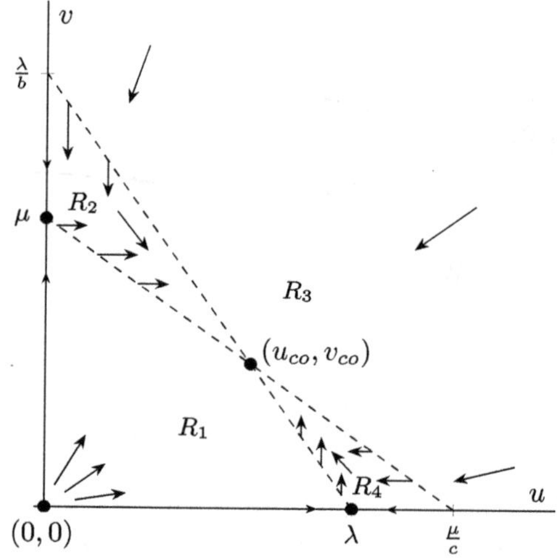

Fig. 11.23. Direction field of (11.72) when $bc < 1$ and (λ, μ) belongs to P_2.

To prove Part (c), we assume that (λ, μ) lies in region P_2. Then, the straight lines $u + bc = \lambda$ and $cu + v = \mu$ cross at the coexistence state (u_{co}, v_{co}), and the direction field of (11.72) looks like the one sketched in Figure 11.23. Now the first quadrant is divided into four regions:

$$R_1 := \{(u, v) \in (0, +\infty)^2 : u + bv < \lambda, \ v + cu < \mu\},$$

$$R_2 := \{(u, v) \in (0, +\infty)^2 : u + bv < \lambda, \ v + cu > \mu\},$$

$$R_3 := \{(u, v) \in (0, +\infty)^2 : u + bv > \lambda, \ v + cu > \mu\},$$

$$R_4 := \{(u, v) \in (0, +\infty)^2 : u + bv > \lambda, \ v + cu < \mu\}.$$

According to Theorem 11.24(ii), $(0,0)$ is an unstable node, $(\lambda, 0)$ and $(0, \mu)$ are saddle points, and the coexistence state, (u_{co}, v_{co}), is a stable node. Moreover, by adapting the analysis of the proof of Part (a), it is easily seen that the unstable manifolds of $(0, \mu)$ and $(\lambda, 0)$ lie in the regions R_2 and R_4, respectively.

Assume that (u_0, v_0) lies in the region R_1. Then, since $u'(t) > 0$ and $v'(t) > 0$ for all $t \in I_{(u_0, v_0)} \cap R_1$, it is apparent that $\Gamma^-_{(u_0, v_0)}$ is well defined and bounded. Actually, by monotonicity, $\alpha(u_0, v_0) = \{(0, 0)\}$. Thus,

$$\lim_{t \downarrow -\infty} (u(t; u_0, v_0), v(t; u_0, v_0)) = (0, 0).$$

Regarding the forward behavior, three possibilities can occur:

(i) either the solution remains in R_1 for all $t > 0$,
(ii) or there exists $t_1 > 0$ such that $cu(t_1) + v(t_1) = \mu$ with $u(t_1) < u_{co}$,
(iii) or there exists $t_2 > 0$ such that $u(t_2) + bv(t_2) = \lambda$ with $u(t_2) > u_{co}$.

Observe that, in case (i), the solution is globally defined in time, and, by monotonicity, it converges to (u_{co}, v_{co}) as $t \uparrow +\infty$. Instead, in case (ii), the direction field forces the solution to enter R_2, while in case (iii) the solution enters R_4.

Similarly, when (u_0, v_0) lies in the region R_3, either the solution enters R_2 or R_4 after a finite time, or it remains in R_3 for all $t > 0$, and it converges to (u_{co}, v_{co}) as $t \uparrow +\infty$.

Finally, if the solution starts in, or enters the region R_2 or R_4, it remains in the interior of that region for all forward times, and it must converge towards (u_{co}, v_{co}) as $t \uparrow +\infty$. Indeed, otherwise, its ω-limit would consist of a non-trivial periodic solution, which is not possible since none of these regions contains other coexistence state.

Therefore, we have shown that the coexistence state (u_{co}, v_{co}) is a global attractor for (11.72) in this case. Figure 11.24 shows the corresponding global phase portrait. $\qquad\square$

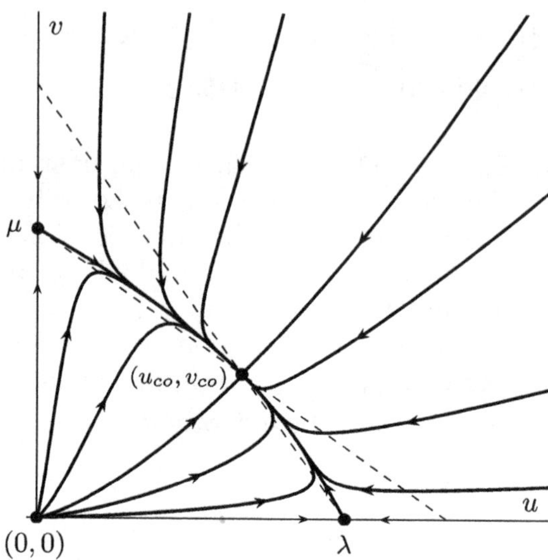

Fig. 11.24. Phase portrait of (11.72) when $bc < 1$ and (λ, μ) belongs to P_2: the coexistence state is a global attractor for the interior of the first quadrant.

Remark 11.26. Subsequently, we give another proof of the fact that the solution of (11.72) is bounded (and thus defined) for all $t \geq 0$. This new proof is different from the one provided in the proof of Theorem 11.25 and is valid for all values of λ, μ, b, c, u_0 and v_0. It is based on a tool that can be applied in more general situations and will be discussed in more detail in Section 11.7.

Let $(u(t), v(t)) := (u(t; u_0, v_0), v(t; u_0, v_0))$ be the maximal solution of (11.72) and set

$$\varphi(t) := \frac{1}{2}\left(u^2(t) + v^2(t)\right) - R^2, \quad t \in I_{(u_0, v_0)},$$

where R is a positive constant to be fixed later. Then, by differentiating,

$$\varphi'(t) = u(t)u'(t) + v(t)v'(t)$$

$$= u(t)\left(\lambda u(t) - u^2(t) - bu(t)v(t)\right)$$

$$\quad + v(t)\left(\mu v(t) - v^2(t) - cu(t)v(t)\right)$$

$$= \lambda u^2(t) + \mu v^2(t) - u^2(t)\left(u(t) + bv(t)\right)$$

$$\quad - v^2(t)\left(cu(t) + v(t)\right).$$

Assume that there exists $t_1 > 0$ such that $\varphi(t_1) = 0$, i.e.

$$u^2(t_1) + v^2(t_1) = 2R^2.$$

Then,

$$\begin{aligned}
(u(t_1) + bv(t_1))^2 &= u^2(t_1) + b^2 v^2(t_1) + 2bu(t_1)v(t_1)\\
&\geq u^2(t_1) + b^2 v^2(t_1) \geq \min\{1, b^2\} 2R^2.
\end{aligned}$$

Similarly,

$$(cu(t_1) + v(t_1))^2 \geq \min\{1, c^2\} 2R^2.$$

Thus, setting

$$\tilde{b} := \sqrt{2\min\{1, b^2\}} > 0, \quad \tilde{c} := \sqrt{2\min\{1, c^2\}} > 0,$$

we obtain that

$$u(t_1) + bv(t_1) \geq \tilde{b}R, \quad cu(t_1) + v(t_1) \geq \tilde{c}R.$$

Therefore,

$$\begin{aligned}
\varphi'(t_1) &\leq \max\{\lambda, \mu\} \left(u^2(t_1) + v^2(t_1)\right) - \left(u^2(t_1)\tilde{b}R + v^2(t_1)\tilde{c}R\right)\\
&\leq \left(\max\{\lambda, \mu\} - R\min\{\tilde{b}, \tilde{c}\}\right) \left(u^2(t_1) + v^2(t_1)\right) < 0,
\end{aligned}$$

provided we take

$$R > \frac{\max\{\lambda, \mu\}}{\min\{\tilde{b}, \tilde{c}\}}.$$

This argument implies that no solution of (11.72) starting in the interior of a disk centered at $(0, 0)$ with any sufficiently large radius can reach the boundary of such a disk in the future. Indeed, the above estimates show that, on the boundary of the disk, the flow of (11.72) points towards its interior.

As a byproduct, for every (u_0, v_0) with $u_0 > 0$ and $v_0 > 0$, the positive semi-orbit $\Gamma^+_{(u_0, v_0)}$ is well defined and bounded, for any value of λ, μ, b and c. This is a general property of a wider class of systems that are referred to as *dissipative*.

11.5.4   *Dynamics of system* (11.72) *when* $bc > 1$

In this section, we consider the case when $bc > 1$. Now, the lines across which the stability of the semitrivial non-negative steady-states $(\lambda, 0)$ and $(0, \mu)$ change, $\mu = c\lambda$ and $\lambda = b\mu$, look like those shown in Figure 11.25. Indeed, since $bc > 1$, we have that $c > \frac{1}{b}$. Note that the relative positions of $\mu = c\lambda$ and $\lambda = b\mu$ are inverted with respect to the case of $bc < 1$, plotted in Figure 11.18.

We observe that the lines $\lambda = b\mu$ and $\mu = c\lambda$ divide the first quadrant of the parameter plane into three regions:

$$P_1 := \{\lambda > 0, \mu > 0 \colon \lambda < b\mu \text{ and } \mu > c\lambda\},$$

$$P_2 := \{\lambda > 0, \mu > 0 \colon \lambda < b\mu \text{ and } \mu < c\lambda\},$$

$$P_3 := \{\lambda > 0, \mu > 0 \colon \lambda > b\mu \text{ and } \mu < c\lambda\}.$$

By Propositions 11.21 and 11.22 and Theorem 11.23, the following result holds.

Theorem 11.27. *Consider* (11.72) *with* $bc > 1$. *Then,* $(0, 0)$ *is an unstable node, and:*

(i) *in the region* P_1, $(0, \mu)$ *is a stable node,* $(\lambda, 0)$ *is a saddle point, and there is no coexistence state;*

(ii) *in* P_2, $(\lambda, 0)$ *and* $(0, \mu)$ *are stable nodes and there is a unique coexistence state,* (u_{co}, v_{co}), *which is a saddle point;*

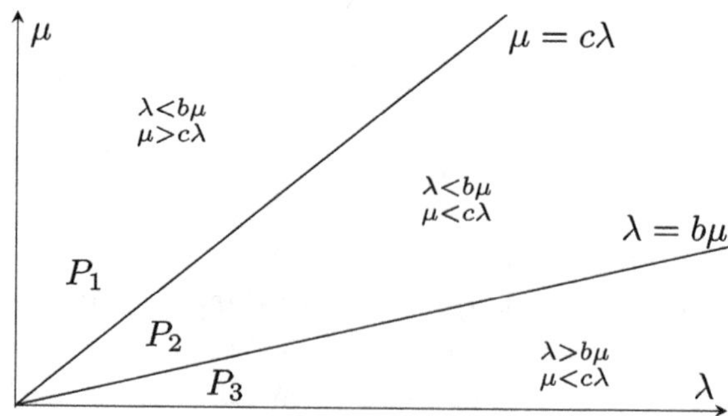

Fig. 11.25. Three parameter regions with different stability of the semitrivial steady states when $bc > 1$.

(iii) *in P_3, $(\lambda, 0)$ is a stable node, $(0, \mu)$ is a saddle point, and there is no coexistence state.*

The main result in this section establishes that, in the region P_1, $(0, \mu)$ is a global attractor for the solutions of (11.72) with $v_0 > 0$, and that, similarly, $(\lambda, 0)$ is a global attractor for the solutions of (11.72) with $u_0 > 0$, in the region P_3. Instead, in the region P_2, i.e. when $(\lambda, 0)$ and $(0, \mu)$ are local attractors, the asymptotic behavior of the solutions of (11.72) with $u_0 > 0$ and $v_0 > 0$ depends on the position of (u_0, v_0) with respect to the stable manifold of the coexistence state (u_{co}, v_{co}), $W^s(u_{co}, v_{co})$. Precisely, the following result holds.

Theorem 11.28. *Assume $bc > 1$. Then:*

(a) *Suppose $\lambda < b\mu$ and $\mu > c\lambda$. Then, for every $u_0 \geq 0$ and $v_0 > 0$, the solution of (11.72), $(u(t; u_0, v_0), v(t; u_0, v_0))$, satisfies*

$$\lim_{t \to +\infty} u(t; u_0, v_0) = 0, \qquad \lim_{t \to +\infty} v(t; u_0, v_0) = \mu. \qquad (11.84)$$

In particular, $(0, \mu)$ is a global attractor for the interior of the first quadrant.

(b) *Suppose $\lambda > b\mu$ and $\mu < c\lambda$. Then, for every $u_0 > 0$ and $v_0 \geq 0$, the solution of (11.72), $(u(t; u_0, v_0), v(t; u_0, v_0))$, satisfies*

$$\lim_{t \to +\infty} u(t; u_0, v_0) = \lambda, \qquad \lim_{t \to +\infty} v(t; u_0, v_0) = 0. \qquad (11.85)$$

In particular, $(\lambda, 0)$ is a global attractor for the interior of the first quadrant.

(c) *Suppose $\lambda < b\mu$ and $\mu < c\lambda$. Then, for each $u_0 > 0$ and $v_0 > 0$:*

(i) *if (u_0, v_0) lies on the stable manifold of (u_{co}, v_{co}), $W^s(u_{co}, v_{co})$, then*

$$\lim_{t \to +\infty} u(t; u_0, v_0) = u_{co} = \frac{\lambda - b\mu}{1 - bc},$$
$$\lim_{t \to +\infty} v(t; u_0, v_0) = v_{co} = \frac{\mu - c\lambda}{1 - bc}; \qquad (11.86)$$

(ii) *if (u_0, v_0) lies above $W^s(u_{co}, v_{co})$ in the (u, v)-plane, (11.84) holds;*

(iii) *if (u_0, v_0) lies below $W^s(u_{co}, v_{co})$ in the (u, v)-plane, (11.85) holds.*

In sum, except when (u_{co}, v_{co}) exists and $(u_0, v_0) \in W^s(u_{co}, v_{co})$, one of the species is driven to extinction by the other.

Proof. The proofs of Parts (a) and (b) follow the same patterns as the proofs of Parts (a) and (b) in Theorem 11.25. So, we do not repeat them.

To prove Part (c), assume that $\lambda < b\mu$, $\mu < c\lambda$, $u_0 > 0$, and $v_0 > 0$. Then, according to Theorem 11.27(ii), $(0, 0)$ is an unstable node, $(\lambda, 0)$ and $(0, \mu)$ are stable nodes, and (u_{co}, v_{co}) is a saddle point. In addition, by Theorem 11.9, the structure of the phase portrait of (11.72) in a neighborhood of these equilibria is equivalent to the one of the corresponding linearized systems. Moreover, under the assumptions of this part, the direction field of (11.72) looks like that shown in Figure 11.26.

In this case, the straight lines $u + bv = \lambda$ and $cu + v = \mu$ cross at (u_{co}, v_{co}) and divide the first quadrant of the (u, v)-plane into

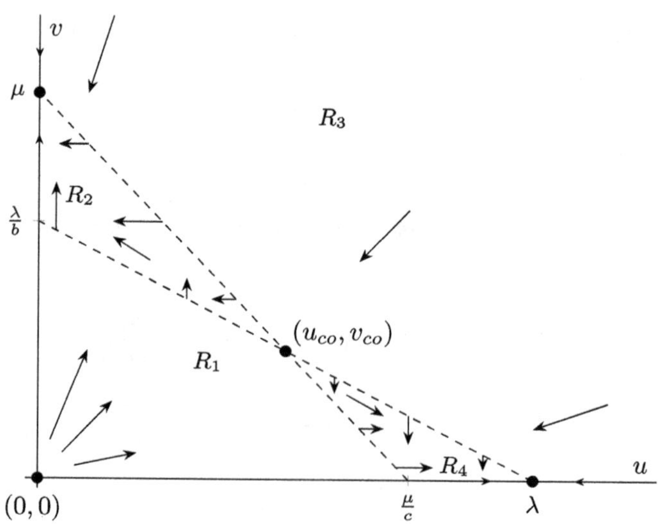

Fig. 11.26. Direction field of (11.72) when $bc > 1$ and (λ, μ) belongs to P_2.

four regions:

$$R_1 := \{(u,v) \in (0,+\infty)^2 : u + bv < \lambda, \ v + cu < \mu\},$$

$$R_2 := \{(u,v) \in (0,+\infty)^2 : u + bv > \lambda, \ v + cu < \mu\},$$

$$R_3 := \{(u,v) \in (0,+\infty)^2 : u + bv > \lambda, \ v + cu > \mu\},$$

$$R_4 := \{(u,v) \in (0,+\infty)^2 : u + bv < \lambda, \ v + cu > \mu\}.$$

Thus, since

$$u' = (\lambda - u - bv)\,u, \quad v' = (\mu - v - cu)\,v,$$

we have that $u' > 0$ and $v' > 0$ in R_1, $u' < 0$ and $v' > 0$ in R_2, $u' < 0$ and $v' < 0$ in R_3, and $u' > 0$, and $v' < 0$ in R_4.

By analyzing the linearized system at (u_{co}, v_{co}), whose matrix is given by (11.81), or, alternatively, by analyzing the direction field, it is possible to show that, in a neighborhood of that equilibrium, the stable manifold $W^s(u_{co}, v_{co})$ lies in the regions R_1 and R_3, while the unstable manifold $W^u(u_{co}, v_{co})$ lies in the regions R_2 and R_4. Moreover, the direction field forces every solution with $(u_0, v_0) \in W^s(u_{co}, v_{co})$ to remain either in R_1 or R_3. In particular, since R_1 is bounded, by reasoning as in the proof of Theorem 11.25, it is possible to show that, if $(u_0, v_0) \in W^s(u_{co}, v_{co}) \cap R_1$, the solution of (11.72) is globally defined and satisfies

$$\lim_{t \downarrow -\infty} (u(t), v(t)) = (0,0) \quad \text{and} \quad \lim_{t \uparrow +\infty} (u(t), v(t)) = (u_{co}, v_{co}).$$

Similarly, if $(u_0, v_0) \in W^u(u_{co}, v_{co}) \cap R_2$, then the solution lies in R_2 for all $t > 0$ and

$$\lim_{t \uparrow +\infty} (u(t), v(t)) = (0, \mu),$$

while, if $(u_0, v_0) \in W^u(u_{co}, v_{co}) \cap R_4$, then the solution lies in R_4 for all $t > 0$ and

$$\lim_{t \uparrow +\infty} (u(t), v(t)) = (\lambda, 0).$$

Thus, the stable and unstable manifolds of (u_{co}, v_{co}) look like those illustrated in Figure 11.27.

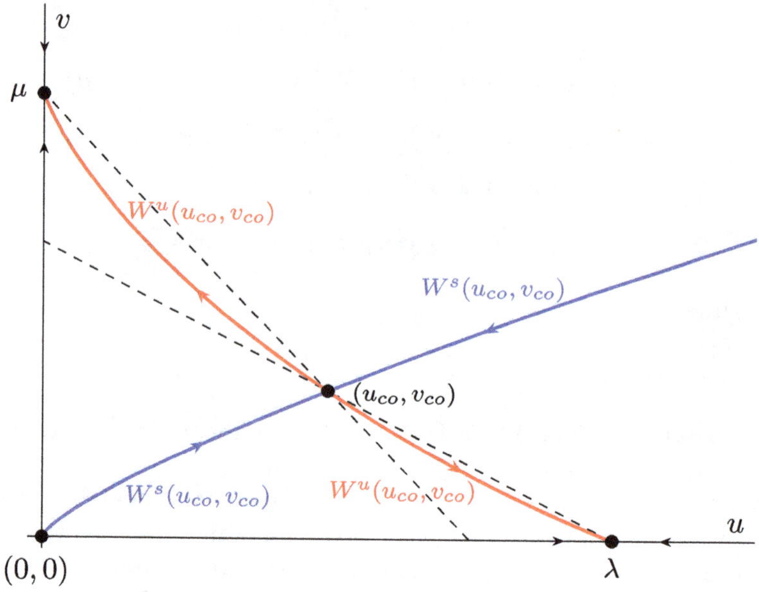

Fig. 11.27. The manifolds $W^s(u_{co}, v_{co})$ and $W^u(u_{co}, v_{co})$ when $bc > 1$ and (λ, μ) belongs to P_2.

In the first quadrant of the (u, v)-plane, the stable manifold $W^s(u_{co}, v_{co})$ consists of a heteroclinic connection from $(0, 0)$ to (u_{co}, v_{co}) plus the graph of a smooth increasing function of u within the region R_3. The unstable manifold consists of two heteroclinic connections between (u_{co}, v_{co}) and each of the semitrivial solutions $(\lambda, 0)$ and $(\mu, 0)$. Obviously, if $(u_0, v_0) \in W^s(u_{co}, v_{co})$, then (11.86) holds true. Moreover, both $W^s(u_{co}, v_{co})$ and $W^u(u_{co}, v_{co})$ divide the interior of the quadrant into two regions: one above and one below each of them. If (u_0, v_0) lies above $W^s(u_{co}, v_{co})$ and $W^u(u_{co}, v_{co})$, then, by reasoning as in the proof of Theorem 11.25, it is possible to show that either the solution remains in R_3 for all $t > 0$ or there exists $t_1 \geq 0$ such that it enters R_2 and remains therein for all $t \geq t_1$. In the former case, (11.84) holds since the solution cannot converge to the saddle point (u_{co}, v_{co}). In the latter one, (11.84) also holds because all solutions in R_2 converge to $(0, \mu)$. If (u_0, v_0) lies above $W^s(u_{co}, v_{co})$ but below $W^u(u_{co}, v_{co})$, then, by reasoning again as in the proof of Theorem 11.25, it is possible to show that there exists $t_2 \geq 0$ such that the solution enters R_2, remains therein for all $t \geq t_2$,

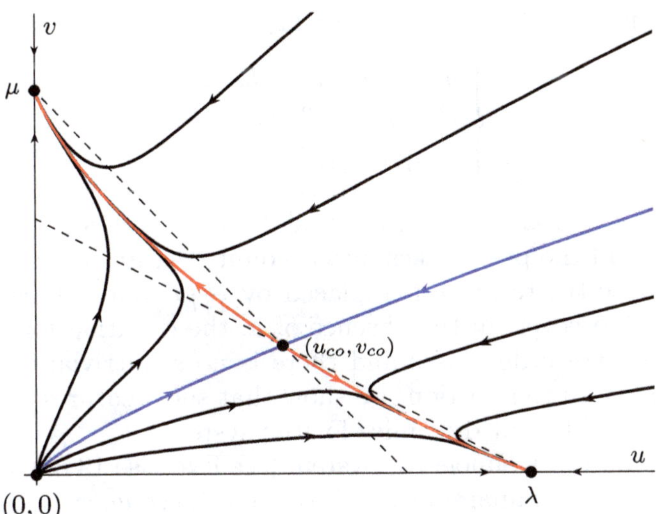

Fig. 11.28. Phase portrait of (11.72) when $bc > 1$ and (λ, μ) belongs to P_2.

and converges to $(0, \mu)$. Summarizing, (11.84) is satisfied whenever (u_0, v_0) lies above $W^s(u_0, v_0)$.

Analogously, one can prove that (11.85) holds if (u_0, v_0) lies below $W^s(u_0, v_0)$, which concludes the proof. Figure 11.28 shows the global phase portrait of (11.72) relative to Part (c). $\qquad\square$

11.6 A Predator-Prey Model with Saturation Effects

In this section, we study the dynamics of the Cauchy problem

$$\begin{cases} u' = 4u - u^2 - 3\frac{uv}{1+u}, \\ v' = -v + 2\frac{uv}{1+u}, \\ u(0) = u_0 \geq 0, \quad v(0) = v_0 \geq 0. \end{cases} \tag{11.87}$$

In Ecology, this system is known as *Holling–Tanner model*, after Holling (1959) and Tanner (1975), and describes the interaction between a predator, v, and its prey, u. With respect to the more classical Lotka–Volterra predator-prey model which is going to be analyzed in Exercise 13 of Chapter 11,

$$\begin{cases} u' = \lambda u - u^2 - buv, \\ v' = \mu v - v^2 + cuv, \\ (u(0), v(0)) = (u_0, v_0), \end{cases}$$

where $\lambda, b, c > 0$ and $\mu < 0$, in (11.87) the interaction between the predators and the preys takes into account a saturation effect of the prey, because the term uv is replaced by $\frac{u}{1+u}v$, and $\frac{u}{1+u}$ is bounded for $u \geq 0$. Moreover, in the absence of u, the equation for v reduces to $v' = -v$, i.e. v decreases and there is no semitrivial equilibrium on the v axis. In this section, we show that such features will lead to a completely different dynamics for the system.

As (11.87) is a Kolmogorov system (see Exercise 13 of Chapter 5), it has a unique maximal solution, $(u(t; u_0, v_0), v(t; u_0, v_0))$, defined for $t \in I := (T_{\min}, T_{\max})$, with

$$-\infty \leq T_{\min} < 0 < T_{\max} \leq +\infty,$$

and, in addition, for every $t \in I$,

$$\operatorname{sign} u(t; u_0, v_0) = \operatorname{sign} u_0, \quad \operatorname{sign} v(t; u_0, v_0) = \operatorname{sign} v_0.$$

In particular, if $u_0 = 0$, then $u(t) = 0$ for all $t \in I$, and $v' = -v$. Thus,

$$(u(t; 0, v_0), v(t; 0, v_0)) = (0, e^{-t}v_0) \quad \text{for all } t \in \mathbb{R}.$$

Consequently, according to this model, in the absence of preys the predators become extinct. Moreover, if $u_0 > 0$ and $v_0 = 0$, then

$$\begin{cases} u' = 4u - u^2, \\ u(0) = u_0 > 0, \end{cases}$$

and, hence,

$$\lim_{t \uparrow +\infty} (u(t; u_0, v_0), v(t; u_0, v_0)) = (4, 0).$$

Therefore, in the absence of predators, the preys follow a logistic equation, and they stabilize, in a monotone way, towards their positive equilibrium 4. Finally, when $u_0 > 0$ and $v_0 > 0$, then $u(t; u_0, v_0) > 0$ and $v(t; u_0, v_0) > 0$ for all $t \in I$.

The equilibria of (11.87) are given by the solutions of the following system

$$\begin{cases} 0 = 4u - u^2 - 3\frac{uv}{1+u} = \left(4 - u - 3\frac{v}{1+u}\right)u, \\[2mm] 0 = -v + 2\frac{uv}{1+u} = \left(-1 + 2\frac{u}{1+u}\right)v. \end{cases}$$

A direct analysis shows that there are three equilibria, all of of them non-negative: the *trivial steady-state* $(0,0)$, the *semitrivial steady-state* $(4,0)$, and the *coexistence state* $(1,2)$.

Setting

$$f(u,v) := \begin{pmatrix} 4u - u^2 - 3\frac{uv}{1+u} \\[2mm] -v + 2\frac{uv}{1+u} \end{pmatrix}, \qquad (u,v) \in (0,+\infty) \times \mathbb{R},$$

we have that

$$Df(u,v) = \begin{pmatrix} 4 - 2u - \frac{3v}{(1+u)^2} & -\frac{3u}{1+u} \\[2mm] \frac{2v}{(1+u)^2} & -1 + \frac{2u}{1+u} \end{pmatrix}, \qquad (u,v) \in (0,+\infty) \times \mathbb{R}.$$

In particular,

$$Df(0,0) = \begin{pmatrix} 4 & 0 \\ 0 & -1 \end{pmatrix}, \qquad Df(4,0) = \begin{pmatrix} -4 & -\frac{12}{5} \\ 0 & \frac{3}{5} \end{pmatrix},$$

$$Df(1,2) = \begin{pmatrix} \frac{1}{2} & -\frac{3}{2} \\ 1 & 0 \end{pmatrix}.$$

Thus, since

$$\sigma(Df(0,0)) = \{4, -1\}, \qquad \sigma(Df(4,0)) = \{-4, \tfrac{3}{5}\},$$

$(0,0)$ and $(4,0)$ are saddle points, while

$$\sigma(Df(1,2)) = \left\{\tfrac{1}{4}(1 - \sqrt{23}i), \tfrac{1}{4}(1 + \sqrt{23}i)\right\}$$

and, hence, $(1,2)$ is an unstable focus.

As usual, we divide the study of the dynamics of (11.87) into several steps.

Step 1. Sketch of the direction field. In order to sketch the direction field of (11.87), note that $u' = 0$ if, and only if, either $u = 0$, or

$$4 - u - 3\frac{v}{1+u} = 0,$$

which can be equivalently expressed as

$$v = \frac{1}{3}(1+u)(4-u)$$

and describes the branch of parabola plotted in Figure 11.29 with a dashed line. Similarly, $v' = 0$ if and only if either $v = 0$, or

$$-1 + 2\frac{u}{1+u} = 0 \quad \text{or, equivalently,} \quad u = 1,$$

which is the straight dashed line plotted in Figure 11.29. Moreover, since

$$\operatorname{sign} u' = \operatorname{sign}\left((1+u)(4-u) - 3v\right), \quad \operatorname{sign} v' = \operatorname{sign}(u-1),$$

if $u > 0$ and $v > 0$, it becomes apparent that

$$u' \begin{cases} > 0 & \text{if } v < \frac{1}{3}(1+u)(4-u), \\ = 0 & \text{if } v = \frac{1}{3}(1+u)(4-u), \\ < 0 & \text{if } v > \frac{1}{3}(1+u)(4-u), \end{cases} \qquad v' \begin{cases} > 0 & \text{if } u > 1, \\ = 0 & \text{if } u = 1, \\ < 0 & \text{if } u < 1. \end{cases}$$

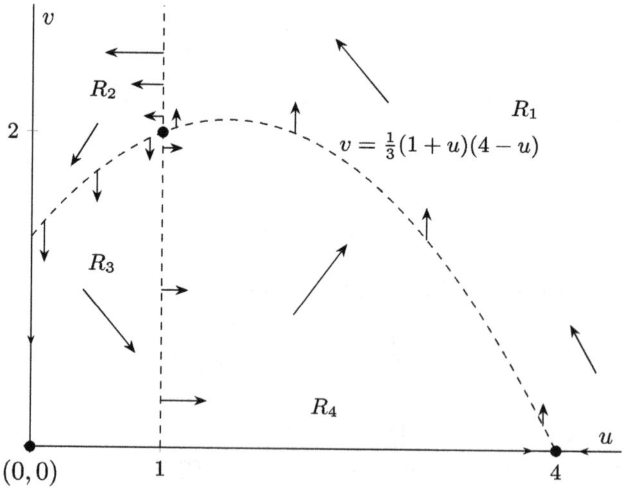

Fig. 11.29.   Direction field of (11.87).

Thus, we can identify the following four regions in the interior of the first quadrant of the phase plane

$$R_1 := \left\{ (u, v) \in (0, +\infty)^2 : u > 1, \; v > \tfrac{1}{3}(1 + u)(4 - u) \right\},$$

$$R_2 := \left\{ (u, v) \in (0, +\infty)^2 : u < 1, \; v > \tfrac{1}{3}(1 + u)(4 - u) \right\},$$

$$R_3 := \left\{ (u, v) \in (0, +\infty)^2 : u < 1, \; v < \tfrac{1}{3}(1 + u)(4 - u) \right\},$$

$$R_4 := \left\{ (u, v) \in (0, +\infty)^2 : u > 1, \; v < \tfrac{1}{3}(1 + u)(4 - u) \right\},$$

and the direction field of (11.87) looks like the one sketched in Figure 11.29. Note that it is compatible with the local structure of the phase portrait at the equilibria $(0, 0)$, $(4, 0)$ and $(1, 2)$. Indeed, $v = 0$ is the stable manifold of $(4, 0)$, and $\{(0, v) : v > 0\}$, is the part of the stable manifold of $(0, 0)$ within the first quadrant. The unstable manifold of $(0, 0)$, within the first quadrant, is the left branch of the stable manifold of $(4, 0)$ on the u-axis. Observe also that, according to the direction field, in a neighborhood of $(4, 0)$ the unstable manifold of such an equilibrium, $W^u(4, 0)$, must lie in the region R_1. Of course, one can obtain this also by computing the tangent to $W^u(4, 0)$ explicitly, as done in the proof of Part (a) in Theorem 11.25.

Step 2. The solutions are globally defined in the future. Subsequently, we show that $T_{\max} = +\infty$ if $u_0 > 0$ and $v_0 > 0$. Indeed, for all $t \in [0, T_{\max})$, we have that

$$u'(t) = \left(4 - u(t) - 3\frac{v(t)}{1 + u(t)} \right) u(t),$$

$$v'(t) = \left(-1 + 2\frac{u(t)}{1 + u(t)} \right) v(t),$$

and, hence,

$$u(t; u_0, v_0) = u_0 e^{\int_0^t \left(4 - u(s) - 3\frac{v(s)}{1 + u(s)} \right) ds},$$

$$v(t; u_0, v_0) = v_0 e^{\int_0^t \left(-1 + 2\frac{u(s)}{1 + u(s)} \right) ds}.$$

Thus, since $u(s) > 0$, $v(s) > 0$ and $\frac{u(s)}{1 + u(s)} < 1$ for all $s \in [0, t]$, we find that, for every $t \in [0, T_{\max})$,

$$0 < u(t; u_0, v_0) < u_0 e^{4t}, \quad 0 < v(t; u_0, v_0) < v_0 e^t.$$

According to Theorem 5.18, $T_{\max} = +\infty$, as claimed.

Step 3. Behavior of the solutions starting in R_1. We now show that, if (u_0, v_0) lies in R_1, then there exists $\tilde{t} > 0$ such that $u(\tilde{t}; u_0, v_0) = 1$. In particular, this will imply that any solution on the unstable manifold of $(4, 0)$ must reach $u = 1$ in a finite time and then pass to the region R_2, as illustrated in Figure 11.30. This fact will be crucial in the analysis of the dynamics of the system, because the arc $\mathcal{A}$ that lies on the unstable manifold of $(4, 0)$ and connects the equilibrium with the line $u = 1$ will act as a barrier for all the solutions as time grows.

For convenience, we denote

$$(u(t), v(t)) := (u(t; u_0, v_0), v(t; u_0, v_0))$$

and assume, by contradiction, that $u(t) > 1$ for all $t \geq 0$. Then, the direction field forces $(u(t), v(t))$ to stay in R_1 for all $t \in [0, +\infty)$, implying that $u'(t) < 0$ and $v'(t) > 0$ for all $t \in [0, +\infty)$.

As a preliminary step, we show that there exists a positive value M, which depends on u_0 and v_0, such that

$$v(t) \leq M \quad \text{for all } t \in [0, +\infty). \tag{11.88}$$

To do so, we proceed by contradiction, assuming that, for every integer $n \geq 1$, there exists a time $t_n > 0$ such that

$$v(t) > n \quad \text{for all } t \geq t_n.$$

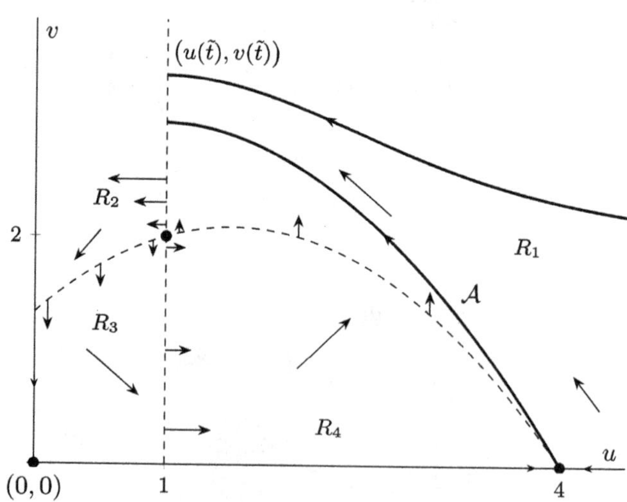

Fig. 11.30. Behavior of the solutions of (11.87) starting in the region R_1.

Then, by integrating

$$\frac{u'(s)}{u(s)} = 4 - u(s) - \frac{3v(s)}{1 + u(s)}, \quad s \geq 0,$$

in $[t_n, t]$, it follows that

$$\ln \frac{u(t)}{u(t_n)} = \int_{t_n}^{t} \frac{u'(s)}{u(s)}\, ds = 4(t-t_n) - \int_{t_n}^{t} u(s)\, ds - 3\int_{t_n}^{t} \frac{v(s)}{1+u(s)}\, ds$$

$$\leq 4(t-t_n) - \frac{3n}{1+u_0}(t-t_n) = \left(4 - \frac{3n}{1+u_0}\right)(t-t_n).$$

Thus, for every $n \geq 1$ and $t \in [t_n, +\infty)$, we have that

$$1 < u(t) \leq u(t_n)e^{\left(4-\frac{3n}{1+u_0}\right)(t-t_n)} \leq u_0 e^{\left(4-\frac{3n}{1+u_0}\right)(t-t_n)},$$

which is impossible because, if we fix a sufficiently large n so that $4 - \frac{3n}{1+u_0} < 0$, then

$$\lim_{t\uparrow+\infty} e^{\left(4-\frac{3n}{1+u_0}\right)(t-t_n)} = 0.$$

Thus, (11.88) holds true and, due to the monotonicity of $u(t)$ and $v(t)$, there exists $(u_\infty, v_\infty) \in [1, u_0] \times (0, M]$ such that

$$\lim_{t\to+\infty} (u(t), v(t)) = (u_\infty, v_\infty).$$

Thanks to Property 2 in Section 9.2, (u_∞, v_∞) must be an equilibrium of the system. Since $v_\infty > 0$, necessarily $(u_\infty, v_\infty) = (1, 2)$, which is impossible, because $(1, 2)$ is an unstable focus. This contradiction shows that there exists $\tilde{t} > 0$ such that $u(\tilde{t}) = 1$.

Step 4. Behavior of the solutions starting in R_2. We now show that, if (u_0, v_0) lies in R_2, then there exists $\hat{t} > 0$ such that

$$(u(t), v(t)) := (u(t; u_0, v_0), v(t; u_0, v_0))$$

leaves R_2 at $t = \hat{t}$, and then enters R_3. Observe that, since the system is autonomous, this analysis also applies to the solutions that, according to Step 3, enter R_2 after leaving R_1.

We argue by contradiction, assuming that the solution stays in R_2 for all $t \geq 0$ (observe that the direction field prevents the solution from leaving R_2 thorough the line $u = 1$). Then, due to the monotonicity of $u(t)$ and $v(t)$, $(u(t), v(t))$ should approach, as $t \uparrow +\infty$, an equilibrium lying in the closure of R_2. However, the only equilibrium with such characteristics is $(1, 2)$, which is an unstable focus. Hence, it cannot be approached by the solution as $t \uparrow +\infty$.

Step 5. Behavior of the solutions starting in R_3. We now show that, if (u_0, v_0) lies in R_3, then there exists $\bar{t} > 0$ such that $(u(t), v(t)) := (u(t; u_0, v_0), v(t; u_0, v_0))$ leaves R_3 at $t = \bar{t}$, and then enters R_4. As above, this analysis also applies to the solutions that, according to Step 4, enter R_3 after leaving R_2.

This can be obtained by reasoning as in Step 4, with the additional observation that $(0, 0)$ is a saddle point whose stable manifold consists of the v-axis, which entails that any solution lying in the interior of the first quadrant cannot converge to it.

Step 6. Behavior of the solutions starting in R_4. By reasoning as in Step 5, and using the fact that v increases, it is possible to prove that, if (u_0, v_0) lies in R_4, there exists $\check{t} > 0$ such that $(u(t), v(t)) := (u(t; u_0, v_0), v(t; u_0, v_0))$ leaves R_4 at $t = \check{t}$, and then enters R_1. As above, this analysis also applies to the solutions that, according to Step 5, enter R_4 after leaving R_3.

In addition, by uniqueness (see Property 1 in Section 9.2), as long as the solution $(u(t), v(t))$ remains in R_1 for $t > \check{t}$, it has to stay below the arc $\mathcal{A} \subset W^u(4, 0)$ that connects $(4, 0)$ with the line $u = 1$, and whose existence has been shown in Step 3.

Step 7. Determination of $\omega(u_0, v_0)$. We now conclude our analysis by showing that, for every $(u_0, v_0) \in (0, +\infty)^2 \setminus \{(1, 2)\}$, $\omega(u_0, v_0)$ consists of a non-trivial periodic solution of (11.87) that attracts the solution

$$(u(t), v(t)) := (u(t; u_0, v_0), v(t; u_0, v_0)).$$

Indeed, if we fix $(u_0, v_0) \neq (1, 2)$, the analysis of the previous steps guarantees that $\Gamma^+_{(u_0, v_0)}$ is bounded and $\omega(u_0, v_0)$ does not contain any equilibrium. Thus, by Theorem 11.15, $\omega(u_0, v_0)$ is the orbit of a non-trivial periodic solution, as claimed. In general, establishing the uniqueness of the non-trivial periodic solution is a real challenge. Some numerical experiments suggest that (11.87) has a unique non-trivial periodic solution. Should it indeed be unique, then it would

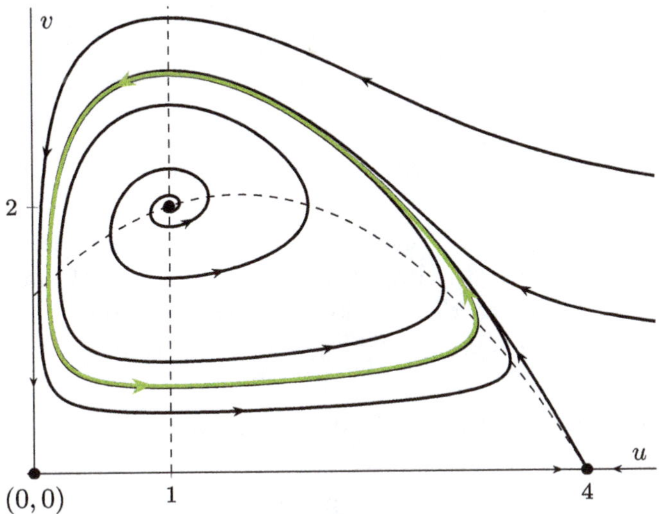

Fig. 11.31. Phase portrait of the Holling–Tanner model (11.87) with a periodic solution plotted in green.

be a *limit cycle*, in the sense that it attracts all solutions from its interior and its exterior, except the equilibrium $(1, 2)$, which is an unstable focus.

Based on this analysis, the phase portrait of (11.87) looks like the one shown in Figure 11.31. The reader should compare this particular dynamics with those analyzed in Section 9.3 and Exercise 13 of Chapter 11, and observe that they are substantially different.

11.7 Lyapunov's Second Method

In this section, we introduce Lyapunov's second method, which relies on analyzing the behavior of a scalar function V, defined in some subset of $\mathbb{R}^N$, along a solution $u(t)$ of $u' = f(u)$, with the goal of establishing some qualitative properties of the system, such as the stability, attractivity, or instability of an equilibrium, or the invariance of a certain set with respect to the underlying flow. In some cases, this method will give information even when Lyapunov's first method, based on the linearization discussed in Section 11.2, is not conclusive.

Actually, previously in this chapter, we have already encountered some examples where this approach was adopted. A first one is the function

$$V(u,v) := u^2 + v^2, \quad (u,v) \in \mathbb{R}^2, \tag{11.89}$$

used in the example in Section 11.4.1 to show that the solutions $(u(t), v(t))$ of (11.60) starting on a circle of sufficiently large radius $R > 0$ cannot leave the disk enclosed by such a circle. Precisely, we have shown that, as soon as $u_0^2 + v_0^2 = R^2$ with $R > \sqrt{2}$, the function

$$\varphi(t) := V(u(t), v(t)), \quad (u(t), v(t)) = (u(t; u_0, v_0), v(t; u_0, v_0)),$$

satisfies $\varphi'(t) < 0$, at least, in a right neighborhood of $t = 0$.

The same function (11.89) has been used in Remark 11.26 to show that the competitive system (11.71) is dissipative because, for sufficiently large $R > 0$, $\varphi(t)$ is decreasing if $\varphi(t) = V(u(t), v(t)) = 2R^2$.

Another function of this kind but of a different nature, Φ, has been constructed in the proof of Theorem 11.4 to establish the instability of the equilibrium x_0 when $Df(x_0)$ possesses an eigenvalue with a positive real part. Essentially, the proof of Theorem 11.4 shows the existence of an initial datum $u_0 \in \mathbb{R}^N$ such that $\varphi(t) := \Phi(u(t; t_0, u_0) - x_0)$, where $u(t; t_0, u_0)$ is the solution of $u' = f(u)$ such that $u(t_0) = u_0$, satisfies $\varphi'(t) > 0$ for all $t \geq t_0$.

These examples motivate the following definition. If $f(x_0) = 0$, a *Lyapunov function* of $u' = f(u)$ at x_0 is any function $V \colon B_\rho(x_0) \to \mathbb{R}$ of class $\mathcal{C}^1(B_\rho(x_0); \mathbb{R})$, for some $\rho > 0$, such that $V(u) > V(x_0)$ for all $u \in B_\rho(x_0) \setminus \{x_0\}$ and, for sufficiently small $\varepsilon \in (0, \rho)$, the function

$$\dot{V}(u) := \langle \nabla V(u), f(u) \rangle, \quad u \in B_\varepsilon(x_0), \tag{11.90}$$

does not change sign in $B_\varepsilon(x_0)$. The quantity (11.90) is often known as the *orbital derivative* of V.

To relate this definition with the above examples, set

$$\varphi(t) := V(u(t)), \quad u(t) = (u_1(t), \ldots, u_N(t)),$$

where V is a Lyapunov function of $u' = f(u)$ at x_0, and $u(t)$ is a solution of the system. Then, by differentiating,

$$\varphi'(t) = \langle \nabla V(u(t)), f(u(t)) \rangle = \dot{V}(u(t)).$$

Thus, a Lyapunov function is monotone (non-decreasing or non-increasing) along the trajectories of the system. Essentially, $V(u) - V(x_0)$ can be seen as a generalized distance function between u and the equilibrium x_0. Naturally, if $\varphi'(t) < 0$, the equilibrium should be an attractor, as the generalized distance between $u(t)$ and x_0 decreases in time, while x_0 should be unstable if $\varphi'(t) > 0$ because in such a case, the generalized distance between $u(t)$ and x_0 increases in time.

The advantage of using Lyapunov functions relies on the fact that, with (11.90), one may try to determine the sign of $\dot{V}(u)$ even without knowing the solutions of the system. Then, one can use the following results, going back to Lyapunov (1907), to make formal the above heuristic discussion and ascertain the stability or instability of the equilibrium x_0.

Theorem 11.29 (Lyapunov stability/instability). *Suppose that $f(x_0) = 0$ and $V \colon B_\rho(x_0) \to \mathbb{R}$ is a Lyapunov function of $u' = f(u)$ at x_0. Then:*

(a) *x_0 is stable if there exists $\varepsilon \in (0, \rho)$ such that*

$$\dot{V}(u) := \langle \nabla V(u), f(u) \rangle \leq 0 \quad \text{for all } u \in B_\varepsilon(x_0).$$

(b) *x_0 is asymptotically stable if there exists $\varepsilon \in (0, \rho)$ such that*

$$\dot{V}(u) := \langle \nabla V(u), f(u) \rangle < 0 \quad \text{for all } u \in B_\varepsilon(x_0) \setminus \{x_0\}.$$

(c) *x_0 is unstable if there exists $\varepsilon \in (0, \rho)$ such that*

$$\dot{V}(u) := \langle \nabla V(u), f(u) \rangle > 0 \quad \text{for all } u \in B_\varepsilon(x_0) \setminus \{x_0\}.$$

When V is a Lyapunov function and $\dot{V}(u) \leq 0$ in a neighborhood of x_0, the previous result only guarantees that x_0 is stable. Nevertheless, some additional information about the global dynamics of the system, particularly about the location of the ω-limit and α-limit sets of the trajectories, can be obtained using the following global invariance principle established by N. N. Krasovskii in 1959 and J. P. LaSalle in 1960, where the only assumption is that $\dot{V}(u) \leq 0$ for all $u \in \mathbb{R}^N$.

Theorem 11.30 (Krasovskii and LaSalle). *Consider the system* $u' = f(u)$, *and suppose that there exists* $V \in C^1(\mathbb{R}^N; \mathbb{R})$ *such that*

$$\dot{V}(u) := \langle \nabla V(u), f(u) \rangle \leq 0 \quad \text{for all } u \in \mathbb{R}^N.$$

Let $\mathcal{T}$ *denote the set of entire trajectories of the system that are contained in the set of critical points of the orbital derivative of* V, *denoted by*

$$\mathcal{C} := \left\{ u \in \mathbb{R}^N : \dot{V}(u) = 0 \right\}.$$

Then,

$$\bigcup_{\substack{u \in \mathbb{R}^N \text{ with} \\ \Gamma_u^- \text{ bounded}}} \alpha(u) \subset \mathcal{T} \quad \text{and} \quad \bigcup_{\substack{u \in \mathbb{R}^N \text{ with} \\ \Gamma_u^+ \text{ bounded}}} \omega(u) \subset \mathcal{T}.$$

In the following section, we illustrate how to use this result to analyze the global dynamics of a relevant example in Physics.

With Lyapunov's second method, some global stability results can also be obtained, such as the following one.

Theorem 11.31 (Global stability). *Suppose that* V *is a global Lyapunov function, in the sense that* $V(u) > V(x_0)$ *for all* $u \in \mathbb{R}^N \setminus \{x_0\}$,

$$\lim_{\|u\| \to +\infty} V(u) = +\infty,$$

and

$$\dot{V}(u) := \langle \nabla V(u), f(u) \rangle < 0 \quad \text{for all } u \in \mathbb{R}^N \setminus \{x_0\}.$$

Then, x_0 *is a global attractor for the solutions of* $u' = f(u)$, *i.e. all the solutions of the system are globally defined in the future and converge to* x_0 *as* $t \to +\infty$.

Based on Theorem 11.30, some sharper refinements of Theorem 11.31 can be obtained (see, e.g. Hale, 1980, Corollary 1.2 on p. 317).

Finally, a slightly stronger result about instability was proved by N. G. Chetaev in 1948.

Theorem 11.32 (Chetaev's instability theorem). *Suppose that x_0 is an equilibrium of $u' = f(u)$, that there exists a continuously differentiable function $V(u)$ such that $x_0 \in \partial G$, where*

$$G := \{u \in \mathbb{R}^N : V(u) > V(x_0)\},$$

and that there exists $\varepsilon > 0$ for which

$$\dot{V}(u) := \langle \nabla V(u), f(u) \rangle > 0 \quad \text{for all } u \in G \cap B_\varepsilon(x_0).$$

Then, x_0 is unstable.

All the results presented in this section rely on the existence of a Lyapunov function in order to ascertain the local or global character of an equilibrium, x_0. Nevertheless, one should keep in mind that constructing suitable Lyapunov functions in applications might be an intricate task, if it is possible at all!

11.8 The Damped Pendulum

In this section, we study the dynamics of the damped pendulum

$$u''(t) + \frac{\kappa}{M} u'(t) + \frac{G}{L} \sin u(t) = 0, \qquad (11.91)$$

where $\kappa \geq 0$ is the damping (or friction) constant. The equivalent first order system associated with (11.91) is

$$\begin{cases} u' = v, \\ v' = -\frac{G}{L} \sin u - \frac{\kappa}{M} v. \end{cases} \qquad (11.92)$$

In this section, for every $(u_0, v_0) \in \mathbb{R}^2$,

$$(u(t), v(t)) := (u(t; u_0, v_0), v(t; u_0, v_0))$$

denotes the unique maximal solution of the Cauchy problem

$$\begin{cases} u' = v, \\ v' = -\frac{G}{L} \sin u - \frac{\kappa}{M} v, \\ u(0) = u_0, \quad v(0) = v_0. \end{cases} \qquad (11.93)$$

The conservative case $\kappa = 0$ has been already analyzed in Section 10.3, where we have introduced the total energy, given by

$$V(u,v) := \frac{v^2}{2} - \frac{G}{L}\cos u, \quad (u,v) \in \mathbb{R}^2, \tag{11.94}$$

and, again for $\kappa = 0$, thanks to (10.7), one has that

$$V(u(t),v(t)) = V(u_0,v_0) \quad \text{for all } t \in \mathbb{R}.$$

Since the nonlinearity of (11.93) is globally Lipschitz in $(u,v) \in \mathbb{R}^2$ also for $\kappa > 0$, $(u(t),v(t))$ is globally defined for all $t \in \mathbb{R}$. Moreover, for every $t \in \mathbb{R}$,

$$-u'(t)\left(\frac{G}{L}\sin u(t) + \frac{\kappa}{M}v(t)\right) = v(t)v'(t).$$

Thus, it follows from (11.94) that

$$\frac{d}{dt}V(u(t),v(t)) = v(t)v'(t) + \frac{G}{L}u'(t)\sin u(t) = -\frac{\kappa}{M}v^2(t) \leq 0. \tag{11.95}$$

Actually, for every $(u,v) \in \mathbb{R}^2$,

$$\dot{V}(u,v) := \left\langle \nabla V(u,v), \left(v, -\frac{G}{L}\sin u - \frac{\kappa}{M}v\right)\right\rangle$$

$$= \left\langle \left(\frac{G}{L}\sin u, v\right), \left(v, -\frac{G}{L}\sin u - \frac{\kappa}{M}v\right)\right\rangle = -\frac{\kappa}{M}v^2 \leq 0.$$

Moreover, the set of critical points of the orbital derivative of V is

$$\mathcal{C} := \{(u,v) \in \mathbb{R}^2 : \dot{V}(u,v) = 0\} = \{(u,v) \in \mathbb{R}^2 : v = 0\},$$

i.e. it consists of the u-axis in the phase plane of (11.92). In particular, $\mathcal{C}$ contains the set of equilibria of (11.92),

$$\mathcal{E} := \{(n\pi, 0): n \in \mathbb{Z}\},$$

and, except for these equilibria, it does not contain any other forward or backward invariant set. Indeed, suppose that, for some $t_0 \in \mathbb{R}$, $v(t_0) = 0$ and $(u(t_0), 0) \notin \mathcal{E}$. Then,

$$v'(t_0) = -\frac{G}{L}\sin u(t_0) - \frac{\kappa}{M}v(t_0) = -\frac{G}{L}\sin u(t_0) \neq 0,$$

and, hence, the solution must abandon $\mathcal{C}$. Therefore, with the notation of Theorem 11.30, $\mathcal{T} = \mathcal{E}$, and, for every $(u_0, v_0) \in \mathbb{R}^2$ such that $\Gamma^+_{(u_0, v_0)}$ is bounded, $\omega(u_0, v_0)$ is an equilibrium of (11.92). Similarly, for every $(u_0, v_0) \in \mathbb{R}^2$ such that $\Gamma^-_{(u_0, v_0)}$ is bounded, $\alpha(u_0, v_0)$ is an equilibrium.

Subsequently, we analyze the local character of the equilibria of (11.92). Setting

$$f(u, v) = \begin{pmatrix} v \\ -\frac{G}{L} \sin u - \frac{\kappa}{M} v \end{pmatrix},$$

we have that

$$Df(u, v) = \begin{pmatrix} 0 & 1 \\ -\frac{G}{L} \cos u & -\frac{\kappa}{M} \end{pmatrix}.$$

Thus, for every integer $n \in \mathbb{Z}$,

$$Df(n\pi, 0) = \begin{pmatrix} 0 & 1 \\ \frac{G}{L} (-1)^{n+1} & -\frac{\kappa}{M} \end{pmatrix},$$

and, hence, the eigenvalues of $Df(n\pi, 0)$ are the roots of the characteristic equation

$$0 = \begin{vmatrix} -z & 1 \\ \frac{G}{L} (-1)^{n+1} & -\frac{\kappa}{M} - z \end{vmatrix} = z \left(\frac{\kappa}{M} + z \right) + \frac{G}{L} (-1)^n,$$

which are

$$z_\pm := \frac{1}{2} \left(-\frac{\kappa}{M} \pm \sqrt{\left(\frac{\kappa}{M} \right)^2 - 4 \frac{G}{L} (-1)^n} \right). \qquad (11.96)$$

Suppose that n is odd. Then, (11.96) becomes

$$z_\pm = \frac{1}{2} \left(-\frac{\kappa}{M} \pm \sqrt{\left(\frac{\kappa}{M} \right)^2 + 4 \frac{G}{L}} \right) \qquad (11.97)$$

which shows that one eigenvalue is negative and the other is positive. Therefore, Theorem 11.9 guarantees that $(n\pi, 0)$ is a saddle point if n is odd.

Suppose that n is even. Then, (11.96) becomes

$$z_\pm = \frac{1}{2}\left(-\frac{\kappa}{M} \pm \sqrt{\left(\frac{\kappa}{M}\right)^2 - 4\frac{G}{L}}\right)$$

and we need to differentiate two cases, according to the size of κ. Precisely, according to Theorem 11.9 again, $(n\pi, 0)$ is a stable focus if n is even and

$$\kappa < 2M\sqrt{\frac{G}{L}}, \tag{11.98}$$

while it is a stable node if n is even and

$$\kappa \geq 2M\sqrt{\frac{G}{L}}. \tag{11.99}$$

From now on, we assume that (11.98) holds. Comparing with the case of the undamped pendulum $\kappa = 0$ treated in Section 10.3 (cf. in particular, Figure 10.9), the centers of the undamped case now become stable foci as $\kappa > 0$ satisfies (11.98). The saddle points of the case $\kappa = 0$ are instead maintained.

In the rest of this section, we analyze the global dynamics of this system. To do so, we fix $n \in \mathbb{Z}$ odd. One of our main goals will be determining the solutions attracted by the stable focus $((n+1)\pi, 0)$.

Step 1. Linearized analysis at saddle-type equilibria. The main goal of this step is determining the local position of the stable and unstable manifolds at the equilibria of saddle type in the phase plane.

First of all, we analyze the contours of the Lyapunov function V introduced above. Thanks to the analysis of Section 10.3 (see also Figure 10.9), we know that $V(u, v) = \frac{G}{L}$ for every $(u, v) = (n\pi, 0)$ and all points (u, v) lying on the heteroclinic connections between $(n\pi, 0)$ and $((n+2)\pi, 0)$ of the undamped counterpart. Let $\mathcal{H}$ denote the set of such points,

$$\mathcal{H} := \left\{(u, v) \in \mathbb{R}^2 : V(u, v) = \frac{G}{L}\right\},$$

which has been represented with a dashed line in Figure 11.32. Since n is odd and

$$V((n+1)\pi, 0) = -\frac{G}{L}\cos((n+1)\pi) = -\frac{G}{L} < \frac{G}{L},$$

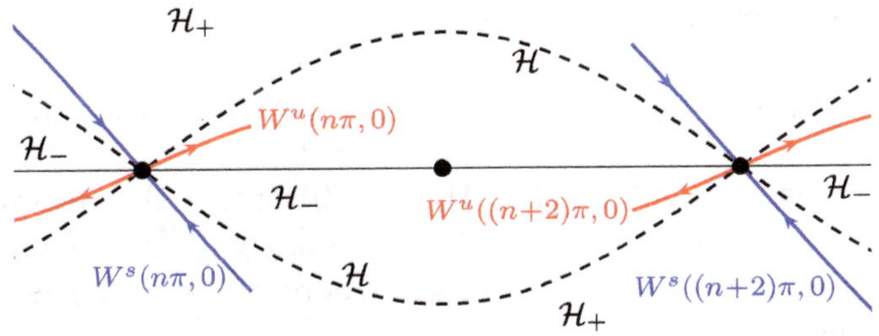

Fig. 11.32. Linearized analysis at the saddle-type equilibria $(n\pi, 0)$ and $((n + 2)\pi, 0)$ of (11.92), with $n \in \mathbb{Z}$ odd, compared with the contour line $\mathcal{H} = \{V(u, v) = \frac{G}{L}\}$.

it becomes apparent that $V < \frac{G}{L}$ in the (infinitely many) bounded components of $\mathbb{R}^2 \setminus \mathcal{H}$. Moreover, since

$$\lim_{v \to +\infty} V(u, v) = \lim_{v \to +\infty} \left(\frac{v^2}{2} - \frac{G}{L} \cos u \right) = +\infty,$$

we find that $V > \frac{G}{L}$ in the unbounded components of $\mathbb{R}^2 \setminus \mathcal{H}$. Subsequently, we set

$$\mathcal{H}_- := \left\{ (u, v) \in \mathbb{R}^2 : V(u, v) < \tfrac{G}{L} \right\},$$

$$\mathcal{H}_+ := \left\{ (u, v) \in \mathbb{R}^2 : V(u, v) > \tfrac{G}{L} \right\}.$$

By construction,

$$\mathbb{R}^2 = \mathcal{H} \cup \mathcal{H}_- \cup \mathcal{H}_+, \quad \partial\mathcal{H}_+ = \partial\mathcal{H}_- = \mathcal{H}.$$

Now, we analyze the linearized system at the saddle point $(n\pi, 0)$ to determine the tangents to its stable and unstable manifolds. According to Theorem 11.9, the unstable manifold of $(n\pi, 0)$ is tangent to the straight line through $(n\pi, 0)$ in the direction of an eigenvector of $Df(n\pi, 0)$ associated with the positive eigenvalue z_+ in (11.97), which is given by any non-zero solution of the linear system

$$\begin{pmatrix} -z_+ & 1 \\ \frac{G}{L} & -\frac{\kappa}{M} - z_+ \end{pmatrix} \begin{pmatrix} u \\ v \end{pmatrix} = \begin{pmatrix} 0 \\ 0 \end{pmatrix},$$

and can be equivalently expressed as $v = z_+ u$. Thus, the unstable manifold of $(n\pi, 0)$ is tangent to the straight line $v = z_+(u - n\pi)$.

By (11.97), $z_+ = z_+(\kappa)$ satisfies

$$z_+(0) = \sqrt{\frac{G}{L}} \quad \text{and} \quad z_+(\kappa) < \sqrt{\frac{G}{L}} \quad \text{for all } \kappa > 0.$$

This allows us to conclude that $W^u(n\pi, 0)$ enters the region $\mathcal{H}_-$ for $u \sim n\pi$, as illustrated in Figure 11.32. Similarly, $W^s(n\pi, 0)$ is tangent to the straight line $v = z_-(u - n\pi)$, where $z_- = z_-(\kappa)$ is given by (11.97). Since

$$z_-(0) = -\sqrt{\frac{G}{L}} \quad \text{and} \quad z_-(\kappa) < -\sqrt{\frac{G}{L}} \quad \text{for all } \kappa > 0,$$

$W^s(n\pi, 0)$ enters $\mathcal{H}_+$ for $u \sim n\pi$, as illustrated in Figure 11.32.

Step 2. Global behavior of $W^s(n\pi, 0)$. The second step consists in showing, by means of Krasovskii–LaSalle's invariance principle (see Theorem 11.30), that both branches of $W^s(n\pi, 0)$ are unbounded in u.

More generally, we take any $(u_0, v_0) \in \mathcal{H}_+$ with $v_0 > 0$ and consider the solution $(u(t), v(t))$ of (11.93); it may be, e.g. any solution on the left branch of $W^s(n\pi, 0)$. By definition of $\mathcal{H}_+$, $V(u_0, v_0) > \frac{G}{L}$, and, due to (11.95), $t \mapsto V(u(t), v(t))$ is non-increasing for all $t < 0$. Thus,

$$V(u(t), v(t)) \geq V(u_0, v_0) > \frac{G}{L} \quad \text{for all } t \leq 0, \tag{11.100}$$

which entails $v(t) > 0$ for all $t \leq 0$ and

$$\Gamma^-_{(u_0, v_0)} \subset \mathcal{H}^+_+ := \mathcal{H}_+ \cap \left\{ (u, v) \in \mathbb{R}^2 : v > 0 \right\}.$$

Suppose that $\Gamma^-_{(u_0, v_0)}$ is bounded. Then, by Theorem 11.30, its α-limit set is an equilibrium of (11.92), say $(\tilde{u}, \tilde{v})$. Since $v(t) > 0$ for all $t \leq 0$, $\tilde{v} \geq 0$. Let $\{t_k\}_{k \geq 1}$ be a sequence of negative times such that

$$\lim_{k \to +\infty} t_k = -\infty, \quad \lim_{k \to +\infty} (u(t_k), v(t_k)) = (\tilde{u}, \tilde{v}).$$

Then, evaluating (11.100) at $t = t_k$ and letting $k \to +\infty$ yields to

$$V(\tilde{u}, \tilde{v}) \geq V(u_0, v_0) > \frac{G}{L}.$$

Thus, $(\tilde{u}, \tilde{v}) \in \mathcal{H}_+$. So, $\tilde{v} > 0$. But this is impossible, because any equilibrium of (11.92) is of the form $(m\pi, 0)$, with $m \in \mathbb{Z}$. This contradiction shows that $\Gamma^-_{(u_0,v_0)}$ is unbounded in the phase plane. Actually, if we denote by $\mathcal{P}_u$ the projection on the first component of $\mathbb{R}^2$, we have that

$$\mathcal{P}_u\left(\Gamma^-_{(u_0,v_0)}\right) = (-\infty, u_0]. \tag{11.101}$$

On the contrary, assume that $\mathcal{P}_u\left(\Gamma^-_{(u_0,v_0)}\right) \subset [u_1, u_0]$ for some $u_1 \in (-\infty, u_0]$. Then, for every $t < 0$,

$$\left| \int_0^t v(s)\,ds \right| = \left| \int_0^t u'(s)\,ds \right| = |u(t) - u(0)| \leq u_0 - u_1.$$

So, the integral on the left-hand side is bounded as $t \to -\infty$. Therefore, since $v(t) > 0$ for all $t < 0$, we obtain that

$$\liminf_{t \to -\infty} v(t) = 0.$$

Consequently, since $u(t) \in [u_1, u_0]$ for all $t \leq 0$, there exists a sequence of negative times, $\{t_k\}_{k \geq 1}$, such that, for some $\hat{u} \in [u_1, u_0]$,

$$\lim_{k \to +\infty} t_k = -\infty, \qquad \lim_{k \to +\infty} u(t_k) = \hat{u}, \qquad \lim_{k \to +\infty} v(t_k) = 0.$$

Hence, $(\hat{u}, 0)$ is an α-limit point of $\Gamma^-_{(u_0,v_0)}$. Furthermore, evaluating (11.100) at t_k and letting $k \to \infty$, it is apparent that $V(\hat{u}, 0) > \frac{G}{L}$, which contradicts

$$V(\hat{u}, 0) = -\frac{G}{L} \cos \hat{u} \leq \frac{G}{L}$$

and completes the proof of (11.101).

By adapting this argument, it readily follows that, for every $(u_0, v_0) \in \mathcal{H}_+$ with $v_0 < 0$, $\Gamma^-_{(u_0,v_0)}$ is unbounded and

$$\mathcal{P}_u\left(\Gamma^-_{(u_0,v_0)}\right) = [u_0, +\infty).$$

As a direct consequence of this analysis, since $W^s(n\pi, 0)$ is a C^1-curve satisfying

$$\mathcal{P}_u\left(W^s(n\pi, 0)\right) = (-\infty, +\infty),$$

$\mathbb{R}^2 \setminus W^s(n\pi, 0)$ consists of two open disjoint regions: $\mathcal{A}^-(n\pi, 0)$, the component containing $(j\pi, 0)$ for all $j \leq n - 1$, and $\mathcal{A}^+(n\pi, 0)$, the

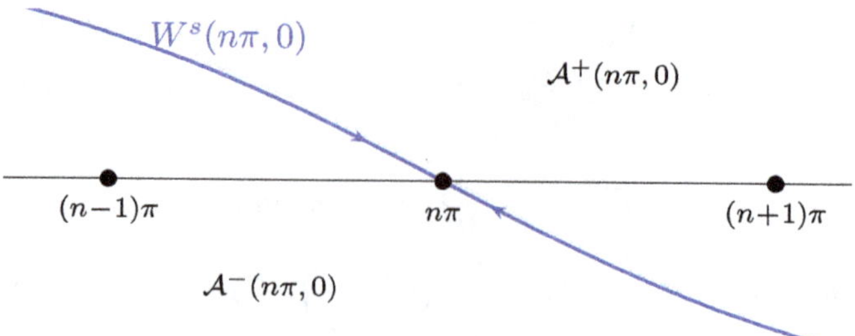

Fig. 11.33. The stable manifold $W^s(n\pi, 0)$, with $n \in \mathbb{Z}$ odd, divides the phase plane of (11.93) into two components: $\mathcal{A}^\pm(n\pi, 0)$.

component containing $(k\pi, 0)$ for all $k \geq n + 1$, as illustrated in Figure 11.33. By construction, setting

$$\mathcal{B}_{n+1} := \mathcal{A}^+(n\pi, 0) \cap \mathcal{A}^-((n + 2)\pi, 0),$$

we have that

$$\{(u, 0) \in \mathcal{B}_{n+1}\} = (n\pi, (n + 2)\pi) \times \{0\}. \tag{11.102}$$

Step 3. Basin of attraction of the foci $((n+1)\pi, 0)$, *with* $n \in \mathbb{Z}$ *odd.* To conclude our analysis, we show that, for every $(u_0, v_0) \in \mathcal{B}_{n+1}$, the unique solution, $(u(t), v(t))$, of (11.93) satisfies

$$\lim_{t \to +\infty} (u(t), v(t)) = ((n + 1)\pi, 0), \tag{11.103}$$

i.e. the focus $((n+1)\pi, 0)$ is a global attractor for all solutions starting in $\mathcal{B}_{n+1}$. Actually, for this reason, $\mathcal{B}_{n+1}$ is called the *basin of attraction* of $((n + 1)\pi, 0)$.

Naturally, (11.103) is obvious if $(u_0, v_0) = ((n+1)\pi, 0)$. Thus, we assume, in addition, that $(u_0, v_0) \neq ((n+1)\pi, 0)$. First of all, observe that $\mathcal{B}_{n+1}$ is an invariant set for (11.92) because

$$\partial \mathcal{B}_{n+1} = W^s(n\pi, 0) \sqcup W^s((n + 2)\pi, 0),$$

where $\sqcup$ indicate the disjoint union, and, hence, no solution starting in $\mathcal{B}_{n+1}$ can escape from it in a finite time, by the uniqueness for problem (11.93).

Next, we show that $(u(t), v(t))$ reaches the axis $v = 0$ in a finite positive time. Indeed, suppose, e.g. $v_0 > 0$, and assume by contradiction that $v(t) > 0$ for all $t > 0$. Then, (11.95) implies that

$V(u(t), v(t))$ is decreasing and, since V is bounded from below by $-\frac{G}{L}$, it becomes apparent that

$$\lim_{t \to +\infty} \frac{d}{dt} V(u(t), v(t)) = -\frac{\kappa}{M} \lim_{t \to +\infty} v^2(t) = 0.$$

Hence,

$$\lim_{t \to +\infty} v(t) = 0.$$

On the other hand, since $u'(t) = v(t) > 0$ for all $t > 0$, $u(t)$ is increasing. Therefore, there exists $\tilde{u} \in [n\pi, (n+2)\pi]$ such that

$$\lim_{t \to +\infty} (u(t), v(t)) = (\tilde{u}, 0).$$

As this entails $(\tilde{u}, 0)$ to be an equilibrium of (11.92), necessarily

$$\tilde{u} \in \{n\pi, (n+1)\pi, (n+2)\pi\}. \tag{11.104}$$

Since

$$(u(t), v(t)) \notin W^s(n\pi, 0) \sqcup W^s((n+2)\pi, 0) \quad \text{for all } t \in \mathbb{R},$$

necessarily, $\tilde{u} = (n+1)\pi$. But this is impossible, because a focus cannot be approached from the half-plane $v > 0$. Therefore, there exists $t_1 > 0$ such that $v(t) > 0$ for all $t \in [0, t_1)$, and $v(t_1) = 0$. Necessarily $u(t_1) \in ((n+1)\pi, (n+2)\pi)$, by the form of the direction field of (11.92). Similarly, also when $v_0 < 0$ there exists $t_1 > 0$ such that $v(t) < 0$ for all $t \in (0, t_1)$, $v(t_1) = 0$, and $u(t_1) \in (n\pi, (n+1)\pi)$.

Iterating the previous argument in $\mathcal{B}_{n+1}$, it becomes apparent that there exists an increasing sequence of times, $\{t_k\}_{k \geq 1}$, such that

$$\lim_{k \to +\infty} t_k = +\infty, \quad v(t_k) = 0 \quad \text{for all } k \geq 1,$$

as well as

$$n\pi < u(t_{2k}) < (n+1)\pi < u(t_{2k-1}) < (n+2)\pi \quad \text{if } v_0 > 0,$$

$$n\pi < u(t_{2k-1}) < (n+1)\pi < u(t_{2k}) < (n+2)\pi \quad \text{if } v_0 < 0.$$

Moreover, since $V(u(t_k), 0)$ is decreasing with respect to k, $|u(t_k) - (n+1)\pi|$ is also decreasing and, actually,

$$\lim_{k \to +\infty} u(t_k) = (n+1)\pi.$$

Therefore, $(u(t), v(t))$ spirals towards the focus $((n+1)\pi, 0)$.

In particular, this provides us with the spiraling behavior of the solutions lying on the right branch of $W^u(n\pi, 0)$, as well as of those lying on the left branch of $W^u((n+2)\pi, 0)$.

These features are summarized in Figures 11.34 and 11.35, where the basin of attraction of $(0,0)$, $\mathcal{B}_0$, and a more global vision of the

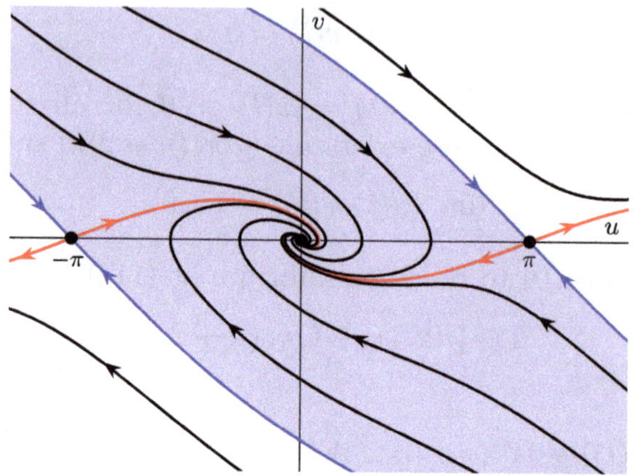

Fig. 11.34. Basin of attraction of the equilibrium $(0,0)$ of (11.92) under condition (11.98), which is given by $\mathcal{B}_0 = \mathcal{A}^+(-\pi, 0) \cap \mathcal{A}^-(\pi, 0)$.

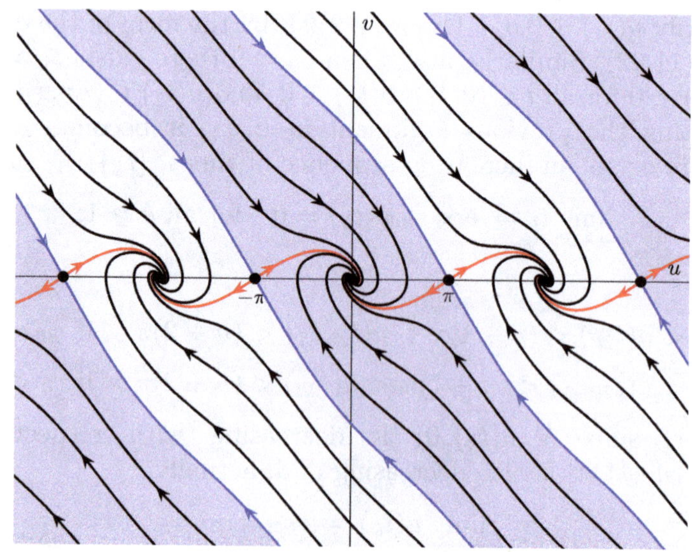

Fig. 11.35. Phase portrait of (11.92) under condition (11.98).

phase portrait of (11.92) under condition (11.98) are represented, respectively.

11.9 Exercises

1. Consider $A(t) = (a_{ij}(t))_{1 \le i,j \le N}$ and $B(t) = (b_j(t))_{1 \le j \le N}$, with $a_{ij}, b_j \in \mathcal{C}(\mathbb{R})$ for all $i, j \in \{1, \dots, N\}$. Prove that the behavior, as $t \uparrow +\infty$, of all the solutions of the linear system

$$u' = A(t)u + B(t)$$

   is the same and that it is independent of t_0, in the sense that all solutions are either unstable, stable, or asymptotically stable. Show that the same happens for $t \downarrow -\infty$.

2. Complete the proof of Lemma 11.5 by checking all the technical details.

3. Consider the system

$$\begin{cases} u_1' = \alpha u_1, \\ u_2' = (\alpha - \gamma)u_2 + \varepsilon u_1 u_3, \\ u_3' = -\gamma u_3, \end{cases} \qquad (11.105)$$

   where $\alpha > \gamma > 0$ and $\varepsilon \ne 0$. Show that there is no map

$$\Psi : (u_1, u_2, u_3) \to (v_1, v_2, v_3)$$

   of class $\mathcal{C}^1$ with non-vanishing Jacobian from a neighborhood of $u = 0$ onto a neighborhood of $v = 0$ that transforms the solutions of (11.105) into those of the linearized system

$$\begin{cases} v_1' = \alpha v_1, \\ v_2' = (\alpha - \gamma)v_2, \\ v_3' = -\gamma v_3. \end{cases}$$

4. Use system

$$\begin{cases} u' = \varepsilon^2 u + v^2, \\ v' = \varepsilon v, \end{cases} \qquad (11.106)$$

   with $\varepsilon > 0$ fixed, to show that, in general, in Theorem 11.9, the function Ψ cannot be taken to be of class $\mathcal{C}^2$, even if f is analytic.

5. Construct the phase portrait of the planar system

$$\begin{pmatrix} u' \\ v' \end{pmatrix} = \begin{pmatrix} 0 & 1 \\ -1 & 0 \end{pmatrix} \begin{pmatrix} u \\ v \end{pmatrix} - \begin{pmatrix} u \sin \frac{1}{u^2 + v^2} \\ v \sin \frac{1}{u^2 + v^2} \end{pmatrix},$$

with $(u, v) \in \mathbb{R}^2 \setminus \{(0,0)\}$.

6. Construct the phase portrait of the planar system

$$\begin{cases} u' = \varepsilon u + v - u \left(u^2 + v^2 \right), \\ v' = -u + \varepsilon v - v \left(u^2 + v^2 \right), \end{cases}$$

according to the values of the parameter $\varepsilon \in \mathbb{R}$.

7. Prove the following characterization of the ω-limit set of x:

$$\omega(x) = \bigcap_{t \geq 0} \overline{\Gamma^+_{u(t;x)}}.$$

Determine an analogous characterization for the α-limit set of x.

8. Prove Theorem 11.20. [Hint: Combine the ingredients of the proof of Theorem 11.15.]

9. Complete the analysis of Section 11.5.4. Specifically, when $\lambda < b\mu$ and $\mu < c\lambda$, determine the tangent lines to $W^s(u_{co}, v_{co})$ and $W^u(u_{co}, v_{co})$, and compare them with the nullclines.

10. Complete the analysis of the dynamics of the competitive system (11.72) for $bc \neq 1$ when the parameters (λ, μ) lie on the lines where the stability of the semitrivial steady-states changes (cf. Figures 11.18 and 11.25).

11. Analyze the dynamics of the competition system (11.72) in the case $bc = 1$.

12. Construct the phase portrait of the damped pendulum (11.91) under condition (11.99).

13. As a continuation of Exercise 8 of Chapter 5, consider the Lotka–Volterra predator-prey model

$$\begin{cases} u' = \lambda u - u^2 - buv, \\ v' = \mu v - v^2 + cuv, \\ (u(0), v(0)) = (u_0, v_0), \end{cases} \tag{11.107}$$

with $u_0 \geq 0$, $v_0 \geq 0$, $\lambda > 0$, $b > 0$, $c > 0$, and $\mu \in \mathbb{R}$. Using the techniques introduced in Section 11.5, for the different values of λ, μ, b, and c:

(a) find the non-negative equilibria of (11.107), as well as the direction-field of the system;

(b) determine the local character of the non-negative equilibria of (11.107);

(c) construct the phase portrait of (11.107).

14. Consider the Lotka–Volterra symbiotic model

$$\begin{cases} u' = \lambda u - u^2 + buv, \\ v' = \mu v - v^2 + cuv, \\ (u(0), v(0)) = (u_0, v_0), \end{cases} \tag{11.108}$$

with $u_0 \geq 0$, $v_0 \geq 0$, $\lambda < 0$, $\mu < 0$, $b > 0$, and $c > 0$. Using the techniques introduced in Section 11.5, for the different values of λ, μ, b, and c:

(a) find the non-negative equilibria of (11.108), as well as direction-field of the system;

(b) determine the local character of the non-negative equilibria of (11.108);

(c) construct the phase portrait of (11.108);

(d) analyze whether the maximal solution of (11.108) is globally defined in the future.

15. Construct the phase portrait of the damped diffusive logistic equation

$$-u'' = ru' + \lambda u - u^3,$$

where $r > 0$ and $\lambda > 0$. [Hint: Adapt the techniques used in Section 11.8.]

16. Given the following parametric version of the model (11.87)

$$\begin{cases} u' = \lambda u - u^2 - b\frac{uv}{1+u}, \\ v' = -\mu v + c\frac{uv}{1+u}, \\ u(0) = u_0 \geq 0, \quad v(0) = v_0 \geq 0, \end{cases} \tag{11.109}$$

where λ, μ, b, and c are positive constants, determine some ranges of the parameters λ, μ, b, and c for which (11.109) has a non-trivial periodic solution.

17. Use polar coordinates to construct the phase portrait of

$$\begin{cases} u' = u - v + \dfrac{uv - u^3 - uv^2}{\sqrt{u^2 + v^2}}, \\[2mm] v' = u + v - \dfrac{u^2 + u^2 v + v^3}{\sqrt{u^2 + v^2}}, \end{cases} \tag{11.110}$$

with $(u, v) \in \mathbb{R}^2 \setminus \{(0,0)\}$, where the direction field is assumed to be $(0,0)$ for $(u, v) = (0,0)$. Observe that this system gives an example of an attractive equilibrium which is Lyapunov unstable.

18. Prove the following result, known as *Bendixson's negative criterion*. Suppose that $u' = f(u)$ is a planar system ($N = 2$) and $\Omega \subset \mathbb{R}^2$ is a simply connected domain such that $\operatorname{div} f(u, v)$ has constant sign in Ω. Then, Ω cannot contain a non-trivial periodic orbit of $u' = f(u)$.

Use this result to show that, for all $g, h \in \mathcal{C}^1(\mathbb{R}; \mathbb{R})$, the system

$$\begin{cases} u' = u + g(v), \\ v' = h(u) - 2v, \end{cases}$$

does not admit any non-trivial periodic solution.

11.10   Final Comments

The modern theory of Dynamical Systems was founded by Poincaré in 1880 and 1892 and by Lyapunov in 1907. Some years later, Birkhoff introduced the concept of *Dynamical Systems* (1917, 1927). In this chapter, we have introduced some of the most fundamental concepts and technical devices of this theory, and we have applied them to analyze the dynamics of the Lotka–Volterra competition system, a Holling–Tanner predator-prey model, and a nonlinear pendulum in the presence of friction.

One of the most important results is the Poincaré–Bendixson theorem, which gives some sufficient conditions for the existence of a periodic solution in a planar autonomous system of ODEs. With respect to the problem of counting the exact number of limit cycles (attractive periodic solutions) of a system, one should be aware that it is one of the 23 problems proposed by D. Hilbert in Paris at the International Conference of Mathematics in 1900. These problems have deeply influenced the development of Mathematics in the

20th century and beyond. Precisely, the second part of Hilbert's 16th problem asks to determine, for planar systems whose nonlinearity is simply polynomial, an upper bound for the number of limit cycles in terms of the degree of the polynomial components of the vector field. To date, this question remains completely open.

Due to the introductory nature of this textbook, we have decided not to include the proofs of the theorems by Grobman and Hartman, stated in Section 11.2, and those by Lyapunov, Chetaev, Krasovskii, and LaSalle, presented in Section 11.7. Some of them will probably be incorporated in a forthcoming volume about selected advanced topics in the theory of Dynamical Systems. In any case, the flavor of the techniques for demonstrating the results related to Lyapunov's second method is given in the analysis of the nonlinear pendulum with friction, presented in Section 11.8, where the energy of the undamped case is employed as a Lyapunov function.

Regarding the Lyapunov stability/instability theorem (see Theorem 11.29), which establishes that the existence of a Lyapunov function with certain properties is sufficient for the stability, asymptotic stability, or instability of an equilibrium, we mention that such conditions are also necessary. Actually, this is the content of the so-called Lyapunov's converse theorems, which have been developed by several authors. The interested reader can refer to Massera (1960), where the author gives a complete review of this topic. Anyway, we stress that, in practice, determining an explicit Lyapunov function might be extremely intricate.

Example (11.105) goes back to Hartman (1960) (see Hartman, 2002, Exercise 7.1 of Chapter IX). Example (11.106) goes back to Sternberg (1957, p. 812) (see Hartman, 2002, Exercise 8.1(c) of Chapter IX). Example (11.110) goes back to Fernández-Pérez and Vegas-Montaner (1996, p. 414). Bendixson's negative criterion (see Exercise 18 of Chapter 11) goes back to Bendixson (1901).

Bibliography

Airy, G. B. (1838). On the intensity of light in the neighbourhood of a caustic. *Trans. Camb. Phil. Soc.*, 6, 379–402.

Amann, H. (1971/72). Existence of multiple solutions for nonlinear elliptic boundary value problems. *Indiana Univ. Math. J.*, 21, 925–935.

Amann, H. (1976). Fixed point equations and nonlinear eigenvalue problems in ordered Banach spaces. *SIAM Rev.*, 18, 620–709.

Amann, H. (1990). *Ordinary Differential Equations: An Introduction to Nonlinear Analysis.* de Gruyter Studies in Mathematics, Vol. 13. Berlin: Walter de Gruyter.

Amann, H. (2005). Maximum principle and principal eigenvalues. In Ferrera, J., López-Gómez, J. and Ruiz del Portal, F. R. (Eds.). *Ten Mathematical Essays on Approximation in Analysis and Topology.* Amsterdam: Elsevier, pp. 1–60.

Amann, H. and López-Gómez, J. (1998). A priori bounds and multiple solutions for superlinear indefinite elliptic problems. *J. Diff. Equ.*, 146, 336–374.

Antón, I. and López-Gómez, J. (2019). Principal eigenvalue and maximum principle for cooperative periodic-parabolic systems. *Nonlinear Anal.*, 178, 152–189.

Apostol, T. M. (1974). *Mathematical Analysis* (2nd ed.). Reading, Massachusetts: Addison Wesley.

Arzelà, C. (1895). Sulle funzioni di linee. *Mem. R. Accad. Bologna*, 5, 225–244.

Ascoli, G. (1883/4). Le curve limiti di una varietà data di curve. *Mem. R. Accad. Lincei*, 18, 521–586.

Banach, S. (1922). Sur les opérations dans les ensembles abstraits et leur application aux équations intégrales. *Fund. Math.*, 3, 133–181.

Begon, M., Harper, J. L. and Townsend, C. R. (1996). *Ecology*. Oxford: Blackwell Science.

Bellman, R. (1943). The stability of solutions of linear differential equations. *Duke Math. J.*, 10, 643–647.

Bendixson, I. (1896). Démostration de l'existence de l'intégrale d'une équation aux dérivées partielles linéaire. *Bull. Soc. Math. Fr.*, 24, 220–225.

Bendixson, I. (1901). Sur les courbes définies par des équations différentielles. *Acta Math.*, 24, 1–88.

Bielecki, A. (1956). Une remarque sur la méthode de Banach-Cacciopoli-Tikhonov dans la théorie des équations différentielles ordinaires. *Bull. Acad. Sci. Polon. Sci.* C1, III, 4, 261–264.

Birkhoff, G. D. (1917). Dynamical systems with two degrees of freedom. *Trans. Am. Math. Soc.*, 18, 199–300.

Birkhoff, G. D. (1927). *Dynamical Systems*. Colloquium Publications. New York: American Mathematical Society.

Cano-Casanova, S., López-Gómez, J. and Takimoto, K. (2012). A quasilinear parabolic perturbation of the linear heat equation. *J. Diff. Equ.*, 252, 323–343.

Casal, A., Eilbeck, J. C. and López-Gómez, J. (1994). Existence and uniqueness of coexistence states for a predator-prey model with diffusion. *Diff. Integral Equ.*, 7, 411–439.

Chetaev, N. G. (1948). Concerning the stability and instability of irregular systems (in Russian). *Akad. Nauk. SSSR. Prikl. Mat. Meh.*, 12, 639–642.

Chicone, C. (2006). *Ordinary Differential Equations with Applications*. New York: Springer-Verlag.

Chpolski, E. (1977). *Physique Atomique*, Tome 1. Moscow: Mir.

Coddington, E. A. and Levinson, N. (1955). *Theory of Ordinary Differential Equations*. New York: McGraw-Hill.

Coppel, W. A. (1977). Ordinary differential equations. *J. Aust. Math. Soc.*, 24, 1–9.

Cubillos, P., López-Gómez, J. and Tellini, A. (2022). Multiplicity of nodal solutions in classical non-degenerate logistic equations. *Electron. Res. Arch.*, 30, 898–928.

Dieudonné, J. (1960). *Foundations of Modern Analysis*. Pure and Applied Mathematics, Vol. X. New York: Academic Press.

Fernández Pérez, C. and Vegas Montaner, J. M. (1996). *Ecuaciones Diferenciales II: ecuaciones no lineales*. Madrid: Pirámide.

Fick, A. (1855). Ueber diffusion. *Ann. Phys.*, XCII, 59–86.

Fisher, R. A. (1937). The wave of advances of advantageous genes. *Ann. Eugen.*, 7, 355–369.

Fourier, J. B. J. (1822). *Théorie Analytique de la Chaleur*. Paris: Firmin Didot.

Fraile, J. M., Koch, P., López-Gómez, J. and Merino, S. (1996). Elliptic eigenvalue problems and unbounded continua of positive solutions of a semilinear equation. *J. Diff. Equ.*, 127, 295–319.

Fraile, J. M., López-Gómez, J. and Sabina, J. C. (1995). On the global structure of the set of positive solutions of some semilinear elliptic boundary value problems. *J. Diff. Equ.*, 123, 180–212.

Friedrichs, K. O. (1956). *Advanced Ordinary Differential Equations*. Lectures Notes of the Institute of Mathematical Sciences. New York: New York University.

Friedrichs K. O. (1965). *Lectures on Advanced Ordinary Differential Equations*. New York: Gordon and Breach Science Publishers.

Frobenius, F. G. (1873). Ueber die Integration der linearen Differentialgleichungen durch Reihen. *J. Reine Angew. Math.*, 76, 214–235.

Fukuhara, M. (1928). Sur les systémes des équations différentielles ordinaires. *Proc. Imp. Acad. Jpn.*, 4, 448–449.

Furter, J. E. and López-Gómez, J. (1997). Diffusion mediated permanence problem for an heterogeneous Lotka-Volterra competition model. *Proc. R. Soc. Edinb.*, 127-A, 281–336.

Grobman, D. M. (1959). Homeomorphisms of systems of differential equations. *Dokl. Acad. Nauk SSSR*, 128, 880–881.

Grobman, D. M. (1962). Topological classification of the neighborhood of a singular point in n-dimensional space. *Mat. Sb. (N.S.)*, 56, 77–94.

Grönwall, T. H. (1919). Note on the derivatives with respect to a parameter of the solutions of a system of differential equations. *Ann. Math. (2)*, 20, 292–296.

Guzmán, M. (1975). *Ecuaciones diferenciales ordinarias*. Madrid: Alhambra.

Hahn, H. (1967). *Stability of motion*. Die Grundlehren der mathematischen Wissenschaften, Band 138. New York: Springer-Verlag.

Hale, J. K. (1980). *Ordinary Differential Equations* (2nd ed.). Malabar, Florida: Robert E. Krieger Publishing Company, Inc.

Hartman, P. (1960). A lemma in the theory of structural stability of differential equations. *Proc. Am. Math. Soc.*, 11, 610–620.

Hartman, P. (1962). On dichotomies for solutions of nth order linear differential equations. *Math. Ann.*, 147, 378–421.

Hartman, P. (1963a). A differential equation with non-unique solutions. *Am. Math. Mon.*, 70, 255–259.

Hartman, P. (1963b). On the local linearization of differential equations. *Proc. Am. Math. Soc.*, 14, 568–573.

Hartman, P. (2002). *Ordinary Differential Equations* (2nd ed.). Philadelphia: Society for Industrial and Applied Mathematics.

Holling, C. S. (1959). The components of predation as revealed by a study of small mammal predation of the European pine sawfly. *Can. Entomol.*, 91, 293–320.

Hopf, E. (1952). A remark on linear elliptic differential equations of the second order. *Proc. Am. Math. Soc.*, 3, 791–793.

John, F. (1978). *Partial Differential Equations* (3rd ed.). Applied Mathematical Sciences, Vol. 1. New York: Springer.

Kamke, E. (1930). *Differentialgleichungen reeller Funktionen*. Leipzig: Akademische Verlagsgesesellschaft.

Kamke, E. (1942). *Differentialgleichungen. Lösungsmethoden und Lösungen I. Gewöhnliche Differentialgleichungen*. Leipzig: Akademische Verlagsgesellschaft.

Kampen, E. R. van (1937). Remarks on systems of ordinary differential equations. *Am. J. Math.*, 59, 144–152.

Keller, H. (1957). On solutions of $\Delta u = f(u)$. *Commun. Pure Appl. Math.*, X, 503–510.

Kneser, H. (1923). Ueber die Lösungen eines Systems gewöhnlicher Differentialgleichungen das der Lipschitzschen Bedingung nicht genügt. *Preuss. Akad. Wiss. Phys.-Math Kl.*, II4, 171–174.

Kolmogorov, A. N., Petrovsky, I. G. and Piskunov, N. S. (1937). Étude de l'équation de la diffusion avec croissance de la quantité de matière et son application à un problème biologique. *Bull. Math. Univ. d'Etat à Moscow (Sèr. Int.)*, A 1, 1–129.

Krasovskii, N. N. (1959). *Problems of the Theory of Stability of Motion* (Russian). English translation: Stanford, CA: Stanford University Press, 1963.

Kuratowski, C. (1922). Une méthode d'élimination des nombres transfinis des raisonnements mathématiques. *Fundam. Math.*, 3, 76–108.

Lang, S. (1994). *Algebraic Number Theory*. Graduate Texts in Mathematics. New York: Springer.

LaSalle, J. P. (1960). Some extensions of Lyapunov's second method. *IRE Trans. Circuit Theory*, CT-7, 520–527.

Laurentiev, M. (1925). Sur une équation différentielle du premier ordre. *Math. Z.*, 23, 197–209.

Lebesgue, H. (1910). Sur l'integration des fonctions discontinues. *Ann. Ec. Norm. Sup.*, XXVII, 361–450.

Lindelöf, E. (1894). Sur l'application des méthodes d'approximations successives à l'étude des intégrales réelles des équations différentielles ordinaires. *J. Math. Pures Appl.*, 10, 117–128.

Lipschitz, R. (1876). Sur la possibilité d'intégrer complètement un système donné d'équations différentielles. *Bull. Sci. Math. Astro.*, 10, 149–159.

López-Gómez, J. (2001a). *Ecuaciones Diferenciales y Variable Compleja.* Madrid: Prentice Hall.

López-Gómez, J. (2001b). *Ecuaciones Diferenciales y Variable Compleja.* Problemas y Ejercicios resueltos. Madrid: Prentice Hall.

López-Gómez, J. (2013). *Linear Second Order Elliptic Operators.* Singapore: World Scientific.

López-Gómez, J. (2015). *Metasolutions of Parabolic Equations in Population Dynamics.* Boca Raton: CRC Press.

López-Gómez, J. and Maire, L. (2018). Multiplicity of large solutions for quasi-monotone pulse nonlinearities. *J. Math. Anal. Appl.*, 459, 490–505.

López-Gómez, J. and Maire, L. (2019). Uniqueness of large solutions for non-monotone nonlinearities. *Nonlinear Anal. Real World Appl.*, 47, 291–305.

López-Gómez, J. and Molina-Meyer, M. (1994). The maximum principle for cooperative weakly coupled elliptic systems and some applications. *Diff. Integral Equ.*, 7, 383–398.

López-Gómez, J., Molina-Meyer, M. and Tellini, A. (2015). Spiraling bifurcation diagrams in superlinear indefinite problems. *Disc. Cont. Dyn. Syst.*, 35, 1561–1588.

López-Gómez, J. and Mora-Corral, C. (2007). *Algebraic Multiplicity of Eigenvalues of Linear Operators.* Operator Theory, Advances and Applications, Vol. 177. Basel: Birkhäuser.

López-Gómez, J., Muñoz-Hernández, E. and Zanolin, F. (2023). Rich dynamics in planar systems with heterogeneous nonnegative weights. *Commun. Pure Appl. Anal.*, 22, 1043–1098.

López-Gómez, J. and Rabinowitz, P. H. (2020). The structure of the set of 1-node solutions of a class of degenerate BVP's. *J. Diff. Equ.*, 268, 4691–4732.

López-Gómez, J. and Tellini, A. (2014). Generating an arbitrarily large number of isolas in a superlinear indefinite problem. *Nonlinear Anal.*, 108, 223–248.

López-Gómez, J., Tellini, A. and Zanolin, F. (2014). High multiplicity and complexity of the bifurcation diagrams of large solutions for a class of superlinear indefinite problems. *Commun. Pure Appl. Anal.*, 13, 1–73.

Lotka, A. J. (1932). The growth of mixed populations: Two species competing for a common food supply. *J. Wash. Acad. Sci.*, 22, 461–469.

Lyapunov, A. M. (1907). Problème général de la stabilité du mouvement. *Ann. Fac. Sci. Univ. Toulouse*, 9, 203–475.

Marsden, J. E. and Tromba, A. J. (2013). *Vector Calculus* (6th ed.). New York: W. H. Freeman and Company.

Massera, J. L. (1960). Converse theorems of Lyapunov's second method. *Bol. Soc. Mat. Mexicana*, 5, 158–163.

Moigno, F. (1840). *Leçons de calcul différentiel et de calcul intégral* (d'après Cauchy). Paris: Mallet-Bachelier.

Molina-Meyer, M. (1995). Existence and uniqueness of coexistence states for some nonlinear elliptic systems. *Nonlinear Anal.*, 25, 279–296.

Molina-Meyer, M. (1996). Global attractivity and singular perturbation for a class of nonlinear cooperative systems. *J. Diff. Equ.*, 128, 347–378.

Müller, M. (1927). Ueber der Fundamentaltheorem in der Theorie der gewöhnlichen Differentialgleichungen. *Math. Z.*, 26, 619–645.

Müller, M. (1928). Beweis eines Satzes des Herrn H. Knesser über die Gesamheit der Lösungen, die ein System gewöhnlicher Differentialgleichungen durch einen Punkt schickt. *Math. Z.*, 28, 349–355.

Nagumo, M. (1926). Eine hinreichende Bedingung für die Unität der Lösung von Differentialgleichungen erster Ordnung. *Jpn. J. Math.*, 3, 107–112.

Nagumo, M. and Fukuhara, M. (1930). Un théorème relatif à l'ensemble des courbes intégrales d'un système d'équations différentielles ordinaires. *Proc. Phys. Math. Soc. Jpn.*, 12, 233–239.

Nelson, E. (1969). *Topics in Dynamics I, Flows*. Princeton, NJ: Princeton University Press.

Olech, C. (1956). On the asymptotic behaviour of the solutions of a system of ordinary nonlinear differential equations. *Bull. Acad. Polo. Sci. Class III*, 4, 555–561.

Oleinik, O. A. (1952). On properties of some boundary problems for equations of elliptic type. *Math. Sb. (N.S.)*, 30, 695–702.

Osgood, W. F. (1898). Beweis der Existenz einer Lösung der Differentialgleichung $dy/dx = f(x, y)$ ohne Hinzunahme der Cauchy-Lipschitz'schen Bedingung. *Monatsh. Math. Phys.*, 9, 331–345.

Osserman, R. (1957). On the inequality $\Delta u \geq f(u)$. *Pac. J. Math.*, 7, 1641–1647.

Peano, G. (1890). Démostration de l'integrabilité des équations différentielles ordinaires. *Math. Ann.*, 37, 182–228.

Peano, G. (1897). Generalità sulle equazioni differenziali ordinarie. *Atti Accad. Sci. Torino*, 33, 9–18.

Pearl, R. and Reed, L. J. (1920). On the rate of growth of the population of the United States since 1790 and its mathematical representation. *Proc. Natl. Acad. Sci.*, 6, 275–288.

Perron, O. (1923). Eine neue Behandlung der Randwertaufgabe für $\Delta u = 0$. *Math. Z.*, 18, 42–54.

Perron, O. (1926), Ueber Ein- und Mehrdeutigkeit des Integrales eines Systems von Differentialgleichungen. *Math. Ann.*, 95, 98–101.

Picard, E. (1890). Mémoire sur la théorie des équations aux dérivées partielles et la méthode des approximations successives. *J. Math. Pures Appl.*, 6, 423–441.

Poincaré, H. (1879). *Sur les propriétés des fonctions définies par les équations aux différences partielles.* Oeuvres, 1. Paris: Gauthier-Villars.

Poincaré, H. (1880). Sur les courbes définies par les équations différentielles. *C. R. Acad. Sci. Paris*, 99, 673–675.

Poincaré, H. (1892). *Les métodhes nouvelles de la mécanique céleste*, Vol. I. Paris: Gauthier-Villars.

Pontriaguin, L. (1975). *Équations Différentielles Ordinaires.* Moscow: Mir.

Puig Adam, P. (1978). *Curso Teórico-Práctico de Ecuaciones Diferenciales Aplicado a la Física y Técnica.* Madrid: Herederos de Pedro Puig Adam.

Riccati, J. F. (1724). Animadversiones in aequationes differentiales secundi gradus. *Actorum Eruditorum, quae Lipsiae publicantur*, Supplementa 8, pp. 66–73.

Rosenblatt, A. (1909). Über die Existenz von Integralen gewöhnlicher Differentialgleichungen. *Ark. Mat. Astro. Fys.*, 5(2) (III 6, 9).

Rothe, F. (1985). The periods of the Volterra-Lotka system. *J. Reine Angew. Math.*, 355, 129–138.

Sattinger, D. (1973). *Topics in Stability and Bifurcation Theory.* Lecture Notes in Mathematics, Vol. 309. Berlin: Springer-Verlag.

Schrödinger, E. (1944). *The Physical Aspect of the Living Cell.* Cambridge: Cambridge University Press.

Sternberg, S. (1957). Local contractions and a theorem of Poincaré. *Am. J. Math.*, 69, 809–824.

Sternberg, S. (1958). On the structure of local homeomorphisms of Euclidean n-space. II. *Am. J. Math.*, 80, 623–631.

Tanner, J. T. (1975). The stability and intrinsic growth rates of prey and predator populations. *Ecology*, 56, 855–867.

Tonelli, L. (1928). Sulle equazioni funzionali del tipo di Volterra. *Bull. Calcutta Math. Soc.*, 20, 31–48.

Tychonoff, A. N. (1930). Über die topologische Erweiterung von Räumen. *Math. Ann.*, 102, 544–561.

Verhulst, P. F. (1838). Notice sur la loi que la population suit dans son accroissement. *Corr. Math. Phys.*, 10, 113–125.

Vinograd, R. È. (1957). Inapplicability of the method of characteristic exponents to the study of non-linear differential equations (in Russian). *Mat. Sb. (N.S.)*, 41(83), 431–438.

Vladimirov, V. S. (1971). *Equations of Mathematical Physics*. New York: Marcel Dekker Inc.

Volterra, V. (1931). *Leçons sur la théorie mathématique de la lutte pour la vie*. Paris: Gauthier-Vilars.

Waldvogel, J. (1986). The period in the Lotka-Volterra system is monotonic. *J. Math. Anal. Appl.*, 114, 178–184.

Wilson, E. O. (1980). *Sociobiology*. Cambridge, MA: Harvard University Press.

Wintner, A. (1945). The nonlocal existence problem of ordinary differential equations. *Am. J. Math.*, 67, 277–284.

Wintner, A. (1946). The infinities in the non-local existence problem of ordinary differential equations. *Am. J. Math.*, 68, 173–178.

Zorn, M. A. (1935). A remark on method in transfinite algebra. *Bull. Am. Math. Soc.*, 41, 667–670.